Engineering Materials and Processes

Series Editor

Brian Derby, Materials Science Center, University of Manchester, Manchester, UK

The Engineering Materials and Processes series focuses on all forms of materials and the processes used to synthesise and formulate them as they relate to the various engineering disciplines. The series deals with a diverse range of materials: ceramics; metals (ferrous and non-ferrous); semiconductors; composites, polymers, biomimetics etc. Each monograph in the series is written by a specialist and demonstrates how enhancements in materials and the processes associated with them can improve performance in the field of engineering in which they are used.

Professor Derby and Springer welcome new book ideas from potential authors. If you are interested in writing for the series please contact: Anthony Doyle (Executive Editor - Engineering, Springer) at anthony.doyle@springer.com.

Jiyang Fan · Paul K. Chu

Silicon Carbide Nanostructures

Fabrication, Structure, and Properties

Second Edition

Springer

Jiyang Fan
Key Laboratory of Quantum Materials and Devices of Ministry of Education
School of Physics
Southeast University
Nanjing, China

Paul K. Chu
Department of Physics, Materials Science and Engineering and Biomedical Engineering
City University of Hong Kong
Hong Kong, China

ISSN 1619-0181 ISSN 2365-0761 (electronic)
Engineering Materials and Processes
ISBN 978-3-032-10985-9 ISBN 978-3-032-10986-6 (eBook)
https://doi.org/10.1007/978-3-032-10986-6

This Springer imprint is published by the registered company Springer Nature Switzerland AG
The registered company address is: Gewerbestrasse 11, 6330 Cham, Switzerland

Preface to the Second Edition

It has been 10 years since the publication of the first edition of this monograph, and silicon carbide nanostructures have continued to be a booming field with more and more intriguing and important new developments. We have largely extended the contents of this book to reflect those newest advances as well as some important correlated points that had been missed in the first edition. We try our best to render this second edition still and even more self-contained. It covers the major areas of diverse silicon carbide nanostructures including zero-dimensional nanoparticles and quantum dots, one-dimensional nanowires and nanotubes, two-dimensional graphene-like monolayers and ultrathin silicon carbide, nanoporous structures, nanostructured films, etc. Topics include synthesis and fabrication approaches, microstructures, unique polytypes, various characteristics, and promising applications, emphasizing the strong correlations between structures and properties, which are the permanent pursuing targets of natural science concerning all scale levels and all kinds of materials and structures existing in nature and in the cosmos. The section of point defects and spin dynamics is highly enlarged to reflect the booming advances in this field, especially the explorations of the applications of the SiC point defects in quantum information. More advances concerning impurities and luminescence as well as presolar SiC grains and explorations of their origins are included. In Chap. 4, we discuss more exciting and important advances toward better understanding the basic characteristics of SiC quantum dots. We add more sections involving preliminary applications of SiC nanoparticles. There is now a section in Chap. 5 devoted to the description of more application potential of SiC nanowires such as microwave absorption and thermal management. The section of two-dimensional honeycomb-structured monolayers and ultrathin layers of SiC is highly enlarged in Chap. 6 to reflect the exciting experimental and theoretical breakthroughs in this important field. In Chap. 8, more interesting advances concerning the biological application potential of SiC nanostructures are summarized and discussed.

We express our great gratitude to our tutors (for Jiyang Fan) or colleagues Prof. Xinglong Wu, Prof. Shining Zhu, Prof. Min Xiao, and others for their always guidance, cooperation, and help. Thank the graduate students Dejian Dai, Xiaoxiao Guo, Yuanyuan Li, Nan Zhang, Xuan Ji, Xueli Sun, and others for their contributions to our explorations of SiC nanostructures. We acknowledge Springer Press, especially senior editor Anthony Doyle, and senior editorial assistant Manju Ramanathan, for their encouragement and patience during the preparation of this book. We also thank our families for their tremendous support without which this book would not have been completed.

Nanjing, China
Hong Kong, China
September 2025

Jiyang Fan
Paul K. Chu

Preface to the First Edition

Although nanostructured silicon carbide is rapidly evolving, the field is still in the infancy stage and full potential of the materials is far from being realized. This book provides the state of the art of the various nanostructures of wide bandgap silicon carbide, including nanoparticles, nanowires, nanotubes, porous structures, nanostructured films, and other complex nanostructures. Topics covered in this book include fabrication methods, polytypes, bulk and surface structures, electronic structures, properties, especially electrical and optical ones, as well as potential applications. The target readership is graduate students and researchers in universities, research institutes, and companies working on SiC and related materials in the fields of materials science, physics, chemistry, electronic engineering, biomedical engineering, and biomedicine.

It is impossible to cover all the aspects of the materials in a single book, and so our intention is to provide an up-to-date review of the relevant results and present a holistic view of this burgeoning field. As a self-contained monograph, Chap. 1 is the brief introduction and Chap. 2 describes the general characteristics of bulk silicon carbide in order to provide necessary and self-contained backgrounds to understand the properties of silicon carbide nanostructures. This chapter emphasizes various unique and core physical characteristics of silicon carbide. It covers bulk crystal structures, introduction of intriguing presolar SiC grains stemming from carbon stars and supernovae, surface reconstruction structures and SiC/SiO_2 interface structures, intrinsic point and extended defects and related photonic and spintronic applications, electronic structures, impurities and luminescence properties, lattice vibration and infrared and Raman characteristics, as well as diverse polytypes and pertinent phase transformations. However, the bulk crystal growth of silicon carbide, general silicon carbide ceramics, and applications of silicon carbide in electric devices are omitted because these fields are far from the topics of this book and too diverse as well as because there are already some excellent books or reviews on these subjects. Chapters 3 and 4 address the porous silicon carbide nanostructures and individual silicon carbide nanoparticles with emphasis on the complex relationship between the structures and luminescence properties and mechanisms. Chapters 5 and 6 describe one-dimensional SiC nanowires and nanotubes as well as two-dimensional SiC monolayers and ultrathin layers concentrating on the synthesis, intrinsic stacking fault defects, unique properties, and applications. Chapter 7 deals with common silicon carbide nanostructured films focusing on those containing silicon carbide nanocrystals. In Chap. 8, the biological applications of silicon carbide nanostructures are discussed. Topics covered include biomorphic silicon carbide ceramics, biological imaging, and cytotoxicity evaluation.

We express our gratitude to our tutors (for Jiyang Fan) or colleagues, Prof. Xinglong Wu, Prof. Shining Zhu, Prof. Min Xiao, Dr. Teng Qiu, Dr. Gaoshan Huang, Dr. Yimin Yang, Ms. Mengxing Liao, and many others, for their cooperation and help on the studies of silicon carbide nanostructures. We acknowledge Springer Press, especially senior editor Anthony Doyle, and senior editorial assistants Simon Rees, Claire Protherough, Gabriella Anderson, and Sathya Subramaniam, for their encouragement and patience during the preparation of this book. We also thank our families for their tremendous support without which this book would not have been completed.

Nanjing, China — Jiyang Fan
Hong Kong, China — Paul K. Chu
May 2014

Competing Interests The authors have no competing interests to declare that are relevant to the content of this manuscript.

Contents

Chapter 1
Introduction

The constituents of the earth are simple in that 92 elements, which occur naturally among all existing 118 elements, make up all the materials around us. However, our world is also very complex because almost unlimited types of matters (especially crystals and molecules) can be formed by combining these elements in different spatial arrangements following different symmetries. In this respect, silicon carbide is unique as it has over two hundred crystalline forms and some polytypes of silicon carbide have even been identified from presolar grains in meteorites. Silicon carbide is a family of wide bandgap semiconductors with excellent electric and electronic characteristics that bode well for applications in high-temperature, high-frequency, and high-power electronic devices. Moreover, silicon carbide is a superhard ceramic having a small density as well as superior thermal and mechanical properties. It is environmentally friendly and biocompatible, thus having many potential applications in energy, biomedical engineering, and medicine. Last but not least, both silicon and carbon are extremely abundant on earth and SiC can be produced economically, a big advantage compared to materials composed of noble metals and rare-earth elements.

On the heels of rapid developments of nanoscience and nanotechnology, researchers have been intrigued by the unique properties of fruitful types of silicon carbide nanostructures and their diverse potential applications. Many silicon carbide nanostructures ranging from quasi-zero-dimensional quantum dots to one-dimensional nanowires and nanotubes, to two-dimensional honeycomb-structured monolayers as well as quasi-two-dimensional nanostructured films, have been synthesized and studied. SiC nanostructures possess properties inherent to nanomaterials such as the quantum confinement effect, surface effects, and other size-sensitive characteristics. In addition, nanostructured SiC has unique electronic structures, outstanding electrical and mechanical properties, superior thermal and chemical stability, and favorable biocompatibility and is suitable for applications to microelectronics, nanoelectronics, photonics, energy storage and conversion, and life science.

J. Fan and P. K. Chu, *Silicon Carbide Nanostructures*, Engineering Materials and Processes, https://doi.org/10.1007/978-3-032-10986-6_1

Researchers began to notice the diversity and potential applications of SiC nanostructures since the 1990s, about the same time when the fabrication and properties of porous Si and Si nanostructures aroused interest. In the early years, researchers were mainly concerned with porous SiC fabricated by electrochemical etching of bulk single-crystalline SiC. Since the invention of solid-state electronic devices, group IV semiconductors have acted as cornerstones of the microelectronics industry which has introduced to modern living the Internet and indispensable electronic gadgets such as personal computers and cell phones. However, in the important application areas involving lighting and display, the roles of group IV semiconductors are rather limited due to their typical indirect bandgaps. As a result, the luminescence arising from the interband transitions of carriers is inefficient compared to direct-bandgap semiconductors such as GaAs. Microscopically, in an ideal indirect-bandgap semiconductor, light absorption or emission occurs via interband transition of an electron, that is, the transition of an electron between two quantum states with different crystal momentum $\hbar k$; in this process, the conservation of energy is readily ensured. However, the small photon momentum cannot ensure conservation of momentum in the transition. Hence, this process typically occurs with absorption or emission of one or a couple of lattice phonons to satisfy conservation of momentum. Such a quantum process involving electrons, photons, and phonons generally has much lower efficiency than the no-phonon electronic transition in a direct-bandgap semiconductor in which only electrons and photons, but no phonons, are involved.

Investigation of a porous Si layer containing nanoparticles reveals that the luminescence efficiency of Si nanoparticles can be improved by several orders of magnitude compared to the bulk materials due to spatial confinement of carriers. In addition, luminescence from porous Si shifts from the near-infrared region of the bulk silicon to the visible regime due to the quantum confinement effect, which predicts that the smaller the quantum dot, the wider is the energy gap. Similar to Si, porous SiC shows visible luminescence with enhanced efficiency. These porous materials are generally fabricated by electrochemical etching of bulk SiC crystals in hydrofluoric acids. However, in most cases, luminescence from porous SiC is characteristic of defect-related luminescence, and hence, light emission is generally attributed to surface defects although the specific defect types are not fully understood.

The nanoparticles and nanowires in porous SiC are not separated but rather they are connected to the SiC substrate and in effect a component of the whole porous SiC crystal. Therefore, their properties are between those of the single crystal and individual nanocrystals. Various types of separate SiC nanocrystals have been fabricated and studied. Gas phase reactions and solution-based chemical reactions can produce SiC nanocrystals, and SiC nanocrystals encapsulated in polymers or other solid matrices have also been produced by techniques such as ion implantation followed by annealing. The luminescence properties of these SiC nanocrystals vary and depend on the fabrication approaches, and as aforementioned, the light emission is generally ascribed to surface defect states. Colloidal SiC nanocrystals suspended in solvents have been prepared by electrochemical or direct chemical etching of SiC bulk crystals followed by ultrasonic treatment. These SiC colloids show intense visible luminescence with tunable peak wavelengths. The light emissions from SiC nanocrystals can have quantum yields higher than that of bulk materials by several orders of magnitude. The origins of their light emissions are rather complex, combining the interband transition luminescence and surface defect states-correlated luminescence. For the luminescence originating from quantum confinement, the emission band shifts monotonically toward the shorter-wavelength region when the particle size diminishes. On the other hand, there are fruitful surface defects for the compound SiC nanocrystals, and they offer diverse localized quantum states, some of which give rise to distinct colors (blue, green, or even red) of luminescence. Therefore, the intrinsic interband transition luminescence of nonlocalized electrons and surface defects luminescence superimpose and give rise to the detected complex and wideband luminescence. Different colors of light emissions, spanning red, yellow, green, blue, and near-UV, are achievable from SiC crystallites with different sizes and distinct surface structures. Blue and quasi-white light-emitting diodes based on SiC nanocrystals have been demonstrated. Furthermore, silicon carbide has high chemical and thermal stability in comparison with other types of semiconductors. This virtue combined with the color-tunable luminescence enables the use of luminescent SiC nanocrystals in harsh environments. In conjunction with the high surface hydrophilicity, excellent biocompatibility, low density, and high hardness, SiC nanocrystals are very useful in biology and medicine. Additionally, SiC has long been used as filling materials to enhance the mechanical properties of ceramics, and it has been shown that incorporation of SiC nanoparticles into metal nanocomposites can boost the mechanical properties. It has also been found that doping with SiC nanoparticles can substantially enhance the performance of some superconductors.

SiC nanowires have been widely investigated. Several methods have been developed to synthesize such materials, and they involve two main growth mechanisms: vapor–liquid–solid growth with the assistance of metal catalysis and vapor–solid growth without catalysts. Some synthesized SiC nanowires have the SiC/amorphous SiO_2 core/shell structure. Carbon-sheathed SiC nanowires have also been observed but are rare. BN-coated SiC nanowires have been prepared. The synthesized SiC nanowires can be as small as several nanometers or as large as several hundred nanometers in diameter. Their lengths are generally on the order of micrometers, although SiC nanowires with lengths of a few centimeters have also been reported. Besides standard nanowires and nanorods, less common quasi-one-dimensional SiC nanostructures with different morphologies of bamboo, bead, helix, belt, flower, and multileg have also been produced. The SiC nanowires and similar structures often exhibit visible or near-UV luminescence, but the luminescence mechanisms are not well understood. SiC nanowires yield stable field electron emissions with favorable characteristics such as low turn-on fields. Large-scale and highly ordered SiC nanowire arrays have been prepared, and some synthesized SiC nanowires have very sharp ends that are beneficial to field emissions. One-dimensional SiC nanostructures are competent for microwave absorption, thermal management and fireproofing, and as building blocks of molecular electronic devices. SiC-based coaxial nanocables such as coaxial SiC/BN nanocables have promising applications in nanoelectronic devices. Silicon carbide nanowires having superior mechanical characteristics and electrical conductivity can be utilized to reinforce composite materials or as nanocontacts in harsh environments.

In comparison with SiC nanowires, SiC nanotubes have been less investigated experimentally. They are generally synthesized using carbon nanotube or semiconductor nanowire templates. Most of the synthesized SiC nanotubes are actually not real nanotubes, because the walls of the nanotubes are composed of cubic SiC nanocrystals with different orientations. In other words, they are polycrystalline. However, SiC nanotubes with structures similar to multiwalled carbon nanotubes have been observed. In comparison, there have been many theoretical studies on the structures, electronic structures, and molecular adsorption properties of SiC nanotubes based on first-principles calculations or molecular dynamics simulations. This is understandable because the structures of SiC nanotubes are simpler than that of most of other SiC nanostructures. It consists of fewer atoms than other SiC nanostructures of the same size, and smaller computational power is required. The theoretical results are generally consistent with experimental data. Two-dimensional honeycomb-structured SiC nanosheets similar to graphene have aroused more and more interest in recent years. They have long been widely investigated and predicted by using density functional theory calculations. However, the fabrication of the 2D SiC in laboratories is challenging. Most of the early observed "2D SiC" materials are either ultrasheets composed of 3C-SiC nanoparticles or small fragments with probable honeycomb structures. Very recent advances in this field have enabled experimental fabrication or synthesis of truly honeycomb-structured SiC monolayers. They are indeed stable. But this field is still in infancy, and more developments are expected.

The research history of SiC films is longer than that of SiC nanoparticles and SiC nanowires. However, in the early investigations of SiC films, little attention had been paid to the nanoscale structures of these films due to the limitations in the preparation methods and characterization capability. However, in the last twenty years, researchers have been able to prepare and identify nanostructured SiC films by many methods. Chemical vapor deposition is the most widely used approach to synthesize nanocrystalline SiC films, and ion implantation and high-temperature sputtering have also been adopted to fabricate nanostructured SiC films. Many thin SiC films comprise embedded SiC nanocrystals, and these nanocrystalline SiC films are potentially useful in electronic devices, solar cells, and sensors. Furthermore, many SiC films yield visible light emissions rendering the materials probably suitable for large-area displays.

As biocompatible materials, SiC nanostructures have potential applications in biological imaging. Silicon carbide is an excellent wear-resistant and benign coating on biomedical implants. Porous SiC films with ultrasmall pores are semi-permeable biomaterials. Biomorphic SiC ceramic materials with a wood-like microstructure have been prepared by infiltrating a wood-derived carbon template with silicon or silicon oxide at high temperatures. The biomorphic SiC ceramics coated with bioactive silica show great promise as dental and orthopedic implants with enhanced mechanical and chemical properties.

Theoretical studies play important roles in understanding the structures and properties of various silicon carbide nanostructures. Common approaches include semi-empirical and first-principles calculations as well as molecular dynamics simulations. Many important properties of SiC nanostructures have been theoretically predicted. For example, the energy gap of a SiC quantum dot depends closely on its dimension and surface composition. Calculations show that the difference between cohesive energy and energy gap of different polytypes of SiC quantum dots vanishes gradually as the size diminishes. With decreasing size, the surfaces of the SiC nanocrystals become more important, and in practice, SiC nanoparticles with sizes close to or below one nanometer resemble giant molecules. As a result, the properties of ultrasmall SiC nanoparticles are distinct from those of bulk materials. For example, they may have complex and surface defects-dominated light emissions. Theoretical investigations show that the electronic structures and mechanical properties of SiC nanowires sensitively depend on the nanowire diameter. SiC nanotubes are generally semiconducting and their bandgap may decrease as the tube diameter decreases. This is contrary to the case of SiC nanocrystals. SiC nanotubes may have an indirect or direct bandgap depending on the chirality, and some SiC nanotubes may even be metallic when the surface is terminated by hydrogen atoms.

In this book, the various fabrication and synthesis techniques, diverse properties, and fruitful potential applications of SiC nanostructures are described and discussed. Containing the latest information and knowledge in this fledgling field, it serves as a valuable reference for graduate students and researchers who are interested in these fascinating nanostructures and potential applications galore.

Chapter 2
General Properties of Bulk SiC

The crystallization and properties of silicon carbide crystals have been extensively studied because as a wide bandgap semiconductor, silicon carbide is ideal for electric and electronic applications requiring high temperature, high frequency, and high power. The electrical and electronic properties as well as device applications of bulk silicon carbide have been discussed in several books [1–3]. This chapter provides a self-contained up to date framework to facilitate understanding of the properties of silicon carbide nanostructures. The various basic characteristics of silicon carbide crystals including updated advances are reviewed and discussed.

2.1 SiC Polytypes

Similar to silicon, silicon carbide is a covalently bonded semiconductor. In the crystalline form, each silicon atom is covalently bonded to four neighboring carbon atoms to form a tetrahedron (Fig. 2.1) and vice versa. There are two types of tetrahedrons in the silicon carbide crystal. The first type is obtained by rotating another tetrahedron along its *c* axis by 180°, and one type of tetrahedron is the mirror image of the other when the *c* axis is parallel to the mirror. Silicon carbide consists of magnetically dilute spin 1/2 isotopes (^{29}Si, 4.7%; ^{13}C, 1.1%) and is ideal for magic angle spinning nuclear magnetic resonance (NMR) spectroscopy study. Silicon carbide polytypes give distinctive ^{29}Si and ^{13}C NMR spectra related to the number and types of lattice sites present, which differ in their next-nearest-neighbor environment. The NMR result shows that only four next-nearest-neighbor silicon (and carbon) environments are possible in crystalline SiC regardless of polytype [4]. The *c* axis denotes the direction normal to the Si–C double atomic layers. In each layer, the Si (or carbon) atoms have a close-packed hexagonal arrangement. There are three types of sites (named A, B, C) in arranging the Si–C double atomic layers, and each layer is normal to the *c* axis (Fig. 2.2).

J. Fan and P. K. Chu, *Silicon Carbide Nanostructures*, Engineering Materials and Processes, https://doi.org/10.1007/978-3-032-10986-6_2

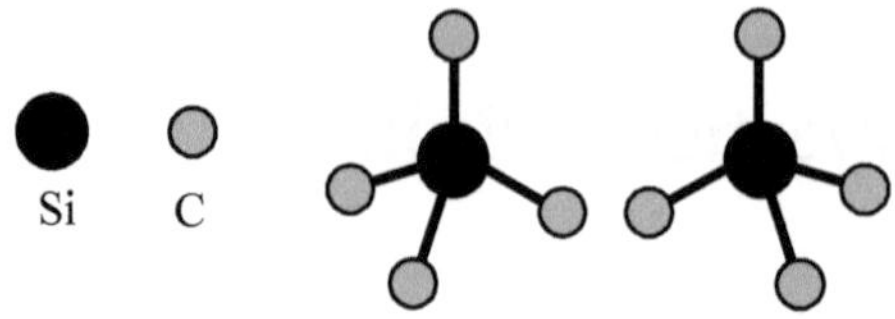

Fig. 2.1 Two types of tetrahedrons forming the building blocks of all SiC crystals with each tetrahedron consisting of one Si atom and four covalently bonded nearest neighboring C atoms (or vice versa). One type is obtained by rotating another type around its *c* axis (vertical here) by 180°

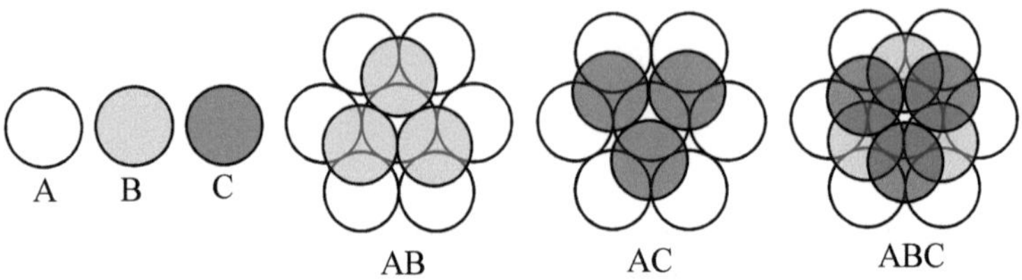

Fig. 2.2 Three types (A, B, C) of Si–C double atomic layer arrangement along the *c* axis (stacking direction) through close-packed spheres. The *c* axis is normal to the paper plane

The three types of double atomic layers A, B, and C form a periodical series with a specified repeating unit along the *c* axis forming a specific polytype of the silicon carbide crystal. For example, in 3C-SiC (Fig. 2.3), the stacking sequence of the double atomic layers is ABCABC... The letter "C" in "3C-SiC" denotes the cubic crystal structure and "3" refers to the number of double atomic layers in one repeating unit (ABC). 3C-SiC is the only polytype with the cubic crystal structure that is also referred to as β-SiC. The dominant polytypes in silicon carbide are hexagonal. The three most common hexagonal polytypes are 2H, 4H, and 6H and their stacking sequences are ABAB..., ABCBABCB..., and ABCACBABCACB..., respectively. Similarly, "H" refers to the hexagonal crystal structure and the number before "H" denotes the number of double atomic layers in one repeating unit. There are other types of silicon carbide crystals, for instance, 15R-SiC where "R" stands for the rhombohedral crystal structure. It should be noted that one Si (or C) atomic layer in the stacking sequence can have a local cubic (k) or hexagonal (h) environment with respect to the immediate neighbors, as shown in Fig. 2.3. 3C-SiC has only k-type atoms and 2H-SiC has only h-type atoms. The atomic number ratio k: h is 1:1 in 4H-SiC and 2:1 in 6H-SiC.

As a result of the different stacking sequence of the double atomic layers, different polytypes of silicon carbide have distinct electronic structures as well as electrical and optical characteristics. For instance, the indirect bandgaps at room temperature of 3C-SiC, 6H-SiC, and 4H-SiC are 2.2 eV, 2.86 (or 3.02) eV, and 3.2–3.3 eV, respectively. The bandgap depends on the temperature and even the measured room temperature bandgap of a specific polytype may vary in a small range depending on the uncertainty

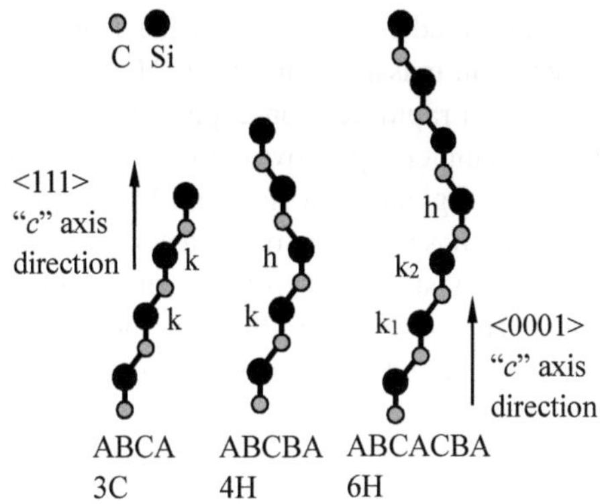

Fig. 2.3 Si–C double atomic layer arrangement in the three most common polytypes of SiC. The *c* axis denotes the stacking direction and *k* and *h* denote atoms with the localized cubic and hexagonal crystal symmetry, respectively

in determining the optical absorption edge and measurement conditions in different laboratories.

2.2 Presolar SiC Grains from Carbon Stars and Supernovae

SiC can form and be synthesized on earth, to one's surprise, in practice SiC exists widely across the whole cosmos. Presolar SiC grains have been frequently found in meteorites. Their morphologies have been extensively studied by using electron microscopy, and their elemental and isotopic compositions have been analyzed by using mass spectrometry.

Including silicon carbide, five types of presolar matters have been identified in chondrites, namely diamond (about 2.6 nm), graphite (0.8–7 μm), silicon carbide (0.05–20 μm), silicon nitride, and aluminum oxide [5]. The anomalous isotopic compositions of Si, C, O, and N demonstrate that the grains have a presolar origin. Studies on presolar diamond, graphite, and silicon carbide in chondrites show that the abundance correlates with the chemical and physical properties of the meteorites [6, 7]. Among the various types of presolar grains, silicon carbide is the most extensively studied because it contains many elements including minor and trace elements. Silicon carbide has long been observed in the interstellar medium and in circumstellar shells. Silicon carbide was identified in two separates from the Murray carbonaceous chondrite that are enriched 20,000-fold in isotopically anomalous neon and xenon [8]. The observed crystalline SiC grains have sizes of 0.1–1 μm. Most grains are cubic and {111}-twinned cubic. The anomalous isotopic composition of its carbon, nitrogen, and silicon indicates the presolar origin of red giants. The strontium isotopic composition determined from individual circumstellar SiC grains extracted from the Murchison meteorite by resonant ionization mass spectrometry confirms the grain origin of slow neutron-capture process (*s*-process) nucleosynthesis at moderate neutron densities [9]. In asymptotic giant branch (AGB) stars, elements heavier than iron observed in nature originate from slow neutron-capture process (*s*-process), whereas in massive stars (> 10 $M_\odot$, with $M_\odot$ the mass of the sun) heavy elements stem from rapid neutron-capture process (*r*-process) or proton-capture process (*p*-process) under supernova explosion conditions. An AGB star is the late evolutionary phase (red giant) of a star with low to intermediate mass of about 0.5–8 $M_\odot$. The AGB star has a carbon–oxygen core and it obtains energy from He fusion in the shell in the early stage and from intermittent burning of the H and He shells in the later stage. The neutrons produced in the He shell are captured by Fe nuclei to generate heavier elements (*s*-process). In the latter stage the heavy mass loss from the star (thermal pulse) would create a planetary nebula.

The polytype distribution of circumstellar SiC has been identified by using transmission electron microscopy (TEM) on presolar silicon carbide in the Murchison carbonaceous meteorite [10]. Only 3C-SiC and 2H-SiC and their intergrowth have been observed (Fig. 2.4). High-resolution TEM (HRTEM) images and electron diffraction of 508 SiC grains demonstrate presence of only cubic 3C (β-SiC) polytype (79.4% of number population) and hexagonal 2H (α-SiC) polytype (2.7%) [11]. Intergrowths of such two polytypes are abundant (17.1%). There are 0.9% of one-dimensionally disordered SiC grains with high density of stacking faults. The secondary ion mass spectroscopy (SIMS) reveals non-solar isotopic compositions of $^{12}C/^{13}C = 64 \pm 4$ and $^{14}N/^{15}N = 575 \pm 24$. These measurements are consistent with mainstream presolar SiC stemming from the expanding atmospheres of AGB carbon stars. The equilibrium condensation calculations employing inferred mineral condensation sequences predict relatively low SiC condensation temperatures in the AGB carbon stars. They agree with the condensation temperatures of 2H- and 3C-SiC, which are the lowest of all SiC polytypes on earth. These explain the occurrence of only 2H and 3C polytypes of SiC in circumstellar outflows. The 2H- and 3C-SiC were supposed to condense at different radii (i.e., temperatures) in the expanding atmospheres of AGB carbon stars.

Much effort has been made to explore the origin of the SiC grains. The investigated all five noble gases (He–Xe) in interstellar SiC grains isolated from the Murchison C2 chondrite show grossly non-solar isotopic and elemental abundances that vary with grain size but are strikingly similar to calculated values for the helium-burning shell of low-mass carbon-rich AGB stars [12]. These grains formed in carbon-star envelopes and were impregnated with noble gas ions from a stellar wind. Therefore, the meteoritic SiC recorded the detailed nuclear and chemical processes in mother carbon stars. The following theoretical analysis based on the models for low-mass thermally pulsed AGB stars of different metallicity [13] explains well a variety of measured abundances, in particular the range of krypton isotope ratios and the excess ^{22}Ne found in the grains. Condensation calculations for C-rich circumstellar envelopes of carbon stars were used to model the condensation sequence of C, TiC, and SiC, and trace-element patterns observed in circumstellar SiC grains [14]. The comparison of calculated patterns for trace-element solid solutions in SiC with the astronomical observations of key elemental abundances (e.g., C, N, O) and isotopic ratios (e.g., $^{12}C/^{13}C$) indicates a correlation between the grains and at least three different groups of carbon stars. The barium isotopic anomaly observed from SiC

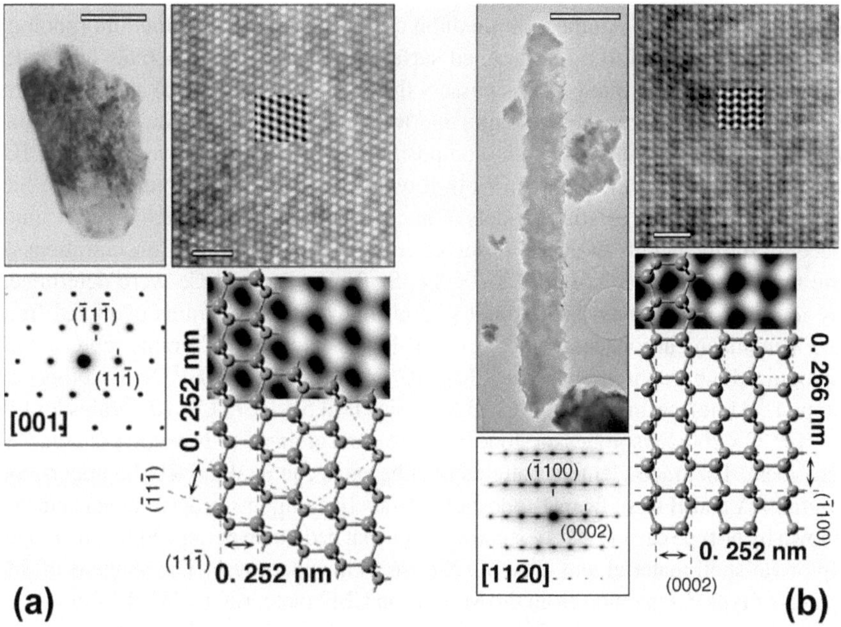

Fig. 2.4 Circumstellar 3C-SiC (**a**) and 2H-SiC (**b**) from Murchison carbonaceous chondrite KJE. Each panel contains the TEM image (top left, scale bar: 0.1 μm), selected-area electron diffraction (SAED) pattern (bottom left), HRTEM image (top right, scale bar: 1 nm), and atomic model of the [011] (for 3C-SiC) or [11$\bar{2}$0] (for 2H-SiC) zone axis superimposed on a simulated lattice image (bottom right). The tetrahedral stacking direction in the HRTEM image and atomic model is horizontal. Reprinted with permission from [10]. Copyright 2003, American Association for the Advancement of Science

meteorite grains was considered to stem from the *s*-process nucleosynthesis in carbon stars that suffer recurrent He shell instability on the asymptotic giant branch [15]. Additionally, the Xe, Nd, and Sm anomalies in the same SiC grains show clearly the *s*-signature and the corresponding interpretation corroborates the carbon-star origin. The analysis of He, Ne, Ar, Kr, and Xe in interstellar SiC isolated from the Murchison C2 chondrite shows highly anomalous isotopic signature of the *s*-process nucleosynthesis [16]. These components resemble predictions for the He-burning shells and envelopes of red giant carbon stars. The grain size dependence of many elemental and isotopic ratios (such as ^{80}Kr/^{82}Kr and ^{86}Kr/^{82}Kr) suggests that the SiC comes from carbon stars of different masses, metallicities, temperatures, and neutron densities. The ^{21}Ne-content of the gas-rich SiC grains suggests a presolar cosmic-ray irradiation of up to 130 Myr (million years), representing the interval between formation of the grains in a circumstellar shell and arrival in the solar system 4.6 Gyr (giga years) ago. Another group has studied the isotopic compositions of the carbon, nitrogen, magnesium, silicon, and titanium in the single interstellar silicon carbide grains (with sizes of 1–10 μm) extracted from the Murchison carbonaceous chondrite [17]. Scanning electron microscopy (SEM) images of these SiC grains are displayed in Fig. 2.5. All the grains exhibit crystalline features but their morphology varies from euhedral to a fine-grained surface, indicating agglomerates of smaller subgrains. Although many grains possess the hexagonal (Fig. 2.5b) and tetrahedral (Fig. 2.5c) shape, electron diffraction and Raman scattering indicate that the grains are β-SiC. The C and N isotopic compositions along with *s*-process Kr, Xe, Ba, and Nd suggest that the AGB or Wolf–Rayet stars are possible sources of the SiC grains. However, the existing models of nucleosynthesis in these stellar sites cannot account for the details of all the observed isotopic compositions. Concentrations of the trace elements Mg, Al, Ca, Ti, V, Fe, Sr, Y, Zr, Ba, and Ce were determined by ion microprobe mass spectrometry in 60 individual SiC grains extracted from the Murchison carbonaceous meteorite [18]. The selected SiC grains represent all recognized subgroups: mainstream ($20 < {}^{12}C/{}^{13}C < 120$; $200 < {}^{14}N/{}^{15}N$; Si isotopes on slope 1.34 line), grains A (${}^{12}C/{}^{13}C < 3.5$), grains B ($3.5 < {}^{12}C/{}^{13}C < 10$), grains Y ($150 < {}^{12}C/{}^{13}C < 260$; Si isotopes on slope 0.35 line), and grains X (${}^{15}N$ excesses, large ${}^{28}Si$ excesses). The trace-element patterns of subgroups can qualitatively be understood if grains A and B arise from J-type carbon stars (lacking in *s*-process enhancements shown by N-type carbon stars) or carbon stars that had not experienced much dredge-up of He-shell material and if grains X arise from supernovae. The analysis of SiC grains of type Z separated from the Murchison CM2 meteorite by IMMS shows ${}^{12}C/{}^{13}C$ ratios from 11 to 120, ${}^{14}N/{}^{15}N$ ratios between 1100 and 19,000, initial ${}^{26}Al/{}^{27}Al$ ratios of less than 0.003, and deficits in ${}^{29}Si$ of up to 150‰ and enrichments in ${}^{30}Si$ of up to 510‰ relative to those of solar system [19]. These isotopic signatures in combination with the predictions from a AGB star model suggest that the Z grains formed in the outflows of low-mass ($< 2.3\ M_{\odot}$) low-metallicity AGB stars that experienced strong cool bottom processing during the red giant phase. The resonant ionization mass spectrometry gives the isotopic composition of Mo in presolar SiC grains from the Murchison meteorite [20]. Relative to terrestrial abundance, the majority of the grains show strong depletions in the *p*-process isotopes ${}^{92}Mo$ and ${}^{94}Mo$ and the *r*-process isotope ${}^{100}Mo$. The observed isotopic patterns of Mo from mainstream SiC grains signify the *s*-process nucleosynthesis and suggest that the grains stem from the low-mass thermally pulsed AGB stars.

The isotopic ratios of presolar SiC grains of types A and B (defined as having ${}^{12}C/{}^{13}C < 10$) in an acid-resistant residue of the Murchison carbonaceous meteorite along with that of 28 previously analyzed A + B grains have been studied [21]. The most likely sources of the A + B grains with solar *s*-process abundance are J-type carbon stars although the origin of these stars is still unclear. Born-again AGB stars are possible sources of the A + B grains with enhanced *s*-process elemental abundance. There are SiC grains of the rare subtype Y, which constitute ~ 1% of Murchison SiC grains larger than ~ 2 μm and are defined as having ${}^{12}C/{}^{13}C > 100$ (89 for solar) and ${}^{14}N/{}^{15}N > 272$ (solar) [22]. The analysis of the isotopic compositions for 27 presolar type-Y SiC grains in an acid-resistant residue of the Murchison (CM2) meteorite indicates an origin in C-rich AGB stars of near-solar metallicity. In a comparative study, the strontium, zirconium, molybdenum, and barium isotopic compositions predicted for the mass losing envelopes of AGB stars of solar metallicity and masses 1.5, 3, and 5 $M_{\odot}$ were compared with the measurements of single presolar SiC

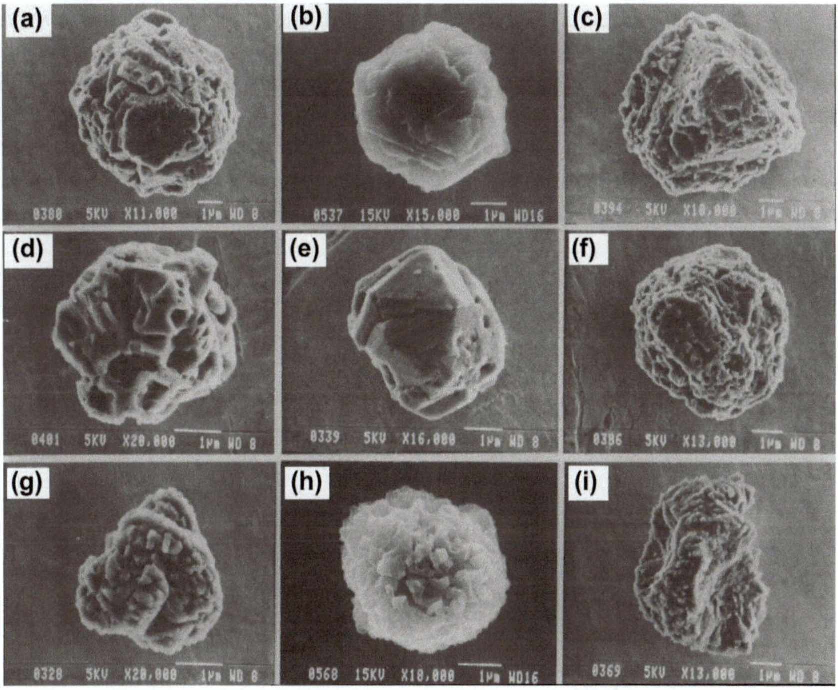

Fig. 2.5 Scanning electron microscopy images of SiC grains extracted from Murchison carbonaceous chondrites KLG (**b**, **h**) and KJH (**a**, **c**, **d**, **e**, **f**, **g**, **i**). The grains show crystalline features. The morphology varies from euhedral (**a**, **b**, **c**, **e**) to fine-grained surface indicating agglomerates of smaller subgrains (**g**, **h**, **i**). Many grains have the hexagonal (**b**) and tetrahedral (**c**) shape. Electron diffraction and Raman scattering studies indicate that the grains are β-SiC. Reprinted with permission from [17]. Copyright 1994, IOP Publishing

grains from the Murchison meteorite [23]. The predicted isotopic compositions in the envelopes of low-mass AGB stars of solar metallicity agree with the isotopic ratios determined from individual presolar SiC grains, whereas the prediction for intermediate mass stars excludes them as the source of these grains. A multiplicity of low-mass AGB stars with metallicity close to that of the sun with different masses and neutron exposure is required to explain the measured spread in the heavy element isotopic compositions among single presolar SiC grains.

The 7.5–13.5 μm infrared spectra of 32 identified or potential carbon stars have been investigated [24]. Figure 2.6 shows the self-absorption fits of the infrared absorption spectra of two carbon stars. Four carbon stars, including AFGL 3068, IRAS 02408+5458, AFGL 2477, and AFGL 5625, show the 11-μm SiC feature in absorption. All of the observed SiC features can be fitted well by α-SiC grains. The observation shows no unambiguous evidence of the presence of β-SiC around these stars. The individual β-SiC nanoparticles with average diameters of 10–20 nm (Fig. 2.7) have been synthesized by laser induced pyrolysis of gas-phase starting materials [25]. The infrared spectra perfectly match the absorption or emission band profile of the 11+ μm feature observed from carbon-rich AGB stars conducted by the *Infrared Space Observatory*. The isotopic composition of Ru in the SiC grains isolated from the Murchison meteorite shows the signature of *s*-process nucleosynthesis in AGB stars and an anomaly in ^{99}Ru that was explained by the in situ decay of ^{99}Tc in the grains [26]. This finding along with the observation of Tc spectral lines in certain stars indicates that the majority of SiC grains come from low-mass AGB stars.

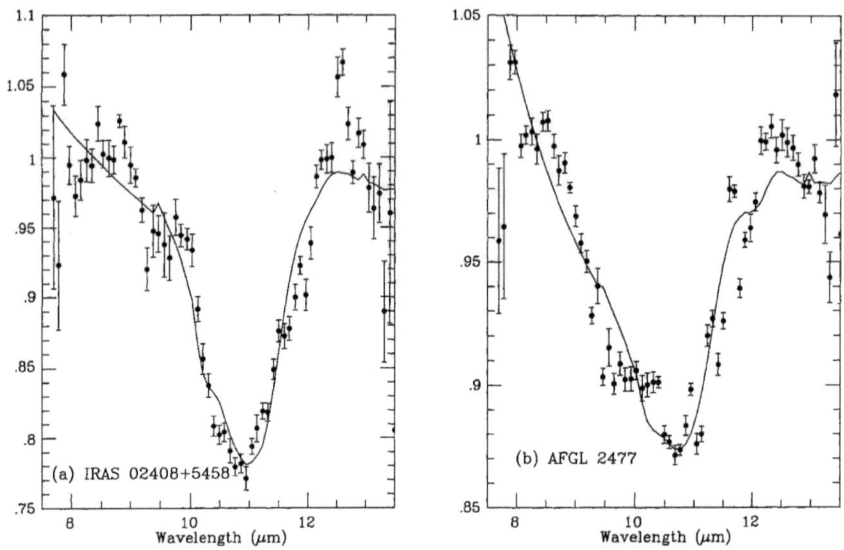

Fig. 2.6 Flux calibrated and normalized infrared spectra of two carbon stars showing absorption features with best χ^2-fits using self-absorbed SiC. Reprinted with permission from [24]. Copyright 1997, IEEE

Fig. 2.7 High-resolution TEM image of one particle showing the β-SiC core as determined by the lattice fringe spacing and electron diffraction pattern (inset). Reprinted with permission from [25]. Copyright 2003, IOP Publishing

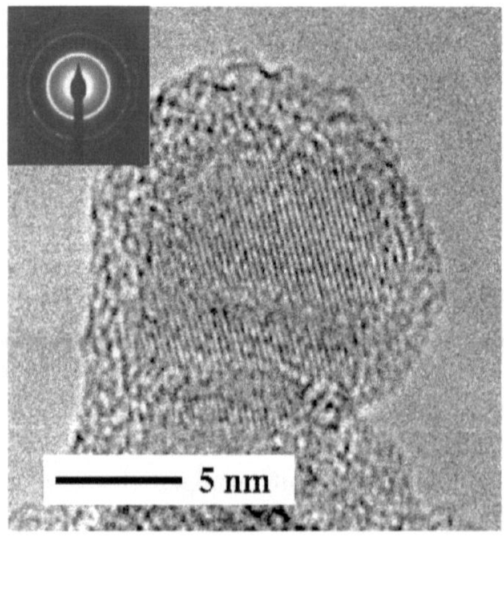

Although many types (M, Y, Z, AB) of presolar SiC grains [27] have been mostly supposed to originate from the carbon-rich AGB stars, several some types of SiC grains (types X and C grains with characteristic Si-isotopic anomalies) may form in the ejecta of supernova explosion. The C, N, Si, Mg–Al, Ca, and Ti isotopic compositions of five single SiC grains 2–9 μm in size extracted from the Murchison carbonaceous meteorite have been determined by ion microprobe [28]. The Ca and Ti anomalies suggest their origin of explosive nucleosynthesis in supernovae and the in situ decay of the radioactive precursors ^{44}Ti and ^{49}V in SiC grains can form in the supernova ejecta. However, no simple formation mechanism can provide a consistent explanation of the isotopic compositions of those grains. Large excess of ^{44}Ca arising from the radioactive decay of short-lived ^{44}Ti, and in ^{46}Ti, ^{49}Ti, and ^{50}Ti has been observed from four graphite grains and five type-X SiC grains extracted from the Murchison meteorite [29]. These grains must have a supernova origin since ^{44}Ti is only produced in supernovae. Furthermore, the Si, C, N, Al, O, and Ti isotopic compositions require a Type II supernova source and suggest heterogeneous mixing of different supernova regions including the nickel core. By using secondary ion mass spectrometry, the isotopic compositions of C, N, O, and Si in 849 small (about 1 μm) individual SiC grains extracted from the Murchison meteorite have been determined [30]. The isotopic compositions of the major elements C and Si of most grains are similar to those observed from larger grains, suggesting an AGB star origin of these grains. Highly anomalous isotopic compositions have been found from a nitride grain which comprises Si_3N_4 with little carbon. The isotopic features of C, N, and Si resemble those of the rare type-X SiC grains, suggesting that these two rare constituents of circumstellar matters formed in type II supernovae. The study of Si, C, N, Mg–Al, Ca, and Ti isotopic data for three micron-sized presolar SiC grains from the Murchison meteorite shows they have very low ^{12}C/^{13}C and ^{14}N/^{15}N ratios [31], similar to grains with a previous assignment of origin of classical ONe novae. The isotopic signatures in one grain (^{28}Si, ^{49}Ti, and ^{44}Ca excesses and a high ^{26}Al/^{27}Al ratio) indicate its actual origin of a supernova. Such result raises the question whether all grains previously attributed to novae might originate from Type II supernovae. The isotopic abundances for C, N, Mg–Al, Si, Ca–Ti, and Fe in presolar SiC type-X grains (0.5–1.5 μm) extracted from the Murchison CM2 meteorite were studied [32]. The isotopic compositions of C, Fe, and Al as well as abundances of extinct ^{44}Ti were well explained both by type Ia and type II supernova models. The isotopic characteristics of C, N, Si and initial ^{26}Al/^{27}Al ratios in small X grains are similar to those of large X grains (2–10 μm). In another study, the type-X SiC particles extracted chemically from meteorites were demonstrated to have chemical and isotopic compositions resembling those from explosive helium burning in ^{14}N-rich matter [33]. It was proposed that these particles are from Type Ia supernovae that explode with a cap of helium atop their CO structure. The enhanced abundance of C and Si suggests that silicon carbide is in fact the natural condensate and it was speculated that the dust grew within the interior of Type Ia supernovae during expansion.

Knowledge of the time span between the formation of presolar grains in stellar outflows and their incorporation into early solar-system solids is essential for better understanding of the processing of grains in the interstellar medium and formation of the solar system. The analysis of the Galactic cosmic rays produced He and Ne in very large presolar SiC grains (diameter ~ 2–50 μm; density ~ 3.2 g cm^{-3}) extracted from the Murchison carbonaceous chondrite suggests interstellar residence times of ~ 3–1100 Myr [34]. The age distribution may be ascribed to starburst activity 1–2 billion years prior to the birth of the sun. Five SiC grains isolated from the Murray and Murchison C2 chondrites were examined by using noble-gas and ion-probe mass spectrometry as well as by SEM–EDX (energy dispersive X-ray spectroscopy) [35]. Their ^{21}Ne-content exceeds up to 40-fold that expected from the recent cosmic ray irradiation, implying a presolar cosmic-ray exposure age of 39 (+ 19, − 13) Myr. The analysis of cosmogenic ^{21}Ne in large presolar SiC grains extracted from the Murchison CM2 meteorite indicates interstellar cosmic-ray exposure ages ranging from 3.9 ± 1.6 Myr to ~ 3 ± 2 Gyr before the start of the solar system ~ 4.6 Gyr ago [36]. A majority of the grains having interstellar lifetimes of < 300 Myr condensed in outflows of AGB stars < 4.9 Gyr ago that possibly formed during an episode of enhanced star formation ~ 7 Gyr ago. A minority of the grains show ages > 1 Gyr.

SiC can also form in the Earth's mantle [37]. Moissanite (natural SiC) has been recovered from podiform chromitites of several ophiolite complexes, including the Luobusa and Donqiao ophiolites in Tibet, the Semail ophiolite in Oman and the United Arab Emirates, and the Ray-Iz ophiolite of the Polar Urals in Russia. Moissanite arises also in diamondiferous kimberlites. Therefore, natural SiC is a widespread mineral in the Earth's mantle which possesses local extremely low redox conditions. Most ophiolite moissanite grains have sizes of 20–150 μm and exhibit colors of light bluish-gray to blue or green. They can exhibit as twinned crystals. Many grains contain native Si and Fe–Si alloys ($FeSi_2$, Fe_3Si_7). Secondary ion mass spectrometric analysis shows that the ophiolite-hosted moissanite has a ^{13}C-depleted isotopic composition much lighter than the main carbon reservoir in the upper mantle. This suggests that moissanite originates from a separate carbon reservoir in the mantle, maybe in the lower mantle.

2.3 Electronic Structures

SiC is an indirect semiconductor with a wide bandgap. The most common polytypes, 3C, 6H, 4H, and 2H, have room-temperature bandgaps of around 2.2, 2.86 (or 3.02), 3.2–3.3, and 3.3 eV (exciton energy gap), respectively. Note that the bandgap subtracted by the exciton binding energy is the exciton energy gap. The electronic structures [electronic dispersion curves or plot of electron single-particle energy $E_n(\boldsymbol{k})$ vs. wavevector $\boldsymbol{k}$ (crystal momentum: $\hbar\boldsymbol{k}$) in the Brillouin zone of a specific crystal, with n of the band number] of silicon carbide crystals have been extensively studied by using empirical-pseudopotential, tight-binding, and first-principles calculations. The first-principles electronic structure calculations start from numerically solving the derivative equations (Hartree–Fock equations or Kohn–Sham Schrödinger-like equations) of the original many-particle Schrödinger equations without using empirical parameters of molecules or crystals, with the aids of modern high-performance computers. Among the ab initio methods, the most widely used one is the density-functional theory (DFT) calculation, which involves numerical iterative solution of the Kohn–Sham variational equations describing the law of the single-particle (Kohn–Sham auxiliary system) energies and wavefunctions. In this Kohn–Sham approach, the original many-body problem is replaced by an auxiliary independent-particle problem, and various physical quantities are expressed as the functionals of the ground-state electron number density $n(\boldsymbol{r})$ of the auxiliary system:

$$n(\boldsymbol{r}) = \sum_{i,\sigma} f_{i,\sigma} \left|\psi_{i,\sigma}(\boldsymbol{r})\right|^2, \tag{2.1}$$

where i is the quantum number of the electron orbital $\psi_{i,\sigma}(\boldsymbol{r})$, and σ is its spin index, $f_{i,\sigma}$ represents the electron occupation number arising from Fermi–Dirac statistics of this orbital at ground state of the auxiliary system at thermal equilibrium ($T = 0\,\mathrm{K}$). In this scenario, the ground-state energy of the actual interacting many-particle system is expressed as:

$$\begin{aligned} E_{\mathrm{KS}} = E_{\mathrm{ion-ion}} + T_{\mathrm{kinetic}}[n(\boldsymbol{r})] + \iiint V_{\mathrm{ext}}(\boldsymbol{r}) n(\boldsymbol{r}) \mathrm{d}^3 r \\ + E_{\mathrm{Hartree}}[n(\boldsymbol{r})] + E_{\mathrm{xc}}[n(\boldsymbol{r})], \end{aligned} \tag{2.2}$$

where $E_{\mathrm{ion-ion}}$ represents the interaction energy between the lattice ions, T_{kinetic} is the sum of the independent-particle kinetic energies of the electrons, $V_{\mathrm{ext}}(\boldsymbol{r})$ is the external potential, which represents the electron–nuclei potential energy in the absence of externally exerted fields, E_{Hartree} is the classical Coulomb energy of the electron density $n(\boldsymbol{r})$ interacting with itself (Hartree energy):

$$E_{\mathrm{Hartree}} = \frac{1}{2} \iiint \frac{e^2 n(\boldsymbol{r}) n(\boldsymbol{r}')}{4\pi\varepsilon_0 \left|\boldsymbol{r} - \boldsymbol{r}'\right|} \mathrm{d}^3 r \mathrm{d}^3 r'. \tag{2.3}$$

In Eq. (2.2), E_{xc} is the exchange–correlation energy of the electrons in the system, which represents the sum of the electron kinetic energies and electron–electron interaction energies of the true interacting many-particle system subtracted by $T_{\mathrm{kinetic}}[n(\boldsymbol{r})]$ plus E_{Hartree}. The Kohn–Sham approach involves solving the single-particle Kohn–Sham equation of the auxiliary system, but which employs an effective single-particle interacting potential energy arising from the last three terms in Eq. (2.2) and an actually interacting electron number density $n(\boldsymbol{r})$. Because there are no accurate analytic expressions of the exchange–correlation functional, various kinds of approximations of the exchange–correlation energy functional has been taken, such as the simplest local density approximation (LDA), more accurate generalized gradient approximation (GGA) (such as PW91, PBE, LYP), and even more accurate and complex approximations such as hybrid functionals (such as HSE06, HSE03, and B3LYP) which partially employ the Fock exchange energy term of the Hartree–Fock

Hamiltonian (another scheme and starting point of ab initio calculations which takes electron wavefunctions rather than electron number density as the core quantity). Note that the LDA or GGA DFT calculations usually (not always) underestimate the bandgaps of the semiconductors or insulators. In contrast, the DFT calculations employing the hybrid exchange–correlation functionals usually give rise to more accurate bandgaps. On the other hand, different kinds of wavefunction basis sets are employed to expand the true single-electron wavefunctions (or more accurately speaking, eigenfunctions of the Kohn–Sham equations) of the auxiliary system. The popular Gaussian software package employs the Gaussian basis sets to cope with mainly molecule systems; in contrast, condensed-matter physicists usually employ the plane-wave basis sets to deal with periodic systems of crystals, such as that used in the VASP simulation package. Furthermore, pseudopotentials (norm-conserving pseudopotential, ultrasoft pseudopotential, and PAW method) are usually employed when dealing with periodic crystal systems to highly reduce the computation time. Pseudopotential is the approximate interaction potential between a valence electron and the ionic core composed of a nucleus and its tightly bound core electrons (other than more mobile valence electrons). When pseudopotentials are employed in a DFT calculation, $n(\boldsymbol{r})$ contains contributions of only the valence electrons, and $V_{\mathrm{ext}}(\boldsymbol{r})$ is the electron–ionic cores potential energy, that is, the sum of pseudopotential energies of all ionic cores of the lattice. More accurate ab initio methods such as GW and GW–BSE (including exciton effects) with considerations of many-particle effects become more and more popular with the rapid increase in computation powers of modern computers.

The electronic band structures of the four polytypes (3C, 2H, 4H, and 6H) of silicon carbide were calculated using local atomic pseudopotentials for carbon and silicon [38]. The calculated indirect bandgaps of the 3C, 2H, 4H, and 6H polytypes are 2.40, 3.35, 2.80, and 2.45 eV, respectively. The indirect bandgap increases with increasing hexagonality of the polytype and the trend is qualitatively consistent with experimental results. However, the calculated bandgaps of the 4H and 6H polytypes are significantly smaller than the reported experimental values. This is because the energy levels in 4H- and 6H-SiC were calculated without further adjustment of the pseudopotentials, unlike the cases of 3C- and 2H-SiC. The calculated direct bandgaps at Γ differ relatively slightly among the four polytypes and they are 5.14, 4.46, 4.60, and 4.40 eV for the 3C, 2H, 4H, and 6H polytypes. The energy band structure of 3C-SiC has been calculated by using the nonlocal empirical-pseudopotential method [39]. The results show that 3C-SiC is an indirect bandgap semiconductor with the valence band maximum at Γ_{15v} and conduction band minimum at X_{1c} (Fig. 2.8). The calculated threshold of the indirect transition is 2.33 eV, corresponding to the Γ_{15v}–X_{1c} transition. This agrees well with the experimental value of 2.39 eV [40]. The threshold of the direct transition is 5.90 eV, which corresponds to the Γ_{15v}–Γ_{1c} transition.

By interpreting the SiC polytypes as natural superlattices consisting of mutually twisted cubic layers, a method has been developed to calculate the electron band structure of any polytype based on an empirical pseudopotential description of cubic SiC [41]. The indirect bandgaps of all SiC polytypes with hexagonality below 50%

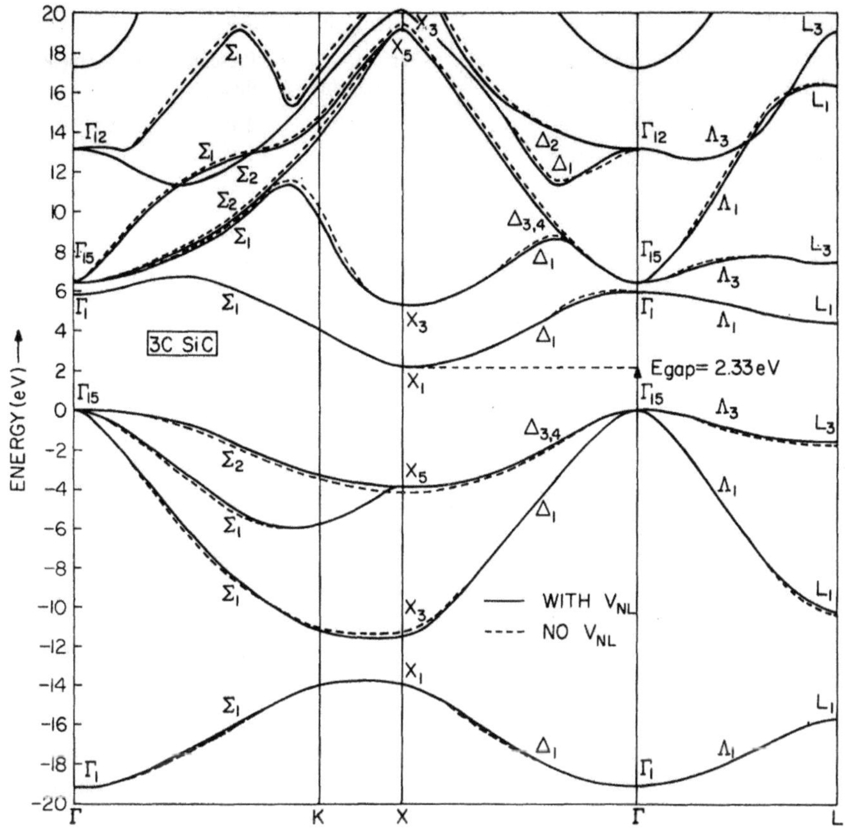

Fig. 2.8 Calculated electronic band structure of 3C-SiC with (solid line) and without (dashed line) inclusion of a nonlocal angular momentum dependent potential in the pseudopotential. Reprinted with permission from [39]. Copyright 1972, American Physical Society

were obtained. The calculated bandgaps of hexagonal and rhombohedral polytypes are in excellent agreement with the experimental data (Table 2.1). The results also reveal the nearly linear relationship between the indirect bandgap and the hexagonality, in agreement with experiments. 2H-SiC is an exception and its bandgap can be reproduced only with the proper choice of the interface position.

The electronic structure and wavefunction of 3C-SiC can be calculated using the first-principles Hartree–Fock–Slater model based on the discrete variational method [42]. The calculated first indirect bandgap is 2.5 eV, close to the experimental value of 2.39 eV. The quasiparticle energy of 3C-SiC was evaluated using the first-principles density-functional theory by adopting the local-density approximation (abbreviated as DFT–LDA) and nonlocal, norm-conserving pseudopotentials [43]. The localized Gaussian orbital basis sets were utilized. The self-energy operator was evaluated utilizing the *GW* approximation and it provided energy correction to the LDA calculation. The calculated indirect bandgap is 2.33–2.35 eV that is in good agreement with

Table 2.1 Hexagonality and energy gaps (eV) of several SiC polytypes

Polytype (Ramsdell)	Polytype ABC	Percent hexagonality	Present theory	Experiment
3C	ABC	0	2.40	2.390
10H	ABCABCBACB	20	2.66	
8H	ABCABACB	25	2.76	2.80
6H	ABCACB	33.3	3.05	3.023
15R	ABCACBCABACABCB	40	3.03	2.986
4H	ABCB	50	3.28	3.263
2H	AB	100	4.04	3.330

Reprinted with permission from [41]. Copyright 1994, American Physical Society

the experimental value. Another theoretical calculation employing the DFT–LDA gave rise to the electronic structures of 3C-, 6H-, 4H-, and 2H-SiC (Fig. 2.9) [44]. The calculated indirect bandgaps of 3C-, 6H-, 4H-, and 2H-SiC are 1.27, 1.96, 2.18, and 2.10 eV, respectively. The energy gap increases with increasing hexagonality with the exception of the 2H polytype. The DFT–LDA calculation generally underestimates the transition energies. This matches the generally accepted notion that the DFT–LDA calculation generally underestimates the fundamental bandgaps of semiconductors and insulators by 30–60%. After introduction of the quasiparticle correction (1.13 eV) to the DFT–LDA band structures, the calculated indirect bandgaps agree with the experimental result. The full-potential linearized augmented-plane-wave method in the DFT–LDA including spin–orbit interaction has been employed to calculate the bandgaps of 3C-, 2H-, 4H-, and 6H-SiC [45, 46]. The calculated bandgaps are smaller than the measured values by about 1.1 eV. The shapes of the bands are consistent with the results obtained from other types of calculations.

The self-consistent ab initio pseudopotential calculation of the electronic properties of 3C, 2H, 4H, and 6H SiC has been performed [47]. The calculated bandgaps were underestimated as well, ranging from 1.04 to 1.28 eV, as a result of the use of the LDA in the calculation. Different polytypes show similar valence band widths and they are 15.33, 15.37, 15.41, and 15.49 eV for 3C, 6H, 4H and 2H polytypes, respectively. The valence band width increases slightly with hexagonality. In contrast, the bandgap and conduction band minimum vary significantly as the hexagonality increases. The electronic excitation properties of various SiC polytypes, 6H, 4H, 2H, and 3C, have been studied by combining the first-principles pseudopotential method in the framework of DFT–LDA and proper calculation of the quasiparticle effect [48]. The quasiparticle shifts added to the density-functional eigenvalues were evaluated using a model dielectric function and an approximate treatment of the electron self-energy concerning local-field effects and dynamic screening. The calculated indirect bandgaps are in good agreement with the measured values, for instance, 2.59 and 3.25 eV for the 3C and 6H polytypes, respectively.

The self-consistent ab initio linear combination of atomic orbital (LCAO) method which includes all electrons and full potentials was used to study the electronic

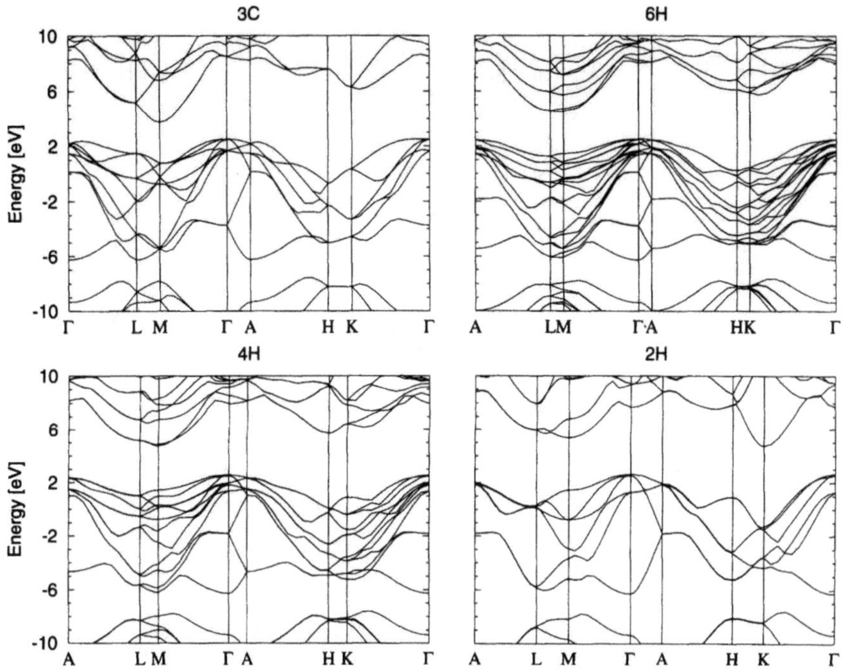

Fig. 2.9 Electronic band structures of 3C-, 6H-, 4H-, and 2H-SiC derived by the first-principles DFT–LDA. Reprinted with permission from [44]. Copyright 1994, American Physical Society

structures of 3C- and 4H-SiC [49]. The total energy calculation for 3C-SiC gave an equilibrium lattice constant of $a = 4.35$ Å, which is in agreement with the experimentally measured value of 4.348 Å. The calculation indicates charge transfer of about 1.4 electrons from a Si atom to a C atom in both 3C- and 4H-SiC. In contrast, the calculation with DFT–LDA-based orthogonalized linear combination of atomic orbitals method gave Mulliken effective charge and shows a charge transfer of the order of 0.85–0.89 electron from Si to C [50].

The dependence of the minimum bandgap and gaps at the relevant k-points on the hexagonality of the SiC polytypes have been summarized [51], as shown in Fig. 2.10. The minimum at K increases from 2H to 3C but shows some non-monotonic oscillations for the lower-hexagonality polytypes. The dashed line in this figure indicates the well-known linear relationship between the minimum indirect bandgap and hexagonality. This law applies to all hexagonalities. This is the absolute minimum gap except for the value very close to 100% hexagonality. This linear relationship and small deviations very close to 3C can be explained [41].

The first-principles pseudopotential calculation in the framework of DFT–LDA shows that the valence-band offset at the interface between SiC polytypes is rather small, with a maximum value of 0.13 eV between the extreme cases of 2H-SiC and 3C-SiC [52]. The wurtzite-SiC valence-band edge is higher in energy. Almost the whole large difference in the bandgaps between these polytypes appears as a

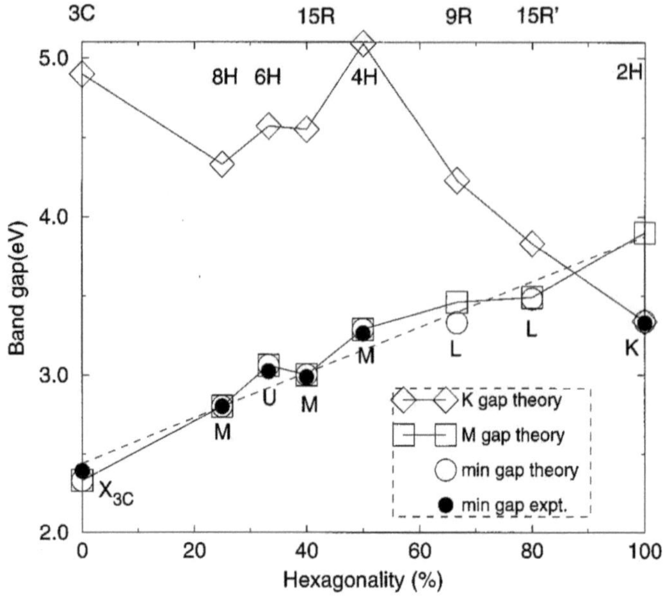

Fig. 2.10 Plot of the minimum and *k*-specific bandgaps of the SiC polytypes as a function of hexagonality. The 1-eV self-energy correction was added to all the LDA calculated gaps. The labels of the minimum gaps indicate their locations in the Brillouin zones and the M, L, U, and K signs refer to the hexagonal Brillouin zones. For 15R, the *M* point corresponds to rhombohedral X_R and for 15R′ and 9R, the hexagonal *L* point corresponds to the L_R point. Reprinted with permission from [51]. Copyright 1997, John Wiley & Sons, Inc.

conduction-band offset. The calculation predicts the crystal-field splitting of the top of the valence band of 2H-SiC to be 0.12 eV. Later, the electronic structure of SiC polytypes was studied with soft X-ray absorption and emission spectroscopy acquiring the local C *p* and Si ($s + d$) partial density of states (PDOS) of *n*-doped 3C-SiC and 4H- and 6H-SiC [53]. The shape and the energetic position of the occupied PDOS of the valence band measured by nonresonantly excited soft X-ray absorption emission spectroscopy are nearly identical for all three polytypes, reflecting the local identity of their crystalline structures. The change of the bandgap from 2.2 eV for 3C-SiC to 3.0 eV for 6H-SiC, and 3.3 eV for 4H-SiC was supposed to owe to changes of the conduction band alone, as reflected by the unoccupied PDOS measured by soft X-ray absorption.

The observation of cyclotron resonance in 6H-SiC layers by optically detected cyclotron resonance (ODCR) spectroscopy at *X*-band microwave frequency was reported [54]. The *n*-type 6H-SiC layers with residual doping concentrations of 10^{14}–10^{15} cm^{-3} were grown by chemical vapor deposition (CVD). Their effective masses were determined as $m^*_\perp = (0.42 \pm 0.02)m_0$ (m_0 is the electron mass) and $m^*_\parallel = (2.0 \pm 0.2)m_0$. The ODCR study was further performed on high-purity *n*-type 4H-SiC epitaxial layers grown by CVD at both *X* band (9.23 GHz) and *Q* band (35.05 GHz) microwave frequencies [55]. They have electron effective masses of $m_\perp^* = 0.42m_0$

and $m_{\parallel}^{*} = 0.29m_0$. A scattering time in the basal plane $\tau_{\perp} \approx 4.3 \times 10^{-11}$ s and the corresponding electron mobility $\mu_{\perp} \approx 1.8 \times 10^5$ cm^2 V^{-1} s^{-1} were obtained from a fit of the ODCR lineshape. Another ODCR measurement indicates that the 4H-SiC epitaxial films grown by CVD with n-type residual doping concentration of 2×10^{14} cm^{-3} and thickness of about 93 μm have m (L) = $(0.33 \pm 0.01)m_0$, m (G) = $(0.58 \pm 0.01)m_0$, and m (K) = $(0.31 \pm 0.01)m_0$ [56]. These values agree well with m (L) = $0.31m_0$, m (G) = $0.57m_0$, and m (K) = $0.28m_0$ derived with the DFT–LDA calculation using linearized augmented plane-wave method. The conduction-band minimum was found to be at the M point of the Brillouin zone. The ODCR at X-band frequency (9.23 GHz) was used to study the hole effective masses in the 4H SiC epitaxial films grown by CVD with an n-type residual doping concentration of 10^{14} cm^{-3} and a thickness of about 8 μm [57]. The samples were slightly compensated, with the concentration of Al acceptors of ~ 10^{13} cm^{-3}. The effective masses of the holes were determined as $m_{\perp}^{*} = (0.66 \pm 0.02)m_0$ and $m_{\parallel}^{*} = (1.75 \pm 0.02)m_0$.

2.4 Defects

There are many intrinsic defects in as-grown or irradiated silicon carbide crystals, for example, point defects and higher-dimensional defects [58]. Point defects mainly consist of silicon vacancies, carbon vacancies, divacancies, antisites, interstitial defects, nitrogen–vacancy (NV) defects, and carbon vacancy–carbon antisite pairs. The higher dimensional intrinsic defects comprise dislocations, micropipes, and stacking faults. There are other types of extrinsic defects in silicon carbide crystals associated with impurities. The presence of intrinsic and extrinsic defects in silicon carbide crystals can significantly degrade the performance of SiC-based devices. Moreover, some defects may serve as luminescence centers or nonradiative recombination centers, thereby affecting the luminescence properties of silicon carbide. Therefore, identification of these defects, knowledge of the characteristics, and finding approaches to mitigate the effects of these defects are crucial to the applications of silicon carbide to electronic devices and light emitting diodes.

2.4.1 Point Defects

Besides intrinsic point defects created during the crystal growth, the particle or ion irradiation can deliberately generate point defects. The electron spin resonance (ESR, also called electron paramagnetic resonance, EPR) spectra of n- and p-type 6H-SiC and n-type 3C-SiC single crystals irradiated at room temperature with either 0.8 or 1.0 MeV electrons or neutrons were tentatively ascribed to uncharged carbon vacancy, negatively charged carbon divacancy, and uncharged silicon vacancy trapped at a boron site [59]. The deep-level transient spectroscopy (DLTS) reveals the defect energy levels in nitrogen-doped 6H-SiC epitaxial layers irradiated with 2-MeV electrons and implanted with 300-keV deuterium or hydrogen at room temperature [60]. It was suggested that the detected E_c–0.51 eV, E_c–0.34 and E_c–0.41 eV, and E_c–0.51 eV levels originate from carbon vacancies, occupation of inequivalent lattice sites, and different charge states of carbon vacancies, respectively. The results for the E_c–0.62 eV and E_c–0.64 eV levels agree with a defect model involving a silicon vacancy on inequivalent sites in 6H-SiC.

The characterization methods for the point defects in SiC mainly contain electric, magnetic, and optical methods. Hall effect measurements and DLTS have been commonly used to determine electrical parameters of impurities or defects such as concentrations, ionization energies or electrical capture cross sections for carriers [61]. The Hall measurements determine the dopants (donors or acceptors) governing the conductivity type. The DLTS monitors the change in charge state of a deep defect in the depletion region by observation of the capacitance transient originating from the return to thermal equilibrium after a perturbation is applied to the system employing a *pn* junction or Schottky barrier. It detects the deep level impurities or defects in semiconductors.

ESR is a powerful technique to detect the point defects with unpaired electrons by measuring signals of their spin-flip induced resonance absorption of electromagnetic quanta when the sample is placed in a static magnetic field [62]. The conventional ESR has a limited sensitivity, so more sensitive optically detected magnetic resonance (ODMR) and electrically detected magnetic resonance (EDMR) techniques have been developed to detect small numbers of spins [63]. The spin state information is converted into that of photons or charges in these techniques. Optical (particularly optical excitation and deexcitation) spectroscopy techniques are very helpful for detecting the structural and electronic properties of the optically and magnetically active point defects through deciphering the light–matter interaction message. Furthermore, the photoluminescence (PL) spectroscopy can combine with theoretical calculations to offer fruitful information concerning the point defects. For example, the optical spectroscopy is capable of revealing the symmetry configurations of the color centers such as divacancy and NV pair in SiC and revealing their sensitivity to the polarization of the excitation light [64]. The quantum mechanical DFT calculations are able to simulate the PL lineshapes of various color centers (single vacancies, double vacancies, and vacancy–impurity pairs) in SiC and yield valuable information such as positions of zero-phonon lines, Huang–Rhys and Debye–Waller factors, and nature of underlying lattice vibrations [65].

2.4.1.1 Vacancies

The vacancies in silicon carbide have been very deeply and extensively studied. Many applications have been conceived. It is formidable to summarize all the works in this fast expanding area and instead, this section gives a skeleton of this field. Besides the above mentioned methods, positron annihilation spectroscopy is a very useful characterization method for the various vacancy defects. First-principles calculations

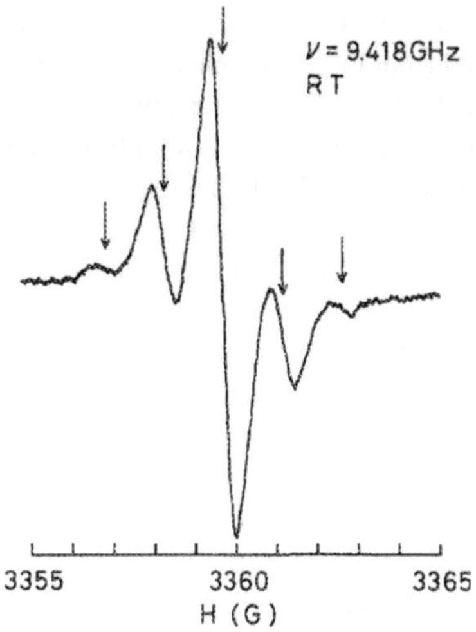

Fig. 2.11 Room temperature electron spin resonance spectrum of 3C-SiC irradiated by 1 MeV electrons with a fluence of 3 $\times$ 10^{18} cm^{-2}. The arrows indicate five lines of the $T1$ center. Reprinted with permission from [68]. Copyright 1989, American Institute of Physics

can be used to study the positron lifetime, affinity, and positron–defect binding energy for different SiC polytypes in perfect and defected states [66]. The results show that the positron lifetime is sensitive to the defect size. The positron lifetime spectroscopy shows that the as-grown *n* and *p*-type 6H-SiC crystals possess V_{Si} vacancy, $V_C V_{Si}$ divacancy, and probably V_C vacancy [67]. The silicon vacancy and carbon vacancy were annealed out in the temperature range of 400–650 °C. The $V_C V_{Si}$ divacancy persisted at an annealing temperature of 1400 °C.

Silicon Vacancies

ESR spectroscopy was used to study the electron irradiation-induced defects in 3C-SiC epitaxial films prepared by CVD [68]. An isotropic ESR center (named $T1$ center) (Fig. 2.11) which consists of five lines equally spaced at about 1.5 G and has a g value of about 2.0029 was observed. This center in electron irradiated 3C-SiC can be annealed in three stages (150, 350, and 750 °C). The $T1$ center is a point defect at the Si site that has 12 second-nearest-neighbor Si atoms. Photoluminescence spectroscopy revealed a dominant PL line at 1.913 eV, and it disappeared after annealing at 100 and 700 °C [69]. The annealing results of the PL line and fractions annealed in each annealing stage agree well with those obtained for the $T1$ ESR center, indicating that the 1.913 eV PL line stems from silicon vacancies (V_{Si}) in 3C-SiC. Positron lifetime and ESR measurements performed for 1-MeV electron-irradiated 3C-SiC identified the annihilation of positrons localized at single-negative silicon vacancies [70]. The positron lifetime at silicon vacancies was determined to be 188 $\pm$ 4 ps, which agrees well with the theoretical positron lifetime [66].

Deep level defects in electron-irradiated 4H-SiC epitaxial layers grown by CVD were studied using DLTS [71]. The measurements performed on electron-irradiated

p^+n junctions within 100–750 K revealed several electron traps and one hole trap with thermal ionization energies of 0.35–1.65 eV. Most of these defects were observed at a dose of irradiation as low as 5×10^{13} cm^{-2}. The capture cross sections of these two defects are temperature independent. The negatively charged Si vacancy in neutron irradiated 4H-SiC has been identified by using EPR and electron nuclear double resonance spectroscopy [72]. The magnetic resonance parameters of the negatively charged Si vacancy are nearly the same for 3C-, 4H-, and 6H-SiC. The ligand hyperfine interactions for the nearest- and next-nearest-neighbor shell of the Si vacancy in 3C-SiC were calculated based on the DFT–LDA method. The calculated result agrees well with the experimental data, thus supporting the assignment of the defect center. Determination of the spin state ($S = 3/2$) also helps to identify the defect. The silicon vacancies (V_{Si}) in 3C- and 2H-SiC in all possible charge states were studied using spin-polarized ab initio local spin-density calculation [73]. For the neutral, single, and double negative charge states, the exchange coupling overcomes the Jahn–Teller energies, and the silicon vacancy was found in the high-spin ground state with T_{d} symmetry. A weak Jahn–Teller effect occurs in the low-spin configuration. The irradiation induced PL spectrum acquired from the electron irradiated 4H- and 6H-SiC epitaxial layers and corresponding optically detected magnetic resonance signals from this band have been investigated [74]. The number of no-phonon lines in the deep PL band is equal to the number of inequivalent sites in the respective polytype. These lines are at 1352 meV ($V2$) and 1438 meV ($V1$) for 4H-SiC and at 1366 meV ($V3$), 1398 meV ($V2$), and 1433 meV ($V1$) for 6H-SiC (Fig. 2.12). The intensity of the PL lines decreased remarkably after short annealing at 750 °C. Electron spin resonance measurement combined with theoretical calculation suggests that the no-phonon lines in the PL originate from internal transitions from an excited state of the neutral Si vacancy at the inequivalent sites in 4H- and 6H-SiC. Moreover, comparison of the characteristics of the triplet states between the polytypes indicates that the $V2$ lines correspond to the hexagonal sites and $V1$ and $V3$ to the cubic sites. The electronic structure of the ground state and lowest-lying excited states of the neutral silicon vacancy in 4H- and 6H-SiC was studied by using high-resolution PL, PL excitation, and Zeeman spectroscopy [75]. The ground and excited states responsible for the defect luminescence with no-phonon lines at 1.438 and 1.352 eV in 4H-SiC and 1.433, 1.398, and 1.368 eV in 6H-SiC were supposed to be singlets. The no-phonon lines show different polarization behaviors for the inequivalent sites. An additional fine structure of the no-phonon lines was observed from the strained samples. The T_{V2a} center, which was previously suggested to be the excited triplet state of the neutral silicon vacancy related defect [74] in the electron-irradiated n-type 4H-SiC was studied by using continuous and pulsed EPR [76]. The spin multiplicity of T_{V2a} was determined to be quartet ($S = 3/2$). Based on the ^{13}C hyperfine spectrum of the nearest-neighbor carbon atoms, it was proposed that the T_{V2a} center is a negatively charged silicon vacancy of $C_{3\text{v}}$ symmetry with the crystal field distorted slightly from regular tetrahedron. The triply degenerate t_2 state of an electronic configuration $a_1^2 t_2^3$ under T_{d} symmetry splits into a_1 and e by the distortion to C_{3v}. The high spin configuration [$a_1 e^2$ or $e^2 a_1$] that reduces the electron repulsion energy is more preferable than the low spin configuration expected from the symmetry-lowering crystal field

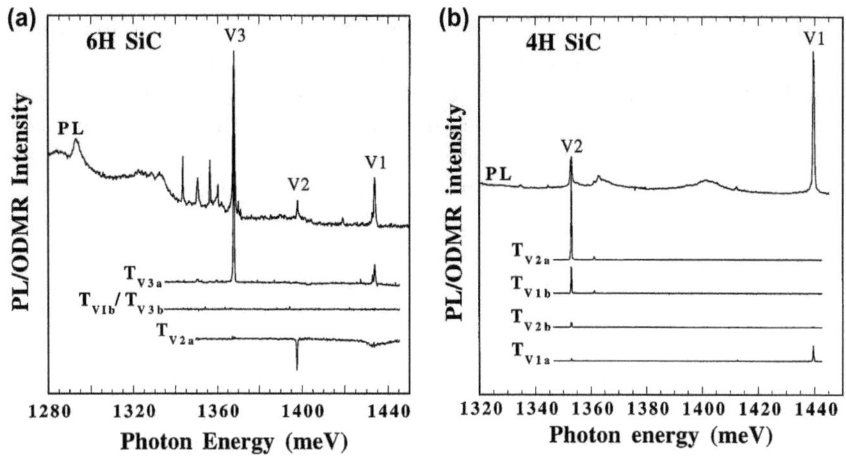

Fig. 2.12 Deep PL band (top) and excitation spectrum (below) of the various optically detected magnetic resonance signals that were detected from the PL band for the **a** 6H-SiC and **b** 4H-SiC. The samples were 2.5 MeV electron-irradiated epitaxial layers prepared by CVD. Reprinted with permission from [74]. Copyright 2000, American Physics Society

alone. The ^{13}C hyperfine spectrum of the nearest-neighbor carbon atoms shows that there are two kinds of V_{Si}^-, which were separately assigned to arise from hexagonal site (h) and quasicubic sites (k in 4H-SiC, k_1 and k_2 in 6H-SiC) [77]. About the assignment of V_{Si}, a proposed model explains the optical transitions and magnetic resonance signals as occurring in the singly negative charge state of the silicon vacancy in 4H- and 6H-SiC [78]. Among two kinds of V_{Si}^-, one type (T_{V2} center) in SiC shows a longer coherence time than the NV center in diamond even though the natural abundance of ^{29}Si is higher than ^{13}C [79]. The multiparticle symmetry-adapted wavefunctions of the negatively charged silicon vacancy defect in hexagonal 4H-SiC were calculated by using group theory and DFT method [80]. The spin–orbit and spin–spin interactions show influence on these states. A later study combining the first-principles calculations and high resolution ESR measurements reliably identifies the Si vacancy related qubits in hexagonal SiC as isolated negatively charged silicon vacancies [81]. The study shows that the $V1$ emission line for the negatively charged silicon vacancy center in 4H-SiC features two well-distinguishable zero-phonon lines at around 1.445 and 1.440 eV [82]. This center combines 40% optical emission into the zero-phonon lines, showing difference in optical properties with varying temperature and polarization, 100% increase in luminescence intensity upon the spin resonance, and long spin coherence time of their spin-3/2 ground states up to 0.6 ms. The negatively charged silicon vacancy centers in 4H-SiC suffer from neither large electron–phonon interaction nor fast spin dephasing [83]. Their $^4A_2$ symmetry in ground and excited states renders optical resonances stable. This in combination with millisecond-long spin coherence times indicates high-fidelity optical initialization and coherent spin control.

The defects in solids can be ideal centers for generating single photon emissions, which have wide applications in quantum communication and quantum computation. This has been demonstrated by the studies of the NV defect centers in diamond and nanodiamond [84]. They showed the virtues of good localization and anti-photobleaching [85, 86]. SiC as a group IV material has excellent characteristics [84] which are similar and even superior in some aspects to diamond and thus can be utilized in single-photon sources. The Si vacancy in SiC has been investigated as a promising quantum system for single-defect and single-photon spectroscopy in the infrared region, by using EPR, ODMR, and high-resolution PL excitation spectroscopy [87]. The PL spectra of Si vacancies acquired from 6H- and 4H-SiC are shown in Fig. 2.13. They comprise zero-phonon lines (no-phonon lines) accompanied by phonon replicas. The number of zero-phonon lines generally agrees with that of the inequivalent lattice sites in the SiC polytypes. These lines arise at 865 nm ($V1$), 887 nm ($V2$), and 906 nm ($V3$) in 6H-SiC and at 862 nm ($V1$) and 917 nm ($V2$) in 4H-SiC. A change of a factor 2–3 was observed in the PL intensity of the zero-phonon lines in zero magnetic field upon absorption of radio waves with energy equal to the fine-structure splitting of spin sublevels of the vacancy ground state. The silicon vacancy related defects in electron-irradiated epitaxial SiC layers showed monoexponential PL decay with a lifetime of 6.1 ns at room temperature [88]. The integrated PL intensity depends linear on the excitation power density up to 20 kW cm^{-2}, suggesting a relatively small absorption cross section of the defects. The optically induced inverse population of the ground-state spin sublevels of the silicon vacancies was realized in 4H- and 6H-SiC upon irradiation with unpolarized light at low temperature and room temperature [89]. The probed silicon vacancy spin ensemble can be prepared in a coherent superposition of the spin states. The Rabi nutation persisted for over 80 μs. Two opposite schemes of optical alignment of the populations between the ground-state spin sublevels of V_{Si} upon illumination with unpolarized light were realized in 4H- and 6H-SiC at room temperature. The silicon vacancy qubits can be selectively addressed and manipulated under the double radio–optical resonance conditions [90]. Their long spin memory was revealed using pulsed magnetic resonance technique.

The study shows that 4H-SiC can host single silicon vacancies with long spin coherence times at room temperature, rendering it promising for long-lived qubits [91]. The single T_{V2} defect centers show narrow PL line ($\ll 1$ nm) at low temperature and wide PL spectrum at room temperature (Fig. 2.14a), similar to the characteristic luminescence of a silicon vacancy defect ensemble at room temperature. The autocorrelation measurement of the photon intensity revealed that at zero delay, the second-order correlation function $g^{(2)}$ was well below 0.5 (Fig. 2.14b), characteristic of single-defect emission. The PL of the single T_{V2} was stable; even 1 mW laser illumination over months did not alter the luminescence rate. The defect PL exhibited no blinking (Fig. 2.14d).

The second-order correlation function is defined as:

$$g^{(2)}(\tau) = \frac{\langle n_1(t) n_2(t+\tau)\rangle}{\langle n_1(t)\rangle\langle n_2(t+\tau)\rangle}, \tag{2.4}$$

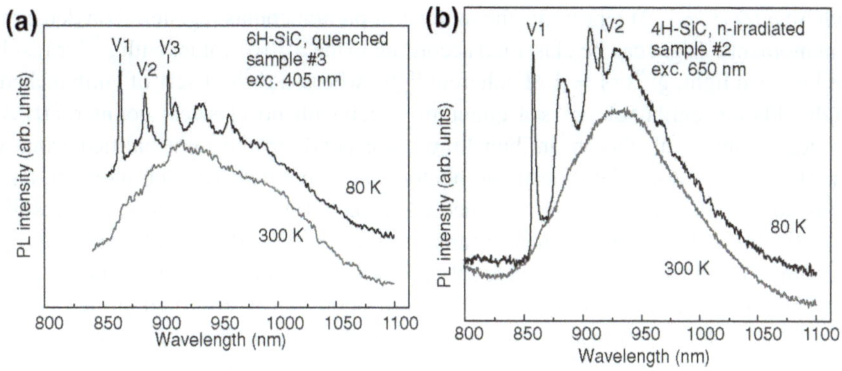

Fig. 2.13 **a** PL spectra acquired from neutron irradiated 6H-SiC at 80 K and room temperature excited by 405 nm light. **b** PL spectra of neutron irradiated 4H-SiC at 80 K and room temperature excited by 650 nm light. Reprinted with permission from [87]. Copyright 2011, American Physical Society

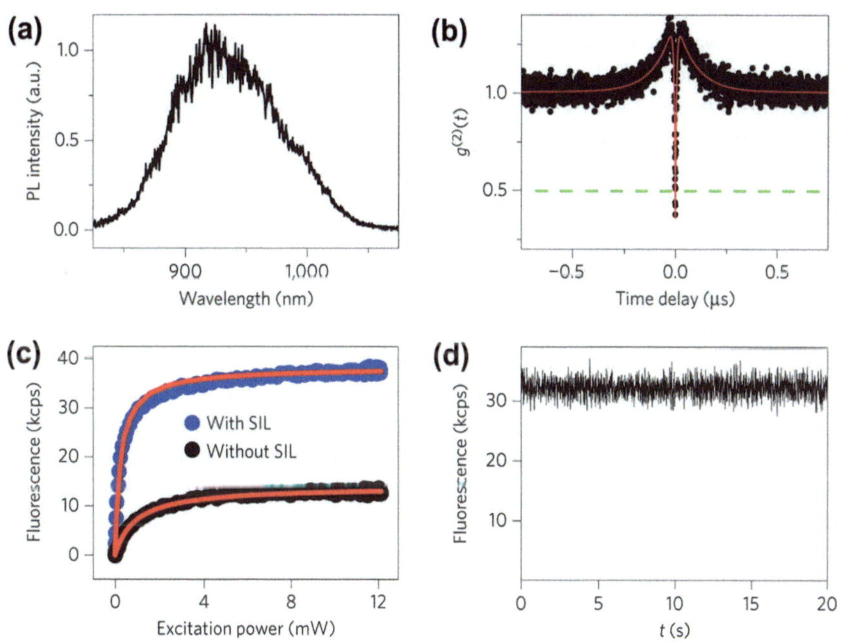

Fig. 2.14 **a** PL spectrum of a single T_{V2} center with 1 mW 730 nm laser excitation. **b** Autocorrelation measurement of a single T_{V2} center measured at 0.1 mW optical power. The red curve is a fit based on the three-states model and the green line indicates $g^{(2)} = 0$. **c** Saturation curves of single T_{V2} centers with and without a solid immersion lens (SIL). Red curves: fits based on the three-states model. **d** Time trace of PL intensity of a single T_{V2} center with 10 ms bins. Reprinted with permission from [91]. Copyright 2015, Springer Nature Limited

where $n_i(t)$ $(i = 1, 2)$ represents the number of photon counts registered on detector i at moment t. Light can be classified according to the $g^{(2)}(0)$ value, with $g^{(2)}(\tau) > 1$ of bunched light, $g^{(2)}(\tau) = 1$ of coherent light, whereas $g^{(2)}(\tau) < 1$ of antibunched light. Photon antibunching is a quantum effect with no classical counterpart. A typical example of photon antibunching corresponds to the light emitted from a single quantum dot, where only one photon can be emitted out each moment, and thus no two photons can be simultaneously detected by two detectors, as a result, $g^{(2)}(0) \approx 0$. Therefore, the measurement of the second-order correlation function becomes a standard approach to identify the isolated luminescent quantum dot.

There have been investigations on the spin–acoustic (or strain) coupling in SiC. The effect of temperature and strain on the optical and spin properties of the silicon-vacancy qubits V_1 and V_2 in 4H-SiC was studied by using DFT calculations [92]. The exhibited polaronic gap of around 5 and 22 meV for the V_1 and V_2 centers caused difference in the temperature-dependent dephasing and zero-field splitting of the excited states. The influence of the crystal deformations on the zero-phonon line of these emitters was discussed. The acoustically driven spin resonances in V_{Si} centers in 4H-SiC were implemented at room temperature [93]. A surface acoustic wave cavity was used to selectively address spin transitions with magnetic quantum number difference of $\pm$ 1 and $\pm$ 2 in the absence of external microwave electromagnetic fields. The spin–acoustic resonances showed nontrivial dependence on the static magnetic field orientation owing to the intrinsic symmetry of the acoustic fields and the peculiar properties of a half-integer spin system. The dynamic strain of surface acoustic waves can efficiently excite magnetic resonances of V_1 centers in 4H-SiC up to room temperature [94]. Based on the width and temperature dependence of the acoustically induced spin resonances of the V_1 centers, they were attributed to transitions between spin sublevels in the excited states. The spin–acoustic control of naturally occurring negatively charged silicon vacancies in a lateral overtone bulk acoustic resonator based on 4H-SiC was demonstrated [95]. The acoustic driving can be used at room temperature to induce coherent population oscillations. The spin–acoustic resonance proved useful as a frequency-tunable probe of bulk acoustic wave resonances, showing the dynamical strain distribution inside a bulk acoustic wave resonator under ambient conditions. The spin–strain interaction characterized by the defect spin-3/2 quadrupole components in SiC coupled to strain fields was theoretically investigated [96]. With consideration of the C_{3v} symmetry of the vacancy site beyond the spherical approximation, the effect of a deviation from the spherical symmetry on spin resonance transition rate was clarified. The ratios of spin–strain coupling parameters were evaluated from the anisotropic field-direction dependence of the transition rate.

The neutron irradiation (0.18–2.5 MeV) of a high-quality 110 mm thick 4H-SiC epitaxial layer in a fission reactor helps to realize control of the V_{Si} density over eight orders of magnitude, from below 10^9 to above 10^{16} cm^{-3} within an accuracy down to a single vacancy level [97]. The defect PL showed no bleaching even after 10^{14} excitation cycles. The vacancy spins can be manipulated using an ODMR technique. The V_1/V_2-type silicon vacancies were controllably generated in 4H-SiC through successive irradiation by 1 and 2 MeV protons without need for pre- or post-irradiation treatment [98]. There was strong increase of near-infrared emission in the irradiation areas (Fig. 2.15). The generation depth and resolution can be predicted by matching the proton energy to the material's stopping power, and the amount of V_{Si} at one specific sample volume is tunable from ensembles of millions to discernible single photon emitters. The generated silicon vacancies showed a long spin–echo coherence time of 42 ± 20 μs at room temperature. The room temperature ODMR and spin echo measurements were performed on an ensemble of vacancies [99]. The spin echo properties depend strongly on magnetic field. The spin echo decay time varied from less than 10 μs at low fields to 80 μs at 68 mT. A strong field-dependent spin echo modulation was observed and attributed to the interaction with nuclear spins. A study shows that 4H-SiC with natural isotope abundance can preserve a coherent spin superposition in silicon vacancies over an unexpectedly long time exceeding 20 ms [100]. The spin-locked subspace with a drastically reduced decoherence rate was attained through suppression of heteronuclear spin crosstalking by applying a moderate magnetic field in combination with dynamic decoupling from the nuclear spin baths. Several phonon-assisted mechanisms of spin–lattice relaxation were identified. The relaxation can be even longer than 10 s at cryogenic temperatures. Molecular dynamics simulations by considering the ionization energy loss for swift ion implantation were used to study the defect distribution and transformation at atomic-scale during hydrogen ion implantation and post-annealing [101]. The results show that low-fluence C pre-implantation is helpful to improve the color center yield for the (carbon and proton) dual-ions implantation. Three-dimensional proton beam writing of optically active coherent vacancy spins in 4H-SiC was implemented through irradiation of the substrate with 2 MeV electron beams [102]. Subsequent lithographic process created 800 nm tall nanopillars with 400–1400 nm diameters. High collection efficiency of up to 22 kilo counts per second from a single silicon vacancy center was realized, for which the single photon emission and optically induced electron-spin polarization properties was preserved. A single V_{Si}-defect emitter array in 4H-SiC was realized through 30-keV carbon ion implantation [103]. V_{Si} defects were generated in predetermined locations. Maskless and targeted generation of single silicon vacancy arrays in 4H-SiC can be realized by controlling the focused He^+ ion beam (implantation energy: 30 keV) with an implantation uncertainty of 60 nm [104]. After optimization, a conversion yield of ~ 6.95% and a generation rate for a single silicon vacancy of ~ 35% were realized.

The silicon vacancies show promising applications in magnetometry. The fine structure of a silicon vacancy in isotopically purified 4H-^{28}SiC and previously unconsidered terms in the spin Hamiltonian originating from the trigonal pyramidal symmetry of this spin-3/2 color center were revealed [105]. These terms give rise to additional spin transitions that lead to a level anticrossing in an external magnetic field. The sharp variation of the PL intensity in the vicinity of this level anticrossing can be used for an all-optical sensing of the magnetic field. The achieved dc magnetic field sensitivity is better than $100\,\mathrm{nT}/\sqrt{\mathrm{Hz}}$ within a volume of 3×10^{-7} mm^3 at room temperature. This contactless method is robust at high temperatures up to 500 K. A compact fiber-integrated SiC silicon-vacancy-based magnetometer at room temperature was demonstrated [106]. The silicon vacancy in a tiny SiC slice was coupled with

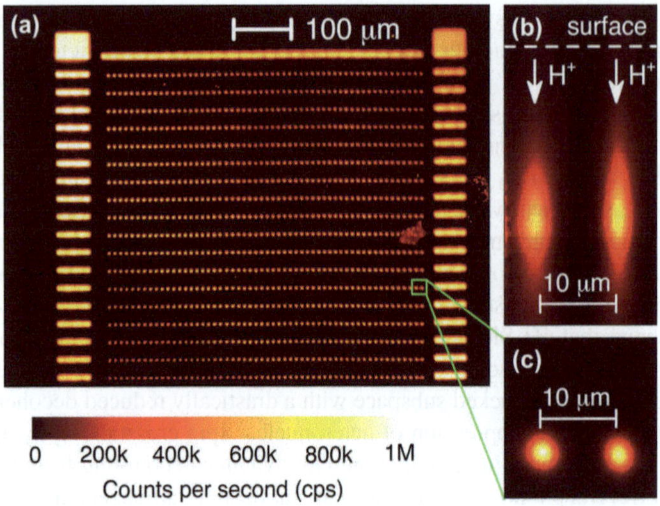

Fig. 2.15 3D proton beam writing of silicon vacancies in 4H-SiC. **a** *XY* confocal PL map of the pattern created using 1.7 MeV protons. The count rate is color-coded using logarithmic scale. **b** Zoom-in of a *XZ* confocal map along the depth direction. **c** Zoom into the *XY* confocal PL map showing two irradiated sites. The color scale codes the count rate in (**b, c**). Reprinted with permission from [98]. Copyright 2017, American Chemical Society

an optical fiber tip, and the readout of the spin signal through the fiber was realized simultaneously. The optimized magnetic field sensitivity reached 12.3 100 $\mu T/\sqrt{\text{Hz}}$. The magnetometer was used to measure the strength and polar angle of an external magnetic field. The optical and spin properties of implanted silicon vacancies in 4H-SiC exhibited single-axis and temperature-independent zero-field splitting [107]. Using this technique, the $Nd_2Fe_{14}B$ showed magnetic phase transition at about 7 GPa. The critical temperature–pressure phase diagram of the superconductor $YBa_2Cu_3O_{6.6}$ was mapped.

The first-principles calculations predict high-spin configurations of the negatively charged V_{Si} defects in 3C-SiC [108]. The coupling is ferromagnetic between the V_{Si} defects at − 2e charge state, whereas the V_{Si} defects at − e charge state prefers to interact antiferromagnetically. Substitution of C with N atoms manipulates the charge states V_{Si} defects and the magnetic interactions between them. The study shows that the spins of a family of homotypic silicon vacancy defects in 6H-SiC can be initialized and read out at room temperature by means of ODMR [109]. The detected two-quantum spin resonances provide strong evidence for the $S = 3/2$ ground state of the silicon vacancies. The optically induced population inversion of these high-spin ground states leads to stimulated microwave emission. This property can be used to realize solid-state masers and extraordinarily sensitive radiofrequency amplifiers. A giant thermal shift of 2.1 MHz K^{-1} related to the excited-state zero-field splitting in the silicon vacancy centers in 4H-SiC was observed [110]. Relative variations of the zero-field splitting for small temperature differences can be detected without application of radiofrequency fields, by simply monitoring the PL intensity in the vicinity of the level anticrossing. This results in an all-optical thermometry technique with sensitivity of 100 mK $\sqrt{\text{Hz}}^{-1}$ for a detection volume of 10^{-6} mm^3. The zero-field splitting in the ground state did not show detectable temperature shift. The waveguide-integrated V_{Si} centers in 4H-SiC maintained spin–optical coherences, enabling nuclear high-fidelity spin qubit operations [111]. Nearly lifetime-limited luminescence and high spin-coherence times for single defects implanted in bulk as well as in nanophotonic waveguides created by reactive ion etching were observed. The high spin–optical coherences of V_{Si} centers in waveguides were used to demonstrate controlled operations on nearby nuclear spin qubits. The entanglement between a silicon vacancy defect in 4H-SiC and a scattered single photon in the zero-phonon line was realized [112]. The spin state was measured by detecting photons scattered in the phonon sideband. The photonic qubit was encoded in the time-bin degree of freedom and measured using an unbalanced Mach–Zehnder interferometer.

There have been also investigations on the silicon vacancies in nanocrystalline SiC. A starting point can be SiC crystals irradiated with neutrons to generate silicon vacancies responsible for the NIR emission; these crystals were milled to give rise to small nanocrystals with sizes as small as 10 nm [113]. The result shows that the silicon vacancies remained thermodynamically stable even in very small SiC nanocrystals. The ODMR on vacancy-related defects in SiC nanocrystals was demonstrated at ambient conditions. The pulsed laser ablation of electron-irradiated bulk 4H-SiC produced 40–50 nm diameter SiC nanoparticles that exhibited PL at 850–950 nm [114]. This PL was attributed to the silicon vacancy centers. The original silicon vacancy centers from the target sample retained in the final colloidal nanoparticles. The V_{Si} from the 6H-SiC micro- and nanoparticles ranging from 100 nm to 5 μm in size exhibited cathodoluminescence [115]. The emitters localized in the particle center and at the edges showed clear shifts in the emission wavelength. By comparing spatial cathodoluminescence maps with strain analysis carried out in transmission electron microscopy, the emission shifts were attributed to compressive strain of 2–3% along the a-axis of the hexagonal structure of the particle. The DFT calculation demonstrates site dependence of properties of bright negatively charged silicon vacancies within a 2H-SiC nanowire [116]. The optical properties of defects show strong dependence on the hybridization of the defect states with the surface states and on the structural changes allowed by proximity to the surfaces. The charge-state conversion and/or migration to thermodynamically favorable undercoordinated surface sites can deteriorate deep-defect properties.

Carbon Vacancies

Positron lifetime measurement and Doppler broadening spectroscopy have been employed to study the vacancies in 6H-SiC formed by irradiation of 2.2- and 10-MeV electrons [117]. Carbon vacancies showed a positron lifetime of 160 ps, 15 ps longer than the bulk lifetime, and the Doppler-broadening S parameter is close to that of bulk. The silicon vacancies showed a positron lifetime of 260 ps and the S parameter is 1.12

times the bulk value. The defects in *n*-type 6H-SiC are neutral and no vacancies were detected in *p*-type materials. Two negative-*U* centers have been observed from 4H-SiC by using capacitance transient techniques [118]. Negative-*U* center refers to the defect which can trap two electrons and there is an effective negative attraction interaction between these two electrons (negative *U*) owing to the compensating energy generated by large lattice relaxation [119]. Each of the observed negative-*U* centers generated two levels in the bandgap, one shallower donor level, and one acceptor level, and the electron ionization energy of the acceptor level is larger than that of the donor level. The two-electron emissions from the two acceptor levels caused the previously reported DLTS peak related to the Z_1 center. The neutral and charged Si and C vacancies in 3C- and 4H-SiC were theoretically studied by using DFT–LDA (or local spin density approximation, LSDA) calculations [120]. A plane-wave-supercell approach was combined with the ultrasoft Vanderbilt pseudopotentials to ensure converged calculation. Generation of the C-site vacancy is generally accompanied by a remarkable Jahn–Teller distortion, whereas only an outward breathing relaxation occurs at the Si site vacancy due to strong localization of the neighboring C dangling bonds. As a result, Si vacancies show high spin configurations and C vacancies show low spin configurations with a negative-*U* behavior. The schematic diagrams of the energy levels of the vacancies are depicted in Figs. 2.16 and 2.17. The defect energy levels in nitrogen-doped 6H-SiC epitaxial layers prepared by CVD and irradiated with 2 MeV electrons and implanted with 300 keV deuterium or hydrogen were investigated by using DLTS at room temperature [121]. Five levels within the energy gap were observed and it was proposed that the E_c–0.51 eV level (E_c referring to the conduction band minimum energy) originates from a carbon vacancy. The two levels at E_c–0.34 and E_c–0.41 eV originate from the occupation of nonequivalent lattice sites. The results for the E_c–0.62 eV and E_c–0.64 eV levels support a defect model involving a silicon vacancy on the nonequivalent sites in the 6H lattice. The intrinsic defects in *p*-type 6H-SiC:Al generated by electron irradiation at 300 keV were studied by using EPR spectroscopy [122]. Two dominant irradiation-induced paramagnetic defects were observed. The silicon-vacancy-related spin $S = 3/2$ defect with a zero-field splitting of 68.7×10^{-4} cm^{-1} was attributed to a Si Frenkel pair aligned parallel to the *c* axis. There is another carbon-vacancy-related spin $S = 1/2$ defect attributed to $V_C{}^+$. The higher electron energy (350 keV) created an additional lower-symmetry $S = 3/2$ spectrum attributed to Si Frenkel pairs, with the interstitial being located at six equivalent off-axis sites. High-energy irradiation (2 MeV) can create only isolated Si vacancies with no zero-field splitting. An EPR spectrum was measured from electron-irradiated *p*-type 4H- and 6H-SiC at a temperature above 25 K [123]. This defect center has the C_{3V} symmetry, with an electron spin of 1/2, and its configuration is depicted in Fig. 2.18. The detailed hyperfine structure due to its interaction with the four nearest silicon neighbors was obtained by using high frequency EPR spectroscopy. In this way, the defect was identified as the positively charged carbon vacancy ($V_C{}^+$). Its *g* values and hyperfine tensor are almost the same in both polytypes and they are independent of the inequivalent lattice sites. Photoexcitation EPR studies were performed on *p*-type 4H-SiC irradiated with 2.5 MeV electrons [124]. At *W*-band frequencies (~ 95 GHz) different EPR spectra could be separated to allow for determination of the ground-state levels of the defects. The results obtained for the positively charged carbon vacancy (V_C^+) were explained by a deep donor model with the (+/0) level located at (1.47+/−0.06) eV above the valence band. First-principles calculation carried out on models for the Z_1/Z_2 defects in 4H-SiC shows that a π-bonded dicarbon interstitial complex next to a nitrogen atom, which is ultra thermally stable and has negative-U character with donor and acceptor levels close to that of the defect, is the most likely candidate of the Z_1/Z_2 defect [125]. EPR and electron-spin-echo studies show that isolated V_{Si}^-, V_{Si}^0, and V_C vacancies are dominant intrinsic paramagnetic defects in 4H-SiC and 6H-SiC treated by neutron irradiation with doses up to 10^{19} cm^{-2} [126]. The 95-GHz EPR spectra at 1.2 K show that the ground state of V_{Si}^0 corresponds to $S = 1$ and that the zero-field splitting parameter D is positive. An anisotropic splitting of the EPR lines of V_{Si}^- in 4H-SiC was observed, which was assumed to stem from small difference in the g tensor of the quasicubic (k) and hexagonal (h) sites. Anisotropic EPR spectra with $S = 1/2$ related to the carbon vacancy were also observed in the neutron-irradiated SiC crystals. The complete set of ^{29}Si hyperfine data reveals that both the EI5 and EI6 centers in radiation-damaged and semi-insulating 4H-SiC should originate from V_C^+ [127]. The locations are quasicubic sites for EI5 and hexagonal sites for EI6. The two types of V_C^+ centers showed remarkable difference in their atomic structures and in the temperature dependence of hyperfine interactions, which depend on the site. The negatively charged carbon vacancies V_C^- were observed in n-type 4H-SiC by using EPR and photoinduced EPR [128]. This EPR center (HEI1) was characterized by an electron spin of 1/2 in a Si–Si antibonding state of V_C. First-principles calculations confirm that the HEI1 center arises from V_C^- at hexagonal sites. The HEI1 spectrum exhibited a transition between C_{1h} and C_{3v} symmetries due to a fast reorientation effect. The photo-EPR data suggest that V_C^{2-} is the dominant form of V_C when the Fermi level lies 1.1 eV below the conduction band. The deep levels in n-type 4H-SiC epilayers fabricated by CVD were studied by using deep level transient spectroscopy [129]. The Z_1/Z_2 and EH_6/EH_7 centers are dominant in the as-grown samples. Their concentrations increased significantly after electron irradiation at 116 keV, by which only carbon atoms may be displaced. They increased with 0.7 power of the electron fluence of irradiation and their concentrations decreased by annealing at 1600–1700 °C. The concentrations of the Z_1/Z_2 and EH_6/EH_7 centers are very close to each other in all kinds of samples (as-grown, as-irradiated, and annealed ones). It has been proposed that both the Z_1/Z_2 and EH_6/EH_7 centers microscopically contain the same defect, possibly the carbon vacancy. Thermal oxidation proved to be able to efficiently eliminate the Z_1/Z_2 and EH_6/EH_7 defects in 4H-SiC [130, 131]. The Z_1/Z_2 and EH_6/EH_7 concentrations were reduced from the initial 10^{12}–10^{13} cm^{-3} to below the detection limit of 1×10^{11} cm^{-3} after thermal oxidation at 1150–1300 °C. However, oxidation produced a high concentration (5–8 × 10^{12} cm^{-3}) of another HK_0 center ($E_v + 0.78$ eV). Further annealing in argon at 1550 °C eliminated this HK_0 center. Therefore, all the major deep defect levels can be eliminated by this two-step thermal treatment and the carrier lifetime in a 98 μm thick n-type 4H-SiC epilayers can be improved from 0.64 to 4.52 μs afterward.

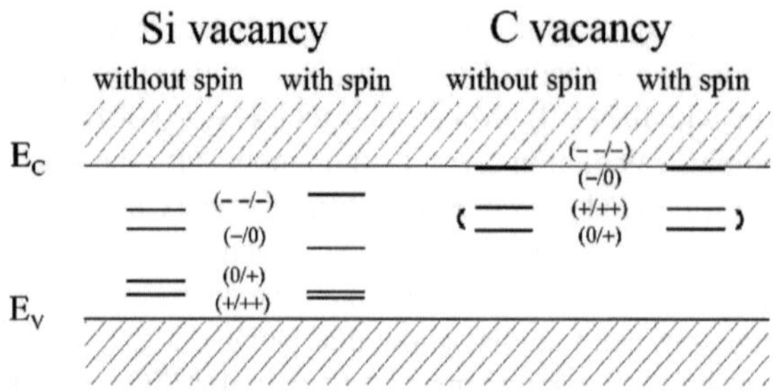

Fig. 2.16 Schematic of the energy levels of the Si and C vacancies at cubic and hexagonal sites in 3C-SiC obtained by local density approximation or local spin density approximation calculation using the 216-atom supercell. Reprinted with permission from [120]. Copyright 1999, American Physical Society

Fig. 2.17 Schematic of the energy levels of the Si and C vacancies in 4H-SiC obtained by local density approximation or local spin density approximation calculation employing the 128-atom supercell. Reprinted with permission from [120]. Copyright 1999, American Physical Society

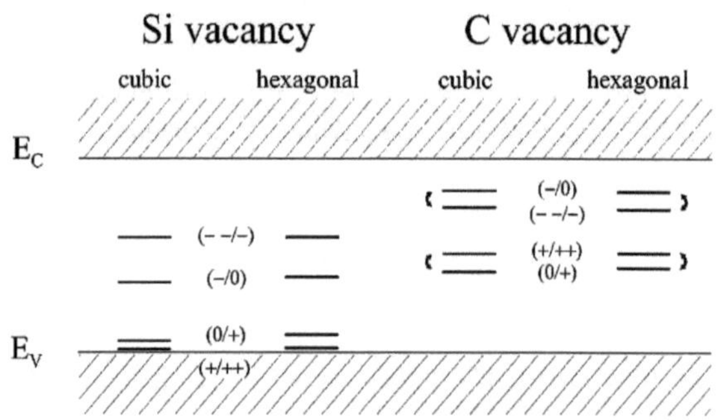

Fig. 2.18 Schematic of an isolated defect ($V_C{}^+$) in SiC with the magnetic field parallel to the *c*-axis. The expected symmetries of the hyperfine tensors representing the interaction with the Si (1) and Si (2, 3, 4) are respectively C_{3V} and C_{1h}. Reprinted with permission from [123]. Copyright 2001, American Physical Society

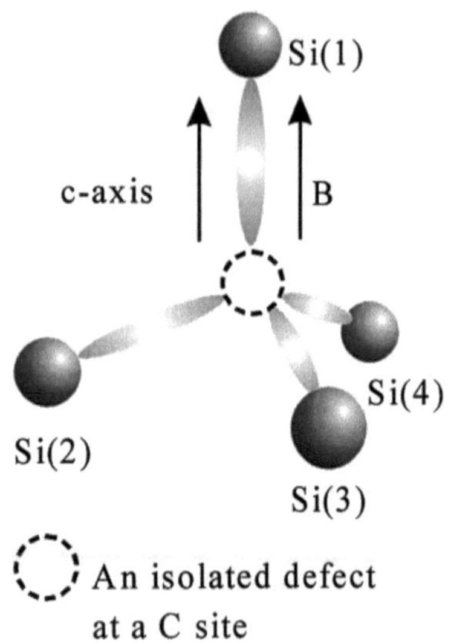

The first-principles G_0W_0–BSE calculations of the absorption spectra show that electron correlations control the photophysics of carbon vacancy in 4H-SiC through both renormalization of the quasiparticle band structure and excitonic effects [132]. Two possible excitation channels for the carbon vacancy that involve conduction and valence bands were considered. Corrections to the Kohn–Sham ionization levels are sensitive to the defect charge state. The BSE excitonic effect leads to red shift of the rapid phase approximation spectrum by 0.23 eV. The EPR spectroscopy determined the energy levels of the carbon vacancy in 4H-SiC and its negative-U properties [133]. The combined EPR and DLTS show that the two most common defects in as-grown 4H-SiC, that is, the $Z_{1/2}$ lifetime-limiting defect and the EH_7 deep defect, are associated with the double acceptor (2−|0) and single donor (0|+) levels of V_C, respectively. The migration mechanism of the carbon vacancy in 3C-SiC and 4H-SiC was investigated using combined theoretical and experimental approaches [134]. The combined nudged elastic band and dimer methods were employed to identify the migration mechanisms, transition state geometries, and energy barriers for V_C migration. In 3C-SiC, V_C migrated with an activation energy of 4.0 eV. In 4H-SiC, the V_C migration is anisotropic and basal-plane selective. Anisotropic migration of V_C in 4H-SiC was demonstrated by using DLTS depth profiling of the $Z_{1/2}$ electron trap in annealed ion-implanted samples. Activation energies of 4.4 ± 0.3 eV and 3.6 ± 0.3 eV were observed for V_C migration along the c and a orientations, respectively.

Divacancies

The exploration of SiC divacancy defects started from the research of irradiation effects on SiC crystals. Many works contribute to the identification of the SiC divacancy defects by employing optical and magnetic resonance methods. Radiation damage caused by the implantation of 200 keV Ge^+ ions into 6H-SiC was studied by using monoenergetic positron Doppler broadening and lifetime techniques [135]. Positron lifetime measurements in combination with theoretical calculations suggest that the main defect produced is the divacancy, although there are also Si monovacancies. In the amorphous surface layer larger agglomerates consisting of at least four but more probably six vacancies were detected. First-principles DFT–LDA (or LSDA) calculation indicates that there is a negative-U behavior at $E_v + 0.7$ eV between the charge states 1+/1− only for nearest-neighbor divacancies on different lattice sites in 4H-SiC, but not for all the other cases in 4H-SiC or for the cubic divacancy in 3C-SiC [136]. Vacancies in electron irradiated 6H-SiC were identified using positron lifetimes and coincident Doppler broadening studies of the positron–electron annihilation radiation in combination with theoretical studies [137]. After 0.3 MeV electron irradiation, carbon vacancies V_C were identified, and after 0.5 MeV electron irradiation, predominantly silicon vacancies V_{Si} were observed. After 2.5 MeV irradiation, divacancies $V_{Si}V_{Si}$ were detected. Positron lifetime spectroscopy was used to detect vacancy-related defects in as-received and 12-MeV proton-irradiated 6H-SiC crystals [138]. The temperature dependence of the lifetime spectra decomposition suggests that neutral and negatively charged vacancy defects existed in crystals before and after proton irradiation at low fluence ($\leq 4 \times 10^{15}$ H$^+$ cm^{-2}). Neutral vacancies were detected after irradiation at high fluence ($\geq 4 \times 10^{16}$ H$^+$ cm^{-2}). Negatively charged silicon monovacancies with 202 ± 8 ps lifetime were detected at low temperature in as-received n-type 6H-SiC and after irradiation at low fluence. Neutral divacancies of (225 ± 5)-ps lifetime were produced by irradiation and dominated the positron trapping at room temperature. The EPR measurement combined with ab initio supercell calculation suggests that the P_6/P_7 (spin resonance signals) defect centers in 3 MeV electron irradiated 4H-SiC originate from the spin-1 triplet ground states of the neutral divacancy in the axial C_{3v} and monoclinic C_{1h} configurations (Fig. 2.19) [139]. The spin density is located mainly on three nearest C neighbors of the Si vacancy and it is negligible on the nearest Si neighbors of the C vacancy. The vacancy model for the P_6/P_7 centers points out that the interaction between V_{Si} and V_C to form divacancies ($V_{Si}V_C$) is significant and so the divacancy is a common defect in SiC. Annealing indicates that the formation of the divacancy is governed mainly by diffusion of V_{Si} and V_C. The annealing behavior of divacancy (P_6/P_7 EPR centers) and a previously unreported EPR center (supposed as $V_C V_{Si} V_C$ trivacancy) in SiC as well as their relationship with the UD1-3 series of PL lines (1.05–1.40 eV) were investigated [140]. The presented evidence supports the assignment of the UD2 PL lines to the P_6/P_7 EPR lines that arise from the divacancy. An ANN1 EPR center and its ^{29}Si and ^{13}C hyperfine spectra support its assignment as a trivacancy. The relationship between the UD1 and UD3 PL signals indicates that both stem from the same impurity or lattice defect. In practice, a later DFT study indicates that the EPR-detected ANN1 center was provisionally identified as the doubly positive trivacancy $V_C V_{Si} V_C$ [141]. The GGA calculations combined with the extended dangling-bond-counting model reveal that V_3 and V_6 are energetically favorable. The most stable form of V_6 is composed of the central Si vacancy surrounded by four C vacancies and the next-neighbor C vacancy, labeled as $V_C{}^k V_{Si}{}^k V_C{}^h V_C{}^h V_C{}^h V_C{}^h$. The divacancy defect formation processes in 3C-SiC were investigated using atomistic models, enhanced sampling simulations, and DFT calculations [142]. The divacancy formation was found to be a thermally activated process that competes with the conversion of V_{Si} to V_C. The divacancy reorientation can occur without dissociation. Increasing the concentration of V_{Si} relative to V_C favors the formation of divacancies.

The fundamental physical properties of the divacancy defects have been deeply investigated. It was demonstrated that several defect spin states in high purity semi-insulating 4H-SiC can be optically addressed and coherently controlled in the time domain at 20–300 K [143]. Using optical and microwave techniques, researchers studied the spin-1 ground state of four inequivalent forms of the neutral carbon–silicon divacancy and a pair of defect spin states of unidentified origin. They lie within 1.09–1.20 eV at 20 K (Fig. 2.20). At low temperatures, PL from a single species of optically active defect usually consists of a sharp zero-phonon line (ZPL) and a broad sideband composed of lower-energy phonon replicas of the ZPL. Four sharp lines (PL1–PL4) in the PL spectrum at 20 K (Fig. 2.20) form a group of lines known as UD-2 line [144] that originates from four distinct forms of neutral divacancy, which is a defect complex consisting of a carbon vacancy adjacent to a silicon vacancy: $[V_{Si}V_C]^0$ [139, 140]. Two types of divacancy (hh and kk) are

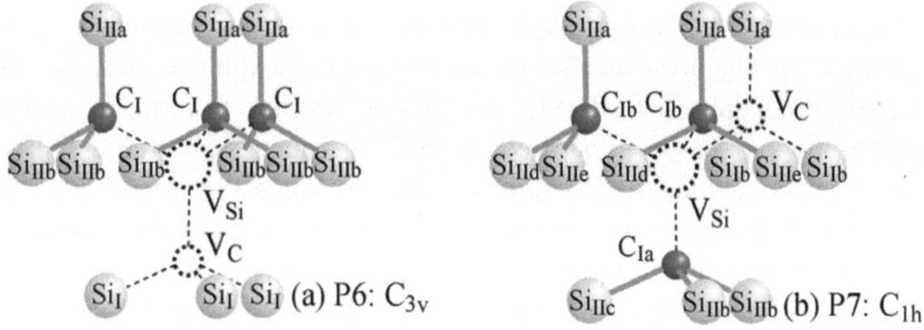

Fig. 2.19 Schematic of the **a** axial (C_{3v}) and **b** monoclinic (C_{1h}) configurations of the divacancy in electron-irradiated 4H-SiC. Reprinted with permission from [139]. Copyright 2006, American Physical Society

oriented along the *c* axis of 4H-SiC. The other two (*hk* and *kh*) are oriented along the basal bond orientations. These four defects give rise to the P_6 (*c*-axis) and P_7 (basal) spin-resonance signals observed in ESR studies of 4H-SiC SiC [139, 145]. PL5 and PL6 are close to one another in energy and share several other features, suggesting they are related. At higher temperatures all six emission lines broaden and diminish, so they are indistinguishable from the phonon sidebands by 200 K. But the PL through these phonon sidebands is still visible at up to 300 K. These defects are optically active near the telecommunication wavelengths and they possess desirable spin coherence properties comparable to those of the NV center in diamond. Further study demonstrates that 4H-, 6H- and 3C-SiC all hosted coherent and optically addressable defect spin states in the 1.08–1.20 eV range, including states with room temperature quantum coherence [146]. It suggests that crystal polymorphism in SiC can be a degree of freedom for engineering spin qubits. The spin relaxation times range from 8 to 24 ms at 20 K. The Hahn-echo coherence times range from 10 to 360 ms at 20 K, with significant dependence on implantation dose and substrate doping type. The spin coherence of the defects of all three polytypes persisted up to room temperature. In as-grown 4H-SiC a neutral divacancy line that persisted up to room temperature and three other ODMR lines of unknown origin all showed Hahn-echo coherence times of 50 ± 10 ms. The combined pulsed spin resonance and ab initio simulations show that spin–spin interactions in 4H-SiC neutral divacancies generate spin states with a strong Stark effect, sub-10^{-6} strain sensitivity, and highly spin-dependent PL with intensity contrasts of 15–36% [147]. These results suggest that SiC color centers are promising for sensing nanoscale electric and strain fields. A later study shows that individual electron spins in high-purity monocrystalline 4H–SiC can be isolated and coherently controlled [148]. These states that are bound to neutral divacancy defects exhibited exceptionally long ensemble Hahn-echo spin coherence times of over 1 ms. A high-fidelity spin–photon interface was demonstrated to exist in isolated divacancies in epitaxial films of 3C-SiC and 4H-SiC [149]. The divacancies in 4H-SiC showed minimal undesirable spin mixing. Linewidths of 100–200 MHz were observed from the 4H-SiC samples, with a particular (*kk*) divacancy showing transitions as narrow as 80 MHz at 5 K. These values are comparable to the 36–483 MHz linewidths of NV centers used in spin–photon and spin–spin entanglement experiments [150–152]. The divacancy's lifetime-limited linewidth is merely 11 MHz, so the observed broadenings in 4H-SiC and 3C-SiC are due to intrinsic interactions with lattice phonons or by impurities. 3C-SiC divacancies showed a millisecond Hahn-echo spin coherence time, which is among the longest measured in a naturally isotopic solid. The charge state control for divacancy and silicon vacancy qubits in 4H-SiC and bidirectional optical charge conversion between the bright and dark states of these defects were implemented [153]. The PL from divacancy ensembles showed increase by up to three orders of magnitude using near-ultraviolet excitation, without degrading the electron spin coherence time. This charge conversion remained stable for hours at cryogenic temperatures. The Hahn-echo coherence time of electron spins related to divacancy defects in 4H-SiC can reach 1.3 ms [154]. A first-principles quantum-bath model reveals two factors determining the robust coherence. In the presence of moderate magnetic fields ($\geq$ 30 mT), the ^{29}Si and ^{13}C paramagnetic nuclear spin baths are decoupled. Homo-nuclear spin pairs in SiC are both diluted and forbidden from forming strongly coupled, nearest-neighbor spin pairs. Longer neighbor distances result in fewer nuclear spin flip-flops, a less fluctuating intra-crystalline magnetic environment, and thus a longer coherence time. A study demonstrated coherent manipulation of single divacancy spins in 4H-SiC with a high readout contrast (– 30%) and a high photon count rate (150 kilo counts per second) under ambient conditions [155]. The arrays of single $V_{Si}V_C$ defects in 4H-SiC were fabricated through carbon ion (C^+) implantation and annealing. The observed performance is competitive with the NV centers in diamond. Coupling between a single defect spin and a nearby nuclear spin was observed.

There have been studies on the temperature dependence of the SiC divacancy defects. A self-protected infrared high-sensitivity thermometry based on the divacancies of electron spins in 4H-SiC was demonstrated [156]. It was concluded that the Ramsey oscillations of the spin sensor are robust against magnetic noise due to a self-protected mechanism from the intrinsic transverse strain of the defect. The ODMR spectrum and Ramsey oscillation of the PL5 divacancy defect spins in 4H-SiC within 300–600 K were investigated [157]. The PL6 defect spin properties in 4H-SiC implanted with N_2^+ showed temperature dependence [158]. The zero-field splitting parameter decreased when the temperature increased. It changed by – 13 MHz from 5 to 300 K. The inhomogeneous dephasing time is stable. The coherence time and depolarization time are longer in the sample implanted with lower ion fluence. The coherence time increased with temperature owing to disappearance of the Jahn–Teller distortion. The coherent control of defect spins was achieved even at 600 K. The high-temperature thermal sensing with the PL5 divacancy defect spins at 450 K was demonstrated.

The excitation state properties of divacancy defect in SiC have been explored. The PL from the divacancy defect in 4H-SiC can be quenched below certain excitation photon energies [159]. The PL decay was supposed to correlate with conversion of the divacancy from neutral to negative charge state due to capture of electrons photoionized from other defects (traps) by the excitation. The quasiparticle optical absorption spectrum of the divacancy defect in 3C-SiC was studied by using *GW–BSE* (*GW*

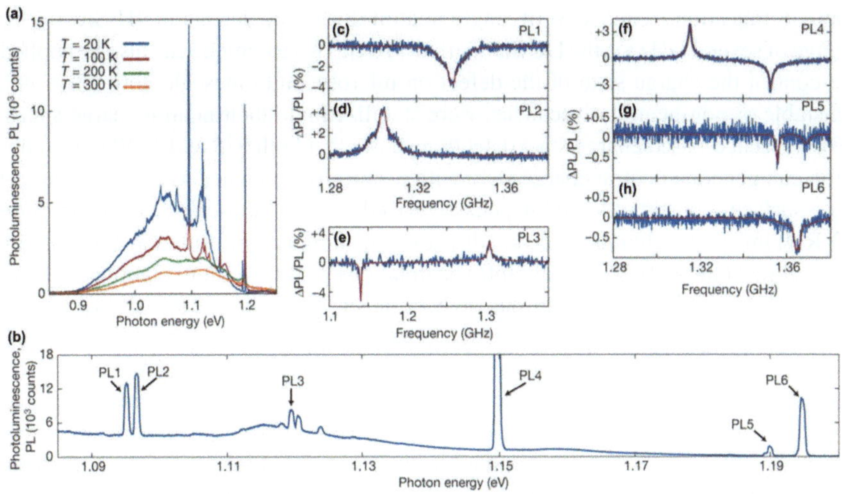

Fig. 2.20 Optical detection of defect spins in 4H-SiC. **a** PL spectra of high-purity semi-insulating 4H-SiC at 20–300 K. **b** Enlarged view of 20-K PL showing six defect lines (PL1–PL6). PL1–PL4 are four distinct forms of neutral divacancy. PL5 and PL6 remain unidentified. **c**–**h** Normalized change in PL (ΔPL/PL) collected from PL1–PL6 as a function of applied microwave frequency at 20 K. Solid red lines in **c**–**h** are fits. Reprinted with permission from [143]. Copyright 2011, Springer Nature Limited

plus Bethe–Salpeter equation) calculations [160]. The low-energy excitonic states are mainly made of transitions from occupied defect states to continuum conduction states from SiC. The mixed character of defect states and bulk states of these low-energy exciton states is distinct from the NV^- center in diamond and the divacancy in 4H-SiC, where the deep defect levels are well separated from bulk states. The dipole moment's orientation of single divacancy defects in 4H-SiC was determined using tightly focused radially and azimuthally polarized laser beams through confocal microscopy [161]. Polar and azimuthal angles of the dipole moment were extracted by comparison with calculated 2D PL intensity distributions.

There are explorations on the interaction between divacancy defect spins, nuclear spins, and phonons as well as external control of divacancy defects. Defect-induced magnetism was observed in neutron irradiated *p*-type 6H-SiC(0001) wafers [162]. The intentionally created defects dominated by divacancies ($V_{Si}V_C$) are responsible for the observed magnetism. First-principles calculations reveal that defect states favor the formation of local moments and the extended tails of defect wavefunctions make long-range spin coupling possible. The study shows that near-infrared light can nearly completely polarize populations of ^{29}Si nuclear spins coupled to paramagnetic color centers in 4H- and 6H-SiC [163]. In this dynamic nuclear polarization process, the optically pumped polarization of electron spins bound to either neutral divacancy or PL6 color centers was transferred to proximate nuclei via the hyperfine interaction. The 99% ± 1% degree of polarization observed at room temperature corresponds to an effective nuclear temperature of 5 μK. Electric fields can be used to tune the optical transition frequencies of single neutral divacancy defects in 4H-SiC over a range of several GHz via the DC Stark effect [164]. This technique can also be applied to control the charge state of the defect on microsecond timescale, through which unstable or non-neutral divacancies were stabilized into their neutral charge states. Fluorescence-based charge state detection shows that both 975 and 1130 nm excitation can generate neutral charge state with near unity efficiency. Gaussian focusing of a surface acoustic wave in SiC, characterized using a nanoscale X-ray diffraction (XRD) imaging technique, delivers direct strain amplitude information at nanoscale spatial resolution [165]. The ab initio calculations provide a more complete picture of spin–strain coupling for various defects in SiC with C_{3v} symmetry. All-optical detection of acoustic paramagnetic resonance without microwave magnetic fields was demonstrated. Mechanically driven Autler–Townes splitting and magnetically forbidden Rabi oscillation were also observed. A study shows the control of isolated ^{29}Si nuclear spins in 4H-SiC to create an entangled state between an optically active divacancy spin and a strongly coupled nuclear register [166]. The authors realized high-fidelity electron spin control ($F = 99.984\%$), alongside extended coherence times (Hahn-echo time of 2.3 ms, dynamical decoupling > 14.5 ms), and a > 40-fold increase in Ramsey spin dephasing time from isotopic purification.

Like other light emitters, the PL of SiC divacancy defects can be enhanced. A single neutral divacancy in 4H-SiC coupled to a photonic crystal cavity showed Purcell enhancement of PL [167]. A combination of nanolithographic technique and a dopant-selective photoelectrochemical etch method was used to produce suspended cavities with quality factors exceeding 5000. Subsequent coupling to a single divacancy led to a Purcell factor of ~ 50, with increased PL of the zero-phonon line and a shortened excited-state lifetime. The off-resonance lifetime is 15.7 ± 0.3 ns and the on-resonance lifetime is 5.3 ± 0.1 ns. The PL can also be enhanced by using surface plasmons. A study shows enhancement of the brightness by 7 times and spin-control strength by 14 times of single divacancy defects in 4H-SiC membranes using surface plasmons generated by gold film coplanar waveguides [168]. Lifetime measurements support the coupling between defects and surface plasmons.

2.4.1.2 Nitrogen–Vacancy Centers

Similar to diamond, SiC can also accommodate NV centers, in which the substitutional N_C sits next to a Si vacancy. First-principles calculations compare the properties of the NV center in diamond to the analogous defect in 4H-SiC [169]. The $N_C V_{Si}{}^{-1}$ center in SiC has a defect-level diagram that is similar to that of the NV^{-1} center in diamond. The splitting between the levels in SiC is much smaller than that in diamond due to the reduced dangling bond overlap caused by the crystal's larger bond lengths. The zero-phonon line has a magnitude that is about half that of the NV^{-1} center in diamond. The DFT calculations employing the screened hybrid functional developed by Heyd, Scuseria, and Ernzerhof (HSE) explore the divacancy $V_C V_{Si}$ in 4H- and 3C-SiC and the NV center in 3C-SiC [170]. The calculated excitation and emission energies of the divacancy in 4H-SiC agree with experimental data. The paramagnetic spin-1 neutral divacancy was identified in 3C-SiC. Unlike the situation in 4H-SiC where semi-insulating material is needed with the Fermi level within a specific range, the desired $S = 1$ states of the $(V_C V_{Si})^0$ and $(N_C V_{Si})^-$ centers are stable in n-type-doped 3C-SiC. The combined EPR spectroscopy and DFT–HSE calculations verified the existence of the NV centers in 4H-SiC in the form of nitrogen–silicon vacancy pairs $N_C V_{Si}$ [171]. Similar to the NV center in diamond, the $^3A_2$ triplet ground state (with a zero-field splitting of 429×10^{-4} cm^{-1}) of the singly negative charge state in axial coordination can be optically spin polarized at room temperature. Evidence of near-infrared PL signature of $(N_C V_{Si})^-$ from 4H-SiC was observed [172]. This center exhibited an $S = 1$ ground state spin similar to the NV^- center in diamond. PL excitation measurements at cryogenic temperature demonstrated efficient photoexcitation of distinct PL from $(N_C V_{Si})^-$ in 4H-SiC. The observed PLX1–PLX4 lines were explained as NV-related zero-phonon lines for four different pairs of orientation (Fig. 2.21). Each of the zero-phonon line was associated to the respective occupation of hexagonal (h) and quasicubic (k) lattice sites in analogy to neutral divacancy centers $(V_C V_{Si})^0$ in that the electronic structure of the $(N_C V_{Si})^-$ centers closely resembles that of the $(V_C V_{Si})^0$ centers in 4H-SiC [171]. In practice, from Fig. 2.21c, e one notes that the spin (magnetization) density is mainly localized at the carbon dangling bonds of the V_{Si} component of the defect pair. A very small part of the density of the unpaired electrons lies at the ^{14}N nucleus. The spin density giving rise to the hyperfine splitting as well as the spin–spin part of the zero-field splitting resembles perfectly that of the divacancy. The electronic properties for both kinds of defects are primarily determined by the $V_{Si}C_3$ core structure of the defect. Therefore, the $(N_C V_{Si})^-$ pairs provide hyperfine splitting with nearest-neighbor C and next-nearest-neighbor Si atoms similar to that of the divacancies. The NV triplet centers can also be generated in all the main SiC polytypes. The axial NV^- pairs in 3C-, 4H-, and 6H-SiC were identified by using EPR spectroscopy and DFT–PBE calculations [173]. The room-temperature spin polarization of the ground state upon near-infrared optical excitation for the NV center in 3C-SiC and axial NV centers in the hexagonal (4H, 6H) polytypes is very efficient (the signals of basal pairs are much weaker). The DFT calculation indicates that the divacancy inevitably forms when the NV center in SiC is produced by implantation or irradiation [174].

The combined EPR and PL spectroscopy shows that the NV^- center in 3C-SiC has a ZPL line at 1.468 μm [175]. This line can be observed up to $T = 100$ K. The negatively charged NV center in 3C-SiC is the structural isomorph of the NV center in diamond and has equally spin $S = 1$ ground and excited states, long spin lattice relaxation times and possesses optically induced ground state spin polarization. The NIR PL properties originating from $N_C V_{Si}{}^{-1}$ centers in high purity semi-insulating and nitrogen-contained 4H-SiC substrates formed by ion beam irradiation and subsequent thermal annealing have been investigated [176]. The NV centers formed exclusively by contained N impurities rather than implanted N ions. The heavier ion irradiation induced the NV center formation effectively. Six NV centers in 6H-SiC were investigated by using PL and EPR spectroscopy combined with DFT–PBE calculations [177]. The absorption ZPLs are very sharp with a half-width of 0.2–0.3 nm. Similar to the case of NV^- centers in 4H-SiC, the Franck–Condon shift between the

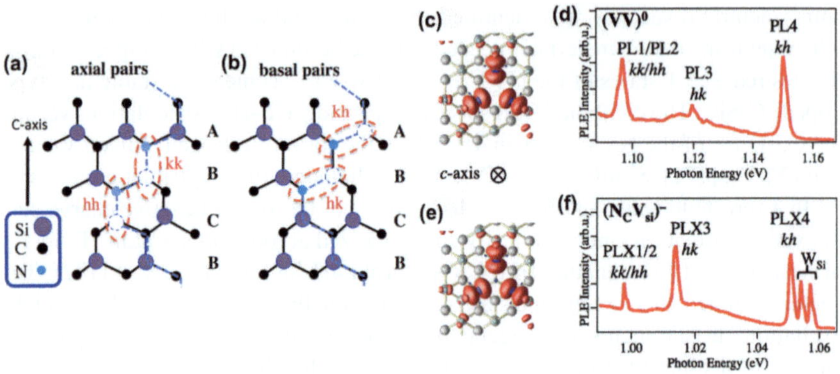

Fig. 2.21 **a, b** Orientation of defect pairs $N_C V_{Si}$ in 4H-SiC. Empty sites are represented by dashed circles. **c, e** DFT-calculated spin densities for the axial (*hh*) divacancy $(V_C V_{Si})^0$ and $(N_C V_{Si})^-$ center, respectively (view along the hexagonal *c* axis). **d** Divacancy PL1–PL4 association to the different configurations (1.409 eV excitation). **f** PLX1–PLX4 lines excited at 1.240 eV and tentatively associated to $(N_C V_{Si})^-$ configurations. Reprinted with permission from [172]. Copyright 2016, American Physical Society

absorption and emission ZPLs of the intracenter transition $^3E \rightarrow {}^3A_2$ is negligible. Time-resolved photo-EPR measurements at 4 K showed that NV^- centers in 6H-SiC have spin lattice relaxation times of several seconds. Coherent control of the NV center (*hh*) ensemble in 4H-SiC was realized at room temperature [178]. The NV concentration increased sixfold through optimization of the implantation conditions. The single NV centers are photostable at room temperature. The PL ZPLs lie within 1150–1250 nm, with a lifetime of ~ 2.1 ns (Fig. 2.22).

Note that the PL decay under excitation of a short light pulse (with time duration of sub-nanoseconds or femtoseconds) usually follows:

$$I(t) = I_0 e^{-\frac{t}{\tau}}, \tag{2.5}$$

where τ is the average lifetime of the excited state of the luminescent system. This equation is derived from another equation describing the decay behavior of the excited state:

$$N(t) = N_0 e^{-\frac{t}{\tau}}, \tag{2.6}$$

where $N(t)$ stands for the number of identical fluorophors (such as luminescent quantum dots) lying at a specific excited quantum state at time t and N_0 is its initial value. We have:

$$\frac{1}{\tau} = \frac{1}{\tau_r} + \frac{1}{\tau_{nr}}, \tag{2.7}$$

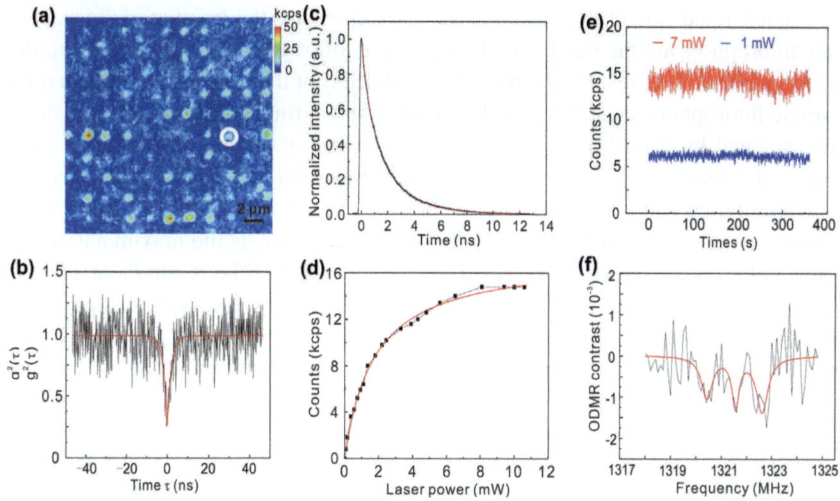

Fig. 2.22 **a** A confocal scan image (20 × 20 μm²) of the single NV center arrays in 4H-SiC with a laser power of 4 mW. The circled point is the studied single NV center. The scale bar is 2 μm. **b** Second-order correlation function of the circled single NV center. **c** Time-resolved PL spectrum of the NV center ensemble at room temperature. **d** Emission saturation curve of a single NV center as a function of laser power. **e** Photostability of the single NV center at two different pump laser powers with a time bin of 100 ms. **f** ODMR spectrum of a single NV center. Reprinted with permission from [178]. Copyright 2020, American Physical Society

where τ_r and τ_{nr} are the radiative and nonradiative transition lifetimes. The spontaneous emission lifetime corresponding to the electron transition between two quantum states without participation of phonons within the electric dipole approximation follows

$$\tau_r = \frac{3h\varepsilon_0 c^3}{16\pi^3 v_0^3 n\mu^2}, \tag{2.8}$$

where h is the Planck's constant, c is the light velocity in vacuum, ν_0 is the emitted-photon frequency, n is the refractive index surrounding the emitter, and μ is the matrix element of the electric dipole moment operator between initial and final quantum states:

$$\mu = \left| \iiint \psi_2^* e\boldsymbol{r}\psi_1 \mathrm{d}V \right|. \tag{2.9}$$

From Eq. (2.6) we further obtain

$$-\frac{\mathrm{d}N(t)}{N(t)\mathrm{d}t} = \frac{1}{\tau}. \tag{2.10}$$

The left-hand side of this equation (or $1/\tau$) represents the fraction of the excited-state fluorophors at time t that would decay at the following unit interval, and another meaning is that it represents the probability of decay at the following unit interval of a single fluorophor that still lies at the excited state at time t. The exponential decay law as stated by Eq. (2.6) is the basic characteristic of quantum mechanics and it reflects the uncertainty in moment of quantum transition of a given excited electron.

The luminescence saturation curve shown in Fig. 2.22d can be fitted using the power dependence model $I(P) = I_s/(1 + P_0/P)$, where I_s is the maximal emission count and P_0 is the saturation power. The fit gives I_s of 17.4 kcps and P_0 of 1.7 mW. The coherence time can reach about 17.1 μs. Three energy levels were supposed to be involved in the luminescence of the NV centers [179]. For *c*-axis NV centers, both excitation and emission polarization degrees exceeded 90%. Photon purity and photostability of the single NV centers were maintained at temperatures up to 400 K. The ODMR spectroscopy with resonant excitation of four NV centers in 4H-SiC was performed and the ground-state energy levels were identified [180]. Coherent manipulation of NV centers in SiC was achieved through Rabi and Ramsey oscillations. The resonantly excited EPR spectroscopy helps to identify the zero-phonon absorption line of the $^3A_2 \rightarrow {}^3E$ transition at 1289 nm and its phonon sideband [181]. The analysis reveals a Huang–Rhys factor of $S = 2.85$ and a Debye–Waller factor of 5.8%. The spin–lattice relaxation time is as long as 17 s at 4 K. The DFT–PBE calculations reveal that the ZPLs of the negatively charged NV center in 4H-SiC are sensitive to the geometrical configurations of the ground and excited states [182]. Supercells with over 2000 atoms are needed to obtain accurate numerical results. The computed decoherence time of the basal NV centers at zero magnetic field is much bigger than that of the axial configurations. Using an ensemble of NV centers in 4H-SiC and the dynamical decoupling technique, a room-temperature quantum sensing of an artificial AC field centered at ~ 900 kHz with a spectral resolution of 10 kHz was realized [183]. The frequency resolution of the sensor was extended to 0.01 kHz using the synchronized readout technique. The photoinduced pulse ESR reveals the room temperature spin coherence of NV defects in 6H-^{28}SiC, which was purified to reduce ^{29}Si concentration to ~ 1%, four times below its natural level [184]. At room temperature, the Hahn-echo coherence time is 23.6 μs, and the spin–lattice relaxation time is 0.1 ms. The inhomogeneous dephasing time is as long as 1.5 μs, about five times greater than that measured for NV defects in SiC with natural isotopic content. The room temperature spin coherence properties of the axial NV (*kk*) center in 4H-SiC was investigated by using pulsed high-frequency ESR and electron–nuclear double resonance techniques [185]. The phase coherence time is 25.3 μs at room temperature for the NV center ensemble. Precise control over NV defect spins was realized through Rabi oscillations. Readout of the ^{14}N nuclear spin states via optically polarized electron spin states of the NV defects was demonstrated.

2.4.1.3 Antisites

The antisite defect was generated by occupation of a Si (C) position by a C (Si) atom. DLTS performed on neutron irradiated 3C-SiC revealed point defect related centers called H_1, H_2, H_3, and E_2 [186]. The electronic energy levels of H centers are located at 0.18 eV (H_1), 0.24 eV (H_2), and 0.514 eV (H_3) above the valence band maximum, whereas the level of E_2 is located at 0.49 eV below the conduction band minimum. 90% of the produced defects were eliminated by annealing at 350 °C, whereas the concentration of the E_2 centers increased gradually with annealing of the H centers. The DFT–LDA calculation suggests that the H centers are ascribed to silicon antisite (Si_C) defects generated by neutron irradiation [187]. In both 3C- and 2H-SiC, the silicon antisites have several ionization levels in the bandgap. These levels in 3C-SiC were calculated by using the plane wave pseudopotential method and the results agree well with the DLTS data (Fig. 2.23).

Ion implantation of SiC may produce damage in the crystal lattice that is hard to repair. Some of these defects remain even after annealing at high temperatures. One of these persistent defects was observed from 3C-SiC after 120 keV He ion implantation or 1 MeV electron bombardment [188]. This defect (D_1) exhibited strong low-temperature luminescence which is independent of the implanted ion, and the luminescence spectrum comprises a zero-phonon line (at 627.98 nm) accompanied by a strong vibronic structure. This luminescence remained after annealing at 1600 °C. The radiation defects in 6H-SiC produced by ion implantation and electron bombardment were investigated [189]. The defects also showed low-temperature luminescence. This D_1 spectrum persisted after annealing at 1700 °C. Its intensity depends strongly on the defect concentration. The defect was proposed to originate from a divacancy. Unlike the case of 3C-SiC, there are three no-phonon lines in the D_1 band (Fig. 2.24) due to the three nonequivalent sites in 6H-SiC. The spectrum showed unusual temperature dependence. The low temperature (1.4 K) spectrum vanished when the high temperature (77 K) form emerged, and this abrupt spectral change was ascribed to lattice distortion at low temperatures. The D_1 centers with similar properties were also observed from 4H-SiC [190]. It is well known that annealing can change the structures and properties of intrinsic defects, especially vacancies.

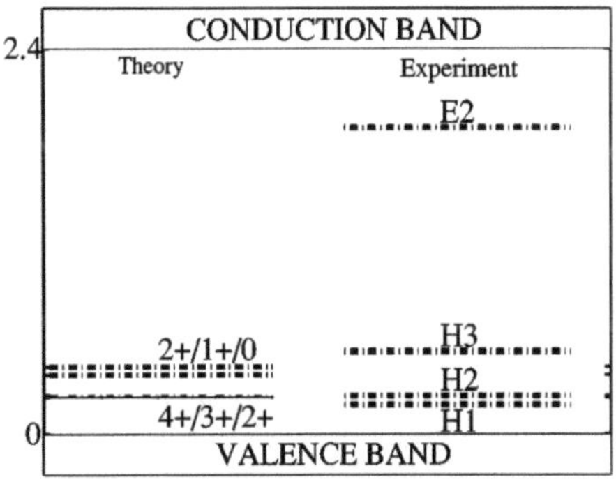

Fig. 2.23 Comparison of the calculated ionization levels of Si antisite in 3C-SiC with the experimental DLTS data obtained from Ref. [186]. Reprinted with permission from [187]. Copyright 1998, American Physical Society

The annealing processes of vacancy-type defects in 3 MeV electron irradiated and as-grown 6H-SiC were investigated by using positron lifetime spectroscopy [191]. A vacancy type defect giving rise to a positron lifetime of 183 ps was detected from the as-grown n-type 6H-SiC. This defect was annihilated after annealing at around 1400 °C. Electron irradiation produced defects related to carbon vacancies, silicon vacancies, and divacancies. The defects related to carbon vacancies and divacancies were annealed at a temperature up to 500 °C, whereas the defects related to silicon vacancies were annealed at around 750 and 1400 °C. The observed defect was suggested to arise from Si vacancies, possibly complexes of silicon vacancies and nitrogen atoms. No vacancy-type defects were detected from p-type 6H-SiC even after electron irradiation. This study shows that the D_1 luminescence center is not related to any vacancy-type defects, such as divacancies, silicon vacancies, and complexes of vacancies and impurities.

There have been further explorations of the nature of the defect D_1. Some luminescence lines with previously unknown origins in silicon carbide were ascribed to bound excitons. The D_1 PL observed from electron irradiated 4H-SiC has been investigated [192]. At temperatures lower than 5 K, the D_1 PL spectrum showed a zero phonon line (L_1, 427.3 nm) and phonon replicas. The L_1 line decreased rapidly with increasing temperature and meanwhile, two higher energy no-phonon lines (M_1, 426.1 nm; H_1, 425.7 nm) emerged. At temperatures higher than 100 K, the D_1 PL

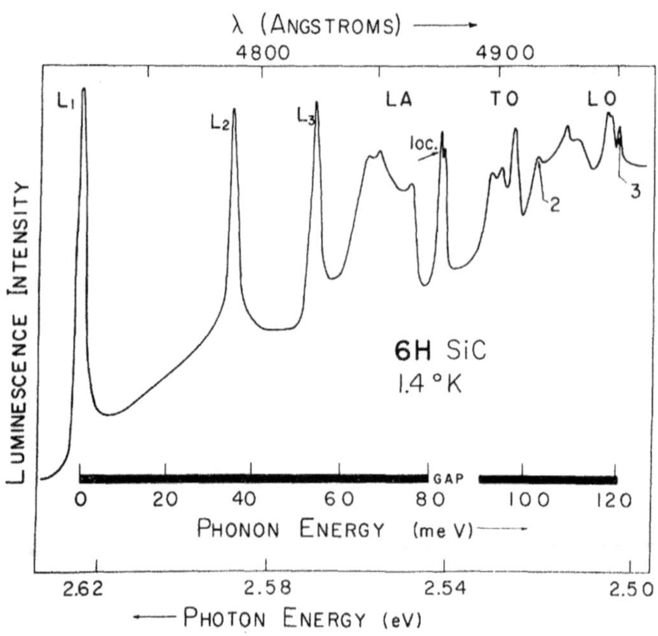

Fig. 2.24 Portion of the 1.4 K D_1 PL spectrum of the 1300 °C annealed and ion bombarded 6H-SiC. L_1, L_2, and L_3 are no-phonon lines and most of the phonon replicas belong to the L_1 peak. Reprinted with permission from [189]. Copyright 1972, American Physical Society

was quenched. The lifetime of the D_1 PL is 450 μs at 1.3 K and 2 μs at 70 K. It was suggested that the D_1 luminescence arises from the exciton recombination at an isoelectronic center, with one of the particles of the exciton weakly bound and the other more tightly bound. The pseudodonor character of the D_1 defect was further studied [193]. The result suggests that the D_1 defect is related to a hole trap at 0.35 eV above the valence band. The minority carrier transient spectroscopy indicates that the D_1 defect indeed correlates with such a hole trap. The study also indicates that the D_1 defect is not correlated to the $Z_{1/2}$ electron trap.

First-principles density-functional calculations were used to investigate antisite pairs in 4H-SiC [194]. The result shows that they are likely form in close proximity under ionizing conditions and they possess thermal stability. The most stable antisite pair may generate the most energetic line in the E_A spectrum (discussed below) or the D_1 band. A subsequent study based on the low-temperature PL and ab initio calculation for 3C- and 4H-SiC shows that the neutral antisite pair could be a candidate for the D_1 defect [195]. The calculated one electron level of the neutral antisite pair is close to the measured transition energy of the zero-phonon line. The calculated hyperfine constants of the antisite pair in its paramagnetic state support this assignment. However, a deeper study based on the local density-functional calculation indicates that the close-by antisite pair is not stable enough to account for D_1 [196]. Instead, D_1 was suggested to be related to an isolated Si antisite.

Another luminescence center called E_A was observed at room-temperature from the 2 MeV electron-irradiated 4H-SiC (n^+- and p^+-type substrates and epilayers grown by CVD with different levels of n- or p-type residual doping) [197]. The spectrum E_A consists of several sharp zero-phonon lines accompanied by a broad phonon sideband (Fig. 2.25). Annealing at around 750 °C induced an abrupt change in the spectrum and further annealing at over 750 °C resulted in the dominance of the strong D_1 PL spectrum (Fig. 2.26). Up to 40 zero-phonon lines (within 2.8–2.9 eV) were resolved in the E_A spectrum. PL excitation spectroscopy shows that the lines come in groups of two to four lines. The characteristics of the line groups suggest that they are attributed to the bound-exciton recombination at isoelectronic defect centers. First-principles DFT calculations show that the lines of the E_A spectrum [194] stem from the recombination of a delocalized electron with a localized hole bound to a Si antisite defect (Si_C) which is in proximity with a C antisite (C_{Si}). Further calculations indicate that the four forms of the close-by antisite pair (Fig. 2.27) in 4H-SiC give rise to the a, b, c, and d members of the E_A PL spectrum [196].

Another famous persistent intrinsic defect in SiC is D_{II}. This defect center was detected from ion implanted and annealed (at 1300 °C) 3C-SiC by low temperature luminescence spectroscopy [198]. The PL spectroscopy revealed a number of high energy localized modes. One mode has an energy of 164.7 meV, which is equivalent to the highest lattice frequency in diamond and far above the 120.5 meV lattice limit of SiC. A carbon di-interstitial was suggested to account for this defect, which appeared after annealing at 1300 °C and persisted after annealing at 1700 °C. The D_{II} center was also observed from the unannealed ion (B, Al, P) implanted 3C-SiC films [199]. The D_{II} zero-phonon line is located at about 2.3 eV. Its intensity increased with annealing temperature, but dropped rapidly for annealing temperatures above

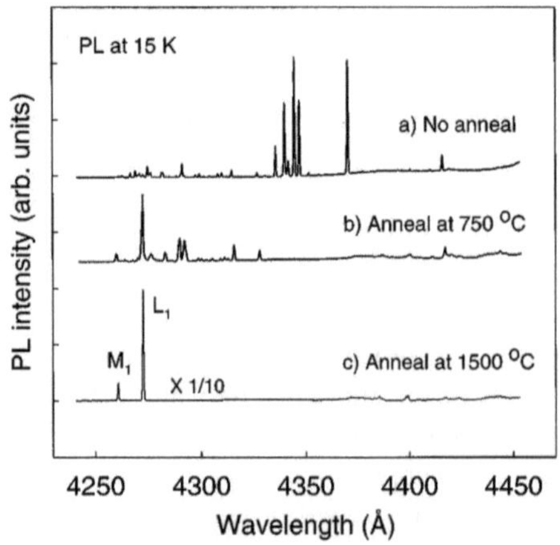

Fig. 2.25 Effects of annealing on the zero-phonon lines in the PL spectrum (at 15 K) of 2 MeV electron-irradiated 4H-SiC. The D_1 lines L_1 and M_1 which dominate the spectrum after annealing at 1500 °C (**c**) are one order of magnitude more intense than the lines of the (**a**) and (**b**) spectra. Reprinted with permission from [197]. Copyright 1999, American Physical Society

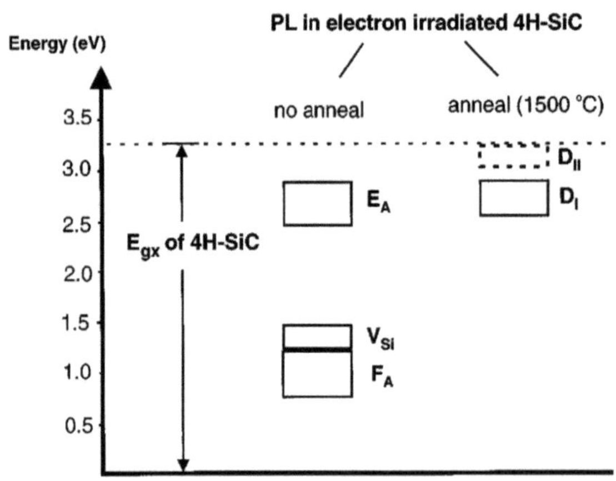

Fig. 2.26 Energy ranges of the main PL spectra of room-temperature electron irradiated 4H-SiC. The excitonic bandgap energy of 4H-SiC is shown on the left, the case right after electron irradiation in the middle, and that after high temperature (about 1500 °C) annealing on the right. Reprinted with permission from [196]. Copyright 1999, American Physical Society

1600 °C. The phonon replicas of the D_{II} zero-phonon line are located in the range of 1.98–2.25 eV. The spectral density of the phonon replicas resembles the phonon density of states in diamond, suggesting that the observed spectrum stems from a carbon-associated defect. The structural, electronic, and vibrational properties of the dicarbon antisite complexes have been investigated by using DFT to check whether they can account for the D_{II} center [200]. Figure 2.28 shows the configuration of the

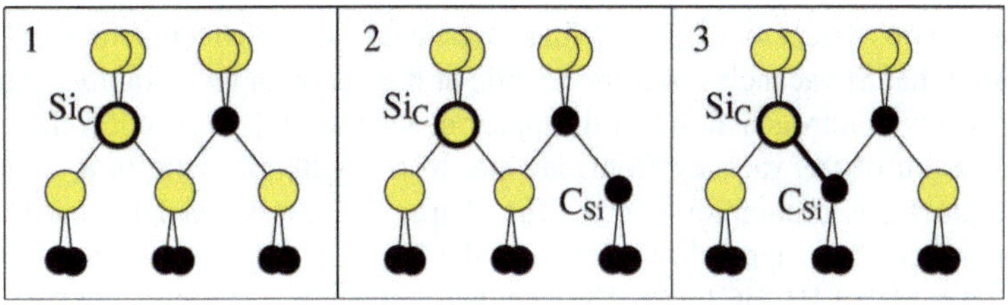

Fig. 2.27 Schematic figure of (1) isolated Si_C, (2) third neighbor antisite pair, and (3) close-by antisite pair in 4H-SiC. Reprinted with permission from [196]. Copyright 2006, American Physical Society

Fig. 2.28 Configuration of the dicarbon antisite $(C_2)_{Si}$ and carbon cluster with four atoms on a silicon site $(C_4)_{Si}$ in 3C-SiC. Reprinted with permission from [200]. Copyright 2004, American Physical Society

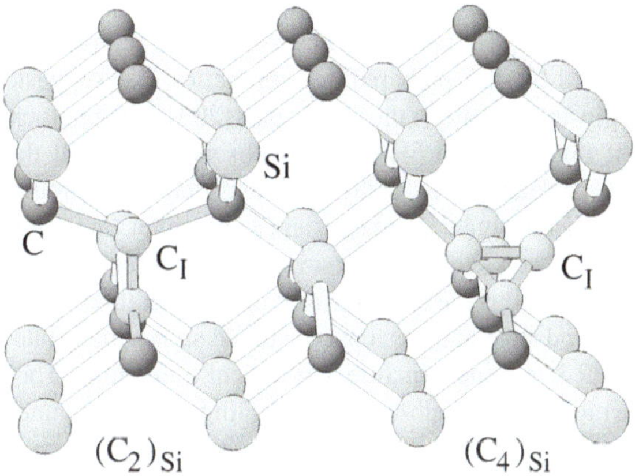

dicarbon antisite $(C_2)_{Si}$, where two carbon atoms sharing a silicon site are surrounded by a tetrahedron of four carbon atoms. The dicarbon antisite possesses a characteristic vibrational pattern in the frequency range of the D_{II} spectrum, but it cannot explain all the experimentally observed phonon replicas. The D_{II}-type centers may be related to larger carbon clusters built on the basis of the dicarbon antisite.

By assuming an antisite defect (Si_C) in near or inside the stacking faults as a color center in the *n*-type region, the performance of the p^+–n–n^+ 4H-SiC single-photon emitting diode was investigated by using 2D numerical simulations [201]. Using a comprehensive theoretical approach and rigorous numerical simulations, the authors indicated that at room temperature, the photon emission rate from the diode can exceed 5 giga counts per second, which is higher than what can be achieved with electrically driven color centers in diamond or epitaxial quantum dots.

2.4.1.4 Carbon Vacancy–Carbon Antisite Pairs

The carbon vacancy–carbon antisite (C_{Si}–V_C) pair is another important point defect in silicon carbide. This presence of $C_{Si}V_C$ pairs is closely related to the evolution of another famous defect in SiC, V_{Si}. What happens to V_{Si} at high temperatures? It was suggested that Si vacancies become mobile at high enough temperatures and form complexes with nitrogen atoms or disappear at a sink [191]. In practice, the motion and diffusion of the vacancy from one site to a neighboring site in a compound semiconductor is a complex matter. The charge self-consistent DFT based tight-binding scheme was applied to investigate the $V_{Si} \rightarrow C_{Si}V_C$ process in a 143 atom supercell model of 4H-SiC [202]. The calculation shows that the energy of the relaxed $C_{Si}V_C$ structure is 1.67 eV lower compared to the V_{Si}. There is an energy barrier of 1.80 eV between these two structures with respect to the Si vacancy (at $T = 0$ K) which can be overcome by thermal excitation. For the neutral C vacancy no such metastability was observed. The corresponding antisite–vacancy complex is instable and relaxes into V_C without barrier.

The *P6*/*P7* spectrum observed from neutron-irradiated and annealed nitrogen-doped 6H-SiC prepared by sublimation was suggested to arise from a photoexcited spin-triplet state of the $C_{Si}V_C$ pair [203]. The photoexcited paramagnetic triplet state of this defect in 6H-SiC was observed from all the inequivalent lattice sites and all orientations by using EPR and magnetic circular dichroism of the absorption. The EPR signal comprises six spin triplet lines (Fig. 2.29). First-principles calculation shows that among several pair defect models, $C_{Si}V_C$ in the charge state $+ 2$ (Fig. 2.30) is the only intrinsic nearest neighbor pair defect that is diamagnetic in its ground state and has a photoexcited spin-triplet state that can generate the observed EPR hyperfine structure and optical transitions. The $C_{Si}V_C$ pair was observed after irradiation and subsequent annealing above the annealing temperature of the silicon vacancy. Therefore, movement of the carbon atom into the neighboring Si vacancy during annealing may result in the formation of the $C_{Si}V_C$ defect. $C_{Si}V_C$ have also been observed from semi-insulating 4H-SiC grown by physical vapor transport and by high-temperature CVD using EPR and PL [204]. All samples exhibited EPR signature of the $C_{Si}V_C$ pair defect in an excited state of its 2+ charge state. The PL signature of this EPR center was supposed to be the UD2 spectrum—a set of four sharp lines within 1.1–1.15 eV. The UD1 line, a pair of sharp PL lines near 1.06 eV and UD3, a single sharp line near 1.36 eV, were also identified. The positively charged $C_{Si}V_C^+$ pairs in electron-irradiated *p*-type 4H-SiC were identified by means of EPR and DFT–LDA calculation [205]. Four EPR centers, *HEI*9*a*/*b* ($S = 1/2$, C_{3v} symmetry) and *HEI*10*a*/*b* ($S = 1/2$, C_{1h} symmetry), were respectively assigned to $C_{Si}V_C^+$ (*hh*)/(*kk*) and $C_{Si}V_C^+$ (*kh*)/(*hk*), based on EPR and theoretical analysis of their ^{13}C hyperfine interactions. Their wavefunctions are highly localized on the C_{Si} atom. An isochronal annealing study suggests that the thermal stability of $C_{Si}V_C^+$ is as high as that of V_C^+. All four configurations of the $C_{Si}V_C^+$ center in ultrapure compensated isotope-enriched 4H-^{28}SiC were observed using EPR and their response to illumination in photo-EPR experiments [206]. The photo-EPR data suggest that the (+|0) energy level of $C_{Si}V_C$ lies at 1.1 eV below the conduction band. Both the positive and neutral charge states of the $C_{Si}V_C$ center can be effectively controlled by optical excitation. The neutral charge state of $C_{Si}V_C$ may also be accessible even in the *p*-type material if the concentration of this defect is bigger than the total concentration of shallow acceptors and divacancy to ensure only a part of it is positively charged.

However, the same spectrum generated from electron-irradiated 4H-SiC has been proposed to arise from the triplet ground states of the neutral divacancy [139]. Instead, another experimentally observed defect (*SI*5 center) in 4H-SiC was assigned to the negatively charged carbon vacancy–carbon antisite pair $C_{Si}V_C^-$ based on electron paramagnetic resonance and first-principles calculation [207]. The vacancy defects

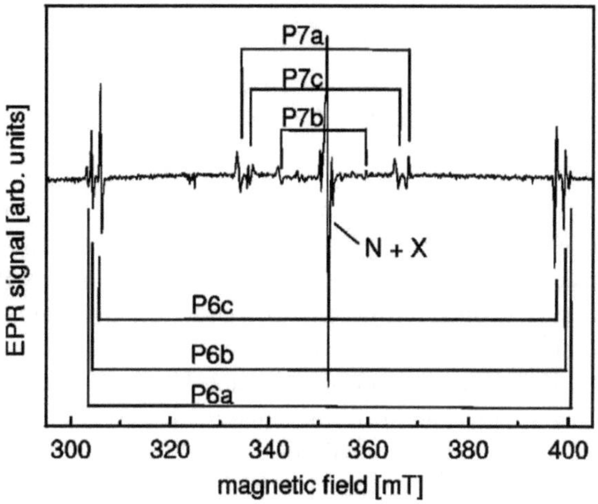

Fig. 2.29 Electron paramagnetic resonance spectrum of neutron irradiated and 1000 °C annealed 6H-SiC acquired with the magnetic field parallel to the *c*-axis of the crystal. Six spin-triplet spectra labeled *P*6a, b, c, and *P*7a, b, c were observed. Reprinted with permission from [203]. Copyright 2001, American Physical Society

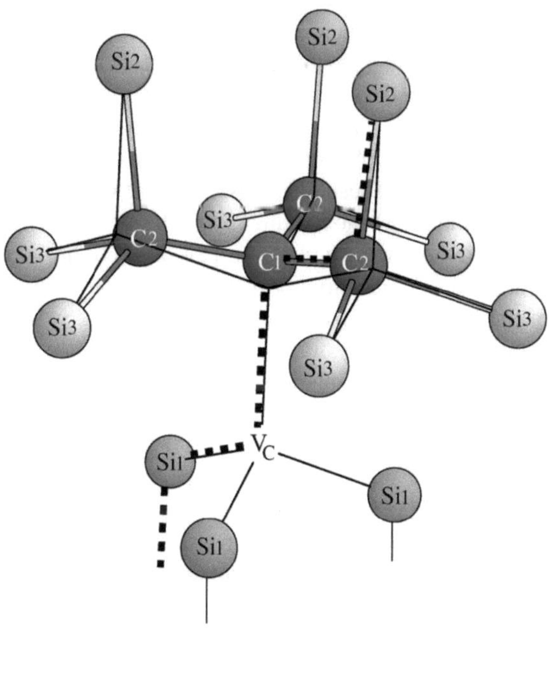

Fig. 2.30 Calculated relaxed structure of $(C_{Si}V_C)^{2+}$ in 6H-SiC. The thin lines mark the position of the bonds in the ideal lattice. The carbon antisite C1 relaxes by 16% of the bond length in the ideal crystal away from the vacancy. Reprinted with permission from [203]. Copyright 2001, American Physical Society

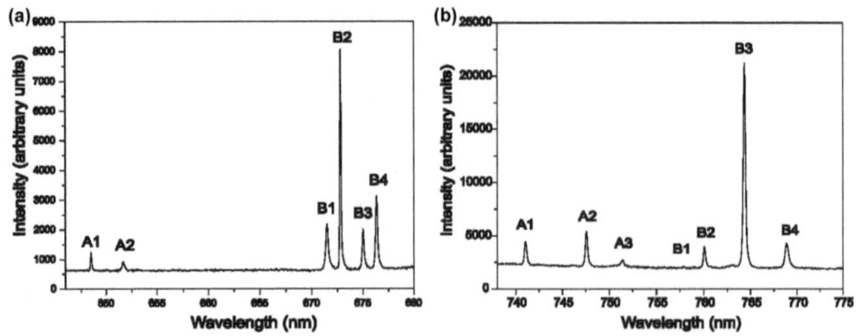

Fig. 2.31 A and B spectra of **a** 4H-SiC and **b** 6H-SiC excited by 488 nm light at 7 K. Reprinted with permission from [208]. Copyright 2009, American Physical Society

and *SI*5 center were found to play an important role in the semi-insulating property of SiC. Two sets of photoluminescence lines (called A and B) were detected from electron irradiated and ion implanted 4H-SiC at 7 K (Fig. 2.31) [208]. The spectral details, temperature dependence, and annealing characteristics suggest that the A and B spectra stem from the neutral on-axis and off-axis carbon antisite–vacancy pairs $C_{Si}V_C{}^0$. The relatively impure and highly compensated-doped samples exhibited very strong A and B PL, whereas the largely uncompensated *n* (N) or *p* (Al) doped samples exhibited mainly PL of the Si vacancy related defects and weak PL of A and B even after annealing at 900 °C.

The carbon antisite–vacancy defect showing *A*/*B* PL can serve as a single-photon source [209]. These defects were produced in high purity and semi-insulating 4H-SiC wafers by irradiation with 2 MeV electrons followed by annealing. Low electron fluences were adopted to give rise to isolated single defects and the annealing temperature was tuned to increase/decrease the number of defects. The sample showed obvious A/B PL lines (Fig. 2.32). The lines are at 648.7 nm (*A*1), 651.8 nm (*A*2), 665.1 nm (*A*3), 668.5 nm (*A*4), 671.7 nm (*B*1), 673.0 nm (*B*2), 675.2 nm (*B*3), and 676.5 nm (*B*4). These lines arise from the $C_{Si}V_C$ defect. In contrast to the previous assignment [208], the ab initio supercell calculation in combination with group theory indicates that the neutral $C_{Si}V_C$ pair does not have visible PL and so the positively charged $C_{Si}V_C$ pair (spin = 1/2) with predicted visible PL was proposed as the origin of the *A*/*B* PL. PL from single defect centers was observed by room-temperature confocal microscopy, as shown in Fig. 2.33a. These spectra are in accordance with the *A*/*B* lines. Single-photon emission was confirmed using a Hanbury Brown–Twiss interferometer based on the observation that the second-order photon correlation function at zero delay is $g^{(2)}(0) \approx 0$ (Fig. 2.33b). The typical acquired excited state lifetime is 1.2 ns. 45% of the defect emitters measured in a sample irradiated and annealed at 300 °C are stable over a long period of time, whereas the other emitters showed PL blinking and some eventually photobleached. The samples annealed at a higher temperature (500 °C) contained nearly no blinking centers.

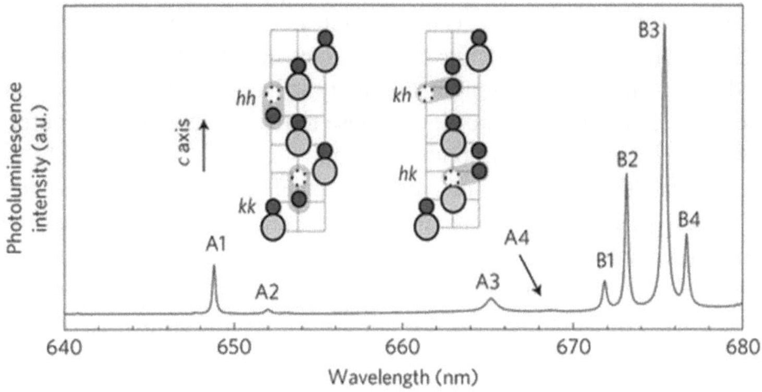

Fig. 2.32 A and B PL lines observed from electron irradiated 4H-SiC acquired at 80 K. The proposed atomic structure of the defects $C_{Si}V_C$ yielding the PL is shown in the (1120) plane (small/large circle representing C/Si atom and small open circles denoting vacancies). The carbon antisite–vacancy pair has four possible configurations: *hh*/*kk* axial configurations with C_{3v} symmetry, and *hk*/*kh* basal plane configurations with C_{1h} symmetry. *h*/*k* denotes the hexagonal/cubic site in the SiC lattice. Reprinted with permission from [209]. Copyright 2014, Springer Nature Limited

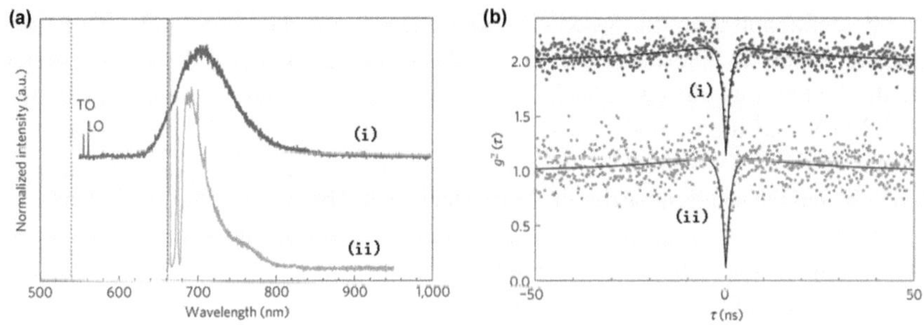

Fig. 2.33 **a** Room-temperature PL spectra of a single photon source in the irradiated 4H-SiC excited by the 532 nm (i) and 660 nm (ii) lines. The dashed lines denote excitation wavelengths. **b** Corresponding anti-bunching curves. Reprinted with permission from [209]. Copyright 2014, Springer Nature Limited

2.4.1.5 Interstitial

Semiconductor materials exhibit few spectroscopically fingerprints of self-interstitials. IR absorption spectroscopy was used to distinguish between amorphous and crystalline structures of excess carbon formed in the region of SiC grain boundaries as an interstitial phase [210]. The amorphous phase produced a compressive strain in the 3C-SiC films, leading to IR active vibration of C–C bonds. The IR absorption spectrum of the interstitial graphite shows that the characteristic absorption of graphite at 1580 cm^{-1} was split into two peaks at 1629 and 1542 cm^{-1} that may be ascribed to crystalline and amorphous graphite phases. DFT calculations were carried out on models for the Z_1/Z_2 defects observed in as-grown and irradiated n-type 4H-SiC [125]. The most likely model involves an interstitial nitrogen–interstitial carbon (N_i-C_i) defect, which is exceptionally thermally stable, bistable, and has negative-U character with donor and acceptor levels close to that attributed to the defect. This defect would disappear when trapping a mobile vacancy which probably explains the simultaneous disappearance of a positron annihilation signal due to vacancies and Z_1/Z_2 defects. DFT–LSDA calculations were carried out to investigate clusters of carbon interstitials in 3C- and 4H-SiC [211]. The formation energies suggest that the complex formation of carbon interstitials or their aggregation to carbon antisites is energetically favored in SiC. The electronic and vibronic properties of the carbon interstitials and their aggregates depend on the polytype. The small carbon interstitial and antisite clusters in 3C- and 4H-SiC were investigated by using DFT–LSDA calculations [212]. All considered defect complexes possess localized vibrational modes whose phonon energies are well above that of SiC bulk phonons. Particularly, the compact antisite clusters show high-frequency localized vibrational modes up to 250 meV. The dicarbon antisite seems to correspond to the D_{II} center, and the carbon split interstitial is a probable origin of the P–T centers. The high-frequency modes suggest that the U center is a carbon antisite-based defect.

A combination of kinetic Activation Relaxation Technique sampling with empirical potential and ab initio based climbing image nudged elastic band calculations was carried out to investigate the $(C_{BC})_3$ cluster in 3C-SiC [213]. It is a very stable small C interstitial cluster in SiC. The fastest migration path has an energy barrier of 4.29 eV and the second-fastest path has a barrier of 4.37 eV. The dissociation barrier of the $(C_{BC})_3$ cluster into a non-interacting $(C_{BC})_2$ defect and $C_{sp<100>}$ is 4.83 eV, and the rotation barrier of the cluster moving between different (111) planes is 4.14 eV. The $(C_{BC})_3$ cluster dissociates at above 1500 K. The M center in 4H-SiC is a bistable defect responsible for a family of electron traps. A bistable defect M was observed from epitaxial 4H-SiC n-type layers implanted at room temperature with a low fluence of MeV protons [214]. It has two DLTS peaks, M_1 and M_3 at E_c–0.42 and E_c–0.75 eV in one configuration and one peak, M_2 at E_c–0.70 eV in another configuration. The thermal activation energies for the transformation between the bistable defect peaks are 0.90 and 1.40 eV. Using DFT–HSE06 calculations and junction spectroscopy, the properties of M, including bistability, annealing, reconfiguration kinetics, and electronic levels, were demonstrated to match those of the carbon self-interstitial [215]. Excess carbon was introduced into low-doped n-type 4H-SiC epilayers by thermal annealing and employing a pyrolyzed carbon source covering the SiC surface [216]. The layers have low initial concentrations of carbon vacancies, and excess carbon would cause complete V_C annihilation and formation of defects due to excess carbon including carbon interstitials C_i and their higher-order complexes. A set of three defect levels were observed in low-temperature DLTS spectra, with activation energies of 0.38, 0.59, and 0.7 eV, which are likely C_i-related. Their properties are different from that found for the M center. The DLTS and Laplace-DLTS study of the low-energy electron irradiated 4H-SiC indicates that the electron irradiation introduced deep level defects of EH_1 and EH_3 [217] that were previously assigned to carbon interstitial-related defects. It was proposed that

EH_1 and EH_3 are identical to M_1 and M_3 that were assigned to carbon interstitial defects. The defects were supposed to be $C_i^{=}$ (*h*) and C_i^0 (*h*), respectively.

2.4.1.6 Studies on Multiple Point Defects

The effects of deviation from stoichiometry on the relative abundances of native defects and electronic properties of 3C-SiC have been investigated by using ab initio calculations [218]. The result shows that the dominant defect in a perfectly stoichiometric SiC is the electrically inactive Si_C–C_{Si} pair. In Si-rich SiC, the concentrations of the stoichiometry compensating defects depend strongly on the Fermi-level position. The Si_C antisite and carbon vacancy are separately the dominant defects in *n*- and *p*-type SiC. As a double donor, the carbon vacancy acts as a charge-compensating defect for lowered Fermi levels and reduces the doping efficiency of acceptors. The electrically inactive C_{Si} antisite dominates in the C-rich SiC regardless of the position of the Fermi level. The slightly C-rich SiC is thus more suitable for *p*-type doping in electronic applications. Two negative-U centers in 4H-SiC were studied by using various capacitance transient techniques [118]. Each center generated one acceptor level (–/0) and one donor level (0/+). The two-electron emissions from the two acceptor levels produced the previously reported DLTS peak associated with the Z_1 center. The levels at E_c–(0.50–0.54) and E_c–(0.43–0.46) eV correspond to the donor levels Z_1^0 and Z_2^0, respectively, and the levels at E_c–0.67 and E_c–0.71 eV correspond to the acceptor levels Z_1^- and Z_2^-, respectively. The electronic and atomic structures of the monovacancies and antisite defects in 4H-SiC in all possible charge states were investigated using a plane-wave pseudopotential method based on DFT–LSDA calculations [219]. The calculation indicates no negative-U behavior for the carbon vacancies. Hence, the single positive charge state of the carbon vacancy V_C^+ is stable, in agreement with previous experimental result. The silicon antisite Si_C^+ was found to be stable at small values of electron chemical potential. These results agree with previous experiments. The capacitance transient techniques indicate low concentrations of electron traps (denoted as $Z_{1/2}$, $EH_{6/7}$ and titanium) and hole traps (denoted as $HS1$ and shallow boron) in the *n*-type 4H-SiC epilayers [220]. The concentration of intrinsic defects increased with growth temperature. The incorporation of shallow B decreased at higher growth temperatures, whereas the Ti concentration is insensitive to the growth temperature. The concentration of shallow B and Ti increased with increasing C/Si ratio. DFT–HSE06 calculations give signs of multiple defect clusters in 4H-SiC by combining divacancies (*VV*) with intrinsic defects and common extrinsic elements of oxygen, nitrogen, and hydrogen [221]. Two configurations of OV^0 (one O atom filling the C vacancy in VV^0) were supposed to be sources of previously reported PL5 and PL6 features in PL spectra, and one configuration of NHV^0 (one H atom attached to the N atom in NV^0) is the source of PL8. Additionally, the OHV^+ (OV with one H atom bonded with the O atom at + 1 charge state) exhibited as potential color centers with $S = 1$.

The concentration of the *EH*6/7 and *HS*1 centers decreased with increasing C/Si ratio. Electron irradiation-induced defects in epitaxially grown 3C-SiC crystals were studied by using ESR measurements [68]. The results reveal an isotropic ESR $T1$ center that consists of five lines equally spaced at about 1.5 G and has a g value of 2.0029 ± 0.0001. Isochronal and isothermal annealing of electron-irradiated 3C-SiC shows three stages (150, 350, and 750 °C) for this center and that the 750 °C stage exhibited first-order reaction with an activation energy of 2.2 ± 0.3 eV. Deep level defects in electron-irradiated 4H-SiC CVD-grown epitaxial layers grown were studied using DLTS [71]. The measurements performed on electron-irradiated p^+n junctions within the temperature range of 100–750 K revealed several electron traps and one hole trap with thermal ionization energies of 0.35–1.65 eV. Most of these defects are observable at a dose of irradiation as low as 5×10^{13} cm^{-2}. The temperature dependence studies show that the capture cross sections of these defects are temperature independent. A type of spin defect in 4H-SiC generated via hydrogen ion implantation followed by high-temperature postannealing was observed [222]. Their PL spectrum and ODMR spectra are different from those of previously discovered defects. The PL peak at 1007 nm (20 K) belongs to ZPL of PL8 defects. Several other peaks were deduced to be unrelated to PL8 defects. The generation of these defects is controllable by optimizing the annealing temperature. Deep centers in n-type 4H-SiC and 6H-SiC irradiated with 8 MeV protons were investigated by using capacitance spectroscopy and EPR [223]. Irradiation led to an increase in the concentration of uncompensated donors in an n-type material. Irradiation generated deep acceptor centers in both polytypes to which electrons from shallower donor levels pass. This shifts the Fermi level downward to render n-type 6H- and 4H-SiC layers semi-insulating at room temperature. 4H-SiC epilayers were irradiated with either protons or electrons and electrically active defects were studied by means of DLTS [224]. Motion of defects occurred at temperatures as low as 350–400 K. Some defects created by the irradiation are unstable and the application of an electric field enhanced recovery and appearance of more stable defects. The irradiation followed by annealing generated two minor levels B and M. Two other levels $S1$ and $S2$ are located at 0.40 and 0.71 eV below E_c with a capture cross section of 8×10^{-17} and 3×10^{-15} cm^2 (335 K), respectively. They were proposed to be two charge states of the same acceptor center S which showed a first order annealing process with an activation energy of 1.8 eV. Electrically active defects in 4H-SiC were produced by irradiation of low energy electrons below the threshold of Si atom displacement [225]. The majority and minority carrier traps induced by radiation were analyzed by using capacitance transient techniques. Four electron traps ($EH1$, Z_1/Z_2, $EH3$, and $EH7$) and one hole trap ($HS2$) were detected (Fig. 2.34) and suggested to be related to the initial displacement of carbon atoms. Their properties are summarized in Table 2.2. The concentrations of these defects increased linearly with irradiation dose, suggesting that no divacancies or di-interstitials were generated. None of the observed defects were found to be an intrinsic defect-impurity complex. The probable candidates for these defects are limited to carbon vacancies, split interstitials, and antisites. No additional defects were detected from samples irradiated by electrons with energies higher than the threshold of Si atom displacement. Deep levels in as-grown and electron-irradiated p-type 4H-SiC epilayers were investigated by using DLTS [226]. The D center and four deep levels were observed from as-grown epilayers. In *p*-type 4H-SiC, reactive ion etching followed by thermal treatment (at 1150 °C) generated the *HK*0 (E_v + 0.79 eV) and *HK*2 (E_v + 0.84 eV) centers. The electron irradiation at 116–400 keV created two deep levels at 0.98 eV (*EP*1) and 1.44 eV (*EP*2), while two additional deep levels (*EP*3 and *EP*4) were observed in the samples irradiated at 400 keV. These centers were annealed out and the *HK*4 (E_v + 1.44 eV) concentration increased after annealing at 950 °C. The electron irradiation at over 160 keV followed by annealing at 950 °C generated three deep levels at 0.30 eV (*UK*1), 0.58 eV (*UK*2), and 1.44 eV (*HK*4). All the centers except for the *D* center were reduced to below the detection limit (1–3 × 10^{11} cm^{-3}) by annealing at 1550 °C for 30 min. Low temperature annealing of electron irradiation-induced deep levels in 4H-SiC was reported [227]. The major DLTS peak S2 associated with the energy level at E_c–0.39 eV vanished within 360–400 K. Some rearrangement of the peak S3 associated with the defect Z_1/Z_2 with energy level at (E_c − 0.5)/(E_c − 0.65) eV occurred within 400–470 K. A net free charge carrier concentration increased along with disappearance of peak S2.

The point defects can strongly affect the carrier lifetimes in SiC. There is a significant reduction in deep level defects and increase of carrier lifetime in 4H-SiC after carrying out carbon or silicon ion implantation into the shallow surface layer of 250 nm and subsequent annealing at ≥ 1600 °C [228]. Reduction of $Z_{1/2}$ and $EH_{6/7}$ traps from 3 × 10^{13} cm^{-3} to below the detection limit 5 × 10^{11} cm^{-3} was observed by using DLTS in the region 4 μm underneath the implanted layer. Minority carrier lifetime increased from 126 to 240 ns after the implantation. It was proposed that the implanted layer acts as a source of carbon interstitials which indiffuse during annealing and accelerate annealing out of grown-in defects in the region underneath the implanted layer. The influence of the carbon-implantation/annealing on

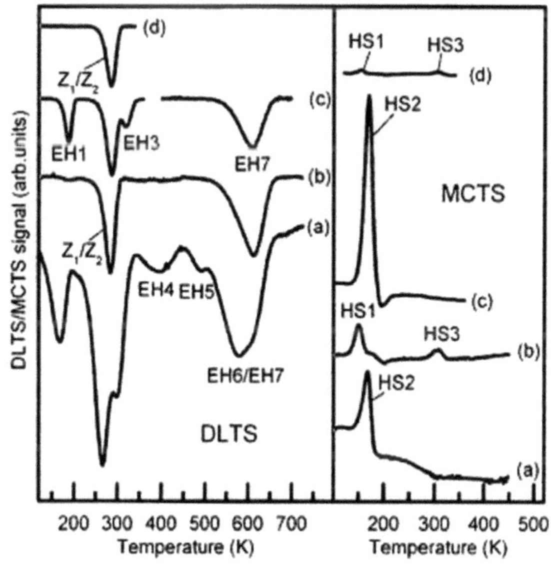

Fig. 2.34 Deep level transient spectroscopy and minority carrier transient spectroscopy spectra of electron irradiated 4H-SiC. **a** Irradiation with 9 MeV electrons. **b** Irradiation with 9 MeV electrons and annealing at 950 °C for 1 h. **c** Irradiation with 210 keV electrons. **d** Irradiation with 210 keV electrons and annealing at 950 °C for 1 h. Reprinted with permission from [225]. Copyright 2004, American Institute of Physics

Table 2.2 Electrical properties of the defects observed in 4H-SiC irradiated by low energy electrons

Position in the bandgap	Label	Peak temperature (K) (for 230 ms rate window)	Capture cross section (cm^2)
E_c–0.41 eV	$EH1$ (SS_1)	188	4×10^{-15}
E_c–0.68 eV	($Z_{1/2}$)	287	2×10^{-14}
E_c–0.71 eV	$EH3$ (S_2)	320	4×10^{-15}
E_c–1.54 eV	$EH7$	607	3×10^{-14}
E_v–0.39 eV	$HS2$	170	2×10^{-14}

The capture cross section is estimated from the intercept of the Arrhenius plot. Reprinted with permission from [225]. Copyright 2004, American Institute of Physics

the main lifetime limiting defect $Z_{1/2}$ in thick 4H-SiC epilayers was studied [229]. The carrier lifetime increased from less than 200 ns to over 1 μs at room temperature in the annealed carbon-implanted layer. The thick 4H-SiC epilayers that experienced carbon-implantation/annealing are applicable for fabrication of high-voltage bipolar devices with improved conductivity modulation. The $Z_{1/2}$ defect has low migration mobility in as-grown samples, and additional mobile C_i limits the amount of $Z_{1/2}$ that is able to be annealed out. The results support association of $Z_{1/2}$ with V_C. C_i seems to indiffuse from the carbon-implanted surface layer into the volume of the epilayer far below the implanted layer and annihilates with V_C in the volume, thus reducing the V_C-related trap density. Carrier lifetimes in *n*-type 4H-SiC epilayers were investigated by using differential microwave photoconductance decay measurements [230]. The $Z_{1/2}$ and $EH_{6/7}$ centers were identified as effective recombination centers. When the $Z_{1/2}$ (and $EH_{6/7}$) concentration is higher than 10^{13} cm^{-3}, the carrier lifetime is inversely proportional to the trap concentration (Fig. 2.35a), and the lifetime increased with increasing excitation light intensity. The energy levels of major deep centers observed in as-grown *n*-type and *p*-type 4H-SiC epilayers are illustrated in Fig. 2.35c. Other recombination processes such as surface recombination and recombination in the substrate limit the lifetime when the $Z_{1/2}$ concentration is lower than 10^{13} cm^{-3}, and the carrier lifetime decreased with increasing light excitation intensity. Carrier lifetime is controllable by controlling the $Z_{1/2}$ and $EH_{6/7}$ concentration by low-energy electron irradiation. The $EH_{6/7}$ concentration is close to the $Z_{1/2}$ concentration for most as-grown, electron-irradiated, and annealed 4H-SiC epilayers (Fig. 2.35b). It is difficult to judge which defect center ($Z_{1/2}$ or $EH_{6/7}$) is the dominant recombination center in the material. In another study, low-injection minority carrier lifetimes and deep trap spectra were investigated in $n-$4H-SiC epilayers in order to enable the separation of bulk lifetimes from surface recombination effects [231]. It was concluded that it is Z_1/Z_2 alone that controls the minority carrier lifetime in this material, as judged from the linear dependence of the inverse bulk minority carrier lifetime on the concentration of Z_1/Z_2 defects and from the behavior of the deep trap spectra in 4H-SiC *p–i–n* diodes under forward bias.

The irradiation effects on defects as well as the defect migration processes in SiC have been extensively studied by using theoretical calculations. In fact, other than

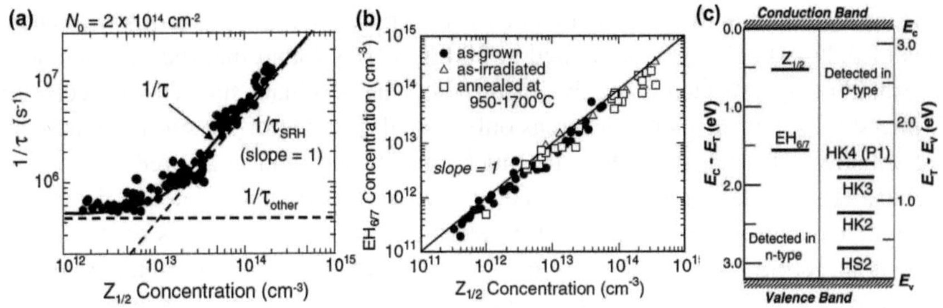

Fig. 2.35 **a** Relation between the inverse of carrier lifetime and the $Z_{1/2}$ concentration in the 4H-SiC epilayers measured with an excitation intensity of 2×10^{14} cm^{-2}. **b** $Z_{1/2}$ concentration versus *EH*6/7 concentration obtained for various as-grown, electron-irradiated, and annealed 4H-SiC epilayers. **c** Energy levels of major deep centers observed in as-grown *n*-type and *p*-type 4H-SiC epilayers. Reprinted with permission from [230]. Copyright 2008, John Wiley & Sons, Inc.

diffusion of the point defects, the C or Si self-diffusion had long been discussed. The ^{14}C self-diffusion coefficients for both lattice (D^*_{lc}) and grain boundary (D^*_{bc}) transport in high purity CVD β-SiC single crystals were studied within 2128–2374 K [232]. D^*_{bc} is 10^5–10^6 faster than D^*_{bc}. A vacancy mechanism was assumed for lattice transport. The crystallography and energetics give support of a path of transport that involves an initial jump to a vacant tetrahedral site succeeded by a jump to a normally occupied C vacancy. The authors further studied the ^{30}Si lattice self-diffusion coefficients within 2283–2547 K [233]. Lattice diffusion of Si was supposed to occur by a mechanism involving a direct jump to the nearest Si vacancy without previous occupation of a normally unfilled position. They also studied self-diffusion of ^{30}Si in high purity and N-doped α-SiC within 2273–2573 K [234]. The diffusion ($D_{\mathrm{Si}}{}^*$) in N-doped crystals exceeded that in pure crystals owing to increase in the concentration of the charged accepter-type Si vacancies in the presence of N species. The diffusion $D_{\mathrm{C}}{}^*$ exceeded $D_{\mathrm{Si}}{}^*$ by 102, mainly because of the greater entropy of migration of C. The threshold energy for atomic displacement and the evolution of high-energy displacement cascades in 3C-SiC were examined using molecular dynamics simulation by employing a modified Tersoff potential and a DFT-derived repulsive potential to describe the inter-atomic interactions [235]. The results indicate that the minimum displacement energy is about 36 eV for Si and 28 eV for C. The cascade lifetime is of the order of 0.1 ps, and the surviving C vacancies and interstitials outnumbered the corresponding Si defects by a factor of ~ 3. Molecular dynamics simulation using a modified Tersoff potential was used to simulate high-energy (50 keV) displacement cascades in 3C-SiC [236]. The results show that the cascade lifetime is 10 times shorter than that in metals, and C interstitials and vacancies are the dominated surviving defects. Antisite defects are generated on both sublattices. Only 19% of the interstitial population forms clusters with small sizes. The largest cluster contains only four interstitial atoms. All clusters are created by a quenched-in mechanism directly from the collisional phase of the cascade to their final arrangements. The simulation further evaluates the primary damage state and defect production in the displacement cascades in 3C-SiC [237]. Recoils with energies of 0.25–50 keV were simulated at 300 K. It was found that the displacement threshold energy surface is highly anisotropic. The dominant surviving defects are C interstitials and vacancies, whereas only a small fraction of the surviving defects are antisites. The defect clusters are much smaller and rarer compared to those in metals. The simulation also shows that the defect production efficiency decreases with increasing recoil energy. The formation energies and properties of the native defects in 3C-SiC were studied by using DFT–LDA based on the pseudopotential plan-wave method combined with the molecular dynamics simulation using Tersoff potentials [238]. It was found that the most favorable configurations for the C interstitials are <100> and <110> dumbbells with formation energies of 3.16–3.59 eV, and the most favorable Si interstitial is Si tetrahedral surrounded by four C atoms with a formation energy of 6.17 eV. The formation energies of the vacancies and antisite defects obtained by molecular dynamics simulation agree with those obtained by DFT calculation. The authors further employed molecular dynamics simulation to investigate cascade overlap, damage accumulation, and amorphization processes in 3C-SiC [239]. A large number of 10 keV displacement cascades were randomly generated in a model crystal to simulate damage and cause amorphization. At low dose the damage state is dominated by point defects and small clusters. The coalescence and growth of clusters at intermediate and higher doses tend to cause amorphization in SiC. The homogeneous nucleation of small clusters at low dose underpins the homogeneous-like amorphization observed in SiC. Both interstitials and antisites play important roles in leading to amorphization in SiC. The topologies (total pair correlation function, bond-angle, and bond-length distributions) of damage accumulation in the crystal suggest occurrence of crystalline-to-amorphous transition at about 0.28 dpa, close to the experimental value of 0.27 dpa under similar irradiation conditions. An empirical many-body interatomic potential was developed by fitting to various equilibrium properties and stable defect configurations in bulk SiC [240]. This parameterized potential was used to calculate defect formation energies and to determine the most stable configurations for interstitials using the molecular dynamics method. Molecular dynamics simulation of displacement cascades in 3C-SiC indicates that most defects created are point defects, where interstitials are displaced only a small distance from the nearest vacancies [241]. 40% of the interstitials produced in SiC survive as freely migrating ones that can contribute to microstructural evolution during irradiation of SiC. The stable Frenkel pairs were created with low-energy recoils using molecular dynamics simulation and they were annealed at different temperatures using simulations. The activation energies for recombination of interstitials with vacancies range from 0.22 to 1.6 eV for C Frenkel pairs and from 0.28 to 0.9 eV for Si Frenkel pairs. The spontaneous recombination distance was estimated to be 0.66 and 0.70 a_0 (lattice constant of SiC) for the C and Si sublattices, respectively. The formation energies of vacancies are smaller than those obtained by ab initio calculations, the properties of antisites and interstitials agree well with those calculated by ab initio methods. DFT–HSE06 calculations clarify the formation energies and charge transition levels of native point defects and carbon clusters in carbon-rich 4H-SiC [242]. The results show that neutral Si antisites are energetically favorable among Si-point defects in a wide range of Fermi level positions, whereas negatively charged Si vacancies and a positively charged Si interstitial on a site surrounded by six Si and four C atoms become favorable under *n*- and *p*-type conditions, respectively. Neutral C antisites are favorable among C-point defects under intrinsic and *n*-type conditions, whereas positively charged C vacancies are favorable under *p*-type conditions. The annealing and defect migration processes in SiC have been extensively studied. The thermal conductivity and effects of point defects of crystalline 3C-SiC were calculated by using molecular dynamics simulation to evaluate the heat current correlation function and thus obtain the conductivity through the Green–Kubo expression in linear response theory [243]. The point defects show little influence on the heat capacity and thermal expansion coefficient, even at a high concentration of 0.5%. The thermal conductivity degrades markedly as a result of the highly enhanced decay of the heat current correlation caused by defect scattering of phonons. The conductivity exhibits as temperature independent. Migration of vacancies at high temperature was studied by the self-consistent charge density functional based tight-binding method [244]. The energy barrier for sublattice migration of V_{Si} is lower than that for V_C and hence, V_{Si} should anneal out at lower temperature as supported by experiments. The calculation shows that the motion of vacancies can mediate the migration of antisites. Vacancy assisted diffusion of carbon antisites is much faster than that of silicon antisites, thereby creating carbon antisite clusters. The result shows that the mobility of carbon antisites can trigger the formation of larger antisite complexes. Diffusion of intrinsic defects in 3C-SiC was studied by using DFT–LDA (or LSDA) calculations [245]. The results show that the vacancies migrate to their own sublattice. The carbon split interstitials (Fig. 2.36) and two silicon interstitials (tetrahedrally carbon-coordinated interstitial in the *p*-type materials and the split interstitial in the compensated and *n*-type materials) are more mobile than the vacancies, thereby playing a prominent role in diffusion. The metastable Si vacancy in *p*-type and compensated SiC is transformed into the more stable vacancy-antisite complex, suppressing its contribution to diffusion processes. The most important migration channels for vacancies and interstitials in 3C-SiC are depicted in Fig. 2.37. The annealing kinetics of mobile intrinsic defects in 3C-SiC was investigated by using the same method [246]. The calculated migration and reaction barriers help to determine the annealing mechanisms. The carbon and silicon interstitials have higher mobility than the vacancies, thus driving the annealing mechanisms at lower temperatures including vacancy-interstitial recombination and formation of interstitial carbon clusters. These clusters emit carbon interstitials at higher temperature. In *p*-type SiC, transformation of the silicon vacancy into the more stable vacancy–antisite complex serves as an annealing mechanism before vacancy migration. The long range migration of point defects in 3C-SiC was subsequently investigated by both molecular dynamics simulation and the nudged-elastic band method [247]. The results show that stable C split interstitials can migrate via the first- or second-nearest neighbor sites, whereas Si interstitials migrate directly from one tetrahedral position to another neighboring equivalent position by a kick-in/kick-out process. C and Si vacancies jump to one of their equivalent sites through a direct migration mechanism. The calculated migration barriers for C and Si interstitials agree with experiments. EPR studies reveal that vacancies (V_C and V_{Si}), $C_{Si}V_C$, and V_CV_{Si} are common defects in high-purity semi-insulating 4H-SiC substrates [248]. The thermal activation energies of ~ 0.8–0.9, ~ 1.1, ~ 1.25–1.3, and ~ 1.5 eV, as determined from the temperature dependence of the resistivity in different types of 4H-SiC substrates, were associated to different deep acceptor levels of V_{Si}, V_C, $C_{Si}V_C$, and V_CV_{Si}. Carrier compensation processes were proposed to explain the changed thermal activation energies due to high temperature annealing. The concentrations of all types of defects decrease after high temperature annealing in all types of samples. The energy transferred to the electron system of pre-damaged 4H-SiC by energetic ions via inelastic ionization can effectively anneal pre-existing defects and restore the structural order [249]. The threshold determined for this recovery process reveals that it can be activated respectively by 750 and 850 keV Si and C self-ions. This technique can optimize the fabrication of SiC-based devices by offering a room-temperature approach to repair atomic lattice structures. The development of defect populations after proton irradiation of *n*-type 4H-SiC and subsequent annealing experiments was studied by using DLTS and PL spectroscopy [250]. A model describing the evolution of irradiation-induced point defects during annealing below 1000 °C proposes that the EH_4 and EH_5 traps found by DLTS originate from the (+/0) charge transition level belonging to different configurations of $C_{Si}V_C$. The transformation channel between V_{Si} and $C_{Si}V_C$ is effectively blocked under *n*-type conditions and is available in samples where the Fermi level moves toward the center of the band gap due to irradiation-induced donor compensation. The annealing of V_{Si} and V_C is dominated by recombination with residual self-interstitials at temperatures up to 400 °C. At higher temperatures, the $C_{Si}V_C$ pair density decreases and V_C density increases, suggesting dissociation of the $C_{Si}V_C$ pair into separate C_{Si} and V_C. Thermal oxidation of 4H-SiC at 1150–1300 °C can reduce the $Z_{1/2}$ and $EH_{6/7}$ concentrations to below 1×10^{11} cm^{-3} [131]. The oxidation also generated a high concentration of *HK*0 center ($E_v + 0.78$ eV). Additional annealing in Ar at 1550 °C eliminated the *HK*0 center. All the major deep levels can be eliminated by the two-step thermal treatment. The carrier lifetime in 4H-SiC epilayers was improved from 0.64 (as-grown) to 4.52 μs by this method. Energy barriers of the key annealing reactions of neutral and charged point defects in SiC were calculated with ab initio DFT methods [251]. In order to search for the lowest energy migration paths, the preliminary path was first established based on ab initio molecular dynamics (AIMD) simulations. It was proposed that the C interstitial and Si antisite annealing reaction provides a critical barrier that governs both the recovery stage III and amorphization processes. Molecular dynamics simulation was employed to study the V_{Si} centers in He and Si ions co-implanted 4H-SiC [252]. The simulation shows that the decrease of the V_{Si} center upon high temperature annealing is partly due to the transformation of some silicon vacancies to $C_{Si}V_C$ defects.

The aggregates of carbon interstitials in 3C- and 4H-SiC have been studied theoretically [211]. Formation of carbon aggregates is found to be energetically favored and all the carbon clusters are electrically active. The electronic and vibronic properties depend strongly on the polytype of SiC. Additionally, comparison of the calculated

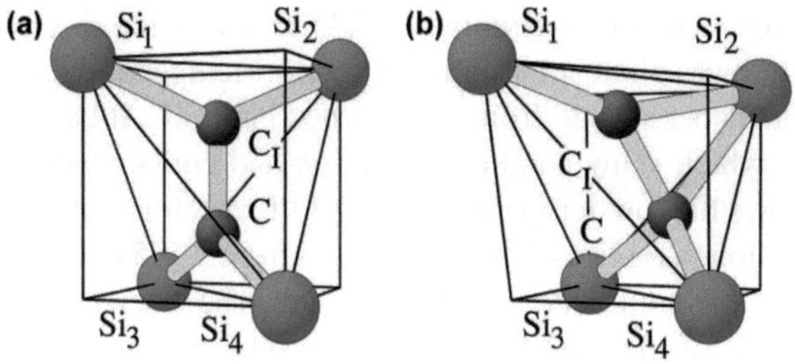

Fig. 2.36 Configuration of the carbon split interstitial $C_{sp<100>}$: **a** <100> orientation with D_{2d}/D_2 symmetry and **b** tilted $C_{sp<100>}$. The $C_{spSi<100>}$ configuration corresponds to the untilted $C_{sp<100>}$ with the C-C_I pair replaced by a Si-C_I pair and C neighbors instead of Si neighbors. Reprinted with permission from [245]. Copyright 2003, American Physical Society

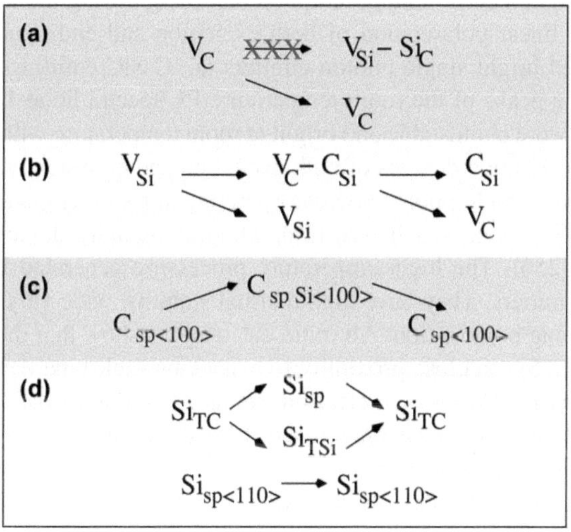

Fig. 2.37 Dominant migration mechanisms of interstitials and vacancies in 3C-SiC: **a** C vacancies (V_C) migrate on the C sublattice only, **b** transformation of the Si vacancy (V_{Si}) into a C vacancy-antisite complex V_C-C_{Si} (*p*-type and compensated) and vacancy migration on the Si sublattice (*n*-type), **c** C-interstitial migration via split-interstitial configurations $C_{sp<100>}$ and $C_{spSi<100>}$, and **d** kick-out mechanism for the C-coordinated Si interstitial (Si_{TC}) via split interstitials ($Si_{sp<100>}$ and $Si_{sp<110>}$) (*p*-type) and direct migration of the Si split interstitial $Si_{sp<110>}$ (compensated and *n*-type). Reprinted with permission from [245]. Copyright 2003, American Physical Society

local vibration modes of all the defect aggregates to those observed from the *P* and D_{II} PL centers suggests that these centers are related to the $(C_2)_{Si}$ and $[(C_2)_{Si}]_2$ defects, respectively.

A single-photon emitter in 4H-SiC was reported [253]. The room-temperature PL spectrum lies in ~ 590–700 nm. Three higher-energy peaks at 584.8 nm (E_1), 597.3 nm (E_2), and 609.6 nm (E_3) were assigned to the room-temperature zero phonon lines. The single-exponential fit of the PL decay curves indicates that there is only one excited state with a lifetime of 3.45 ± 0.1 ns for E_1 and 4.65 ± 0.3 ns for E_3. The emitter was photostable at room and low temperatures, emitting photons over 2 × 10^6 counts per second from unpatterned bulk SiC. It has two orthogonally polarized states, which show parallel absorption and emission dipole orientations. Figure 2.38a shows the low-temperature (18 K) PL spectra comprising sharp ZPLs and low-energy phonon replicas. The ZPL ranges from ~ 575 to 610 nm (Fig. 2.38b). The ZPL and polarization state are independent. The linewidth of the ZPL that accounts for > 30% of the total PL spectrum is smaller than 0.1 nm. The relation of the linewidth of the E_1 peak at 581.4 nm with temperature follows T^3 behavior (Fig. 2.38c). The emitter density increases twice if irradiating the sample with electrons (density: 10^{17} cm^{-2}, energy: 10 MeV) before annealing. Single photon emitters emitting in a visible region of 600–800 nm were also observed from 3C-SiC crystals [254]. The emitters show PL intensity reaching up to 2.45 × 10^5 counts per second. There was no significant blinking and photobleaching during the measuring time. The degrees of linear polarization of both excitation and emission reached 80%. A study revealed bright single-photon emitters in 3C-SiC emitting in the telecom range [255]. The peaks of the room-temperature PL spectra lie at 1085, 1188, and 1250 nm. They were photostable and bright at room temperature with a count rate of ~ MHz. The polarization degrees of both excitation and emission reach up to 97%. Electrically driven single-photon sources operating at room temperature were integrated into 4H-SiC devices, achieved using photolithography, dopant implantation, and annealing [256]. The high-temperature processing generated highly efficient single-photon emitters. They have high thermal stability, wide varying ZPL wavelengths, and strong polarization. Ab initio calculations show that the luminescence may stem from a Si_C in close proximity to a stacking fault-type defect. New color centers were observed from both electron irradiated and non-irradiated SiC lateral *p–i–n* diodes [257]. One type is only observable by charge carrier injection into the intrinsic layer, while two other types can be observed by optical excitation in both doped and intrinsic layers. The surface oxide is responsible for the origin of the PL emitters. These optically and electrically driven emitters exhibited clear signatures of nonclassical light within the visible to near-infrared spectral region. Both emitter types in the irradiated samples generated linearly polarized light. Room temperature electrical control of single photon emission from color centers in 4H-SiC surface was implemented [258]. Defects as single photon sources formed on the surface of *n*-type epilayers of 4H-SiC p^+–n–n^+ diodes by field oxidation process. The confocal microscopy shows that the electroluminescence intensity of single photon source can be controlled by minority carrier injection under forward bias voltages, PL intensity can be controlled by reverse bias voltages. Both electroluminescence and PL spectra exhibited no significant variations due to applied bias voltages. A single-photon emitter in 3C-SiC was verified, which can operate as an optical switch at room temperature under pulsed green laser illumination with a near-infrared pulsed laser as the control gate [259]. An ultrafast and reversible optical modulation with a high PL intensity suppression ratio up to 97.9% and a response time as short as 287.9 ± 5.7 ps was demonstrated.

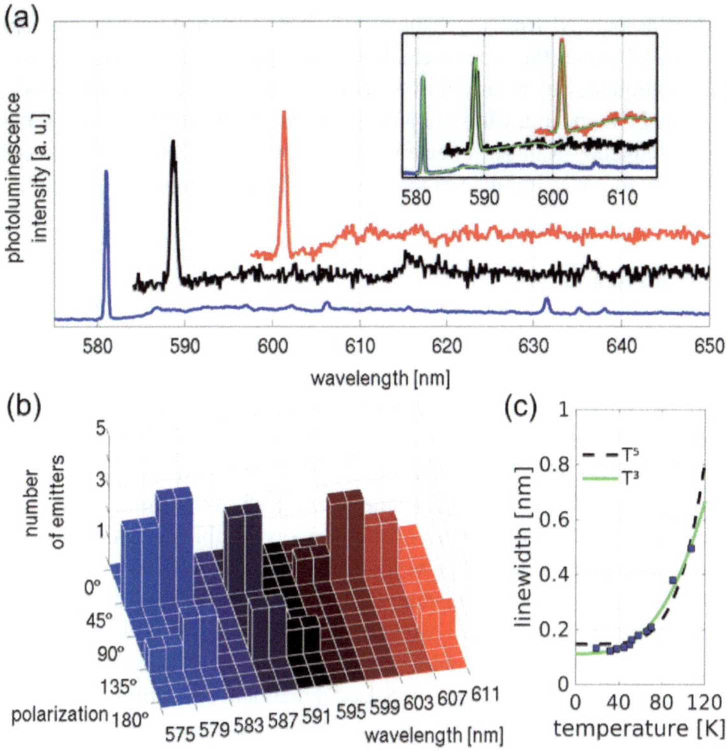

Fig. 2.38 **a** PL spectrum of E_1, E_2, and E_3 (from left to right) at 18 K. Inset: magnified spectral range containing ZPL (E_1, 581.2 nm; E_2, 588.9 nm; E_3, 601.5 nm) and first phonon modes, highlighted with Gaussian fits. **b** Histogram of ZPLs with polarization of various emitters at different wavelengths recorded at cryogenic temperatures. **c** Plot of the linewidth broadening with temperature of the peak at 581.4 nm (E_1, squares) and a fit ($\propto T^{3.02}$) (solid line). The dashed line shows fit with T^5. Reprinted with permission from [253]. Copyright 2016, Optica Publishing Group

The spin of optically addressable point defects can be coherently driven with alternating current (AC) electric fields [260]. Based on magnetic-dipole forbidden spin transitions, this scheme enables spatially confined spin control, the imaging of GHz-frequency electric fields, and the characterization of defect spin multiplicity. This technique can be used to control ensembles of QL1 defect in 6H-SiC that belongs to a UD-2 family of intrinsic defects. Positron annihilation characteristics (lifetime and Doppler broadening) in six basic vacancy-type defects of 6H-SiC and two NV complexes were determined using ab initio calculations [261]. They show that the coincidence Doppler broadening technique is the most relevant for observing the NV complexes, $N_C V_{Si}$, and $C_{Si} V_C$. The calculated positron characteristics for the other studied defects are too close for the defects to be distinguished using a single positron annihilation technique. A concept of operating V_{Si} centers as qudits was developed, involving room-temperature coherent control of high-dimensional quantum bits (qudits) associated with V_{Si} spins in 6H-^{28}SiC enriched with nuclear spin-free isotopes [262]. Besides the experimental excitation of a spectrally narrow qudit mode at the pump frequency by means of two-frequency ODMR, several other modes were excited in the ESR spectra, whose relative positions depend on the external magnetic field. Two-frequency Ramsey interferometry performed demonstrated absolute dc magnetometry that was immune to thermal noise and strain inhomogeneity.

2.4.1.7 Spin Readout

A new self-calibrating, solid-state based magnetometer was demonstrated [263], which measured magnetic field induced changes in current within a SiC *pn* junction caused by the interaction of external magnetic fields with the intrinsic defects. A deep-learning-enhanced method was proposed to extract effective parameters from continuous-wave ODMR spectra and Rabi oscillations with less time consumption [264]. A well-trained convolutional neural network shows promise of predicting the resonance peaks of ODMR spectra or the period of Rabi oscillations. Single-shot readout of single defects in 4H-SiC via spin-to-charge conversion was demonstrated [265], whereby the defect's spin state was mapped onto a long-lived charge state. This technique helps to achieve over 80% readout fidelity without pre- or post-selection, resulting in a high signal-to-noise ratio that enables measurement of long spin coherence times. Combined with pulsed dynamical decoupling sequences in an isotopically purified host material, single-spin $T_2 > 5$ s was realized. Coherent spin control and all-electrical readout of a small ensemble of spins in a SiC junction diode were demonstrated using pulsed electrically detected magnetic resonance [266]. A lock-in detection scheme based on a three stage modulation cycle enhanced the signal-to-noise ratio. This technique enables observation of coherent spin dynamics such as Rabi spin nutation, spin dephasing, and spin decoherence. Single shot readout for a nuclear spin in SiC was reported [267]. A deterministic nuclear spin initialization and readout fidelity of 94.95% with measurement duration of 1 ms were realized. Readout fidelity of 99.03% within 0.28 ms by sacrificing the success efficiency was realized by a dual-step readout scheme. A high-fidelity quantum state readout of a V_2 center (single silicon vacancy at the quasicubic site) nuclear spin in 4H-SiC was demonstrated based on a repetitive readout technique [268]. Up to 99.5% readout fidelity and 99% for state preparation were realized. This efficient readout helps to initialize the nuclear spin by measurement and its Rabi and Ramsey nutation was demonstrated. The nuclear spin was used as a long-lived memory for quantum sensing application of a weakly coupled diatomic nuclear-spin bath. Photo-electrical detection of single spins of single V_2 center in 4H-SiC was demonstrated [269]. With efficient photocarrier collection, a 1.7–2 times better signal-to-noise ratio was achieved for single spins of silicon vacancies with electrical detection than with optical detection suffering from saturating luminescence and internal reflection. Single-defect-like features were observed from the photocurrent image where PL is absent in the spectrum range of silicon vacancies.

2.4.2 Extended Defects

There are various extended defects in SiC, including 1D dislocations and micropipes commonly existing in general crystals and 2D stacking faults that occur owing to the polytypes of SiC [270]. Like point defects, these extended defects have also significant influence on the performances of the SiC-based electric and optical devices. Micropipes were found to severely limit the performance of 4H-SiC Schottky diodes, whereas carrot-like defects did not influence the value of breakdown voltage [271]. The screw dislocation density as identified by X-ray topography under the active area of the diode exhibited direct influence on the breakdown voltage. Only the diodes with low density of screw dislocations and free from micropipes could block 2 kV or higher. There have been extensive studies regarding the origins and properties of such extended defects. X-ray topography reveals there are a number of structural defects in α-SiC crystals grown in the $[000\bar{1}]$ and $[1\bar{1}00]$ directions by the modified-Lely method [272]. Micropipes and screw dislocations were observed in the $[000\bar{1}]$ grown crystal, which contained misoriented subgrains with tilt and twist components. Stacking faults and edge dislocations were observed in the $[1\bar{1}00]$ grown crystal, which had only tilted subgrain boundaries. Most of the defects formed at the very initial stage of growth.

2.4.2.1 Dislocations and Micropipes

Dislocations

Dislocations have long been observed from SiC crystals and have been widely investigated [273]. There are different types of dislocations in SiC crystals, including deformation-induced basal plane dislocations, and usual grown-in threading edge dislocations and threading screw dislocations. The correlations between dislocations and performances of the SiC-based electric devices had been explored. The nanoscale pits above the threading dislocations were found to be in the same location as the leakage current points in 4H-SiC Schottky barrier diodes and junction barrier Schottky diodes [274]. The study compared the leakage current of 1.2 kV and 200 A diodes with and without nanoscale pits. The leakage current in diodes without nanoscale pits is lower than those in diodes with pits. In the diodes without nanoscale pits the leakage currents were generated at step-bunching area and the leakage current at the threading dislocations was not observed.

X-ray topography and PL spectroscopy serve as major techniques to detect the dislocations in SiC. The various dislocations and glide elements in 6H-SiC have been investigated by etching in conjunction with optical microscopy and X-ray diffraction microscopy [275]. The conventional etching technique is suitable for identification of the intersection of dislocations on the (0001) surfaces. High-density dislocations in the (0001) planes were observed by diffraction microscopy and dislocations with $[11\bar{2}0]$ vectors were observed in addition to evidence of slip on both the basal planes and "puckered" pyramidal plane. Pileups formed by slip and dislocation walls formed by climb were identified. The dislocations in the parallel glide planes interact with each other (Fig. 2.39) and they line up in a direction perpendicular to the glide planes. A 6H-SiC crystal exhibiting three large growth spirals on its (0001) face was examined [276]. X-ray topography revealed a giant screw dislocation with Burgers vector parallel to [0001] at the center of each spiral and that these dislocations had hollow cores running their lengths. Diameters (~ 10 μm) of the hollow cores within the crystal are an order of magnitude bigger than the diameters of the microscopically visible holes at the surface outcrops of the dislocations. 6H-SiC single crystals were plastically deformed by uniaxial compression at a constant strain rate at temperatures of 1100–1800 °C, and their critical resolved shear stress τ_c and strain-rate sensitivity were measured [277]. The flow rate follows $(\tau_c - \tau_i)^n \exp(-U/kT)$, where the activation energy U, the stress exponent n and the stress τ_i were evaluated to be 3.4 ± 0.7 eV, 3.1 ± 0.4 and 3 MPa, respectively. The study shows that plastic strain of SiC crystals under the present deformation conditions is due to dislocation glide motion on the basal slip system. In the studied temperature range the basal dislocation motion is dominated by the Peierls mechanism rather than by the solution-hardening mechanism. Propagation of basal plane dislocations from the off-axis 4H-SiC substrate into the homo-epitaxial layer was investigated by chemical etching, optical microscopy, and transmission electron microscopy [278]. The basal plane dislocations in the substrates were converted into threading edge dislocations in the epilayer at the epilayers/substrate interface. TEM reveals that the converted dislocations are inclined from the *c*-axis toward the down-step direction by about 15°. The in-plane orientation of the inclination is toward the step flow direction within ± 10°. The conversion can be interpreted as a result of the image force in the epilayers between flowing growth steps and basal plane dislocations. Screw dislocation, threading edge dislocation, and basal plane dislocation in 4H-SiC substrates and epitaxial layers were observed by using X-ray topography [279]. Basal plane dislocations in the substrate are arc-shaped and have no distribution direction, whereas those in the epitaxial layer are straight and directed toward the off-orientation. The propagation of dislocations from the substrate to the epitaxial layer was observed (Fig. 2.40). Black arc-shaped lines marked A are basal plane dislocations in the substrate. Screw dislocations marked B and threading edge dislocations marked C observed at the same arrangement in three figures suggest that threading edge dislocations C in Fig. 2.40a, b were propagated from threading edge dislocations in the substrate in the same manner as screw dislocations. Threading edge dislocations D and basal plane dislocation E in the epitaxial layer were propagated from two basal plane dislocations A in the substrate. Threading edge dislocations in the substrate propagated into the epitaxial layer. Most of the basal plane dislocations in the substrate deflected to threading edge dislocations, and the rest propagated into the epitaxial layer.

High-geometrical-resolution imaging of dislocations in 4H-SiC was demonstrated by using monochromatic synchrotron topography [280]. In back-reflection topographs, 1*c* screw dislocation images are "magnified" to appear as circular white

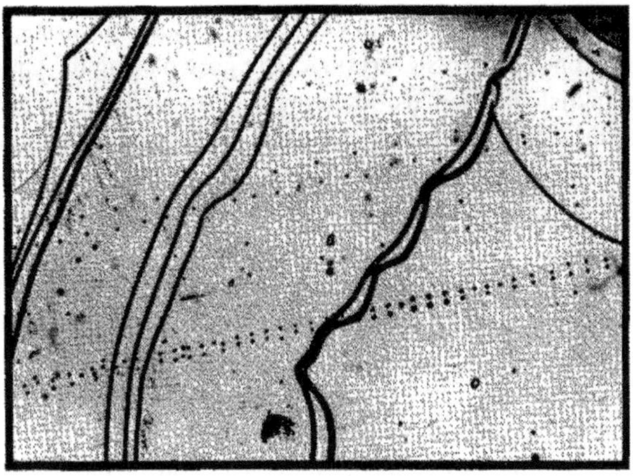

Fig. 2.39 Interaction between dislocations in parallel glide planes (180×). Reprinted with permission from [275]. Copyright 1960, American Institute of Physics

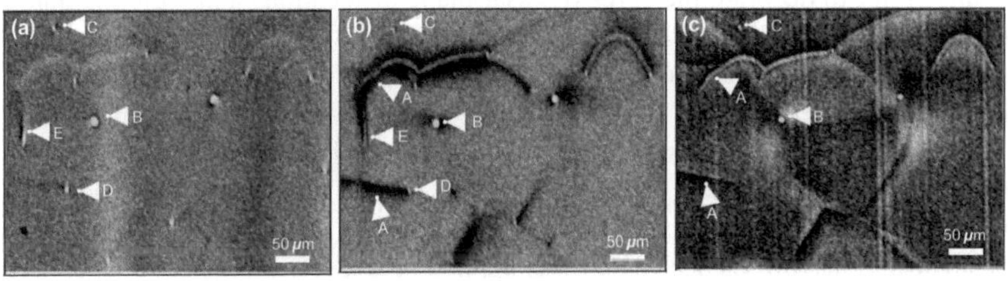

Fig. 2.40 Various dislocations in grazing incident X-ray reflection topographs of **a**, **b** 4H-SiC epitaxial layer, and **c** 4H-SiC substrate before epitaxial growth. The epitaxial layer is 11 μm thick. These pictures were taken from the same position in the sample. The X-ray wavelength and diffraction vector are **a** 1.455 Å (penetration depth: 5 μm), $\mathbf{g} = 11\bar{2}8$; and **b**, **c** 0.726 Å (penetration depth: 30 μm), $\mathbf{g} = 22\bar{4}{\cdot}16$. Reprinted with permission from [279]. Copyright 2004, Elsevier B.V.

spots, while basal plane dislocations with opposite edge Burgers vector components exhibit two distinct kinds of contrast images. The conversion of threading screw dislocations in 4H-SiC crystals that grow in an induction-heating furnace by the top-seeded solution growth method provides an efficient method of obtaining high-quality SiC crystals [281]. X-ray topography revealed that almost all threading screw dislocations in the off-axis Si-face 4H-SiC{0001} seed crystal were converted into Frank-type stacking faults on the basal planes. The conversion ratio is highly influenced by the surface polarity of the seed crystal. X-ray topography was used to evaluate the observable depth of dislocations in 4H-SiC(0001) wafers with a 4° off-axis angle toward the [11$\bar{2}$0]-direction [282]. The observable depths of the basal plane dislocations were determined to be 6.66–21.10 μm. The amorphization and dislocation evolution mechanisms of 6H-SiC were investigated by using nano-indentation, TEM, and molecular dynamics simulations [283]. Two major plastic deformation mechanisms of 6H-SiC under nano-indentation are amorphization near the residual indentation mark and dislocation nucleation and propagation below the amorphization region. The 4H-SiC crystals were grown using the gas-source method and the changes in dislocation densities along the growth direction were investigated [284]. The densities of threading and basal plane dislocations in the crystal grown at ~ 3 mm h^{-1} were reduced to be less than 1/10 and 1/20, respectively. A surface morphological defect in 4H-SiC epilayers was observed [285]. The defect appeared on the epilayer surface as an inclined line at an angle of $\pm$ 80° to the step-flow direction $[11\bar{2}0]$. It originated from a threading screw dislocation intersecting the surface and its orientation was controlled by the sign of the Burgers vector of the dislocation. The defect formed through the interaction of local spiral growth associated with threading screw dislocations and step-flow growth related to the substrate offcut. Basal-plane dislocations greatly challenge the reliability of bipolar power devices based on the 4H-SiC. Nitrogen doping can advance the conversion of basal-plane dislocations to threading edge dislocations and improve the reliability of 4H-SiC-based power devices. Kelvin probe force microscopy and first-principles calculations indicate that basal-plane dislocations may create acceptor-like states in undoped 4H-SiC, N atoms prefer to decorate the cores of basal-plane dislocations and yield the donor-like behavior of basal-plane dislocations in N-doped 4H-SiC [286]. Threading edge, screw, and mixed dislocations as well as basal plane dislocations were observed in $(11\bar{2}0)$-oriented 6H-SiC wafers by using synchrotron X-ray topography [287]. Threading edge dislocations were observed to be more likely off *c*-axis compared to threading screw and mixed dislocations. X-ray section topography is an advanced imaging technique enabling 3D observation of crystal defects. An incident X-ray beam with a narrow vertical width was applied, and a series of sheet images were collected by step-scanning a translational stage, then the 3D structure was constructed by stacking and connecting the sheet images using "ImageJ" software. 3D micro-X-ray topography captured the structure of threading screw dislocations deep in the 4H-SiC(0001) crystals [288]. A 3D image was constructed by stacking the X-ray topographs of the collected sheets. The propagation of the threading screw dislocation core-lines toward the inside of the sample was captured. The threading screw dislocations parallel to and inclined from the $\langle 0001 \rangle$ axis were observed, and the inclined ones are predominant and have different orientations and angles (Fig. 2.40a, b). The linear dislocations marked by white arrows are basal plane dislocations near the surface. Dark contrast lines propagating inside the sample were observed. They appeared as slim cones with a strong strain field at the surface and a weak field at the deep location. The strain field of the surface was spread by the surface relaxation effect [289] (wherein the contrast features of X-ray topographic images of screw-type basal plane dislocations in 4H-SiC wafers were studied, as shown in Fig. 2.40c, d). The contrast weakens at the deep location owing to stronger absorption of penetrated X-rays. Therefore, the dark contrast lines projected in the planar X-ray topograph of Fig. 2.41b may reflect the strain fields of the threading screw dislocation core-lines penetrating inside the sample. Black arrows in Fig. 2.41a mark the individual threading screw dislocations. The dislocation core-line A (visible only near the terminal due to overlap of surrounding strain fields) is parallel to the $\langle 0001 \rangle$ axis, whereas the dislocation core-lines B, C, and D are inclined from $\langle 0001 \rangle$ axis.

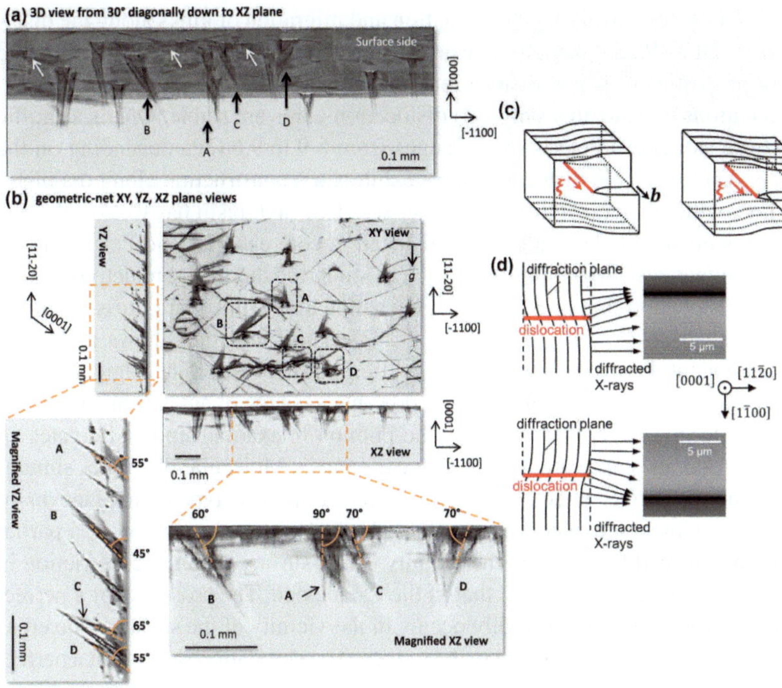

Fig. 2.41 3D micro-X-ray topography image at **g** = 00012 (**g**: diffraction vector) of 4H-SiC(0001): **a** 3D view from 30° diagonally down to the XZ plane and **b** geometric-net XY, XZ, and YZ plane views of (**a**). Upper part represents the surface shown in (**a**). The magnified XZ and YZ views are displayed. In the magnified views of threading screw dislocations—A, B, C, and D the dashed lines denote inclined angles from the (0001) plane. Reprinted with permission from [288]. Copyright 2024, IOP Publishing. **c** Schematic showing the displacement field around right-handed and left-handed screw dislocations. ξ represents the dislocation line direction. **d** The contrast formation mechanism and simulated images for $11\bar{2}8$ reflection for X-rays diffracted 1 μm from right-handed or left-handed basal plane dislocations in 4H-SiC. Reprinted with permission from [289]. Copyright 2021, Elsevier B.V.

Theoretical simulations or DFT calculations can also give insights into the properties of dislocations. The structures of straight 90° glide partial dislocations in cubic $Si_{43}C_{43}H_{68}$ were calculated using an ab initio local density functional cluster method [290]. Si partials containing core Si atoms show strong reconstruction with Si–Si bond. C partials possessing core C atoms show weak reconstruction. The calculated formation and migration energies of kinks on the partials indicate that C partials are more mobile. The calculations predict that *n*-type and *p*-type doping separately increases the mobility of C partials and Si partials. A theoretical study has been conducted on the 30 and 90° Shockley partial dislocations lying in the {111} and basal planes of cubic and hexagonal silicon carbide [291]. The core structure, energetics, and electronic structure of the dislocations were determined. The thermal activation barriers to glide motion of 30 and 90° Shockley partials were calculated in terms of a process involving the formation and migration of kinks along the dislocation line. DFT–PBE calculations were performed to determine the stability and mobility properties of screw dislocations in 4H-, 2H- and 3C-SiC [292]. Numerical simulations indicate that only two dislocation cores are stable. One is a shuffle core, whose associated Peierls stress ranges from 8.9 to 9.6 GPa depending on the polytype. Another stable dislocation core exhibits a reconstruction along the dislocation line, with greater stability, but is also sessile. Polytypism has weak influence on these results, particularly regarding dislocation core energies and Peierls stress. Molecular dynamics simulations were used to study the thermal conductivity along the screw dislocation line in 3C-SiC <110> nanowires with square cross-sectional areas ranging from 4×4 to 7×7 nm^2 [293]. The simulation shows a Burgers vector dependent thermal conductivity reduction in SiC. The effect originates from the highly deformed core region, which represents a source of anharmonic phonon–phonon scattering. High strain reduces the phonon relaxation time and creates an internal thermal resistance around the dislocation axis. Molecular dynamics simulations explore the surface effect on the conversion from screw-type basal plane dislocation to threading edge dislocation in 4H-SiC [294]. The constriction of a partial dislocation pair readily occurs in the vicinity of the surface, and the constriction in the Si-face substrate is easier than that in the C-face one. The cross slip of a perfect screw basal plane dislocation readily occurs in the vicinity of the surface. The cross slip in the Si-face is easier than that in the C-face. The simulation shows that a perfect screw basal plane dislocation-to-threading edge dislocation conversion in an off-cut substrate occurs even at a low temperature (500 K). The authors further investigated the effect of surface on the contraction of partial dislocations pair in 4H-SiC [295]. The contraction of the partial dislocation parallel to the surface occurs if the distance from the surface is smaller than 0.25 nm. For the partial dislocations intersected to the surface, the dislocation segment located less than 0.25 nm below the surface contracts readily. Because the cross slip of the perfect screw dislocation readily occurs in the vicinity of the surface, the basal plane dislocation-to-threading edge dislocation conversion is controllable by the contraction of partial dislocations located less than 0.25 nm below the surface. Nanoindentations of 3C-SiC at temperatures of 300–2000 K were studied using molecular dynamics simulations [296]. The load–displacement curves indicate weak temperature dependence of elastic responses and strong temperature dependence of plastic deformation. The critical mean contact pressure as a criterion for the plasticity onset decreases with temperature. Plastic deformation at low temperatures starts with nucleation and expansion of perfect dislocations, whereas the perfect dislocations dissociate into Shockley partial dislocations with stacking faults at high temperatures. DFT–PBE calculations show that perfect screw dislocations in 3C-SiC can change the spin configuration of a divacancy located inside its core, without changing the localized nature of its defect states [297].

Micropipes

Micropipes which constitute a major defect in mass-produced SiC wafers are hollow tubular defects penetrating the SiC single crystals. Their radii range from a few tens of nanometers to several tens of micrometers and the density at the growth surface ranges from several hundred to less than one micropipe per cm^2 [298]. These micropipe defects can severely hamper the performance of SiC power devices [299]. Micropipe defects in 4H- and 6H-SiC substrates have been found to lead to preavalanche reverse-bias point failure in many epitaxially grown *pn* junction devices with an area of 1 mm^2 or larger. The SiC power device ratings will be restricted to several amps or less if the defect density is on the order of 100's of micropipes cm^{-2}.

Many studies indicate that the micropipes in SiC are hollow core dislocations [291]. The dependence of the micropipe radius on the Burgers vector content of the micropipes in different SiC has been analyzed and the relationship can be explained by Frank's model of a hollow core dislocation when a mixed dislocation is assumed [300]. The origin of micropipes and nanopipes in SiC are hollow superscrew dislocations formed by the attraction of unit screw dislocations to a nanopipe formed at the axis of some triple junctions [301].

A model has been proposed to explain the formation of screw dislocations (including micropipe nucleation) in SiC [302]. Micropipes have been demonstrated to nucleate at the sites of foreign materials inclusions by using synchrotron white beam X-ray topography and transmission optical microscopy (Fig. 2.42). Incorporation of the inclusions into the growing crystal can lead to deformation of the protruding ledge which constitutes the overgrowing layer. Deformation is accommodated by the crystal lattice creating pairs of opposite sign screw dislocations (Fig. 2.43) which then propagate with the growing crystal. Using a series of synchrotron imaging techniques in the reflection geometrics, micropipes in 6H-SiC crystals have been mapped [303]. The combined synchrotron white beam X-ray topography techniques and corresponding simulation (Fig. 2.44) demonstrate explicitly that the micropipes are pure superscrew dislocations. The details of the superscrew dislocations, including the spatial distribution of the strain fields, magnitude of the Burgers vectors, dislocation senses, and surface relaxation effects, can also be accurately determined by these techniques.

Synchrotron X-ray topography and Nomarski optical microscopy were carried out on 6H-SiC crystals grown by the sublimation physical vapor transport technique [304]. Two kinds of dislocations were observed: dislocations exhibiting bimodal images of various widths and line directions approximately parallel to the [0001] axis and dislocations confined to the basal plane. The larger bimodal image width dislocations are hollow core micropipes. Contrast analysis of topographic images obtained in transmission and back-reflection geometries indicates that micropipes are Frank-type hollow-core screw dislocations with Burgers vectors typically equal to 3–7 times the *c* lattice constant. X-ray topography revealed many line defects approximately parallel to the [0001] axis that are screw dislocations with Burgers

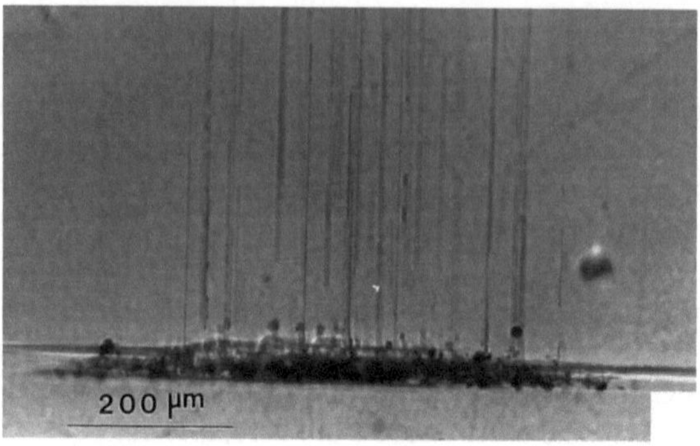

Fig. 2.42 Optical micrograph showing nucleation of micropipes at a group of inclusions during growth in a 4H-SiC wafer. Reprinted with permission from [302]. Copyright 1999, American Institute of Physics

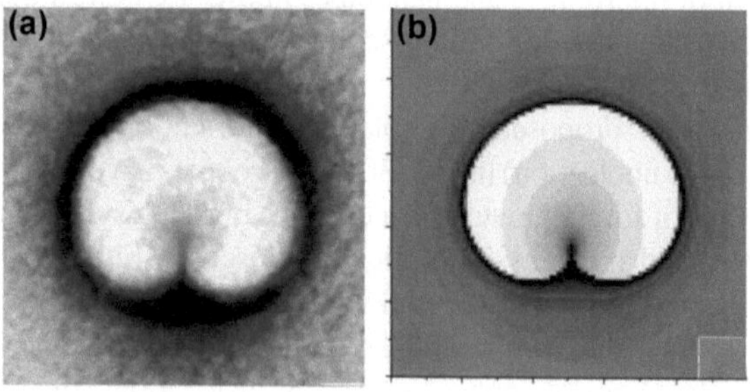

Fig. 2.43 **a** Synchrotron white beam X-ray topography image recorded in the back-reflection geometry from a (0001) Lely platelet showing a pair of opposite-sign screw dislocations. **b** Simulated image of two opposite-sign screw dislocations with 30 µm spacing. The image dimensions are 200 $\times$ 200 µm^2. Reprinted with permission from [302]. Copyright 1999, American Institute of Physics

vectors equal to the *c* lattice constant and there are no micropipes associated with them. A new morphology defect related to micropipes was observed in 4H-SiC thick homoepitaxial layers [305]. Raman and PL characterizations reveal that this defect is a pure morphology defect without polytype transition. Molten KOH etching reveals that this defect originates from micropipes. A non-destructive method for detecting sub-micrometer-sized micropipes on an entire wafer surface of 4H-SiC epitaxial layers was investigated [306]. The location of line-shaped depressions was identified by using optical inspection. The sub-micrometer-sized micropipes were observed at the sites of leakage current. Device simulation of the blocking-voltage characteristics of *pn* diodes indicates that the increase in leakage current depends on the electric field at the sub-micrometer-sized micropipes, regardless of their core size.

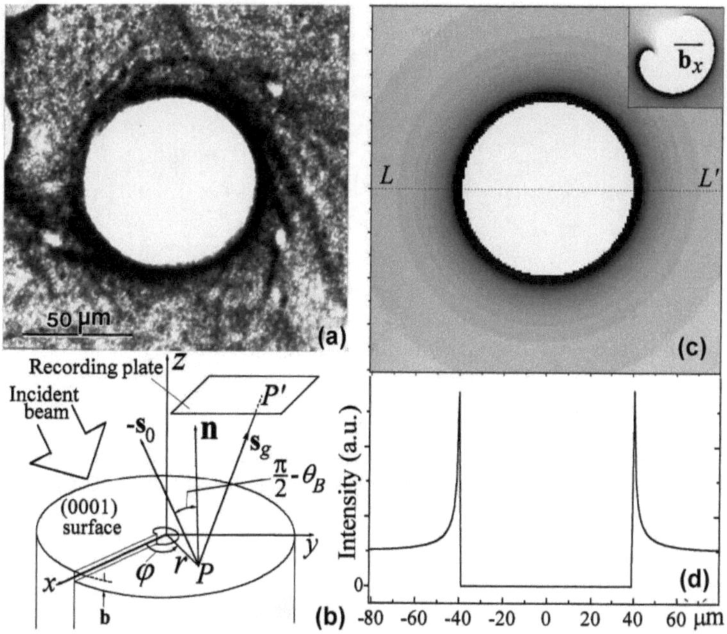

Fig. 2.44 **a** Synchrotron back-reflection topographic image of an 8*c* micropipe under diffraction conditions $\mathbf{g} = 00024$ (and its harmonics), $\theta_B = 81.4°$, and sample-to-film distance equal to 20 cm. **b** Diffraction geometry. **c** Simulation of an 8*c* superscrew dislocation. The inset corresponds to the simulation of a mixed dislocation. **d** Radial diffraction intensity profile of (**c**). Reprinted with permission from [303]. Copyright 1999, American Institute of Physics

2.4.2.2 Stacking Faults

Basic Properties of SiC Stacking Faults

Stacking faults (SFs) are segments in the SiC crystal where rows of Si–C bilayers deviate from the perfect stacking sequence (along *c* axis) of the mother crystal. These 2D defects are actually quantum wells sandwiched between two bulk-size mother polytype layers, forming a double-heterostructure like crystal. Therefore, they are intriguing and unique crystal systems for SiC or similar crystals that possess a great many polytypes. There has been great interest in the stacking faults in SiC.

The defects in single crystalline and polycrystalline SiC were examined by using TEM, and the result shows that dislocations are not as common as stacking faults [307]. Dislocations have been observed from nearly all the materials examined (Fig. 2.45). The observed stacking faults are either "stress-induced" extrinsic or "grown-in" intrinsic ones. The stacking fault energy was measured to be 1.9 ergs cm^{-2} by the extended node method. The structure of the stacking faults in β-SiC has been investigated by using high-resolution TEM [308]. The HRTEM images of the stacking faults in a specimen sintered at 2000 °C along with their structure

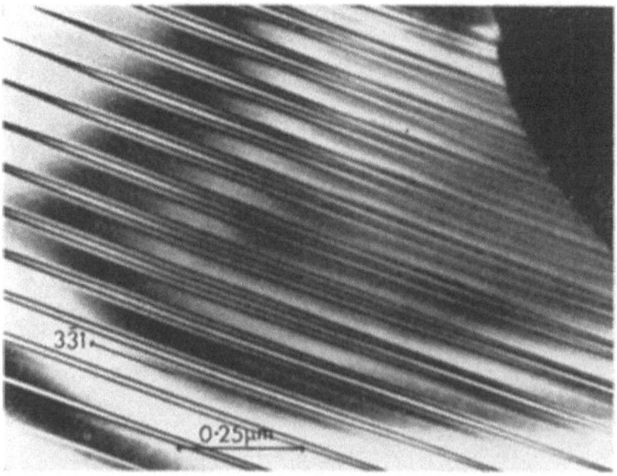

Fig. 2.45 TEM image of the regularly spaced stacking faults in hot-pressed SiC. Reprinted with permission from [307]. Copyright 1972, Springer Science + Business Media, LLC, part of Springer Nature

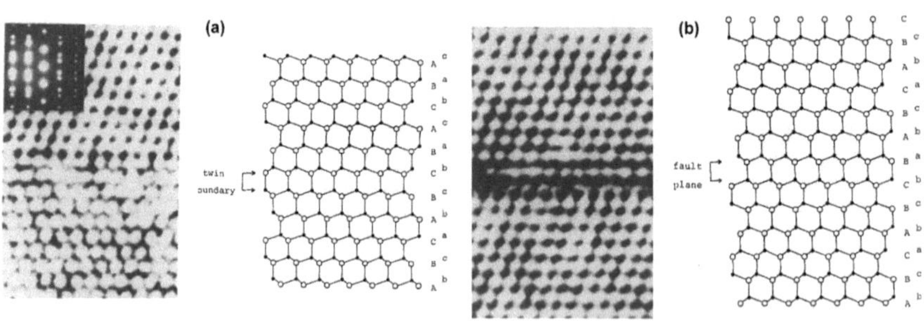

Fig. 2.46 High-resolution TEM images of **a** twin stacking fault and **b** deformation stacking fault found in β-SiC ceramics sintered at 2000 °C. The right panel shows the structural models based on the stacking sequence. Reprinted with permission from [308]. Copyright 1989, John Wiley & Sons, Inc.

models are shown in Fig. 2.46. There are many twin boundaries with nearly perfect atomic arrangement in each twinned domain. The result shows that the stacking faults initially present in the β-SiC powders were eliminated as grain growth proceeded at elevated temperatures.

The effects of stacking faults on the XRD profiles of β-SiC powders have been studied by using simulation with supercells of up to 100 times the volume of a conventional unit cell [309]. Stacking faults in β-SiC powders give rise to all the features observed from the diffraction pattern, that is, an additional peak at $d = 0.266$ nm, a high background intensity around the peak at $d = 0.252$ nm, and broadening of the 3C-SiC peaks. The simulation also indicates that the distribution of the faults is not uniform and random, although a unique description of the fault population is not available. Factors affecting SF formation during the synthesis of β-SiC were

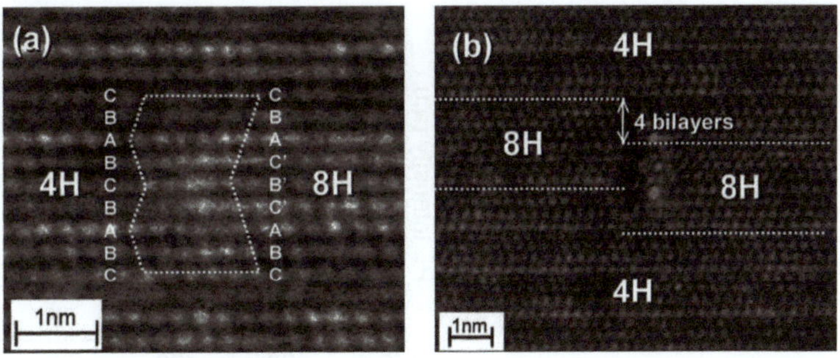

Fig. 2.47 **a, b** High-resolution TEM images of the stacking faults in the 4H-SiC epilayer. Reprinted with permission from [312]. Copyright 2005, American Institute of Physics

investigated [310]. Two main reaction routes in the carbothermal reduction of SiO_2 (solid–solid and solid–gas reactions) led to the formation of β-SiC, depending on the vapor pressure of the SiO gas. The β-SiC formed by the solid–gas (carbon–SiO gas) reaction exhibited a whisker morphology with a high SF density, whereas that formed by the solid–solid (Si–carbon) reaction exhibited spherical particles with a low SF density. The average size of the synthesized particles decreased with decreasing reaction temperature and time. The deformed 4H- and 6H-SiC single crystals by compression at 1300 °C have been investigated [311]. All the deformation-induced dislocations were dissociated into two partials bounding a ribbon of intrinsic stacking fault. Using TEM, the stacking fault energy was determined from the separation width of the two partials of dissociated dislocations. It is 14.7 ± 2.5 mJ m^{-2} for 4H-SiC and 2.9 ± 0.6 mJ m^{-2} for 6H-SiC. The calculated stacking fault energies based on the axial next-nearest-neighbor Ising spin model are within 5% and 40% of the experimental values for 6H- and 4H-SiC, respectively.

The structure of in-grown stacking faults in 4H-SiC(0001) epilayers prepared by CVD has been investigated [312]. The in-grown stacking faults nucleated near the substrate/epilayer interface and expanded with increasing epilayer thickness in a triangular shape. TEM confirms that the in-grown stacking fault is the 1*c* of 8H polytype (Fig. 2.47). The low-temperature PL spectrum obtained from the same area (absent in other areas) comprises four peaks at 463.1, 468.3, 471.8, and 473.1 nm (Fig. 2.48) corresponding to phonon replicas with a bandgap of 2.710 eV, confirming the 8H polytype. The epilayers prepared under the conditions of slow growth, high temperature, and improved substrate surface have reduced densities of in-grown stacking faults.

The density, shape, and structure of the in-grown stacking faults in 4H-SiC(0001) epitaxial layers were investigated by using cathodoluminescence, PL, and TEM [313]. The in-grown stacking faults have the 8H structure (Fig. 2.49) and exhibited a PL peak at 2.56–2.70 eV. The stacking faults in the Schottky barrier diode lowered the barrier height and decreased the breakdown voltage. The stacking faults in 4H-SiC epilayers were studied by using micro PL spectroscopy and PL intensity mapping [314]. Three types of stacking faults, namely intrinsic Frank, double

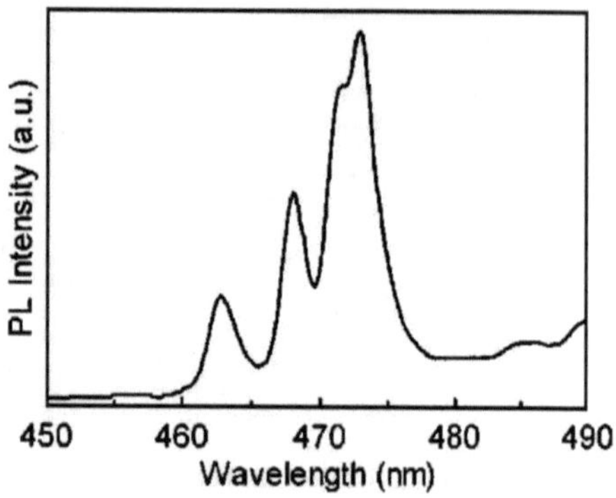

Fig. 2.48 PL spectrum of an in-grown stacking fault in the 4H-SiC epilayer acquired at 9 K excited by the 325 nm line. The four peaks are at 463.1, 468.3, 471.8, and 473.1 nm. Reprinted with permission from [312]. Copyright 2005, American Institute of Physics

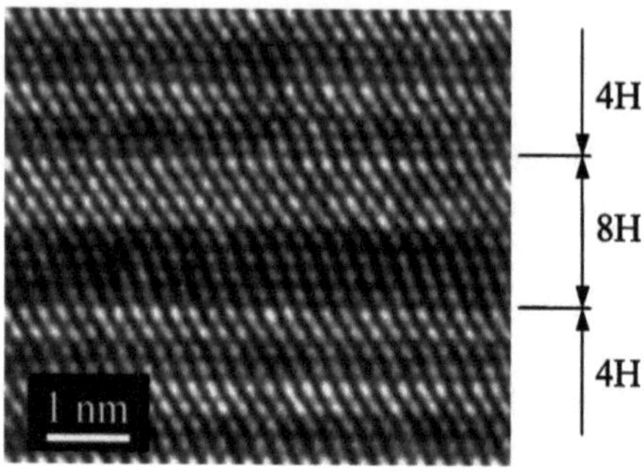

Fig. 2.49 Cross-sectional HRTEM image of an in-grown stacking fault in the 4H-SiC epilayer prepared by chemical vapor deposition. Reprinted with permission from [313]. Copyright 2005, American Institute of Physics

Shockley, and in-grown stacking faults, were identified. Each kind of stacking fault showed distinct PL located at 420 nm (intrinsic Frank), 500 nm (double Shockley), and 455 nm (in-grown), respectively (Fig. 2.50). The region free of stacking faults showed PL with a peak at 386 nm (close to 4H-SiC bandgap) accompanied by a tail on the low energy side. The shape and distribution of the stacking faults can be profiled by micro-PL intensity mapping (Fig. 2.51). 4H-SiC epilayers were characterized by micro-PL spectroscopy and micro-PL intensity mapping at room temperature [315]. A type of SF with a peak emission wavelength at 480 nm (2.58 eV) was identified. The micro-PL intensity mapping reveals triangular shape of this SF. High-resolution TEM spectroscopy shows that the SF stacking sequence is (3, 5) in Zhdanov's notation, consistent with that of the triple Shockley SF.

Frank-type defects on the basal plane in thick 4H-SiC epitaxial layers were studied by using PL spectroscopy and PL imaging microscopy [316]. The Frank-type defect consisting of intrinsic SF (single missing layer SF) was determined to have PL at ~ 488 nm at room temperature. The Frank-type defect consisting of the extrinsic SF (single extra layer SF) exhibited PL at ~ 424 nm. The Frank-type defect comprising multiple layers SF had emission at ~ 457 nm. The high-resolution PL imaging confirmed that the Franck partial dislocations on the basal plane exhibited PL in the near-infrared region (> 700 nm). The Franck partial dislocations on the missing-layer side of the triangular shaped SF showed stronger PL compared to those on the

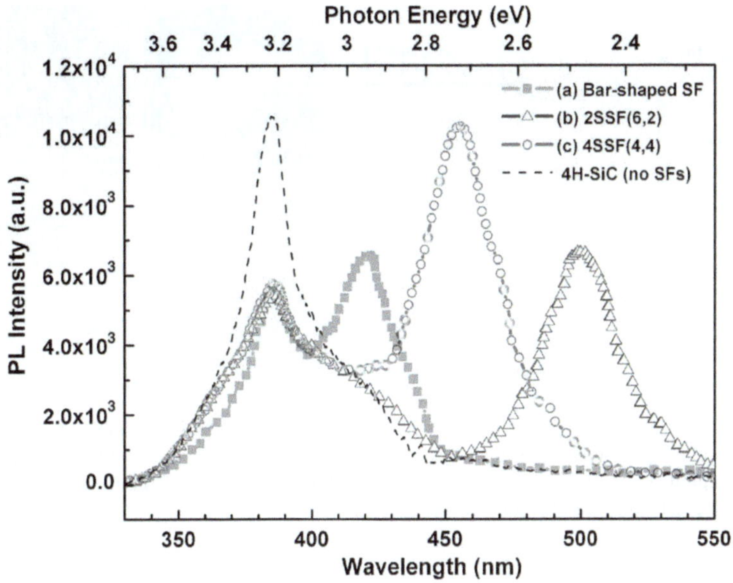

Fig. 2.50 PL spectra obtained from the 4H-SiC epilayer in the areas of: bar-shaped (intrinsic Frank) stacking fault, double Shockley stacking fault (2SSF), and in-grown stacking fault (4SSF). The spectrum acquired from the area without SFs is shown by the dashed line. Reprinted with permission from [314]. Copyright 2008, American Institute of Physics

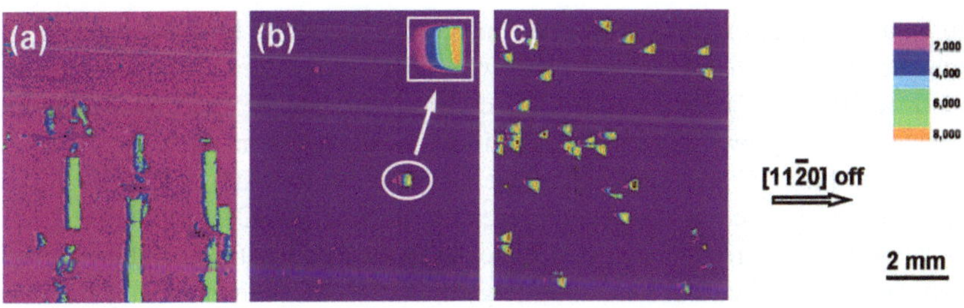

Fig. 2.51 PL intensity mapping of the 4H-SiC epilayer at **a** 420 nm, **b** 500 nm, and **c** 455 nm. Reprinted with permission from [314]. Copyright 2008, American Institute of Physics

extra-layer side. Frank-type SFs in 4H-SiC epitaxial layers were investigated using room-temperature PL mapping and high-angle annular dark-field (HAADF) scanning TEM (STEM) [317]. The intrinsic Frank-type SF (5, 2) and intrinsic multilayer Frank-type SF (4, 2), which were previously reported to have an elongated triangular shape, were found to have a broader triangular shape (Fig. 2.52) that is typical of Shockley-type SFs. The shaping of SFs could be explained based on the directions of the partial dislocations bounding the SFs. Frank-type SFs (3, 2, 2, 3) and (3, 3) as well as Shockley-type SFs (3, 3, 3, 2, 3, 2) and (3, 2, 3, 3, 3, 2) were observed. All of

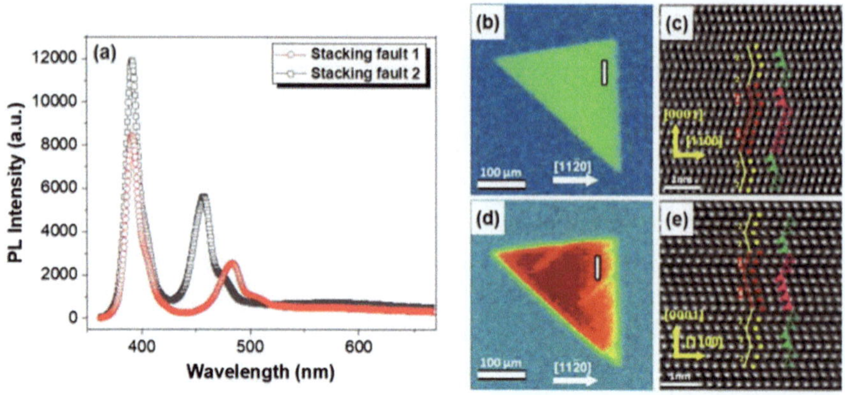

Fig. 2.52 **a** PL spectra of SFs 1 and 2. **b** PL mapping and **c** HAADF HR-STEM images of SF 1. **d** PL mapping and **e** HAADF high-resolution STEM images of SF 2. Positions of focused ion beam sampling are marked with rectangles in (**b**) and (**d**). Reprinted with permission from [317]. Copyright 2024, American Institute of Physics

them are composed of stacking numbers 2 and 3 in the Zhdanov notation and have characteristic PL peaked at 427–428 nm.

Expansion Mechanism of Stacking Faults

The formation and expansion mechanism of stacking faults have been explored. Spontaneous formation of stacking faults was observed from heavily nitrogen-doped ($n > 1 \times 10^{19}$ cm^{-3}) 4H-polytype SiC crystals by using TEM [318]. Faults are present in the as-grown boules and additional faults were generated by annealing in argon at 1150 °C. All the faults have an identical structure consisting of six layers stacked in a cubic sequence embedded in the 4H-SiC crystal (Fig. 2.53). The stacking faults result from two Shockley partial dislocations gliding on two neighboring basal planes of SiC. The energy of the 4H-SiC with faults may be lower than that of the perfect heavily doped crystal at typical processing temperatures and this explains the structural transformation. The generation and evolution of defects in 4H–SiC *pn* junctions due to carrier injection under forward bias were investigated by using synchrotron X-ray topography, electroluminescence imaging, and KOH etching [319]. The defects are Shockley SFs with rhombic or triangular shapes bound by partial dislocation loops with dislocation lines along Peierls valleys $<11\bar{2}0>$ or along the intersection of the basal plane containing the fault and diode surface. The Burgers vector of all bounding partials belongs to 1/3 $<10\bar{1}0>$-type. Only two among six possible types of partial dislocations with these properties were observed in the volume of the epitaxial layer. One was identified as 30° carbon-core and another as 30° silicon-core partial dislocation. Only silicon-core partial dislocation exhibited movement and emitting light under forward bias. Low angle grain boundaries were observed to act as one of some possible nucleation sites of SFs. Detailed study shows that the structural transformation is ascribed to the quantum well action [320], the electrons in highly *n*-type 4H-SiC enter stacking fault-induced quantum well states to lower the system energy. The net energy gain derived from the charge neutrality equation shows dependence on the temperature and nitrogen doping concentration. Doping concentrations in excess of 3×10^{19} cm^{-3} should result in spontaneous formation of double layer stacking faults at the device processing temperatures, in agreement with observations. Charge buildup in the stacking faults can produce an electrostatic potential which increases the forward voltage drop in SiC *p–i–n* diodes. The annealing effect on single Shockley faults in *n*-type 4H-SiC epitaxial layers was investigated by using PL mapping in combination with high-power laser illumination [321]. Comparison before and after annealing at 350–550 °C shows that annealing caused shrinking of the faulted area of single Shockley faults. High-power laser illumination on the same area annealed at 550 °C led to recovery of the right-angled triangular single Shockley faults, but the isosceles triangular single Shockley faults did not reform. PL spectroscopy revealed carrier recombination from single Shockley SFs and partial dislocations in 4H-SiC [322]. The PL from single Shockley SFs and partial dislocations has a fast decay component compared with that from the band edge, suggesting faster carrier recombination in single Shockley SFs and at partial dislocations than that in regions without them. As a result of slower recombination at Si-core partial dislocations compared with that in single Shockley SFs and at C-core partial dislocations, the velocity of single Shockley SFs expansion is limited by the carrier recombination at Si-core partial dislocations. The decay time at the Si-core partial dislocation decreased with increasing temperature. The expansion rate of single Shockley-type stacking faults in 4H-SiC was examined under UV illumination in various basal plane dislocation structures [323]. In the case of 30° Si-core partial dislocations at the expansion front, some basal plane dislocations exhibited very slow expansion. PL imaging revealed that the basal plane dislocations are accompanied by threading edge dislocations. Additional high-resolution STEM analysis revealed that the leading 30° C-core partial dislocation might account for the very slow expansion of the single Shockley-type stacking faults. The structures of SFs and inverted domain boundaries in 3C-SiC epitaxial layer were investigated by using STEM and molecular dynamic simulations [324, 325]. The domain boundaries propagate in a complex way through the crystal, forming "complex-domain boundaries" that interact strongly with SFs. Domain boundaries can terminate and generate SFs. Threading partial dislocations that limit the SFs appear to have non-conventional line directions such as <112>, <123>, and <134>.

Stacking Faults in *p-i-n* and Schottky Diodes

The structure of the stacking faults formed in forward-biased 4H- and 6H-SiC pn^- diodes was investigated by using TEM [326]. The typical fault densities are between 10^3 and 10^4 cm^{-1}. The analysis of the stacking sequences (Fig. 2.54) and partial dislocations bounding the faults indicates that the faults are isolated single-layer

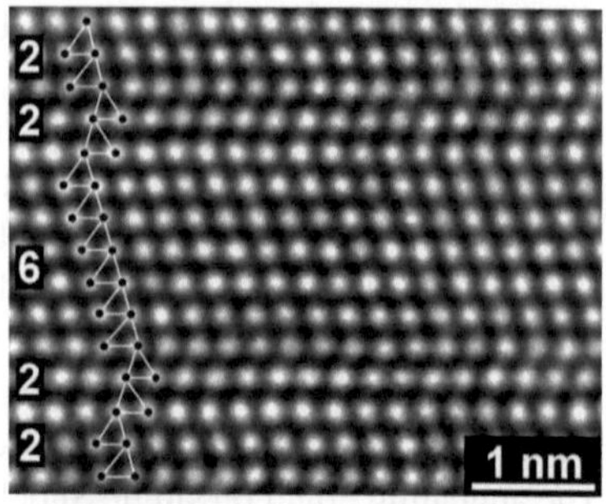

Fig. 2.53 HRTEM image of a stacking fault in the heavily doped *n*-type 4H-SiC crystal. Reprinted with permission from [318]. Copyright 2002, American Institute of Physics

Fig. 2.54 High-resolution TEM image of a stacking fault (marked by arrow) in a 4H-SiC diode. Reprinted with permission from [326]. Copyright 2002, American Institute of Physics

Shockley ones bounded by partial dislocations. The stacking faults formed during forward biasing of 4H-SiC *p–i–n* diodes were identified using light emission imaging and transmission electron microscopy [327]. These faults may have either faulted dislocations or faulted dislocation loops and the leading partial dislocations of the faulted loops may have either carbon cores or silicon cores, whereas the partials of faulted dislocations are silicon cores. These stacking faults seem to relieve both tension and compression during *p–i–n* diode operation, indicating the presence of a complex inhomogeneous strain field within the diode. The formation and growth of the stacking faults causing degradation in bipolar *p–i–n* diodes were studied by using electrical, optical, and structural techniques [328]. The stacking fault has at least three different origins, resulting in two different types of stacking faults. SF growth in stressed SiC *p–i–n* diodes was examined using electroluminescence imaging at 80 and 160 A cm^{-2} [329]. Dark areas in the emission developed because of SFs and the current capability of the diode drops. The low-current images are bright lines at dislocations bounding the SFs and at or near the SF intersection with the surface. SFs nucleated 1–2 μm below the surface. Most continued to grow until they spanned the diode.

The driving force of stacking fault expansion in 4H-SiC *p–i–n* diodes was investigated by using optical emission microscopy and TEM [330]. The stacking-fault expansion and properties of partial dislocations are inconsistent with stress as the driving force. Instead, the thermodynamic free energy difference between the perfect and fault structure was suggested as the driving force for stacking fault formation in the diodes. This implies that the hexagonal polytypes of SiC may be metastable at room temperature and may transform to a fault structure if the activation energy of the partial dislocation glide is available. This activation energy can be provided by the electron–hole recombination during operation of the diodes. The morphology and nucleation sites of SFs formed during the forward operation of 4H-SiC *p–i–n* diodes were studied using optical emission microscopy and TEM [331]. Partial dislocations bounding the SFs are mostly aligned to the $<11\bar{2}0>$ orientations with Burgers vectors of 1/3 $<1\bar{1}00>$ type. Dislocation half-loop arrays in the blocking layer act as nucleation sites of double-rhombic SFs. The SF morphology indicates that short basal plane segments associated with threading dislocations act as the origin of rhombic SFs. All dislocations in a half-loop array have the same Burgers vector and nucleate on a single basal plane. Most pre-existing basal plane dislocations within the blocking layer dissociate to form SFs during the degradation. Basal plane dislocations aligned along the off-cut direction form rectangular SFs, while others break up into partial dislocation segments along the $<11\bar{2}0>$ direction. All nucleation sites of the SFs correspond to pre-existing dislocation segments residing in basal planes. The extension of SFs in a forward-biased 4H-SiC *p–i–n* diodes by the recombination-enhanced motion of leading partial dislocations was investigated by using optical emission microscopy [332]. The results are consistent with electrically active SFs in 4H-SiC with energy levels of ~ 0.2 eV below the conduction band bottom that can trap electrons. These levels as radiative recombination centers make the SFs luminescent in the presence of an e–h plasma. There are two sites acting as recombination centers on leading (and possibly trailing) partial dislocations in 4H-SiC, a radiative one with an energy level of ~ 1.8 eV and a nonradiative one at an energy level of ~ 2.2 eV. The nucleation and expansion of SFs induced by recombination in 4H- and 6H-SiC epitaxial films have been studied [333]. The activation energy for partial dislocation glide under optical excitation decreased to about 0.25 eV, about 2 eV lower than that for pure thermal activation. The low activation threshold is in good agreement with the observation in forward operating *pn*-junction devices, implying that radiation enhanced dislocation glide is a general phenomenon in hexagonal SiC. The elementary process controlling the expansion of the stacking faults was supposed to be kink pair nucleation aided by the phonon-kick mechanism, based on thermal activation and below-gap excitation spectroscopy of dislocation glide. A detailed model was presented to describe a possible driving force responsible for SF expansion and contraction in 4H-SiC [334]. It is based on the relationship between the electron quasi-Fermi energy and the energy of the highest occupied electronic state within the SF 2D density of states. The SF expansion was supposed to be energetically driven by the electronic energy gain caused by trapping of conduction electrons within the SF quantum wells, as this energy gain significantly outweighs the relatively few structural energy cost associated with incorporating a planar 3C-SiC SF within the 4H host lattice. The expansion behavior of a single Shockley SF arising from a basal plane dislocation in a 4H-SiC epitaxial layer under electron beam irradiation was observed [335]. The width of the SF is proportional to the electron beam current. Electron beam irradiation at a fixed spot outside a SF expanded the

SF as effectively as direct SF irradiation. The increase in the width of an SF under selective excitation of a Si-core partial dislocation is much stronger than that under direct SF excitation or under both indirect SF excitation and indirect Si-core partial dislocation excitation. Electron–hole recombination at the Si-core partial dislocation could drive the SF expansion. The supply of a threshold number of electron–hole pairs causes mobility of the C-core partial dislocation.

Electrical properties of stacking faults and bounding partial dislocations in 4H-SiC Schottky diode were studied by using electron-beam-induced current (EBIC) and cathodoluminescence techniques [336]. Basal plane dislocation shows easy dissociation into stacking fault-spaced two partial dislocations [$Si_{(g)}$ 30° and $C_{(g)}$ 30° partials]. The EBIC contrast of $C_{(g)}$ 30° partial is several percent higher than that of $Si_{(g)}$ 30° partial. The stacking fault in EBIC image is brighter than the background, being related to its quantum-well state. Cathodoluminescence spectroscopy shows a new peak (417 nm) of emission appearing at the stacking fault position. The influence of four types of SFs on electron conduction in an *n*-type SiC epilayer was investigated by examining Schottky barrier diodes which completely contain the SF plane in the active area [337]. The forward current conductance is significantly limited by the SF and depends on the SF type, whereas the reverse leakage current of all Schottky barrier diodes is lower than the detection limit at reverse voltages up to − 40 V. The limitation of the forward current conduction due to SFs was moderated by increasing temperature. The impact of SFs on the characteristics of 4H-SiC Schottky diodes was investigated under both forward and reverse biases [338]. In the presence of SFs, no significant effect on the ideality factor and barrier height was observed under forward bias, whereas an anomalous increase in the leakage current occurred under reverse bias. The reverse leakage was explained by a space-charge limited current model, consistent with the presence of a trap distribution in the band gap of 4H-SiC. The weak temperature dependence of the reverse current suggests that carrier tunneling sets in and may coexist with space-charge limited current.

Quantum Well Models for Stacking Faults

Stacking faults can be regarded as a quantum well sandwiched between the normal silicon carbide crystals, and the quantum states and energy levels of the stacking faults can thus be evaluated. The calculated and experimental data provide the basis to understand the formation and properties of stacking faults. A unique luminescence band belonging to the stacking fault was observed from the 4H-SiC bipolar diode after extended forward voltage operation [339]. As shown in Fig. 2.55, a series of lines located within 410–435 nm (E_{SF}) are present in the PL spectrum of the stressed diode but absent in the non-stressed one. The panchromatic cathodoluminescence image (Fig. 2.56a) of the stressed diode contains triangular features that are assigned to the stacking faults. The triangular regions are brighter in the image acquired with a monochromator at 425 nm (Fig. 2.56b), indicating that the E_{SF} emission stems from the entire stacking faults. The band-edge luminescence at 390 nm within the stacking faults was suppressed (Fig. 2.56c), implying competition between two recombination

Fig. 2.55 PL spectra of a stressed (top) and non-stressed (bottom) 4H-SiC diode acquired at 2 K. P_0, Q_0, and P_{77} are the nitrogen bound exciton lines. A new set of lines labeled E_{SF} appear in the PL spectrum of the stressed diode between 410 and 435 nm. Reprinted with permission from [339]. Copyright 2001, American Institute of Physics

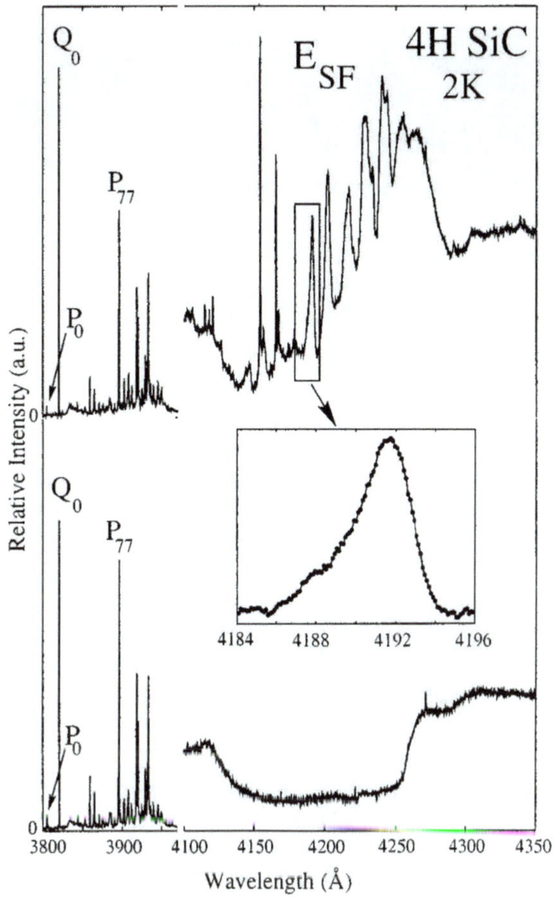

processes. The measured E_{SF} decay time is within 7–470 ns and has the same order as that of the intrinsic exciton in 4H-SiC. The E_{SF} emission was attributed to the exciton recombination at the quantum well created by the stacking fault.

The stacking fault-like quantum wells in SiC crystals can be created intentionally. Multi-quantum well structures with 3C-SiC wells between 4H-SiC barriers (consisting of more than 20 heterojunctions) were fabricated by a two-step procedure using solid source molecular beam epitaxy [340]. The structures were produced by selective nucleation of wire-like 3C-SiC nuclei on the terraces of well-prepared hexagonal SiC(0001) substrates followed by a step-flow of both 3C wires and surrounding hexagonal SiC materials (Fig. 2.57). The thickness of the 3C-SiC quantum well is about 2.3 nm (corresponding to three unit cells in <111> direction), thus giving rise to spatial quantization of electrons. The PL spectrum obtained at 6 K comprises dominant L_1 emission line related to the D_1 center (attributed to 4H-SiC) as well as an additional band at around 2.26 eV (Fig. 2.58). These new lines can be explained by a model of a triangular electron quantum well in 3C-SiC.

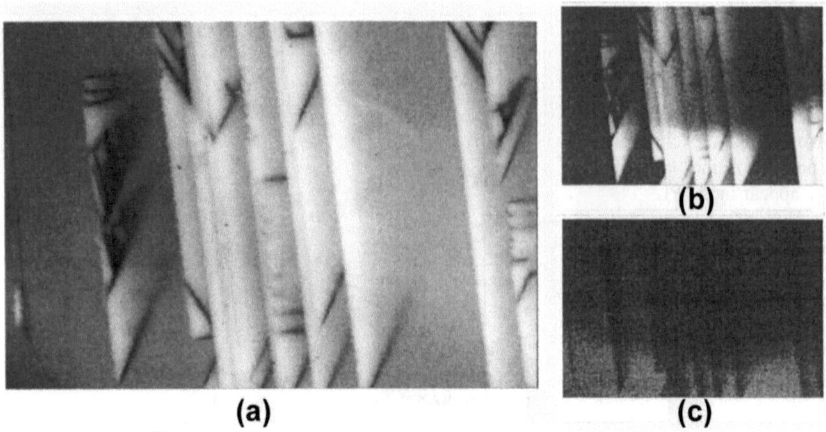

Fig. 2.56 **a** Panchromatic cathodoluminescence image of the stressed 4H-SiC diode showing stacking faults. Cathodoluminescence image of the same diode at **b** 425 nm and **c** 390 nm. Reprinted with permission from [339]. Copyright 2001, American Institute of Physics

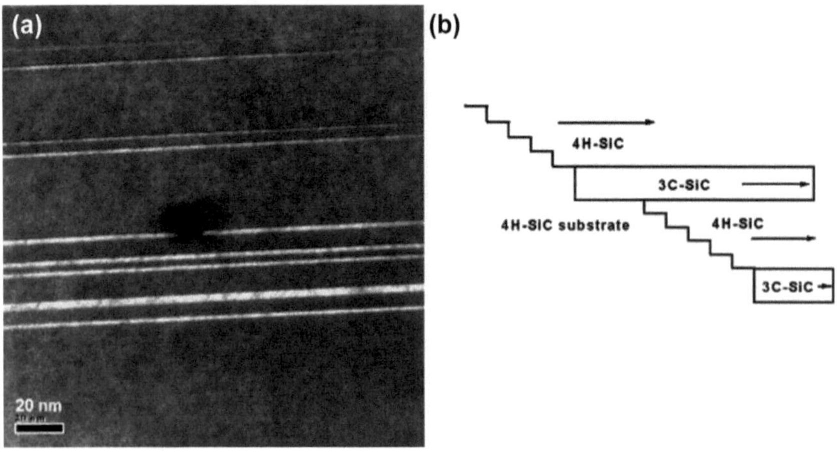

Fig. 2.57 **a** TEM image of the 3C/4H-SiC multi-quantum well structure fabricated on 4H-SiC(0001) by step-controlled epitaxy. **b** Schematic illustration of the formation process of the structure on a tilted substrate. Reprinted with permission from [340]. Copyright 2001, Elsevier B.V.

Low-temperature PL in 4H/3C/4H-SiC single quantum wells in the 20–50 μm thick 4H-SiC epitaxial film has been investigated [341]. The quantum well is 3.25 nm thick, consisting of thirteen 3C-SiC bilayers as revealed by HRTEM (Fig. 2.59). Figure 2.60 shows a portion of the PL spectrum taken from the same area where the TEM image was taken. In this region, four well resolved peaks between 580 and 600 nm are observable, but they are absent in a normal 4H-SiC film. The energy separation between these peaks approximately equals the energy difference between

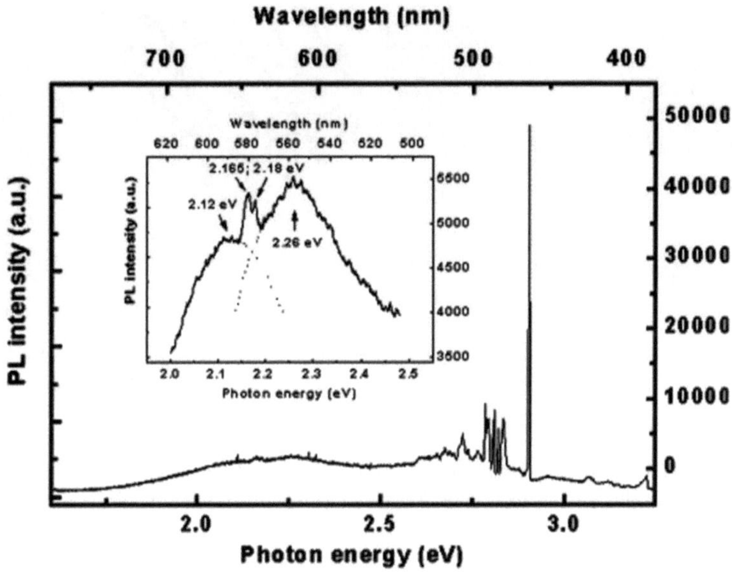

Fig. 2.58 PL spectrum of the 3C/4H-SiC multi-quantum well structure. The inset shows the magnified region between 2 and 2.5 eV. Reprinted with permission from [340]. Copyright 2001, Elsevier B.V.

the four major phonons in 3C-SiC, and thus these peaks belong to the phonon replicas. An effective exciton energy gap of about 2.177 eV (for the 3C quantum well) was derived and it is more than 200 meV below the usual exciton bandgap of bulk 3C-SiC as a result of Stark effect due to the built-in electric field in the quantum well in conjunction with the quantum confinement effect. Calculation shows a strong internal electric field on the order of 1 MV cm^{-1} to satisfy the observed large red shift of PL and this field was proposed to arise from the difference in spontaneous polarization between 3C- and 4H-SiC.

The study using ballistic electron emission microscopy in ultrahigh vacuum indicates that annealing of 4H-SiC wafer in dry oxygen at 1150 °C for 90 min can produce double stacking faults 3C-SiC inclusions [342]. Distinctive quantum well structures corresponding to individual inclusions were observed. The two-dimensional quantum well conduction band minimum was determined to be approximately 0.53 eV below the conduction band minimum of bulk 4H-SiC. Macroscopic diode *I–V* and microscopic ballistic electron emission microscopy measurements indicate that there is little variation in the interface state density or fixed charge across the inclusions. The PL of SFs with a complicated stacking sequence in 4H-SiC epitaxial layer has been investigated [343]. Three types of SF with complicated structures were identified, whose stacking sequences are expressed in Zhdanov notation as (3, 3, 4), (3, 3, 3, 3), and (3, 3, 2, 2, 4). PL imaging and X-ray topography show that these SFs were propagated from *p*-type substrates. The PL photon energy decreased with

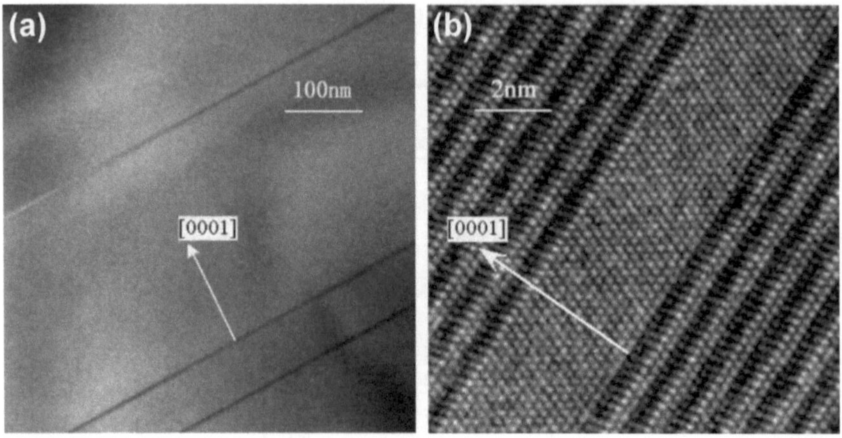

Fig. 2.59 **a** TEM image and **b** high-resolution TEM image of the cross section of an undoped 4H-SiC epitaxial film with 3C inclusions. Reprinted with permission from [341]. Copyright 2003, American Institute of Physics

increasing maximum number of layers of 3C-type sequence in the SFs. The quantum-well model with the Stark effect explains this. It suggests that SFs with complicated structures have multiple quantum wells. The authors extended this theory from 3C-type SFs to SFs with complex structures. The study of the electronic properties of single- and double-layer stacking faults in 4H-SiC clarifies the apparent distinction of recombination-enhanced defect reactions at these faults [344]. PL imaging spectroscopy and DLTS revealed the key constituents of radiative recombination and provide evidence of the nonradiative centers at $E_v + 0.38$ eV responsible for the recombination-enhanced mobility of Si-core partial dislocations. This energy level model allows for qualitative description of the recombination activity at different types of stacking faults and bounding partial dislocations.

Other Characterization Methods for Stacking Faults

Besides conventional PL spectroscopy, cathodoluminescence, and nonlinear spectroscopy, Raman scattering and infrared absorption spectroscopy can also offer information about the characteristics of the SFs in SiC. In-grown SFs in *n*-type 4H-SiC epitaxial layers were investigated by real-color cathodoluminescence imaging and spectroscopy at room temperature and liquid helium temperatures [345]. SFs with 8H stacking order as well as double layer and multilayer 3C-SiC structures were observed. 8H SFs and triangular surface defects can be generated from similar nucleation sources. Time-resolved measurements revealed that compared to defect-free regions, the carrier lifetimes decreased significantly in the presence of SFs corresponding to triangular surface defects and 3D 3C-SiC inclusions. 3D defect imaging in 4H-SiC epilayers was demonstrated by using optical second-harmonic generation

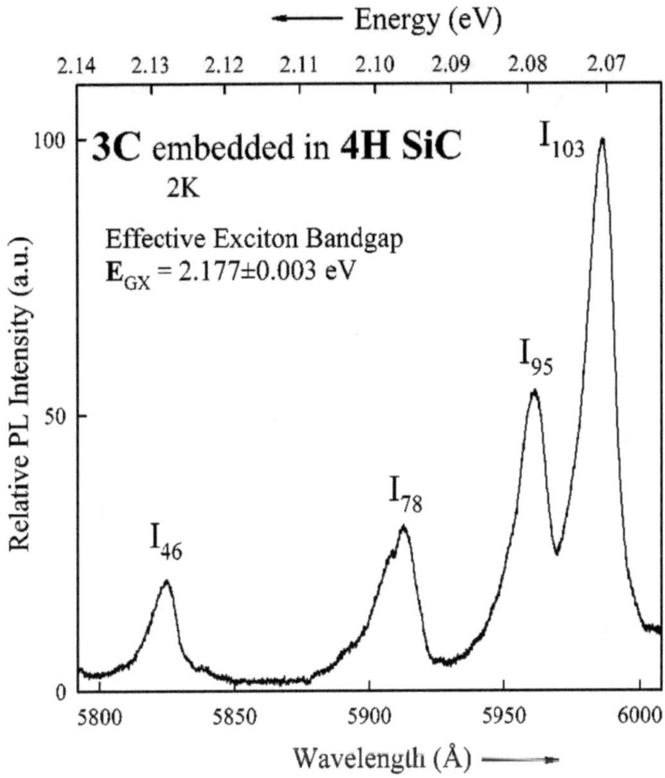

Fig. 2.60 PL spectrum of the 4H-SiC epitaxial film with 3C-SiC inclusions acquired at 2 K. The subscript of the marked lines denotes the momentum-conserving phonon energy in meV. Reprinted with permission from [341]. Copyright 2003, American Institute of Physics

(SHG) and two-photon-excited PL [346]. The SHG method offers clear 3D images of 3C inclusions because 3C-SiC is SHG-active, but 4H-SiC host crystal is not SHG-active in the case of *c*-axis incidence. The two-photon-excited PL approach also gives rise to 3D images of 3C inclusions, 8H SFs, and single Shockley SFs in the epilayers. Temperature dependence of the Raman spectra of microcrystalline 3C-SiC free-standing films was studied [347]. The measurements were conducted under direct laser heating of the thin films for temperatures up to 1700 K. The temperature dependence of the TO- and LO-phonon frequencies agrees well with that of 3C-SiC single crystals [348], but the Raman lines are significantly broader (Fig. 2.61). The influence of SFs on the linewidth was examined by comparing the experimental data with computer simulated Raman intensity profiles of 3C-SiC structures having randomly distributed SFs. Good agreement regarding the linewidth and disorder-induced peak shift was observed by assuming an average SF distance of 6 Å. The observed irreversible narrowing of the Raman lines at temperatures over 1900 K was attributed to annealing of SFs in 3C-SiC. The temperature dependence of the Raman

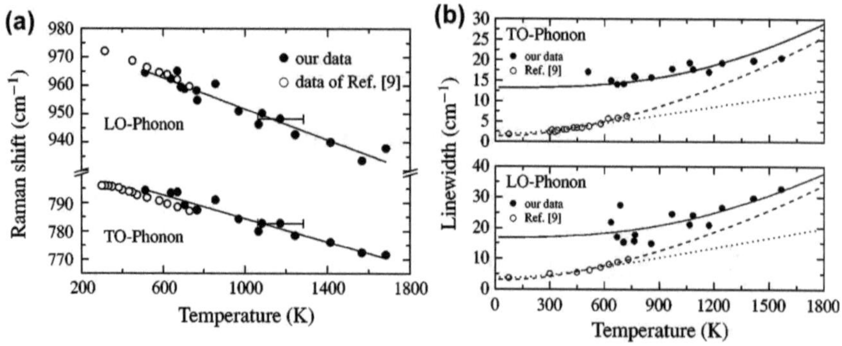

Fig. 2.61 Temperature dependence of **a** LO and TO Raman shifts and **b** linewidths for microcrystalline 3C-SiC free-standing films (filled circles) and for 3C-SiC single crystals (open circles) from Ref. [348]. Dotted line in (**b**) is a least-squares fit to the data of Ref. [348] considering three-phonon processes, whereas the solid and dashed lines are least-squares fits considering four-phonon processes. Reprinted with permission from [347]. Copyright 1998, American Physical Society

spectra of microcrystalline 3C-SiC free-standing films was examined. The measurements were performed under direct laser heating of the thin films for temperatures up to 1700 K. The temperature dependence of the TO- and LO-phonon frequencies agrees well with that of single-crystal 3C-SiC, but the Raman lines are considerably broader. The influence of stacking faults on the linewidth was studied by comparing the results with computer simulated Raman intensity profiles of 3C-SiC structures having randomly distributed stacking faults. Good agreement with respect to the linewidth and disorder-induced peak shift was found if the average stacking fault distance was assumed to be 6 Å. An irreversible narrowing of the Raman lines was observed at temperatures above 1900 K, which was attributed to annealing of the stacking faults in 3C-SiC.

Raman spectra of 6H-SiC crystals containing SFs were measured using the back scattering geometry for the *c* face [349]. The intensity of the transverse optical (TO) phonon band at 796 cm^{-1}, which corresponds to the phonon mode at the Γ point in 3C-SiC, is sensitive to the SFs. The intensity of this band depends on the SF density. It was explained by the bond polarizability model. The TO mode at 796 cm^{-1} in 6H- and 4H-SiC should be forbidden for the back scattering geometry using the (0001) face. However, the study shows that this band arose weakly for these polytypes containing SFs [350]. This was supposed to be ascribed to the breakdown of the wavevector selection rule by SFs. Nano Fourier-transform infrared (FTIR) spectroscopy was used for the characterization of in-grown SFs in a 4H-SiC epitaxial layer [351]. A local spectral shift of the mid-infrared near-field response was observed, in agreement with the identification of the defect stacking order as 3C-SiC from simulations based on finite dipole model.

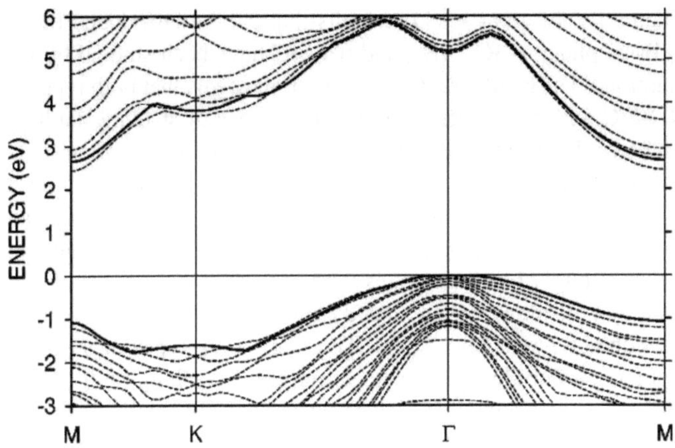

Fig. 2.62 Band edges of 4H-SiC near the fundamental gap projected on the Γ–M–K basal plane calculated by the first-principles method (solid lines). The dashed lines show the band structure results of the model containing the stacking faults. Reprinted with permission from [352]. Copyright 2001, American Institute of Physics

Theoretical Calculations of Stacking Faults

Several groups have theoretically studied stacking faults by first-principles calculations. The first-principles calculation was performed on the stacking faults in 4H-SiC [352]. An interface band in the gap with a maximum depth of 0.2–0.3 eV below the conduction band minimum at the $\overline{M}$ point was obtained (Fig. 2.62) because the local environment resembles that of the lower gap 6H polytype. The formation energies of the stacking faults in 3C-, 4H-, and 6H-SiC are on the order of a few meV per pair. Hence, the thermodynamic driving force results in growth of the stacking faults in an *n*-type sample. Nonradiative recombination of electrons trapped at the stacking fault with holes provides the necessary energy to overcome the partial dislocation motion barriers required to increase the stacking fault area in a device under forward bias. This explains the observed increase in stacking faults in SiC devices under long time forward bias operation and degradation of the devices by the negative effects of trapping electrons on transport. Trapping of electrons in the stacking fault interface states in 4H-SiC serves as driving force for growth of stacking faults, because the energy to create the additional stacking fault area is offset by the energy gained by capturing electrons in an interface localized state below the conduction band [353]. Screening significantly reduces the potential barrier (for electrons) adjacent to the quantum well.

The structurally different stacking faults that can be introduced by glide along the (0001) basal plane in 3C-, 4H-, and 6H-SiC have been studied by using DFT–LDA calculations. The band-structure calculation reveals that both types of stacking faults in 4H-SiC and two of the three different types of stacking faults in 6H-SiC induce quasi-two-dimensional energy states in the band gap at about 0.2 eV below the conduction band minimum, thus being electrically active in *n*-type SiC [354]. The third type of stacking faults in 6H-SiC and the only stacking fault in 3C-SiC do not give rise to states clearly separated from the band edges, but instead produce rather strongly localized states with energies very close to the band edges [355]. Figure 2.63 displays the calculated band structure of the perfect 6H-SiC and the 6H-SiC containing stacking faults. There are three types of stacking faults in 6H-SiC that can be introduced by dislocation glide: stacking fault (42), defined as (A|CABAC) inside the faulted cell, stacking fault (3111), defined as (AB|ABAC), and stacking fault (24), defined as (ABC|BAC). Herein "|" denotes the slip plane. Strong localization of the stacking fault gap state wavefunction is within roughly 10–15 Å along the direction perpendicular to the stacking fault plane. This quantum well like feature of certain stacking faults in SiC can be understood in terms of the remarkable conduction-band offsets between cubic and hexagonal polytypes. The stacking fault energies were also calculated using the supercell method and axial next-nearest neighbor Ising model method. Both calculated results agree well with the measured low stacking fault energies determined experimentally. The 4H-SiC which contains two, three, and four stacking faults in neighboring glide planes was investigated [356]. The structural defects can be viewed as thin 3C-like inclusions in the 4H-SiC matrix. The stacking fault energy of two stacking faults in adjacent basal planes is reduced by approximately a factor of 4 relative to that of the isolated stacking fault. This implies that double stacking faults (3C-like inclusions 1.25 nm in thickness) may be quite common in 4H-SiC. Electronic structures of extended planar defects in 3C-, 4H-, 6H-, and 15R-SiC, including twin boundaries, SFs, and polytype inclusions were calculated by using the DFT method [357]. SF energies were calculated using both supercell method and axial next-nearest-neighbor Ising model. The dependence of 4H-SiC SF energy on temperature was studied using calculation of the free energy of phonons based on the DFT–GGA method [358]. Both harmonic vibration approximation and quasi-harmonic approximation were used in order to account for the effects of thermal expansion. The SF energies of a single Shockley-type SF at room temperature and at 1500 K are 14.5 and 12.8 mJ m^{-2}. They are insensitive to temperature. The SF energy of a double Shockley-type SF is 8 mJ m^{-2} at room temperature and 11 mJ m^{-2} at 1500 K, and the energy increases with temperature. By applying the obtained SF energy to the quantum well action model, the critical minority carrier density was estimated to be 1.0×10^{16}–1.0×10^{17} cm^{-3}.

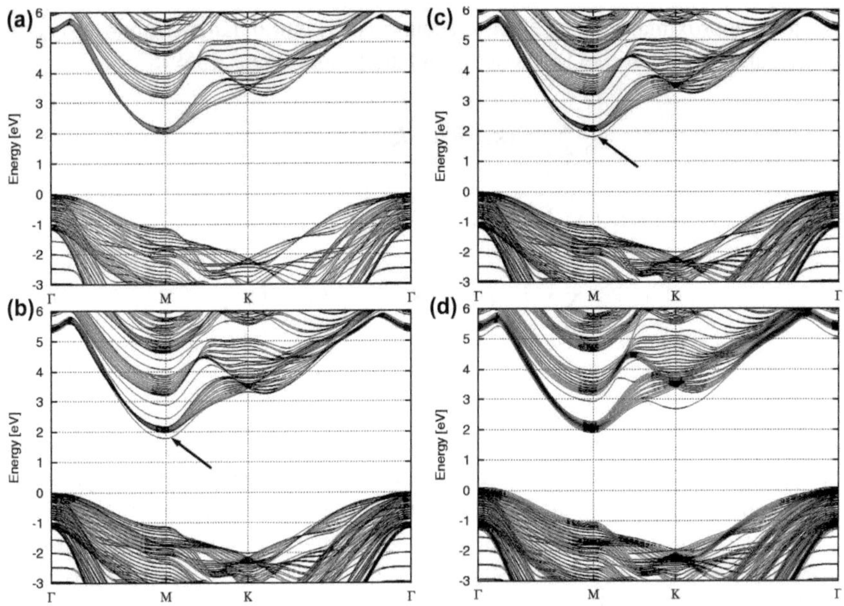

Fig. 2.63 Kohn–Sham band structure of the 96 atom supercells for **a** perfect 6H-SiC, **b** with stacking fault (42), **c** with stacking fault (24), and **d** with stacking fault (3111). The arrows point to the split-off bands below the conduction band caused by the stacking fault. Reprinted with permission from [355]. Copyright 2003, American Physical Society

2.5 Lattice Vibration and Infrared/Raman Spectra

The lattice dynamical vibration properties of silicon carbide and associated infrared and Raman scattering characteristics have been extensively studied. Under the small-amplitude vibration approximation, the overall lattice vibration can be decomposed into different amplitudes and frequencies of harmonic oscillations. These harmonic oscillations comprise a series of higher-frequencies transverse optical (TO) and longitudinal optical (LO) modes, and a series of lower-frequencies transverse acoustic (TA) and longitudinal acoustic (LA) modes. In the more accurate scenario of quantum mechanics, each vibrational mode represents a quantum harmonic oscillator, whose energy is composed of a constant energy of $\frac{\hbar\omega}{2}$ plus integer (n) number of discrete energy packets called "phonons", that is, $E_n = \frac{\hbar\omega}{2} + n\hbar\omega$. Phonon dispersion curve refers to plot of the relationship between phonon frequency (y coordinate) and wave vector (x coordinate) in a crystal, describing wave vector dependence of the vibration frequency within the Brillouin zone. The phonon dispersion curve depends on the crystal structure including compositional elements and their ways of spatial arrangement (symmetries). It reflects the complete set of vibrational modes in a crystal. Harmonic oscillators and corresponding phonons are basic thermal quantities of crystals. However, it should be mentioned that some thermal phenomena such as thermal expansion only arises when the anharmonic-vibration effect beyond harmonic approximation is taken into account. The Hamiltonian of a vibrating crystal lattice can be expressed in terms of normal mode coordinates [359]:

$$\begin{aligned}\mathcal{H} = &\frac{1}{2}\sum_{\boldsymbol{k},n}\dot{Q}(\boldsymbol{k},n)\dot{Q}(-\boldsymbol{k},n)\\ &+\frac{1}{2}\sum_{\boldsymbol{k},n}\omega_0^2(\boldsymbol{k},n)Q(\boldsymbol{k},n)Q(-\boldsymbol{k},n)\\ &+\sum_{m>2}\frac{1}{m!}\sum_{\boldsymbol{k}_1,n_1}\cdots\sum_{\boldsymbol{k}_m,n_m}V_m\begin{pmatrix}\boldsymbol{k}_1\cdots\boldsymbol{k}_m\\ n_1\cdots n_m\end{pmatrix}\\ &Q(\boldsymbol{k}_1,n_1)\cdots Q(\boldsymbol{k}_m,n_m)\Delta(\boldsymbol{k}_1+\cdots\boldsymbol{k}_m),\end{aligned} \tag{2.11}$$

where $\boldsymbol{k}$ and n are lattice wavevector and branch index of the phonon dispersion curve, $Q(\boldsymbol{k}, n)$ is the normal coordinate and $\omega_0(\boldsymbol{k}, n)$ is the circular frequency of the vibration mode, V_m is the coupling constant, and $\Delta(\boldsymbol{G})$ equals unity if $\boldsymbol{G}$ is a reciprocal lattice vector and zero otherwise. The first two terms on the right-hand side are the kinetic and potential energies of the harmonic oscillators, whereas the third term contains all anharmonic interaction energies.

The infrared absorption technique measures the "fingerprint" phonon frequencies characteristic of a solid by recording the discrete light absorption peaks, each of these peaks corresponds to a crystal phonon frequency, where an incident infrared photon would be resonantly absorbed, along with creation of a phonon. For the Raman measurement, an incident visible or near-UV photon inelastically collides with the lattice, outgoing as an energy-diminished (or increased) photon, accompanied by creation (or annihilation) of a lattice phonon with energy equal to the energy difference between the incident and outgoing photons. This process is called Stokes (or anti-Stokes) Raman scattering. From the microscopic quantum mechanical viewpoint, the infrared light absorption and Raman scattering processes can be regarded as "photon–phonon" collision processes. During such processes, the conservation laws such as energy conservation and momentum conservation must be satisfied. The infrared absorption and Raman scattering processes can be allowed (active) or forbidden (inactive) by selection rules of quantum transitions. The selection rules correlate intimately with conservation laws and specific symmetries of the crystal. The infrared absorption or Raman scattering process allowed by selection rules (infrared or Raman active) can be different, rendering these two characterization methods sometimes complementary to some extent depending on the symmetries of the specific crystal. Infrared light absorption and Raman scattering are also very powerful tools for detecting the surface vibration (phonon) modes or surface functional groups of bulk solids and nanostructures.

2.5.1 Lattice Vibration and Phonon Dispersion Curves

The phonon dispersion curves in the various symmetry directions in cubic SiC were derived using a rigid ion model consisting of short range central and noncentral interactions and long range Coulombic interactions among ions with the appropriate effective ionic charges [360]. The calculated dispersion curve in the <111> direction (Fig. 2.64) was compared to Raman data acquired from various polytypes of SiC [361, 362], and the agreement between calculation and experiments is very good. The phonon density of states and Debye characteristic temperature of SiC were also derived. The adiabatic bond-charge model was applied to the 3C, 6H, 4H, and 2H polytypes of SiC [363]. The short range elastic forces were described by two-body and three-body interactions among silicon ions, carbon ions, and bond charges, whereas the long range Coulomb forces were taken into account by the Ewald technique. The model parameters were fitted to the frequencies obtained from Raman and luminescence measurements. The resulting phonon dispersion curves along the high symmetry lines in the face-centered cubic and hexagonal Brillouin zone agree with experimental results. In the 3C case, the calculation yields the same eigenvectors as ab initio calculation. The independent elastic constants of the different polytypes were derived using the resulting sound velocities and they agree reasonably well with experimental ones. The plane wave pseudopotential approach of the DFT employing the LDA was used to study the ground-state properties of the 3C, 2H, and 4H polytypes of SiC [364]. The linear response theory within the DFT was employed to determine the dynamic lattice properties of 3C-SiC. The phonon dispersion curves of cubic SiC are intermediate between those of Si and diamond but not average and consistent with measured values (Fig. 2.65). The phonon dispersion curves of 2H- and 4H-SiC were also derived. The calculated ground-state properties of the cubic 3C and hexagonal 2H and 4H polytypes of SiC are quite similar, especially concerning ground state energies and volumes as well as the valence charge densities along the bonding and stacking directions. In contrast to elemental crystals like Si and diamond, SiC has noticeable ionicity, that is, charge asymmetry. The employed calculation procedure has limitations in that it neglects the effects of the vibrational energy and crystal entropy. Phonon dispersion relations in 3C-SiC were obtained for the entire Brillouin zone by inelastic X-ray scattering using a synchrotron radiation source [365]. The DFT calculations reproduced the experimental phonon dispersion remarkably well along high-symmetry paths in the Brillouin zone of Γ–K [110], X–Γ [100], and Γ–L [111]. Inelastic X-ray scattering data for 4H-SiC along the *c* axis were also examined.

2.5.2 Infrared Spectroscopy of Phonons

Infrared spectroscopy imparts information about the lattice vibration properties of various polytypes of silicon carbide. The infrared transmission and reflectivity of

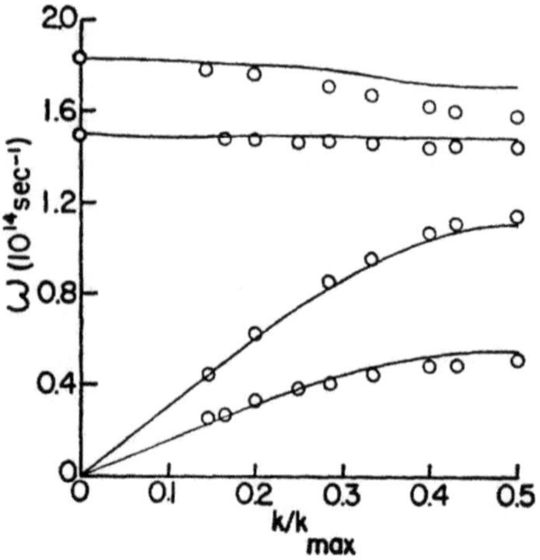

Fig. 2.64 Calculated phonon dispersion curves (solid lines) of 3C-SiC in the <111> direction. The circles denote measured values obtained from Ref. [361] ($k_{max} = 2\pi/a$). Reprinted with permission from [360]. Copyright 1969, American Physical Society

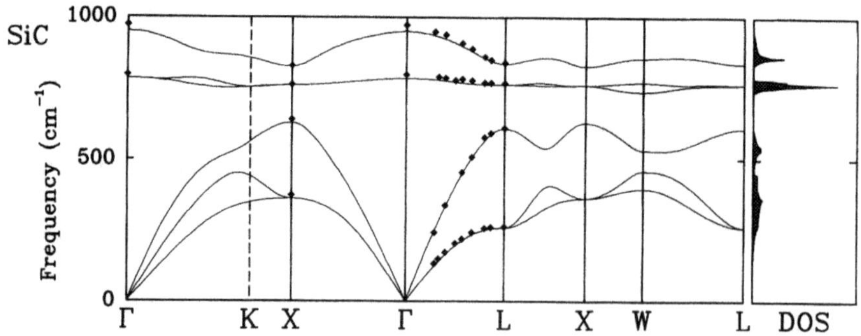

Fig. 2.65 Calculated phonon dispersion curves and density of states (DOS) of 3C-SiC (Γ − K: [110], X − Γ: [100], Γ − L: [111]). The diamond signs denote the experimental first-order Raman scattering data obtained from Ref. [362]. Reprinted with permission from [364]. Copyright 1994, American Physical Society

several hexagonal SiC crystals in the 1–25 μm spectral region have been measured [366] and the reflectance spectra of both ordinary and extraordinary rays have been obtained (Figs. 2.66 and 2.67). Resonance occurs at 12.60 and 12.73 μm, respectively, and based on the reflectivity data, the high frequency dielectric constant was calculated to be 6.7. The reflectance spectra can be fitted within experimental errors by the classical dispersion theory using the properly selected dispersion parameters, and the low frequency dielectric constant was derived to be 10.0 accordingly. The study of the surface effect suggests that reflectivity is an intrinsic property of the crystal. The absorption coefficient was derived as a function of wavelength in the range 1–10 μm based on the transmission measurement data.

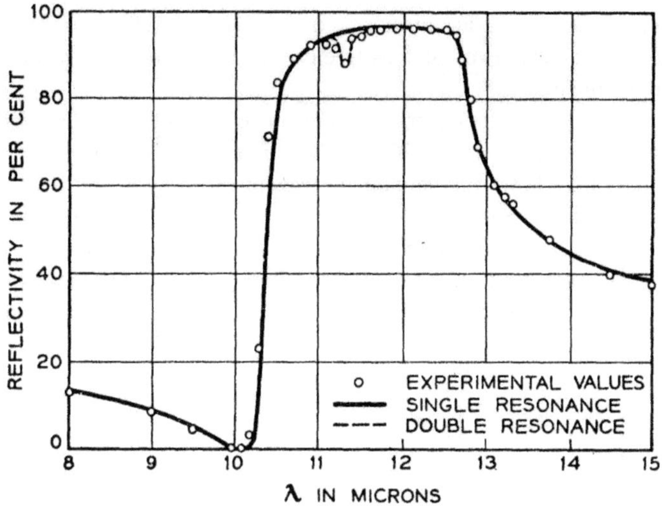

Fig. 2.66 Reflectivity versus wavelength of the extraordinary ray in α-II SiC. The circles denote measured values and the solid line is the calculated curve. Reprinted with permission from [366]. Copyright 1959, Americal Physical Society

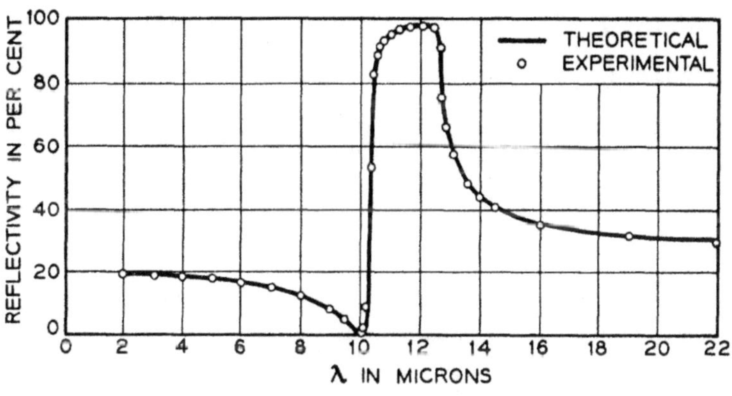

Fig. 2.67 Reflectivity versus wavelength of the ordinary ray in α-II SiC. The circles denote measured values and the solid line shows the calculated curve. Reprinted with permission from [366]. Copyright 1959, Americal Physical Society

The transmission and reflection of 3C-SiC films were investigated in the 1–15 μm range [367] and the results were analyzed by using the classical dispersion theory. The dispersion parameters for the fundamental resonance at 12.60 μm are essentially the same as those of the ordinary ray in the hexagonal α-II SiC. The measured and calculated transmission and reflection spectra are displayed in Fig. 2.68. In the immediate region of resonance, the agreement between theory and experiments is very good for both transmission and reflection. However, at the shorter wavelengths, the theory predicts higher transmission values and smaller reflection values because other absorption mechanisms such as free carrier absorption were neglected in the theoretical treatment.

Multiple phonon absorption processes also occur in SiC. The featured infrared absorption bands obtained from SiC have been interpreted as arising from multiple phonon absorption [368]. The phonon energies obtained by measuring the indirect interband absorption in SiC are 0.045 eV (TA), 0.067 eV (LA), 0.0955 eV (TO), and 0.1055 eV (LO). The ten infrared absorption peaks in the energy range of 0.130–0.300 eV stem from multiple phonon absorption, and Fig. 2.69 depicts a part of the infrared absorption spectrum. The four predicted summation bands show good

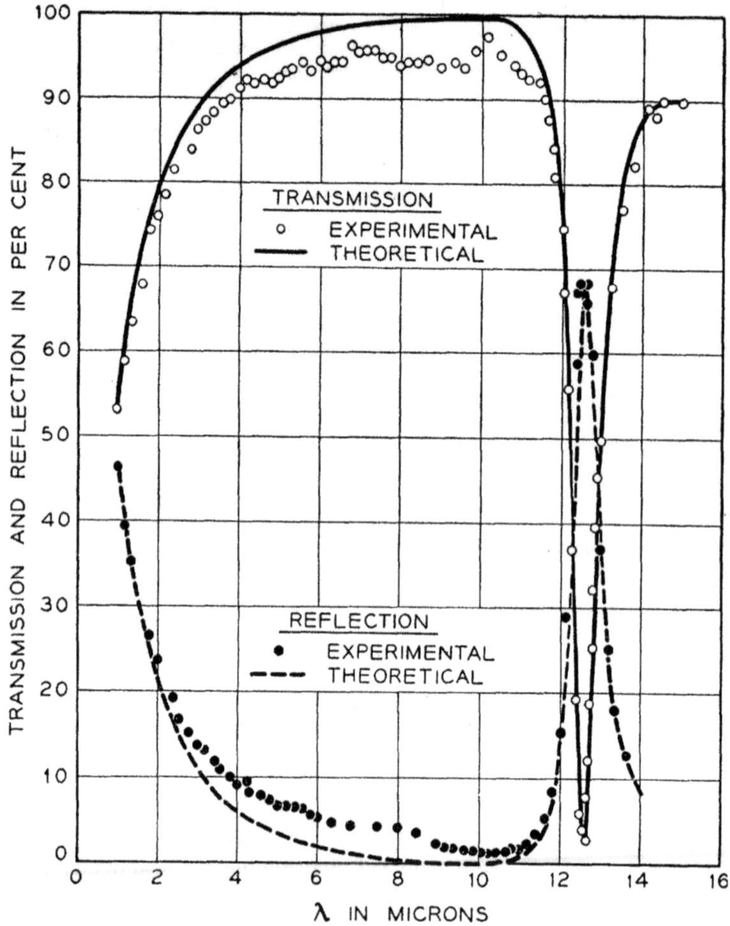

Fig. 2.68 Transmission and reflection of a 0.06-μm thick β-SiC film. Reprinted with permission from [367]. Copyright 1959, American Physical Society

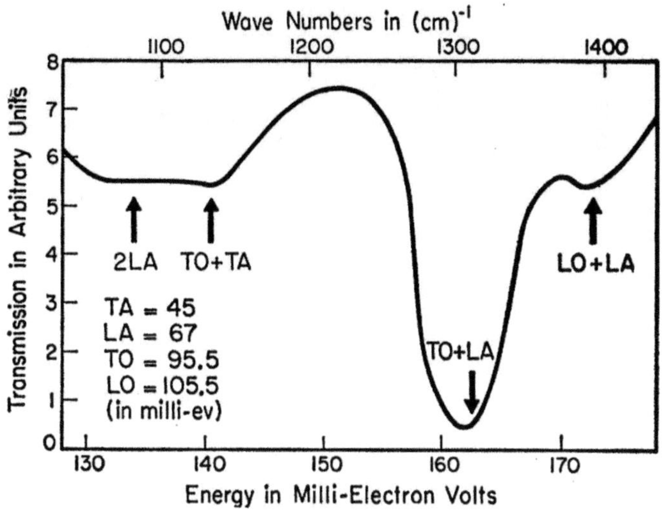

Fig. 2.69 Two-phonon optical acoustical summation bands in the infrared spectrum of SiC. The arrows point to the energy values calculated using the four assigned phonon energies. Reprinted with permission from [368]. Copyright 1961, American Physical Society

agreement with the corresponding peaks of the measured spectrum. Infrared absorption in several types of SiC crystals was studied [369]. In most polytypes, additional one-phonon absorption lines are permitted by group-theoretical selection rules. A procedure was proposed that makes it possible to classify all the additional weak lines. Two lines at 62 and 110 meV in 6H-SiC and one line at 70 meV in 15R-SiC were observed. The classification procedure applies also to Raman lines.

Since the infrared reflection spectrum depends on many sample parameters, the infrared spectra of SiC films can provide useful information about the status of the films. Several unusual features are present in the infrared reflection spectra of the CVD β-SiC films concerning the surface roughness and possible conducting layers at the film-substrate interface [370]. The free carrier concentration and thickness of the film can be determined from the reflectance extrema. Generally, the complex dielectric function describing the optical responses including light reflectance of a semiconductor to infrared light follows [371]:

$$\varepsilon = (n + i\kappa)^2 = \varepsilon_\infty \left(1 - \frac{\omega_p^2}{\omega(\omega + i\gamma)} + \frac{\omega_{LO}^2 - \omega_{TO}^2}{\omega_{TO}^2 - \omega^2 - i\Gamma\omega}\right), \tag{2.12}$$

where n is the refractive index, κ is the extinction coefficient, ω is the photon frequency, ε_∞ is the high-frequency dielectric constant, ω_{TO} and ω_{LO} are the TO and LO phonon frequencies, Γ is the phonon damping constant, $\omega_p = \left(\frac{4\pi Ne^2}{m^* \varepsilon_\infty}\right)^{1/2}$ is the bulk plasma frequency of free carriers with a density of N and an effective mass of m^*, and γ is the free carrier damping constant. The light reflectance follows:

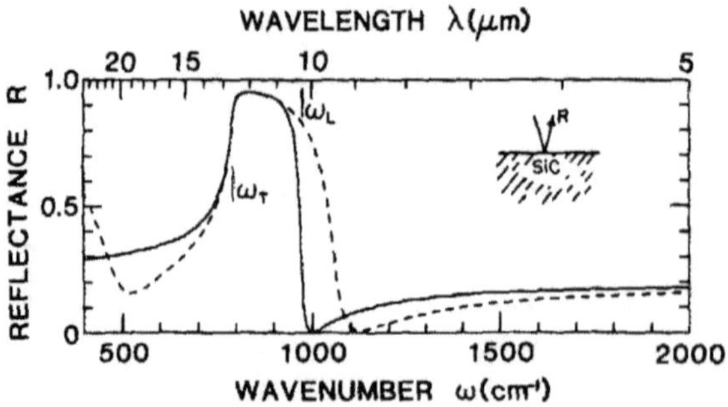

Fig. 2.70 Calculated reflectance at normal incidence for semi-infinite *n*-type β-SiC. The solid and dashed lines correspond to $N = 0$ and $N = 1 \times 10^{19}\ \text{cm}^{-3}$, respectively. The inset shows the sample configuration. The positions of ω_{TO} and ω_{LO} are marked. Reprinted with permission from [370]. Copyright 1986, American Institute of Physics

$$R = \frac{(n-1)^2 + \kappa^2}{(n+1)^2 + \kappa^2}. \tag{2.13}$$

Comparison of the calculated reflectance spectra (Fig. 2.70) for the intrinsic and *n*-type β-SiC shows clearly the influence of the free carriers on the profile of the reflectance spectrum. These values show good agreement with the values obtained by using other methods and hence, infrared reflectance measurement can nondestructively assess the fabricated films.

Spectral analysis may also reveal the crystal anisotropy, as demonstrated by the polarized infrared reflectivity spectra of 6H-SiC single crystals at different incident angles [372]. A sharp line in the Reststrahlen band was observed from the extraordinary ray and its line shape depends on the incident angle. Numerical calculation based on a Lorentz oscillator model demonstrates that this sharp line arises from the crystal anisotropy of 6H-SiC, suggesting that the infrared reflectivity of many anisotropic solids may depend on the incident angle. Infrared ellipsometry was used to determine the free electron concentrations in bulk and epitaxial nitrogen-doped 4H- and 6H-SiC [373] as well as epitaxial layer thickness. Data obtained from 6H-SiC can be modeled well if the large effective mass anisotropy is taken into account, whereas effective mass anisotropy has no noticeable effects on the 4H-SiC spectra. The two-phonon interaction influences the line shape in the region between 1300 and 1700 cm^{-1} and should be considered in calculating the free carrier concentration. To interpretate the 11.3 μm absorption feature observed in the spectra of carbon-rich stars, infrared data on some SiC particulate samples were presented [374]. The spectra show a variety of SiC phonon features within 10–13 μm. The main parameters determining the band profile are grain size and shape as well as impurities in the material. Free charge carriers generated by such as nitrogen doping were found to be a very common characteristic of many SiC particles. These free charge carriers produce strong plasmon absorption in the near and middle infrared, which may also influence the 10–13 μm feature profile via plasmon–phonon coupling. Experiments and theoretical calculations show that there is no systematic dependence of the infrared absorption band profile on the crystal type (α- vs. β-SiC). Films of a-SiC:H grown on Si single crystal (100) by plasma-enhanced CVD (PECVD) using monomethylsilane (CH_3SiCl_3) source and hydrogen carrier gas were examined by using FTIR spectroscopy [375]. The quantity of the hydrogen bonds of the films is proportional to the square of the plasma power density. The shift of the SiH_n ($n = 1, 2$) peak reveals that the nearest atoms to Si in the film are replaced by Si or H atoms from C in CH_n with increasing hydrogen content. In another study, the FTIR Spectroscopy was utilized to analyze the changes in chemical bonding that occurs in PECVD a-SiC:H thin films [376]. The decrease in mass density and dielectric constant of the films is accompanied by both increase in terminal SiH_x and CH_x bonding and decrease in Si–C network bonding. At densities of 1.85 g cm^{-3}, the concentration of terminal SiH_x bonding peaks and subsequent hydrogen incorporation were achieved via terminal CH_3 groups. The dielectric properties of the 4H- and 6H-SiC in the 0.1–19 THz range (below the fundamental TO phonons) were investigated [377]. Folding of the Brillouin zone due to the specific superlattice structure of the two polytypes causes activation of acoustic phonon modes. The combined ultrabroadband terahertz time-domain spectroscopy and simulations based on density-functional perturbation theory were used to observe these modes, including band splitting due to the dissimilar carbon and silicon sublattices of the structures. The weakest mode of the infrared active folded phonons in 4H-SiC, folded TA mode, was studied [378]. The measurement of far-infrared transmission and theoretical simulation using an oscillator model determined the lifetime of the folded TA phonons to be 58 ± 4 ps at room temperature and it increased to 140 ± 7 ps at 5 K. Oscillator strength of the folded TA phonon resonance was determined to be $(6.0 \pm 0.5) \times 10^{-4}$ and is independent of temperature.

2.5.3 *Raman Spectroscopy of Phonons*

Raman scattering conveys fruitful information about the lattice dynamical vibration properties of solids determined by their various aspects of microstructures including crystal structures, defects, degrees of amorphization, surfaces and interfaces, and strains. Examination of Raman scattering in 6H-SiC crystals [361] revealed fifteen phonon lines (Fig. 2.71). Polarized light was used to identify the mode symmetry. The modes were classified by large zone analysis, yielding dispersion-like curves (Fig. 2.72). All observed narrow lines can be explained and only two of the expected lines were not observed. Analysis of the dependence of phonon energy on propagation direction indicates that certain infrared and Raman active modes have extremely small infrared strength due to the specific polytype structure of 6H-SiC.

The first-order Raman spectra of more polytypes of SiC crystals, including 3C, 4H, 6H, 15R, and 21 R, have been acquired [362], and the phonon dispersion curves were established based on the Raman measurements. Overall 9 lines were observed from 4H-SiC, 16 lines from 15R-SiC, and 14 lines from 21R-SiC. The symmetry

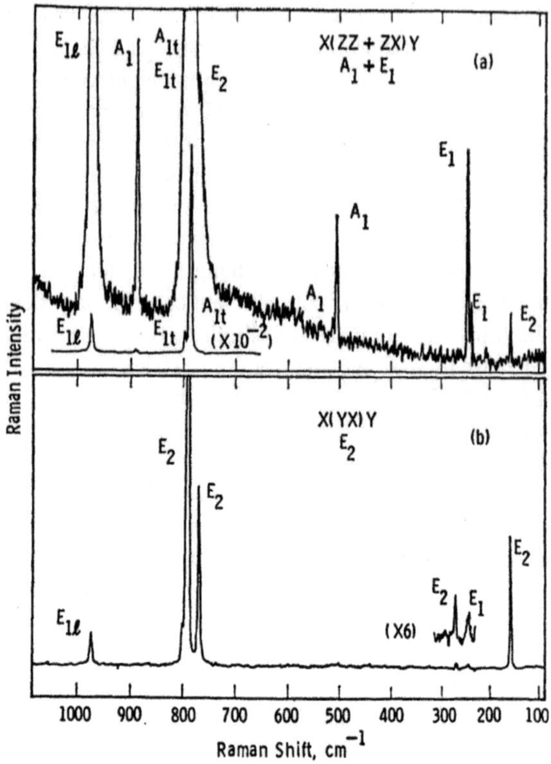

Fig. 2.71 Full one-phonon emission Raman spectrum of 6H-SiC crystal with arrangements permitting **a** A_1 and E_1 modes, and **b** E_2 modes. Reprinted with permission from [361]. Copyright 1968, American Physical Society

of each phonon mode was determined by analyzing the polarization. The modes were classified using a standard large zone and then the positions of the three groups of one-phonon lines and 15 previously observed lines in 6H-SiC were assembled in a single large zone (Fig. 2.73). This assignment results in a set of SiC phonon dispersion curves, confirming the existence of a common phonon spectrum for all SiC polytypes in the axial direction. The velocities of the longitudinal and transverse acoustic phonon waves derived from the dispersion curves are in good agreement with experimental results.

A theoretical method based on bond polarizability was adopted to calculate the Raman intensity profiles of the folded modes of SiC [379]. The relative Raman intensity of the folded modes was calculated using bond Raman polarizability together with the eigenvectors of the linear chain model. In this method, the bonds in a unit cell are classified into groups in which the individual Raman polarizability is equivalent. The Raman polarizability is expressed as the sum of the product of the bond Raman polarizability and relative displacement of the end atoms linked by the bond. The calculated Raman intensity profiles are in qualitative agreement with the experimental Raman spectra of the folded modes arising from TA and TO phonon branches along the c direction (Fig. 2.74). The calculated dispersion curves are compared to the measured ones in Fig. 2.75. The calculated frequencies of the TA branch are slightly

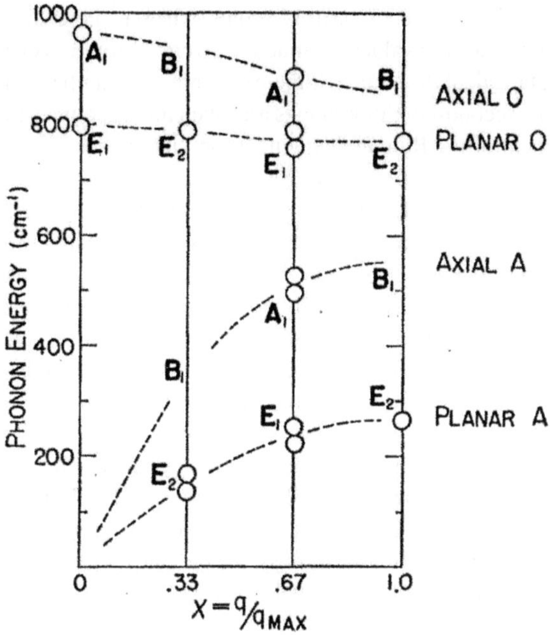

Fig. 2.72 Plot of experimental data of 6H-SiC in the form of dispersion curves in the large zone. The doublet separation is exaggerated. This figure is valid for weak modes for any phonon propagation direction and appropriate for strong modes only when the phonon propagation direction is along the *c* axis. Reprinted with permission from [361]. Copyright 1968, American Physical Society

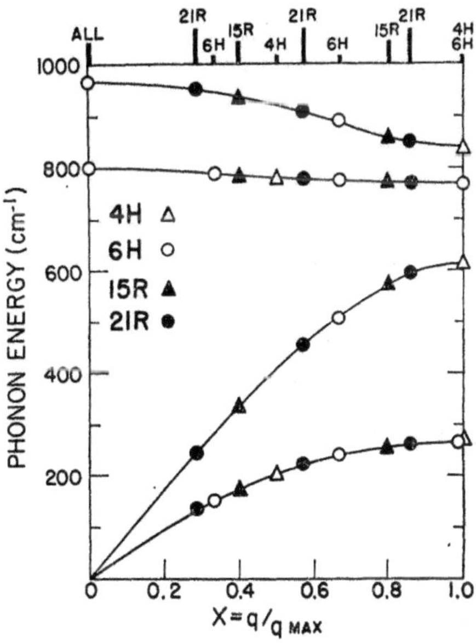

Fig. 2.73 Combined dispersion curves based on measured Raman data obtained from four polytypes of SiC. The Raman accessible values of $x = q/q_{max}$ are marked at the top of the figure. Reprinted with permission from [362]. Copyright 1968, American Physical Society

smaller than the measured values, whereas there are slightly larger deviations from the measured values (as much as 20 cm^{-1}) for the TO branch at around the zone edge. The calculation may fit the experimental data better if the force constants between the second-neighbor planes are taken into account. The results indicate that measurement of the Raman frequencies and intensities is a useful technique for identifying the polytype structure of SiC.

Raman backscattering studies for 3C-SiC single-crystal films grown by CVD with film thicknesses of 600 Å–17 μm were conducted [380]. Raman spectra of samples with thicknesses > 4 μm show a sharp and strong feature obeying the selection rule for 3C-SiC LO(Γ) (LO phonon at the Γ point) phonon line. The Raman signals from the SiC film and the Si substrate show the same polarization behavior, confirming their same crystalline orientations. Although there is a big lattice mismatch of 20% between 3C-SiC and Si, the Raman shifts between 3C-SiC/Si and free 3C-SiC films are ≤ 2 cm^{-1} and reflect strains of 0.1–0.2% in films thicker than 4 μm. The Si 522 cm^{-1} phonon from a Si wafer was enhanced in intensity by a factor of 2–3 due to a CVD overlayer of 3C-SiC. The 3C-SiC LO(Γ) from SiC/Si samples was enhanced by a factor of 2 or 3 after removal of the Si substrate. The forward scattering has different selection rules from the back scattering and thus could cause arising of the forbidden TO phonon and depolarization of the 3C-SiC Raman phonons from the SiC

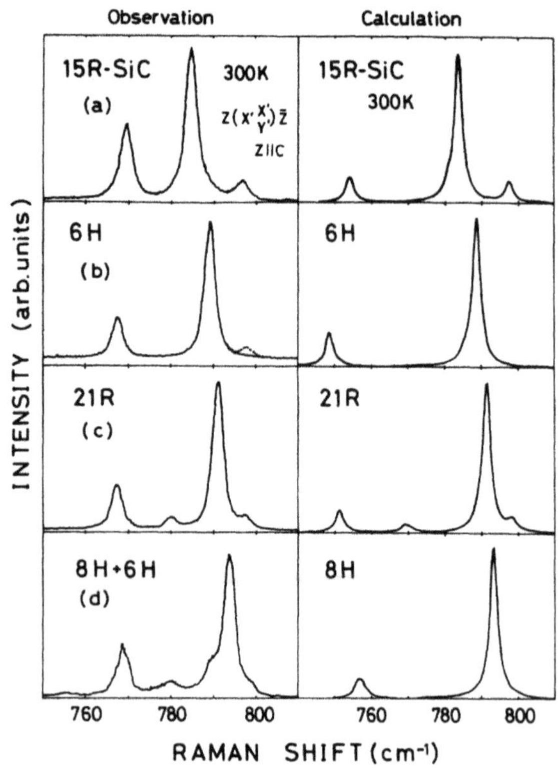

Fig. 2.74 Comparison between the measured and calculated Raman spectra of the folded TO modes of the four polytypes of SiC: **a** 15R, **b** 6H, **c** 21R, and **d** 8H. Reprinted with permission from [379]. Copyright 1986, American Physical Society

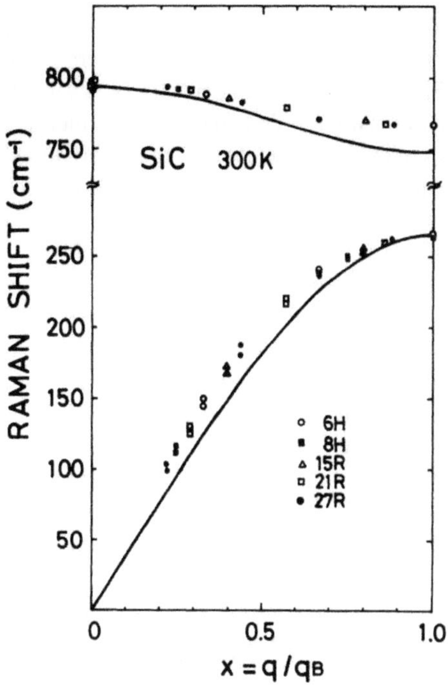

Fig. 2.75 Assembled phonon dispersion curves in the basic zone based on the data acquired from five polytypes of SiC. The solid lines are the calculated curves by the linear-chain model. Reprinted with permission from [379]. Copyright 1986, American Physical Society

free films. Raman scattering was utilized to evaluate SiC heteroepitaxial thin layers [381]. The heteroepitaxial layer of 3C-SiC can grow on the (0001) surface of 6H-SiC platelets using conditions similar to those used to fabricate 3C-SiC on Si. The Raman spectra were acquired in the backscattering geometry using the 514.5 nm laser line at room temperature. The incident light is normal to the 3C-SiC(111) and 6H SiC(0001) surfaces and polarized in a vertical plane. Figure 2.76 shows the TO and LO regions of the Raman spectra of the 6H-SiC sample before and after growth of the epitaxial layers. The intensity is normalized to the 6H-SiC main peaks at 789 cm^{-1} in the TO region and 967 cm^{-1} in the LO region. In the TO region, the intensity of the peak corresponding to the 3C-SiC TO phonon mode (796 cm^{-1}) increased after epitaxy, whereas in the LO region, the peak corresponding to 3C-SiC LO phonon mode (972 cm^{-1}) and the diminished peak of 6H-SiC LO phonon were observed. The layer thickness is about 10–15 μm and the surface exhibited a single-crystal reflection high energy electron diffraction pattern. The results indicate that the 3C-SiC(111) layer was heteroepitaxially grown on the 6H-SiC(0001) surface. This is understandable because that the atomic arrangements on the (111) plane of 3C-SiC and the (0001) plane of 6H-SiC are identical and the only structural difference between 3C in the [111] direction and 6H-SiC in the [0001] direction is the stacking order. 3C-, 4H-, and 6H-SiC epitaxial layers were investigated by Auger electron spectroscopy (AES), high-resolution electron energy loss spectroscopy (EELS), and Raman spectroscopy to determine the excitation energies of the optical Fuchs–Kliewer surface phonons and their relation to bulk phonon frequencies [382]. The surfaces were treated in a buffered HF solution. Loss structures attributed to excitation of Fuchs–Kliewer phonons were resolved. Their energies are at 115.9 ± 1 meV irrespective of the SiC polytype. Raman spectra of several SiC polytypes (3C, 6H, 15R) made from natural silicon (^{28}Si) and ^{30}Si were measured [383]. The isotope shifts of the phonon frequencies show characteristic variations of the effective wave vectors in 3C-SiC which arise from Brillouin zone backfolding. This allows for determination of the phonon eigenvectors of 3C-SiC for the dispersion branches along the [111] direction. The TO and LO phonon frequencies at Γ point were determined to be 796 and 971 cm^{-1}, respectively. The observed magnitudes of the Si and C ion displacements as well as their relative phase confirm the bond charge model and the ab initio calculation results. Phonon anisotropic properties of 4H-, 6H- and 15R-SiC were investigated by using polarization Raman scattering [384]. The E_1, E_2 and A_1 modes are anisotropic in the (1120) and the (1100) planes of 4H- and 6H-SiC, while E_2 modes are isotropic in the (0001) plane. The E_1 mode of 799 cm^{-1} in the (0001) plane exhibited anisotropic properties due to the stacking faults in SiC.

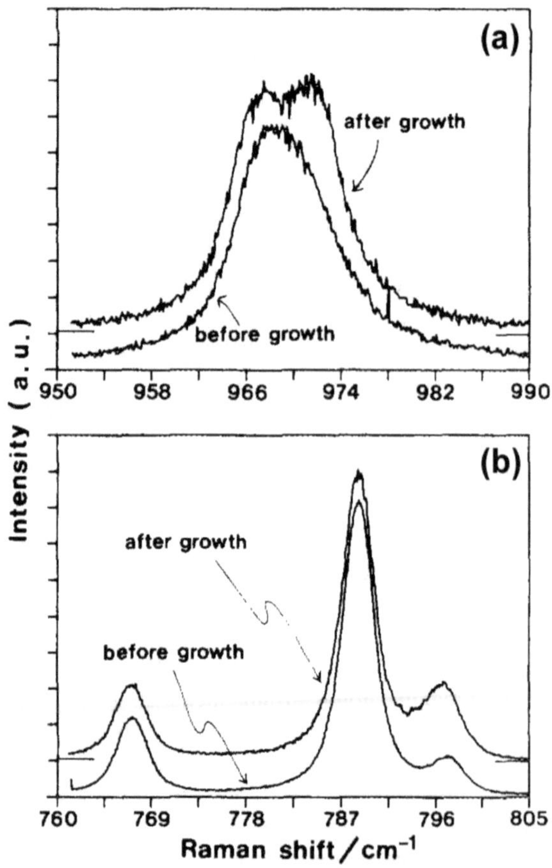

Fig. 2.76 **a** LO and **b** TO regions of the Raman spectra of 6H-SiC before and after growth of the epitaxial layer of cubic SiC on the (0001) surface. The intensities are normalized using the 6H-SiC main peaks at 789 cm^{-1} in the TO region and 967 cm^{-1} in the LO region. Reprinted with permission from [381]. Copyright 1987, American Institute of Physics

A simple realspace scheme for deriving atomic force constants was developed [385]. It was used to predict the phonon dispersion relation of SiC and to calculate the atomic force constants between a pair of nearest-neighbor atoms around a carbon vacancy. Using a revised lattice dynamics model and the bond polarizability concept, the Raman intensity profiles in SiC polytypes were examined [386]. The force constant parameters of a linear chain model were such determined that the calculated phonon dispersion and Raman intensity profiles fit the experimental results. The employed lattice dynamics model takes into account the third-neighbor plane forces and the difference between the forces in the hexagonal and cubic configurations. Quantitative agreement between the calculated and observed Raman profiles was obtained for the folded modes of the TO and TA branches for a number of polytypes. This agreement indicates that the set of force constants can be commonly used for any polytype of SiC and the bond polarizability concept is applicable to folded modes of both acoustic and optical branches.

Low-temperature second-order Raman spectra of 3C-SiC were measured [387]. The Γ_1 spectrum is strong, whereas the Γ_{12} and Γ_{15} spectra are faint. The ab initio calculated phonon eigensolutions and phenomenological polarizability coefficients agree well with experiment. The Γ_1 spectrum exhibits three distinct peaks at 1302, 1400, and 1619 cm^{-1} which occur in the gaps of the overtone density of states. This exemplifies the importance of taking into account the coupling matrix elements rather than simply relying on the one-to-one correspondence between the overtone density of states and the Γ_1 Raman spectrum. The first- and second-order Raman spectra of semi-insulating 4H-SiC have been investigated at room temperature by nonresonant excitation [388]. The results were compared to Raman spectra obtained from *n*-type doped 4H- and 6H-SiC. The second order Raman spectra (Fig. 2.77) of *n*-type doped and semi-insulating 4H-SiC are quite similar. The optical branches of the second-order Raman spectra are different between 4H- and 6H-SiC and more complicated than the two-phonon Raman spectra of 3C-SiC. The two-phonon spectra were assigned by comparison with the calculation, and both overtones and combination bands are present in the second-order Raman spectra of 4H- and 6H-SiC.

2.5.4 *Pressure and Stress Effects*

The lattice compression (or tension) can affect lattice dynamical properties of silicon carbide and also the Raman spectra. The pressure dependence of the optical phonons and transverse effective charge in 3C-SiC has been experimentally investigated [389]. The first-order Raman spectra of 3C-SiC were acquired at different hydrostatic pressures up to 22.5 GPa (Fig. 2.78). The measured LO(Γ, at the Brillouin-zone center) and TO(Γ) phonon energies can be determined within experimental errors by the following least square fits:

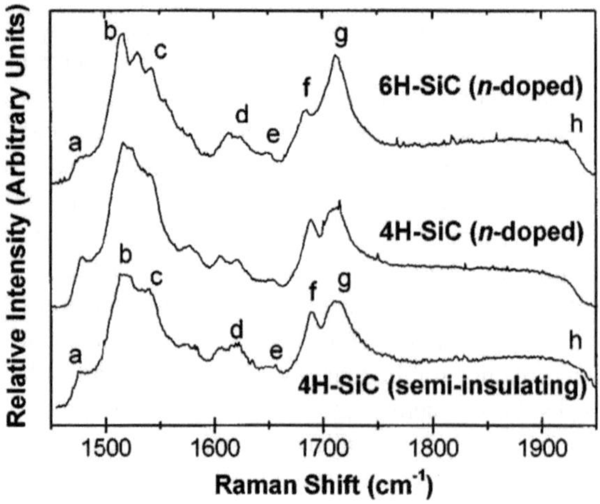

Fig. 2.77 Second-order Raman spectra acquired from semi-insulating 4H-SiC and *n*-type (nitrogen concentration: 2.1×10^{18} cm^{-3}) 4H- and 6H-SiC. Reprinted with permission from [388]. Copyright 1999, American Physical Society

$$\omega_{\mathrm{TO}} = (796.5 \pm 0.3) + (3734 \pm 30)\left[-\frac{\Delta a}{a_0}\right],$$
$$\omega_{\mathrm{LO}} = (973 \pm 0.3) + (4532 \pm 30)\left[-\frac{\Delta a}{a_0}\right], \tag{2.14}$$

where ω_{TO} and ω_{LO} are in cm^{-1} and a_0 is the lattice constant at room temperature. The fitted curve is shown in Fig. 2.78. The quadratic fit to the measured phonon energies versus pressure gives rise a sublinear relationship:

$$\omega_{\mathrm{TO}} = (796.2 \pm 0.3) + (3.88 \pm 0.08)p - (2.2 \pm 0.4) \times 10^{-2}p^2,$$
$$\omega_{\mathrm{LO}} = (972.7 \pm 0.3) + (4.75 \pm 0.09)p - (2.5 \pm 0.4) \times 10^{-2}p^2, \tag{2.15}$$

with ω_{TO} and ω_{LO} given in cm^{-1} and p in GPa. The mode-Grüneisen parameters $\gamma_{\mathrm{i}} = -\partial \ln \omega_{\mathrm{i}}/\partial \ln V$ for the TO(Γ) and LO(Γ) phonons were evaluated from the coefficients of the linear terms in $\Delta a/a_0$ of Eq. (2.1). They are $\gamma_{\mathrm{LO}} = 1.55 \pm 0.01$ and $\gamma_{\mathrm{TO}} = 1.56 \pm 0.01$. The LO(Γ)–TO(Γ) splitting increases linearly with lattice compression (Fig. 2.79) and also shows a sublinear dependence on pressure. The corresponding mode-Grüneisen parameter is $\gamma_{\mathrm{LO\text{-}TO}} = 1.52 \pm 0.05$. This increasing splitting implies an increase in the transverse effective charge under pressure. The transverse effective charge is defined as the macroscopic dipole moment per unit displacement of the sublattice (Si or C) when the macroscopic electric field in the crystal is kept zero. The relationship between the transverse effective charge and lattice compression was explained by the microscopic pseudopotential calculation but cannot be explained by the semi-empirical models such as the bond orbital model (Fig. 2.80).

The linewidth of the long wavelength optical phonons of 3C-SiC increases with pressure at above 10 GPa (Fig. 2.81) [390]. This was supposed to arise from the increase in the decay rate when Γ optical phonons decay into two acoustical phonons.

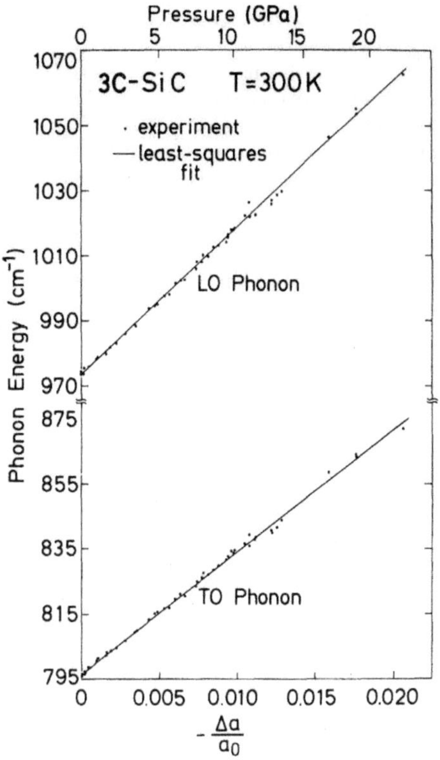

Fig. 2.78 Plot of the measured LO(Γ) and TO(Γ) phonon energies versus relative lattice compression (linear lower scale) and pressure (upper scale) in 3C-SiC crystals. The solid lines are least-square fits of the experimental points. Reprinted with permission from [389]. Copyright 1982, American Physical Society

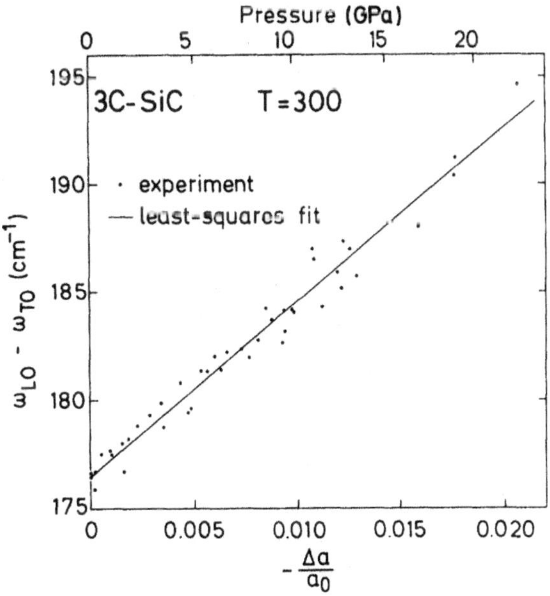

Fig. 2.79 Plot of the LO(Γ)–TO(Γ) splitting versus lattice compression and pressure in 3C-SiC crystals. The solid line through the experimental points is the least-square fit. Reprinted with permission from [389]. Copyright 1982, American Physical Society

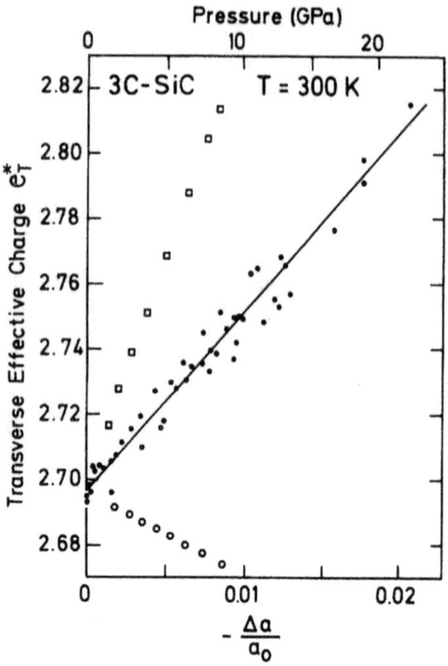

Fig. 2.80 Plot of transverse effective charge versus lattice compression and pressure in 3C-SiC crystals. The data are acquired from the measured pressure dependence of the LO and TO phonons in Figs. 2.78 and 2.79. The solid line is the least-square fit. The squares represent the full pseudopotential calculation result and the open circles are the calculated results based on the bond-orbital model. The calculated transverse effective charges for $\Delta a = 0$ are renormalized to the experimental values in both cases. Reprinted with permission from [389]. Copyright 1982, American Physical Society

Different results were obtained from the low temperature (6 K) experimental and theoretical study of the pressure dependence of the linewidths of the TO and LO phonons in 3C-SiC [391]. While the linewidth of the TO mode remained unchanged at pressures up to 35 GPa, that of the LO mode exhibited monotonic increase up to 26 GPa, followed by an abrupt decrease above this pressure. The experimental results agree well with first-principles calculations and this anomalous behavior was proposed to originate from the anharmonic effect. The high-pressure behavior of the direct and indirect bandgaps of 3C-SiC was examined using ab initio pseudopotential calculations [392]. The fundamental bandgap from Γ_{15}^{v} to X_1^{c} decreased linearly with pressure up to 200 kbar. The linear pressure coefficients of the fundamental bandgap and the TO phonon frequency at the Γ point in the Brillouin zone agree with measured values.

The pressure effect in the Raman spectra of 6H-SiC shows features different from 3C-SiC. The TO and LO modes of 6H-SiC were observed at pressures up to 95 GPa, indicating the structural stability of the 6H polytype [393]. The sample remained optically transparent at 95 GPa and there was no sign of the expected metallic phase. The LO(Γ) and TO(Γ) Raman frequencies increased with pressure. The LO–TO splitting is 181 cm^{-1} at the ambient pressure and increased rapidly below 60 GPa, whereas it tended to saturate at 205 cm^{-1} under over 60 GPa (Fig. 2.82). The transverse effective charge increased with pressure at low pressures until reaching a maximum at about 40 GPa and then decreased (Fig. 2.83). This suggests increased covalent binding at high pressures and this variation agrees with the proposed theory. The mode-Grüneisen parameters γ of the SiC LO and TO modes were derived from

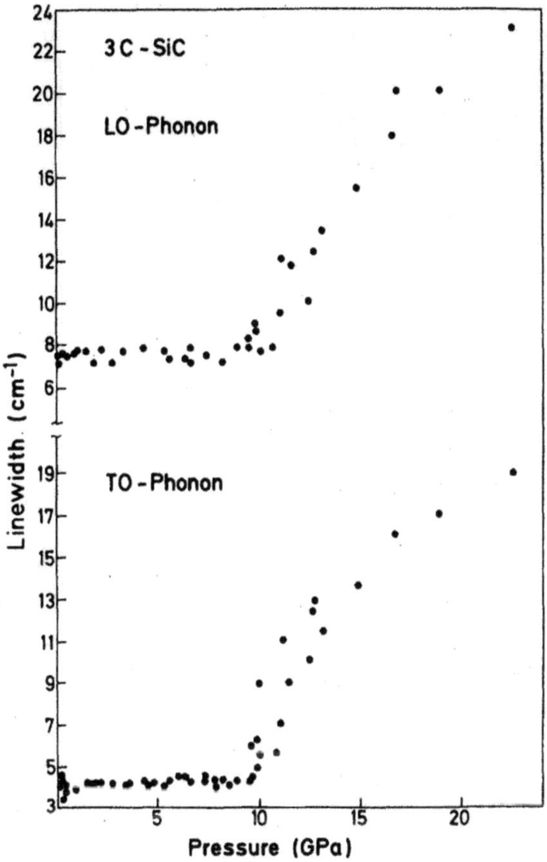

Fig. 2.81 Plot of LO(Γ) and TO(Γ) Raman linewidths versus pressure in 3C-SiC crystals. Reprinted with permission from [390]. Copyright 1982, American Physical Society

the experimental data. Comparison of the density variation of γ with diamond and Si leads to the conclusion that γ of the LO mode in 6H-SiC is softer than that of the TO mode and they both exhibit anomalous decrease. The pressure-dependent lattice dynamics of 4H-SiC was investigated using diamond anvil cell [394]. Both the zone-center LO and TO modes shifted to higher frequencies with increasing pressure. The LO–TO splitting is described by the cubic function with respect to pressure and it decreased under over 33 GPa. The change of transverse effective charge and the ionic character of 4H-SiC exhibited cubic dependence on pressure due to the nonequivalent lattice dynamics along orientations parallel to and perpendicular to the *c*-axis of 4H-SiC. The LO mode is softer than the TO mode.

Both the residual strain and strain relaxation effect in heteroepitaxial 3C-SiC/Si system were investigated by using micro-Raman spectroscopy [395]. A theory of inhomogeneous shift and broadening for optical phonons was developed, which takes into account the phonon interaction with the static strain fluctuations. The Raman scattering cross section was calculated, including both disorder and spatial dependence of the average strain with distance from the interface. Three regimes of

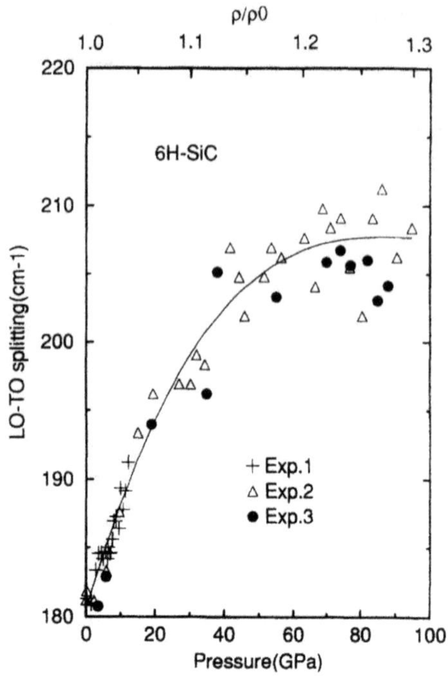

Fig. 2.82 LO–TO splitting in 6H-SiC as a function of pressure obtained by three different experiments. The upper scale is the density variation from the equation of state of 6H-SiC and the solid curve represents the quadratic fit of the data. Reprinted with permission from [393]. Copyright 1994, American Physical Society

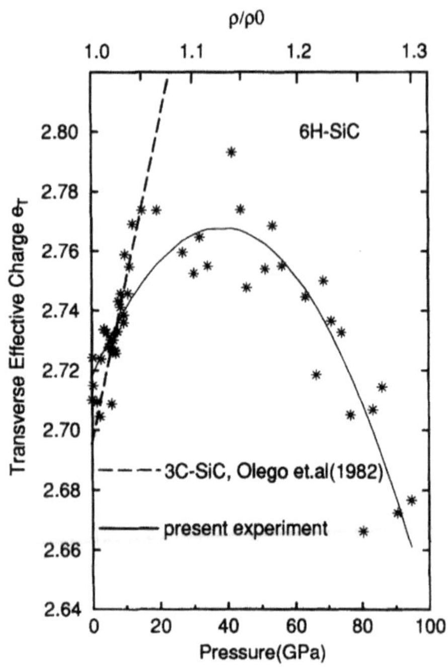

Fig. 2.83 Transverse effective charge as a function of pressure. Reprinted with permission from [393]. Copyright 1994, American Physical Society

short- and long-range disorder as well as anisotropic (2D-like) disorder with different lineshapes can be observed. From the change in coupling constant plotted versus distance from the interface, one concludes that the mean-squared strain relaxes in the bulk of the epitaxial samples according to approximate z^{-1} dependence. Micro-Raman spectroscopy was employed to measure the residual stress in the SiC fibers at various stages of the fabrication process [396]. Samples of the SiC fiber-reinforced ceramic matrix composites were analyzed after mechanical braiding and at 33, 66 and 100% chemical vapor infiltration densification of the composites. Raman spectra revealed significant band shifts of the SiC peaks and these shifts correspond to − 716 MPa of residual stress following the braiding process and − 1075 MPa after the densification process.

2.5.5 Temperature Effects

The temperature effect on the lattice dynamical properties has been investigated. The temperature dependence of the long wavelength optical phonons of 3C-SiC up to 750 K shows that the TO and LO phonon Raman lines shift to lower frequencies and broaden with temperature (Fig. 2.84) [348]. The shift in the LO line is stronger than that of the TO line while the LO–TO splitting decreased with temperature. The line width also increased with temperature and this was explained by assuming that the lifetime of the optical phonon is determined by its decay into two LA phonons, similar to the cases of Si and Ge. The transverse effective charge decreased slightly with temperature. The temperature and pressure data elucidate the explicit temperature contribution to the temperature dependence of the transverse effective charge. The results also agree with calculations based on the pseudopotential expression for charges. The temperature dependence of the Raman spectra of 3C-SiC microcrystalline films consisting of randomly oriented crystallites of about 100 nm is quite different from that of 3C-SiC single crystals [347]. The Raman spectra were acquired at elevated temperatures up to 1700 K and the temperature dependence of the TO- and LO-phonon frequencies is consistent with that of the 3C-SiC single crystals, but the Raman lines are much broader. This can be attributed to the effects of stacking faults by comparing the experimental results with simulated Raman intensity profiles of 3C-SiC comprising randomly distributed stacking faults. The simulated linewidth and disorder-induced peak shift show good agreement with experiments provided that the average stacking fault distance is assumed to be 6 Å. The Raman lines show irreversible narrowing at temperatures above 1900 K, possibly arising from annealing of the stacking faults in 3C-SiC. Temperature-resolved infrared spectral emissivity of 3C-SiC in the 2–20 μm range for temperatures up to 900 °C was investigated [397]. SiC showed significant temperature dependence of emittance in the absorption region from 10 to 14 μm and exhibited a crossover of emittance isotherms at individual wavelengths. The shifts of phonon wavenumbers of undoped 6H–SiC as well as in undoped and nitrogen-doped 4H–SiC were investigated by using Raman spectroscopy in the temperature range of 3–112 °C [398]. All spectra were acquired

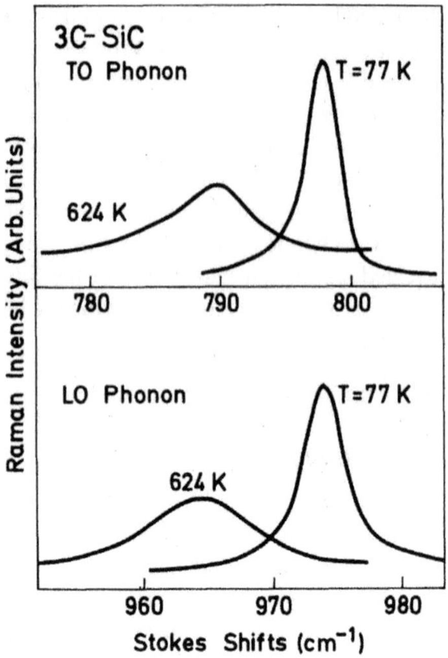

Fig. 2.84 First-order Raman lines of 3C-SiC recorded at two temperatures. Reprinted with permission from [348]. Copyright 1982, American Physical Society

with the incident laser beam oriented parallel to as well as perpendicular to the *c*-axis to account for the anisotropy of the phonon dispersion. Nearly all individual peaks shifted linearly toward smaller wavenumbers with increasing temperature. Only the peak of the A_1 (LO) in nitrogen-doped 4H-SiC shifted to bigger wavenumbers. The Raman peaks of all phonons depend linearly on temperature and phonon frequency. Temperature evolution of the far-infrared transmission spectra of 6H-SiC crystal slab was experimentally examined within 5–320 K [399]. Temperature dependence of the phonon lifetimes for long-lived phonons, that is, the IR active folded TA phonon doublet, was determined. It was interpreted using first-principles anharmonic lattice dynamics calculations. Three-phonon anharmonic processes dominated in the folded TA phonon decay at temperatures above 160 K, resulting in rapid decrease of the phonon lifetimes with increasing temperature. Contributions of four-phonon anharmonic processes were negligible at below 320 K.

2.5.6 Phonon–Plasmon Coupled Modes

The phonon–plasmon coupled modes in SiC and Raman scattering from these coupled modes have been investigated. The plasmon arises when a polar semiconductor is highly doped. The plasmon oscillation refers to long-wavelength electron-density oscillation. It is a longitudinal vibrational mode of the electron gas and the energy quantum is called plasmon. The plasmon oscillation can couple to the electric field of the long wavelength LO phonon to give rise to the phonon–plasmon coupled mode. The Raman data acquired from nitrogen-doped 6H- and 15R-SiC revealed the valley–orbit transitions for donors on nonequivalent sites in two types of SiC [400]. The transitions can be grouped into two widely separated energy regions, suggesting that the donor wavefunctions are sensitive to the impurity potential beyond the central cell. The Raman symmetry of the valley–orbit transitions was found to be E_2, which provides understanding of the position of the conduction band minima in the Brillouin zone. The Raman spectrum shows a shifted asymmetric LO phonon line (Fig. 2.85) interpreted as the LO phonon–plasmon coupled mode (L^+). This suggests a metallic system, and the calculated mobility is in agreement with measured values in heavily doped samples. The asymmetry of the mode and absence of the L^- LO phonon–plasmon coupled mode may be caused by the short electron collision lifetime. The efficiency of light scattering from a LO phonon coupled to a single component plasmon in a semiconductor was derived by using the semiclassical theory [401]. Two terms contribute to the efficiency. The first one arises from the modulation of polarizability by atomic displacements and the macroscopic longitudinal field obeying the ordinary Raman polarization selection rules. The second term comes from free electron charge density fluctuation that obeys different selection rules and is proportional to the square of the wave vector transferred. This theory combined with experimental data provides insight to coupled plasmon–phonon scattering in heavily nitrogen-doped 6H-SiC. The dominant light scattering mechanism is polarizability modulation, whereas the density mechanism was predicted to be 3500 times weaker. The absolute efficiency of the coupled mode spectrum in doped materials and A_1 (LO) phonon in lightly doped crystals have also been derived.

The transport properties including free carrier density and mobility of SiC are usually achieved by the Hall effect measurements. Raman spectroscopy also proves the ability to give rise to transport parameters. Particularly, Raman scattering can offer information regarding carrier concentrations. LO phonon-overdamped plasmon coupled modes in n-type β-SiC epitaxial films were measured for the carrier concentrations ranging from 6.9×10^{16} to 2×10^{18} cm^{-3} [402]. The carrier concentrations and damping constants were determined by lineshape fitting of the coupled modes and compared with the values derived from the Hall measurements. The concentrations achieved from the two approaches are in agreement. The fitting-derived Faust–Henry coefficient is 0.35. The lineshape analysis of the coupled mode indicates that the dominant scattering mechanisms in β-SiC are deformation-potential and electrooptic mechanisms. The Raman profiles of the axial and planar LO phonon–plasmon coupled modes in n-type 6H- and 4H-SiC bulk crystals with free carrier concentrations of 10^{16}–10^{18} cm^{-3} have been studied [403]. In the axial mode, the direction of plasmon oscillation and atomic displacement are parallel to the c axis, whereas in the planar mode, they are perpendicular to the c axis. In 6H-SiC, the planar mode exhibits more dramatic variations in the peak intensity, width, as well as shift with carrier concentration than the axial mode (Fig. 2.86). In contrast, the spectra of 4H-SiC show no noticeable difference in the peak intensity and width between the axial and planar modes, but the planar mode shows slightly bigger peak shift

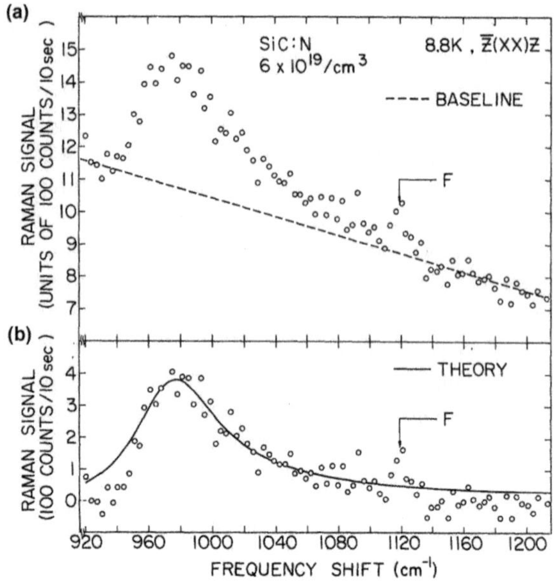

Fig. 2.85
a Photon-counting spectrum of the shifted A_1 (LO) phonon of nitrogen-doped 6H-SiC measured at 8.8 K. **b** Shifted A_1 (LO) phonon with the baseline of part (**a**) (dashed line) subtracted and theoretical fit (solid line). Reprinted with permission from [400]. Copyright 1972, American Physical Society

than the axial mode (Fig. 2.87). The anisotropic plasmon frequencies and carrier damping constants were derived using a fit of the calculated band shape with the experimental Raman band profiles, and the agreement between the experimental and calculated profiles is rather good. Using the plasmon frequencies in combination with the carrier concentrations obtained from Hall measurements, the longitudinal ($m_\parallel$) and transverse ($m_\perp$) effective masses of electrons can be derived, namely $m_\parallel = 1.4m_0$ and $m_\perp = 0.35m_0$ for 6H-SiC and $m_\parallel = 0.48m_0$ and $m_\perp = 0.30m_0$ for 4H-SiC. The carrier mobility obtained from the Raman analysis is also anisotropic.

The relationship between the LO phonon–plasmon coupled mode position and carrier concentration can be used to map the carrier concentration spatially. This was

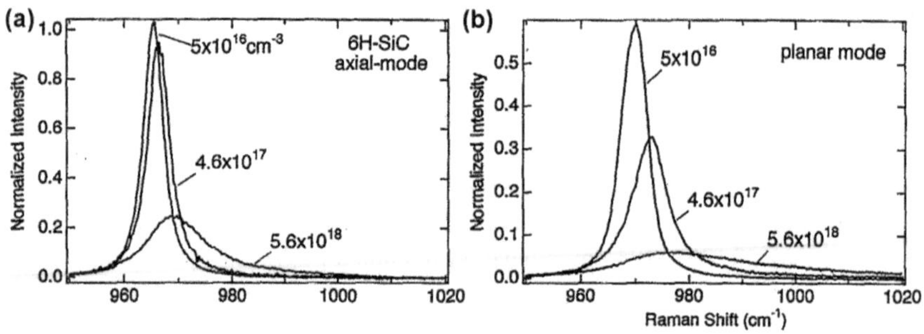

Fig. 2.86 Raman profiles of **a** axial and **b** planar LO phonon–plasmon coupled modes of *n*-type 6H-SiC with different carrier concentrations. Reprinted with permission from [403]. Copyright 1995, American Institute of Physics

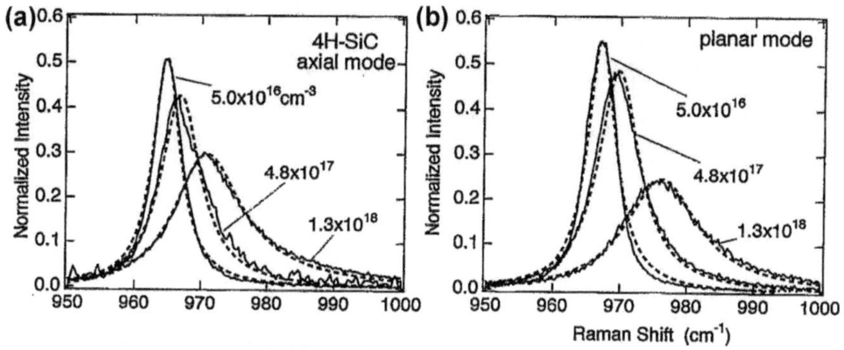

Fig. 2.87 Comparison between measured (solid line) and calculated (dotted line) Raman spectra of **a** axial and **b** planar LO phonon–plasmon coupled modes in *n*-type 4H-SiC with different concentrations. Reprinted with permission from [403]. Copyright 1995, American Institute of Physics

demonstrated using the semi-insulating and *n*-type nitrogen-doped (concentration between 2.1×10^{18} and 1.2×10^{19} cm^{-3}) 4H-SiC and 6H-SiC [404]. Significant coupling between the A_1 LO phonon and plasmon mode was observed and there is direct correlation between the position of the coupled mode peak and carrier concentration. Figure 2.88a shows the nitrogen concentration estimated from the coupled LO phonon frequency across the wafer. The spatial dependence is not the same in different SiC wafers and the concentration profile is generally not symmetrical with respect to wafer rotation. The wafer center has a large carrier concentration and the far edges have doping levels much smaller than the nominal value specified for the wafer. Figure 2.88b shows the resistivity map of a doped SiC wafer, confirming the spatially nonuniform dopant concentration. The resistivity variation in the most heavily doped wafer is consistent with the observed carrier concentration variation obtained by Raman measurement of the same wafer. The results indicate that Raman scattering can be employed as a noncontact, nondestructive, and in situ probe for the spatial distribution of carriers in SiC wafers. The LO phonon–plasmon coupled modes in doped SiC wafers were analyzed [405]. Changes in the carrier concentration of *n*-type 4H-SiC caused frequency shift of the coupled mode that is close to the A_1 (LO) phonon mode. The free carrier density can be estimated from the frequency shift of the coupled mode for low carrier concentrations. The plasmon damping constant shows no noticeable influence. At higher concentrations, the plasmon damping constant affects the Raman frequency shift of the coupled mode and reliable transport data is only available by a full lineshape analysis. The experimental results for *n*-type 4H–SiC bulk samples with free electron concentrations of 1×10^{17}–5×10^{18} cm^{-3} confirm such conclusions. The photoexcited carrier concentrations of laser-irradiated semi-insulating 4H-SiC as extracted from the Raman lineshape of LO phonon–plasmon coupled mode were studied [406]. The LO peaks of high-purity and V-doped semi-insulating 4H-SiC exhibited no shift with increasing (532 nm) laser power owing to low concentrations of photoexcited

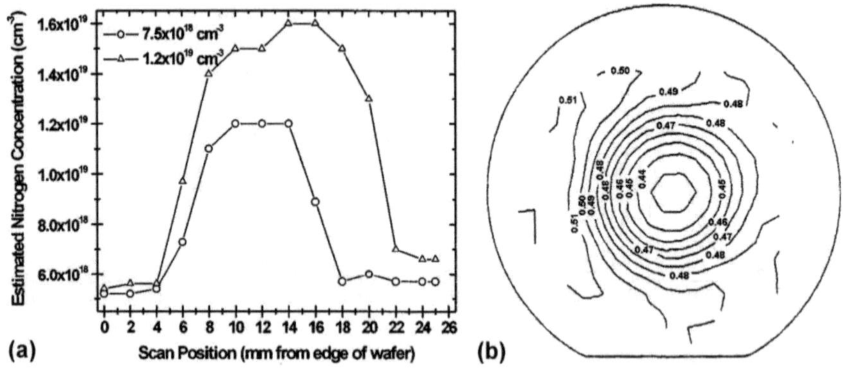

Fig. 2.88 **a** Estimated carrier concentration profile across two heavily doped 4H-SiC wafers determined by Raman spectroscopy with nominal nitrogen concentrations denoted in the legend. **b** Resistivity map for the 4H-SiC wafer nominally doped with 1.2×10^{19} cm^{-3}. Reprinted with permission from [404]. Copyright 1998, American Institute of Physics

carriers. The photoexcited carrier concentration varied nonlinearly with (355 nm) laser power because of trap-assisted Auger recombination.

2.5.7 Electron–Phonon Coupling in SiC

Electron–phonon coupling is one of the critical characteristics of solids. Renormalization of the electronic bandgap due to electron–phonon coupling in the tetrahedral semiconductors diamond, silicon, and 3C-SiC was studied by using DFT–LDA calculations [407]. There is significant difference between the coupling of the vibrational state to the valence band maximum and to the conduction band minimum. The strength of phonon coupling to the valence band maximum is similar between the three systems and is dominated by vibrations that change the bond length. The coupling strength to the conduction band minimum differs significantly in three types of crystals. The lattice thermal conductivity in intrinsic and doped 3C-SiC with charge carrier concentrations of 10^{17}–10^{21} cm^{-3} was studied by using DFT and density functional perturbation theory calculations [408]. The combined phonon dispersion, group velocities, and phonon–phonon scattering rates yield a thermal conductivity of 491 W m^{-1} K^{-1} at 300 K for undoped bulk 3C-SiC. The lattice thermal conductivity of doped 3C-SiC decreases highly at high carrier concentrations. The effects of electron–phonon interaction are much stronger for hole- than electron-doped material owing to the features of the electronic band structure near the band edges. At a high carrier concentration of 10^{21} cm^{-3}, the thermal conductivity drops by 57% for hole and 32% for electron doping. Electron–phonon coupling matrix elements were calculated using DFT and density-functional perturbation theory methods [409].

The phonon limited carrier mobility of 3C-SiC was quantified by solving the Boltzmann transport equation. The mobilities of electrons and holes are dominated by the intra-valley LA phonon scattering at 300 K due to the low occupation number of high-frequency polar LO phonons. At a high temperature (800 K), LO phonons govern the scattering instead. The maximum mean-free paths of electrons and holes at room temperature were calculated to be 40 and 15 nm, respectively. The drift and Hall mobility tensors of 4H-SiC were calculated from first-principles [410]. The calculated GW electronic structure and ab initio electron–phonon interactions help to solve the Boltzmann transport equation without fitting parameters. The electron Hall factor depends on the direction of external magnetic field, and the hole Hall factor exhibits different temperature dependency for magnetic field parallel to or perpendicular to the crystal *c* axis.

2.5.8 Surface Phonon Polaritons in SiC

Like surface plasmons in metals, the surface phonons in silicon carbide can strongly couple with incident photons, producing surface phonon polaritons. This is an interesting phenomenon and has potential applications. The reflectivity and coherent thermal emission properties of a one-dimensional SiC grating were investigated both experimentally and theoretically [411]. The study was conducted in a spectral region of 10–11.5 μm where the real part of the permittivity of SiC is negative. The reflectivity spectra in *p* polarization exhibit obvious dips due to surface phonon polariton excitations as shown in Fig. 2.89a. The reflectivity is different from that of the flat surface. Three dips appear at 917, 947, and 962 cm^{-1} and these peaks are assigned to surface phonon polaritons. Peaks in the same directions and at the same frequencies were observed from thermal emission (Fig. 2.89b). The peaks at 916, 945, and 961 cm^{-1} for *p* polarization were ascribed to light emissions due to thermally excited surface phonon polaritons. Similar effects were not observed for *s* polarization. In both cases, experiments and calculation agree and the dispersion relationship obtained theoretically and experimentally (Fig. 2.90) are consistent. The experimental dispersion relationship was derived from the reflectivity data, whereas the theoretical data were obtained from the simulated reflectivity spectra.

The boundary element method was adopted to study small particle phonon polariton excitation in 6H-SiC [412]. Phonon polaritons were observed from the 10–12 μm region in the far-infrared spectrum. For incident light in near resonance with the phonon polariton, the scattered electromagnetic fields in the far field and near-field around the structure were all enhanced. Calculation shows that the circular and elliptical cylinders support phonon polaritons at distinguishable wavelengths that are related to the axis ratio of the particle (Fig. 2.91). The scattering cross section curve of larger particles broadens and shifts to red. The single phonon polariton is split into a band with geometries other than elliptical. Two phonon polaritons are excited for the rectangular cylinders and five phonon polaritons are excited for the triangular cylinders. The phonon polaritons in SiC were more efficiently excited than

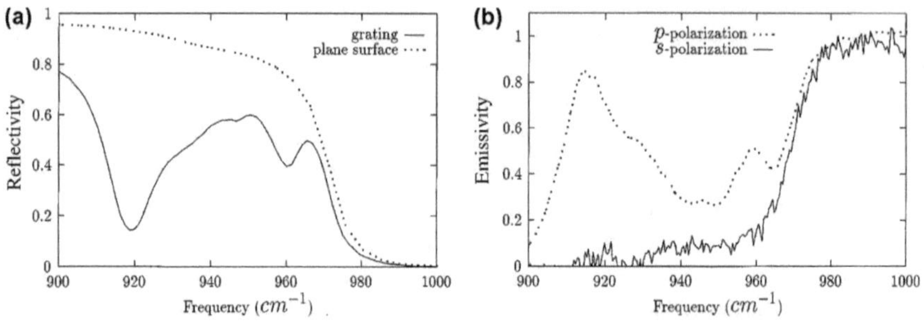

Fig. 2.89 **a** Reflection spectra of the SiC grating and SiC plane surface for p polarization at near normal incidence (incidence angle: 6°). **b** Emission spectra of the SiC grating for both polarizations at near normal incidence (incidence angle: 6°). Reprinted with permission from [411]. Copyright 1997, American Physical Society

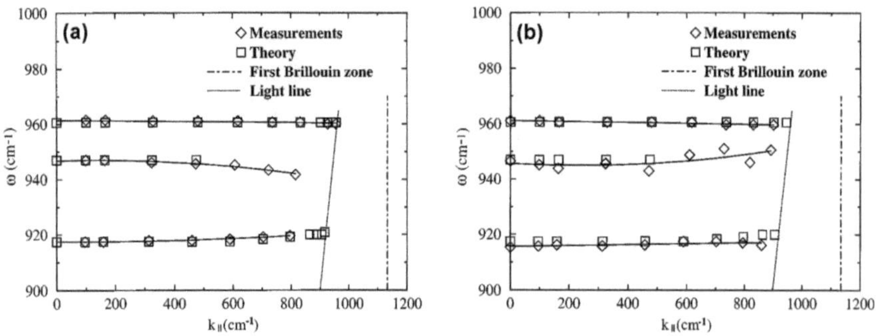

Fig. 2.90 Dispersion relation of the surface phonon polariton on a SiC grating obtained from reflectivity data (**a**) and from emissivity data at 100° (**b**). Reprinted with permission from [411]. Copyright 1997, American Physical Society

plasmon polaritons excited in silver due to the smaller imaginary part of the dielectric constant of SiC.

Silicon carbide particles exhibited electric and magnetic resonance in the infrared region [413]. Experimental extinction spectra and Mie theory calculation of a single SiC rod reveal three mid-infrared resonant modes. The measured extinction spectrum of a typical SiC rod with a diameter of 1 μm is depicted in Fig. 2.92a. With respect to TM polarization (E field parallel to the whisker's long axis), a sharp peak at 774 cm^{-1} and a broader low frequency peak were observed; for TE polarization (E field perpendicular to long axis), a sharp peak arises at 785 cm^{-1}, whereas there is no featureless extinction at low frequencies. The calculation was performed on the basis of Mie theory for infinitely long cylinders (1 μm in diameter) for normal incidence illumination where the incident, scattered, and interior fields can be considered purely TE or TM. The calculation results (Fig. 2.92b) agree with experimental ones. For TM polarization, the calculated extinction consists of a broad low-frequency peak and a sharp peak at 777 cm^{-1}. The red shift of the low frequency peak relative to the

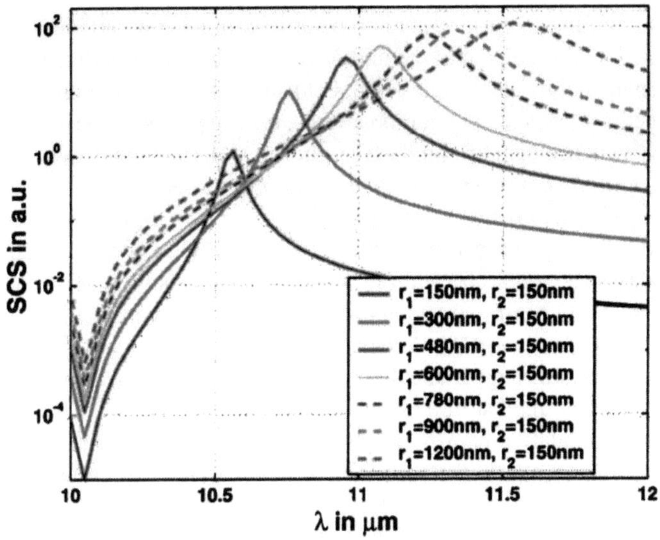

Fig. 2.91 Scattering cross sections of an elliptical 6H-SiC cylinder as a function of the axis ratio (from 1 to 8). Reprinted with permission from [412]. Copyright 2005, Optica Publishing Group

experimental spectrum may result from off-normal rays by the microscope objective. For TE polarization, a sharp peak also arises at 777 cm^{-1} and it is superimposed on a flat background. The experimentally observed 774 cm^{-1} peak (TM polarization) and 785 cm^{-1} peak (TE polarization) are degenerate and the actual separation may be caused by experimental complications. The existence of both electric and magnetic resonances may enable design of metamaterials.

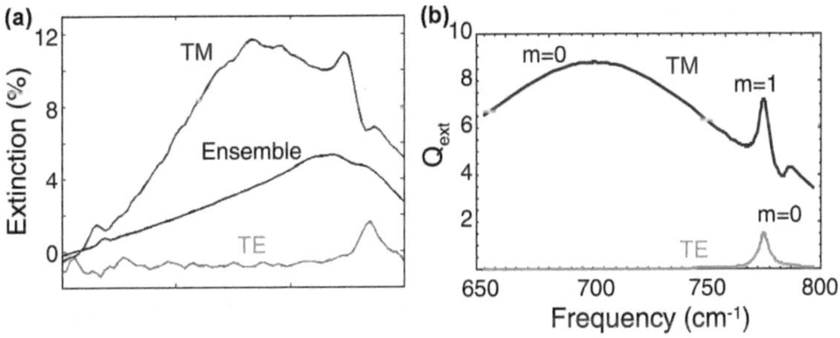

Fig. 2.92 **a** Ensemble, TE (*E* field perpendicular to long axis), and TM (*E* field parallel to long axis) polarized extinction spectra of a typical 1 μm diameter SiC whisker (length: 40–100 μm). **b** Calculated extinction efficiency of an infinitely long 1 μm diameter SiC cylinder for both TE and TM illumination. The spectrum is dominated by three modes: TM_0, TM_1, and TE_0, where the subscript denotes the *m* number of the mode. Reprinted with permission from [413]. Copyright 2007, American Physical Society

Critical coupling to surface phonon polaritons in silicon carbide was observed by attenuated total reflection of the mid-infrared radiation through a double scan of the wavelength and incident angle [414]. Critical coupling means complete coupling of laser energy to surface phonon polaritons, resulting in disappearing reflectivity and largest enhancement in the surface phonon polariton field over the incident laser field. Critical coupling occurs when the prism coupling loss is equal to loss in silicon carbide and the substrate. In this process, the surface phonon polaritons were launched using evanescent light coupled to the Otto configuration. pL-scale analyte index sensing can be realized on the basis of analysis of surface phonon polaritons in the mid-infrared region excited at the silicon carbide/analyte interface in the Otto configuration [415]. Attenuated total reflectance measurement provides the analyte indexes through a double scan of the wavelength and incident angle for analyte volumes as small as 100 pL.

Numerical simulation shows that the incident infrared optical field (10.6 μm) in gold microstructures on the surface of silicon carbide can be enhanced due to excitation of surface phonon polaritons [416]. The enhancement arises from two contributions: resonant coupling of the conventional gold antenna and excitation of surface phonon polaritons of SiC. The amplitude of the local field is enhanced by up to 106 times. Another study considers the thermal emission of cross-slit SiC grating in the Reststrahlen region over all emission angles [417]. The experiment shows that the thermal excitation of surface phonon polaritons on the surface of 2D grating allows for getting a high emissivity in both polarizations, which was collimated in *p* polarization for a specific wavelength determined by the periodicity of the grating.

A theoretical study shows that the silicon carbide spheres embedded in potassium bromide can act as metamaterials [418]. Near-field radiative heat is transferred between these isotropic and dielectric based metamaterials. The electric and magnetic surface polaritons are excited and mediate the near-field radiative heat transfer. Metamaterials showing negative index in the visible (blue) were realized using co-sputtered SiC and Ag nanoparticles on a glass substrate [419]. The real part of the negative index of the metamaterials at 405 nm obtained by the double Michelson interferometry setup shows a negative phase delay. The simulation reveals that the metamaterials can yield near-field super-resolution imaging for both TE and TM polarizations.

Continuous multispectral imaging of surface phonon polaritons on 6H-SiC in a wavelength range of 943–896 cm^{-1} was observed under excitation by an external cavity quantum cascade laser using scattering-type scanning near-field optical microscopy [420]. The theoretical dispersion relation and measured in-plane wave vectors of the surface phonon polaritons agree well in the entire measurement range. A nonlinear shift in the dispersion relation of a surface phonon polariton in 4H-SiC was observed with grating-coupled pump–probe reflection spectroscopy [421]. Upon excitation of the surface phonon polariton, an instantaneous frequency shift of the surface phonon polariton mode at a constant wavevector was observed. This pump-induced frequency shift is equivalent to a nonlinear dispersion shift and to a Kerr-like nonlinear phase shift. A nonlinear oscillator model indicates that the primary origin is either the Born effective charge or the phonon anharmonicity depending on the frequency within the Reststrahlen band. The instantaneous shift was followed by a picosecond recovery, reflecting the energy relaxation and dissipation of the excited surface phonon polariton. Surface phonon polariton dispersion in SiC free-standing membranes of a few hundred nanometers in thickness was acquired through near-field spectroscopy [422]. Several branches of dispersion curves were observed, which were rationalized as multiple reflections of tip and edge launched surface phonon polaritons. Polaritonic waves at the surface of SiC can be excited under irradiation of a CO_2 laser (~ 10 μm) [423]. They were visualized using apertureless near-field interference scanning probe microscopy. The Bragg scattering (diffraction) outside the grating was identified from the near-field scans in the vicinity of gold film periodical strip structures, with coherently summed contributions from separate strips.

2.5.9 Raman Scattering in Specially Structured SiC

Raman scattering can give rise to information about the structural properties of some specially structured SiC. The evolution of Raman spectra reveals structural and compositional transformations of SiC fibers made from polycarbosilane caused by heat treatment [424]. Raman spectra indicate that the SiC fiber consists of amorphous or microcrystalline SiC, carbon microcrystals, and silicon oxide. The amount of microcrystalline carbon in the fiber increased with heat treatment temperature up to 1400 °C and decreased abruptly in the fibers heated above 1500 °C. The tensile strength of the fiber dropped to zero after heating at 1500 °C. Carbon microcrystals precipitated from the Si–C random network with excess carbon and they suppressed the growth of SiC crystals. Carbon microcrystals play an important role in maintaining the high mechanical strength of the SiC fiber. Raman spectroscopy was used to investigate the structure of ion-irradiated α-SiC single crystals at room temperature and 400 °C [425]. Irradiation induced decrease of the Raman line intensities related to crystalline SiC, arising of several new Si–C vibration bands ascribed to breakdown of the Raman selection rules, and formation of homonuclear Si–Si and C–C bonds within the SiC network. At low doses, the overall sp^3 bond structure and chemical order were almost completely conserved. The amorphous state showed strong randomization of the Si–Si, Si–C, and C–C bonds. The relative Raman intensity decreased exponentially versus increasing dose due to absorption of the irradiated layer. The total disorder follows a sigmoidal curve, which follows the direct impact/defect stimulated model. The chemical disorder expressed as the ratio of C–C bonds to Si–C bonds increased exponentially versus the dose. The neutron irradiation effects of 3C-SiC at temperatures of 235–750 °C and neutron doses of 0.01–11.8 displacements per atom were evaluated by using Raman spectroscopy [426]. The SiC optical phonon lines shifted to lower wavenumbers after irradiation of the samples. The irradiation-induced peak shift was explained by combinations of lattice strain, reduction of the elastic modulus, and other factors including decreasing coherent domain size. Amorphous SiC samples prepared by plasma CVD and reactive sputtering methods were investigated [427]. In the CVD samples, the x (a-Si_xC_{1-x}) dependence of the intensity ratio $I_{Si\text{–}Si}/I_{C\text{–}C}$ was determined to follow $x^3/(1-x)^3$. The Raman efficiency of the Si–C band is much smaller than that of the Si–Si and C–C bands and that of the C–C bands is the largest. The samples prepared by the reactive sputtering method showed amorphous structure or a structure similar to glassy carbon. The C^+ and He^+ irradiation-induced amorphization of 3C-SiC nanograins in an amorphous SiC matrix was examined by using Raman spectroscopy and Rutherford backscattering spectrometry [428]. SiC grains with an average size of ~ 6 or ~ 20 nm were observed to be fully amorphized at a lower dose at room temperature compared with their monocrystalline counterpart under the identical irradiation condition. Phonon confinement effect and surface optical mode in SiC nanocrystals were investigated by using Raman spectroscopy [429]. Effect of anisotropy of phonon dispersion curves was analyzed. The additional 875 cm^{-1} band was attributed to defects and amorphous SiC. SiC grains were regarded as column array by taking grain boundaries as surrounding media, which explains surface optical phonon mode at 915 cm^{-1}.

2.6 Light Absorption, Luminescence, and Impurities

Light absorption and emission are very important optical processes in solids. On the one hand, they can serve as probes of the electronic structures and defect characteristics of solids. On the other hand, luminescence from solids has applications in many areas including displays, lighting, information communication, etc. As a wide bandgap semiconductor, the luminescence properties of silicon carbide have technological and commercial interest. The light absorption and emission spectra of silicon carbide often contain features and fruitful information concerning the impurities that naturally form during growth of SiC crystals or are intentionally doped to give rise to *n*- or *p*-type SiC crystals.

2.6.1 Light Absorption

As a direct probe method for energy band structures, light absorption by some common polytypes of silicon carbide has been extensively investigated. The UV–Vis absorption spectra from hexagonal SiC were acquired at temperatures ranging from 77 to 717 K [430]. Figure 2.93 shows the absorption spectrum [$\alpha^{1/2}$ vs. E (photon energy)] at 700 K along with the fit. The fit is the sum of the following two terms representing phonon absorption and phonon emission, respectively:

$$\alpha^{\frac{1}{2}}(\text{abs.}) = \left[\frac{A}{e^{\theta/T}-1}\right]^{1/2}\left[h\nu - \left(E_g - k_B\theta\right)\right], \tag{2.16}$$

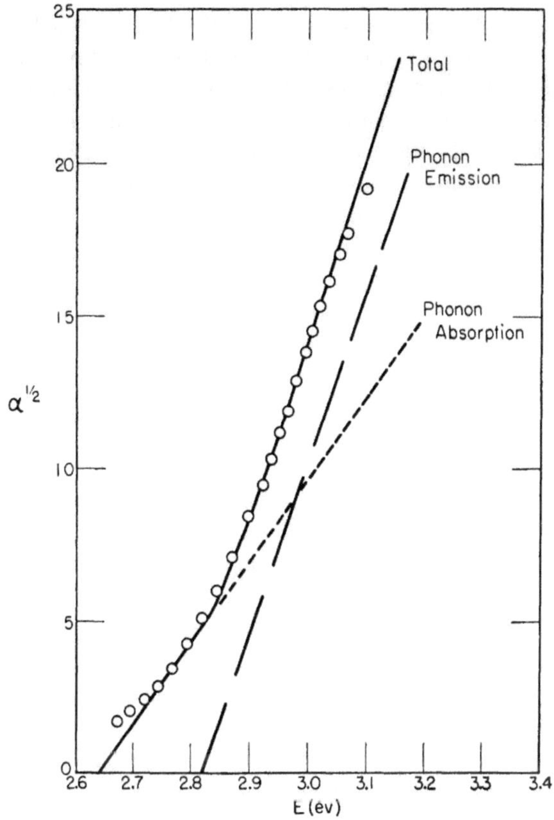

Fig. 2.93 Comparison between observed and calculated absorption by hexagonal silicon carbide at 700 K. Reprinted with permission from [430]. Copyright 1957, American Physical Society

$$\alpha^{\frac{1}{2}}(\text{emiss.}) = \left[\frac{A}{1 - e^{-\theta/T}}\right]^{1/2} \left[h\nu - \left(E_g + k_B\theta\right)\right], \tag{2.17}$$

where α is the photon absorption coefficient corresponding to absorption or emission of a phonon, h is Planck's constant, ν is the photon frequency, E_g is the bandgap of silicon carbide, k_B is Boltzmann's constant, and A is a constant depending on the properties of the materials and electron–phonon interaction. To obtain the best fit based on Eqs. (2.16) and (2.17), $k_B\theta$ (phonon energy) = 0.09 eV and $A = 2500$. The good fit indicates that the interband transitions in SiC are indirect. The minimum energy gap at 300 K is 2.86 eV and above this temperature, $dE_g/dT = -3.3 \times 10^{-4}$ eV per degree. Optical experiments on 15R SiC were examined [431]. The absorption edge is due to indirect exciton transitions (2.986 eV) at 6 K. The sample exhibited two distinct PL spectra arising from two types of nitrogen–exciton complexes, from which 18 phonon energies were obtained. Only four series of lines are observable in each spectrum, although there are five inequivalent nitrogen sites in 15R SiC. The four nitrogen donor ionization energies are 0.14, 0.16, 0.16, and 0.20 eV. There are evidences for exciton hopping, localized vibrations, and spin–orbit splitting of

4.8 meV. The optical absorption of 4H, 6H, 8H, and 15R SiC crystals were measured within 0.35–2.5 μm with polarization of light parallel to and perpendicular to the *c*-axis of the crystals [432]. The observed absorption bands were regarded to arise from electron excitation from the conduction band minimum to other sites of increased density of states in the higher empty bands. All polytypes exhibited strong anisotropy of free carrier absorption. Four higher energy absorption edges were observed by studying absorption of 6H-SiC with a thickness of 1.8 μm prepared by grinding and polishing [433]. Three indirect absorption edges (Fig. 2.94) were observed at 3.0, 3.7, and 4.1 eV (from fitted lines 1, 2, and 3), whereas the fourth absorption edge is at 4.6 eV (from line 4) but cannot be definitively identified as direct or indirect. The following simplified equation was used to fit the experimental data:

$$\alpha = \sum_i \alpha_i = \sum_i A_i (h\nu - E_{gi})^2, \tag{2.18}$$

where α is the absorption constant in cm^{-1}, A_i is a constant, and E_{gi} is the energy threshold. A_i and E_{gi} were chosen to yield the best fit. In this equation, the effects of excitons and phonons were neglected since they were not observed experimentally.

A study shows that nitrogen doping of 3C-SiC can enhance its optical absorption [434]. A narrow band near 3.1 eV was supposed to stem from direct interband transition from the X_{1c} conduction band to the higher X_{3c} band. Wavelength modulated absorption spectra of the free excitons in 6H, 15R, and 3C SiC were measured [435]. The spin–orbit splitting values of the valence bands of the uniaxial and cubic polytypes are 7 and 10 meV, respectively. With consideration of the exciton binding energy of 27 meV, the fundamental bandgap of 3C-SiC was determined to be 2.417 eV (2 K).

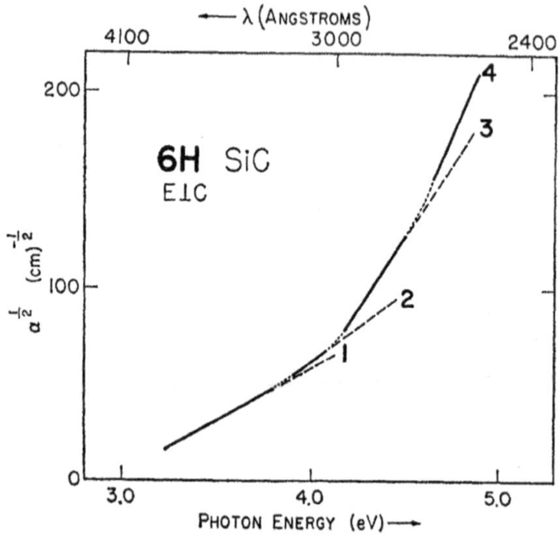

Fig. 2.94 Plot of $\alpha^{1/2}$ versus $h\nu$ for the UV–Vis absorption spectrum of 6H-SiC with data obtained from many samples. The measured values are on the straight-line segments (solid) or in the transitional regions (dotted). The four straight-line segments are extrapolations (dashed). Reprinted with permission from [433]. Copyright 1968, American Physical Society

The valence bands are highly non-parabolic at low energies owing to the small spin–orbit splitting. UV reflectivity (4–10 eV) for an electric field perpendicular to the *c*-axis was examined for 4H, 15R, 6H, and 3C SiC polytypes [436]. The positions of all major features agree well with theoretical results obtained by the scalar-relativistic full-potential linear muffin-tin orbital (LMTO) method in the local-density approximation except for a polytype and energy independent upward shift of the conduction band. The major peaks in the reflectivity are associated with relatively extended areas of nearly parallel bands at general *k* points. The authors further investigated the validity of the interpretation of the 1–3-eV absorption features in heavily doped *n*-type SiC polytypes in terms of transitions from the lowest conduction band to higher bands [437]. The result shows that this model agrees well with the energy band difference calculated using the LMTO–LDA method for 3C, 6H, 4H, 8H, and 15R SiC polytypes. It also explains the polarization dependence of the spectra in terms of selection rules and the symmetry of the bands, and explains qualitatively the relative intensities of the main absorption features for different polarization in terms of the relative magnitudes of the matrix elements.

The absorption coefficients of 4H-SiC at room temperature in the wavelength range between 325 and 390 nm were determined [438]. The absorption coefficients from 335 to 390 nm were measured using a Xe arc lamp as the light source and light propagating along the *c* axis, whereas the 325 nm absorption coefficient was determined using a He–Cd laser source. The results are displayed in Fig. 2.95. At high energy, the square root of the absorption coefficient is proportional to the photon energy subtracted by the bandgap, as required by the indirect bandgap nature. By using the known relationship between the bandgap and temperature, the absorption coefficients of 4H-SiC at 2 K can be derived. The absorption coefficients of 4H-, 6H- and 3C-SiC at room temperature between 297 and 390 nm were measured using xenon and tungsten lamps together with lasers of several wavelengths [439]. On the basis of the temperature dependence of the bandgaps, the absorption coefficients at 2 K were estimated.

The effect of doping on absorption for *n*-type (N) 6H-SiC, 4H-SiC and 15R-SiC as well as *p*-type (Al) 6H-SiC, (Al) 4H-SiC, and (B) 6H-SiC was investigated [440]. The bandgap narrowing at higher doping concentration was observed. For *n* type doping below bandgap absorption bands at 464 nm for 4H-SiC, at 623 nm for 6H-SiC, and at 422 and 734 nm for 15R-SiC were observed. Hall measurements reveal linear relation between the peak intensities of these absorption bands and the charge carrier concentrations. A purely optical wafer mapping of the spatial variation of the charge carrier concentration was demonstrated. Fundamental band edge absorption was investigated in nominally undoped ($n < 10^{14}$ cm^{-3}) and heavily doped ($n \approx 8 \times 10^{18}$ cm^{-3}) 4H-SiC by a spectroscopy technique based on spatially and time-resolved free-carrier absorption [441]. The spectra lie in 0.02–500 cm^{-1} at 75–450 K. The experimental data are supported by an indirect transition theory with a unique set of momentum-conserving phonons. The data fit at 75 K gives an exciton binding energy of 30 $\pm$ 10 meV. The detected polarization anisotropy of absorption with respect to *c* axis agrees with the selection rules for the corresponding phonon branches.

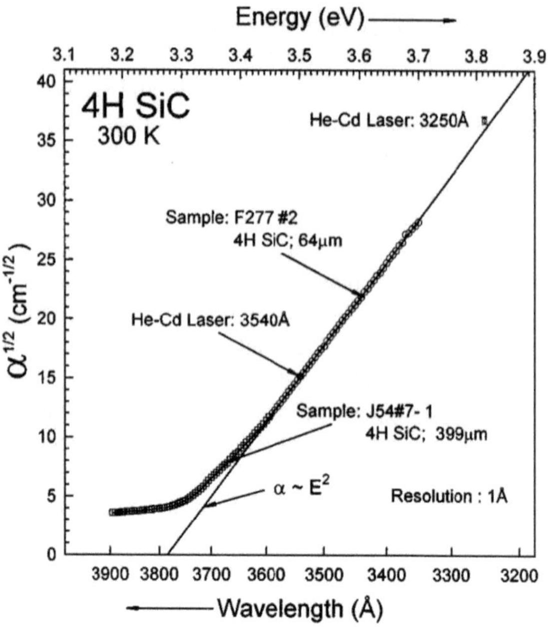

Fig. 2.95 Square root of the absorption coefficient versus wavelength of 4H-SiC. Reprinted with permission from [438]. Copyright 1998, American Institute of Physics

Transient infrared pump–probe spectroscopy with variable levels of photocarrier injection was employed to explore the effects of free carriers in the Reststrahlen band region of 4H-SiC, where the permittivity is negative due to charge oscillations of the polar optical phonons in the midinfrared [442]. The experiments provide evidence for the LO phonon–free carrier plasma coupling-mediated carrier-tunable permittivity of 4H-SiC. By fitting the transient reflectance spectra with a model of permittivity that includes phononic and free-carrier terms, the free-carrier densities and carrier-induced damping under various optical pumping levels were extracted. By constructing a cone–pillar double-structure surface on SiC, a high absorption ($>97\%$) of normally incident electromagnetic waves in the whole Reststrahlen band was realized experimentally [443]. Simulations reveal that this high absorption is ascribed to the efficient coupling of incident electromagnetic waves into bulk evanescent waves within 10.33–10.55 μm and the efficient coupling of incident electromagnetic waves into surface phonon polaritons within 10.55–12.6 μm. Wavelength-modulated absorption spectra of 4H-SiC over 3500–3800 Å were measured at a resolution of $<$ 0.1 Å [444]. By comparing the measurements with the low-temperature PL spectra in ultrapure 4H-SiC, several features were identified (Fig. 2.96), which were ascribed to a 56 ± 3 meV crystal-field splitting of the valence band maximum or a 136 ± 3 meV separation between the two lowest conduction band minima. The spin–orbit splitting of the valence band contributes to nonparabolic dispersion near the valence band maximum and generates several previously misidentified features in the wavelength-modulated absorption spectra. Fine-structure splitting in the free exciton ground state

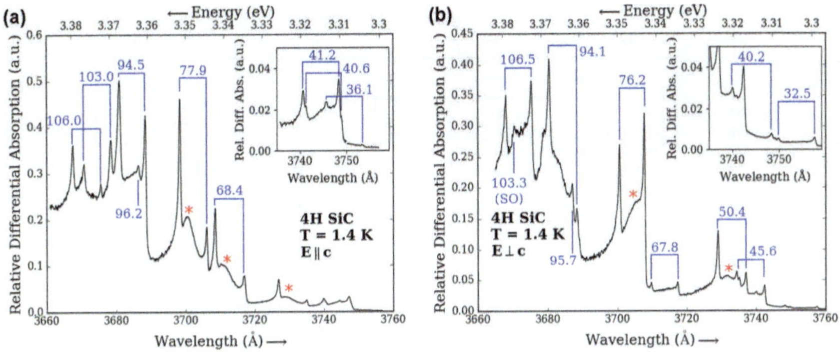

Fig. 2.96 Wavelength-modulated absorption spectra within ~ 100 meV of the fundamental absorption edge in 4H-SiC at 1.4 K. The incident light is polarized with **a** **E** ∥ c or **b** **E** ⊥ c. The spin–orbit splitting of the valence band generates pairs of peaks labeled by brackets and the energy (in meV) of momentum-conserving phonon participating in the indirect transition from two topmost valence band maxima to the lowest conduction band minimum. Asterisks indicate features due to nonparabolicity in the valence bands. Insets show absorption onsets due to the lowest energy principal phonons. Reprinted with permission from [444]. Copyright 2020, American Physical Society

was identified (Fig. 2.97), which manifests as four small (0.7 ± 0.1 meV) splittings in the wavelength-modulated absorption spectrum due to mass anisotropy and electron–hole exchange interaction. Phonon-assisted temperature-dependent optical absorption spectra of the SiC polytypes (3C, 2H, 4H, 6H, and 15R) were calculated using first-principles approaches based on DFT and related techniques. The results agree with experimentally determined absorption coefficients in the spectral region between direct and indirect bandgaps. Consideration of the temperature-dependent bandgaps predicts well temperature dependence of the spectra [445].

2.6.2 *Impurities and Luminescence*

2.6.2.1 N and P Impurities

Some specific luminescence lines observed from bulk silicon carbide were ascribed to recombination of excitons bound to impurities or free-to-bound recombination. The exciton–impurity complex was discussed [446] and various donors and acceptors in Si have been investigated leading to the observation of luminescence arising from the decay of the exciton bound to impurities [447]. The phonon assisted indirect transitions involving the excitons–impurity complexes in SiC generally give rise to a series of emission lines, from which the energies of the phonons involved can be derived. Photoluminescence from 6H-SiC exhibited 50 lines at low temperatures [448]. Figure 2.98 depicts the PL spectrum collected at 6 K. It shows three no-phonon lines, P_0, R_0, and S_0 caused by exciton recombination at the three nonequivalent

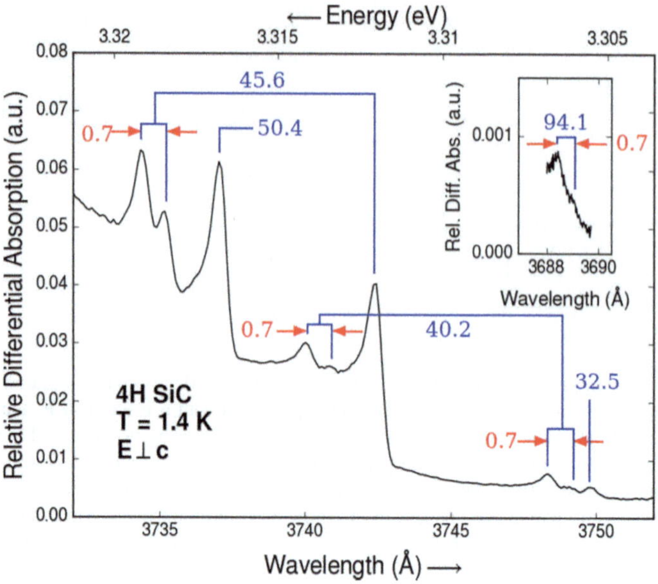

Fig. 2.97 Expanded view of the wavelength-modulated absorption spectrum shown in Fig. 2.96b. The peaks are labeled by the energy (in meV) of the principal phonon participating in the indirect transition. The 45.6 and 40.2 meV phonons give rise to a spin–orbit split pair with further, 0.7 ± 0.1 meV fine-structure splitting in either one or both of the spin–orbit split peaks. The inset shows an additional fine-structure splitting for the 94.1 meV phonon. These splittings are due to a fourfold splitting in the free exciton ground state owing to electron mass anisotropy and electron–hole exchange interaction. Reprinted with permission from [444]. Copyright 2020, American Physical Society

unionized nitrogen donors along with the phonon replicas. For example, P_0 is related to a series of lines. The P series are due to recombination at the P nitrogen center with emission of one of 17 phonons. The displacement of a P line from P_0 corresponds to the energy of a phonon. The 17 measured phonon energies can be assembled into an appropriate extended Brillouin zone. The spin–orbit splitting of the valence bands is only 4.8 meV, suggesting that the hole is confined to the carbon sublattice. Luminescence due to recombination of free excitons involving phonons has also been observed (Fig. 2.99).

Besides the excitons bound to neutral nitrogen impurity atoms, the exciton bound to the singly ionized nitrogen impurity can yield luminescence in 6H-SiC [449]. Photoluminescence from the ionized nitrogen–exciton complexes in N-doped 6H-SiC was observed at 6 K and the PL spectrum is shown in Fig. 2.100. The ionization energies of the three nonequivalent nitrogen donors are 0.17, 0.20, and 0.23 eV, respectively. The observed quenching of edge emission in the impure samples was ascribed to exciton hopping. Figure 2.101 displays the energy ranges of the two distinct luminescence spectra arising separately from the exciton–neutral nitrogen complex (four particle complex) or exciton–ionized nitrogen complex (three particle

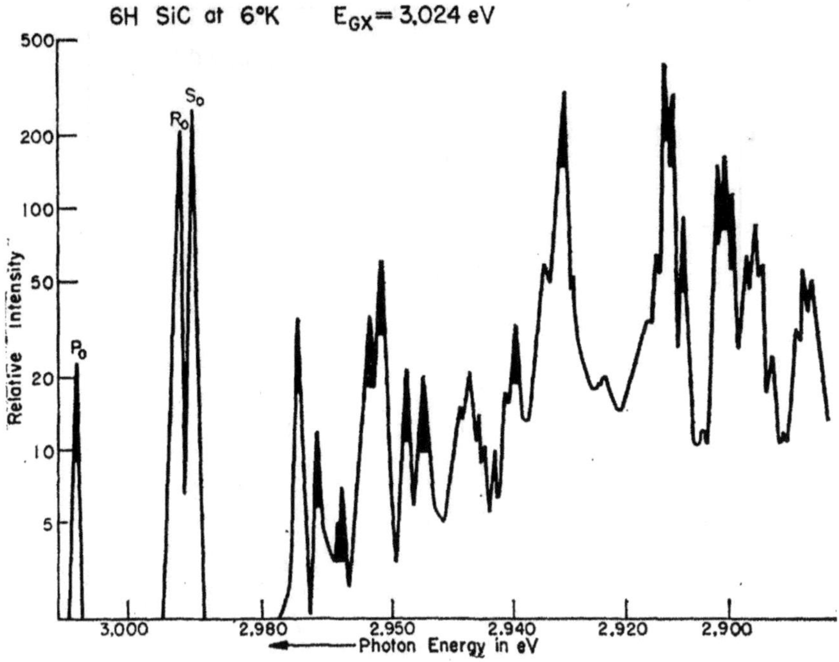

Fig. 2.98 Photoluminescence spectrum obtained from 6H-SiC at 6 K. The black peaks belong to the *P* series. Reprinted with permission from [448]. Copyright 1962, American Physical Society

complex). The three-particle spectrum extends into the lower energy region with multi-phonon emission and becomes gradually a continuum. Four-particle emissions have higher photon energies than three-particle emissions because the exciton is more tightly bound to ionized nitrogen atoms than neutral ones. In 15R-SiC, the photoluminescence spectra due to two kinds of nitrogen–exciton complexes are very similar to those of 6H-SiC, and similar phonon spectra were acquired [40]. The four nitrogen ionization energies are approximately 0.14, 0.16, 0.16, and 0.20 eV, which are smaller than those in 6H-SiC. Only four of the five nonequivalent nitrogen atoms in 15R-SiC were detected from the PL spectra. 15R-SiC also exhibits indirect absorption at the absorption edges. The edges have a similar shape as 6H-SiC due to the similarity in the phonon spectrum. 15R-SiC has an exciton energy gap of 2.986 eV at 6 K which is smaller than that of 6H-SiC. According to the luminescence from thermally excited states of the exciton–neutral nitrogen complexes, it was proposed that 15R-SiC has a second valence band that splits from the first one by 4.8 meV due to the spin–orbit interaction; the holes were supposed to be largely confined to the carbon sublattice. The result shows evidence of the presence of six conduction band minima.

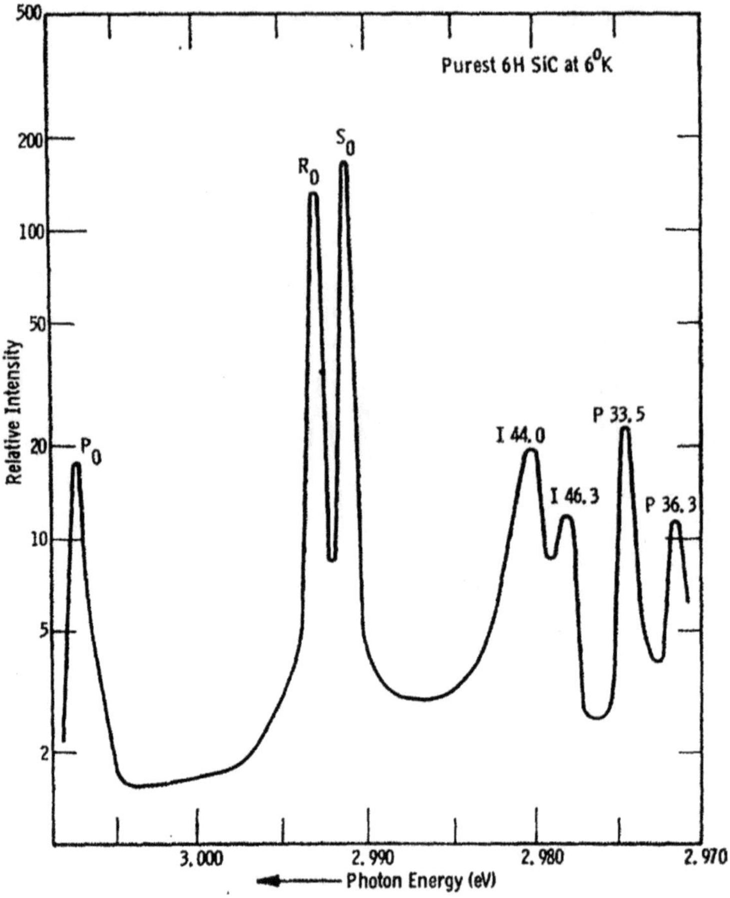

Fig. 2.99 Photoluminescence spectrum of the purest 6H-SiC sample at 6 K. Two additional intrinsic lines (*I* 44.0 and *I* 46.3) due to recombination of free excitons are shown and the numbers are the phonon energies in meV. Reprinted with permission from [448]. Copyright 1962, American Physical Society

Luminescence from the excitons–nitrogen complex and interband absorption in cubic SiC have been investigated [40]. Absorption measurements at 4.2 K demonstrate that 3C-SiC has an indirect absorption edge associated with exciton transitions and the exciton energy gap is 2.390 eV. The phonon energies determined from the absorption spectrum are 46, 79, 94, and 103 meV. The exciton–neutral nitrogen complex luminescence spectrum at 6 K provides accurate phonon energies. The thermally excited states of the complex give rise to additional lines in the luminescence spectrum at a higher temperature. Comparison shows that 3C-, 6H-, and 15R-SiC have very close phonon energies and the values are 46.3 (TA), 79.4 (LA), 94.4 (TO), and 102.8 (LO) meV for 3C-SiC. An important empirical correlation between the energy gap and hexagonality (percent "hexagonal", Table 2.3) for seven polytypes

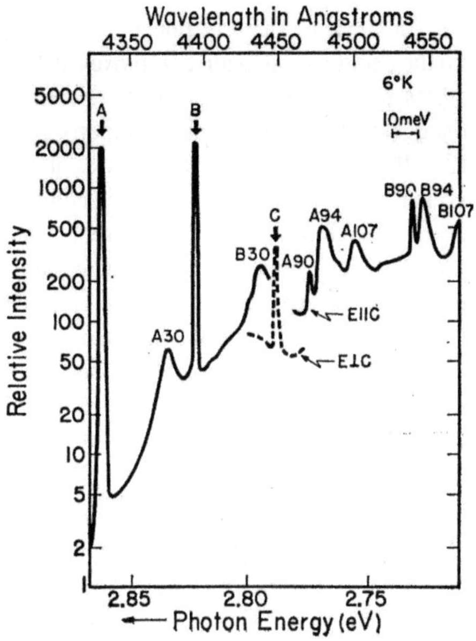

Fig. 2.100 Photoluminescence spectrum of 6H-SiC showing the portion of the ionized nitrogen–exciton complex band, including the no-phonon lines of *A*, *B*, and *C* along with some peaks arising from phonon emission. The peaks are labeled with their series letter and the number indicates the phonon energy in meV. Reprinted with permission from [449]. Copyright 1963, American Physical Society

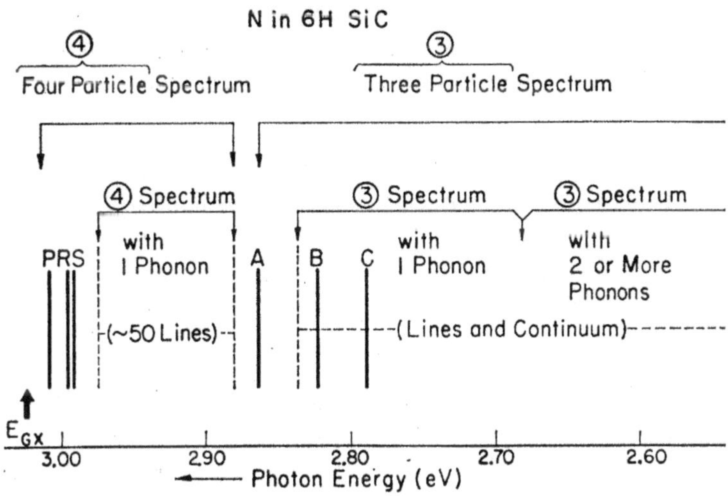

Fig. 2.101 Energies of no-phonon lines *P*, *R*, and *S* of the exciton–neutral nitrogen complex photoluminescence spectrum and *A*, *B*, and *C* of the exciton–ionized nitrogen complex photoluminescence spectrum as well as energy ranges of photons emitted with simultaneous phonon emission. E_{Gx} is the exciton energy gap. Reprinted with permission from [449]. Copyright 1963, American Physical Society

of SiC (Fig. 2.102) was found. It is a nearly linear relationship. The low temperature luminescence from the excitons–nitrogen complexes in 4H-SiC was investigated [450]. 18 phonon energies were derived from the PL spectra. The interpretation of the phonon spectrum suggests positions of low symmetry for the conduction-band minima, requiring a 12-valley model for electrons. It was argued that such a method can be generalized to other SiC polytypes to locate the conduction-band minima. It explains the observed similarity in the phonon spectra of seven SiC polytypes. The relatively large and pure 2H-SiC crystals were fabricated and their absorption and luminescence properties at 2–8 K were investigated [451]. 2H-SiC shows an indirect energy gap of 3.330 eV that is the largest among all the observed SiC polytypes. The group theory selection rules were utilized to analyze the polarized luminescence in order to determine the active phonons. The spectrum is consistent with the selection rules by assuming that there are conduction band minima at the K points of the Brillouin zone. SiC resembles diamond more than Si concerning the "trace variable" forces.

Nitrogen donor is an important impurity in silicon carbide and it has been intensively studied. Infrared absorption and Hall effect measurements were utilized to study nitrogen-doped 6H-SiC single crystals [452]. Three sets of infrared absorption lines were observed and attributed to transitions from the $1s$ (A_1) and $1s$ (E) ground states to $2p_0$, $2p_\pm$, $3p_0$, and $3p_\pm$ excited states of the nitrogen donors on the h lattice site of 6H-SiC and from $1s$ (A_1) to $2p_\pm$ and $3p_\pm$ states of the nitrogen donors on k_1 and k_2 lattice sites. The ground state binding energies of these donor species are 81.0, 137.6, and 142.4 meV, respectively. The h donor center exhibits valley–orbit splitting of 12.6 meV. Figure 2.103 shows the schematic of the energy levels of the nitrogen donors residing at h, k_1, and k_2 lattice sites as determined from infrared spectra. The transverse and longitudinal effective electron masses are 0.24 $\pm$ 0.01m_0 and 0.34 $\pm$ 0.02m_0, respectively, as obtained by fitting the effective-mass formalism to the observed transition energies. The energy positions of the ground and excited states of the nitrogen donor in the nitrogen-doped 4H-SiC were also investigated [453]. The two electrically active levels (Hall effect) and three series of absorption lines (infrared spectra) were assigned to two nitrogen donor species that

Table 2.3 E_{Gx} (exciton bandgap) of seven SiC polytypes and corresponding "hexagonal" percentage

Polytype (Ramsdell notation)	Exciton bandgap E_{Gx}	Jagodzinski notation	Hexagonality h
3C	2.390	k	0
8H	2.80	hkkk	25
21R	2.86	hkkhkkk	29
15R	2.986	hkhkk	40
33R	3.01	hkkhkkhkkhk	36
6H	3.023	hkk	33
4H	3.263	hk	50

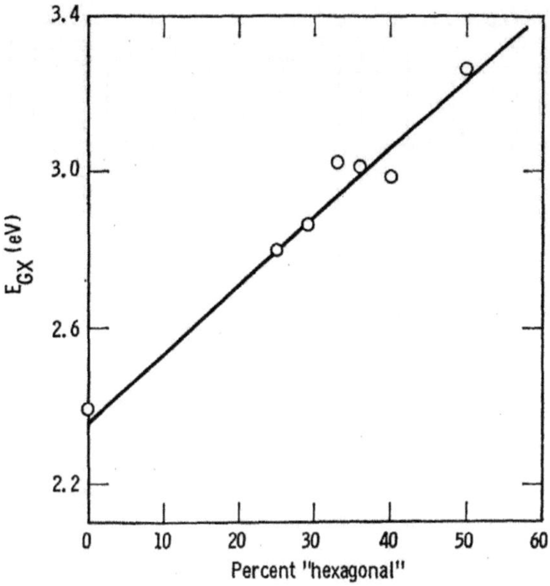

Fig. 2.102 Exciton bandgap E_{Gx} versus hexagonal percentage (hexagonality) of seven SiC polytypes. The values of E_{Gx} were obtained from absorption measurements at 4.2 K or luminescence measurements at 6 K. These values and the hexagonality values are summarized in Table 2.3. Reprinted with permission from [40]. Copyright 1964, American Physical Society

substitute on the two inequivalent lattice sites (h, k) in 4H-SiC. The nitrogen donors on hexagonal sites (h) exhibit a valley–orbit splitting of the ground-state level of 7.6 meV. The energies of the excited states of both nitrogen donors were identified by effective mass approximation assuming the transverse effective mass of $0.18m_0$ and longitudinal effective mass of $0.22m_0$.

Nitrogen-bound exciton luminescence can be used to determine the nitrogen dopant concentration in silicon carbide, as demonstrated by low temperature photoluminescence spectroscopy conducted on uncompensated 4H- and 6H SiC [454]. This was realized by comparing the intensity of the nitrogen bound exciton lines to that of the free exciton (as an internal reference). A line-fitting method with the proper line shape was employed to separate the contribution of the bound exciton and free exciton lines in the PL spectrum. The ratio of the bound exciton zero phonon lines to the free exciton most intensive phonon replica (around 77 meV) shows good linear dependence on the dopant concentration as determined by using capacitance–voltage measurements on both polytypes. When the real line shape, influence of the spectrometer transfer function, and structure of the PL spectrum in the vicinity of the free exciton replica were taken into account, the fit allows for determination of the N dopant concentration in the ranges between 1×10^{14} and 3×10^{16} cm^{-3} in 4H-SiC and 1×10^{14} and 10^{17} cm^{-3} in 6H-SiC. The free exciton emission was not observed above these concentrations.

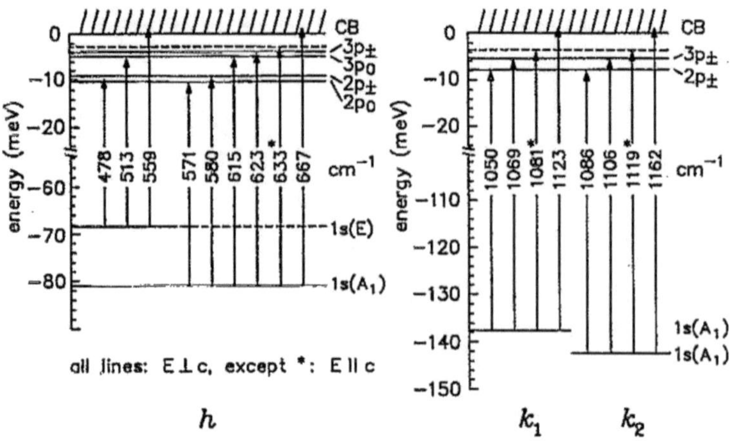

Fig. 2.103 Energy levels of nitrogen donors residing at the *h*, k_1, and k_2 lattice sites of nitrogen-doped 6H-SiC. Reprinted with permission from [452]. Copyright 1992, American Institute of Physics

Phonons at the *M* point of the Brillouin zone (at the position of the conduction-band minimum) of 4H-SiC was studied both experimentally and theoretically [455]. 24 phonons were classified by symmetry into $8M_1 + 4M_2 + 4M_3 + 8M_4$, where M_i are the irreducible representations of the group of the wavevector at the *M* point. The polarization selection rules imply that among 24 phonon replicas (lines), there are 12 with parallel polarization (M_1 and M_3 symmetry) and 12 with polarization (M_2 and M_4 symmetry) perpendicular to the crystal *c* axis. The phonons of each symmetry are equally distributed above and below the phonon energy gap, leading to six replicas of each polarization above and below the gap. 22 of the 24 phonons at the *M* point in 4H-SiC were experimentally observed. The free-exciton PL spectrum of 4H-SiC comprises a detailed fine structure due to different phonons involved in the indirect gap transition. First-principles calculations of the phonons, their symmetry labeling, and their contribution to the PL spectrum were conducted [456]. The electron–phonon coupling matrix elements were computed via density-functional perturbation theory, and the energy bands and optical phonon matrix elements were calculated by using the DFT method. The results agree well with the observed phonon energies and polarization dependence of the spectrum. The discrepancy in PL intensity may arise from competition between phonon-assisted PL and electron trapping in the unoccupied hexagonal site N-donor shallow level with 53-meV binding energy.

Aluminum-doped 6H-SiC epilayers were implanted with phosphorus and subsequently annealed at 1400–1700 °C [457]. Combined Hall effect, admittance spectroscopy, and PL characterizations reveal that phosphorus acts as a shallow donor. Two determined ionization energies of (80 ± 5) meV and (11 ± 5) meV were assigned to phosphorus atoms residing at hexagonal and cubic lattice sites, respectively. The annealing process caused an activation energy of the phosphorus donors of 2.5 eV. The low-temperature PL spectra comprise a set of four phosphorus-related lines at about 420/421 nm, whose intensity increased in parallel with the electrical activation of phosphorus donors by raising the annealing temperature. The Hall effect investigation was conducted on N- and P-implanted as well as on (N + P)-coimplanted (at 500 °C) 4H-SiC epilayers [458]. Post-implantation anneal at 1700 °C for 30 min was conducted to electrically activate the implanted N^+ and P^+ ions. The Hall effect investigation indicates that there is a critical donor concentration of (2–5) $\times 10^{19}$ cm^{-3}. Below this value, N- and P-donors result in comparable sheet resistances. The critical concentration represents an upper limit for electrically active N donors; P donors can be activated at concentrations above 10^{20} cm^{-3}. This accounts for the observed low sheet resistance of 35 Ω per square. In another study, high-dose P^+ implantation into 4H-SiC(0001) followed by high-temperature annealing was examined [459]. Annealing highly suppressed surface roughening of implanted SiC. The surface stoichiometry of implanted SiC was examined by using X-ray photoelectron spectroscopy. Hall effect measurement shows no significant difference in the sheet resistance between SiC annealed with a graphite cap and without a graphite cap. The sheet resistance has weak temperature dependence, and it has a minimum value of 45 Ω per square at an implant dose of 6.0×10^{16} cm^{-2}. Spontaneous formation of SFs in heavily nitrogen-doped 4H-SiC crystals was observed by using TEM [318]. SFs lie in as-grown boules, and annealing in argon at 1150 °C generated more SFs. All SFs have identical structure consisting of six layers stacked in a cubic sequence and were supposed to arise from two Shockley partial dislocations gliding on two neighboring basal planes of SiC. The energy of faulted 4H-SiC has a lower energy than the perfect heavily doped ($n > 1 \times 10^{19}$ cm^{-3}) crystal at typical processing temperatures, which offers driving force for transformation.

2.6.2.2 B and Al Impurities

Deep impurity centers in 6H-SiC epitaxial layers doped with boron were studied by using DLTS, admittance spectroscopy, and PL spectroscopy [460]. Two electrically active defect centers were identified: the isolated boron acceptor at $E_B = E_v + 0.3$ eV and the boron-related D-center at $E_D = E_v + 0.58$ eV. The yellow luminescence observed in these SiC layers was supposed to arise from pair recombination via D-center and nitrogen donor. Carrier recombination mechanisms in heavily aluminum doped *p*-type 4H-SiC were investigated by using PL and time-resolved PL spectroscopy [461]. The dominant recombination mechanism varies with injection conditions for the samples with Al concentration less than 10^{19} cm^{-3}. Under low injection conditions, recombination via the Al acceptor level appears dominant, exhibiting weak temperature dependence. Under high injection conditions, Shockley–Read–Hall recombination via deep recombination centers becomes predominant, leading to shorter carrier lifetimes with increasing temperature. Low (keV) and high (MeV) energy Al and B implanted *n*-type 6H- and 3C-SiC at both room temperature and 850 °C were investigated [462]. The implanted wafer was annealed at 1100, 1200, or 1400 °C for 10 min. Secondary ion mass spectrometry, Rutherford backscattering, PL, Hall effect, and capacitance–voltage techniques were employed. The samples implanted at 850 °C were not amorphized and the damage peak vanished after 1400 °C annealing. For the samples that were amorphized by the implantation, a distinct damage peak remained in the Rutherford backscattering spectrum even after 1400 °C annealing. P-type conduction was only observed in samples implanted by Al at 850 °C and annealed at 1400 °C in Ar, with 1% dopant electrical activation. Implantation of Al and B (at room temperature or elevated temperatures of 500–700 °C) in combination with high temperature (1500–1750 °C) annealing proves to be a successful approach to achieve *p*-type conductivity in 4H/6H SiC crystals [463]. The generated electrically active acceptors increased with increasing annealing temperature. Hot implantation was required in the case of high Al fluences to avoid amorphization of the implanted SiC layers. Rutherford backscattering and cross-sectional TEM investigations show that randomized SiC layers cannot completely be recrystallized even at annealing temperatures of 1800 °C. Severe outdiffusion of B atoms at the crystal surface occurred at temperatures over 1750 °C. DLTS spectra of implanted SiC samples show a series of peaks belonging to intrinsic defect centers. No levels were observed from the Al^+-implanted *p*-type SiC, and only the D-center was detected in the B^+-implanted *p*-type SiC. The generation of the D-center and B acceptor is controllable by co-implantation of Si^+/B^+ and C^+/B^+ according to the site-competition effect.

Surface roughening of implanted and annealed 4H-SiC was examined by using atomic force microscopy (AFM) [464]. *n*-type 4H-SiC epitaxial layers (5×10^{15} cm^{-3}, 4 μm thick) were implanted with B or Al at a dose of 4×10^{14} cm^{-2} or 2×10^{15} cm^{-2}, respectively. Apparent activation energies for roughening following implantation with Al and B (annealed in argon) are 1.1 and 2.2 eV, respectively. The implant-activation depends sublinearly on time. The activation percentage after 5 min anneal following boron implantation is about a factor of two less than that after 40 min anneal. Roughness values for anneal in argon at 1750 °C are ~ 8.0 nm. The roughness decreased to 1.0 nm (root-mean square) after anneal of boron or aluminum implants in silane at 1100 °C, with respect to 8 and 11 nm for anneal of B and Al implants in argon at the same temperature.

The degree of incomplete ionization of Al doped 4H-SiC was studied [465]. Analysis of published measurements of ionization energy, resistivity, and Hall mobility for varying Al concentration was used to construct two consistent models to yield calculations of the fraction of mobile holes to dopant atoms p/N_A. A theoretical model includes the effects of doping-dependent ionization energy, quantum–mechanical spreading of the acceptor density of states, and density of states smearing due to disorder effects. The second model was derived from experimental Hall and resistivity data. The compensation of native defects and self-compensation in Al-doped 4H-SiC were studied [466]. The positively charged carbon vacancies (V_C^{2+}) were found to be the dominant compensating centers in Al-doped 4H-SiC. For an Al concentration of 10^{16}–10^{19} cm^{-3}, the concentration of holes is smaller by one order of magnitude than the Al concentration because of the compensation of V_C^{2+}. For an Al concentration of over 10^{20} cm^{-3}, the concentration of holes is only in the order of 10^{19} cm^{-3} owing to the dominant compensation of V_C^{2+} and supplementary self-compensation of interstitial Al (Al_i^{3+}). The local structure around boron doped in

6H-SiC was investigated using neutron holography [467]. 3D atomic images reconstructed from multiple-wavelength holograms confirmed boron substitution for both silicon and carbon. Boron atoms were found to be selectively located at the Si–C-cubic site layer. Depth profiles of deep levels in the tail region of Al ion-implanted *n*-type 4H-SiC were investigated by using DLTS [468]. Implantation and subsequent activation annealing at 1750 °C for 20 min generated deep levels located at E_c–0.55 eV, E_c–0.64 eV, and E_c–1.50 eV (E_c: conduction band bottom). Their densities are 20–40 times lower than the density of the implanted Al atoms. The densities of these defects and Al atoms exhibited an exponential decay along the depth direction with a decay length of 140–190 nm. Another deep level located at E_c–1.30 eV was detected and the density of this trap has a decay length of 62 nm.

Superconductivity was observed in heavily boron-doped bulk 3C-SiC [469]. The compound exhibited zero resistivity and diamagnetic susceptibility below a critical temperature T_c of ~ 1.4 K and at an effective boron doping concentration of over 10^{21} cm^{-3}. The H–T phase diagram of this superconducting compound was determined from AC susceptibility. In finite DC magnetic fields a hysteresis was observed between cooling and subsequent warming runs. Therefore, unlike the type-II superconductivity in boron-doped diamond and silicon, a type-I superconductivity with a critical field $H_c(0)$ of about 100 Oe was realized in boron-doped SiC. An obvious jump of the specific heat at T_c demonstrates bulk nature of the superconductivity. Typical glassy ferromagnetism was observed from Al-doped 4H-SiC [470]. The coeffect of sp^2/sp^3 configuration along with the structural defects was proposed to account for the origin of this ferromagnetic order. The aluminum was thus demonstrated to be a novel dopant in SiC that not only stabilizes the crystal structure but also induces ferromagnetic order.

2.6.2.3 Donar–Acceptor Pair Emission

Donor–acceptor pair luminescence plays an important role in silicon carbide. Boron and nitrogen impurities in 6H-SiC single crystals have been studied by using ESR spectroscopy [471]. Both impurities substitute for carbon and occupy the three nonequivalent carbon sites with equal probability. The hyperfine structure was well resolved for both species. The nitrogen hyperfine structure is interpretable in terms of some *s* character for the unpaired electron, whereas the boron hyperfine structure indicates predominantly *p* character. Photoluminescence excited from the donor–acceptor pairs in cubic SiC by a laser at 1.8 K has been studied [472]. The spectrum consists of a series of lines and some of them are displayed in Fig. 2.104. The lines arise from the nitrogen (N)–aluminum (Al) pairs (donor N in C sites and acceptor Al in Si sites) based on the crystal growth and freedom from extraneous lines. The following relationship [473] was employed to analyze the origin of the observed luminescence lines:

$$h\nu = E_g - (E_D + E_A) - E_C + E_{vdW}, \tag{2.19}$$

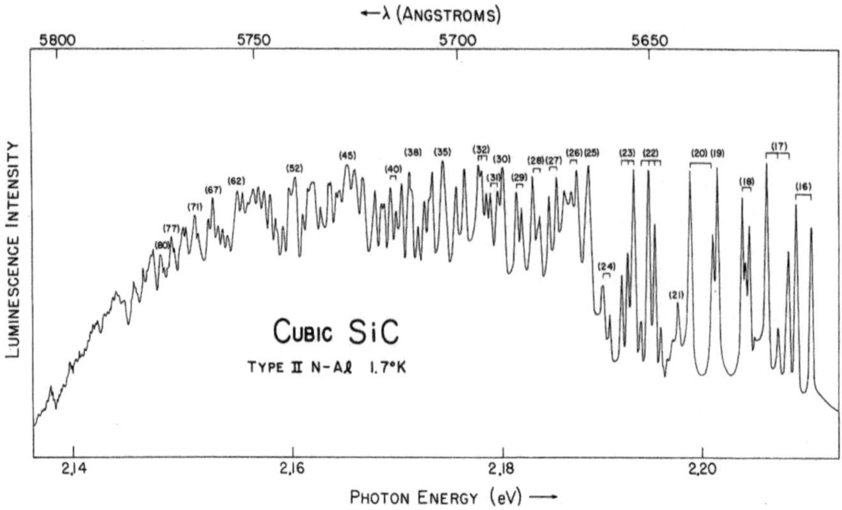

Fig. 2.104 Photoluminescence lines for shell numbers bigger than 15. The shell numbers (*m*) are displayed in brackets and shell substructure components are identified for small *m*. Reprinted with permission from [472]. Copyright 1970, American Physical Society

where E_g is the bandgap, E_D and E_A are the donor and acceptor ionization energies, E_C is the Coulombic interaction energy between the donor and acceptor ions after electron–hole recombination, and E_{vdW} is the interaction energy between the neutral donor and acceptor atoms before recombination. E_C has discrete values allowed by the lattice structure and so for a given donor, E_g was evaluated for each shell (denoted by an integer m) of equidistant acceptors. The same table of donor–acceptor shell substructure components of GaP was used to determine the shell numbers in cubic SiC considering the identical (zinc blende) crystal structures. In the N–Al spectrum of cubic SiC, the lines were well resolved up to the 80th shell and extrapolation yields the photon energy for distant donor–acceptor pairs, $h\nu_\infty$, of 2.0934 eV. With the aid of the bandgap of cubic SiC (2.39 eV), the donor and acceptor ionization energies can be derived, $E_D = 118$ meV for N, and $E_A = (179 \text{ meV}) + E_x$, for Al. Here, E_x represents the exciton binding energy. The donor–acceptor spectrum exhibits as an envelope at low resolution (Fig. 2.105) and the low resolution spectrum shows several phonon replicas from which the phonon energies were derived to be 69, 94, and 118 meV.

Donor and acceptor related luminescence from 4H-, 6H-, and 3C-SiC single crystals has been investigated in the temperature range between 2 and 450 K [474]. The SiC crystals containing A1, Ga, or B acceptors were studied and the luminescence color varies from violet to near infrared depending on the energy level of the acceptor and polytype. All the luminescence at low temperature was proposed to be due to pair recombination between the unintentionally introduced nitrogen donor and respective acceptor. Each PL spectrum comprises a zero phonon line and several phonon

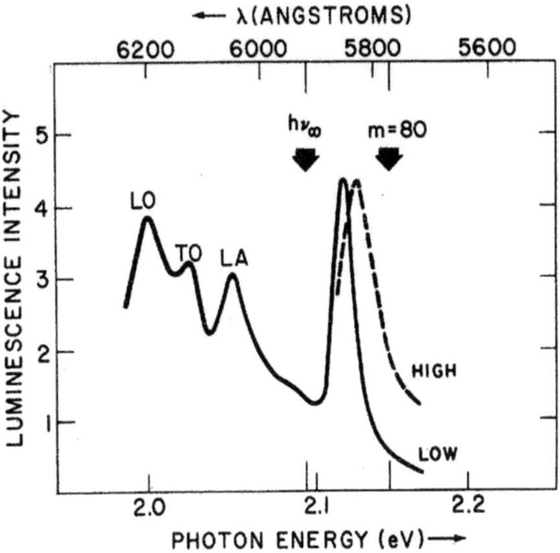

Fig. 2.105 Low-resolution envelope of the donor–acceptor PL spectrum (strongest peak) and phonon replicas (LA, TO, and LO). The solid and dashed lines were obtained at small and large excitation intensities, respectively. The position of the $m = 80$ line shown in Fig. 2.104 is marked and the photon energy of the distant pair, $h\nu_\infty = 2.0934$ eV, is also marked. Reprinted with permission from [472]. Copyright 1970, American Physical Society

replicas. The higher temperature spectrum consists of another emission band stemming from recombination of free electrons with bound holes at the acceptors. The ionization energies of the acceptors were determined from the luminescence spectra and thermal quenching of the emission intensity. The B acceptor has higher ionization energy than Al and Ga, and so it shows a different luminescence spectrum; it is intense even at room temperature. The ionization energies of A1, Ga, and B in 4H-SiC are 168 meV + E_x, 249 meV + E_x, and 628 meV + E_x, respectively, where E_x is the exciton binding energy on the order of 10 meV. It has been demonstrated that the site dependent impurity levels are caused by nonequivalent sites in 4H, 6H, and 15R SiC [475]. Figure 2.106 shows the luminescence spectra of these three polytypes of SiC at 77 K. These spectra show three series of peaks, A_0, B_0, and C_0 along with the respective phonon replicas. B and C arise from the donor–acceptor pair transitions and A is attributed to free to-acceptor transition. The peak intensity ratios of C_0 to B_0 under high intensity excitation, where both the B and C series appearing with full intensity, are respectively 0.96 for 4H, 1.9 for 6H, and 1.4 for 15R SiC. These values agree well with the ratios of the number of cubic to hexagonal like sites (1 for 4H, 2 for 6H, and 1.5 for 15R). Hence, the B and C series of emission lines arise from donor–acceptor pair transitions between acceptors and donors in hexagonal-like and cubic-like sites, respectively, as suggested previously [476]. This implies that every impurity atom enters each type of site with the same equal probability. Further analysis based on the configuration coordinate phonon spectra supports the notion that all the impurities in cubic sites occupy deeper levels than those in hexagonal sites (Table 2.4). The ratios of the ionization energies between the cubic and hexagonal sites, which are roughly constant and independent of the polytypes and impurity types, are 1.0–1.08 for acceptors and 1.55–1.88 for donors. The origin of the site effect on the impurity level was explained by assuming a local

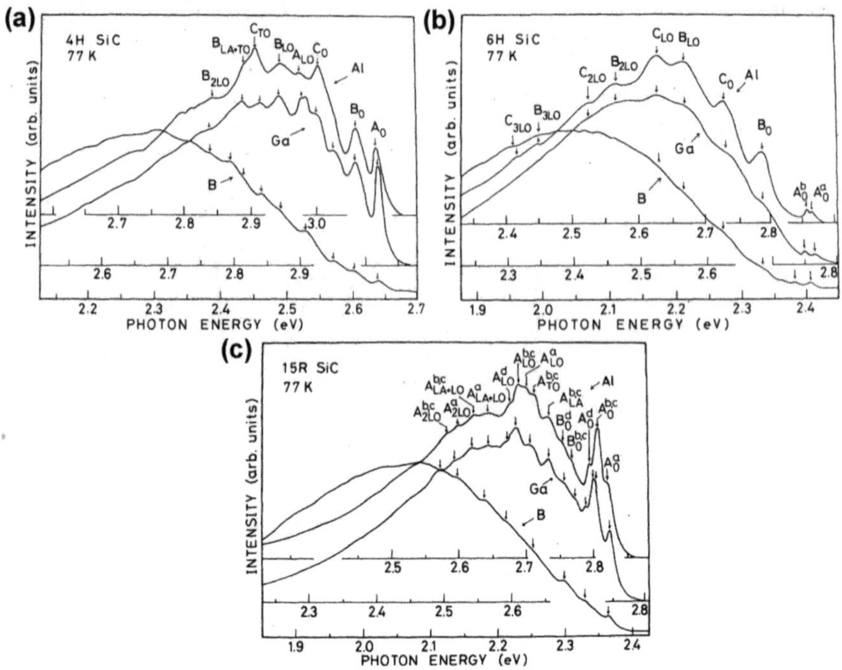

Fig. 2.106 Photoluminescence spectra of Al-doped, Ga-doped, and B-doped **a** 4H, **b** 6H, and **c** 15R SiC single crystals at 77 K. The subscripts indicate the types of phonons and superscripts indicate the types of sites which acceptors substitute for. Reprinted with permission from [475]. Copyright 1980, American Physical Society

dielectric constant and local effective mass based on the quantum defect model. The exciton binding energy depends almost linearly on the ionization energy of nitrogen donors in 3C, 4H, 6H, and 15R SiC.

The donor–acceptor pair light emission from silicon carbide can persist at elevated temperatures. The highly efficient visible luminescence from N and B doped 6H-SiC epilayers was observed at room temperature [477]. Figure 2.107 displays the PL spectra of two samples with B concentrations of (a) 2×10^{17} cm^{-3} (sample 1) and (b) 2×10^{18} cm^{-3} (sample 2). The N concentration is approximately 4×10^{18} cm^{-3} in both samples. The temperature varies from 10 to 250 K in the PL measurements. The sample with a small B concentration shows both N–B donor–acceptor emission at approximately 630 nm and N–Al (unintentionally introduced) donor–acceptor emission at 460 nm at temperatures below 100 K. However, the N–Al emission was quenched above 100 K. The peak wavelength of the N–B emission gradually shifted to 576 nm with increasing temperature. The N–B emission band is broad, with a full width at half maximum (FWHM) of 110 nm at 250 K. The sample with a large B concentration exhibited only the N–B emission even at low temperatures. The PL intensity from the highly doped sample at high temperature is significantly larger (Fig. 2.108). A high internal quantum efficiency of 95% was estimated for the highly

Table 2.4 Ionization energies of donors and acceptors in cubic-like sites (E^C) and hexagonal-like sites (E^H), their average ratios, and free-exciton binding energies (E_X) in 3C, 4H, 6H, and 15R SiC

Ionization energy (meV)

	Site	N	Al	Ga	B	E_X
3C	*C*	56.5	254	343	735	13.5
6H	*C*	155	249	333	723	78
	H	100	239	317	698	
	E^C/E^H	1.55	1.040	1.050	1.036	
15R	*C*	112	236	320	700	40
			230	311		
			223	305		
	H	64	221	300	666	
			206	282		
	E^C/E^H	1.75	1.076	1.072		
4H	*C*	124				20
			191	267	647	
	H	66				
	E^C/E^H	1.88	1.0	1.0	1.0	

B-doped sample at 250 K. The donor–acceptor recombination lifetimes are 2.5 and 5.0 ms for samples 1 and 2, respectively, as determined from the time-resolved photoluminescence measurement, and in good agreement with the calculated value based on the rate equation. It has also been demonstrated that when the N and B concentrations are in a range of 10^{18} cm^{-3}, 6H-SiC showed the most intense luminescence when the concentration difference between the donors and acceptors was about 4.6×10^{18} cm^{-3} [478]. The PL spectrum comprises a maximum at about 587 nm with a FWHM of 120 nm. However, the obtained lifetimes are 0.5 ns, 30 ns, 0.5 μs, 1.2 μs, and 1.1 μs for different samples and much smaller than those reported in Ref. [477].

Low-temperature PL spectroscopy was used to characterize as-grown and ion-implanted 3C-SiC films grown by CVD on Si(100) substrates [199]. The D_1 and D_2 defect PL bands were observed in the as-grown CVD films. The defect PL bands in the two forms of 3C-SiC exhibited similar dependence on post-implantation annealing temperature. The spectral details of the defect PL bands and their annealing characteristics are independent of the particular implanted-ion species. The effects of boron and phosphorus doping on structural, optical, and electrical properties of amorphous and microcrystalline SiC films were investigated [479]. The doped microcrystalline SiC films exhibited an optical gap as high as 2.1 eV and a conductivity of 10^{-3} Ω^{-1} cm^{-1}. The excitation power and temperature dependence of PL in 6H-SiC was investigated [480]. A sublinear power-function dependence of the donor–acceptor pair emission intensity on the excitation power was explained by a rate-equation model. The small

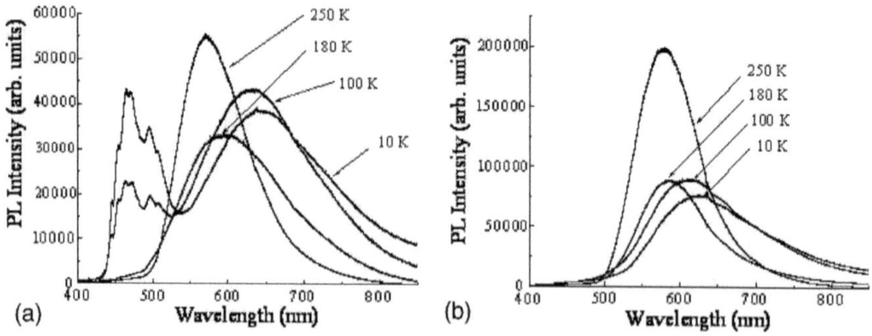

Fig. 2.107 Photoluminescence spectra of N and B doped 6H-SiC epilayers excited by 351 nm at different temperatures. The B concentrations are **a** 2×10^{17} cm^{-3} (sample 1) and **b** 2×10^{18} cm^{-3} (sample 2), respectively. Reprinted with permission from [477]. Copyright 2006, American Institute of Physics

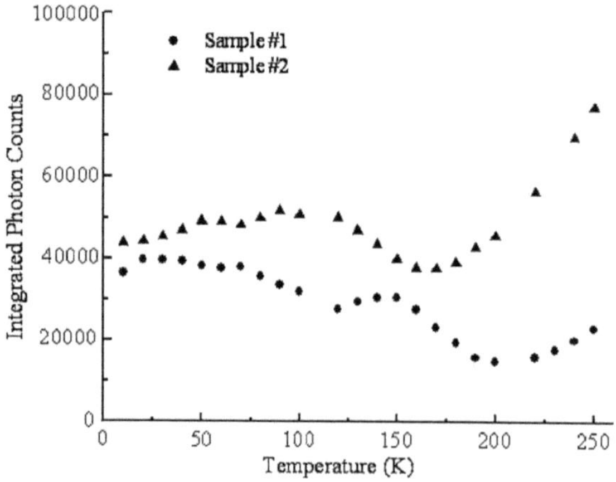

Fig. 2.108 Integrated photon count in the PL spectrum as a function of temperature for low and highly doped 6H-SiC. Reprinted with permission from [477]. Copyright 2006, American Institute of Physics

power exponent ($k = 0.3$) in a power-law dependence of PL intensity on the excitation power at room temperature for B and N codoped 6H-SiC suggests that Auger recombination is the main carrier recombination pathway. In comparison, the Auger recombination in an Al and N codoped SiC is weaker.

Optical detection of magnetic resonance was reported for the blue donor–acceptor pair PL in 6H-SiC comprising nitrogen and aluminum [481]. The nitrogen donor resonance is isotropic, with $g = 2.004 \pm 0.002$. Resonances for the aluminum acceptor in three inequivalent silicon lattice sites were detected with $g_{\parallel} = 2.412 \pm 0.002$, 2.400 $\pm$ 0.002, and 2.325 $\pm$ 0.002; and $g_{\perp} = 0$. This suggests an effective-mass-like character for the acceptor, although with reduced orbital contribution due to localization. Models for doping-induced band edge displacements and bandgap narrowing in both *n*-type and *p*-type 3C-, 4H-, and 6H-SiC were presented [482]. The models take into account three different electron effective mass components associated with hexagonal lattices. A careful treatment of minority carrier correlation energy was made by adopting a two-band model for the dielectric function of a hole gas in the plasmon-pole approximation. The band edge displacements follow a simple analytical function of doping concentration.

2.6.2.4 Other Impurity-Correlated Luminescence Models

Besides donor–acceptor pair, exciton bound to donor complex, there are other models proposed to explain the low-temperature luminescence lines in silicon carbide. A series of bound exciton lines just below the luminescence of single excitons bound to the shallow nitrogen donors [483] can be attributed to recombination in the multiple bound exciton complexes. This model was supported by the spectral positions, Zeeman effect, and behavior under optical pumping. Two series of near-infrared emission bands were observed from boron-doped 3C-SiC crystals [484]. The dominant series of luminescence bands (at about 1.6 eV) at low temperature were assigned to nitrogen donor–boron acceptor pair recombination. The other series of bands was explained by the free-to-bound transition model concerning the spectral shape of the zero phonon line and temperature dependence. In this scenario, the electrons in the conduction band recombine with the holes trapped at the boron acceptors to produce luminescence. The ionization energies of the nitrogen donor and boron acceptor were estimated to be 0.05 and 0.73 eV, respectively. The exciton can also be bound to the neutral aluminum acceptor complex in 3C-, 6H-, and 4H-SiC, as confirmed by low-temperature photoluminescence and cathodoluminescence experiments [485]. The correlated lines were observed only from lightly doped *p*-type SiC films, and the intensity increased with aluminum concentration. The multiplicity of the observed lines agrees with the prediction from the symmetry-based models. In another study on low temperature photoluminescence from the 6H, 4H, and 15R polytypes of SiC, a couple of new features different from the well-known emission lines close to the bandgap were observed from the three polytypes and attributed to recombination of the bound excitons at aluminum acceptors [486]. There are two distinct boron-related centers in silicon carbide polytypes, one shallow with an ionization energy of 300 meV and the other deep with an ionization energy of 650 meV. 4H-SiC homoepitaxial films can be intentionally doped with the shallow boron center by controlling the silicon to carbon source gas ratio during CVD [487]. For samples produced using a small Si/C ratio, the shallow boron centers tend to be incorporated into the silicon sublattice, as confirmed by the temperature-dependent Hall effect, admittance spectroscopy, and DLTS. A luminescence peak at around 383.8 nm was observed from the samples at low temperature and this peak stems from recombination of an exciton–neutral shallow boron acceptor complex. The first evidence for assignment of this luminescence peak is that its intensity increased with boron concentration and secondly, the observed momentum-conserving phonon replicas have energies consistent with other previously observed exciton–neutral acceptor complexes in SiC. Lastly, the excited states were observed at elevated temperatures and this is similar to the cases of excitons bound to neutral aluminum and gallium acceptors.

2.6.2.5 Transition Metal Impurities

A variety of kinds of transition metal impurities in SiC have also been investigated. Among them, vanadium has attracted great interest owing to its excellent properties when it is doped in SiC. Trace impurities of vanadium in the SiC crystals prepared by a modified Lely process were detected based on the strong and polytype-dependent photoluminescence in the 1.3–1.5 μm near-infrared region as well as by infrared absorption [488]. The spectra originate from the intra-3d-shell transitions $^2E(3d^1) \rightarrow {}^2T_2(3d^1)$ of $V^4{}_{Si}{}^+(3d^1)$. ESR spectroscopy reveals that V_{Si} in SiC serves as a deep acceptor and possibly also as a deep donor. The vanadium impurity was suggested to be able to reduce the minority-carrier lifetime in SiC-based optoelectronic devices. By using ESR and photo-ESR, vanadium was identified as an electrically amphoteric deep level impurity introducing both donor D^0/D^+ states and acceptor A^0/A^- states in the band gap of 4H- and 6H-SiC [489]. Photo-ESR indicates that the vanadium donor level in 6H-SiC lies near midgap, at $E_v + 1.6$ eV. Omnipresent titanium impurities form complexes with nitrogen donors, and the corresponding donor level of the (TiN) pair in 6H-SiC lies at E_c–0.6 eV. Hall effect, DLTS, and optical absorption measurements were used to determine the position of the vanadium acceptor level in vanadium and nitrogen-doped 6H- and 4H-SiC [490]. Hall effect data indicate that the acceptor level lies at 0.80 and 0.66 eV beneath the conduction band edge for 4H-SiC and 6H-SiC, respectively. The DLTS signature of the defect in 4H-SiC shows an ionization energy of 0.80 eV and a capture cross section of 1.8×10^{-16} cm^{-2}. The optical absorption measurement reveals existence of the vanadium acceptor level. The DLTS measurement and secondary ion mass spectroscopy indicate that the maximum solubility of vanadium in SiC is 3.0×10^{17} cm^{-3}. The vanadium acceptor level could be used for producing semi-insulating silicon carbide. The optical properties of tetravalent-vanadium impurities in 4H-SiC were investigated [491]. Two crystalline sites showed PL at 1.28 and 1.33 μm (Fig. 2.109), with lifetimes of 163 and 43 ns, respectively, and the PL was separately stable up to 50 and 20 K. The doublet at emission wavelengths of 1278.0 and 1279.9 nm (α_3 and α_2) was attributed to the hexagonal site in 4H-SiC, and the single emission line at 1334 nm (β) was attributed to the quasicubic site in 4H-SiC. Spectrally broad PL was observed at up to room temperature.

The magneto-optical properties of the V^{3+} center in 4H-SiC was investigated by using EPR and photo-EPR spectroscopy [492]. The defect proves to possess a high-spin $S = 1$ ground state, optically induced ground-state spin polarization of more than 70%, long spin–lattice relaxation times of the order of a second, and associated zero-phonon PL lines at 1.8 μm. The zero-field splitting parameter D increased linearly with temperature between 4 and 292 K with a rate of 440 kHz K^{-1}. These properties make the vanadium acceptor a promising candidate for qubit applications with optical properties in the telecommunication band. Controlled defect introduction using 100 MeV Ag ion irradiation and quenching of PL from neutral vanadium V^{4+} atoms in vanadium-doped semi-insulating 4H-SiC were reported [493]. With increasing Ag ion fluence, intensity decrease from V^{4+} related alpha and beta emissions and correlated increase in thermally stimulated current were observed. Single

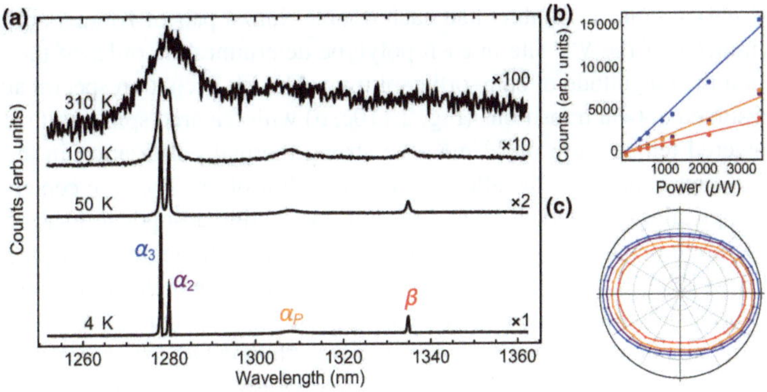

Fig. 2.109 **a** PL spectra of V-doped 4H-SiC at denoted temperatures showing a 2-nm split doublet associated with the hexagonal site. **b** Power dependence of the PL at 3.5 K, showing no saturation in measured regime. **c** Polarization dependence of the four different features (the axes are with respect to the input field). Reprinted with permission from [491]. Copyright 2019, American Physical Society

V^{4+} vanadium centers in 4H-SiC with telecom-wavelength emission and a coherent $S = 1/2$ spin state were investigated [494]. Spin-dependent optical transitions were observed. Engineering the isotopic composition of the SiC matrix helps to reduce the inhomogeneous spectral distribution of different emitters down to 100 MHz. The dopant concentration was tuned to stabilize the telecom-wavelength V^{4+} charge state. In isotopically enriched $4H^{28}Si^{12}C$ layer (the isotope purity is 99.85% for ^{28}Si and 99.98% for ^{12}C, as confirmed by secondary ion mass spectroscopy), the V^{4+} charge state decayed rapidly with a timescale of 129 ± 6 ms. In contrast, it survived for several seconds in the sample featuring a natural abundance of Si and C isotopes. The spin relaxation lifetime of an ensemble of vanadium (V^{4+}) centers with the neutral charge state and occupation of the α sites in 4H- and 6H-SiC was studied [495]. These defects have optical transitions at 1278.78 and 1308.56 nm. They exhibited a large spin contrast exceeding 90% and a long spin-relaxation time of up to 25 s at 100 mK, and of order 1 s at moderate field strengths ($0.1\ \text{T} < \boldsymbol{B} < 0.5\ \text{T}$) and at 1.3 K. These measurements were complemented by a characterization of the ensemble charge state dynamics. The β-site defects showed spin relaxation lifetimes below 1 ms. Vanadium (V^{4+}) in 4H- and 6H-SiC was used to establish a potential telecom spin–photon interface [496]. A vanadium atom substitutes for a silicon atom in the SiC lattice and occupies one of two inequivalent sites in 4H-SiC [a hexagonal (h) and a quasicubic (k) site] and one of three inequivalent sites in 6H-SiC [a hexagonal (h) and two quasicubic (k_1, k_2) sites] (Fig. 2.110a). The neutral charge state of vanadium (V^{4+}) possesses an active electron (spin 1/2) in its $3d_1$ orbital, forming a free-ion ^2D state. The crystal field of the SiC lattice induces splitting of the orbital levels into an orbital doublet ground state and a triplet excited state, and the latter further splits into an orbital singlet and doublet owing to C_{3v} symmetry of 4H- and 6H-SiC (Fig. 2.110b). The spin–orbit interaction further turns the orbital singlets into a Kramers doublet and each doublet into a pair of Kramers doublets. The symmetry of the V^{4+} site in each polytype determines the order of the orbital levels and the magnitude of spin–orbit splitting. The PL excitation spectra at 2.7 K show resolved optical transitions (Fig. 2.110c, d) with featured spin–orbit splitting. At a lowered temperature of 22 mK, the strong thermal polarization in the 4H-β and 6H-β sites suppresses the allowed GS2–ES1 transition. The spin center shows efficient optical spin polarization and readout facilitating all-optical measurements of temperature-dependent spin relaxation times. When the temperature decreased from 2 K to 100 mK, a 4-order-of-magnitude increase in spin relaxation time across all measured sites was observed, with site-specific values ranging from 57.1 ms to 27.9 s. The relaxation mechanism involves a two-phonon Orbach process.

Bulk polycrystalline $Si_{1-x}Mn_xC$ (x: 0–0.1) samples were prepared by the solid-state reaction method [497]. XRD analysis shows that the appropriate Mn doping did not change the crystal structure of 3C-SiC. Ferromagnetism can be developed when Mn substitution is as low as $x = 0.01$. The Curie temperature increases monotonously with increasing Mn-doped quantity up to $x = 0.05$, and then keeps nearly constant when x increases. The effective magnetic moments are 2.05, 1.19, 0.90, and 0.89 μ_B

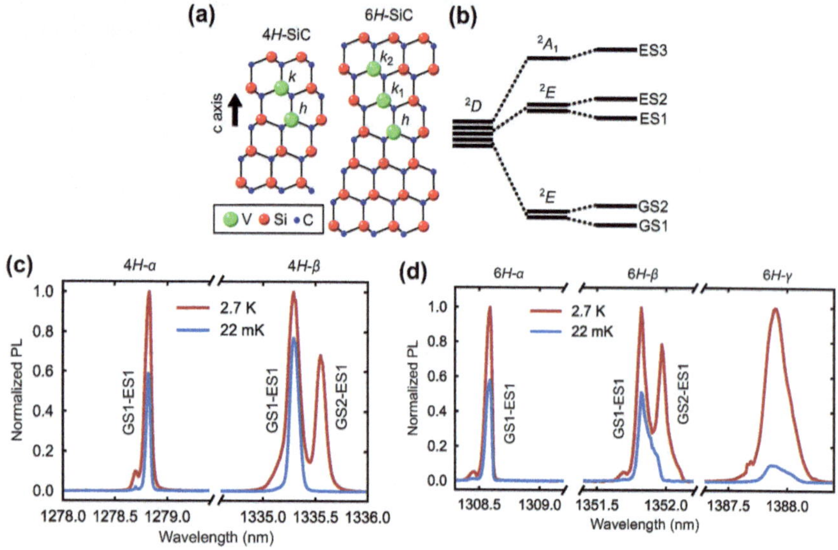

Fig. 2.110 Optical spectroscopy of V^{4+} in 4H-SiC and 6H-SiC. **a** A vanadium defect can substitute one of the inequivalent Si sites in 4H-SiC and 6H-SiC lattices. **b** Orbital level structure of V^{4+} in SiC. The crystal field from the SiC lattice splits the orbital d^1 states into an orbital doublet ground state and an excited state consisting of an orbital singlet and doublet state. The order of the orbital singlet and doublet depends on the site. The spin–orbit interaction induces further splitting in the orbital doublets. **c, d** Resonant PL excitation spectra at 2.7 K (red) and 22 mK (blue) for 4H-SiC and 6H-SiC. The optical transitions in the 6H-γ site are within the inhomogeneous linewidth and selective excitation on the GS1–ES1 transition is challenging. Reprinted with permission from [496]. Copyright 2024, American Physical Society

per Mn atom for the samples with $x = 0.01$, 0.03, 0.05, and 0.1, respectively. Low Mn-doped (< 10^{-3} molar fraction) 6H-SiC was synthesized by reaction of high-purity silicon, graphite, and manganese powder [498]. Raman scattering shows an unusual shift in Raman peak with altering Mn contents. The typical ferromagnetic order was established at as low Mn-doped concentration as 10^{-4} molar fraction at ~ 250 K. The defects-related effects other than the Mn content were supposed to play a more important role in determining the magnetic ordering.

Implantation of Fe, Ni or Mn at doses of 3–5 × 10^{16} cm^{-2} into *p*-type 6H-SiC was carried out at a temperature of ~ 350 °C followed by annealing at 700–1000 °C for 5 min [499]. To the sensitivity of both XRD and selected-area diffraction pattern analysis, no secondary phases were detected. Signatures of ferromagnetism were observed in all the highest dose samples, with Curie temperatures of 50 K (Ni), 250 K (Mn), and 270 K (Fe). *p*-type 6H-SiC crystals were implanted with ^{57}Fe ions with a nominal dose of 1.0 × 10^{16}, 2.0 × 10^{16}, 3.0 × 10^{16} or 2.0 × 10^{17} cm^{-2} at 100 or 200 keV ion energy to produce diluted magnetic semiconductors [500]. The implanted samples experienced thermal annealing at 1000 °C for 2 min. XRD, cross-sectional TEM, and sputter-Auger depth profiling were used for the structural characterizations. The magnetic properties were acquired by using superconducting quantum interference device magnetometry and ^{57}Fe conversion electron Mössbauer spectroscopy at room temperature and 4.2 K. The results show that ferromagnetism in ^{57}Fe-implanted SiC for Fe concentrations above 3% stems mostly from epitaxial superparamagnetic Fe_3Si (and probably a small fraction of Fe nanoparticles) in the SiC matrix. Blocking temperatures range from 400 K for a dose of 2.0 × 10^{16} cm^{-2} to 220 K for 3.0 × 10^{16} cm^{-2}. For the lowest dose of 1.0 × 10^{16} cm^{-2} at 200 keV, ferromagnetism via weak magnetic hyperfine interaction was observed below 20 K. d^0 magnetism can be strengthened through hole doping by 3d-transition metal substitution at the Si site of 3C-SiC [501]. The valence state of Mn, Co, and Ni was found to be 2+ while Cr takes a 3+ valence state. The Mn-doped sample exhibited the highest Curie temperature, saturation magnetization, and spontaneous magnetization value. Analysis of magnetization data and ESR spectra suggests the enhancement of the d^0 magnetism due to the generation of holes caused by the transition metal doping at the Si site. The room temperature ferromagnetic response in the doped samples could be explained using modified Bound Magnetic Polaron model.

Ti$^+$-implanted 4H- and 6H-SiC epilayers were studied by DLTS [502]. Two Ti centers with energy levels at E_c–(117 ± 8) meV and E_c–(160 ± 10) meV were observed in 4H-SiC. These levels were assigned to the ionized Ti acceptor $Ti^{3+}(3d_1)$ residing at hexagonal and cubic Si lattice sites. The Ti acceptor levels were supposed to be resonant in the conduction band for 6H-SiC. Ensembles of molybdenum (Mo) defect centers in 6H- and 4H-SiC with 1.1521 and 1.1057 eV transition energies were reported [503]. The ground-state and excited-state spin of both defects is $S = 1/2$, with large g-factor anisotropy. This agrees with theoretical predictions that Mo resides at a hexagonal lattice site in 4H-SiC and 6H-SiC. The *p*-type host environment suggests that this occurs for Mo at a silicon substitutional site. The ZPL for Mo in 6H-SiC lies at 1.1057 eV (1121.3 nm) and at 1.1521 eV (1076.2 nm) for 4H-SiC ($T = 3.5$ K). The results show optical lifetimes of ~ 60 ns and inhomogeneous spin dephasing times of ~ 0.3 μs. The optical measurement of an ensemble of Mo impurities in 6H-SiC shows a spin lifetime of 2.4 s at 2 K [504]. The slow spin relaxation was ascribed to suppression of spin–lattice and spin–spin interactions owing to a combination of crystal-field with axial symmetry and spin–orbit coupling. Intense erbium luminescence around 1.54 μm at 2–300 K was observed from Er^{3+}-implanted 4H-, 6H-, 15R-, and 3C-SiC samples [505]. The ion implant fluence is about 10^{13} erbium ions cm^{-2} and the implanted samples were annealed at 1700 °C. The most striking feature about the spectra is that the relative integrated intensity of the erbium luminescence is nearly constant from 2 to 400 K. This is different from the narrower gap semiconductors in which the 1.54 μm luminescence was observed to drop significantly. No major difference exists in the spectra of the hexagonal and rhombohedral polytypes, but there is a difference for cubic SiC.

There have been some theoretical investigations of the transition metals in SiC. The magnetic properties of 3C-SiC doped by first-row transition metals were studied by using the linearized muffin-tin orbital method within both the atomic sphere approximation and full potential implementations [506]. The magnetic properties depend on the doping site. The substitution on the Si site shows a nonmonotonic behavior of the magnetic moments with *d* state filling owing to the larger crystal field than spin splitting caused by large hybridization of the transition metal *d* states with neighboring dangling bonds. This generates low-spin states. For the unrelaxed structures, substitution on the C site leads to large-spin states, obeying an atomic-like Hund's rule owing to weak hybridization with surrounding dangling bonds. The relaxation effect is large for the C site and small for the Si site, leading to suppression of high magnetic moments for the C substitution. The formation energies favor Si substitution. For neighboring pairs of transition metal dopants, there is generally suppression of the magnetic moments, except for Cr and Mn on Si sites. Ferromagnetic coupling is preferred for Cr. Mn shows ferromagnetic coupling. Thermodynamic considerations suggest that Cr and Mn doping on Si is favorable. The strong energetic preference for ferromagnetic over antiferromagnetic ordering suggests high Curie temperatures. The authors further investigated the first row transition-metal-doped 3C- and 4H-SiC [507]. The calculation shows that the Ti impurity is active for 4H-SiC but not for 3C-SiC, whereas V and Cr impurities are active for both polytypes. Cr shows a very local exchange interaction that decays rapidly after second-nearest neighbors, which is similar for different polytypes and different sites. Mn has a long range of exchange interaction, which is very sensitive to the location of the Mn pairs.

The band structure and magnetic properties of 3C- and 6H-SiC doped with 3d transition metals were studied by using ab initio calculations [508]. For 3C-SiC, Cr and Mn generate half-metallic ferromagnetism at both Si and C substitution sites, but with different magnetic moments. Fe-doped SiC remains paramagnetic. 6H-SiC shows similar behaviors, but Fe on the Si site at low concentrations generates ferromagnetic ordering. Ab initio calculations can reproduce the anisotropy in the g tensor of the V and Mo defects in 4H- and 6H-SiC with Kramers doublet spin ground state, leading to reduced and vanishing interaction with the magnetic field in parallel and transverse directions, respectively [509]. The role of dopants on deformation and mechanical properties of 4H-SiC was studied using nanoindentation [510]. The hardness, elastic modulus, and fracture toughness of 4H-SiC decreased with the order of vanadium (V) doping, undoping, and nitrogen doping. Basal plane dislocations, threading edge dislocations, and cracks formed during the nanoindentation. 4H-SiC turned amorphous SiC and 3C-SiC when the depth of the ident penetration from the surface increased. N doping weakens the SiC bond strength and enhanced the glide and piling up of basal plane dislocations. V doping hindered the glide of basal plane dislocations, which accumulated high-stress field and facilitated the polymorph transition.

The transition metal V or Mo spin-1/2 defects in SiC, where neutral V^0 or positively charged Mo^+ substitutes for a silicon atom, were studied [511]. The spin structure as well as magnetic and optical resonance properties of the active electron emerges from the interplay of the crystal potential and spin–orbit coupling. The spin–orbit coupling results in additional allowed transitions and modification of the g-tensor. An effective Hamiltonian was derived to describe the dependence of the Rabi frequency on the magnitude and direction of the static and driving fields. The authors developed a theory for the interaction of an active electron in the D shell of a transition metal defect in SiC with the transition metal nuclear spin and derived the effective hyperfine tensor within the Kramers doublets formed by the spin–orbit coupling [512]. They showed that nuclear-spin polarization enables the use of well-known methods for initialization and long time coherent storage of quantum states [513]. The authors further studied the main effects of strain on the electronic levels and optical electric-dipole transitions of the V defect in SiC [514]. Strain shows the ability of engineering the g tensor, electronic selection rules, and hyperfine interaction.

2.6.2.6 H and He Impurities

Hydrogen and helium can also act as impurities in SiC. Hydrogen impurities in 3C-SiC have been studied theoretically based on DFT–LDA calculations [515]. The stable configurations and formation energies of various hydrogen and dihydrogen defects were calculated. The interstitial H is readily incorporated into the p-type 3C-SiC during growth and acts as a shallow donor compensating the acceptors. In n-type 3C-SiC, interstitial H_2 is more stable but its formation energy is too high to allow for significant incorporation. (V_C + H) behaves as a hole trap and (V_{Si} + H) as an electron trap. The implanted hydrogen is favorably trapped by V_{Si}. The electron trap V_{Si} can be passivated by four hydrogen atoms, but V_C remains a hole trap. Low-temperature hydrogen annealing of SiC introduces a sufficient amount of interstitial hydrogen for complete passivation of the p-type materials. In the n-type materials, the dominant defects are (V_{Si} + nH) electron traps which can reduce the free carrier concentration. High temperature annealing in hydrogen is effective in introducing hydrogen even to the n-type materials. It was suggested that some of the paramagnetic signals which have been assigned to pure vacancies may arise from (V + nH) complexes. The energy position and thermal stability of deep levels in nitrogen-doped 4H-SiC epitaxial layers after 1.2 MeV proton implantation and

15 MeV electron irradiation were studied by using DLTS [516]. Isochronal annealing was performed at 100–1200 °C in steps of 50 °C. The DLTS measurements carried out within 120–630 K after each annealing step revealed the presence of ten electron traps lying at 0.45–1.6 eV below the conduction band edge (E_c). Three traps located at 0.69, 0.73, and 1.03 eV below E_c were observed only after proton implantation, and they were proposed to be hydrogen related. The E_c–0.73 eV level shows a very narrow depth distribution, confined within the implantation profile; it may stem from a defect involving only one H atom.

Room temperature helium implantation into 4H-SiC below the threshold amorphization dose can induce strain, which arises from both point defects and He-related defects [517]. For a helium concentration lower than 0.5%, the strain profile follows the point defect profile, whereas at higher concentrations the He ions show a dominant effect on the strain. The near surface strain gradually relaxes upon annealing at up to 1500 °C, whereas the maximum strain relaxation stops at a temperature where helium ions agglomerate into platelets. When the vacancies turn mobile, these platelets become bubble clusters which expel dislocation loops whose migration is enhanced by the strain. The energetics and stability of helium atoms and intrinsic point defects in 3C-SiC crystal were investigated by using first-principles calculations [518]. The interstitial He shows smaller formation energy than that of point defects. The He–C interaction is stronger than He–Si interaction. He interstitials may form He bubbles without involving other types of structural defects. A Si (C) monovacancy can accommodate up to 11 (9) He atoms. The He clusters trapped in the Si or C monovacancy cause large pressure on the order of GPa. The microstructure evolution of hydrogen-implanted 6H-SiC at different temperatures and fluences was investigated [519]. For H-implanted samples with low fluence at room temperature, dense blister cavities were observed after annealing at 1100 °C, whereas no cavities arose in the sample implanted with high fluence due to loss of elastic energy during the crystalline-to-amorphous transition. For higher implantation temperatures of 450–900 °C, amorphization did not occur and H-containing microcracks grew laterally below the surface. Blisters appeared on the surface of the samples implanted at 900 °C even without annealing. For He implantation, blisters did not form.

2.6.2.7 Carrier Recombination Lifetimes

The room-temperature minority carrier lifetimes were measured for 6H-SiC epitaxial layers with n-type doping of 10^{14}–10^{17} cm^{-3} [520]. The samples were grown by CVD using silane and propane or silane and methane in a hydrogen ambient. The minority carrier lifetimes were measured by monitoring the decay of the near-bandgap room-temperature PL of the samples after excitation with a short laser pulse. Lifetimes of the thick low-doped samples reached as high as 0.45 μs. The lifetime decreased after introduction of compensating acceptors. The authors further measured the minority carrier lifetimes of n-type 4H-SiC epitaxial layers [521]. A map of the minority carrier lifetime from a 45 μm thick 4H-SiC layer was made, which reveals that the lifetime is correlated with the morphology of the layer. The minority carrier lifetime reached as high as 2.1 μs at room temperature. At 200 °C, the minority carrier lifetime was about 5 μs. Carrier lifetimes in 4H-SiC epilayers were investigated by using differential microwave photoconductivity decay measurements [522]. When the $Z_{1/2}$ concentration is higher than 10^{13} cm^{-3}, the $Z_{1/2}$ center acts as a recombination center. The carrier lifetime shows positive dependence on the injection level (number of irradiated photons). When the $Z_{1/2}$ concentration is smaller than 10^{13} cm^{-3}, other recombination processes such as surface recombination limit the lifetime. In this case the carrier lifetime decreased with increasing injection level. Therefore, control of the $Z_{1/2}$ concentration by low-energy electron irradiation controlled the lifetime.

The effects of measurement techniques and conditions on measured carrier lifetimes in n^- 4H-SiC epilayers were investigated both experimentally and through carrier dynamics simulations to understand differences between reported lifetimes [523]. The common time-resolved PL, transient free carrier absorption, and microwave photoconductivity decay approaches were compared. For the limits of high or low injection, these techniques can reflect very different lifetimes. The effect of measurement conditions on the carrier lifetime was considered by simulating the carrier dynamics assuming a dominant Z_1/Z_2 defect. The electron and hole capture cross sections for the 0/+ transition of Z_1/Z_2 were found to be 2–4 × 10^{-15} cm^2 and 1–2 × 10^{-14} cm^2, respectively. The temperature dependence of the lifetime is relatively weak below 500 K and thermally activated above this temperature due to electron emission from the Z^0 state. Electron emission from Z^- and hole emission become important at higher temperatures. Simulations with both Z_1/Z_2 and EH_6/EH_7 defects suggest the latter does not contribute significantly to the lifetime in as-grown epilayers owing to a small capture cross section for holes. The carrier lifetime of ~ 265 μm thick n-type 4H-SiC epilayers prepared using carbon-implantation/annealing was examined [524]. A very long minority carrier lifetime of 18.5 μs and a high injection lifetime of 19.2 μs were acquired from time-resolved PL and microwave photoconductivity decay measurements, respectively. The influence of surface recombination on the carrier lifetime was eliminated.

The interband Auger recombination in 4H-SiC was studied using a time-resolved photoinduced absorption technique [525]. The Auger recombination coefficient was derived from the kinetics of electron–hole plasma in heavily doped n-type 4H-SiC and in low-doped 4H-SiC epitaxial layers within 300–565 K. The value decreased from $(7 \pm 1) \times 10^{-31}$ to $(3 \pm 1) \times 10^{-32}$ cm^6 s^{-1}. The remarkable decrease in Auger recombination rate with increasing temperature is related to temperature-dependent threshold energy of Auger process. Excited carrier recombination was observed from 4H-SiC under the high injection condition using time-resolved free carrier absorption measurements [526]. The Auger recombination coefficient was found to depend on the excited carrier concentration, whereas the traps have negligible effects on the Auger recombination coefficient.

2.7 Surface Structures

Silicon carbide has many polytypes and each polytype has its own crystal structure and specific surface structures. The surface atoms in silicon carbide alter the bonding structures to terminate partial dangling bonds by forming Si–Si bonds, C–C bonds, double or triple bonds, or bonding with foreign atoms. On the other hand, silicon carbide is a compound semiconductor that is composed of two elements and both silicon and carbon atoms on the surface can react with various atoms such as hydrogen and oxygen ones to produce different surface chemical structures. Therefore, silicon carbide has rather complex surface structures, which affect its electrical and optical properties. Some surface defects can significantly alter the normal performance of SiC-based electronic devices and so there have been extensive investigations on the surface structures of silicon carbide. Here, the major phenomena involving the surfaces of silicon carbide are reviewed and discussed.

2.7.1 Surface Reconstruction

In general, on the semiconductor surface, owing to the absence of atoms on the air side of the semiconductor/air interface, the surface atoms experience different forces compared to the interior ones. As a result, the surface structure relaxes and reconstructs to have the minimum surface energy. Assume that the reconstructed surface has a two-dimensional lattice with unit vectors of $\boldsymbol{b}_1$ and $\boldsymbol{b}_2$, the substrate surface has two-dimensional lattice with unit vectors of $\boldsymbol{a}_1$ and $\boldsymbol{a}_2$, and on a certain substrate crystal surface $X\{hkl\}$, reconstruction is given by $(\boldsymbol{b}_1 \| \boldsymbol{a}_1, \boldsymbol{b}_2 \| \boldsymbol{a}_2)$ and then we have:

$$\boldsymbol{b}_1 = p\boldsymbol{a}_1, \quad \boldsymbol{b}_2 = q\boldsymbol{a}_2, \tag{2.20}$$

where p and q are two proportionality factors. In general, the reconstructed surfaces are denoted by the following notations [527]:

$$X\{hkl\}(p \times q), \quad X\{hkl\}c(p \times q), \ \text{or} \ X\{hkl\}(p \times q) - R^\circ, \tag{2.21}$$

where c represents a possible center, and R° denote the angles by which the reconstructed surface rotates relative to the substrate surface.

It is very difficult to determine the atomic positions of the reconstructed surface experimentally. Investigators generally utilize scanning tunneling microscopy (STM) and low energy electron diffraction (LEED) to experimentally study SiC surfaces. First-principles calculations and ab initio molecular dynamics simulations have also been adopted to theoretically study SiC. Based on the experimental and theoretical results, various structural models have been proposed. The (100) surface on cubic SiC is one of the most widely studied surfaces of SiC. This surface has several types

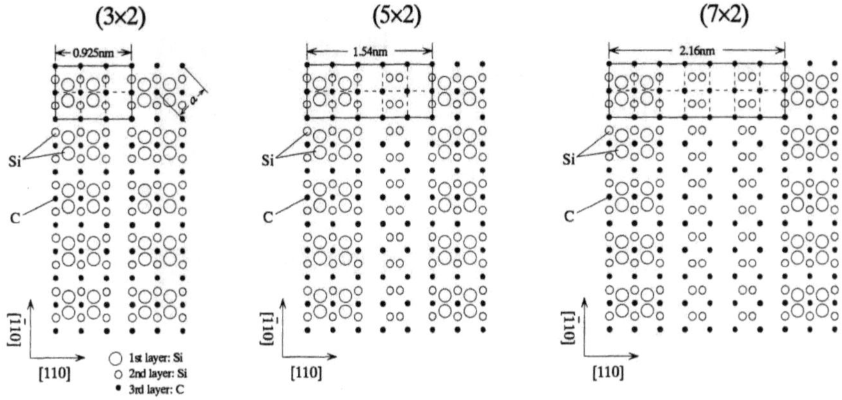

Fig. 2.111 3C-SiC(001)-(3 × 2), (5 × 2), and (7 × 2) structural models. Reprinted with permission from [529]. Copyright 1994, American Physical Society

of surface reconstruction such as the C-terminated $c(2 \times 2)$ surface, Si-terminated 2×1 surface, and Si-rich 3×2 surface.

The β-SiC(100) surface was studied by LEED, AES, high-resolution electron energy loss spectroscopy, and core level excitation EELS [528]. Two Si-terminated phases with (3 × 2) symmetry and (2 × 1) symmetry were discovered. The atomic resolution images obtained by STM reveal that the 3C-SiC(001) surfaces have Si adlayers on the Si-terminated surface [529]. An additional dimer-row model for the surface with extra Si atoms shows structural features consistent with STM and LEED. The unit cell of the Si-saturated (3 × 2) surface observed by LEED comprises a pair of additional Si dimers on the Si-terminated surface, forming a straight line along the direction through the centers of two constituent atoms of a dimer (Fig. 2.111). There are also reconstruction models for the (5 × 2) and (7 × 2) surfaces in which the unit cell of the uppermost atomic layer also consists of a pair of additional Si dimers.

Clear atom images of individual Si dimers were observed from the Si-rich 3C-SiC(100)-(3 × 2) surfaces by using STM [530]. The sample has a high-quality surface with a low defect density (Fig. 2.112). The STM image shows that the Si–Si dimers form rows perpendicular to the dimer direction in a (3 × 2) atomic arrangement with clear evidence of asymmetric dimers tilted in the same direction (not alternate tilting). This is a very open atomic surface arrangement. On the top atomic layers, three types of defects are identified, namely missing dimers (type A), dimer pairs (type B), and dimers having different tilt angles (type C). The proposed structure model is illustrated in Fig. 2.113. This model is based on the clear identification of individual Si atoms from the STM topography (filled and empty electronic states). This model gives a Si-coverage of 1/3, different from conclusions drawn in the aforementioned STM study [529] where there is a coverage of 2/3 with pairs of Si dimers oriented parallel to the rows (Fig. 2.111).

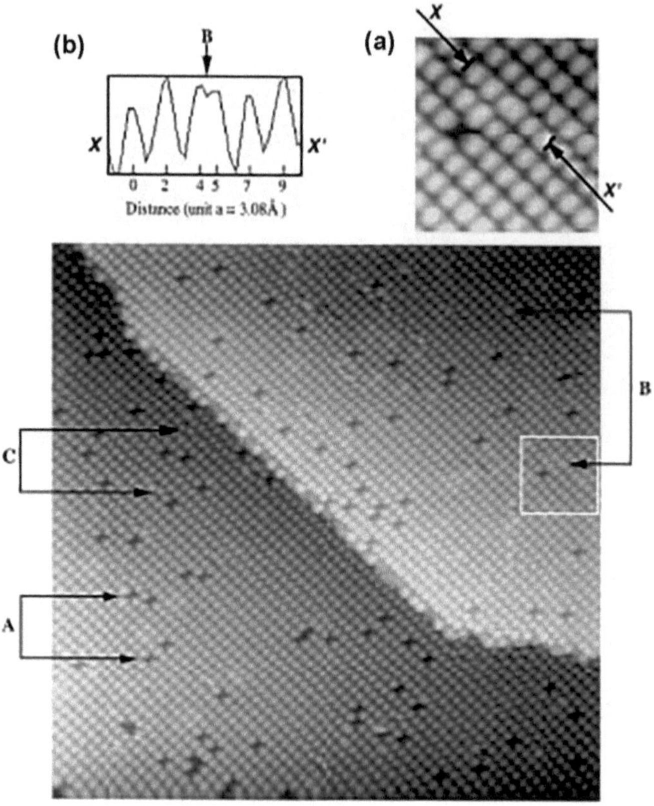

Fig. 2.112 40 × 40 nm STM image (filled states) of the 3C-SiC(100)-(3 × 2) surface with A, B, and C denoting three types of defects. **a** Magnified image of the area having type A (missing dimer) and B (dimer pair) defects. **b** Atomic profile of dimers along the xx' axis displaying the two components of a dimer pair (marked by B). Reprinted with permission from [530]. Copyright 1996, American Physical Society

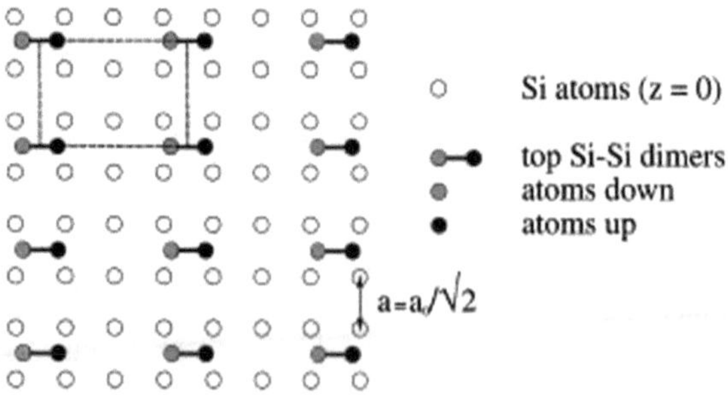

Fig. 2.113 Schematic of the 3 × 2 Si–Si dimers on a nonreconstructed 3C-SiC(100) surface. The surface primitive unit cell parameter is $a = a_0/\sqrt{2}$, and a_0 (4.36 Å) is the 3C-SiC lattice constant. The dashed line refers to the (3 × 2) unit cell. Reprinted with permission from [530]. Copyright 1996, American Physical Society

Surface compositions ranging from Si-rich to Si-deficient for 3C-SiC(001) were prepared by cleaning it in ultrahigh vacuum via heating within a Si atomic flux [531]. Samples were characterized by using AES, EELS, X-ray photoelectron spectroscopy (XPS), and ultraviolet photoelectron spectroscopy (UPS) as well as LEED. Several reconstructions were observed: (3 × 1), *c*(4 × 2), (2 × 1), *c*(2 × 2), and (1 × 1). Surface models were proposed, which include ordered arrays of surface Si dimers for Si-rich and Si-terminated surfaces and graphite formation for Si-deficient surfaces. The structural and electronic properties of five different configurations of polar 3C-SiC(001) surfaces were studied by using the self-consistent ab initio calculation [532]. No dramatic Si dimers on the Si-terminated (2 × 1) surfaces were found, whereas strong carbon dimers were found from various C-terminated surfaces serving as the reconstruction building blocks. The 3C-SiC(100)-*c*(4 × 2) surface reconstruction was studied by using atomic-resolution STM [533]. The STM images agree well with the corresponding STM image calculation (Fig. 2.114). A model of dimer rows was proposed. In this picture, the dimer row consists of alternatively up and down dimers within the row in an "undulating" type of arrangement so as to reduce the surface stress. The 3C-SiC(100) and Si(100) have different *c*(4 × 2) surface reconstruction. The straight, long, and highly stable Si atomic lines self-organized on the 3C-SiC(100) surfaces were observed by using STM [534]. The Si atomic lines are composed of Si dimers formed by annealing of the primary 3C-SiC(100)-(3 × 2) structure at above 1000 °C (Fig. 2.115). The number and spacing of the Si atomic lines can be controlled by the annealing time and temperature. The density of the Si atomic lines decreased with temperature until the isolated Si atomic lines were observed. The phase transition was observed with annealing. The 3C-SiC(100)-(3 × 2) surface transforms to the 3C-SiC(100)-*c*(4 × 2) surface on which the Si atomic lines sit. These atomic channels form upon the phase transition between the (3 × 2) and *c*(4 × 2) reconstruction by selective Si atom organization. Figure 2.116 describes these two surface reconstructions. The Si atomic lines result in a lateral mismatch between the *c*(4 × 2) Si-dimer rows. Formation of the carbon atomic chains were also observed from the C-terminated 3C-SiC(100) surfaces by using atom-resolved STM [535]. These carbon atom lines were thermally stable up to 1200 °C. They comprise buckled single bond sp^3 C–C dimers with the direction perpendicular to the sp C≡C triple bond dimers, forming a *c*(2 × 2) surface reconstruction. Their density increased with annealing time and they did not cause or come from surface antiphase boundaries. The anisotropic tensile or compressive surface stress was proposed as leading driving force in this carbon sp → sp^3 diamond like transformation. Combined variable temperature STM and angle-resolved photoemission spectroscopy were used to study transition between two 3C-SiC(100) surface structures [536]. There is reversible temperature-dependent phase transition from a semiconducting *c*(4 × 2) surface at 25 °C to a metallic 2 × 1 structure at 400 °C. This transition arises from temperature-induced disruption of the *c*(4 × 2) structure composed of alternately up and down dimers into a structure having all dimers at the same height giving a 2 × 1 symmetry. This configuration favors electronic orbital overlap between Si top surface atoms, leading to surface metallization.

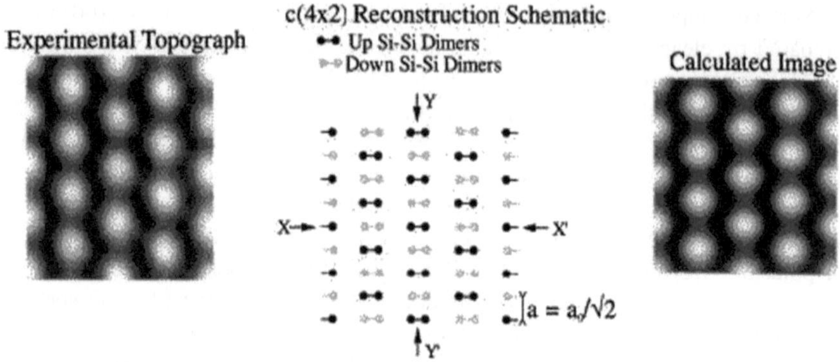

Fig. 2.114 Experimental STM topographical image (filled electronic states) of the Si-terminated 3C-SiC(100)-$c(4 \times 2)$ surface showing the Si dimers (left panel) and schematic of this surface showing the alternating up and down dimers (middle panel). The unreconstructed surface primitive lattice parameter is $a = a_0/\sqrt{2}$, where $a_0 = 4.36$ Å is the 3C-SiC lattice constant. The right panel shows the calculated STM image. Reprinted with permission from [533]. Copyright 1997, American Physical Society

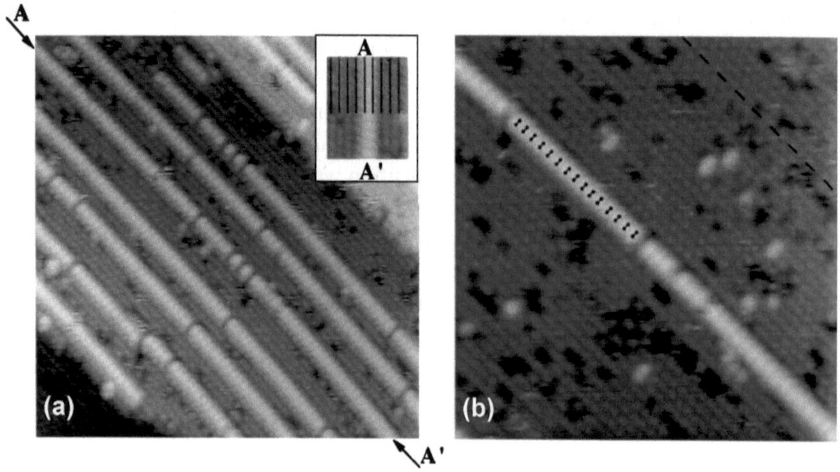

Fig. 2.115 **a** 32 × 32 nm STM topographical image of the 3C-SiC(100)-(3 × 2) surface after annealing at 1100 °C. There are Si atomic lines along with $c(4 \times 2)$ surface reconstruction between them. **b** 20 × 20 nm^2 STM topographical image of 3C-SiC(100)-(3 × 2) surface after annealing at 1150 °C. A single Si-dimer line is displayed. The dark dashed line in the top right corner denotes the $c(4 \times 2)$ dimer row direction. Reprinted with permission from [534]. Copyright 1997, American Physical Society

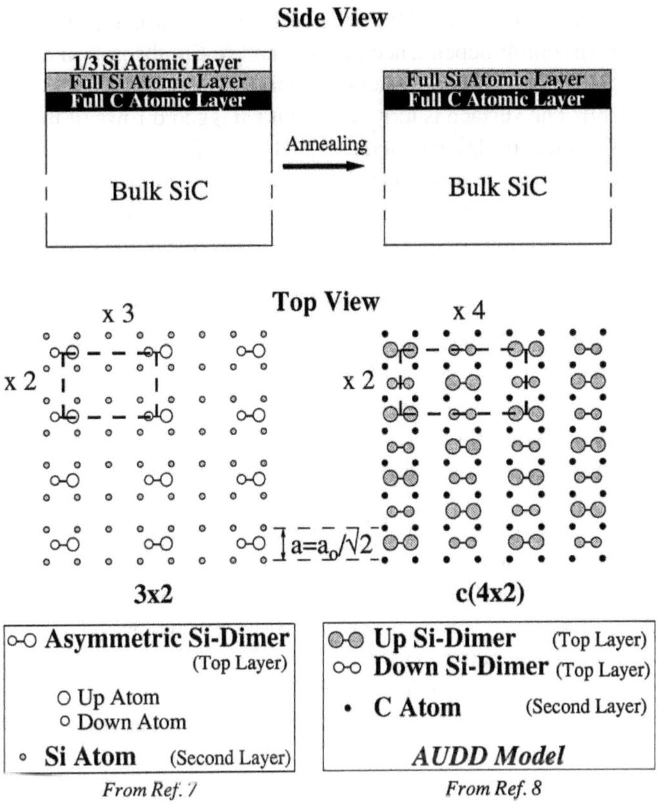

Fig. 2.116 Schematic of the 3C-SiC(100)-(3 × 2) and 3C-SiC(100)-*c*(4 × 2) reconstructed surfaces. Annealing induces phase transition between them. Reprinted with permission from [534]. Copyright 1997, American Physical Society

The atomic structure of the 3C-SiC(100)-*c*(2 × 2) surface was studied using dynamic calculation of LEED intensities [537]. The *c*(2 × 2) surface was prepared in an ultrahigh vacuum by removal of surface silicon in high temperature annealing or deposition of surface carbon by exposing the stoichiometric (2 × 1) surface at 1125 K to C_2H_4. Both samples contain surfaces terminated with C_2 groups in staggered silicon bridge sites (Fig. 2.117). The difference is that the surface created by silicon sublimation contains weak silicon dimer bonds in the second atomic layer of the *c*(2 × 2) surface, whereas the C_2H_4-generated *c*(2 × 2) surface has no such dimer bonds. Thermal decomposition of C_2H_4 yields hydrogen which saturates silicon dangling bonds in the second atomic layer, thus prohibiting the formation of the Si dimers. Ab initio calculation of the reconstruction of the 3C-SiC(100) surfaces confirms that the lowest energy *c*(2 × 2) reconstructed surface comprises triply bonded carbon dimers in a bridging position between neighboring underlying silicon dimers [538]. Additionally, the most stable structure of the (3 × 2) reconstructed surface with a 1/3 monolayer excess of silicon is an alternate dimer row structure. In another study, the single-domain 3C-SiC(100)-$c(2 \times 2)$ surface structure was investigated utilizing the polarization dependence of C-1s and Si-2p absorption as revealed by near edge X-ray absorption fine structure combined with molecular orbital cluster calculation [539]. The surface is terminated with staggered rows of unusual, nearly triply bonded C dimers bridging underlying Si dimers. The electronic structure of the C dimers was determined by resonant valence-band photoemission accompanying C-1s exciton auto-ionization. The ab initio molecular dynamics simulation indicates that under different preparation conditions and thermal treatment, the C-terminated (001) surfaces of 3C-SiC can have different $c(2 \times 2)$ and $p(2 \times 1)$ reconstructions [540]. The atomic structure of the C-terminated 3C-SiC(100)-$c(2 \times 2)$ reconstructed surface was investigated by using STM and surface-core-level-resolved photoelectron diffraction [541]. This surface consists of anomalous bridge-bonded C dimers with a C–C bond length of 1.22 Å. The angle-resolved photoemission study identified two occupied π state bands attributed to the normal and parallel p orbitals of the triple-bonded C dimers, and this explains the stability of the triple-bonded surface reconstruction.

Scanning tunneling microscopy images show a 6×6 reconstruction of the 3C-SiC(111) surface annealed at 1150–1200 °C [542]. Contrast reversal was observed under reversed tunneling voltage bias. Spectroscopic I/V data show presence of a graphite layer on the top surface. A surface model was proposed, where an incommensurate graphite monolayer grows on a (1×1) Si-terminated 3C-SiC(111) surface. It explains the discrepancy between the $6\sqrt{3} \times 6\sqrt{3}$ geometry observed by LEED and the 6×6 geometry observed by using STM on the same surface. There are fewer reports on the 3C-SiC(111)-(3×3) reconstructed surfaces, and this surface structure

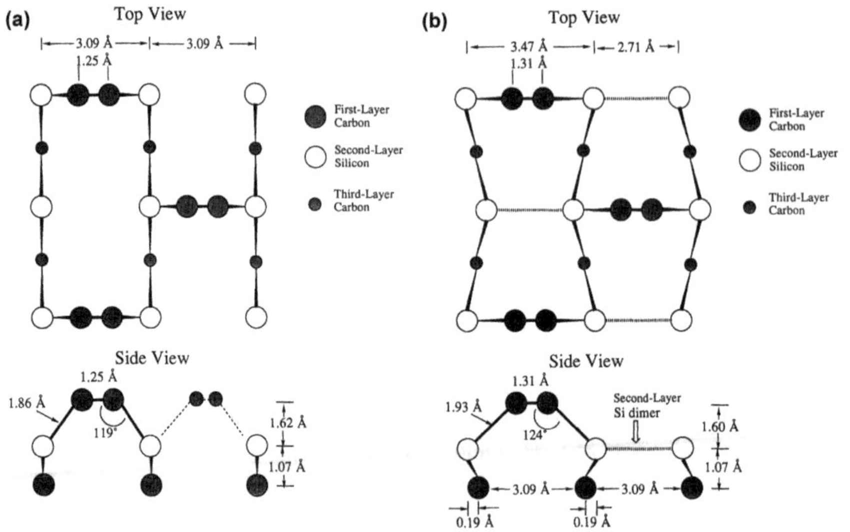

Fig. 2.117 Schematic of the best fit structure of the 3C-SiC(100)-$c(2 \times 2)$ surface produced by **a** C_2H_4 exposure and **b** Si sublimation. Reprinted with permission from [537]. Copyright 1991, American Physical Society

has been investigated by using STM, conventional LEED, LEED holography, and DFT calculation [543]. STM reveals a single adatom per unit cell. LEED and DFT results reveal a Si tetramer on a twisted Si adlayer with clover-like rings. This twist model with one dangling bond left per unit cell is a novel ($n \times n$)-reconstruction mechanism for group-IV (111) surfaces.

Hexagonal SiC may also exhibit the (3 × 3) reconstruction. The drastic (3 × 3) reconstructions of 3C-SiC(111), 4H-SiC(0001), and 6H-SiC(0001) surfaces were identified by using quantitative LEED and holographic interpretation of diffraction intensities, STM, and Auger electron spectroscopy [544]. This observed (3 × 3) reconstruction is a new type of semiconductor ($n \times n$) surface reconstruction, showing the feature of considerable reduction of surface dangling bonds. A modified version of the (3 × 3) reconstruction, that is, the 6H-SiC(0001)-(3 × 3)-Si reconstruction, was observed by using STM [545]. The surface has a hexagonal structure with one Si adatom per unit cell. A reconstruction model in which the unit cell comprises 11 Si atoms was proposed. There are three dimers and three rest atoms in the first layer and one adatom in the second layer. In the first layer, the dimers and other atoms form six- and nine-atom rings which surround unoccupied sites of the outermost silicon atoms of the substrate. For the three dangling bonds in a unit cell, one belongs to the adatom and the other two belong to the silicon atoms of the substrate. STM conducted on the Si-terminated (0001) and C-terminated ($000\bar{1}$) surfaces of 6H-SiC [546] shows that the (3 × 3) reconstruction arises after annealing at 850 °C followed by transformation to the ($\sqrt{3} \times \sqrt{3}$) reconstruction after further annealing at 950 °C. The (3 × 3) surface has a 4/9 monolayer of adatoms and the ($\sqrt{3} \times \sqrt{3}$) surface has a 1/3 monolayer of adatoms. A new (9 × 9) reconstruction on the (0001) surface was observed after further annealing the (3 × 3) surface at 900 °C under a silicon flux. A graphitized surface with a (6 × 6) reconstruction was observed from the (0001) surface when the temperature was suddenly raised to 1150 °C without a silicon flux.

Oxidation and HF etching of the 6H-SiC(0001) surface can produce an atomically flat surface on a 100 nm scale with bulk like periodicity and stoichiometry [547]. The surface dangling bonds are terminated by hydroxyl groups. Thermal treatment can remove the remaining oxygen atoms from the surface and yield the carbon $\sqrt{3} \times \sqrt{3}R30°$ reconstructed surface where carbon–carbon bonds are present on the surface. The $\sqrt{3} \times \sqrt{3}$ reconstruction of the Si-terminated 6H-SiC(0001) surface was investigated by using STM [548]. The STM images are consistent with a structural model composed of 1/3 layer of Si or C adatoms in threefold symmetry sites on top of the outermost Si–C bilayer. Several types of defects were observed from the $\sqrt{3} \times \sqrt{3}$ surfaces. The first-principles total energy calculation determines the chemical identity and adsorption site of the adatoms on the 6H-SiC(0001) $\sqrt{3} \times \sqrt{3}$ surface [549]. Over the entire range of thermodynamically allowed chemical potentials, Si adatoms are preferred over C adatoms. The adatoms prefer the T_4 site over the H_3 site for both Si and C. The calculation shows that in the Si-T_4 model, there is a half-filled adatom dangling-bond band at 1.2 eV above the valence-band maximum with a bandwidth of 0.35 eV. This 6H-SiC(0001) $\sqrt{3} \times \sqrt{3}$ surface was studied by using angle-resolved photoemission spectroscopy [550]. The 6H and 4H polytypes show similar results. A surface state band with semiconducting occupation at around 1.0 eV above the valence band maximum with a width of about 0.2 eV was observed.

LEED and AES experiments of the SiC(0001) surfaces indicate that upon heat-treatment these surfaces were readily "covered" with a layer of graphite by evaporation of silicon [551]. The graphite layer has a distinct crystallographic relation to the SiC crystal. It is monocrystalline on the Si-face and mostly polycrystalline on the C-face. The graphite monolayer already formed at a very low temperature of 800 °C. The 6H-SiC(0001)$\sqrt{3} \times \sqrt{3}R30°$ and $6\sqrt{3} \times 6\sqrt{3}R30°$ reconstructed surfaces were studied by using photoemission spectroscopy [552]. The reconstructed surfaces were fabricated by heating the unreconstructed 1×1 surface to 950 and 1150 °C, respectively. These surfaces show no oxygen related signals. The spectral features for the $\sqrt{3}$ reconstruction cannot be explained by a structural model composed of Si or C adatoms on top of a Si–C bilayer. The $6\sqrt{3}$ reconstructed surface contains a larger amount of carbon atoms. The carbon is not graphitic, and graphitization was observed only after the surface was heated to a high temperature. The atomic structure of the carbon nanomesh template (so-called $6\sqrt{3} \times 6\sqrt{3}R30°$ reconstruction) on the 6H-SiC(0001) surface (Fig. 2.118) was investigated by using STM, LEED, synchrotron photoemission spectroscopy (PES) and DFT calculations [553]. The carbon nanomesh template was proposed to result from self-organization of excess carbon atoms, forming a novel honeycomb arrangement atop the 6H-SiC(0001) surface after annealing at 1100 °C. Angle-resolved synchrotron PES experiments revealed two carbon nanomesh-related C 1s components at binding energies of 285.1 eV (S_1) and 283.8 eV (S_2). The S_2 component was assigned to carbon atoms that bond to one underlying Si atom, and the S_1 component to carbon atoms bonded to other carbon atoms without Si–C bond formation. Further annealing at 1300 °C for 5 min gave a C 1s spectrum dominated by the signal of the graphite layer. It was concluded that graphitization of the SiC surface only occurs at temperatures higher than that required for the formation of the pure carbon nanomesh surface. These distinguish the carbon nanomesh surface and graphitization of the SiC surface. The graphitization of the SiC surface at certain high temperature was accompanied by evaporation of silicon atoms from the bulk.

The relaxed and $\sqrt{3} \times \sqrt{3}R30°$ reconstructed 6H-SiC(0001) surfaces were investigated by using ab initio calculation in the framework of DFT–LDA [554]. Both the Si- and C-terminated reconstructed surfaces were examined. Eight structural models were optimized for the reconstructed surfaces with Si or C adatoms and Si or C trimers adsorbed in threefold-symmetric T_4 or H_3 positions. For the Si-terminated surface, the optimal surface structure involves Si adatoms in T_4 sites and it is consistent with experimental results. For the C-terminated surface, none of the investigated $\sqrt{3} \times \sqrt{3}$ adatom or trimer configurations is the optimal configuration. STM conducted on the 6H-SiC(0001) $\sqrt{3} \times \sqrt{3}$ surfaces of both *n*- and *p*-type materials with low tunnel currents reveals distinct bands of empty and filled states [555]. The density of states with semiconductor features contains a state about 1 eV above the Fermi level and another state 1 eV below the Fermi level. STM images of the surface taken at positive and negative polarities indicate that these two states are localized

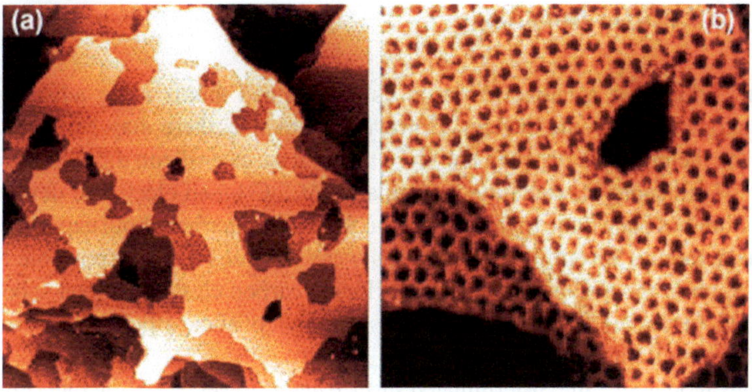

Fig. 2.118 **a** $150 \times 150\,\text{nm}^2$ and **b** enlarged $43 \times 43\,\text{nm}^2$ STM filled state images of SiC honeycomb template after prolonged annealing at 1100 °C for 20 min. $V_T = 2.5$ V, $I = 0.2$ nA. Reprinted with permission from [553]. Copyright 2005, Elsevier B.V.

in the same spatial position, thus supporting a silicon adatom model which predicts a Mott–Hubbard-type density of states.

The (001) and (111) surfaces of β-SiC as well as the (0001) surface of 6H-SiC were studied using LEED and Auger and energy loss spectroscopy [556]. The (001) orientation exhibited five phases: 3×2, $c(4 \times 2)$, 2×1, $c(2 \times 2)$, and 1×1. It was proposed that ordering of asymmetric dimers is responsible for the $c(4 \times 2)$ and 3×2 structures. The $c(2 \times 2)$ exhibited altered bonding at the surface, implying absence of the usual dimerization. Results for (111) and (0001) surfaces are indistinguishable. Three distinct phases were observed for these orientations: 3×3, $\sqrt{3} \times \sqrt{3}R30°$, and 1×1. The 3×3 was identified with an adsorbed Si bilayer. The electronic structures of Si-rich SiC(111)/(0001)3×3 and $\sqrt{3} \times \sqrt{3}$ surfaces oriented parallel to the c axis were studied by using first-principles calculations and ARPES [557]. Independent of the reconstruction model, but in particular for the most favorable Si adtetramer-adlayer and T_4-site Si adatom models, the DFT calculations give rise to half-filled dangling bond bands within the fundamental gap, suggesting metallic surfaces. ARPES showed only completely occupied surface-state bands but no density of states at the Fermi energy. This discrepancy was explained by a Mott–Hubbard picture. The reconstruction geometry of the $\sqrt{3} \times \sqrt{3}R30°$ phase on 4H-SiC(0001) was identified as a T_4 site Si adatom model [558]. The reconstruction consists of Si adatoms in T_4 sites as found by LEED crystallography. A rearrangement of the surface stacking different from that of normal 4H-SiC was identified, which was kinetically triggered when the superstructure was prepared from a sufficiently Si enriched surface. The new stacking sequence resembles a part of the 3C or 6H polytype unit cell.

2.7.2 *SiC–SiO$_2$ Interface*

Silicon carbide has a native oxide on its surface and this is a big advantage for applications in electronic devices. However, in practice, there may be many active defects at the interface between the SiC wafer and surface silicon dioxide layer. The defects significantly degrade the performance of SiC-based electronic devices and much work has been done to investigate the SiC/SiO$_2$ interfaces both experimentally and theoretically. Several types of defects were supposed to be responsible for the degradation of the SiC-based devices, including carbon clusters, silicon suboxides, and Si–C–O clusters. Researchers have also thought of some efficient approaches to mitigate these unfavorable defects.

In hexagonal SiC, the Si-terminated surface has a slower oxidation rate compared to the C-terminated surface and this difference is caused by an unknown interface compound. The angle-resolved XPS study of the oxidized Si-(0001) and C-(000$\bar{1}$) surfaces of 6H-SiC revealed the interfacial oxide $Si_4C_{4-x}O_2$ ($x < 2$) [559]. The slow oxidizing Si-(0001) surface has this oxide with a large thickness (1 nm), whereas the C-(000$\bar{1}$) surface has smaller one and it diminished quickly upon oxidation above 1000 K. Amorphous and single-crystal α-SiC exposed to various oxygen sources (residual gas in an ultrahigh vacuum environment, ambient air, ozone, and oxygen plasma) at room temperature were studied [560]. XPS shows that silicon oxycarbide species formed when these SiC materials were exposed to oxygen. With extended exposure to ambient air, a SiO$_2$ layer subsequently formed over the silicon oxycarbide. The native oxide on the SiC crystal mainly consists of silicon oxycarbide species. The native oxide on the amorphous SiC contained about 6 Å of SiO_xC_y and 26 Å of SiO$_2$ after exposures to ambient air for over a month. The SiC crystal contained a 9.3 Å thick native oxide that mainly consisted of silicon oxycarbide species after extended exposures to air. The exposures to O$_3$ and oxygen plasma enhanced the oxidation, leading to formation of thicker SiO$_2$ layers on both investigated samples. XPS analysis shows there is an interface layer near the SiO$_2$/SiC boundary, where SiO$_2$ stoichiometry collapses and there exists the bondings other than Si–O$_2$ and Si–C, including Si–Si, Si–O–C, and C–O bonds [561]. C–O bonds were regarded as gaseous CO and/or a part of Si–O–C, and the extra O bonds at the interface possibly belong to gaseous single oxygen atoms.

Using TEM and EELS, the structure and chemistry of the interface of the thermally grown SiO$_2$/4H-SiC heterostructure were analyzed [562]. Distinct layers several nanometers thick on each side of the interface as well as partial amorphization of the SiC surface were observed (Fig. 2.119). These interfacial layers were ascribed to the formation of a ternary Si–C–O phase during thermal oxidation. In practice, simple consideration of the topology and geometry suggests that neither the (0001) surface of the hexagonal SiC nor the (001) surface of the cubic SiC could accommodate a suitable geometry and bond lengths for the abrupt (no suboxide layer) SiC–SiO$_2$ interface [563]. This suggests that suboxide bonding at the SiC–SiO$_2$ interface is inevitable. Charge transition levels of various defects at the 4H- and 6H-SiC/SiO$_2$

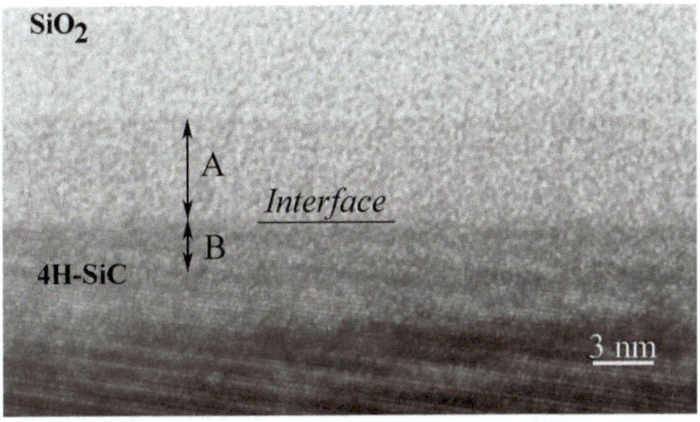

Fig. 2.119 HRTEM image of the SiO_2/4H-SiC interface showing the transition layers A (4.8 nm) and B (3.3 nm) on either side of the interface. Reprinted with permission from [562]. 2008, American Institute of Physics

interfaces were discussed by using DFT calculations employing hybrid density functionals [564]. Carbon-related defects involving single atoms and dimers in both the crystalline SiC substrate and the amorphous SiO_2 oxide were examined. Oxygen- and hydrogen-related defects including Si–Si bond (O vacancy), Si_2–C–O structure, peroxy linkage, and hydrogen bridge (Si–H–Si) were also calculated. The calculation indicates that the Si_2–C–O structure is the best candidate for the high defect density measured near the conduction band of SiC.

The electronic structure of the SiC/SiO_2 interfaces for different SiC polytypes (3C, 4H, 6H, 15R) was investigated by using internal photoemission of electrons [565]. The SiC valence band maximum is located at 6 eV below the oxide conduction band edge for all polytypes. The interface states were found in the energy range up to 1.5 eV above the top of the SiC valence band. The electron spectrum is similar to that of sp^2-bonded carbon clusters, demonstrating the existence of elemental carbon at the SiC/SiO_2 interface. Using electrical analysis and internal photoemission spectroscopy, the energy distribution of the electron states at the SiC/SiO_2 interfaces produced by oxidation of various SiC polytypes (3C, 4H, 6H) was determined [566]. Different polytypes showed similar distributions of interface traps over the SiC bandgap, indicative of a common nature of the interfacial defects. The major defects were proposed to be the carbon clusters at the SiC/SiO_2 interface and near interfacial defects in SiO_2 (Fig. 2.120). In contrast, no signals from the dopant-related defects and dangling bonds at the SiC surface were observed. A large density of interface states at around 0.1 eV below the conduction band of 4H-SiC at the SiC/SiO_2 interface was revealed in low-temperature electrical measurements and photon-stimulated electron tunneling experiments [567]. These states were ascribed to the defects in the near-interfacial oxide layer and they can significantly trap the electrons from SiC. These states may cause degradation of the electron mobility observed from the surface channel of the 4H-SiC/SiO_2 electronic devices.

EELS conducted on the thermally grown SiO_2/6H-SiC(0001) interfaces revealed high carbon concentrations at distinct interface regions [568]. These C-rich regions have a thickness of about 1.0–1.5 nm (Fig. 2.121). The oxides were grown on clean *n*-type 6H-SiC at 1100 °C in a wet oxygen ambient. EELS shows that the CVD deposited SiO_2/SiC interfaces have no C-rich regions. The interface state density in

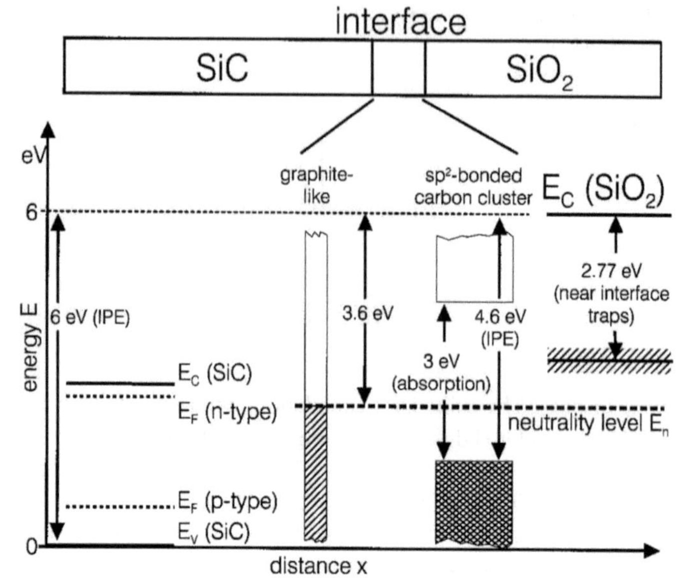

Fig. 2.120 Schematic of the carbon cluster model of the interface states in the SiC/SiO_2 metal–oxide–semiconductor structures. The dominant interface states are the sp^2-bonded carbon clusters with wide energy gaps (3 eV) and graphite-like carbon clusters. The near-interface oxide traps are located at 2.77 eV below the conduction band minimum of SiO_2. Reprinted with permission from [566]. Copyright 1997, John Wiley & Sons, Inc.

the metal–oxide–SiC diodes with thermally grown SiO_2 is about 9×10^{11} cm^{-2} eV^{-1} at $E - E_v = 2.0$ eV. This is consistent with the experimental value for the SiC metal–oxide–semiconductor (MOS) diodes without post-oxidation annealing. These C-rich regions may contribute to the low channel mobility in the SiC-based metal–oxide–semiconductor field-effect transistors (MOSFETs). DFT calculation was employed to study the possible defect structures arising from the interaction between O_2 molecules and an ideal portion of the SiC/SiO_2 interface [569]. The defect structures produced along the calculated oxidation routes remain at the interface in a significant concentration as oxidation ceases. The C–C bonds can explain most of the observed interfacial states but not the large density near the conduction band of 4H-SiC. These electron traps with very high density near the conduction band of 4H-SiC in the metal–silicon dioxide–SiC structures are called near-interface traps. By means of capacitance–voltage and thermal dielectric relaxation current analysis, the near interfacial traps close to the 4H-SiC conduction band edge in the 4H-SiC-based MOS structures with either thermal or deposited oxide were investigated [570]. The signals corresponding to two groups of trap levels were detected and the total trapped charge exceeded 1×10^{13} cm^{-2}. The observed density and energy distribution of these traps are nearly identical in all thermal and deposited oxides, suggesting that they are intrinsic defects at the SiO_2/SiC interface. The high-resolution electron microscopy combined with EELS show no sign of graphitic region at the SiO_2/SiC interface or in the bulk SiO_2

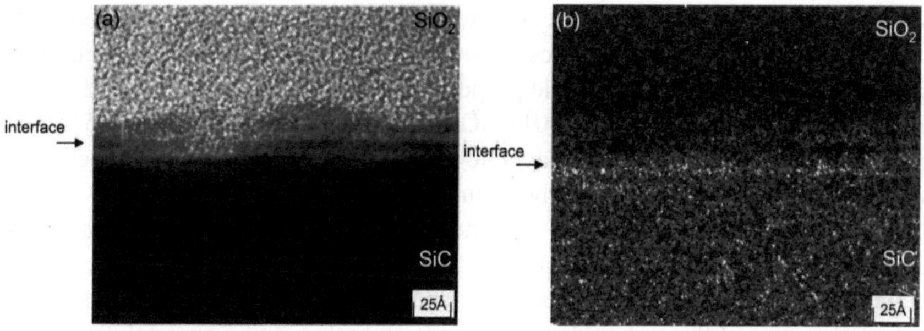

Fig. 2.121 **a** HRTEM image and **b** corresponding electron energy loss spectroscopy map of C at a thermally grown SiO_2/6H-SiC(0001) interface. Reprinted with permission from [568]. Copyright 2000, American Institute of Physics

at a detection limit of 0.7 nm in samples after re-oxidation. No amorphous carbon was observed near the SiO_2/SiC interface, suggesting that the near interface traps close to the 4H-SiC conduction band are not related to the carbon structures at the interface. Sometimes, the near interface traps were assumed to originate from oxides near the interface.

The energy level positions of the intrinsic and carbon defects in SiO_2 were investigated by using DFT calculations [571]. The vacancies and oxygen interstitials can be excluded as the origin of the near interface traps, whereas silicon interstitials and carbon dimers generate gap levels in the energy range inferred from experiments. The correlation between atomic structure and the electrical properties of thermally grown SiO_2/4H-SiC(0001) interfaces was investigated by using synchrotron XPS and electrical measurements of SiC–MOS capacitors [572]. The oxide interface was found to be dominated by Si–O bonds. There existed no distinct C-rich layer beneath the SiC substrate. Electrical characterization of SiC–MOS capacitors indicated accumulation of both negative fixed charges and interface defects, which correlates well with the structural change in the oxide interface. The atomic structures of the SiO_2/4H-SiC(0001) interface states were studied using hard XPS, extended X-ray absorption fine structure, and ab initio molecular dynamics simulations [573]. Continuous interface states in the whole gap and interface states with discrete energy levels near the conduction band minimum were observed. Hard XPS shows that carbon clusters and carbon vacancies formed at the SiO_2/4H-SiC(0001) interface. Simulations indicate that the interface states in the whole gap were attributed to the various atomic geometries of the CßSi3 species and carbon clusters with various sizes and surrounding atoms. The interface states with discrete energy levels near the conduction band minimum could be ascribed to carbon vacancies formed on the steps at the SiO_2/4H-SiC(0001) interface.

A paramagnetic interface defect in thermally oxidized porous *n*-type doped 4H-SiC/SiO_2 was observed [574]. Its axial symmetry and resolved hyperfine interactions suggest it is ascribed to a sp^3 carbon dangling bond center situated at the SiC side of the interface. This defect is electrically active and pins the Fermi level in the oxidized samples. Its formation seems to correlate with interstitial oxygen diffusion at the interface during the oxidation process. No silicon related paramagnetic dangling bond centers were observed. The electric properties of the carbon dangling bond centers at thermally oxidized 4H-SiC(0001)/SiO_2 interfaces were investigated [575]. By combining electrically detected magnetic resonance spectroscopy and capacitance–voltage measurements, the two carbon dangling bond electronic levels [(0/−) and (+/0)] were separately determined as ~ 1.2 and 0.6 eV from the valence band maximum. The effective correlation energy of the carbon dangling bond is 0.6 eV, 1.5 times bigger than that of the silicon dangling bond at Si/SiO_2 interfaces. This suggests potential impact of the carbon dangling bonds on both *p*- and *n*-channel devices.

Some efficient methods have been exploited to reduce the surface defects at the SiC/SiO_2 interface so as to improve the performance of the SiC-based MOS devices. Pre-oxidation cleaning of SiC (3C, 4H, 6H polytypes) surfaces by explosion to ultraviolet radiation and oxygen was found to result in significant improvement in the electronic properties of SiC/SiO_2 interfaces [576]. This treatment removed the defect species that were otherwise present at the interface after thermal oxidation of SiC. Carbon clusters were suggested as the diminished defect species responsible for a substantial part of the SiC/SiO_2 interface states. Nitridation is another efficient way to mitigate the defects at the SiC/SiO_2 interface.

Interfacial characteristics of Al/SiO_2/n-type 6H-SiC MOS capacitors fabricated by rapid thermal processing with N_2O and NO annealing were investigated [577]. The conductance technique yielded interface state density values at room temperature. Rapid thermal processing oxidation in pure O_2 produced an excellent SiO_2/*n*-type 6H-SiC interface with interface state density in the order of 10^{10}–10^{11} eV^{-1} cm^{-2}. NO annealing improved the SiO_2/*n*-type 6H-SiC interface, whereas N_2O annealing increased the interface state density. XPS analysis of NO and Ar annealed SiO_2/4H-SiC MOS structures revealed significant difference between NO and Ar annealed SiO_2/SiC interfaces [578]. This owes to accumulation of nitrogen and formation of stronger Si≡N bonds at the interface. NO annealing improved the SiO_2 /SiC interface. NO annealing produced a less complex SiO_2/SiC interface free of the oxide/carbon compounds compared to the Ar annealing.

Results of capacitance–voltage measurements were reported for MOS capacitors fabricated using the 4H-SiC doped with either nitrogen or aluminum [579]. Annealing in nitric oxide after oxidation/reoxidation process led to slight increase in the defect state density in the lower portion of the band gap for *p*-SiC and significant decrease in the density of states in the upper half of the gap for *n*-SiC. Theoretical calculations account for these results in terms of N passivating C and C clusters at the oxide–semiconductor interface. Analysis of fast and slow traps at the interface of 4H-SiC with oxide grown in O_2, N_2O, and NO indicates that the major positive effect of nitridation is the dramatic reduction of the slow electron trap density [580]. These traps may be the defects in the near-interfacial oxides. This study confirms that nitridation also reduces the fast interface states associated with clustered carbon. Although nitridation has a positive effect in improving the SiC/SiO_2 interface quality, this effect can be saturated. Nitridation with NO significantly reduces the interfacial trap density near the conduction band, but this effect was saturated at 2.5 ×

10^{14} cm^{-2} of nitrogen, leaving a trap density that is still higher than that in silicon [581]. Nitrogen passivation caused dissolution of the defects, leading to lowered trap energy in the bandgap. This behavior is consistent with a model in which the energy of the trap of carbon clusters or silicon suboxide depends on the defect size. The nitrogen and hydrogen effects on the SiC–SiO_2 interfaces have been investigated both experimentally and theoretically [582]. The system studied is the Si-terminated SiC faces in metal–oxide–semiconductor structures. H_2 showed no effect on the density of electrically active interface defects but in contrast, nitridation followed by hydrogenation yielded complementary passivation and substantial reduction in the interface defects. First-principles calculation indicates that isolated C dangling bonds can be passivated by H_2, whereas correlated C dangling bonds on neighboring threefold-coordinated C atoms can only be passivated by monatomic H. Different from previous viewpoints regarding the carbon clusters as the SiC–SiO_2 interface defects, carbon was proposed to exist as a bonded Si–C–O interlayer (one to two atomic layers) and the correlated C dangling bonds were supposed to be the major origin of the observed interface defects. The effect of nitridation was proposed to convert the Si–C–O bond into a Si–C–O–N bond along with reduction in threefold C atoms (N replaces threefold C). The hydrogen atoms result in conversion of Si–C–O–N to Si–C–O–N–H, thus reducing the interfacial states. The residual interface states were ascribed to the Si–Si defects at the interface. Detailed DFT calculation performed on the 4H-SiC/SiO_2 interface [583] reveals that the hole traps near the valence band edge stem from C–C bonds, whereas the amphoteric states deep in the bandgap arise from dangling bonds. The chemical disorder contributes interfacial states with energies close to both band edges. However, the observed electron traps with a large density near the conduction band were proposed to be defects in the oxide near the interface. Silicon interstitials were found to yield a broad distribution of electron traps slightly farther from the conduction band edge, whereas the carbon dimers substituting for two nearby oxygen atoms in the silica network may give rise to the sharp peak of electron traps with critically high density at about 0.1 eV below the conduction band edge of 4H-SiC. The analysis also shows that NO or N atoms attack the carbon dimers at the interface, thereby eliminating carbon. The evolution of the density of interface states, electron traps, and hole traps in 4H-SiC MOS capacitors was examined as a function of the nitrogen content (adjusting the gate oxide NO annealing time) [584]. The result shows that the accumulation of nitrogen at the oxide/semiconductor interface after nitridation of the SiO_2/4H-SiC interface is responsible for reduction in interface state density, immunity to electron injection, and increased hole trapping yields. The generation of acceptor interface states upon electron injection was suppressed after minimum N incorporation, however, the density of oxide hole traps increases linearly with the amount of nitrogen. It was proposed that the nitridation causes formation of a silicon oxynitride layer. Annealing in NO can generate fast states at SiO_2/4H-SiC interfaces [585]. The interface state density was evaluated from the difference between quasi-static and theoretical capacitances. Very fast states that were absent in the as-oxidized samples were generated by NO annealing; meanwhile, the states present at the as-oxidized interface diminished by 90%. The response frequency (measured by the conductance method) of the very fast states is higher than 1 MHz and increases when the energy level approaches the conduction band edge. It is 100 MHz (5 ns) at $E_c - E_T = 0.4$ eV at room temperature. The SiO_2/SiC interface annealed in NO at 1250 °C exhibited the lowest interface state density, and NO annealing at over 1250 °C is ineffective owing to increase in the very fast states.

Hydrogen annealing can also reduce the interface-state density, as demonstrated for the 4H-SiC MOS structures [586]. The density was reduced to as low as 1×10^{11} eV^{-1} cm^{-2} at $E_c - E = 0.6$ eV using hydrogen annealing at over 800 °C, where $E_c - E$ represents the energy level from the conduction band edge. Secondary ion mass spectroscopy revealed that the interface-state density decreases with increasing hydrogen concentration at the SiO_2/4H-SiC interface. The interface states were supposed to stem from the C and Si dangling bonds in that the interface-state density decreased when the hydrogen annealing temperature increased and saturated at ~ 800 °C. Effective removal of near-interface traps in SiO_2/4H-SiC(0001) structures through phosphorus incorporation was reported [587]. Low-temperature capacitance–voltage and thermal dielectric relaxation current measurements were used to investigate near-interface traps in oxides prepared by dry oxidation, NO annealing, and POCl3 annealing. The density of electrons trapped in near-interface traps in POCl3-annealed oxide is smaller than that in dry and NO-annealed oxides. The significant elimination of near-interface traps lowers the interface state density and improves the channel mobility in 4H-SiC MOSFETs.

Effects of wet atmosphere during oxidation and anneal on properties of thermally oxidized *p*-type and *n*-type 4H- and 6H-SiC MOS interfaces were investigated [588]. The wet atmosphere effectively reduced a negative flatband shift caused by deep donor-type interface states in *p*-type SiC MOS capacitors. Wet reoxidation anneal yielded negative fixed charges near the interface. In *n*-type SiC MOS capacitors, the flatband shift has a positive value when using wet atmosphere. The interface properties of *p*-type MOS capacitors and the channel mobility of MOSFET's exhibited little correlation. The threshold voltage of MOSFET's processed by wet reoxidation anneal is higher than that without reoxidation anneal. The transconductance of 4H-SiC MOSFETs is typically much lower compared to the 6H-SiC-based devices. Using a modified capacitance–voltage technique, the large area 4H-SiC MOSFETs were studied and the relationship between the SiC/SiO_2 interface trap density and energy in the bandgap was investigated [589]. The interfacial trap density increases toward the conduction band edge in both polytypes and it is much higher in 4H-SiC compared to 6H-SiC for devices fabricated in the same lot. As a result, the 4H-SiC-based devices will trap more inversion electrons, thus resulting in the lower transconductance with respect to the 6H-SiC-based devices. The electron mobility and free electron density in 4H- and 6H-SiC MOS inversion layers were measured using the Hall effect technique [590]. The 4H-SiC inversion layers exhibited very poor conductance caused by severe trapping of electrons at the Si/SiO_2 interface. The trapping and drop in mobility due to increased Coulombic scattering result in reduced conductance through decrease in the number of free electrons in the inversion layer. Various processing variables have influence on the inversion layer electron mobility of (0001)-oriented 4H-SiC n-channel MOSFETs [591]. The process variables include implant anneal temperature, oxidation procedure, postoxidation annealing in nitric oxide (NO), type of gate material, and high-temperature Ohmic contact anneal. Postoxidation anneal in NO yielded significant increase in electron mobility, whereas other processes have only minor effects on the channel mobility.

There is a question on whether the SiC/SiO_2 interface can be prepared without many interface defects. Lowering the oxidation temperature of 6H-SiC to 1050 °C can significantly reduce the effective oxide charge and interface state densities in oxides grown on *p*-type 6H-SiC [592]. An even lower temperature re-oxidation anneal after the oxidation yielded better results. This anneal dramatically improved the electrical properties of the interface and lowered the interface state density. The net oxide charge density on *p*-type 6H-SiC was lowered, but remained high (1.0×10^{12} cm^{-2}). The interface state density of 1.0×10^{11} cm^{-2} eV^{-1} approached acceptable MOS device levels. The breakdown fields of the oxides were substantially improved owing to both lower oxidation temperature and re-oxidation anneal. The low temperature oxidation followed by re-oxidation anneal for MOSFETs yielded a room temperature mobility of 72 cm^2 V^{-1} s^{-1}. The highly ordered monolayers of silicon dioxide have been prepared on the hexagonal 4H- and 6H-SiC(0001) surfaces by using hydrogen plasma or etching in hydrogen [593]. LEED and AES revealed that the bond angles and distances of the surfaces are in agreement with those of bulk SiO_2. The semiconductor surface was passivated as a result of saturation of all the dangling bonds and it preserved the perfect order also in air. These ideal oxide monolayers may act as a seed for fabricating epitaxial oxides with a small defect density. An effective method was developed to reduce defects at a SiC/SiO_2 interface [594]. To avoid the usage of oxidation of SiC which inevitably leads to defect creation, this scheme includes four successive steps: H_2 etching of SiC, Si deposition, low-temperature (750 °C) oxidation of Si to form SiO_2, and high-temperature (1600 °C) N_2 annealing to introduce nitrogen atoms. The interface state density was estimated to be on the order of 10^{10} cm^{-2} eV^{-1}, two orders of magnitude smaller than that of an interface formed by SiC oxidation. Structural and electronic properties of 4H(0001)-SiC/SiO_2 interface was studied by sequential use of classical and ab initio simulation calculations [595]. An abrupt model structure describing the transition between crystalline SiC and amorphous SiO_2 without showing any coordination defect was demonstrated. In the atomically abrupt interface model, the full oxide bandgap was recovered at a distance of ~ 5 Å from the interface. This agrees with estimates from internal photoemission measurements. The best agreement between calculated band offsets and experimental data was achieved for a hybrid functional in which a screened Coulomb potential was used in the Hartree–Fock exchange term, and the calculated band offsets underestimate the experimental values by ~ 25%, but agree with experiment within a few percent when expressed with respect to the oxide bandgap.

Confocal microscopy can be used to address single 4H-SiC/SiO_2-related defects [596]. A systematic investigation of the defect density of the SiC/SiO_2 interface was conducted via this technique by varying the parameters of a nitric oxide passivation anneal after oxidation. Standard capacitance-based characterization techniques were used to evaluate optical emission rates and densities of the optically active

SiC/SiO_2-related defects. The oxidation temperature influences formation of single photon-emitting defects located at the SiO_2/4H-SiC interface [597]. Thermal oxidation at 900–1300 °C was followed by cooling down in N_2 or O_2 ambient. Single photon emission was confirmed. For the N_2 cooled samples, the density of interface single photon emitters increased with decreasing oxidation temperature, with a controllable density within 10^5–10^8 cm^{-2}. For the O_2 cooled samples, many interface single photon emitters formed irrespective of the oxidation temperature owing to low-temperature oxidation during cooling after oxidation. The impact of oxidation and annealing on the density and optical properties of color centers at the SiO_2/4H-SiC interface was investigated [598]. Post-oxidation CO_2 annealing at 1200 °C led to decrease in the color center density by two orders of magnitude. One third of the color centers in the as-oxidized sample at 1200 °C showed second-order correlation function values exceeding 0.5 and did not exhibit single photon emission. All investigated defects behaved as single photon emitters in the sample annealed in CO_2 at 1200 °C. Hence CO_2 annealing reduced the density of color centers and yielded well-isolated single photon emitters at the SiO_2/SiC interface. The color centers might be interface defects with deep trap levels.

2.7.3 *Other Types of Surface Treatment and Properties*

Surface functionalization of silicon carbide with species other than oxygen has been pursued. Hydrides on the 6H-SiC(000$\bar{1}$) and (0001) surfaces before and after hydrogen annealing were investigated by using infrared attenuated total reflection spectroscopy [599, 600]. The absorption signals of CH_2 and CH_3 arose from the 6H-SiC(000$\bar{1}$) and (0001) surfaces before hydrogen annealing. A sharp C–H stretching vibration assigned to a monohydride appeared for the hydrogen annealed 6H-SiC(000$\bar{1}$) surface, whereas a Si–H stretching vibration signal was observed from the 6H-SiC(0001) surface. Atomically flat surfaces were fabricated by hydrogen etching of 6H- and 4H-SiC(0001) surfaces at 1600–1700 °C [601]. Etching eliminated scratches on the polished substrates. Atomic force microscopy images disclose periodic arrays of atomically flat terraces that are a few hundred nanometers wide. The terraces are separated by 1.5 nm tall steps in the <1100> directions. Hydrogenated 6H-SiC(0001) and (000$\bar{1}$) surfaces were also obtained by high-temperature hydrogen treatment [602]. The samples exhibited unreconstructed surfaces with a high crystallographic order as revealed by LEED. Sharp Si–H stretching modes were observed from the SiC(0001) surfaces. The *n*- and *p*-type samples showed no surface band bending, indicating that the surfaces were electronically passivated by charged surface states with densities below 7×10^{10} cm^{-2} for *p*-type and 1.7×10^{12} cm^{-2} for *n*-type samples. No surface oxide was detected by X-ray photoelectron spectroscopy after the samples were kept in air for two days. Atomic hydrogen can induce surface metallization on the SiC surface [603]. This was realized on the surface of the single-domain β-SiC(100) thin film prepared on a silicon

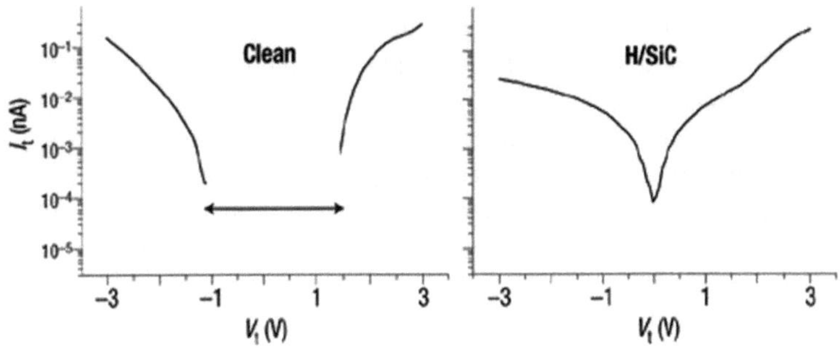

Fig. 2.122 Electronic structure of the clean and H-exposed β-SiC(100) 3 × 2 surface obtained by surface tunneling spectroscopy. The spectra show gap closing of the H-treated β-SiC(100) 3 × 2 surface. Reprinted with permission from [603]. Copyright 2003, Springer Nature Limited

substrate (Fig. 2.122). Photo-electron and photo-absorption spectroscopies and scanning tunneling techniques show that metallization takes place in the subsurface region of the 3 × 2 reconstruction. Metallization was proposed to arise from the competition between hydrogen termination of surface-dangling bonds and hydrogen-generated steric hindrance below the surface.

Ab initio molecular dynamic simulation shows that water molecules interact differently with Si and C terminated 3C-SiC(001) surfaces [604, 605]. The water coverage was selected to be 1/4 to 1 monolayer and water molecules tend to dissociate on the Si-terminated surface, leading to formation of Si–H and Si–OH bonds on the surface. In contrast, the C terminated surface which does not react with water molecules is hydrophobic. The effect of surface polarity on the water wetting behavior of 4H- and 6H-SiC crystals was explored [606]. The contact angle of deionized water on the carbon face is much bigger than that on the silicon face for both polytypes. The contact angle of tetrachloromethane is nearly the same on these two faces. Molecular dynamics simulations indicate that a bigger contact angle on the C face results from different charge agglomeration on the two faces.

Organic molecules can be attached to the silicon carbide surface and this type of passivation is often necessary for biological applications of silicon carbide. The alkyl monolayers can be covalently bound to the 3C-SiC surface via a thermal reaction with 1-alkene [607]. The SiC wafer was first treated in a diluted HF solution to remove the native SiO_2 layer and provided a reactive hydroxyl surface and finally a hydrophobic methyl-terminated surface. Attachment of ω-functionalized 1-alkene also produced a well-defined functionalized surface. The attached monolayers are covalently attached and the resulting surface is stable under both acidic and alkaline conditions. The alkyl monolayers can also form on 3C-SiC by UV irradiation in the presence of alkenes at room temperature [608]. Methyl undec-10-enoate and 2,2,2-trifluoroethyl undec-10-enoate can be bonded to the SiC surfaces and the ester terminated surfaces allow for further attachment of amine compounds. Using the density-functional simulation, molecular sensitization of the nonpolar SiC(110) surface was studied [609]. Aromatic fragments up to a full porphyrin molecule were considered and the result shows that there is a relatively low affinity between aromatic rings and SiC surface. The only exception is the pyrrole ring which produced an exothermic reaction, leading to the stable hybrid interfaces.

Organosilanes can offer covalent functionalization of *n*-type 6H-SiC [610]. Wet-chemically processed self-assembled layers of octadecyltrimethoxysilane and amino-propyldiethoxymethylsilane were investigated by using contact angle measurements, AFM, thermal desorption, and XPS. The organic layers are smooth, with wetting angles up to 100°. Desorption at 830 K proves covalent bonding of the organic molecules to the SiC surface. Functionalization of *n*-type (100) and (111) 3C-SiC surfaces with organosilanes was further realized [611]. The wet chemical technique produced self-assembled monolayers of amino-propyldiethoxymethylsilane and octadecyltrimethoxysilane. The reaction on the hydroxylated surfaces with organosilanes diminishes the surface band bending, suggesting that functionalization results in partial passivation of electrically active surface states. Two types of silane coupling agents, namely 3-aminopropyl triethoxysilane (KH550) and 3-mercaptopropyl trimethoxysilane (KH590), were used to modify the surfaces of SiC powder [612]. KH590 has a better effect than KH550 for hydrophobic modification of SiC. Compared with using KH590 alone, SiC powder modified by KH590 and hexadecyl iodiele exhibited better water resistance with increase of contact angle from 106.8 to 127.5°. The FTIR, XPS, and XRD analyses show that KH550/KH590 and hexadecyl iodiele can be covalently bonded to the surface of SiC powder, without altering its crystal configuration.

Electron spectroscopic comparison of the C-rich SiC(000$\overline{1}$) and Si-rich SiC(0001) surfaces after cleaning and disordering by Ar^+ ion sputtering and subsequent annealing was reported [613]. Three distinct temperature regions with different carbon surface segregation kinetics on SiC(000$\overline{1}$) were observed. Only one temperature region for C-segregation on SiC(0001) was observed. Two crystal surfaces exhibited no spectroscopic difference below 900 K. Both faces are terminated by a surface graphite layer between 900 and 1300 K, and the C-rich face has an additional carbon surface segregation process. The C-terminated surface graphitized at a higher rate than the Si-terminated surface at over 1300 K. Massive graphitization on both surfaces above 1300 K was ascribed to Si(*g*) sublimation from the SiC surfaces. Molecular dynamics simulations give insight into the phase stability of 3C-SiC upon heating and the threshold displacement energy surfaces for C and Si primary knock-on atoms in 3C-SiC and 6H-SiC [614]. The simulations indicate strong anisotropy in the threshold displacement energy surface for both Si and C primary knock-on atoms. The two polytypes exhibit much similarity in properties of the threshold displacement energy surface. The minimum displacement energies of 21 eV for C and 35 eV for Si agree with experimental observations. Growth of SiC on off-oriented 6H-SiC{0001} substrates was quantitatively analyzed [615], on the basis of the theory proposed by Burton, Cabrera, and Frank [616]. With measurement of the critical growth conditions where the growth mode changes from step-flow to 2D nucleation, the model yields the surface diffusion lengths of adatoms on 6H-SiC(0001)Si and (000$\overline{1}$)C faces. The (000$\overline{1}$)C faces possess a much higher nucleation rate and a much longer surface diffusion length. This results in little difference in step-flow growth condition on both faces. The analysis reveals negligible desorption of adatoms. Step bunching on 6H-SiC(0001)-vicinal face etched by HCl at 1390–1500 °C was investigated by using AFM [617]. Continuous parallel and periodic microsteps with six-bilayer height are laid perpendicular to the off direction when the substrate has the inclination toward near <01$\overline{1}$0> or <11$\overline{2}$0>, although those perpendicular to <11$\overline{2}$0> are apt to decompose into three bilayer or less. AFM and high-resolution TEM revealed interesting behaviors of vicinal 4H and 6H-SiC(0001) surfaces [618]. Each surface showed self-ordering of nanofacets, regardless of polytypes and vicinal angles, after HCl (0–0.1%)/H_2 gas etching at 1430 °C at 760 Torr for 15–30 min. Equilibrium surface phase separation induces two typical facet planes: (0001) and high index $(11\overline{2}n)$ (Fig. 2.123a). The $(11\overline{2}n)$ plane may have a minimal free energy due to attractive step–step interaction. Figure 2.123b shows a typical lattice image in 3.5° off the 6H-SiC surface region. A facet consists of several bi-monolayer steps and (0001) terraces of ~ 1 nm width. This surface configuration is promoted energetically by step–step interactions, leading to formation of the $(11\overline{2}n)$ facet. The differing ordering distances in 4H and 6H polytypes imply polytype dependence of nanofaceting. The Casimir force between a gold-coated microsphere and doped SiC samples was measured by using AFM [619]. The result demonstrates a large contact potential (0.67 V) and a relatively small density of charges trapped in SiC.

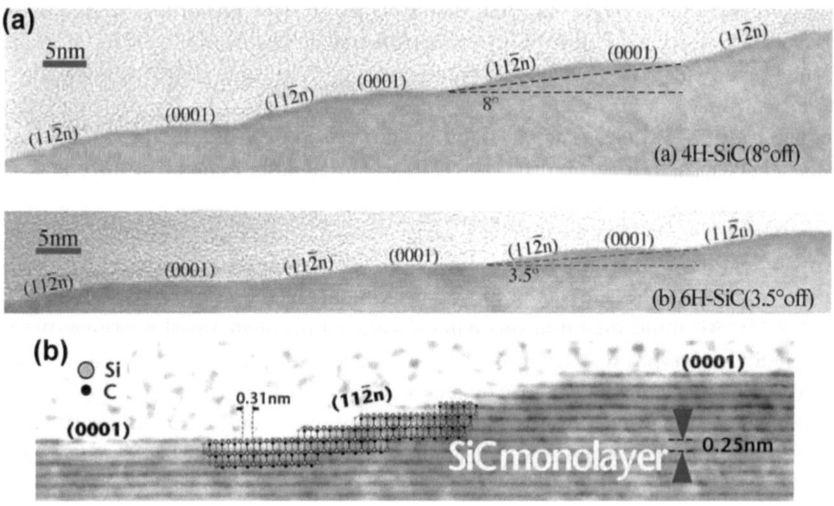

Fig. 2.123 **a** Cross-sectional HRTEM images of 8° off 4H-SiC and 3.5 off 6H-SiC surfaces. The ordering of facets in nanometer scale (nanofaceting) is apparent. The vicinal angle (3.5° or 8°) is conserved. **b** A cross-sectional HRTEM image of $(11\overline{2}n)$ facet observed in a 3.5° off 6H-SiC surface. The atomic drawing of the lattice is superimposed. Reprinted with permission from [618]. Copyright 2003, American Physical Society

2.8 Polytypic Phase Transformation

Silicon carbide has many polytypes with different stacking sequences of Si–C double atomic layers. This property renders silicon carbide a very special semiconductor, but what is so special about the material that so many crystal structures exist in nature? Much work has been done with the intention to clarify this mystery. Some insights were extracted from a detailed survey of 200 SiC crystals grown by the modified Lely technique [620]. The crystals were examined by using X-ray oscillation photography and synchrotron topography. Over half (~ 114) of the crystals are multi-polytypic. 37 crystals show complex polytype combinations, 16 of which include a long-period phase. 9 crystals show a sandwich structure, in which a longer-period polytype layer inside the crystal is surrounded each side by shorter-period polytypes. The reverse structures or pure long-period polytype layers were not observed.

The studies have revealed that one polytype of silicon carbide can transform into another polytype under appropriate conditions such as under pressure. There have been many investigations concerning the phase transformation between different polytypes of silicon carbide. In practice, SiC polytype transformation had long been observed experimentally. For example, in 1971, the phase transitions and habit changes of SiC crystals of 2H, 3C, 4H and 15R polytypes had been studied [621]. All types transformed to tabular 6H-SiC crystals at 1500–2830 °C. The 2H whiskers transformed to 6H crystals via a metastable 3C phase. The transition rates of normal crystals of 3C, 4H and 15R to 6H are much lower than the rate of the 2H → 3C transformation of whiskers. The activation energies were derived to be about 140 kcal per mole.

The factors that determine the stability and growth of SiC polytypes still remain debated. Proposed governing factors include temperature-dependent difference in bulk total energy, biaxial strain caused by point defects, and surface characteristics. Theoretical calculations have been widely utilized to explore the origins of polytypism as well as structural and electronic difference among different polytypes of silicon carbide. A model was proposed to account for the origin of polytypism in SiC [622]. Ab initio quantum mechanical calculation of the total energies of five simple SiC polytypes with norm-conserving pseudopotentials shows that the system is extremely close to a multiphase degeneracy between all the considered structures. Two mechanisms can result in split of the infinite multi-phase degeneracy and give an infinite number of regular SiC structures. The first is a result of the presence of phonons. The second depends on two aspects, an electronically mediated interaction between planes which does not need to be long ranged but depends on their separation, thus giving forces on the layers, and the longitudinal relaxation of the interlayer spacings in response to the above forces. There is a paradox regarding the growth of SiC polytypes from the vapor or the melt [623]. Since 3C-SiC is not the structure with the lowest energy as shown by many calculations, it is perplexing why the cubic structure is experimentally preferred. This was explained by calculations as due to the constrained equilibrium when adding one atomic double layer at a time to the growing crystal in the hexagonal direction without allowing rearrangement of the lower layers. The different roles of donor and acceptor impurities have been discussed. The electronic structure calculation indicates that the valence band offset between different polytypes is small and therefore, there should be large offsets between the conduction bands due to the different bandgap values. As a result, the acceptors do not favor any particular polytype and donors favor the cubic form since it has the smallest bandgap. An atomistic simulation based on an empirical potential and first-principles calculation explores the effects of various process variables on the formation of polytypes during growth of SiC single crystals [624]. The main role of process variables (temperature, surface type, growth rate, atmospheric condition, dopant type) are not to directly change the relative stability of SiC polytypes but to change the formation tendency of point detects. The biaxial local strain caused by point detects shows influence on the relative stability of SiC polytypes and was proposed as a governing factor determining the selective growth of SiC polytypes.

The structural and electronic properties of various polytypes (3C, 2H, 4H, 6H) of SiC have been determined by using ab initio pseudopotential calculations [47]. All the polytypes show similar equilibrium lattice constants and bulk moduli. The calculated total energies of the polytypes are very close and the difference is within 4.3 meV per atom. The small difference may account for the polytypism of SiC. 2H-SiC was found to be the most unstable, whereas 4H-SiC has the lowest energy due to the attractive interactions between the alternating cubic and hexagonal stacking layers. The asymmetric charge distribution for a Si–C bond was found to be on the boundary separating the zinc-blende and wurtzite phases and suggested to be related to the polytypism of silicon carbide. The ground-state properties of cubic and hexagonal SiC polytypes were investigated by using softened ab initio pseudopotentials and a total-energy minimization procedure [625]. Atomic relaxation within the hexagonal unit cells plays an important role in yielding the correct structural properties. Calculations using unrelaxed and relaxed geometries give rise to distinct orders of cohesive energies versus percentage of hexagonality in SiC structures. Calculation based on the relaxed structures yields similar results to those in Ref. [47]. The largest difference between the cohesive energies is 3 meV per atom and the differences in the cohesive energies of different SiC polytypes are smaller than the thermal energies under reasonable crystal growth conditions. Therefore, nonequilibrium effects related to the details of the growth and effects of the vibrating lattice were proposed to be the major forces driving polytypism during growth of SiC crystals. The zero-temperature total energy differences between more polytypes of SiC (3C, 6H, 4H, 2H, 15R, and 9R) were determined by well converged all electron density-functional calculations [626]. The order of total energies for different polytypes is consistent with previous calculated results and the results obtained by different groups are summarized in Fig. 2.124, and these different groups of data show qualitative agreement. The total energies of a variety of SiC polytypes were calculated using the nonorthogonal tight binding method [627]. The energies of the polytypes computed with up to 62 atoms per unit cell show that the hexagonal wurtzite structure has the highest energy and the 4H structure has the smallest energy while the cubic zinc-blende structure has intermediate energy. DFT calculations with a correction of the long-range dispersion force (van der Waals force, or vdw force) give a viewpoint into the stability of cubic

SiC [628]. The result suggests that the vdW interaction is more sensitive to the cubic site than the hexagonal site; hence its influence increases with decreasing hexagonality of the polytype, implying more stability of cubic polytype. However, another comprehensive DFT study of thermodynamic stability of 3C, 2H, 4H, and 6H polytypes of SiC gives distinct conclusions [629]. The authors employed and compared various DFT codes (VASP, CP2K, and FHI-AIMS) and diverse exchange–correlation functionals (LDA, PBE, PBEsol, PW91, HSE06, SCAN, and RTPSS) as well as nine different vdW corrections. The result shows that at $T = 0$ K, 4H-SiC is more stable than 3C-SiC, and the stability further increases with temperature by including entropic effects from lattice vibrations. Specifically, at 0 K, the 3C, 4H, and 6H polytypes have very similar energies, with 4H and 6H being more stable than 3C by − 1 meV per Si–C. The 2H polytype is less stable than 3C by 5 meV per S–iC. At the CVD processing temperature of ~ 1800 K, the phonon vibrational energy stabilizes the hexagonal polytypes 4H (6H) over 3C by about − 3.0 (− 2.8) meV per Si–C. Neither the most advanced vdW corrections nor strain on the lattice have significant influence on the relative polytype stability (Fig. 2.125). Calculations of the energies of the SiC(0001) surfaces that are commonly exposed during epitaxial growth show that the Si-terminated surfaces of 3C-SiC are significantly more stable than that of 4H-SiC. It was concluded that the difference in surface energy is likely the driving force for 3C-nucleation, whereas the difference in bulk thermodynamic stability slightly favors 4H and 6H polytypes.

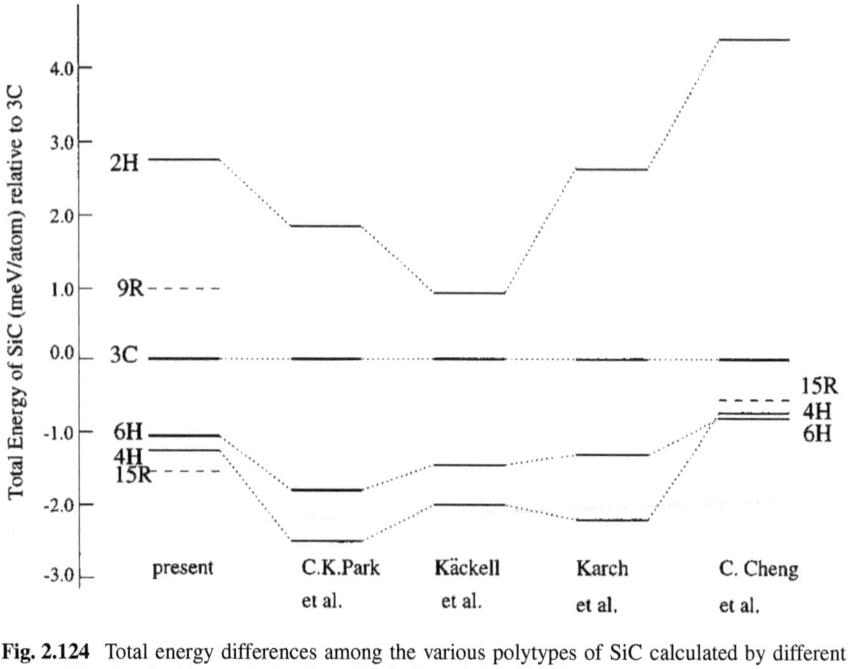

Fig. 2.124 Total energy differences among the various polytypes of SiC calculated by different groups. Reprinted with permission from [626]. Copyright 1998, American Physical Society

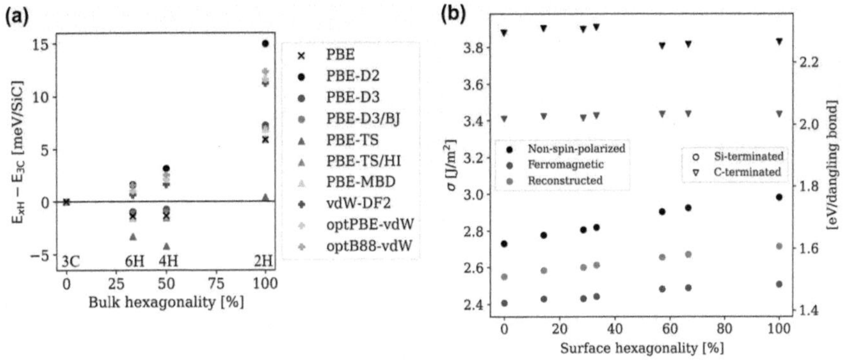

Fig. 2.125 **a** Energy differences for 3C, 6H, 4H, and 2H-SiC as a function of hexagonality, as calculated either with VASP–PBE and a vdW correction or with a vdW exchange–correlation functional. **b** Absolute energies of the Si- and C-terminated surfaces versus hexagonality for the non-spin-polarized and ferromagnetic ideal surface, and a 2 × 1 (nonmagnetic) reconstruction in the case of Si termination. All results were obtained with VASP–PBE. Reprinted with permission from [629]. Copyright 2022, American Physical Society

The DFT calculations including the vdW correction predict a temperature-dependent hierarchy of SiC polytypes in agreement with experimental results [630]. The result shows that 3C-SiC has the highest cohesive energy but the lowest entropy (Fig. 2.126). At high temperatures, the higher entropic contribution to the free energy of the hexagonal polytypes stabilizes their crystal structures. As a result, 6H- and 4H-SiC become thermodynamically more stable than 3C-SiC. An analysis reveals that the positive formation energies for both extrinsic and intrinsic stacking faults in 3C-SiC decrease substantially with increasing temperature; particularly, the extrinsic stacking faults become negative at temperatures above 1750 K. This indicates that too high deposition temperatures should be avoided in order to reduce stacking faults in 3C-SiC. The formation energies of stacking fault in 6H- and 4H-SiC only increase slightly with temperature.

Since the total energy varies only slightly among different polytypes of SiC, as indicated by the aforementioned theoretical calculations, the polytypic transformation between different polytypes takes place relatively readily under various conditions. Synthesis of SiC from silica and carbon as well as silicon and carbon under different conditions was investigated [631]. The primary synthesized SiC is always β-SiC that can form at as low as 525 °C when carbon and silicon are contained in a low melting alloy of aluminum and zinc. β-SiC is stable up to 2100 °C and starts to morph into α-SiC slowly and monotropically at 2100 °C until completion at 2400 °C. The 3C-SiC single crystals in an argon atmosphere transformed into 6H-SiC at temperatures above 2150 °C [632]. Raman scattering was performed to determine the spatial distribution and depth profile of the phase-transformed 6H-SiC region in the annealed cubic samples. The phase transformation was explained by the displacement of the atomic double layers. Cubic SiC crystals can transform partially into noncubic stacking faults at 1400 °C [633]. The 2H-SiC crystals undergo phase

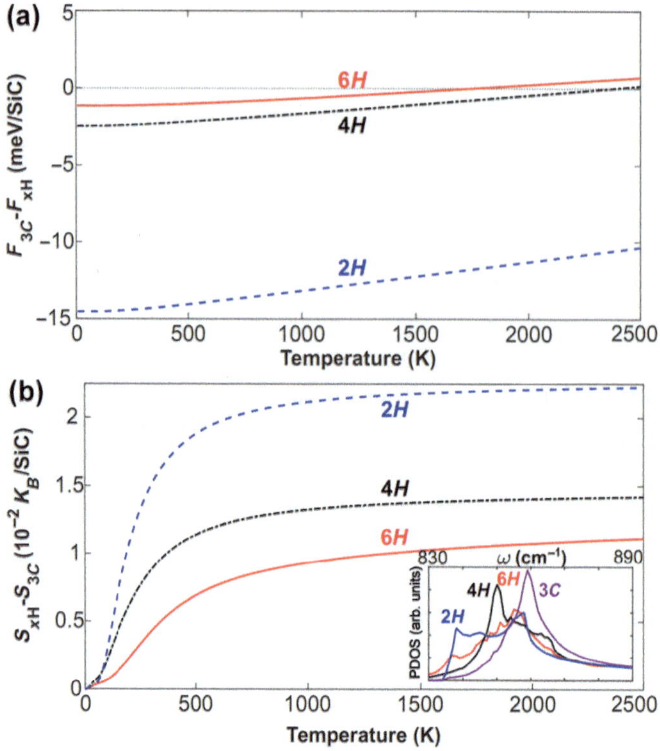

Fig. 2.126 **a** Difference between the Helmholtz free energy of 3C-SiC with respect to the values of 6H, 4H, and 2H polytypes. **b** Difference between the entropy of xH and 3C polytypes. The inset shows the phonon density of states in the region near the LO band. Reprinted with permission from [630]. Copyright 2019, American Physical Society

transformation at temperatures as low as 400 °C, some form a structure with one-dimensional disorder along the crystal *c* axis and others become the faulted cubic or 6H structure. Dislocations significantly enhance this transformation which takes place through a slip process perpendicular to the *c* axis. The β to α transformation in conventionally sintered, hot-pressed, and reaction-sintered polycrystalline SiC has been studied. Optical and transmission electron microscopies reveal that the transformation takes place via the rapid growth of composite grains comprising α-SiC plates sandwiched between recrystallized β-SiC envelopes [634]. Growth of these composite grains into the fine-grained β matrix occurs more quickly than thickening of the α plates into the β envelopes. The composite plates are formed by the extreme anisotropy of the interfacial energy between the two polytypes of SiC. That is, the $(111)_{\beta}||(0001)_{\alpha}$ interfaces have energies several orders of magnitude smaller than the random β/α interfaces [635]. The interface between the β and α regions of partially transformed grains always contains thin α lamellae adjacent and parallel to coherent β twins or at stacking faults [636]. These planar defects are nucleation sites for the transformation, and growth of α phase results from the motion of partial dislocations nucleated at the intersection of coherent and incoherent twin boundaries. However, the initial stages of β to α transformation have a different mechanism, dominated by elimination of high-energy β–α interfaces [637]. The phase transformation from β to α SiC in sintered SiC shows dependence on the initial β ratio and sintering atmosphere [638]. The transformation rate decreases with increasing initial β ratio in the starting powder and in the presence of nitrogen. No transformation occurs for the initial pure β-SiC powders with 10.34 wt% Y_2O_3 and 2.95 wt% AlN as additives. The materials without phase transformation have a homogeneous microstructure with equiaxed grains. In contrast, the SiC powders with a large initial *α/β*-ratio experience phase transformation containing elongated grains. Neutron irradiation with 9 dpa (displacements per atom) at ≈ 1440 °C can drive transformation of β-SiC to α-SiC, with radiation-induced Frank dislocation loops serving as the nucleation sites [639]. The results indicate incipient β → α phase transformation, in the form of small (5–10 nm) pockets of α-SiC forming in the β matrix.

The α to β transformation in silicon carbide has been observed, although the latter may have a little higher cohesive energy. On account of Si enrichment on the 6H-SiC(0001) surfaces during formation of the $\sqrt{3} \times \sqrt{3}R30°$ reconstructed phase, a cubic stacking sequence appears at the surface [558]. LEED crystallography shows that the ultimate reconstructed surface consists of Si adatoms in T_4 sites and STM discloses mesa-like structures with various atomic periodicities before formation of the $\sqrt{3} \times \sqrt{3}R30°$ phase. This Si enriched state provides atoms for the formation of the modified stacking sequence. Dry thermal oxidation of a highly doped (1.7 × 10^{19} cm^{-3}) *n*-type 4H-SiC epilayer can induce formation of single and multiple stacking faults which sometimes give rise to 3C-SiC bands [640]. The single stacking faults and bands of 3C-SiC in a 4H-SiC matrix in the oxidized 4H-SiC epilayer were observed by using transmission electron microscopy (Fig. 2.127). These bands are parallel to the 4H-SiC(0001) basal plane, but the unoxidized control sample did not show the bands. The cathodoluminescence spectrum (at 300 K) obtained from the oxidized materials comprises a new peak at 2.5 eV in addition to the 3.22 eV one belonging to 4H-SiC. This 2.5 eV peak is absent from the sample without oxidation, indicating its origin of stacking faults. The motion of Shockley partial dislocations on the parallel (0001) slip planes was supposed to be responsible for the polytypic transformation. The appearance and displacement of the partials were suggested to be caused by stresses induced by the heavy doping in the epilayer or nucleation from defects. However, a study of the 4H-SiC *p–i–n* diodes by emission microscopy and TEM indicates that the properties of the partial dislocations and stacking fault expansion are inconsistent with stress as the driving force [330]. The thermodynamic free energy difference between the perfect and faulted structures was suggested to be the driving force in the diodes, implying that the hexagonal polytypes of SiC are metastable at room temperature and would transform to a faulted structure when the activation energy of the partial dislocation slip is achieved. The electron–hole recombination during operation of the diode may offer this activation energy. Calculations predict that 4H is the most stable polytype at least among 3C, 2H, 4H, 6H,

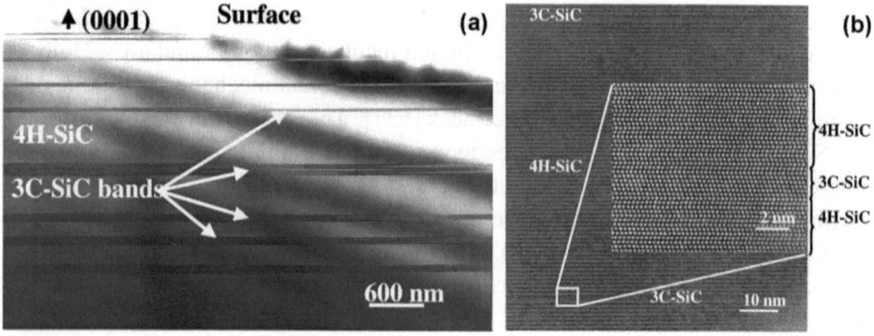

Fig. 2.127 **a** Cross-sectional TEM image of the oxidized 4H-SiC epilayer. The 3C-SiC bands are parallel to the 8°-tilted basal plane. **b** HRTEM image of 4H-SiC and 3C-SiC within a band showing two 3C-SiC subbands spaced by 4H-SiC. Higher magnification image in the inset indicates that this 3C-SiC subband is composed of seven Si–C bilayers. Reprinted with permission from [640]. Copyright 2001, American Institute of Physics

and 15R polytypes, but the energy differences among them are so small that 4H may also convert to other polytypes under certain conditions.

Annealing at 2400 °C in argon atmosphere can result in phase transformation of 2H-SiC crystals into 6H-SiC crystals [641]. The first step in the process of the 2H to 6H as well as the 3C to 6H transformation in SiC is the insertion of a large number of random stacking faults. Both transformations are time-dependent and neither leads to a change of the crystal shape. This indicates that the transformation proceeds through the nucleation and propagation of stacking faults, causing suitable layer displacements to cause the change in structure. The resulting structure is never fully ordered, suggesting that the forces causing the ordering are weak.

At high pressures, silicon carbide crystals may morph into the rocksalt-type structure. The high pressure properties of SiC were calculated by using the ab initio pseudopotential method [642]. Figure 2.128 displays the calculated total energies of the zinc-blende, rocksalt, and β-Sn structures of silicon carbide. The zinc-blende structure is stable up to 660 kbar, above which it transforms into the rocksalt phase. The rocksalt structure of SiC is more stable than the β-Sn structure. Other possible phases like the B-8 structure were not calculated and therefore, the calculated threshold transition pressure represents only an upper limit for the transition pressure from zinc-blende to another structure. The calculation also indicates that the indirect bandgap of cubic SiC decreases with increasing pressure. This tendency is similar to that of silicon but not diamond. X-ray diffraction was employed to study the 3C and 6H polytypes of SiC at pressures up to 105 and 95 GPa, respectively [643]. 3C-SiC underwent phase transition into the rocksalt phase at above 100 GPa together with a volume reduction of 20.3%, whereas 6H-SiC remained stable up to the highest applied pressure but exhibited a premonition of phase transition above 90 GPa. The ab initio density-functional theory is also applicable to the determination of the ultra-high pressure dependence of materials. The pressure dependent properties of cubic and hexagonal polytypes of SiC were calculated based on the density-functional theory using the plane-wave pseudopotential approach [644]. NaCl, NiAs, CsCl, and β-Sn structures of SiC were considered as potential high-pressure phases and the calculation predicted a pressure induced phase transition from zinc blende to rock-salt at a critical pressure of 0.67 Mbar. However, the calculated transition pressure is lower than the experimental value. The high pressure transition to the NaCl structure was predicted for 3C-, 2H-, and 4H-SiC.

Isothermal-isobaric molecular dynamics simulation was performed to investigate the pressure induced structural transformation in SiC [645]. The simulation reproduced the reversible transformation between the fourfold coordinated zinc-blende structure and six-fold coordinated rocksalt structure. The molecular dynamics cell of the system changes from cubic to monoclinic after the onset of phase transformation from zinc-blende to rocksalt (Fig. 2.129). During the cubic-to-monoclinic conversion, the silicon and carbon sublattices shift relatively to each other along the (100) direction in the zinc-blende structure (Fig. 2.130). This direction is the (110) direction in the resultant rocksalt structure. The calculated volume changes at the transition and hysteresis agree well with the experimental values. The pressure-induced zinc blende to rocksalt phase transition of SiC was investigated by using periodic linear

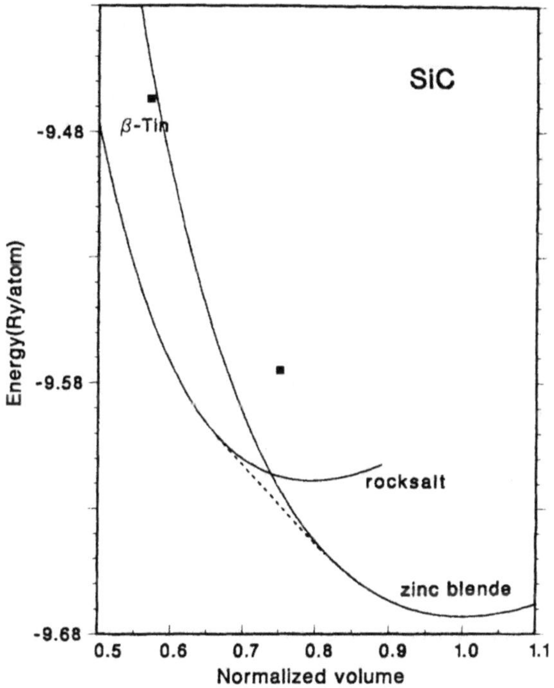

Fig. 2.128 Calculated crystal total energy as a function of volume normalized to the calculated equilibrium volume of 1.0369 nm^3 per atom of the zinc-blende phase of SiC. The dashed line denotes the common tangent of the two curves and solid squares are the energies of the β-Sn phase. Reprinted with permission from [642]. Copyright 1987, American Physical Society

combination of atomic orbitals-DFT least-enthalpy calculations [646]. A transformation pathway based on a *Pmm*2 orthorhombic intermediate state with two SiC units per cell is energetically favored over the *R*3*m* mechanism. The activation enthalpy is 0.75 eV per SiC unit at the predicted transition pressure of 92 GPa. A unified mechanism of high pressure transformation of the various tetrahedrally bonded polytypes of SiC to the rocksalt structure was proposed [647]. It involves several strains and intersublattice motions. Calculation of the energy barriers associated with these transition paths using first-principles pseudopotential method indicates that the 3C transition has a lower enthalpy barrier than the other polytypes at pressures above the equilibrium transition pressure. An ab initio simulation reproduced the reversible phase transition between the zinc blende structure and rocksalt structure [648]. The transition was supposed to be based on a tetragonal and an orthorhombic intermediate states with space groups of $I\bar{4}m2$ and *Imm*2. Using the proposed effective interatomic interaction potential for SiC, molecular dynamics simulations give rise to the structural energies in several SiC polytypes [649]. The cubic structure has the lowest energy, followed by the wurtzite and rocksalt structures. The pressure for the 3C-to-rocksalt transformation from the common tangent was derived to be 90 GPa. A DFT–GGA calculation shows that the pressures of transition of SiC from zinc blende structure to rocksalt structure are 74.6 GPa from total energy–volume data and 75.4 GPa from the enthalpy calculations [650]. The zinc-blende to rocksalt transition of SiC can occur at a lower pressure of ~ 40 GPa when laser-heated to moderately high temperatures [651]. There is a large volume decrease across the transition of ~ 17%. The time-resolved pump–probe XRD can monitor the structures of SiC single crystals and polycrystals laser-shocked to as high as 206 GPa [652]. Experiments revealed transformation from a low-pressure tetrahedral phase to the high-pressure rocksalt structure. The low- and high-pressure phases coexisted in a mixed-phase region and completely transformed to the rocksalt phase above 200 GPa. Time-resolved measurements during shock loading and release revealed a large hysteresis upon unloading, with the rocksalt phase retained to as low as 5 GPa. The sample finally reverted to a mixture of polytypes of the low-pressure phase.

By means of variable temperature scanning tunneling microscopy and angle-resolved photoemission spectroscopy, the transition between two β-SiC(100) surface structures has been observed [653]. There is a reversible temperature-dependent phase transition from a semiconducting $c(4 \times 2)$ surface at 25 °C to a metallic (2×1) structure at 400 °C (Fig. 2.131). This transition stems from the temperature induced disruption of the $c(4 \times 2)$ structure composed of alternately up and down dimers into the (2×1) structure with all dimers of the same height. This (2×1) structure gives rise to electronic orbital overlap between the top surface Si atoms, leading to surface metallization. Xenon chloride excimer laser-induced melt-mediated phase separation and surface reconstruction of 4H-SiC crystals were observed by using high-resolution TEM and a time-resolved reflectance method [654]. A single-pulse laser irradiation induced melting of the SiC surface, leading to phase separation into a disordered carbon layer with partially graphitic domains (~ 2.5 nm) and polycrystalline silicon

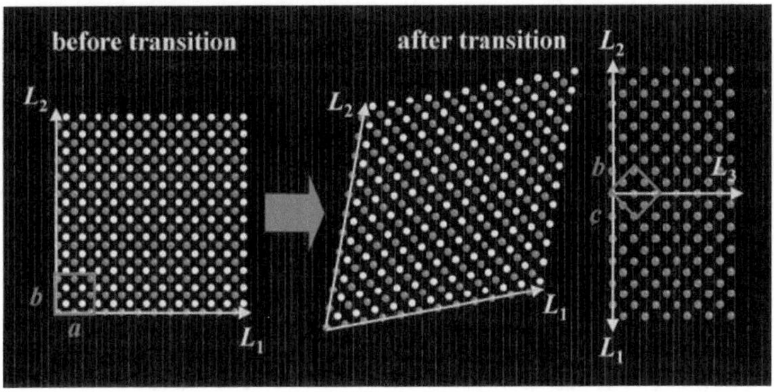

Fig. 2.129 Average atom positions in the molecular dynamics cell prior to and after the zinc-blende to rocksalt transition in silicon carbide. The gray and white balls separately mark the positions of the silicon and carbon atoms. The red box shows the unit cell of each structure. Reprinted with permission from [645]. Copyright 2000, American Physical Society

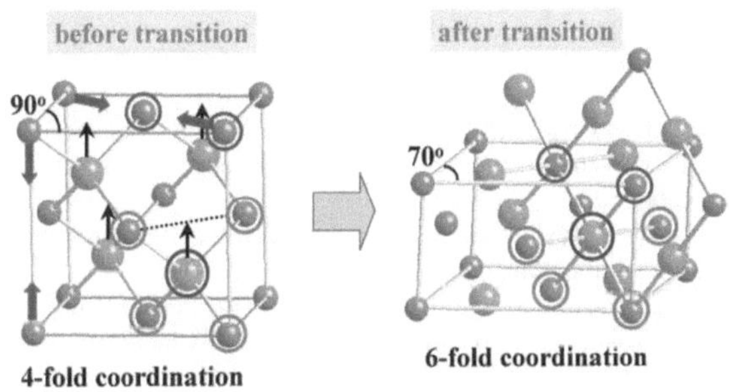

Fig. 2.130 Schematic of mechanism of the SiC phase transformation from the zinc-blende structure to rocksalt structure. The big and small balls show separately the positions of silicon and carbon atoms. The arrows mark the directions along which the atoms move during the phase transformation. Reprinted with permission from [645]. Copyright 2000, American Physical Society

(~ 5 nm). Additional pulse irradiation caused sublimation of the separated silicon element and transformation of the disordered carbon layer into multilayer graphene.

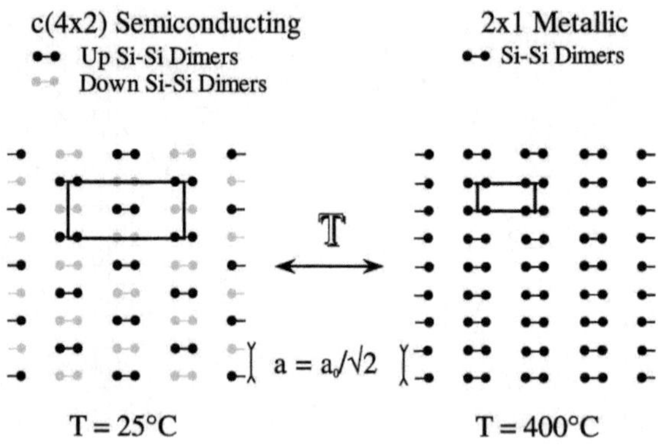

Fig. 2.131 Schematic of the reversible 3C-SiC(100) surface phase transition from a semiconducting $c(4 \times 2)$ reconstruction with alternatively up and down dimers to a (2×1) metallic reconstruction with all dimers of the same height. The surface unit cells are marked. Reprinted with permission from [653]. Copyright 1997, American Physical Society

References

1. Harris GL (ed) (1995) Properties of silicon carbide. INSPEC, London
2. Saddow SE, Agarwal A (eds) (2004) Advances in silicon carbide processing and applications. Artech House, Norwood
3. Shur M, Rumyantsev S, Levinshtein M (eds) (2006/2007) SiC materials and devices (volumes 1 and 2). World Scientific, Singapore
4. Hartman JS, Richardson MF, Sherriff BL, Winsborrow BG (1987) Magic angle spinning NMR studies of silicon carbide: polytypes, impurities, and highly inefficient spin-lattice relaxation. J Am Chem Soc 109:6059–6067
5. Nicolussi GK, Davis AM, Pellin MJ, Lewis RS, Clayton RN, Amari S (1997) s-Process zirconium in presolar silicon carbide grains. Science 277:1281–1283
6. Huss GR, Lewis RS (1995) Presolar diamond, SiC, and graphite in primitive chondrites: abundances as a function of meteorite class and petrologic type. Geochim Cosmochim Acta 59:115–160
7. Huss GR, Meshik AP, Smith JB, Hohenberg CM (2003) Presolar diamond, silicon carbide, and graphite in carbonaceous chondrites: implications for thermal processing in the solar nebula. Geochim Cosmochim Acta 67:4823–4848
8. Bernatowicz T, Fraundorf G, Ming T, Anders E, Wopenka B, Zinner E, Fraundorf P (1987) Evidence for interstellar SiC in the Murray carbonaceous meteorite. Nature 330:728–730
9. Nicolussi GK, Pellin MJ, Lewis RS, Davis AM, Clayton RN, Amari S (1998) Strontium isotopic composition in individual circumstellar silicon carbide grains: a record of *s*-process nucleosynthesis. Phys Rev Lett 81:3583–3586
10. Daulton TL, Bernatowicz TJ, Lewis RS, Messenger S, Stadermann FJ, Amari S (2003) Polytype distribution in circumstellar silicon carbide. Science 296:1852–1855
11. Daulton TL, Bernatowicz TJ, Lewis RS, Messenger S, Stadermann FJ, Amari S (2003) Polytype distribution of circumstellar silicon carbide: microstructural characterization by transmission electron microscopy. Geochim Cosmochim Acta 67:4743–4767
12. Lewis RS, Amari S, Anders E (1990) Meteoritic silicon carbide: pristine material from carbon stars. Nature 348:293–298
13. Gallino R, Busso M, Picchio G, Raiteri CM (1990) On the astrophysical interpretation of isotope anomalies in meteoritic SiC grains. Nature 348:298–302

14. Lodderasn K, Fegle B (1995) The origin of circumstellar silicon carbide grains found in meteorites. Meteontics 30:661478
15. Gallino R, Raiteri CM, Busso M (1993) Carbon stars and isotopic Ba anomalies in meteoritic SiC grains. Astrophys J 410:400–411
16. Lewis RS, Amari S, Anders E (1994) Interstellar grains in meteorites: II. SiC and its noble gases. Geochim Cosmochim Acta 58:471–494
17. Hoppe P, Amari S, Zinner E, Ireland T, Lewis RS (1994) Carbon, nitrogen, magnesium, silicon, and titanium isotopic compositions of single interstellar silicon carbide grains from the Murchison carbonaceous chondrite. Astrophys J 430:870–890
18. Amari S, Hoppe P, Zinner E, Lewis RS (1995) Trace-element concentrations in single circumstellar silicon carbide grains from the Murchison meteorite. Meteoritics 30:679–693
19. Hoppe P, Annen P, Strebel R, Eberhardt P, Gallino R, Lugaro M, Amari S, Lewis RS (1997) Meteoritic silicon carbide grains with unusual Si-isotopic compositions: evidence for an origin in low-mass, low-metallicity asymptotic giant branch stars. Astrophys J 487:L101–L104
20. Nicolussi GK, Pellin MJ, Lewis RS, Davis AM, Amari S, Clayton RN (1998) Molybdenum isotopic composition of individual presolar silicon carbide grains from the Murchison meteorite. Geochim Cosmochim Acta 62:1093–1104
21. Amari S, Nittler LR, Zinner E, Lodders K, Lewis RS (2001) Presolar SiC grains of type A and B: their isotopic compositions and stellar origins. Astrophys J 559:463–483
22. Amari S, Nittler LR, Zinner E, Gallino R, Lugaro M, Lewis RS (2001) Presolar SiC grains of type Y: origin from low-metallicity asymptotic giant branch stars. Astrophys J 546:248–266
23. Lugaro M, Davis AM, Gallino R, Pellin MJ, Straniero O, Käppeler F (2003) Isotopic compositions of strontium, zirconium, molybdenum, and barium in single presolar SiC grains and asymptotic giant branch stars. Astrophys J 593:486–508
24. Speck AK, Barlow MJ, Skinner CJ (1997) The nature of the silicon carbide in carbon star outflows. Mon Not R Astron Soc 288:431–456
25. Clément D, Mutschke H, Klein R, Henning T (2003) New laboratory spectra of isolated β-SiC nanoparticles: comparison with spectra taken by the *Infrared Space Observatory*. Astrophys J 594:642–650
26. Savina MR, Davis AM, Tripa CE, Pellin MJ, Gallino R, Lewis RS, Amari S (2004) Extinct technetium in silicon carbide stardust grains: implications for stellar nucleosynthesis. Science 303:649–652
27. Stephan T, Trappitsch R, Hoppe P, Davis AM, Bose M, Boujibar A, Gyngard F, Hynes KM, Liu N, Nittler LR, Ogliore RC (2024) The presolar grain database. I. Silicon carbide. Astrophys J Suppl S 270:27
28. Amari S, Hoppe P, Zinner E, Lewis RS (1992) Interstellar SiC with unusual isotopic compositions: grains from a supernova. Astrophys J 394:L43–L46
29. Nittler LR, Amari S, Zinner E, Woosley SE, Lewis RS (1996) Extinct ^{44}Ti in presolar graphite and SiC: proof of a supernova origin. Astrophys J 462:L31–L34
30. Hoppe P, Strebel R, Eberhardt P, Amari S, Lewis RS (1996) Small SiC grains and a nitride grain of circumstellar origin from the Murchison meteorite: implications for stellar evolution and nucleosynthesis. Geochim Cosmochim Acta 60:883–907
31. Nittler LR, Hoppe P (2005) Are presolar silicon carbide grains from novae actually from supernovae. Astrophys J 631:L89–L92
32. Hoppe P, Strebel R, Eberhardt P, Amari S, Lewis RS (2000) Isotopic properties of silicon carbide X grains from the Murchison meteorite in the size range 0.5–1.5 µm. Meteorit Planet Sci 35:1157–1176
33. Clayton DD, Arnett D, Kane J, Meyer BS (1997) Type X silicon carbide presolar grains: type Ia supernova condensates. Astrophys J 486:824–834
34. Heck PR, Gyngard F, Ott U, Meier MMM, Ávila JN, Amari S, Zinner EK, Lewis RS, Baur H, Wieler R (2009) Interstellar residence times of presolar SiC dust grains from the Murchison carbonaceous meteorite. Astrophys J 698:1155–1164
35. Zinner E, Ming T, Anders E (1989) Interstellar SiC in the Murchison and Murray meteorites: isotopic composition of Ne, Xe, Si, C, and N. Geochim Cosmochim Acta 53:3273–3290

36. Hecka PR, Greera J, Kööpa L, Trappitschd R, Gyngarde F, Busemanng H, Madeng C, Ávilah JN, Davisa AM, Wielerg R (2020) Lifetimes of interstellar dust from cosmic ray exposure ages of presolar silicon carbide. Proc Natl Acad Sci USA 117:1884–1889
37. Trumbull RB, Yang J-S, Robinson PT, Di Pierro S, Vennemann T, Wiedenbeck M (2009) The carbon isotope composition of natural SiC (moissanite) from the earth's mantle: new discoveries from ophiolites. Lithos 113:612–620
38. Junginger HG, van Haeringen W (1970) Energy band structures of four polytypes of silicon carbide calculated with the empirical pseudopotential method. Phys Stat Sol 37:709–719
39. Hemstreet LA Jr, Pong CY (1972) Electronic band structure and optical properties of 3C-SiC, BP, and BN. Phys Rev B 6:1464–1480
40. Choyke WJ, Hamilton DR, Patrick L (1964) Optical properties of cubic SiC: luminescence of nitrogen-exciton complexes, and interband absorption. Phys Rev 133:A1163–A1166
41. Backes WH, Bobbert PA, van Haeringen W (1994) Energy-band structure of SiC polytypes by interface matching of electronic wave functions. Phys Rev B 49:7564–7568
42. Lubinsky AR, Ellis DE, Painter GS (1975) Electronic structure and optical properties of 3C-SiC. Phys Rev B 11:1537–1546
43. Rohlfing M, Krüger P, Pollmann J (1993) Quasiparticle band-structure calculations for C, Si, Ge, GaAs, and SiC using Gaussian-orbital basis sets. Phys Rev B 48:17791–17805
44. Käckell P, Wenzien B, Bechstedt F (1994) Electronic properties of cubic and hexagonal SiC polytypes from *ab initio* calculations. Phys Rev B 50:10761–10768
45. Persson C, Lindefelt U (1996) Detailed band structure for 3C-, 2H-, 4H-, 6H-SiC, and Si around the fundamental band gap. Phys Rev B 54:10257–10260
46. Persson C, Lindefelt U (1997) Relativistic band structure calculation of cubic and hexagonal SiC polytypes. J Appl Phys 82:5496–5508
47. Park CH, Cheong B-H, Lee K-H, Chang KJ (1994) Structural and electronic properties of cubic, 2H, 4H, and 6H SiC. Phys Rev B 49:4485–4493
48. Wenzien B, Käckell P, Bechstedt F, Cappellini G (1995) Quasiparticle band structure of silicon carbide polytypes. Phys Rev B 52:10897–10905
49. Zhao GL, Bagayoko D (2000) Electronic structure and charge transfer in 3C- and 4H-SiC. New J Phys 2:16.1–16.12
50. Ching WY, Xu Y-N, Rulis P, Ouyang L (2006) The electronic structure and spectroscopic properties of 3C, 2H, 4H, 6H, 15R and 21R polymorphs of SiC. Mater Sci Eng A Struct 422:147–156
51. Lambrecht WRL, Limpijumnong S, Rashkeev SN, Segall B (1997) Electronic band structure of SiC polytypes: a discussion of theory and experiment. Phys Stat Sol (b) 202:5–33
52. Qteish A, Heine V, Needs RJ (1992) Polarization, band lineups, and stability of SEC polytypes. Phys Rev B 45:6534–6542
53. Lüning J, Eisebitt S, Rubensson J-E, Ellmers C, Eberhardt W (1999) Electronic structure of silicon carbide polytypes studied by soft X-ray spectroscopy. Phys Rev B 59:10573–10582
54. Son NT, Kordina O, Konstantinov AO, Chen WM, Sörman E, Monemar B, Janzén E (1994) Electron effective masses and mobilities in high-purity 6H–SiC chemical vapor deposition layers. Appl Phys Lett 65:3209–3211
55. Son NT, Chen WM, Kordina O, Konstantinov AO, Monemar B, Janzén E, Hofman DM, Volm D, Drechsler M, Meyer BK (1995) Electron effective masses in 4H SiC. Appl Phys Lett 66:1074–1076
56. Volm D, Meyer BK, Hofmann DM, Chen WM, Son NT, Persson C, Lindefelt U, Kordina O, Sörman E, Konstantinov AO, Monemar B, Janzén E (1996) Determination of the electron effective-mass tensor in 4H SiC. Phys Rev B 53:15409–15412
57. Son NT, Hai PN, Chen WM, Hallin C, Monemar B, Janzén E (2000) Hole effective masses in 4H SiC. Phys Rev B 53:R10544–R10546
58. Dalibor T, Pensl G, Matsunami H, Kimoto T, Choyke WJ, Schöner A, Nordell N (1997) Deep defect centers in silicon carbide monitored with deep level transient spectroscopy. Phys Stat Sol (a) 162:199–225

59. Balona LAS, Loubser JHN (1970) ESR in irradiated silicon carbide. J Phys C Solid State Phys 3:2344–2351
60. Aboelfotoh MO, Doyle JP (2000) Defect energy levels in electron-irradiated and deuterium-implanted 6H silicon carbide. Phys Rev B 59:10823–10829
61. Pensl G, Choyke WJ (1993) Electrical and optical characterization of SiC. Phys B 185:264–283
62. Isoya J, Umeda T, Mizuochi N, Son NT, Janzén E, Ohshima T (2008) EPR identification of intrinsic defects in SiC. Phys Stat Sol (b) 245:1298–1314
63. Carrington A, McLachlan AD (1967) Introduction to magnetic resonance with applications to chemistry and chemical physics. Harper & Row, Publishers
64. Shafizadeh D, Davidsson J, Ohshima T, Abrikosov IA, Son NT, Ivanov IG (2024) Selection rules in the excitation of the divacancy and the nitrogen-vacancy pair in 4H- and 6H-SiC. Phys Rev B 109:235203
65. Hashemi A, Linderälv C, Krasheninnikov AV, Ala-Nissila T, Erhart P, Komsa H-P (2021) Photoluminescence line shapes for color centers in silicon carbide from density functional theory calculations. Phys Rev B 103:125203
66. Brauer G, Anwand W, Nicht E-M, Kuriplach J, Šob M, Wagner N, Coleman PG, Puska MJ, Korhonen T (1996) Evaluation of some basic positron-related characteristics of SiC. Phys Rev B 54:2512–2517
67. Ling CC, Beling CD, Fung S (2000) Isochronal annealing studies of n-type 6H-SiC with positron lifetime spectroscopy. Phys Rev B 62:8016–8022
68. Itoh H, Hayakawa N, Nashiyama I, Sakuma E (1989) Electron spin resonance in electron-irradiated 3C-SiC. J Appl Phys 66:4529–4531
69. Itoh H, Yoshikawa M, Nashiyama I, Okumura H, Misawa S, Yoshida S (1995) Photoluminescence of radiation induced defects in 3C-SiC epitaxially grown on Si. J Appl Phys 77:837–842
70. Kawasuso A, Itoh H, Morishita N, Yoshikawa M, Ohshima T, Nashiyama I, Okada S, Okumura H, Yoshida S (1998) Silicon vacancies in 3C-SiC observed by positron lifetime and electron spin resonance. Appl Phys A 67:209–212
71. Hemmingsson C, Son NT, Kordina O, Bergman JP, Janzén E, Lindström JL, Savage S, Nordell N (1997) Deep level defects in electron-irradiated 4H SiC epitaxial layers. J Appl Phys 81:6155–6159
72. Wimbauer T, Meyer BK, Hofstaetter A, Scharmann A, Overhof H (1997) Negatively charged Si vacancy in 4H SiC: a comparison between theory and experiment. Phys Rev B 56:7384–7388
73. Torpo L, Nieminen RM, Laasonen KE, Pöykkö S (1999) Silicon vacancy in SiC: a high-spin state defect. Appl Phys Lett 74:221–223
74. Sörman E, Son NT, Chen WM, Kordina O, Hallin C, Janzén E (2000) Silicon vacancy related defect in 4H and 6H SiC. Phys Rev B 61:2613–2620
75. Wagner M, Magnusson B, Chen WM, Janzén E, Sörman E, Hallin C, Lindström JL (2000) Electronic structure of the neutral silicon vacancy in 4H and 6H SiC. Phys Rev B 62:16555–16560
76. Mizuochi N, Yamasaki S, Takizawa H, Morishita N, Ohshima T, Itoh H, Isoya J (2002) Continuous-wave and pulsed EPR study of the negatively charged silicon vacancy with $S = 3/2$ and C_{3v} symmetry in n-type 4H-SiC. Phys Rev B 66:235202
77. Mizuochi N, Yamasaki S, Takizawa H, Morishita N, Ohshima T, Itoh H, Isoya J (2003) EPR studies of the isolated negatively charged silicon vacancies in n-type 4H- and 6H-SiC: identification of C_{3v} symmetry and silicon sites. Phys Rev B 68:165206
78. Janzén E, Gali A, Carlsson P, Gällström A, Magnusson B, Son NT (2009) The silicon vacancy in SiC. Phys B 404:4354–4358
79. Yang L-P, Burk C, Widmann M, Lee S-Y, Wrachtrup J, Zhao N (2014) Electron spin decoherence in silicon carbide nuclear spin bath. Phys Rev B 90:241203(R)
80. Soykal ÖO, Dev P, Economou SE (2016) Silicon vacancy center in 4H-SiC: electronic structure and spin-photon interfaces. Phys Rev B 93:081207(R)

81. Ivády V, Davidsson J, Son NT, Ohshima T, Abrikosov IA, Gali A (2017) Identification of Si-vacancy related room-temperature qubits in 4H silicon carbide. Phys Rev B 96:161114(R)
82. Nagy R, Widmann M, Niethammer M, Dasari DBR, Gerhardt I, Soykal ÖO, Radulaski M, Ohshima T, Vučković J, Son NT, Ivanov IG, Economou SE, Bonato C, Lee S-Y, Wrachtrup J (2018) Quantum properties of dichroic silicon vacancies in silicon carbide. Phys Rev Appl 9:034022
83. Nagy R, Niethammer M, Widmann M, Chen Y-C, Udvarhelyi P, Bonato C, Ul Hassan J, Karhu R, Ivanov IG, Son NT, Maze JR, Ohshima T, Soykal ÖO, Gali Á, Lee S-Y, Kaiser F, Wrachtrup J (2019) High-fidelity spin and optical control of single silicon-vacancy centres in silicon carbide. Nat Commun 10:1954
84. Fan J, Chu PK (2010) Group IV nanoparticles: synthesis, properties, and biological applications. Small 6:2080–2098
85. Kurtsiefer C, Mayer S, Zarda P, Weinfurter H (2000) Stable solid-state source of single photons. Phys Rev Lett 85:290–293
86. Mizuochi N, Makino T, Kato H, Takeuchi D, Ogura M, Okushi H, Nothaft M, Neumann P, Gali A, Jelezko F, Wrachtrup J, Yamasaki S (2012) Electrically driven single-photon source at room temperature in diamond. Nat Photonics 6:299–303
87. Baranov PG, Bundakova AP, Soltamova AA, Orlinskii SB, Borovykh IV, Zondervan R, Verberk R, Schmidt J (2011) Silicon vacancy in SiC as a promising quantum system for single-defect and single-photon spectroscopy. Phys Rev B 83:125203
88. Hain TC, Fuchs F, Soltamov VA, Baranov PG, Astakhov GV, Hertel T, Dyakonov V (2014) Excitation and recombination dynamics of vacancy-related spin centers in silicon carbide. J Appl Phys 115:133508
89. Soltamov VA, Soltamova AA, Baranov PG, Proskuryakov II (2012) Room temperature coherent spin alignment of silicon vacancies in 4H- and 6H-SiC. Phys Rev Lett 108:226402
90. Riedel D, Fuchs F, Kraus H, Väth S, Sperlich A, Dyakonov V, Soltamova AA, Baranov PG, Ilyin VA, Astakhov GV (2012) Resonant addressing and manipulation of silicon vacancy qubits in silicon carbide. Phys Rev Lett 109:226402
91. Widmann M, Lee S-Y, Rendler T, Son NT, Fedder H, Paik S, Yang L-P, Zhao N, Yang S, Booker I, Denisenko A, Jamali M, Momenzadeh SA, Gerhardt I, Ohshima T, Gali A, Janzén E, Wrachtrup J (2015) Coherent control of single spins in silicon carbide at room temperature. Nat Mater 14:164–168
92. Udvarhelyi P, Thiering G, Morioka N, Babin C, Kaiser F, Lukin D, Ohshima T, Ul-Hassan J, Son NT, Vučković J, Wrachtrup J, Gali A (2020) Vibronic states and their effect on the temperature and strain dependence of silicon-vacancy qubits in 4H-SiC. Phys Rev Appl 13:054017
93. Hernández-Mínguez A, Poshakinskiy AV, Hollenbach M, Santos PV, Astakhov GV (2020) Anisotropic spin-acoustic resonance in silicon carbide at room temperature. Phys Rev Lett 125:107702
94. Vasselon T, Hernández-Mínguez A, Hollenbach M, Astakhov GV, Santos PV (2023) Acoustically induced spin resonances of silicon-vacancy centers in 4H-SiC. Phys Rev Appl 20:034017
95. Dietz JR, Jiang B, Day AM, Bhave SA, Hu EL (2023) Spin-acoustic control of silicon vacancies in 4H silicon carbide. Nat Electron 6:739–745
96. Koga M, Matsumoto M (2024) Theory of spin-acoustic resonance for spin-3/2 Si vacancy with C_{3v} site symmetry in silicon carbide. J Phys Soc Jpn 93:014703
97. Fuchs F, Stender B, Trupke M, Simin D, Pflaum J, Dyakonov V, Astakhov GV (2015) Engineering near-infrared single-photon emitters with optically active spins in ultrapure silicon carbide. Nat Commun 6:7578
98. Kraus H, Simin D, Kasper C, Suda Y, Kawabata S, Kada W, Honda T, Hijikata Y, Ohshima T, Dyakonov V, Astakhov GV (2017) Three-dimensional proton beam writing of optically active coherent vacancy spins in silicon carbide. Nano Lett 17:2865–2870
99. Carter SG, Soykal ÖO, Dev P, Economou SE, Glaser ER (2015) Spin coherence and echo modulation of the silicon vacancy in 4H-SiC at room temperature. Phys Rev B 92:161202(R)

100. Simin D, Kraus H, Sperlich A, Ohshima T, Astakhov GV, Dyakonov V (2017) Locking of electron spin coherence above 20 ms in natural silicon carbide. Phys Rev B 95:161201(R)
101. Sun T, Xu Z, Wu J, Fan Y, Ren F, Song Y, Yang L, Tan P (2023) Divacancy and silicon vacancy color centers in 4H-SiC fabricated by hydrogen and dual ions implantation and annealing. Ceram Int 49:7452–7465
102. Radulaski M, Widmann M, Niethammer M, Zhang JL, Lee S-Y, Rendler T, Lagoudakis KG, Son NT, Janzén E, Ohshima T, Wrachtrup J, Vučković J (2017) Scalable quantum photonics with single color centers in silicon carbide. Nano Lett 17:1782–1786
103. Wang J, Zhou Y, Zhang X, Liu F, Li Y, Li K, Liu Z, Wang G, Gao W (2017) Efficient generation of an array of single silicon-vacancy defects in silicon carbide. Phys Rev Appl 7:064021
104. He Z-X, Li Q, Wen X-L, Zhou J-Y, Lin W-X, Hao Z-H, Xu J-S, Li C-F, Guo G-C (2023) Maskless generation of single silicon vacancy arrays in silicon carbide by a focused He^+ ion beam. ACS Photonics 10:2234–2240
105. Simin D, Soltamov VA, Poshakinskiy AV, Anisimov AN, Babunts RA, Tolmachev DO, Mokhov EN, Trupke M, Tarasenko SA, Sperlich A, Baranov PG, Dyakonov V, Astakhov GV (2016) All-optical dc nanotesla magnetometry using silicon vacancy fine structure in isotopically purified silicon carbide. Phys Rev X 6:031014
106. Quan W-K, Liu L, Luo Q-Y, Liu X-D, Wang J-F (2023) Fiber-integrated silicon carbide silicon-vacancy-based magnetometer. Opt Lett 48:1423–1426
107. Wang J-F, Liu L, Liu X-D, Li Q, Cui J-M, Zhou D-F, Zhou J-Y, Wei Y, Xu H-A, Xu W, Lin W-X, Yan J-W, He Z-X, Liu Z-H, Hao Z-H, Li H-O, Liu W, Xu J-S, Gregoryanz E, Li C-F, Guo G-C (2023) Magnetic detection under high pressures using designed silicon vacancy centres in silicon carbide. Nat Mater 22:489–494
108. Zhao M, Pan F, Mei L (2010) Ferromagnetic ordering of silicon vacancies in N-doped silicon carbide. Appl Phys Lett 96:012508
109. Kraus H, Soltamov VA, Riedel D, Väth S, Fuchs F, Sperlich A, Baranov PG, Dyakonov V, Astakhov GV (2014) Room-temperature quantum microwave emitters based on spin defects in silicon carbide. Nat Phys 10:157–162
110. Anisimov AN, Simin D, Soltamov VA, Lebedev SP, Baranov PG, Astakhov GV, Dyakonov V (2016) Optical thermometry based on level anticrossing in silicon carbide. Sci Rep 6:33301
111. Babin C, Stöhr R, Morioka N, Linkewitz T, Steidl T, Wörnle R, Liu D, Hesselmeier E, Vorobyov V, Denisenko A, Hentschel M, Gobert C, Berwian P, Astakhov GV, Knolle W, Majety S, Saha P, Radulaski M, Son NT, Ul-Hassan J, Kaiser F, Wrachtrup J (2022) Fabrication and nanophotonic waveguide integration of silicon carbide colour centres with preserved spin-optical coherence. Nat Mater 21:67–73
112. Fang R-Z, Lai X-Y, Li T, Su R-Z, Lu B-W, Yang C-W, Liu R-Z, Qiao Y-K, Li C, He Z-G, Huang J, Li H, You L-X, Huo Y-H, Bao X-H, Pan J-W (2024) Experimental generation of spin-photon entanglement in silicon carbide. Phys Rev Lett 132:160801
113. Muzha A, Fuchs F, Tarakina NV, Simin D, Trupke M, Soltamov VA, Mokhov EN, Baranov PG, Dyakonov V, Krueger A, Astakhov GV (2014) Room-temperature near-infrared silicon carbide nanocrystalline emitters based on optically aligned spin defects. Appl Phys Lett 105:243112
114. Castelletto S, Almutairi AFM, Thalassinos G, Lohrmann A, Buividas R, Lau DWM, Reineck P, Juodkazis S, Ohshima T, Gibson BC, Johnson BC (2017) Fluorescent color centers in laser ablated 4H-SiC nanoparticles. Opt Lett 42:1297–1300
115. Vásquez GC, Bathen ME, Galeckas A, Bazioti C, Johansen KM, Maestre D, Cremades A, Prytz Ø, Moe AM, Kuznetsov AY, Vines L (2020) Strain modulation of Si vacancy emission from SiC micro- and nanoparticles. Nano Lett 20:8689–8695
116. Joshi T, Dev P (2022) Site-dependent properties of quantum emitters in nanostructured silicon carbide. PRX Quant 3:020325
117. Dannefaer S, Craigen D, Kerr D (1995) Carbon and silicon vacancies in electron-irradiated 6H-SiC. Phys Rev B 51:1928–1930
118. Hemmingsson CG, Son NT, Ellison A, Zhang J, Janzén E (1998) Negative-U centers in 4H silicon carbide. Phys Rev B 58:R10119–R10122

119. Watkins GD, Troxell JH (1980) Negative-U properties for point defects in silicon. Phys Rev Lett 44:593–596
120. Zywietz A, Furthmüller J, Bechstedt F (1999) Vacancies in SiC: influence of Jahn–Teller distortions, spin effects, and crystal structure. Phys Rev B 59:15166–15180
121. Aboelfotoh MO, Doyle JP (1999) Defect energy levels in electron-irradiated and deuterium-implanted 6H silicon carbide. Phys Rev B 59:10823–10829
122. von Bardeleben HJ, Cantin JL, Henry L, Barthe MF (2000) Vacancy defects in p-type 6H-SiC created by low-energy electron irradiation. Phys Rev B 62:10841–10846
123. Son NT, Hai PN, Janzén E (2001) Carbon vacancy-related defect in 4H and 6H SiC. Phys Rev B 63:201201(R)
124. Son NT, Magnusson B, Janzén E (2002) Photoexcitation electron paramagnetic resonance studies of the carbon vacancy in 4H-SiC. Appl Phys Lett 81:3945–3947
125. Eberlein TAG, Jones R, Briddon PR (2003) Z_1/Z_2 defects in 4H-SiC. Phys Rev Lett 90:225502
126. Orlinski SB, Schmidt J, Mokhov EN, Baranov PG (2003) Silicon and carbon vacancies in neutron-irradiated SiC: a high-field electron paramagnetic resonance study. Phys Rev B 67:125207
127. Umeda T, Isoya J, Morishita N, Ohshima T, Kamiya T (2004) EPR identification of two types of carbon vacancies in 4H-SiC. Phys Rev B 69:121201(R)
128. Umeda T, Ishitsuka Y, Isoya J, Son NT, Janzén E, Morishita N, Ohshima T, Itoh H, Gali A (2005) EPR and theoretical studies of negatively charged carbon vacancy in 4H-SiC. Phys Rev B 71:193202
129. Danno K, Kimoto T (2006) Investigation of deep levels in n-type 4H-SiC epilayers irradiated with low-energy electrons. J Appl Phys 100:113728
130. Hiyoshi T, Kimoto T (2009) Reduction of deep levels and improvement of carrier lifetime in n-type 4H-SiC by thermal oxidation. Appl Phys Express 2:041101
131. Hiyoshi T, Kimoto T (2009) Elimination of the major deep levels in n- and p-type 4H-SiC by two-step thermal treatment. Appl Phys Express 2:091101
132. Bockstedte M, Marini A, Pankratov O, Rubio A (2010) Many-body effects in the excitation spectrum of a defect in SiC. Phys Rev Lett 105:026401
133. Son NT, Trinh XT, Løvlie LS, Svensson BG, Kawahara K, Suda J, Kimoto T, Umeda T, Isoya J, Makino T, Ohshima T, Janzén E (2012) Negative-U system of carbon vacancy in 4H-SiC. Phys Rev Lett 109:187603
134. Bathen ME, Coutinho J, Ayedh HM, Ul Hassan J, Farkas I, Öberg S, Frodason YK, Svensson BG, Vines L (2019) Anisotropic and plane-selective migration of the carbon vacancy in SiC: theory and experiment. Phys Rev B 100:014103
135. Brauer G, Anwand W, Coleman PG, Knights AP, Plazaola F, Pacaud Y, Skorupa W, Störmer J, Willutzki P (1996) Positron studies of defects in ion-implanted SiC. Phys Rev B 54:3084–3092
136. Torpo L, Staab RM, Nieminen TEM (2002) Divacancy in 3C- and 4H-SiC: an extremely stable defect. Phys Rev B 65:085202
137. Rempel AA, Sprengel W, Blaurock K, Reichle KJ, Major J, Schaefer H-E (2002) Identification of lattice vacancies on the two sublattices of SiC. Phys Rev B 89:185501
138. Henry L, Barthe M-F, Corbel C, Desgardin P, Blondiaux G, Arpiainen S, Liszkay L (2003) Silicon vacancy-type defects in as-received and 12-MeV proton-irradiated 6H-SiC studied by positron annihilation spectroscopy. Phys Rev B 67:115210
139. Son NT, Carlsson P, Hassan J, Janzén E, Umeda T, Isoya J, Gali A, Bockstedte M, Morishita N, Ohshima T, Itoh H (2006) Divacancy in 4H-SiC. Phys Rev Lett 96:055501
140. Carlos WE, Garces NY, Glaser ER (2006) Annealing of multivacancy defects in 4H-SiC. Phys Rev B 74:235201
141. Iwata J-I, Shinei C, Oshiyama A (2016) Density-functional study of atomic and electronic structures of multivacancies in silicon carbide. Phys Rev B 93:125202
142. Lee EMY, Yu A, de Pablo JJ, Galli G (2021) Stability and molecular pathways to the formation of spin defects in silicon carbide. Nat Commun 12:6325
143. Koehl WF, Buckley BB, Heremans FJ, Calusine G, Awschalom DD (2011) Room temperature coherent control of defect spin qubits in silicon carbide. Nature 479:84–88

144. Magnusson B, Janzén E (2005) Optical characterization of deep level defects in SiC. Mater Sci Forum 483–485:341–346
145. Baranov PG, Il'in IV, Mokhov EN, Muzafarova MV, Orlinskii SB, Schmidt J (2005) EPR identification of the triplet ground state and photoinduced population inversion for a Si-C divacancy in silicon carbide. JETP Lett 82:441–443
146. Falk AL, Buckley BB, Calusine G, Koehl WF, Dobrovitski VV, Politi A, Zorman CA, Feng PX-L, Awschalom DD (2013) Polytype control of spin qubits in silicon carbide. Nat Commun 4:1819
147. Falk AL, Klimov PV, Buckley BB, Ivády V, Abrikosov IA, Calusine G, Koehl WF, Gali Á, Awschalom DD (2014) Electrically and mechanically tunable electron spins in silicon carbide color centers. Phys Rev Lett 112:187601
148. Christle DJ, Falk AL, Andrich P, Klimov PV, Ul Hassan J, Son NT, Janzén E, Ohshima T, Awschalom DD (2015) Isolated electron spins in silicon carbide with millisecond coherence times. Nat Mater 14:160–163
149. Christle DJ, Klimov PV, de las Casas CF, Szász K, Ivády V, Jokubavicius V, Ul Hassan J, Syväjärvi M, Koehl WF, Ohshima T, Son NT, Janzén E, Gali Á, Awschalom DD (2017) Isolated spin qubits in SiC with a high-fidelity infrared spin-to-photon interface. Phys Rev X 7:021046
150. Togan E, Chu Y, Trifonov AS, Jiang L, Maze J, Childress L, Dutt MVG, Sørensen AS, Hemmer PR, Zibrov AS, Lukin MD (2010) Quantum entanglement between an optical photon and a solid-state spin qubit. Nature 466:730
151. Buckley BB, Fuchs GD, Bassett LC, Awschalom DD (2010) Spin-light coherence for single-spin measurement and control in diamond. Science 330:1212
152. Bernien H, Childress L, Robledo L, Markham M, Twitchen D, Hanson R (2012) Two-photon quantum interference from separate nitrogen vacancy centers in diamond. Phys Rev Lett 108:043604
153. Wolfowicz G, Anderson CP, Yeats AL, Whiteley SJ, Niklas J, Poluektov OG, Heremans FJ, Awschalom DD (2017) Optical charge state control of spin defects in 4H-SiC. Nat Commun 8:1876
154. Seo H, Falk AL, Klimov PV, Miao KC, Galli G, Awschalom DD (2016) Quantum decoherence dynamics of divacancy spins in silicon carbide. Nat Commun 7:12935
155. Li Q, Wang J-F, Yan F-F, Zhou J-Y, Wang H-F, Liu H, Guo L-P, Zhou X, Gali A, Liu Z-H, Wang Z-Q, Sun K, Guo G P, Tang J-S, Li H, You L-X, Xu J-S, Li C-F, Guo G-C (2022) Room-temperature coherent manipulation of single-spin qubits in silicon carbide with a high readout contrast. Natl Sci Rev 9:nwab122
156. Zhou Y, Wang J, Zhang X, Li K, Cai J, Gao W (2017) Self-protected thermometry with infrared photons and defect spins in silicon carbide. Phys Rev Appl 8:044015
157. Yan F-F, Wang J-F, Li Q, Cheng Z-D, Cui J-M, Liu W-Z, Xu J-S, Li C-F, Guo G-C (2018) Coherent control of defect spins in silicon carbide above 550 K. Phys Rev Appl 10:044042
158. Lin WX, Yan F-F, Li Q, Wang J-F, Hao Z-H, Zhou J-Y, Li H, You L-X, Xu J-S, Li C-F, Guo G-C (2021) Temperature dependence of divacancy spin coherence in implanted silicon carbide. Phys Rev B 104:125305
159. Magnusson B, Son NT, Csóré A, Gällström A, Ohshima T, Gali A, Ivanov IG (2018) Excitation properties of the divacancy in 4H-SiC. Phys Rev B 98:195202
160. Gao W, da Jornada FH, Ben MD, Deslippe J, Louie SG, Chelikowsky JR (2022) Quasiparticle energies and optical excitations of 3C-SiC divacancy from GW and GW plus Bethe–Salpeter equation calculations. Phys Rev Mater 6:036201
161. Zhou J-Y, Li Q, Hao Z-Y, Yan F-F, Yang M, Wang J-F, Lin W-X, Liu Z-H, Liu W, Li H, You L, Xu J-S, Li C-F, Guo G-C (2021) Experimental determination of the dipole orientation of single color centers in silicon carbide. ACS Photonics 8:2384–2391
162. Liu Y, Wang G, Wang S, Yang J, Chen L, Qin X, Song B, Wang B, Chen X (2011) Defect-induced magnetism in neutron irradiated 6H-SiC single crystals. Phys Rev Lett 106:087205
163. Falk AL, Klimov PV, Ivády V, Szász K, Christle DJ, Koehl WF, Gali Á, Awschalom DD (2015) Optical polarization of nuclear spins in silicon carbide. Phys Rev Lett 114:247603

164. de las Casas CF, Christle DJ, Ul Hassan J, Ohshima T, Son NT, Awschalom DD (2017) Stark tuning and electrical charge state control of single divacancies in silicon carbide. Appl Phys Lett 111:262403
165. Whiteley SJ, Wolfowicz G, Anderson CP, Bourassa A, Ma H, Ye M, Koolstra G, Satzinger KJ, Holt MV, Heremans FJ, Cleland AN, Schuster DI, Galli G, Awschalom DD (2019) Spin–phonon interactions in silicon carbide addressed by Gaussian acoustics. Nat Phys 15:490–495
166. Bourassa A, Anderson CP, Miao KC, Onizhuk M, Ma H, Crook AL, Abe H, Ul-Hassan J, Ohshima T, Son NT, Galli G, Awschalom DD (2020) Entanglement and control of single nuclear spins in isotopically engineered silicon carbide. Nat Mater 19:1319–1325
167. Crook AL, Anderson CP, Miao KC, Bourassa A, Lee H, Bayliss SL, Bracher DO, Zhang X, Abe H, Ohshima T, Hu EL, Awschalom DD (2020) Purcell enhancement of a single silicon carbide color center with coherent spin control. Nano Lett 20:3427–3434
168. Zhou J-Y, Li Q, Hao Z-H, Lin W-X, He Z-X, Liang R-J, Guo L, Li H, You L, Tang J-S, Xu J-S, Li C-F, Guo G-C (2023) Plasmonic-enhanced bright single spin defects in silicon carbide membranes. Nano Lett 23:4334–4343
169. Weber JR, Koehl WF, Varley JB, Janotti A, Buckley BB, Van de Walle CG, Awschalom DD (2011) Defects in SiC for quantum computing. J Appl Phys 109:102417
170. Gordon L, Janotti A, Van de Walle CG (2015) Defects as qubits in 3C- and 4H-SiC. Phys Rev B 92:045208
171. von Bardeleben HJ, Cantin JL, Rauls E, Gerstmann U (2015) Identification and magneto-optical properties of the NV center in 4H-SiC. Phys Rev B 92:064104
172. Zargaleh SA, Eble B, Hameau S, Cantin J-L, Legrand L, Bernard M, Margaillan F, Lauret J-S, Roch J-F, von Bardeleben HJ, Rauls E, Gerstmann U, Treussart F (2016) Evidence for near-infrared photoluminescence of nitrogen vacancy centers in 4H-SiC. Phys Rev B 94:060102(R)
173. von Bardeleben HJ, Cantin JL, Csóré A, Gali A, Rauls E, Gerstmann U (2016) NV centers in 3C, 4H, and 6H silicon carbide: a variable platform for solid-state qubits and nanosensors. Phys Rev B 94:121202(R)
174. Csóré A, von Bardeleben HJ, Cantin JL, Gali A (2017) Characterization and formation of NV centers in 3C, 4H, and 6H SiC: an ab initio study. Phys Rev B 96:085204
175. Zargaleh SA, Hameau S, Eble B, Margaillan F, von Bardeleben HJ, Cantin JL, Gao W (2018) Nitrogen vacancy center in cubic silicon carbide: a promising qubit in the 1.5 μm spectral range for photonic quantum networks. Phys Rev B 98:165203
176. Sato S-I, Narahara T, Abe Y, Hijikata Y, Umeda T, Ohshima T (2019) Formation of nitrogen-vacancy centers in 4H-SiC and their near infrared photoluminescence properties. J Appl Phys 126:083105
177. Khazen K, von Bardeleben HJ, Zargaleh SA, Cantin JL, Zhao M, Gao W, Biktagirov T, Gerstmann U (2019) High-resolution resonant excitation of NV centers in 6H-SiC: a matrix for quantum technology applications. Phys Rev B 100:205202
178. Wang J-F, Yan F-F, Li Q, Liu Z-H, Liu H, Guo G-P, Guo L-P, Zhou X, Cui J-M, Wang J, Zhou Z-Q, Xu X-Y, Xu J-S, Li C-F, Guo G-C (2020) Coherent control of nitrogen-vacancy center spins in silicon carbide at room temperature. Phys Rev Lett 124:223601
179. Wang J-F, Liu Z-H, Yan F-F, Li Q, Yang X-G, Guo L, Zhou X, Huang W, Xu J-S, Li C-F, Guo G-C (2020) Experimental optical properties of single nitrogen vacancy centers in silicon carbide at room temperature. ACS Photonics 7:1611–1616
180. Mu Z, Zargaleh SA, von Bardeleben HJ, Fröch JE, Nonahal M, Cai H, Yang X, Yang J, Li X, Aharonovich I, Gao W (2020) Coherent manipulation with resonant excitation and single emitter creation of nitrogen vacancy centers in 4H silicon carbide. Nano Lett 20:6142–6147
181. von Bardeleben HJ, Cantin J-L, Gerstmann U, Schmidt WG, Biktagirov T (2021) Spin polarization, electron–phonon coupling, and zero-phonon line of the NV center in 3C-SiC. Nano Lett 21:8119–8125
182. Zhu Y, Kovos B, Onizhuk M, Awschalom D, Galli G (2021) Theoretical and experimental study of the nitrogen-vacancy center in 4H-SiC. Phys Rev Mater 5:074602

183. Jiang Z, Cai H, Cernansky R, Liu X, Gao W (2023) Quantum sensing of radio-frequency signal with NV centers in SiC. Sci Adv 9:eadg2080
184. Murzakhanov F, Mamin G, Sadovnikova M, Mokhov E, Nagalyuk S, Gafurov M, Soltamov V (2024) Coherence of NV defects in isotopically enriched 6H-^{28}SiC at ambient conditions. Appl Phys Lett 125:204004
185. Murzakhanov F, Sadovnikova M, Mamin G, Sannikov K, Shakirov A, von Bardeleben HJ, Mokhov E, Nagalyuk S (2024) Room temperature coherence properties and ^{14}N nuclear spin readout of NV centers in 4H–SiC. Appl Phys Lett 124:034001
186. Nagesh V, Farmer JW, Davis RF, Kong HS (1987) Defects in neutron irradiated SiC. Appl Phys Lett 50:1138–1140
187. Torpo L, Pöykkö S, Nieminen RM (1998) Antisites in silicon carbide. Phys Rev B 57:6243–6246
188. Choyke J, Patrick L (1971) Photoluminescence of radiation defects in cubic SiC: localized modes and Jahn–Teller effect. Phys Rev B 4:1843–1847
189. Patrick L, Choyke J (1972) Photoluminescence of radiation defects in ion-implanted 6H-SiC. Phys Rev B 5:3253–3259
190. Haberstroh C, Helbig R (1994) Some new features of the photoluminescence of SiC(6H), SiC(4H), and SiC(15R). J Appl Phys 76:509–513
191. Kawasuso A, Itoh H, Okada S, Okumura H (1996) Annealing processes of vacancy-type defects in electron-irradiated and as-grown 6H-SiC studied by positron lifetime spectroscopy. J Appl Phys 80:5639–5645
192. Egilsson T, Bergman JP, Ivanov IG, Henry A, Janzén E (1999) Properties of the D_1 bound exciton in 4H-SiC. Phys Rev B 59:1956–1963
193. Storasta L, Carlsson FHC, Sridhara SG, Bergman JP, Henry A, Egilsson T, Hallén A, Janzén E (2001) Pseudodonor nature of the DI defect in 4H-SiC. Appl Phys Lett 78:46–48
194. Eberlein TAG, Fall CJ, Jones R, Briddon PR, Öberg S (2002) Alphabet luminescence lines in 4H-SiC. Phys Rev B 65:184108
195. Gali A, Deák P, Rauls E, Son NT, Ivanov IG, Carlsson FHC, Janzén E, Choyke WJ (2003) Correlation between the antisite pair and the D_I center in SiC. Phys Rev B 67:155203
196. Eberlein TAG, Jones R, Öberg S, Briddon PR (2006) Density functional theory calculation of the D_I optical center in SiC. Phys Rev B 74:144106
197. Egilsson T, Henry A, Ivanov IG, Lindström JL, Janzén E (1999) Photoluminescence of electron-irradiated 4H-SiC. Phys Rev B 59:8008–8014
198. Patrick L, Choyke WJ (1973) Localized vibrational modes of a persistent defect in ion-implanted SiC. J Phys Chem Solids 34:565
199. Freitas JA, Bishop SG, Edmond JA, Ryu J, Davis RF (1987) Photoluminescence spectroscopy of ion-implanted 3C-SiC grown by chemical vapor deposition. J Appl Phys 61:2011–2016
200. Mattausch A, Bockstedte M, Pankratov O (2004) Carbon antisite clusters in SiC: a possible pathway to the D_{II} center. Phys Rev B 69:045322
201. Khramtsov IA, Vyshnevyy AA, Fedyanin DY (2018) Enhancing the brightness of electrically driven single-photon sources using color centers in silicon carbide. NPJ Quant Inform 4:15
202. Rauls E, Lingner T, Hajnal Z, Greulich-Weber S, Frauenheim T, Spaeth J-M (2000) Metastability of the neutral silicon vacancy in 4H-SiC. Phys Stat Sol (b) 217:R1
203. Lingner T, Greulich-Weber S, Spaeth J-M, Gerstmann U, Rauls E, Hajnal Z, Frauenheim T, Overhof H (2001) Structure of the silicon vacancy in 6*H*-SiC after annealing identified as the carbon vacancy-carbon antisite pair. Phys Rev B 64:245212
204. Carlos WE, Glaser ER, Shanabrook BV (2003) Optical and magnetic resonance signatures of deep levels in semi-insulating 4H SiC. Phys B 340–342:151–155
205. Umeda T, Ishoya J, Ohshima T, Morishita N, Itoh H, Gali A (2007) Identification of positively charged carbon antisite-vacancy pairs in 4H-SiC. Phys Rev B 75:245202
206. Son NT, Stenberg P, Jokubavicius V, Abe H, Ohshima T, Ul Hassan J, Ivanov IG (2019) Energy levels and charge state control of the carbon antisite-vacancy defect in 4H-SiC. Appl Phys Lett 114:212105

207. Umeda T, Son NT, Isoya J, Janzén E, Ohshima T, Morishita N, Itoh H, Gali A, Bockstedte M (2006) Identification of the carbon antisite-vacancy pair in 4*H*-SiC. Phys Rev Lett 96:145501
208. Steeds JW (2009) Photoluminescence study of the carbon antisite-vacancy pair in 4H- and 6H-SiC. Phys Rev B 80:245202
209. Castelletto S, Johnson BC, Ivády V, Stavrias N, Umeda T, Gali A, Ohshima T (2014) A silicon carbide room-temperature single-photon source. Nat Mater 13:151–156
210. Sun Y, Miyasato T, Wigmore JK (1999) Characterization of excess carbon in cubic SiC films by infrared absorption. J Appl Phys 85:3377–3379
211. Gali A, Deák P, Ordejón P, Son NT, Janzén E, Choyke WJ (2003) Aggregation of carbon interstitials in silicon carbide: a theoretical study. Phys Rev B 68:125201
212. Mattausch A, Bockstedte M, Pankratov O (2004) Structure and vibrational spectra of carbon clusters in SiC. Phys Rev B 70:235211
213. Jiang H, Jiang C, Morgan D, Szlufarska I (2014) Accelerated atomistic simulation study on the stability and mobility of carbon tri-interstitial cluster in cubic SiC. Comput Mater Sci 89:182–188
214. Martin DM, Nielsen HK, Lévêque P, Hallén A, Alfieri G, Svensson BG (2004) Bistable defect in mega-electron-volt proton implanted 4H silicon carbide. Appl Phys Lett 84:1704–1706
215. Coutinho J, Gouveia JD, Makino T, Ohshima T, Pastuović Ž, Bakrač L, Brodar T, Capan I (2021) M center in 4H-SiC is a carbon self-interstitial. Phys Rev B 103:L180102
216. Karsthof R, Bathen ME, Kuznetsov A, Vines L (2022) Formation of carbon interstitial-related defect levels by thermal injection of carbon into n-type 4H-SiC. J Appl Phys 131:035702
217. Knežević T, Hadzipasić A, Ohshima T, Makino T, Capan I (2022) M-center in low-energy electron irradiated 4H-SiC. Appl Phys Lett 120:252101
218. Wang C, Bernholc J, Davis RF (1988) Formation energies, abundances, and the electronic structure of native defects in cubic SiC. Phys Rev B 38:12752–12755
219. Torpo L, Marlo M, Staab TEM, Nieminen RM (2001) Comprehensive *ab initio* study of properties of monovacancies and antisites in 4H-SiC. J Phys Condens Matter 13:6203–6231
220. Zhang J, Storasta L, Bergman JP, Son NT, Janzén E (2003) Electrically active defects in n-type 4H–silicon carbide grown in a vertical hot-wall reactor. J Appl Phys 93:4708–4714
221. Bai R, Liu D, Chen S, Wu Y-N (2025) Origin of the unidentified color centers in 4H-SiC from first principles. Phys Rev B 111:014106
222. Yan F-F, Yi A-L, Wang J-F, Li Q, Yu P, Zhang J-X, Gali A, Wang Y, Xu J-S, Ou X, Li C-F, Guo G-C (2020) Room-temperature coherent control of implanted defect spins in silicon carbide. npj Quant Inform 6:38
223. Lebedev AA, Veinger AI, Davydov DV, Kozlovski VV, Savkina NS, Strel'chuk AM (2000) Doping of n-type 6H–SiC and 4H–SiC with defects created with a proton beam. J Appl Phys 88:6265–6271
224. David ML, Alfieri G, Monakhov EM, Hallén A, Blanchard C, Svensson BG, Barbot JF (2004) Electrically active defects in irradiated 4H-SiC. J Appl Phys 95:4728–4733
225. Storasta L, Bergman JP, Janzén E, Henry A, Lu J (2004) Deep levels created by low energy electron irradiation in 4H-SiC. J Appl Phys 96:4909–4915
226. Dannoa K, Kimoto T (2007) Deep level transient spectroscopy on as-grown and electron-irradiated p-type 4H-SiC epilayers. J Appl Phys 101:103704
227. Castaldini A, Cavallini A, Rigutti L, Nava F (2004) Low temperature annealing of electron irradiation induced defects in 4H-SiC. Appl Phys Lett 85:3780–3782
228. Storastaa L, Tsuchida H (2007) Reduction of traps and improvement of carrier lifetime in 4H-SiC epilayers by ion implantation. Appl Phys Lett 90:062116
229. Storasta L, Tsuchida H, Miyazawa T, Ohshima T (2008) Enhanced annealing of the $Z_{1/2}$ defect in 4H–SiC epilayers. J Appl Phys 103:013705
230. Kimoto T, Danno K, Suda J (2008) Lifetime-killing defects in 4H-SiC epilayers and lifetime control by low-energy electron irradiation. Phys Stat Sol (b) 245:1327–1336
231. Klein PB, Shanabrook BV, Huh SW, Polyakov AY, Skowronski M, Sumakeris JJ, O'Loughlin MJ (2006) Lifetime-limiting defects in n^- 4H-SiC epilayers. Appl Phys Lett 88:052110

232. Hong JD, Davis RF (1979) Self-diffusion of ^{14}C in polycrystalline β-SiC. J Mater Sci 14:2411–2421
233. Hong JD, Davis RF, Newbury DE (1980) Self-diffusion of ^{30}Si in polycrystalline β-SiC. J Mater Sci 15:2073–2080
234. Hong JD, Davis RF, Newbury DE (1981) Self-diffusion of silicon-30 in α-SiC single crystals. J Mater Sci 16:2485–2494
235. Devanathan R, Weber WJ, de la Rubia TD (1998) Computer simulation of a 10 keV Si displacement cascade in SiC. Nucl Instrum Methods Phys Res Sect B 141:118–122
236. Gao F, Weber WJ (2000) Atomic-scale simulation of 50 keV Si displacement cascades in β-SiC. Phys Rev B 63:054101
237. Devanathan R, Weber WJ, Gao F (2001) Atomic scale simulation of defect production in irradiated 3C-SiC. J Appl Phys 90:2303–2309
238. Gao F, Bylaska EJ, Weber WJ, Corrales LR (2001) Ab initio and empirical-potential studies of defect properties in 3C-SiC. Phys Rev B 64:245208
239. Gao F, Weber WJ (2002) Cascade overlap and amorphization in 3C-SiC: defect accumulation, topological features, and disordering. Phys Rev B 66:024106
240. Gao F, Weber WJ (2002) Empirical potential approach for defect properties in 3C-SiC. Nucl Instrum Methods Phys Res Sect B 191:504–508
241. Gao F, Weber WJ (2003) Recovery of close Frenkel pairs produced by low energy recoils in SiC. J Appl Phys 94:4348–4356
242. Kobayashi T, Harada K, Kumagai Y, Oba F, Matsushita Y-I (2019) Native point defects and carbon clusters in 4H-SiC: a hybrid functional study. J Appl Phys 125:125701
243. Li J, Porter L, Yip S (1998) Atomistic modeling of finite-temperature properties of crystalline b-SiC II. Thermal conductivity and effects of point defects. J Nucl Mater 255:139–152
244. Rauls E, Frauenheim T, Gali A, Deák P (2003) Theoretical study of vacancy diffusion and vacancy-assisted clustering of antisites in SiC. Phys Rev B 68:155208
245. Bockstedte M, Mattausch A, Pankratov O (2003) *Ab initio* study of the migration of intrinsic defects in 3*C*-SiC. Phys Rev B 68:205201
246. Bockstedte M, Mattausch A, Pankratov O (2004) *Ab initio* study of the annealing of vacancies and interstitials in cubic SiC: vacancy-interstitial recombination and aggregation of carbon interstitials. Phys Rev B 69:235202
247. Gao F, Weber WJ, Posselt M, Belko V (2004) Atomistic study of intrinsic defect migration in 3C-SiC. Phys Rev B 69:245205
248. Son NT, Carlsson P, Ul Hassan J, Magnusson B, Janzén E (2007) Defects and carrier compensation in semi-insulating 4H-SiC substrates. Phys Rev B 75:155204
249. Zhang Y, Sachan R, Pakarinen OH, Chisholm MF, Liu P, Xue H, Weber WJ (2015) Ionization-induced annealing of pre-existing defects in silicon carbide. Nat Commun 6:8049
250. Karsthof R, Bathen ME, Galeckas A, Vines L (2020) Conversion pathways of primary defects by annealing in proton-irradiated n-type 4H-SiC. Phys Rev B 102:184111
251. Zheng M-J, Swaminathan N, Morgan D, Szlufarska I (2013) Energy barriers for point-defect reactions in 3C-SiC. Phys Rev B 88:054105
252. Fan Y, Song Y, Xu Z, Dong B, Wu J, Rommel M, Zhang K, Zhao J, Zhu R, Li B, Li Q, Fang F (2021) Molecular dynamics simulation of color centers in silicon carbide by helium and dual ion implantation and subsequent annealing. Ceram Int 47:24534–24544
253. Lienhard B, Schröder T, Mouradian S, Dolde F, Tran TT, Aharonovich I, Englund D (2016) Bright and photostable single-photon emitter in silicon carbide. Optica 3:768–774
254. He M, Ju Z, Xue Y, Lin J, Cao Y, Wu B, Wu E (2024) Bright single-photon emitters in cubic silicon carbide. IEEE J Sel Top Quant 30:0100106
255. Wang J, Zhou Y, Wang Z, Rasmita A, Yang J, Li X, von Bardeleben HJ, Gao W (2018) Bright room temperature single photon source at telecom range in cubic silicon carbide. Nat Commun 9:4106
256. Lohrmann A, Iwamoto N, Bodrog Z, Castelletto S, Ohshima T, Karle TJ, Gali A, Prawer S, McCallum JC, Johnson BC (2015) Single-photon emitting diode in silicon carbide. Nat Commun 6:7783

257. Widmann M, Niethammer M, Makino T, Rendler T, Lasse S, Ohshima T, Ul Hassan J, Son NT, Lee S-Y, Wrachtrup J (2018) Bright single photon sources in lateral silicon carbide light emitting diodes. Appl Phys Lett 112:231103
258. Sato S-I, Honda T, Makino T, Hijikata Y, Lee S-Y, Ohshima T (2018) Room temperature electrical control of single photon sources at 4H-SiC surface. ACS Photonics 5:3159–3165
259. He M, Cao Y, Lin J, Ju Z, Wu B, Wu E (2024) Ultrafast optical modulation of the fluorescence from a single-photon emitter in silicon carbide. Photonics Res 12:941–946
260. Klimov PV, Falk AL, Buckley BB, Awschalom DD (2014) Electrically driven spin resonance in silicon carbide color centers. Phys Rev Lett 112:087601
261. Linez F, Makkonen I, Tuomisto F (2016) Calculation of positron annihilation characteristics of six main defects in 6H-SiC and the possibility to distinguish them experimentally. Phys Rev B 94:014103
262. Soltamov VA, Kasper C, Poshakinskiy AV, Anisimov AN, Mokhov EN, Sperlich A, Tarasenko SA, Baranov PG, Astakhov GV, Dyakonov V (2019) Excitation and coherent control of spin qudit modes in silicon carbide at room temperature. Nat Commun 10:1678
263. Cochrane CJ, Blacksberg J, Anders MA, Lenahan PM (2016) Vectorized magnetometer for space applications using electrical readout of atomic scale defects in silicon carbide. Sci Rep 6:37077
264. Liao Y-W, Li Q, Yang M, Liu Z-H, Yan F-F, Wang J-F, Zhou J-Y, Lin W-X, Tang Y-D, Xu J-S, Li C-F, Guo G-C (2022) Deep-learning-enhanced single-spin readout in silicon carbide at room temperature. Phys Rev Appl 17:034046
265. Anderson CP, Glen EO, Zeledon C, Bourassa A, Jin Y, Zhu Y, Vorwerk C, Crook AL, Abe H, Ul-Hassan J, Ohshima T, Son NT, Galli G, Awschalom DD (2022) Five-second coherence of a single spin with single-shot readout in silicon carbide. Sci Adv 8:eabm5912
266. Lew CT-K, Sewani VK, Iwamoto N, Ohshima T, McCallum JC, Johnson BC (2024) All-electrical readout of coherently controlled spins in silicon carbide. Phys Rev Lett 132:146902
267. Lai X-Y, Fang R-Z, Li T, Su R-Z, Huang J, Li H, You L-X, Bao X-H, Pan J-W (2024) Single-shot readout of a nuclear spin in silicon carbide. Phys Rev Lett 132:180803
268. Hesselmeier E, Kuna P, Knolle W, Kaiser F, Son NT, Ghezellou M, Ul-Hassan J, Vorobyov V, Wrachtrup J (2024) High-fidelity optical readout of a nuclear-spin qubit in silicon carbide. Phys Rev Lett 132:180804
269. Nishikawa T, Morioka N, Abe H, Murata K, Okajima K, Ohshima T, Tsuchida H, Mizuochi N (2025) Coherent photoelectrical readout of single spins in silicon carbide at room temperature. Nat Commun 16:3405
270. Tsuchida H, Ito M, Kamata I, Nagano M (2009) Formation of extended defects in 4H-SiC epitaxial growth and development of a fast growth technique. Phys Stat Sol B 246:1553–1568
271. Wahab Q, Ellison A, Henry A, Janzén E, Hallin C, Di Persio J, Martinez R (2000) Influence of epitaxial growth and substrate-induced defects on the breakdown of 4H–SiC Schottky diodes. Appl Phys Lett 76:2725–2727
272. Takahashi J, Ohtani N, Kanaya M (1996) Structural defects in a-SiC single crystals grown by the modified-Lely method. J Cryst Growth 167:596–606
273. Li J, Yang G, Liu X, Luo H, Xu L, Zhang Y, Cui C, Pi X, Yang D, Wang R (2022) Dislocations in 4H silicon carbide. J Phys D Appl Phys 55:463001
274. Fujiwara H, Naruoka H, Konishi M, Hamada K, Katsuno T, Ishikawa T, Watanabe Y, Endo T (2012) Impact of surface morphology above threading dislocations on leakage current in 4H-SiC diodes. Appl Phys Lett 101:042104
275. Amelinckx S, Strumane G, Webb WW (1960) Dislocations in silicon carbide. J Appl Phys 31:1359–1370
276. Krishna P, Jiang SS, Lang AR (1985) An optical and X-ray topographic study of giant screw dislocations in silicon carbide. J Cryst Growth 71:41–56
277. Fujita S, Maeda K, Hyodo S (1987) Dislocation glide motion in 6H SiC single crystals subjected to high-temperature deformation. Philos Mag A 55:203–215
278. Ha S, Mieszkowski P, Skowronski M, Rowland LB (2002) Dislocation conversion in 4H silicon carbide epitaxy. J Cryst Growth 244:257–266

279. Ohnoa T, Yamaguchib H, Kurodac S, Kojimaa K, Suzukia T, Arai K (2004) Direct observation of dislocations propagated from 4H-SiC substrate to epitaxial layer by X-ray topography. J Cryst Growth 260:209–216
280. Huang XR, Black DR, Macrander AT, Maj J, Chen Y, Dudley M (2007) High-geometrical-resolution imaging of dislocations in SiC using monochromatic synchrotron topography. Appl Phys Lett 91:231903
281. Yamamoto Y, Harada S, Seki K, Horio A, Mitsuhashi T, Ujihara T (2012) High-efficiency conversion of threading screw dislocations in 4H-SiC by solution growth. Appl Phys Express 5:115501
282. Ishiji K, Kawado S, Hirai Y, Nagamachi S (2017) Determination of observable depth of dislocations in 4H-SiC by X-ray topography in back reflection. Jpn J Appl Phys 56:106601
283. Wu Z, Liu W, Zhang L, Lim S (2020) Amorphization and dislocation evolution mechanisms of single crystalline 6H-SiC. Acta Mater 182:60–67
284. Hoshino N, Kamata I, Kanda T, Tokuda Y, Kuno H, Tsuchida H (2020) Reduction in dislocation densities in 4H-SiC bulk crystal grown at high growth rate by high-temperature gas-source method. Appl Phys Express 13:095502
285. Karhu R, Ghezellou M, Ul Hassan J (2022) The origin and formation mechanism of an inclined line-like defect in 4H-SiC epilayers. Phys Stat Sol B 259:2100512
286. Li J, Luo H, Yang G, Zhang Y, Pi X, Yang D, Wang R (2022) Nitrogen decoration of basal-plane dislocations in 4H-SiC. Phys Rev Appl 17:054011
287. Peng H, Liu Y, Chen Z, Cheng Q, Hu S, Raghothamachar B, Dudley M, Sampayan K, Sampayan S (2022) Synchrotron X-ray topographic characterization of dislocations in 6H-SiC axial samples. J Cryst Growth 579:126459
288. Ishiji K, Yoneyama A, Inaba M, Fukuda K, Sakaki A, Ohmagari S, Sugie R (2024) 3D structure of threading screw dislocation at a deep location in 4H-SiC using 3D micro-X-ray topography. Jpn J Appl Phys 63:02SP25
289. Fujie F, Peng H, Ailihumaer T, Raghothamachar B, Dudley M, Harada S, Tagawa M, Ujihara T (2021) Synchrotron X-ray topographic image contrast variation of screw-type basal plane dislocations located at different depths below the crystal surface in 4H-SiC. Acta Mater 208:116746
290. Sitch PK, Jones R, Öberg S, Heggie MI (1995) Ab initio investigation of the dislocation structure and activation energy for dislocation motion in silicon carbide. Phys Rev B 52:4951–4955
291. Blumenau AT, Fall CJ, Jones R, Öberg S, Frauenheim T, Briddon PR (2003) Structure and motion of basal dislocations in silicon carbide. Phys Rev B 68:174108
292. Pizzagalli L (2014) Stability and mobility of screw dislocations in 4H, 2H and 3C silicon carbide. Acta Mater 78:236–244
293. Ni Y, Xiong S, Volz S, Dumitrică T (2014) Thermal transport along the dislocation line in silicon carbide. Phys Rev Lett 113:124301
294. Tamura Y, Sakakima H, Takamoto S, Hatano A, Izumi S (2019) Reaction pathway analysis for the conversion of perfect screw basal plane dislocation to threading edge dislocation in 4H-SiC. Jpn J Appl Phys 58:081005
295. Hirano A, Sakakima H, Hatano A, Izumi S (2021) Reaction pathway analysis for the contraction of 4H-SiC partial-dislocations pair in the vicinity of surface. Jpn J Appl Phys 60:085502
296. Nishimura K, Saitoh K-I (2023) Temperature dependence of mechanical properties and defect formation mechanisms in 3C-SiC: a molecular dynamics study. Comput Mater Sci 227:112281
297. Barragan-Yani D, Wirtz L (2024) Assessing the potential of perfect screw dislocations in SiC for solid-state quantum technologies. Phys Rev Res 6:L022055
298. Heindl J, Strunk HP, Heydemann VD, Pensl G (1997) Micropipes: hollow tubes in silicon carbide. Phys Stat Sol (a) 162:251–262
299. Neudeck PG, Powell JA (1994) Performance limiting micropipe defects in silicon carbide wafers. IEEE Electron Device Lett 15:63–65

300. Heindl J, Dorsch W, Strunk HP, Müller G, Eckstein R, Hofmann D, Winnacker A (1998) Dislocation content of micropipes in SiC. Phys Rev Lett 80:740–741
301. Pirouz P (1998) On micropipes and nanopipes in SiC and GaN. Philos Mag 78:727–736
302. Dudley M, Huang XR, Huang W, Powell A, Wang S, Neudeck P, Skowronski M (1999) The mechanism of micropipe nucleation at inclusions in silicon carbide. Appl Phys Lett 75:784–786
303. Huang XR, Dudley M, Vetter WM, Huang W, Wang S, Carter CH (1999) Direct evidence of micropipe-related pure superscrew dislocations in SiC. Appl Phys Lett 74:353–355
304. Dudley M, Wang S, Huang W, Carter CH, Tsvetkov VF, Fazi C (1995) White-beam synchrotron topographic studies of defects in 6H-SiC single crystals. J Phys D Appl Phys 28:A63–A68
305. Yang J, Song H, Jian J, Wang W, Chen X (2021) Characterization of morphological defects related to micropipes in 4H-SiC thick homoepitaxial layers. J Cryst Growth 568–569:126182
306. Kobayashi K, Mori Y, Konishi K, Hasegawa M, Kobayashi K, Shima A (2023) Non-destructive detection of sub-micrometer-sized micropipes in silicon carbide using mirror electron microscope. J Appl Phys 134:195701
307. Stevens R (1972) Defects in silicon carbide. J Mater Sci 7:517–521
308. Koumoto K, Takeda S, Pai CH, Sato T, Yanagida H (1989) High-resolution electron microscopy observations of stacking faults in β-SiC. J Am Ceram Soc 72:1985–1987
309. Pujar VV, Cawley JD (1995) Effects of stacking faults on the X-ray diffraction profiles of β-SiC powders. J Am Ceram Soc 78:774–782
310. Seo WS, Koumoto K (1996) Stacking faults in beta-SiC formed during carbothermal reduction of SiO_2. J Am Ceram Soc 79:1777–1782
311. Hong MH, Samant AV, Pirouz P (2000) Stacking fault energy of 6H-SiC and 4H-SiC single crystals. Philos Mag 80:919–935
312. Izumi S, Tsuchida H, Kamata I, Tawara T (2005) Structural analysis and reduction of in-grown stacking faults in 4H-SiC epilayers. Appl Phys Lett 86:202108
313. Fujiwara H, Kimoto T, Tojo T, Matsunami H (2005) Characterization of in-grown stacking faults in 4H-SiC (0001) epitaxial layers and its impacts on high-voltage Schottky barrier diodes. Appl Phys Lett 87:051912
314. Feng G, Suda J, Kimoto T (2008) Characterization of stacking faults in 4H-SiC epilayers by room-temperature microphotoluminescence mapping. Appl Phys Lett 92:221906
315. Feng G, Suda J, Kimoto T (2009) Triple Shockley type stacking faults in 4H-SiC epilayers. Appl Phys Lett 94:091910
316. Kamata I, Zhang X, Tsuchida H (2010) Photoluminescence of Frank-type defects on the basal plane in 4H–SiC epilayers. Appl Phys Lett 97:172107
317. Na M, Bahng W, Jung H, Oh C, Jang D, Hong S-K (2024) Observation of broad triangular Frank-type stacking faults and characterization of stacking faults with emission wavelengths below 430 nm in 4H–SiC epitaxial layers. Appl Phys Lett 124:152109
318. Liu JQ, Chung HJ, Kuhr T, Li Q, Skowronski M (2002) Structural instability of 4H–SiC polytype induced by n-type doping. Appl Phys Lett 80:2111–2113
319. Skowronski M, Liu JQ, Vetter WM, Dudley M, Hallin C, Lendenmann H (2002) Recombination-enhanced defect motion in forward-biased 4H–SiC p-n diodes. J Appl Phys 92:4699–4704
320. Kuhr TA, Liu JQ, Chung HJ, Skowronski M, Szmulowicz F (2002) Spontaneous formation of stacking faults in highly doped 4H–SiC during annealing. J Appl Phys 92:5863–5871
321. Miyanagi T, Tsuchida H, Kamata I, Nakamura T, Nakayama K, Ishii R, Sugawara Y (2006) Annealing effects on single Shockley faults in 4H-SiC. Appl Phys Lett 89:062104
322. Kato M, Katahira S, Ichikawa Y, Harada S, Kimoto T (2018) Observation of carrier recombination in single Shockley stacking faults and at partial dislocations in 4H-SiC. J Appl Phys 124:095702
323. Nishio J, Ota C, Iijima R (2024) Effect of basal plane dislocation structures on single Shockley-type stacking fault expansion rate in 4H-SiC. Jpn J Appl Phys 63:020802

324. Zimbone M, Barbagiovanni EG, Bongiorno C, Calabretta C, Calcagno L, Fisicaro G, La Magna A, La Via F (2020) Generation and termination of stacking faults by inverted domain boundaries in 3C-SiC. Cryst Growth Des 20:3104–3111
325. Zimbone M, Sarikov A, Bongiorno C, Marzegalli A, Scuderi V, Calabretta C, Miglio L, La Via F (2021) Extended defects in 3C-SiC: stacking faults, threading partial dislocations, and inverted domain boundaries. Acta Mater 213:116915
326. Liu JQ, Skowronski M, Hallin C, Söderholm R, Lendenmann H (2002) Structure of recombination-induced stacking faults in high-voltage SiC p-n junctions. Appl Phys Lett 80:749–751
327. Twigg ME, Stahlbush RE, Fatemi M, Arthur SD, Fedison JB, Tucker JB, Wang S (2003) Structure of stacking faults formed during the forward bias of 4*H*-SiC *p-i-n* diodes. Appl Phys Lett 82:2410–2412
328. Jacobson H, Bergman JP, Hallin C, Janzén E, Tuomi T, Lendenmann H (2004) Properties and origins of different stacking faults that cause degradation in SiC PiN diodes. J Appl Phys 95:1485–1488
329. Stahlbush RE, Fatemi M, Fedison JB, Arthur SD, Rowland LB, Wang S (2002) Stacking-fault formation and propagation in 4H-SiC PiN diodes. J Electron Mater 31:370–375
330. Ha S, Skowronski M, Sumakeris JJ, Paisley MJ, Das MK (2004) Driving force of stacking-fault formation in SiC *p-i-n* diodes. Phys Rev Lett 92:175504
331. Ha S, Skowronski M, Lendenmann H (2004) Nucleation sites of recombination-enhanced stacking fault formation in silicon carbide *p-i-n* diodes. J Appl Phys 96:393–398
332. Galeckas A, Linnros J, Pirouz P (2002) Recombination-enhanced extension of stacking faults in 4H-SiC p-i-n diodes under forward bias. Appl Phys Lett 81:883–885
333. Galeckas A, Linnros J, Pirouz P (2006) Recombination-induced stacking faults: evidence for a general mechanism in hexagonal SiC. Phys Rev Lett 96:025502
334. Caldwell JD, Stahlbush RE, Ancona MG, Glembocki OJ, Hobart KD (2010) On the driving force for recombination-induced stacking fault motion in 4H–SiC. J Appl Phys 108:044503
335. Ishikawa Y, Sudo M, Yao Y-Z, Sugawara Y, Kato M (2018) Expansion of a single Shockley stacking fault in a 4H-SiC (1120) epitaxial layer caused by electron beam irradiation. J Appl Phys 123:225101
336. Chen B, Chen J, Sekiguchi T, Ohyanagi T, Matsuhata H, Kinoshita A, Okumura H, Fabbri F (2008) Electron-beam-induced current study of stacking faults and partial dislocations in 4H SiC Schottky diode. Appl Phys Lett 93:033514
337. Asada S, Murata K, Tsuchida H (2022) Limited current conduction due to various types of stacking faults in n-type 4H-SiC epilayers. Appl Phys Express 15:045502
338. Vivona M, Fiorenza P, Scuderi V, La Via F, Giannazzo F, Roccaforte F (2023) Space charge limited current in 4H-SiC Schottky diodes in the presence of stacking faults. Appl Phys Lett 123:072101
339. Sridhara SG, Carlsson FHC, Bergman JP, Janzén E (2001) Luminescence from stacking faults in 4H SiC. Appl Phys Lett 79:3944–3946
340. Fissela A, Kaisera U, Schrötera B, Richtera W, Bechstedt F (2001) MBE growth and properties of SiC multi-quantum well structures. Appl Surf Sci 184:37–42
341. Bai S, Devaty RP, Choyke WJ, Kaiser U, Wagner G, MacMillan MF (2003) Determination of the electric field in 4H/3C/4H-SiC quantum wells due to spontaneous polarization in the 4H SiC matrix. Appl Phys Lett 83:3171–3173
342. Ding Y, Park K-B, Pelz JP, Palle KC, Mikhov MK, Skromme BJ, Meidia H, Mahajan S (2004) Quantum well state of self-forming 3C-SiC inclusions in 4H SiC determined by ballistic electron emission microscopy. Phys Rev B 69:041305(R)
343. Suo H, Yamashita T, Eto K, Miyasaka A, Osawa H, Kato T, Okumura H (2022) Photoluminescence wavelength from stacking fault with complicated structure in 4H-SiC epitaxial layer. Jpn J Appl Phys 61:105502
344. Galeckas A, Hallén A, Majdi S, Linnros J, Pirouz P (2006) Combined photoluminescence-imaging and deep-level transient spectroscopy of recombination processes at stacking faults in 4H-SiC. Phys Rev B 74:233203

345. Maximenko SI, Freitas JA, Klein PB, Shrivastava A, Sudarshan TS (2009) Cathodoluminescence study of the properties of stacking faults in 4H-SiC homoepitaxial layers. Appl Phys Lett 94:092101
346. Tanuma R, Tsuchida H (2014) Three-dimensional imaging of extended defects in 4H-SiC by optical second-harmonic generation and two-photon-excited photoluminescence. Appl Phys Express 7:021304
347. Rohmfeld S, Hundhausen M, Ley L (1998) Raman scattering in polycrystalline 3C-SiC: influence of stacking faults. Phys Rev B 58:9858–9862
348. Olego D, Cardona M (1982) Temperature dependence of the optical phonons and transverse effective charge in 3C-SiC. Phys Rev B 25:3889–3896
349. Nakashima S, Nakatake Y, Harima H, Katsuno M, Ohtani N (2000) Detection of stacking faults in 6H-SiC by Raman scattering. Appl Phys Lett 77:3612–3614
350. Nakashima S, Nakatake Y, Ishida Y, Talkahashi T, Okumura H (2001) Detection of defects in SiC crystalline films by Raman scattering. Phys B 308–310:684–686
351. Criswell SG, Mahadik NA, Gallagher JC, Barnett J, Kim L, Ghorbani M, Kamaliya B, Bassim ND, Taubner T, Caldwell JD (2024) Nanoscale infrared spectroscopic characterization of extended defects in 4H-silicon carbide. Nano Lett 24:114–121
352. Miao MS, Limpijumnong S, Lambrecht WRL (2001) Stacking fault band structure in 4H-SiC and its impact on electronic devices. Appl Phys Lett 79:4360–4362
353. Lambrecht WRL, Miao MS (2006) Electronic driving force for stacking fault expansion in 4*H*-SiC. Phys Rev B 73:155312
354. Iwata H, Lindefelt U, Öberg S, Briddon PR (2001) Localized electronic states around stacking faults in silicon carbide. Phys Rev B 65:033203
355. Lindefelt U, Iwata H, Öberg S, Briddon PR (2003) Stacking faults in 3C-, 4H-, and 6H-SiC polytypes investigated by an ab initio supercell method. Phys Rev B 67:155204
356. Iwata H, Lindefelt U, Öberg S, Briddon PR (2003) Cubic polytype inclusions in 4H-SiC. J Appl Phys 93:1577–1585
357. Iwata H, Lindefelt U, Öberg S, Briddon PR (2002) Theoretical study of planar defects in silicon carbide. J Phys Condens Matter 14:12733–12740
358. Sakakima H, Takamoto S, Hatano A, Izumi S (2020) Temperature-dependent stacking fault energies of 4H-SiC: a first-principles study. J Appl Phys 127:125703
359. Dove MT (1993) Introduction to lattice dynamics. Cambridge University Press, Cambridge
360. Vetelino JF, Mitra SS (1969) Lattice dynamics of cubic SiC. Phys Rev 178:1349–1352
361. Feldman DW, Parker JH, Choyke WJ, Patrick L (1968) Raman scattering in 6H SiC. Phys Rev 170:698–704
362. Feldman DW, Parker JH, Choyke WJ, Patrick L (1968) Phonon dispersion curves by Raman scattering in SiC, polytypes 3C, 4H, 6H, 15R, and 21R. Phys Rev 173:787–793
363. Hofmann M, Zywietz A, Karch K, Bechstedt F (1994) Lattice dynamics of SiC polytypes within the bond-charge model. Phys Rev B 50:13401–13411
364. Karch K, Pavone P, Windl W, Schütt O, Strauch D (1994) *Ab initio* calculation of structural and lattice-dynamical properties of silicon carbide. Phys Rev B 50:17054–17063
365. Serrano J, Strempfer J, Cardona M, Schwoerer-Böhning M, Requardt H, Lorenzen M, Stojetz B, Pavone P, Choyke WJ (2002) Determination of the phonon dispersion of zinc blende silicon carbide by inelastic X-ray scattering. Appl Phys Lett 80:4360–4362
366. Spitzer WG, Kleinman D, Walsh D (1959) Infrared properties of hexagonal silicon carbide. Phys Rev 113:127–132
367. Spitzer WG, Kleinman D, Frosch CJ (1959) Infrared properties of cubic silicon carbide films. Phys Rev 113:133–136
368. Patrick L, Choyke WJ (1961) Lattice absorption bands in SiC. Phys Rev 123:813–815
369. Patrick L (1968) Infrared absorption in SiC polytypes. Phys Rev 167:809–813
370. Holm RT, Klein PH, Nordquist PER (1986) Infrared reflectance evaluation of chemically vapor deposited β-SiC films grown on Si substrates. J Appl Phys 60:1479–1485
371. Palik ED (ed) (1985) Handbook of optical constants of solids. Academic, New York

372. Engelbrecht F, Helbig R (1993) Effect of crystal anisotropy on the infrared reflectivity of 6H-SiC. Phys Rev B 48:15698–15707
373. Tiwald TE, Woollam JA, Zollner S, Christiansen J, Gregory RB, Wetteroth T, Wilson SR, Powell AR (1999) Carrier concentration and lattice absorption in bulk and epitaxial silicon carbide determined using infrared ellipsometry. Phys Rev B 60:11464–11474
374. Mutschke H, Andersen AC, Clément D, Henning T, Peiter G (1999) Infrared properties of SiC particles. Astron Astrophys 345:187–202
375. Kaneko T, Nemoto D, Horiguchi A, Miyakawa N (2005) FTIR analysis of a-SiC:H films grown by plasma enhanced CVD. J Cryst Growth 275:e1097–e1101
376. King SW, French M, Bielefeld J, Lanford WA (2011) Fourier transform infrared spectroscopy investigation of chemical bonding in low-k a-SiC:H thin films. J Non-Cryst Solids 357:2970–2983
377. Tarekegne AT, Zhou BB, Kaltenecker K, Iwaszczuk K, Clark S, Jepsen PU (2019) Terahertz time-domain spectroscopy of zone-folded acoustic phonons in 4H and 6H silicon carbide. Opt Express 27:3618–3628
378. Shalygin VA, Adamov RB, Moldavskaya MD, Vinnichenko MY, Firsov DA (2020) Far-infrared spectroscopy of folded transverse acoustic phonons in 4H–SiC. Appl Phys Lett 117:202105
379. Nakashima S, Katahama H, Nakakura Y, Mitsuishi A (1986) Relative Raman intensities of the folded modes in SiC polytypes. Phys Rev B 33:5721–5729
380. Feng ZC, Mascarenhas AJ, Choyke WJ, Powell JA (1988) Raman scattering studies of chemical-vapor-deposited cubic SiC films of (100) Si. J Appl Phys 64:3176–3186
381. Okumura H, Sakuma E, Lee JH, Mukaida H, Misawa S, Endo K, Yoshida S (1987) Raman scattering of SiC: application to the identification of heteroepitaxy of SiC polytypes. J Appl Phys 61:1134–1136
382. Nienhaus H, Kampen TU, Mönch W (1995) Phonons in 3C-, 4H-, and 6H-SiC. Surf Sci 324:L328–L332
383. Widulle F, Ruf T, Buresch O, Debernardi A, Cardona M (1999) Raman study of isotope effects and phonon eigenvectors in SiC. Phys Rev Lett 82:3089–3092
384. Qin X, Li X, Chen X, Yang X, Zhang F, Xu X, Hu X, Peng Y, Yu P (2019) Raman scattering study on phonon anisotropic properties of SiC. J Alloy Compd 776:1048–1055
385. Lee DH, Joannopoulos JD (1982) Simple scheme for deriving atomic force constants: application to SiC. Phys Rev Lett 48:1846–1849
386. Nakashima S, Tahara K (1989) Raman scattering determination of structures for SiC polytypes: quantitative evaluation with a revised model of lattice dynamics. Phys Rev B 40:6339–6344
387. Windl W, Karch K, Pavone P, Schutt O, Strauch D, Weber WH, Hass KC, Rimai L (1994) Second-order Raman spectra of SiC: experimental and theoretical results from ab initio phonon calculations. Phys Rev B 49:8764–8767
388. Burton JC, Sun L, Long FH, Feng ZC, Ferguson IT (1999) First- and second-order Raman scattering from semi-insulating 4H-SiC. Phys Rev B 59:7282–7284
389. Olego D, Cardona M, Vogl P (1982) Pressure dependence of the optical phonons and transverse effective charge in 3C-SiC. Phys Rev B 25:3878–3888
390. Olego D, Cardona M (1982) Pressure dependence of Raman phonons of Ge and 3C-SiC. Phys Rev B 25:1151–1160
391. Debernardi A, Ulrich C, Syassen K, Cardona M (1999) Raman linewidths of optical phonons in 3C-SiC under pressure: first-principles calculations and experimental results. Phys Rev B 59:6774–6783
392. Cheong BH, Chang KJ, Cohen ML (1991) Pressure dependences of band gaps and optical-phonon frequency in cubic SiC. Phys Rev B 44:1053–1056
393. Liu J, Vohra YK (1994) Raman modes of 6H polytype of silicon carbide to ultrahigh pressures: a comparison with silicon and diamond. Phys Rev Lett 72:4105–4108
394. Zhang J, Liang T, Lu Y, Xu B, Deng T, Zhang Y, Zeng Z, Pi X, Yang D, Wang R (2022) Effect of hexagonality on the pressure-dependent lattice dynamics of 4H-SiC. New J Phys 24:113015

395. Falkovsky LA, Bluet JM, Camassel J (1998) Strain relaxation at the 3C-SiC/Si interface: Raman scattering experiments. Phys Rev B 57:11283–11294
396. Nance J, Subhash G, Sankar B, Haftka R, Kim NH, Deck C, Oswald S (2021) Measurement of residual stress in silicon carbide fibers of tubular composites using Raman spectroscopy. Acta Mater 217:117164
397. Cagran CP, Hanssen LM, Noorma M, Gura AV, Mekhontsev SN (2007) Temperature-resolved infrared spectral emissivity of SiC and Pt–10Rh for temperatures up to 900 °C. Int J Thermophys 28:581–597
398. Bauer M, Gigler AM, Huber AJ, Hillenbrand R, Stark RW (2009) Temperature-depending Raman line-shift of silicon carbide. J Raman Spectrosc 40:1867–1874
399. Pshenay-Severin DA, Adamov RB, Vinnichenko MY, Moldavskaya MD, Shalygin VA (2023) Decay processes of long-lived phonons in 6H-SiC. J Phys Condens Matter 35:175701
400. Colwell PJ, Klein MV (1972) Raman scattering from electronic excitations in *n*-type silicon carbide. Phys Rev B 6:498–515
401. Klein MV, Ganguly BN, Colwell PJ (1972) Theoretical and experimental study of Raman scattering from coupled LO-phonon-plasmon modes in silicon carbide. Phys Rev B 6:2380–2388
402. Yugami H, Nakashima S, Mitsuishi A, Uemoto A, Shigeta M, Furukawa K, Suzuki A, Nakajima S (1987) Characterization of the free-carrier concentrations in doped β-SiC crystals by Raman scattering. J Appl Phys 61:354–358
403. Harima H, Nakashima S, Uemura T (1995) Raman scattering from anisotropic LO-phonon-plasmon-coupled mode in *n*-type 4H- and 6H-SiC. J Appl Phys 78:1996–2005
404. Burton JC, Sun L, Pophristic M, Lukacs SJ, Long FH, Feng ZC, Ferguson IT (1998) Spatial characterization of doped SiC wafers by Raman spectroscopy. J Appl Phys 84:6268–6273
405. Chafai M, Jaouhari A, Torres A, Antón R, Martín E, Jiménez J, Mitchel WC (2001) Raman scattering from LO phonon-plasmon coupled modes and Hall-effect in n-type silicon carbide 4H–SiC. J Appl Phys 90:5211–5215
406. Gao M-M, Fan L-Y, Gong X-Y, You J-L, Chen Z-Z (2022) Photoexcited carrier dynamics in semi-insulating 4H-SiC by Raman spectroscopy. J Appl Phys 132:135702
407. Monserrat B, Needs RJ (2014) Comparing electron-phonon coupling strength in diamond, silicon, and silicon carbide: first-principles study. Phys Rev B 89:214304
408. Wang T, Gui Z, Janotti A, Ni C (2017) Strong effect of electron-phonon interaction on the lattice thermal conductivity in 3C-SiC. Phys Rev Mater 1:034601
409. Meng F, Ma J, He J, Li W (2019) Phonon-limited carrier mobility and temperature-dependent scattering mechanism of 3C-SiC from first principles. Phys Rev B 99:045201
410. Deng T, Yang D, Pi X (2023) Phonon-limited carrier mobilities and Hall factors in 4H-SiC from first principles. Phys Rev B 107:235203
411. Le Gall J, Olivier M, Greffet J-J (1997) Experimental and theoretical study of reflection and coherent thermal emission by a SiC grating supporting a surface-phonon polariton. Phys Rev B 55:10105–10114
412. Rockstuhl C, Salt MG, Herzig HP (2005) Analysis of the phonon-polariton response of silicon carbide microparticles and nanoparticles by use of the boundary element method. J Opt Soc Am B 22:481–487
413. Schuller JA, Zia R, Taubner T, Brongersma ML (2007) Dielectric metamaterials based on electric and magnetic resonances of silicon carbide particles. Phys Rev Lett 99:107401
414. Neuner B III, Korobkin D, Fietz C, Carole D, Ferro G, Shvets G (2009) Critically coupled surface phonon-polariton excitation in silicon carbide. Opt Lett 34:2667–2669
415. Neuner B III, Korobkin D, Fietz C, Carole D, Ferro G, Shvets G (2010) Midinfrared index sensing of pL-scale analytes based on surface phonon polaritons in silicon carbide. J Phys Chem C 114:7489–7491
416. Kim HC, Cheng X (2010) Surface phonon polaritons on SiC substrate for surface-enhanced infrared absorption spectroscopy. J Opt Soc Am B 27:2393–2397
417. Arnold C, Marquier F, Garin M, Pardo F, Collin S, Bardou N, Pelouard J-L, Greffet J-J (2012) Coherent thermal infrared emission by two-dimensional silicon carbide gratings. Phys Rev B 86:035316

418. Francoeur M, Basu S, Petersen SJ (2011) Electric and magnetic surface polariton mediated near-field radiative heat transfer between metamaterials made of silicon carbide particles. Opt Express 19:18774–18788
419. Nehmetallah G, Aylo R, Powers P, Sarangan A, Gao J, Li H, Achari A, Banerjee PP (2012) Co-sputtered SiC + Ag nanomixtures as visible wavelength negative index metamaterials. Opt Express 20:7095–7100
420. Dougakiuchi T, Kawada Y, Takebe G (2018) Continuous multispectral imaging of surface phonon polaritons on silicon carbide with an external cavity quantum cascade laser. Appl Phys Express 11:032001
421. Kitade S, Yamada A, Morichika I, Yabana K, Ashihara S (2021) Nonlinear shift in phonon-polariton dispersion on a SiC surface. ACS Photonics 8:152–157
422. Mancini A, Nan L, Wendisch FJ, Berté R, Ren H, Cortés E, Maier SA (2022) Near-field retrieval of the surface phonon polariton dispersion in free-standing silicon carbide thin films. ACS Photonics 9:3696–3704
423. Ivchenko VS, Kazantsev DV, Ievleva VA, Kazantseva EA, Kuntsevich AY (2024) Phonon-polariton Bragg generation at the surface of silicon carbide. Appl Phys Lett 125:171601
424. Sasaki Y, Nishina Y, Sato M, Okamura K (1987) Raman study of SiC fibres made from polycarbosilane. J Mater Sci 22:443–448
425. Sorieul S, Costantini J-M, Gosmain L, Thomé L, Grob J-J (2006) Raman spectroscopy study of heavy-ion-irradiated α-SiC. J Phys Condens Matter 18:5235–5251
426. Koyanagi T, Katoh Y, Lance MJ (2018) Raman spectroscopy of neutron irradiated silicon carbide: correlation among Raman spectra, swelling, and irradiation temperature. J Raman Spectrosc 49:1686–1692
427. Inoue Y, Nakashima S, Mitsuishi A, Tabata S, Tsuboi S (1983) Raman spectra of amorphous SiC. Solid State Commun 48:1071–1075
428. Zhang L, Jiang W, Pan C, Fadanelli RC, Ai W, Chen L, Wang T (2019) Raman study of amorphization in nanocrystalline 3C–SiC irradiated with C^+ and He^+ ions. J Raman Spectrosc 50:1197–1204
429. Xiao Z, Yang Y, Ouyang S, Kou Z, Huang B, Luo X (2015) Raman investigation of defective SiC nanocrystals. J Raman Spectrosc 46:1225–1229
430. Choyke WJ, Patrick L (1957) Absorption of light in alpha SiC near the band edge. Phys Rev 105:1721–1723
431. Patrick L, Hamilton DR, Choyke WJ (1963) Optical properties of 15R SiC: luminescence of nitrogen-exciton complexes, and interband absorption. Phys Rev 132:2023–2031
432. Biedermann E (1965) The optical absorption bands and their anisotropy in the various modifications of SiC. Solid State Commun 3:343–346
433. Choyke WJ, Patrick L (1968) Higher absorption edges in 6H-SiC. Phys Rev 172:769–772
434. Patrick L, Choyke WJ (1969) Optical absorption in n-type cubic SiC. Phys Rev 186:775–777
435. Humphreys RG, Bimberg D, Choyke WJ (1981) Wavelength modulated absorption in SiC. Solid State Commun 39:163–167
436. Lambrecht WRL, Segall B, Yoganathan M, Suttrop W, Devaty RP, Choyke WJ, Edmond JA, Powell JA, Alouani M (1994) Calculated and measured UV reflectivity of SiC polytypes. Phys Rev B 50:10722–10726
437. Limpijumnong S, Lambrecht WRL, Rashkeev SN, Segall B (1999) Optical-absorption bands in the 1–3 eV range in n-type SiC polytypes. Phys Rev B 59:12890–12899
438. Sridhara SG, Devaty RP, Choyke WJ (1998) Absorption coefficient of 4H silicon carbide from 3900 to 3250 Å. J Appl Phys 84:2963–2964
439. Sridhara SG, Eperjesi TJ, Devaty RP, Choyke WJ (1999) Penetration depths in the ultraviolet for 4H, 6H and 3C silicon carbide at seven common laser pumping wavelengths. Mater Sci Eng B 61–62:229–233
440. Weingärtner R, Wellmann PJ, Bickermann M, Hofmann D, Straubinger TL, Winnacker A (2002) Determination of charge carrier concentration in n- and p-doped SiC based on optical absorption measurements. Appl Phys Lett 80:70–72

441. Grivickas P, Grivickas V, Linnros J, Galeckas A (2007) Fundamental band edge absorption in nominally undoped and doped 4H-SiC. J Appl Phys 101:123521
442. Spann BT, Compton R, Ratchford D, Long JP, Dunkelberger AD, Klein PB, Giles AJ, Caldwell JD, Owrutsky JC (2016) Photoinduced tunability of the reststrahlen band in 4H-SiC. Phys Rev B 93:085205
443. Chen D, Dong J, Yang J, Hua Y, Li G, Guo C, Xie C, Liu M, Liu Q (2018) Realization of near-perfect absorption in the whole reststrahlen band of SiC. Nanoscale 10:9450–9454
444. Klahold WM, Choyke WJ, Devaty RP (2020) Band structure properties, phonons, and exciton fine structure in 4H-SiC measured by wavelength-modulated absorption and low-temperature photoluminescence. Phys Rev B 102:205203
445. Zhang X, Kioupakis E (2023) Phonon-assisted optical absorption of SiC polytypes from first principles. Phys Rev B 107:115207
446. Lampert MA (1958) Mobile and immobile effective-mass-particle complexes in nonmetallic solids. Phys Rev Lett 1:450–453
447. Haynes JR (1960) Experimental proof of the existence of a new electronic complex in silicon. Phys Rev Lett 4:361–363
448. Choyke WJ, Patrick L (1962) Exciton recombination radiation and phonon spectrum of 6H SiC. Phys Rev 127:1868–1877
449. Hamilton DR, Choyke WJ, Patrick L (1963) Photoluminescence of nitrogen-exciton complexes in 6H SiC. Phys Rev 131:127–133
450. Patrick L, Choyke WJ, Hamilton DR (1965) Luminescence of 4H SiC, and location of conduction-band minima in SiC polytypes. Phys Rev 137:A1515–A1520
451. Patrick L, Hamilton DR, Choyke WJ (1966) Growth, luminescence, selection rules, and lattice sums of SiC with wurtzite structure. Phys Rev 143:526–536
452. Suttrop W, Pensl G, Choyke WJ, Stein R, Leibenzeder S (1992) Hall effect and infrared absorption measurements on nitrogen donors in 6H-silicon carbide. J Appl Phys 72:3708–3713
453. Götz W, Schöner A, Pensl G, Suttrop W, Choyke WJ, Stein R, Leibenzeder S (1993) Nitrogen donors in 4H-silicon carbide. J Appl Phys 73:3332–3338
454. Ivanov IG, Hallin C, Henry A, Kordina O, Janzén E (1996) Nitrogen doping concentration as determined by photoluminescence in 4H- and 6H-SiC. J Appl Phys 80:3504–3508
455. Ivanov IG, Lindefelt U, Henry A, Kordina O, Hallin C, Aroyo M, Egilsson T, Janzén E (1998) Phonon replicas at the M point in 4H-SiC: a theoretical and experimental study. Phys Rev B 58:13634–13647
456. Lyu S, Lambrecht WRL (2020) First-principles study of the phonon replicas in the photoluminescence spectrum of 4H-SiC. Phys Rev B 101:045203
457. Troffer T, Peppermüller C, Pensl G, Rottner K, Schöner A (1996) Phosphorus-related donors in 6H-SiC generated by ion implantation. J Appl Phys 80:3739–3743
458. Laube M, Schmid F, Pensl G, Wagner G, Linnarsson M, Maier M (2002) Electrical activation of high concentrations of N^+ and P^+ ions implanted into 4H–SiC. J Appl Phys 92:549–554
459. Negoro Y, Katsumoto K, Kimoto T, Matsunami H (2004) Electronic behaviors of high-dose phosphorus-ion implanted 4H–SiC (0001). J Appl Phys 96:224–228
460. Suttrop W, Pensl G, Lanig P (1990) Boron-related deep centers in 6H-SiC. Appl Phys A 51:231–237
461. Tanaka K, Kato M (2024) Carrier recombination in highly Al doped 4H-SiC: dependence on the injection conditions. Jpn J Appl Phys 63:011002
462. Rao MV, Griffiths P, Holland OW, Kelner G, Freitas JA Jr, Simons DS, Chi PH, Ghezzo M (1995) Al and B ion-implantations in 6H- and 3C-SiC. J Appl Phys 77:2479–2485
463. Troffer T, Schadt M, Frank T, Itoh H, Pensl G, Heindl J, Strunk HP, Maier M (1997) Doping of SiC by implantation of boron and aluminum. Phys Stat Sol (a) 162:277
464. Capano MA, Ryu S, Cooper JA, Melloch MR, Rottner K, Karlsson S, Nordell N, Powell A, Walker DE (1999) Surface roughening in ion implanted 4H-silicon carbide. J Electron Mater 28:214–218

465. Darmody C, Goldsman N (2019) Incomplete ionization in aluminum-doped 4H-silicon carbide. J Appl Phys 126:145701
466. Huang Y, Wang R, Zhang Y, Yang D, Pi X (2022) Compensation of p-type doping in Al-doped 4H-SiC. J Appl Phys 131:185703
467. Hayashi K, Lederer M, Fukumoto Y, Goto M, Yamamoto Y, Happo N, Harada M, Inamura Y, Oikawa K, Ohoyama K, Wellmann P (2022) Determination of site occupancy of boron in 6H–SiC by multiple-wavelength neutron holography. Appl Phys Lett 120:132101
468. Fujii H, Kaneko M, Kimoto T (2024) Depth profiles of deep levels generated in the tail region of Al ion implantation into n-type 4H-SiC. Jpn J Appl Phys 63:061003
469. Ren ZA, Kato J, Muranaka T, Akimitsu J, Kriener M, Maeno Y (2007) Superconductivity in boron-doped SiC. J Phys Soc Jpn 76:103710
470. Song B, Bao H, Li H, Lei M, Peng T, Jian J, Liu J, Wang W, Wang W, Chen X (2009) Observation of glassy ferromagnetism in Al-doped 4H-SiC. J Am Chem Soc 131:1376–1377
471. Woodbury HH, Ludwig GW (1961) Electron spin resonance studies in SiC. Phys Rev 124:1083–1089
472. Choyke WJ, Patrick L (1970) Luminescence of donor-acceptor pairs in cubic SiC. Phys Rev B 2:4959–4965
473. Hopfield JJ, Thomas DG, Gershenzon M (1963) Pair spectra in GaP. Phys Rev Lett 10:162–164
474. Suzuki A, Matsunami H, Tanaka T (1977) Photoluminescence due to Al, Ga, and B acceptors in 4H-, 6H-, and 3C-SiC grown from a Si melt. J Electrochem Soc 124:241–246
475. Ikeda M, Matsunami H, Tanaka T (1980) Site effect on the impurity levels in 4H, 6H, and 15R SiC. Phys Rev B 22:2842–2854
476. Hagen SH, Van Kemenade AWC, van der Does de Bye JAW (1973) Donor-acceptor pair spectra in 6H and 4H SiC doped with nitrogen and aluminium. J Lumin 8:18–31
477. Kamiyama S, Maeda T, Nakamura Y, Iwaya M, Amano H, Akasaki I, Kinoshita H, Furusho T, Yoshimoto M, Kimoto T, Suda J, Henry A, Ivanov IG, Bergman JP, Monemar B, Onuma T, Chichibu SF (2006) Extremely high quantum efficiency of donor-acceptor-pair emission in N- and-B-doped 6H-SiC. J Appl Phys 99:093108
478. Ou Y, Jokubavicius V, Kamiyama S, Liu C, Berg RW, Linnarsson M, Yakimova R, Syväjärvi M, Ou H (2011) Donor-acceptor-pair emission characterization in N-B doped fluorescent SiC. Opt Mater Express 1:1439–1446
479. Demichelis F, Pirri CF, Tress E (1992) Influence of doping on the structural and optoelectronic properties of amorphous and microcrystalline silicon carbide. J Appl Phys 72:1327–1333
480. Tarekegne AT, Shi X, Gan Y, Chen Y, Ou H (2020) Dependence of photoluminescence emission on excitation power and temperature in highly doped 6H-SiC. Phys Rev Appl 13:064002
481. Dang LS, Lee KM, Watkins GD, Choyke WJ (1980) Optical detection of magnetic resonance for an effective-mass-like acceptor in 6H-SiC. Phys Rev Lett 45:390–394
482. Lindefelt U (1998) Doping-induced band edge displacements and band gap narrowing in 3C-, 4H-, 6H-SiC, and Si. J Appl Phys 84:2628–2637
483. Dean PJ, Herbert DC, Bimberg D, Choyke WJ (1976) Donor exciton satellites in cubic silicon carbide: multiple bound excitons revisited. Phys Rev Lett 37:1635–1638
484. Kuwabara H, Yamada S (1975) Free-to-bound transition in β-SiC doped with boron. Phys Stat Sol (a) 30:739–746
485. Clemen LL, Devaty RP, MacMillan MF, Yoganathan M, Choyke WJ, Larkin DJ, Powell JA, Edmond JA, Kong HS (1993) Aluminum acceptor four particle bound exciton complex in 4H, 6H, and 3C SiC. Appl Phys Lett 62:2953–2955
486. Haberstroh C, Helbig R, Stein RA (1994) Some new features of the photoluminescence of SiC(6H), SiC(4H), and SiC(15R). J Appl Phys 76:509–513
487. Sridhara SG, Clemen LL, Devaty RP, Choyke WJ, Larkin DJ, Kong HS, Troffer T, Pensl G (1998) Photoluminescence and transport studies of boron in 4H SiC. J Appl Phys 83:7909–7919
488. Schneider J, Müller HD, Maier K, Wilkening W, Fuchs F, Dörnen A, Leibenzeder S, Stein R (1990) Infrared spectra and electron spin resonance of vanadium deep level impurities in silicon carbide. Appl Phys Lett 56:1184–1186

489. Maier K, Schneider J, Wilkening W, Leibenzeder S, Stein R (1992) Electron spin resonance studies of transition metal deep level impurities in SiC. Mater Sci Eng B 11:27–30
490. Jenny JR, Skowronski J, Mitchel WC, Hobgood HM, Glass RC, Augustine G, Hopkins RH (1996) Deep level transient spectroscopic and Hall effect investigation of the position of the vanadium acceptor level in 4H and 6H SiC. Appl Phys Lett 68:1963–1965
491. Spindlberger L, Csóré A, Thiering G, Putz S, Karhu R, Ul Hassan J, Son NT, Fromherz T, Gali A, Trupke M (2019) Optical properties of vanadium in 4H silicon carbide for quantum technology. Phys Rev Appl 12:014015
492. von Bardeleben HJ, Zargaleh SA, Cantin JL, Gao WB, Biktagirov T, Gerstmann U (2019) Transition metal qubits in 4H-silicon carbide: a correlated EPR and DFT study of the spin S = 1 vanadium V^{3+} center. Phys Rev Mater 3:124605
493. Chakravorty A, Kabiraj D (2022) Role of Fermi-level depinning in quenching of V^{4+} related photoluminescence in semi-insulating 4H-SiC. Semicond Sci Technol 37:095024
494. Cilibrizzi P, Arshad MJ, Tissot B, Son NT, Ivanov IG, Astner T, Koller P, Ghezellou M, Ul-Hassan J, White D, Bekker C, Burkard G, Trupke M, Bonato C (2023) Ultra-narrow inhomogeneous spectral distribution of telecom-wavelength vanadium centres in isotopically-enriched silicon carbide. Nat Commun 14:8448
495. Astner T, Koller P, Gilardoni CM, Hendriks J, Son NT, Ivanov IG, Ul Hassan J, van der Wal CH, Trupke M (2024) Vanadium in silicon carbide: telecom-ready spin centres with long relaxation lifetimes and hyperfine-resolved optical transitions. Quant Sci Technol 9:035038
496. Ahn J, Wicker C, Bitner N, Solomon MT, Tissot B, Burkard G, Dibos AM, Zhang J, Heremans FJ, Awschalom DD (2024) Extended spin relaxation times of optically addressed vanadium defects in silicon carbide at telecommunication frequencies. Phys Rev Appl 22:044078
497. Ma SB, Sun YP, Zhao BC, Tong P, Zhu XB, Song WH (2007) Magnetic properties of Mn-doped cubic silicon carbide. Phys B 394:122–126
498. Song B, Bao H, Li H, Lei M, Jian J, Han J, Zhang X, Meng S, Wang W, Chen X (2009) Magnetic properties of Mn-doped 6H-SiC. Appl Phys Lett 94:102508
499. Theodoropoulou N, Hebard AF, Chu SNG, Overberg ME, Abernathy CR, Pearton SJ, Wilson RG, Zavada JM, Park YD (2002) Magnetic and structural properties of Fe, Ni, and Mn implanted SiC. J Vac Sci Technol A 20:579–582
500. Stromberg F, Keune W, Chen X, Bedanta S, Reuther H, Mücklich A (2006) The origin of ferromagnetism in ^{57}Fe ion-implanted semiconducting 6H-polytype silicon carbide. J Phys Condens Matter 18:9881
501. Moharana GP, Singh SK, Narayanan HK (2023) Enhancement of weak magnetism of 3C–SiC by 3d-transition metal doping. Solid State Commun 369:115217
502. Dalibor T, Pensl G, Nordell N, Schöner A (1997) Electrical properties of the titanium acceptor in silicon carbide. Phys Rev B 55:13618–13624
503. Bosma T, Lof GJJ, Gilardoni CM, Zwier OV, Hendriks F, Magnusson B, Ellison A, Gällström A, Ivanov IG, Son NT, Havenith RWA, van der Wal CH (2018) Identification and tunable optical coherent control of transition-metal spins in silicon carbide. npj Quant Inform 4:48
504. Gilardoni CM, Bosma T, van Hien D, Hendriks F, Magnusson B, Ellison A, Ivanov IG, Son NT, van der Wal CH (2020) Spin-relaxation times exceeding seconds for color centers with strong spin–orbit coupling in SiC. New J Phys 22:103051
505. Choyke WJ, Devaty RP, Clemen LL, Yoganathan M, Pensl G, Hässler C (1994) Intense erbium-1.54-μm photoluminescence from 2 to 525 K in ion-implanted 4H, 6H, 15R, and 3C SiC. Appl Phys Lett 65:1668–1670
506. Miao MS, Lambrecht WRL (2003) Magnetic properties of substitutional 3d transition metal impurities in silicon carbide. Phys Rev B 68:125204
507. Miao MS, Lambrecht WRL (2006) Electronic structure and magnetic properties of transition-metal-doped 3C and 4H silicon carbide. Phys Rev B 74:235218
508. Shaposhnikov VL, Sobolev NA (2004) The electronic structure and magnetic properties of transition metal-doped silicon carbide. J Phys Condens Matter 16:1761–1768
509. Csóré A, Gali A (2020) Ab initio determination of pseudospin for paramagnetic defects in SiC. Phys Rev B 102:241201(R)

510. Liu X, Zhang J, Xu B, Lu Y, Zhang Y, Wang R, Yang D, Pi X (2022) Deformation of 4H-SiC: the role of dopants. Appl Phys Lett 120:052105
511. Tissot B, Burkard G (2021) Spin structure and resonant driving of spin-1/2 defects in SiC. Phys Rev B 103:064106
512. Tissot B, Burkard G (2021) Hyperfine structure of transition metal defects in SiC. Phys Rev B 104:064102
513. Tissot B, Trupke M, Koller P, Astner T, Burkard G (2022) Nuclear spin quantum memory in silicon carbide. Phys Rev Res 4:033107
514. Tissot B, Udvarhelyi P, Gali A, Burkard G (2024) Strain engineering for transition-metal defects in SiC. Phys Rev B 109:054111
515. Aradi B, Gali A, Deák P, Lowther JE, Son NT, Janzén E, Choyke WJ (2001) *Ab initio* density-functional supercell calculations of hydrogen defects in cubic SiC. Phys Rev B 63:245202
516. Alfieri G, Monakhov EV, Svensson BG, Hallén A (2005) Defect energy levels in hydrogen-implanted and electron-irradiated n-type 4H silicon carbide. J Appl Phys 98:113524
517. Leclerc S, Beaufort MF, Declémy A, Barbot JF (2008) Evolution of defects upon annealing in He-implanted 4H-SiC. Appl Phys Lett 93:122101
518. Sun D, Li R, Ding J, Zhang P, Wang Y, Zhao J (2017) Interaction between helium and intrinsic point defects in 3C-SiC single crystal. J Appl Phys 121:225111
519. Daghbouj N, Li BS, Callisti M, Sen HS, Lin J, Ou X, Karlik M, Polcar T (2020) The structural evolution of light-ion implanted 6H-SiC single crystal: comparison of the effect of helium and hydrogen. Acta Mater 188:609–622
520. Kordina O, Bergman JP, Henry A, Janzén E (1995) Long minority carrier lifetimes in 6H SiC grown by chemical vapor deposition. Appl Phys Lett 66:189–191
521. Kordina O, Bergman JP, Hallin C, Janzén E (1996) The minority carrier lifetime of n-type 4H- and 6H-SiC epitaxial layers. Appl Phys Lett 69:679–681
522. Danno K, Nakamura D, Kimoto T (2007) Investigation of carrier lifetime in 4H-SiC epilayers and lifetime control by electron irradiation. Appl Phys Lett 90:202109
523. Klein PB (2008) Carrier lifetime measurement in n^- 4H-SiC epilayers. J Appl Phys 103:033702
524. Miyazawa T, Ito M, Tsuchida H (2010) Evaluation of long carrier lifetimes in thick 4H silicon carbide epitaxial layers. Appl Phys Lett 97:202106
525. Galeckas A, Linnros J, Grivickas V, Lindefelt U, Hallin C (1997) Auger recombination in 4H-SiC: unusual temperature behavior. Appl Phys Lett 71:3269–3271
526. Tanaka K, Nagaya K, Kato M (2023) 4H-SiC Auger recombination coefficient under the high injection condition. Jpn J Appl Phys 62:SC1017
527. Lüth H (2010) Solid surfaces, interfaces and thin films, 5th edn. Springer-Verlag, Berlin
528. Dayan M (1986) The β-SiC(100) surface studied by low energy electron diffraction, Auger electron spectroscopy, and electron energy loss spectra. J Vac Sci Technol A 4:38–45
529. Hara S, Misawa S, Yoshida S, Aoyagi Y (1994) Additional dimer-row structure of 3C-SiC(001) surfaces observed by scanning tunneling microscopy. Phys Rev B 50:4548–4553
530. Semond F, Soukiassian P, Mayne A, Dujardin G, Douillard L, Jaussaud C (1996) Atomic structure of the β-SiC(100)-(3 × 2) surface. Phys Rev Lett 77:2013–2016
531. Parrill TM, Chung YW (1991) Surface analysis of cubic silicon carbide (001). Surf Sci 243:96–112
532. Sabisch M, Krüger P, Mazur A, Rohlfing M, Pollmann J (1996) First-principles calculations of β-SiC(001) surfaces. Phys Rev B 53:13121–13132
533. Soukiassian P, Semond F, Douillard L, Mayne A, Dujardin G, Pizzagalli L, Joachim C (1997) Direct observation of a β-SiC(100)-$c(4 \times 2)$ surface reconstruction. Phys Rev Lett 78:907–910
534. Soukiassian P, Semond F, Mayne A, Dujardin G (1997) Highly stable Si atomic line formation on the β-SiC(100) surface. Phys Rev Lett 79:2501–2798
535. Derycke V, Soukiassian P, Mayne A, Dujardin G, Gautier J (1998) Carbon atomic chain formation on the β-SiC(100) surface by controlled $sp \rightarrow sp^3$ transformation. Phys Rev Lett 81:5868–5871

536. Aristov VY, Douillard L, Fauchoux O, Soukiassian P (1997) Temperature-induced semiconducting c(4 × 2) double left right arrow metallic (2 × 1) reversible phase transition on the beta-SiC(100) surface. Phys Rev Lett 79:3700–3703
537. Powers JM, Wander A, Rous PJ, Van Hove MA, Somorjai GA (1991) Structural analysis of the β-SiC(100)-*c*(2 × 2) surface reconstruction by automated tensor low-energy electron diffraction. Phys Rev B 44:11159–11166
538. Yan H, Smith AP, Jónsson H (1995) Atomic structure of β-SiC(100) surfaces: an ab initio study. Surf Sci 330:265–275
539. Long JP, Bermudez VM, Ramaker DE (1996) Structural determination of β-SiC(100)-*c*(2 × 2) from C-1*s* surface-core-exciton and Si-2*p* absorption. Phys Rev Lett 76:991–994
540. Catellani A, Galli G, Gygi F (1996) Reconstruction and thermal stability of the cubic SiC (001) surfaces. Phys Rev Lett 77:5090–5093
541. Yeom HW, Shimomura M, Kitamura J, Hara S, Tono K, Matsuda I, Mun BS, Huff WAR, Kono S, Ohta T, Yoshida S, Okushi H, Kajimura K, Fadley CS (1999) Atomic and electronic-band structures of anomalous carbon dimers on 3C-SiC(001)-*c*(2 × 2). Phys Rev Lett 83:1640–1643
542. Chang CS, Tsong IST, Wang YC, Davis RF (1991) Scanning tunneling microscopy and spectroscopy of cubic β-SiC(111) surfaces. Surf Sci 256:354–360
543. Starke U, Schardt J, Bernhardt J, Franke M, Reuter K, Wedler H, Heinz K, Furthmüller J, Käckell P, Bechstedt F (1998) Novel reconstruction mechanism for dangling-bond minimization: combined method surface structure determination of SiC(111)-(3 × 3). Phys Rev Lett 80:758–761
544. Schardt J, Bernhardt J, Starke U, Heinz K (2000) Crystallography of the (3 × 3) surface reconstruction of 3C-SiC(111), 4H-SiC(0001), and 6H-SiC(0001) surfaces retrieved by low-energy electron diffraction. Phys Rev B 62:10335–10344
545. Kulakov MA, Henn G, Bullemcr B (1996) SiC(0001)3 × 3-Si surface reconstruction—a new insight with a STM. Surf Sci 346:49–54
546. Li L, Tsong IST (1996) Atomic structures of 6H-SiC (0001) and ($000\overline{1}$) surfaces. Surf Sci 351:141–148
547. Starke U, Bram C, Steiner P-R, Hartner W, Hammer L, Heinz K, Müller K (1995) The (0001)-surface of 6H-SiC: morphology, composition and structure. Appl Surf Sci 89:175–185
548. Owman F, Mårtensson P (1995) STM study of the SiC(0001)$\sqrt{3} \times \sqrt{3}$ surface. Surf Sci 330:L639–L645
549. Northrup JE, Neugebauer J (1995) Theory of the adatom-induced reconstruction of the SiC(0001)$\sqrt{3} \times \sqrt{3}$ surface. Phys Rev B 52:R17001–R17004
550. Johansson LI, Owman F, Mårtensson P (1996) Surface state on the SiC(0001)-$\sqrt{3} \times \sqrt{3}$ surface. Surf Sci 360:L478–L482
551. Van Bommel AJ, Crombeen JE, Van Tooren A (1975) LEED and Auger electron observations of the SiC(0001) surface. Surf Sci 48:463–472
552. Johansson LI, Owman F, Mårtensson P (1996) High-resolution core-level study of 6H-SiC(0001). Phys Rev B 53:13793–13802
553. Chen W, Xu H, Liu L, Gao XY, Qi DC, Peng GW, Tan SC, Feng YP, Loh KP, Wee ATS (2005) Atomic structure of the 6H–SiC(0001) nanomesh. Surf Sci 596:176–186
554. Sabisch M, Krüger P, Pollmann J (1997) *Ab initio* calculations of structural and electronic properties of 6H-SiC(0001) surfaces. Phys Rev B 55:10561–10570
555. Ramachandran V, Feenstra RM (1999) Scanning tunneling spectroscopy of Mott–Hubbard states on the 6H-SiC(0001)$\sqrt{3} \times \sqrt{3}$ surface. Phys Rev Lett 82:1000–1003
556. Kaplan R (1999) Surface structure and composition of β-SiC and 6H-SiC. Surf Sci 215:111–134
557. Furthmüller J, Bechstedt F, Hüsken H, Schröter B, Richter W (1998) Si-rich SiC(111)/(0001)3 × 3 and $\sqrt{3} \times \sqrt{3}$ surfaces: a Mott–Hubbard picture. Phys Rev B 58:13712–13716
558. Starke U, Schardt J, Bernhardt J, Franke M, Heinz K (1999) Stacking transformation from hexagonal to cubic SiC induced by surface reconstruction: a seed for heterostructure growth. Phys Rev Lett 82:2107–2110

559. Hometz B, Michel H-J, Halbritter J (1994) ARXPS studies of SiO_2-SiC interfaces and oxidation of 6H SiC single crystal Si-(001) and C-(00$\bar{1}$) surfaces. J Mater Res 9:3088–3094
560. Önneby C, Pantano CG (1997) Silicon oxycarbide formation on SiC surfaces and at the SiC/SiO_2 interface. J Vac Sci Technol A 15:1597–1602
561. Hijikata Y, Yaguchi H, Yoshikawa M, Yoshid S (2001) Composition analysis of SiO_2/SiC interfaces by electron spectroscopic measurements using slope-shaped oxide films. Appl Surf Sci 184:161–166
562. Zheleva T, Lelis A, Duscher G, Liu F, Levin I, Das M (2008) Transition layers at the SiO_2/SiC interface. Appl Phys Lett 93:022108
563. Buczko R, Pennycook SJ, Pantelides ST (2000) Bonding arrangements at the Si-SiO_2 and SiC-SiO_2 interfaces and a possible origin of their contrasting properties. Phys Rev Lett 84:943–946
564. Devynck F, Alkauskas A, Broqvist P, Pasquarello A (2011) Charge transition levels of carbon-, oxygen-, and hydrogen-related defects at the SiC/SiO_2 interface through hybrid functionals. Phys Rev B 84:235320
565. Afanas'ev VV, Bassler M, Pensl G, Schulz MJ, von Kamienski ES (1996) Band offsets and electronic structure of SiC/SiO_2 interfaces. J Appl Phys 79:3108–3114
566. Afanas'ev VV, Bassler M, Pensl G, Schulz M (1997) Intrinsic SiC/SiO_2 interface states. Phys Stat Sol (a) 162:321–337
567. Afanas'ev VV, Stesmans A, Bassler M, Pensl G, Schulz MJ (2000) Shallow electron traps at the 4H–SiC/SiO_2 interface. Appl Phys Lett 76:336–337
568. Chang KC, Nuhfer NT, Porter LM, Wahab Q (2000) High-carbon concentrations at the silicon dioxide-silicon carbide interface identified by electron energy loss spectroscopy. Appl Phys Lett 77:2186–2188
569. Knaup JM, Deák P, Frauenheim T, Gali A, Hajnal Z, Choyke WJ (2005) Theoretical study of the mechanism of dry oxidation of 4H-SiC. Phys Rev B 71:235321
570. Pippel E, Woltersdorf J, Ólafsson HÖ, Sveinbjörnsson EÖ (2005) Interfaces between 4H-SiC and SiO_2: microstructure, nanochemistry, and near-interface traps. J Appl Phys 97:034302
571. Knaup JM, Deák P, Frauenheim T, Gali A, Hajnal Z, Choyke WJ (2005) Defects in SiO_2 as the possible origin of near interface traps in the SiC/SiO_2 system: a systematic theoretical study. Phys Rev B 72:115323
572. Watanabe H, Hosoi T, Kirino T, Kagei Y, Uenishi Y, Chanthaphan A, Yoshigoe A, Teraoka Y, Shimura T (2011) Synchrotron X-ray photoelectron spectroscopy study on thermally grown SiO_2/4H-SiC(0001) interface and its correlation with electrical properties. Appl Phys Lett 99:021907
573. Yamashita Y, Nara J, Indari ED, Yamasaki T, Ohno T, Hasunuma R (2022) Experimental and theoretical studies on atomic structures of the interface states at SiO_2/4H-SiC(0001) interface. J Appl Phys 131:215303
574. Cantin JL, von Bardeleben HJ, Shishkin Y, Ke Y, Devaty RP, Choyke WJ (2004) Identification of the carbon dangling bond center at the 4H-SiC/SiO_2 interface by an EPR study in oxidized porous SiC. Phys Rev Lett 92:015502
575. Sometani M, Nishiya Y, Kondo R, Inohana R, Zeng H, Hirai H, Okamoto D, Matsushita Y-I, Umeda T (2023) Energy levels of carbon dangling-bond center (PbC center) at 4H-SiC(0001)/SiO_2 interface. APL Mater 11:111119
576. Afanas'ev VV, Stesmans A, Bassler M, Pensl G, Schulz MJ, Harris CI (1996) Elimination of SiC/SiO_2 interface states by preoxidation ultraviolet-ozone cleaning. Appl Phys Lett 68:2141–2143
577. Li H-F, Dimitrijev S, Harrison HB, Sweatman D (1997) Interfacial characteristics of N_2O and NO nitrided SiO_2 grown on SiC by rapid thermal processing. Appl Phys Lett 70:2028–2030
578. Li H-F, Dimitrijev S, Sweatman D, Harrison HB, Tanner P, Feil B (1999) Investigation of nitric oxide and Ar annealed interfaces by X-ray photoelectron spectroscopy. J Appl Phys 86:4316–4321
579. Chung GY, Tin CC, Williams JR, McDonald K, Ventra MD, Pantelides ST, Feldman LC, Weller RA (2000) Effect of nitric oxide annealing on the interface trap densities near the band edges in the 4H polytype of silicon carbide. Appl Phys Lett 76:1713–1715

580. Afanas'ev VV, Stesmans A, Ciobanu F, Pensl G, Cheong KY, Dimitrijev S (2003) Mechanisms responsible for improvement of 4H-SiC/SiO_2 interface properties by nitridation. Appl Phys Lett 82:568–570
581. McDonald K, Weller RA, Pantelides ST, Feldman LC, Chung GY, Tin CC, Williams JR (2003) Characterization and modeling of the nitrogen passivation of interface traps in SiO_2/4H-SiC. J Appl Phys 93:2719–2722
582. Wang S, Dhar S, Wang S, Ahyi AC, Franceschetti A, Williams JR, Feldman LC, Pantelides ST (2007) Bonding at the SiC-SiO_2 interface and the effects of nitrogen and hydrogen. Phys Rev Lett 98:026101
583. Deák P, Knaup JM, Hornos T, Thill C, Gali A, Frauenheim T (2007) The mechanism of defect creation and passivation at the SiC/SiO_2 interface. J Phys D Appl Phys 40:6242–6253
584. Rozen J, Dhar S, Zvanut ME, Williams JR, Feldman LC (2009) Density of interface states, electron traps, and hole traps as a function of the nitrogen density in SiO_2 on SiC. J Appl Phys 105:124506
585. Yoshioka H, Nakamura T, Kimoto T (2012) Generation of very fast states by nitridation of the SiO_2/SiC interface. J Appl Phys 112:024520
586. Fukuda K, Suzuki S, Tanaka T (2000) Reduction of interface-state density in 4H–SiC n-type metal–oxide–semiconductor structures using high-temperature hydrogen annealing. Appl Phys Lett 76:1585–1587
587. Okamoto D, Yano H, Hatayama T, Fuyuki T (2010) Removal of near-interface traps at SiO_2/4H-SiC (0001) interfaces by phosphorus incorporation. Appl Phys Lett 96:203508
588. Yano H, Katafuchi F, Kimoto T, Matsunami H (1999) Effects of wet oxidation/anneal on interface properties of thermally oxidized SiO/SiC MOS system and MOSFET's. IEEE Trans Electron Devices 46:504–510
589. Saks NS, Mani SS, Agarwal AK (2000) Interface trap profile near the band edges at the 4H-SiC/SiO_2 interface. Appl Phys Lett 76:2250–2252
590. Saks NS, Agarwal AK (2000) Hall mobility and free electron density at the SiC/SiO_2 interface in 4H–SiC. Appl Phys Lett 77:3281–3283
591. Lu C-Y, Cooper JA, Tsuji T, Chung G, Williams JR, McDonald K, Feldman LC (2003) Effect of process variations and ambient temperature on electron mobility at the SiO_2/4H-SiC interface. IEEE Trans Electron Devices 50:1582–1588
592. Lipkin LA, Palmour JW (1996) Improved oxidation procedures for reduced SiO_2/SiC defects. J Electron Mater 25:909–915
593. Bernhardt J, Schardt J, Starke U, Heinz K (1999) Epitaxially ideal oxide-semiconductor interfaces: silicate adlayers on hexagonal (0001) and ($000\bar{1}$) SiC surfaces. Appl Phys Lett 74:1084–1086
594. Kobayashi T, Okuda T, Tachiki K, Ito K, Matsushita Y-I, Kimoto T (2020) Design and formation of SiC (0001)/SiO_2 interfaces via Si deposition followed by low-temperature oxidation and high-temperature nitridation. Appl Phys Express 13:091003
595. Devynck F, Giustino F, Broqvist P, Pasquarello A (2007) Structural and electronic properties of an abrupt 4H-SiC(0001)/SiO_2 interface model: classical molecular dynamics simulations and density functional calculations. Phys Rev B 76:075351
596. Johnson BC, Woerle J, Haasmann D, Lew CT-K, Parker RA, Knowles H, Pingault B, Atature M, Gali A, Dimitrijev S, Camarda M, McCallum JC (2019) Optically active defects at the SiC/SiO_2 interface. Phys Rev Appl 12:044024
597. Kaneko M, Takashima H, Shimazaki K, Takeuchi S, Kimoto T (2023) Impact of the oxidation temperature on the density of single-photon sources formed at SiO_2/SiC interface. APL Mater 11:091121
598. Nakanuma T, Tahara K, Kutsuki K, Shimura T, Watanabe H, Kobayashi T (2023) Control on the density and optical properties of color centers at SiO_2/SiC interfaces by oxidation and annealing. Appl Phys Lett 123:102102
599. Tsuchida H, Kamata I, Izumi K (1997) Si–H bonds on the 6H–SiC(0001) surface after H2 annealing. Jpn J Appl Phys 36:L699–L702

600. Tsuchida H, Kamata I, Izumi K (1997) Infrared spectroscopy of hydrides on the 6H-SiC surface. Appl Phys Lett 70:3072–3074
601. Ramachandran V, Brady MF, Smith AR, Feenstra RM, Greve DW (1998) Preparation of atomically flat surfaces on silicon carbide using hydrogen etching. J Electron Mater 27:308–312
602. Sieber N, Mantel BF, Seyller T, Ristein J, Ley L, Heller T, Batchelor DR, Schmeißer D (2001) Electronic and chemical passivation of hexagonal 6H-SiC surfaces by hydrogen termination. Appl Phys Lett 78(1084–1086):1216–1218
603. Derycke V, Soukiassian PG, Amy F, Chabal YJ, D'angelo MD, Enriquez HB, Silly MG (2003) Nanochemistry at the atomic scale revealed in hydrogen-induced semiconductor surface metallization. Nat Mater 2:253–258
604. Cicero G, Catellani A, Galli G (2004) Atomic control of water interaction with biocompatible surfaces: the case of SiC(001). Phys Rev Lett 93:016102
605. Cicero G, Galli G, Catellani A (2004) Interaction of water molecules with SiC(001) surfaces. J Phys Chem B 108:16518–16524
606. Zhong WW, Huang YF, Gan D, Xu JY, Li H, Wang G, Meng S, Chen XL (2016) Wetting behavior of water on silicon carbide polar surfaces. Phys Chem Chem Phys 18:28033
607. Rosso M, Arafat A, Schroën K, Giesbers M, Roper CS, Maboudian R, Zuilhof H (2008) Covalent attachment of organic monolayers to silicon carbide surfaces. Langmuir 24:4007–4012
608. Rosso M, Giesbers M, Arafat A, Schroën K, Zuilhof H (2009) Covalently attached organic monolayers on SiC and Si_xN_4 surfaces: formation using UV light at room temperature. Langmuir 25:2172–2180
609. Catellani A, Calzolari A (2012) Functionalization of SiC(110) surfaces via porphyrin adsorption: ab initio results. J Phys Chem C 116:886–892
610. Schoell SJ, Hoeb M, Sharp ID, Steins W, Eickhoff M, Stutzmann M, Brandt MS (2008) Functionalization of 6H-SiC surfaces with organosilanes. Appl Phys Lett 92:153301
611. Schoell SJ, Sachsenhauser M, Oliveros A, Howgate J, Stutzmann M, Brandt MS, Frewin CL, Saddow SE, Sharp ID (2013) Organic functionalization of 3C-SiC surfaces. ACS Appl Mater Interfaces 5:1393–1399
612. Shang X, Zhu Y, Li Z (2017) Surface modification of silicon carbide with silane coupling agent and hexadecyl iodiele. Appl Surf Sci 394:169–177
613. Muehlhoff L, Choyke WJ, Bozack MJ, Yates JT (1986) Comparative electron spectroscopic studies of surface segregation on SiC(0001) and SiC(000$\bar{1}$). J Appl Phys 60:2842–2853
614. Devanathan R, Weber WJ (2000) Displacement energy surface in 3C and 6H SiC. J Nucl Mater 278:258–265
615. Kimoto T, Matsunami H (1994) Surface kinetics of adatoms in vapor phase epitaxial growth of SiC on 6H-SiC{0001} vicinal surfaces. J Appl Phys 75:850–859
616. Burton WK, Cabrera N, Frank FC (1951) Philos Trans R Soc Lond A 243:299
617. Nakamura S-I, Kimoto T, Matsunami H, Tanaka S, Teraguchi N, Suzuki A (2000) Formation of periodic steps with a unit-cell height on 6H–SiC (0001) surface by HCl etching. Appl Phys Lett 76:3412–3414
618. Nakagawa H, Tanaka S, Suemune I (2003) Self-ordering of nanofacets on vicinal SiC surfaces. Phys Rev Lett 91:226107
619. Sedighi M, Svetovoy VB, Palasantzas G (2016) Casimir force measurements from silicon carbide surfaces. Phys Rev B 93:085434
620. Fisher GR, Barnes P (1990) Towards a unified view of polytypism in silicon carbide. Philos Mag B 61:217–236
621. Bootsma GA, Knippenberg WF, Verspui G (1971) Phase transformations, habit changes and crystal growth in SiC. J Cryst Growth 8:341–353
622. Cheng C, Needs RJ, Heine V (1988) Inter-layer interactions and the origin of SiC polytypes. J Phys C Solid State Phys 21:1049–1063
623. Heine V, Cheng C, Needs RJ (1991) The preference of silicon carbide for growth in the metastable cubic form. J Am Ceram Soc 74:2630–2633

624. Kang K-H, Eun T, Jun M-C, Lee B-J (2014) Governing factors for the formation of 4H or 6H-SiC polytype during SiC crystal growth: an atomistic computational approach. J Cryst Growth 389:120–133
625. Käckell P, Wenzien B, Bechstedt F (1994) Influence of atomic relaxations on the structural properties of SiC polytypes from *ab initio* calculations. Phys Rev B 50:17037–17046
626. Limpijumnong S, Lambrecht WRL (1998) Total energy differences between SiC polytypes revisited. Phys Rev B 57:12017–12022
627. Bernstein N, Gotsis HJ, Papaconstantopoulos DA, Mehl MJ (2005) Tight-binding calculations of the band structure and total energies of the various polytypes of silicon carbide. Phys Rev B 71:075203
628. Kawanishia S, Mizoguchi T (2016) Effect of van der Waals interactions on the stability of SiC polytypes. J Appl Phys 119:175101
629. Ramakers S, Marusczyk A, Amsler M, Eckl T, Mrovec M, Hammerschmidt T, Drautz R (2022) Effects of thermal, elastic, and surface properties on the stability of SiC polytypes. Phys Rev B 106:075201
630. Scalise E, Marzegalli A, Montalenti F, Miglio L (2019) Temperature-dependent stability of polytypes and stacking faults in SiC: reconciling theory and experiments. Phys Rev Appl 12:021002
631. Baumann HN (1952) The relationship of alpha and beta silicon carbide. J Electrochem Soc 99:109–114
632. Yoo WS, Matsunami H (1991) Solid-state phase transformation in cubic silicon carbide. Jpn J Appl Phys 30:545–553
633. Powell JA, Will HA (1972) Low-temperature solid-state phase transformations in 2H silicon carbide. J Appl Phys 43:1400–1408
634. Heuer AH, Fryburg GA, Ogbuji LU, Mitchell TE (1978) $\beta \rightarrow \alpha$ transformation in polycrystalline SiC: I, microstructural aspects. J Am Ceram Soc 61:406–412
635. Mitchell TE, Ogbuji LU, Heuer AH (1978) $\beta \rightarrow \alpha$ transformation in polycrystalline SiC: II, interfacial energetics. J Am Ceram Soc 61:412–413
636. Ogbuji LU, Mitchell TE, Heuer AH (1981) The $\beta \rightarrow \alpha$ transformation in polycrystalline SiC: III, the thickness of α plates. J Am Ceram Soc 64:91–99
637. Ogbuji LU, Mitchell TE, Heuer AH, Shinozaki S (1981) The $\beta \rightarrow \alpha$ transformation in polycrystalline SiC: IV, a comparison of conventionally sintered, hot-pressed, reaction-sintered, and chemically vapor-deposited samples. J Am Ceram Soc 64:100–105
638. Nader M, Aldinger F, Hoffmann MJ (1999) Influence of the α/β-SiC phase transformation on microstructural development and mechanical properties of liquid phase sintered silicon carbide. J Mater Sci 34:1197–1204
639. Parish CM, Koyanagi T, Kondo S, Katoh Y (2017) Irradiation-induced β to α SiC transformation at low temperature. Sci Rep 7:1198
640. Okojie RS, Xhang M, Pirouz P, Tumakha S, Jessen G, Brillson LJ (2001) Observation of 4H-SiC to 3C-SiC polytypic transformation during oxidation. Appl Phys Lett 79:3056–3058
641. Krishna P, Marshall RC (1971) Direct transformation from the 2H to the 6H structure in single-crystal silicon carbide. J Cryst Growth 11:147–150
642. Chang KJ, Cohen ML (1987) *Ab initio* pseudopotential study of structural and high-pressure properties of SiC. Phys Rev B 35:8196–8201
643. Yoshida M, Onodera A, Ueno M, Takemura K, Shimomura O (1993) Pressure-induced phase transition in SiC. Phys Rev B 48:10587–10590
644. Karch K, Bechstedt F, Pavone P, Strauch D (1996) Pressure-dependent properties of SiC polytypes. Phys Rev B 53:13400–13413
645. Shimojo F, Ebbsjö I, Kalia RK, Nakano A, Rino JP, Vashishta P (2000) Molecular dynamics simulation of structural transformation in silicon carbide under pressure. Phys Rev Lett 84:3338–3341
646. Catti M (2001) Orthorhombic intermediate state in the zinc blende to rocksalt transformation path of SiC at high pressure. Phys Rev Lett 87:035504

647. Miao MS, Lambrecht WRL (2003) Unified path for high-pressure transitions of SiC polytypes to the rocksalt structure. Phys Rev B 68:092103
648. Durandurdu M (2004) Pressure-induced phase transition of SiC. J Phys Condens Matter 16:4411–4417
649. Vashishta P, Kalia RK, Nakano A, Rino JP (2007) Interaction potential for silicon carbide: a molecular dynamics study of elastic constants and vibrational density of states for crystalline and amorphous silicon carbide. J Appl Phys 101:103515
650. Lu Y-P, He D-W, Zhu J, Yang X-D (2008) First-principles study of pressure-induced phase transition in silicon carbide. Phys B 403:3543–3546
651. Daviau K, Lee KKM (2017) Zinc-blende to rocksalt transition in SiC in a laser-heated diamond-anvil cell. Phys Rev B 95:134108
652. Tracy SJ, Smith RF, Wicks JK, Fratanduono DE, Gleason AE, Bolme CA, Prakapenka VB, Speziale S, Appel K, Fernandez-Pañella A, Lee HJ, MacKinnon A, Tavella IF, Eggert JH, Duffy TS (2019) In situ observation of a phase transition in silicon carbide under shock compression using pulsed X-ray diffraction. Phys Rev B 99:214106
653. Aristov VY, Douillard L, Fauchoux O, Soukiassian P (1997) Temperature-induced semiconducting $c(4 \times 2) \Leftrightarrow$ metallic (2×1) reversible phase transition on the β-SiC(100) surface. Phys Rev Lett 79:3700–3703
654. Choi I, Jeong HY, Shin H, Kang G, Byun M, Kim H, Chitu AM, Im JS, Ruoff RS, Choi SY, Lee KJ (2016) Laser-induced phase separation of silicon carbide. Nat Commun 7:13562

Chapter 3
Nanoporous SiC

There are three motives to investigate surface etching of silicon carbide. The first is the creation of patterns on SiC for electronic applications, and the second is that porous silicon carbide obtained by etching yields stronger luminescence than the bulk materials. Last but not least, porous silicon carbide has promising applications in sensing.

3.1 Fabrication Using Various Etching Methods

Silicon carbide is a polar compound semiconductor, and hence, the opposite faces of {001} wafers have different chemical properties and etched patterns. Theoretical studies show that there are small differences in the X-ray reflection intensity from opposite crystal faces [1]. These differences were determined experimentally to allow for identification of the crystallographic polarity under conditions of careful surface preparation which assures identical diffraction conditions on both sides.

Photoelectrochemical etching of silicon carbide has been extensively investigated. Silicon carbide is chemically inert; thus electrochemical etching is commonly adopted to efficiently etch SiC. In 1992, Laser-assisted photoelectrochemical etching of *n*-type 3C–SiC in a HF-based aqueous electrolyte was reported [2]. The measured etch rates of 1–100 μm min^{-1} in this process are much faster than that with previous etching methods. Much higher etch rates can be obtained when (257 nm) UV rather than (514 nm) visible light was used. UV radiation offers efficient photogenerated holes near the SiC surface. These holes are transported in the presence of an external bias to the SiC/electrolyte interface, where dissolution occurs through holes-participated anodic oxidation of the SiC wafer and the removal of the generated oxide by F^- ions present in the electrolyte. By practicing photoelectrochemistry, *n*-type 3C–SiC can be selectively etched while *p*-type 3C–SiC was used as an etch stop [3]. Porous silicon carbide was prepared by anodizing *n*-type 6H–SiC in HF

J. Fan and P. K. Chu, *Silicon Carbide Nanostructures*, Engineering Materials and Processes, https://doi.org/10.1007/978-3-032-10986-6_3

under UV-light illumination [4]. Transmission electron microscopy revealed that the obtained porous SiC network is composed of 10–30 nm pores with interpore spacing ranging from 5 to 150 nm. The selected-area electron diffraction patterns show that the porous layers are single-crystal SiC. Photoelectrochemical etching has the advantages of high etching rates and dopant selectivity compared to other etching methods such as reactive ion etching [5]. However, a rough surface morphology results from etching and may be ascribed to the large defect density. The suggested reactions in anodization are:

$$SiC + 4H_2O + 8h^+ -> SiO_2 + CO_2 + 8H^+, \tag{3.1}$$

$$SiC + 2H_2O + 4h^+ -> SiO + CO + 4H^+. \tag{3.2}$$

The overall reaction is:

$$2SiC + 6H_2O + 12h^+ -> SiO_2 + CO_2 + SiO + CO + 12H^+. \tag{3.3}$$

Anodic dissolution of 6H–SiC in HF occurs via oxidation of the surface described by the above equations, followed by removal of the oxide by HF:

$$6HF + SiO_2 -> H_2SiF_6 + 2H_2O. \tag{3.4}$$

The generated H_2SiF_6 is soluble in the aqueous electrolyte.

In a two-step etching method, the 6H–SiC was first anodized to form a deep porous layer which was then removed by thermal oxidation followed by dipping in HF [6]. The *n*-type and *p*-type materials were etched by photoelectrochemical and dark electrochemical techniques, respectively. Etching rates as high as 400 nm min^{-1} for *n*-SiC and 2.2 μm min^{-1} for *p*-SiC were achieved in anodization, which produced nearly mirror surfaces.

To overcome the poor etching efficiency and low aspect ratio of wet etching of SiC, a strategy exploiting anisotropic charge carrier flow to enable wet etching of 6H–SiC micro/nanostructures was developed [7]. Specifically, an etchant containing H_2O_2, HF, and deionized water was adopted; the bottom surface of a SiC wafer was coated with catalytic Pt (16 nm), an ultraviolet light illuminated the top SiC surface patterned with Au (65 nm) mask. Spatial charge separation across the wafer was achieved, and photogenerated electrons flowed to the bottom to participate in the reduction reaction of an oxidant in the etchant solution, whereas holes flowed to the top to drive oxidation of SiC to trigger subsequent etching (Fig. 3.1). This design suppresses recombination-caused charge loss. When a top metal catalyst mask was used, this strategy yielded an impressive vertical etching rate of 0.737 μm min^{-1} and an aspect ratio of 3.2.

Electrochemical anodization has been applied to 3C–SiC [8]. The porous 3C–SiC layer was produced on a 3C–SiC single crystal by anodic dissolution in HF under

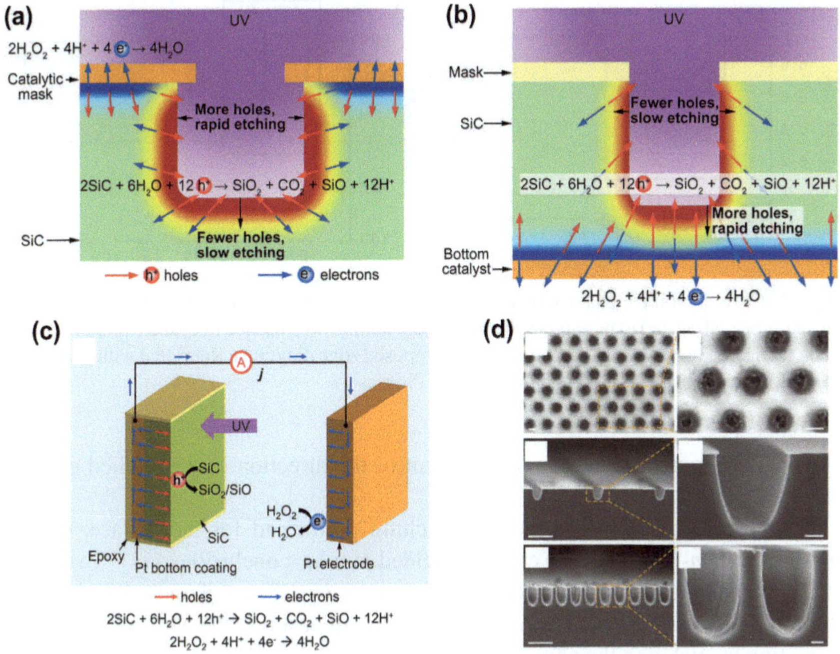

Fig. 3.1 Schematic illustration of **a** a common photoelectrochemical etching method and **b** modified etching method employing a bottom metal catalyst coating. **c** Schematic illustration of the electrochemical characterization. **d** Demonstration of etched 6H–SiC wafers. Top-view and cross-sectional view SEM images of nanopore array (scale bar = 200 nm), 3-μm-wide trench array (scale bar = 10/1 μm), and dense fin/trench array (scale bar = 10/1 μm) as well as corresponding magnified images. Reprinted with permission from [7]. Copyright 2022, John Wiley & Sons, Inc

UV-light illumination. Scanning electron microscopy revealed that the as-prepared 3C–SiC has a porous structure with a minimal size of 20 nm.

A model was presented to explain the formation mechanism of the porous silicon carbide [9]. The porous *n*-type 6H–SiC produced by electrochemical etching with (photo-assisted mode) and without (dark current mode) UV-light illumination was investigated. Photo-assisted etching at a low anode bias produced semi-insulating porous materials with an average fiber thickness below 50 nm. Etching in darkness and at a high bias voltage gave rise to porous SiC with a fiber thickness of 0.5–1 μm and lower resistivity. Self-regulation of fiber size during electrochemical etching arises from Fermi-level pinning to the surface states in thin fibers (Fig. 3.2). Fermi-level pinning results in the increase in resistivity, thereby blocking fiber dissolution and ensuring stability of thin fibers in the electrolyte.

Photoelectrochemical etching of highly doped *n*-type 4H–SiC in HF along different crystallographic orientations under low voltage and/or low current condition has been studied [10]. Scanning electron microscopy revealed that the pores formed by anodization are anisotropic (Fig. 3.3). The Si-terminated crystallographic planes are more resistant to electrolytic attack than the C or mixed C–Si-terminated ones.

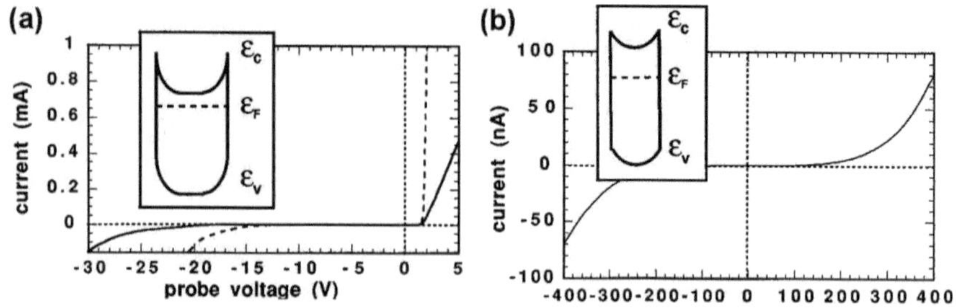

Fig. 3.2 Current–voltage characteristics of a point probe contact on porous 6H–SiC for **a** dark-current-mode fabricated materials (solid line) and crystalline SiC (dashed line) and **b** photo-assisted-mode fabricated materials. The insets show the proposed zone diagrams. Reprinted with permission from [9]. Copyright 1994, American Institute of Physics

The resultant pore structure is independent of the direction of the applied external electric field.

Comparative photoelectrochemical etching performed for the Si-face, C-face, and a-face of *n*-type 6H–SiC crystals yielded similar conclusions [11]. Two kinds of columnar pores were observed: nanocolumnar ($\sim$ 20 nm diameter) and macro-columnar ($\sim$ 1 μm diameter) pores. The results suggest that the nanocolumnar pore formation is related to the anisotropic electrochemical oxidation reaction rate on the different crystal faces. The electrochemical oxidation rate increases with the surface C atom density of SiC. It was proposed that for the Si-face SiC, the preferred etching directions are $\langle 11\bar{2}0\rangle$, and the etching rate on the Si-face in the $[000\bar{1}]$ direction is slower; hence, the pores constantly branched and grew, forming dendritic patterns (Fig. 3.4a). For the C-face etching, the straight columnar pores (Fig. 3.4b) suggest

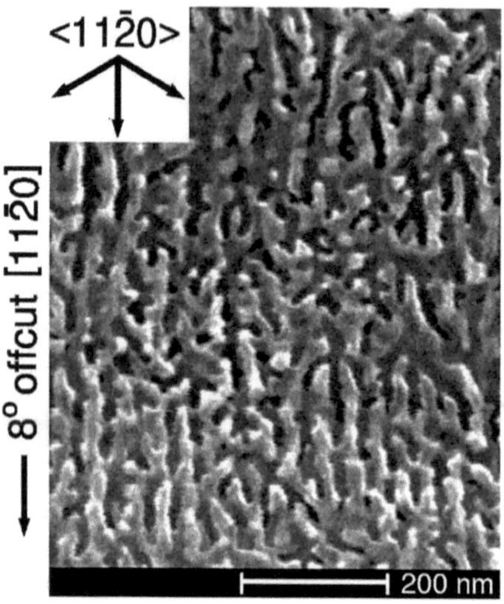

Fig. 3.3 SEM image of an anodized 4H–SiC sample exhibiting the triangular porous morphology. 2 μm of the material was removed by reactive ion etching before imaging. Reprinted with permission from [10]. Copyright 2004, American Institute of Physics

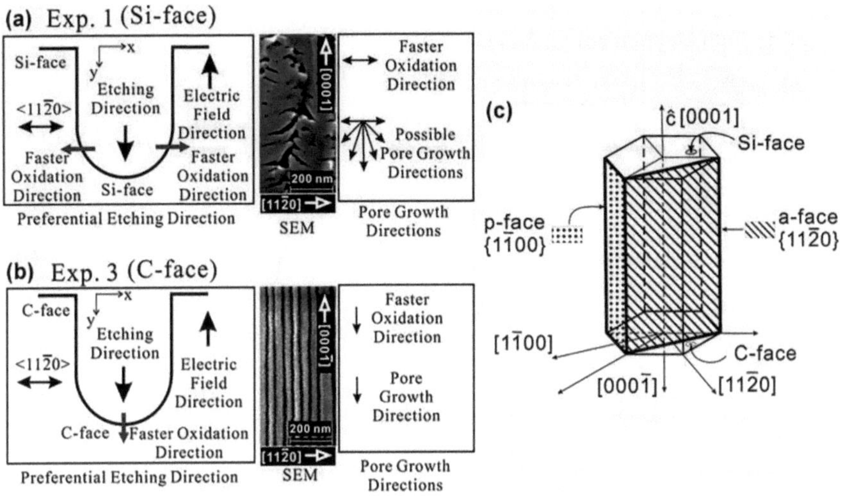

Fig. 3.4 Schematic of preferred oxidation direction, cross-sectional SEM image, and schematic of actual pore growth for the case of **a** Si-face etching and **b** C-face etching in 6H–SiC. **c** Hexagonal pyramid illustrating hexagonal polytypes of SiC. Four important surfaces are indicated. A p-face belongs to the family of {1$\bar{1}$00} planes, whereas an a-face belongs to the family of {11$\bar{2}$0} planes. Reprinted with permission from [11]. Copyright 2008, John Wiley & Sons, Inc

that the preferred etching direction is parallel to the etching direction. This suggests that the oxidation rate in [0001] direction (C-face etching direction, Fig. 3.4c) is faster than the rate in the [11$\bar{2}$0] direction (a-face etching direction) and the [000$\bar{1}$] Si-face etching direction. A study shows that for the electrochemical etching of 4H–SiC crystals by employing HF and H_2O_2 in the electrolyte, the cycle period of the pulsed voltage plays a crucial role in the formation of highly uniform nanohole arrays [12]. The diameters, wall thickness, and depth of the nanoholes can be modulated through modulating the cycle times.

The wetting characteristics of polished or polished and thermally oxidized, on- and off-axis (0001)Si surfaces 6H–SiC in selected acids and bases were determined and compared with that of Si (111) surfaces [13]. AES, XPS, and LEED were used to characterize the chemical state and order of the surfaces. The oxidized SiC surfaces were hydrophilic after oxide removal with a 10: 1 HF solution and were terminated with a monolayer containing OH, CO, CH, and F species. Similarly treated (000$\bar{1}$) C, (11$\bar{2}$0), and (10$\bar{1}$0) surfaces showed the same effects. In contrast, the as-polished SiC surfaces were hydrophobic and covered with a thin (5–10 Å) contamination layer composed mainly of C–C, C–F, and Si–F bonded species. Removal of this layer led to a disordered hydrophilic SiC surface. A 20 Å amorphous Si capping layer passivated the SiC surfaces and offered hydrophobic surfaces. A set of analytical measurements along with first-principles calculations (with DMol3) give insight into the effect of HF etching at SiO_2/SiC interfaces [14]. Unlike the well-known hydrophobic H-terminated surfaces derived from oxidized Si surfaces, HF etching

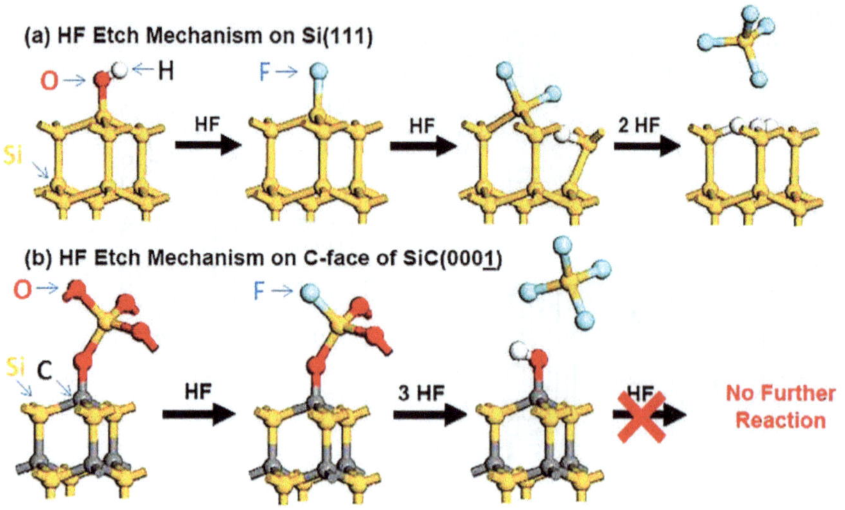

Fig. 3.5 Schematic illustration of mechanisms for **a** H passivation of Si(111) surface and **b** OH termination of C-terminated (000$\bar{1}$) SiC. Reprinted with permission from [14]. Copyright 2009, American Chemical Society

of oxidized SiC yielded hydrophilic, hydroxyl-terminated surfaces (Fig. 3.5). The stability and hydrophilicity of the –OH layer on the C-face of SiC are crucial for subsequent surface engineering. The hydroxyl termination of SiC surfaces renders it possible to graft silane molecules onto HF-etched SiC surfaces. The higher stability of the OH-terminated C–SiC surfaces offers a higher barrier for the process, but usual C–SiC surfaces possess a high enough step density and atomic roughness to present enough Si surface atoms for grafting. This is important for the biomedical applications of SiC in that its hydroxyl terminating groups readily bind large biological molecules such as proteins.

Special hierarchical structures of silicon carbide can be prepared by electroless chemical etching. The nanoplatelets or ordered nanostructures were created by preferential etching of SiC whiskers in a mixture of hydrofluoric acid and nitric acid (3: 1 ratio) at 100 °C [15]. The SEM and TEM images of the pagoda-like SiC nanostructures are shown in Fig. 3.6. A model based on the high-resolution TEM and Raman data (Fig. 3.7) was proposed to explain the etching mechanism. In this model, the whiskers are composed of β-SiC and periodically distributed stacking faults (α-SiC). Preferential etching of β-SiC with respect to α-SiC produced pagoda-shaped nanostructures. In addition to large area etching, a limited area etching on the silicon carbide was also realized. The *n*-type 4H–SiC was electrochemically etched by HF with a glassy-carbon needle electrode (130 μm in diameter) [16], and pores with the same diameter as that of the needles formed at the SiC/glass-carbon contact region. The rate of pore formation increases when sulfuric acid was added to the HF solution.

Some methods other than conventional chemical or electrochemical etching by hydrofluoric acid have been developed to prepare porous silicon carbide. Step

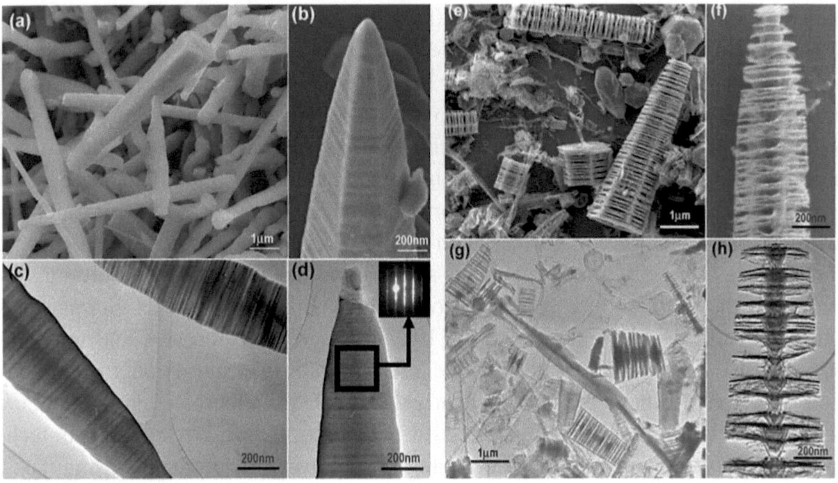

Fig. 3.6 SEM (**a**, **b**) and TEM (**c**, **d**) images of pristine β-SiC whiskers. The inset in **d** is selected-area diffraction pattern of the denoted area. SEM (**e**, **f**) and TEM (**g**, **h**) images of etched SiC whiskers [15]. Reprinted with permission from. Copyright 2006, American Chemical Society

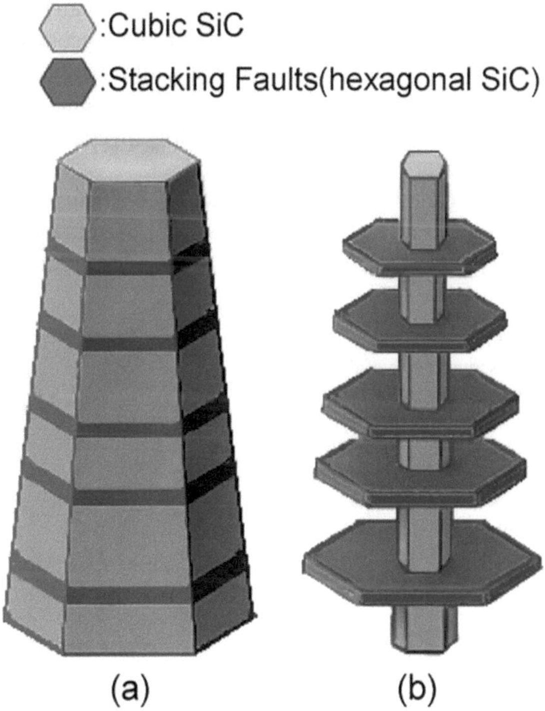

Fig. 3.7 Schematic showing the transformation of a pristine SiC whisker, **a** to a nanostructured pagoda-like structure, **b** by selective etching of β-SiC. Reprinted with permission from [15]. Copyright 2006, American Chemical Society

bunching was achieved on the 6H–SiC (0001) vicinal face by etching with HCl [17]. When the substrate was inclined toward < $01\bar{1}0$ > or < $11\bar{2}0$ >,, continuous parallel and periodic microsteps with six bilayer heights were produced perpendicular to the off direction (Fig. 3.8). The combination of Pt-assisted chemical etching (in HF and $K_2S_2O_8$ solutions) and reactive ion etching of *n*-type 6H–SiC gave rise to large-scale porous materials [18]. The surface roughness increased dramatically in comparison with individual metal-assisted etching or reactive ion etching. Subsequent functionalization of the porous SiC surface with perfluorooctyl trichlorosilane resulted in a superhydrophobic SiC surface which is stable in both acidic and basic solutions. A similar method can produce 3D submicron features with nonporous 4H–SiC surfaces and near-vertical sidewalls (Fig. 3.9) [19]. The etch rate, porosity, and undercut depend on the Pt catalyst pattern, UV excitation intensity, and oxidant concentration. The etch rate increased nearly linearly with the UV excitation intensity. Patterns as small as 150 nm were demonstrated using nanosphere lithography. This technique has advantage in etching nanoscale features over conventional plasma-based methods. Mesoporous silicon carbide was fabricated by a modified sol–gel method [20]. Tetraethoxysilane and phenolic resin were used to generate a binary carbonaceous silicon xerogel, and nickel nitrate was used as the pore adjusting reagent. Carbothermal reduction of the binary xerogel at 1250 °C in argon produced the SiC which has a surface area of 112 $m^2\ g^{-1}$ and mean pore size of 10 nm (Fig. 3.10). The macroporous silicon carbide monoliths with geometric surface areas of 10^5–10^8 m^2 per m^3 were synthesized by capillary filling of packed beds of polystyrene or silica spheres with low viscosity preceramic polymers [21]. Subsequent curing, pyrolysis, and removal of spheres resulted in SiC inverted beaded monoliths that were stable at temperature as high as 1200 °C.

KOH can also etch SiC crystals. The etching mechanism of SiC single crystals by molten KOH was investigated [22]. A nitrogen gas purge largely reduced the etching rate. The etching rate of SiC($000\bar{1}$)C is about four times larger than that of

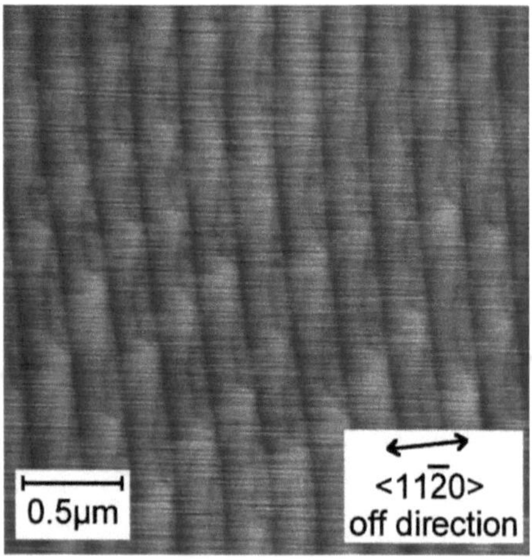

Fig. 3.8 Atomic force microscopy image of the etched 6H–SiC (0001) surface with off direction toward < $11\bar{2}0$ > showing microsteps with triangular depressions. Reprinted with permission from [17]. Copyright 2000, American Institute of Physics

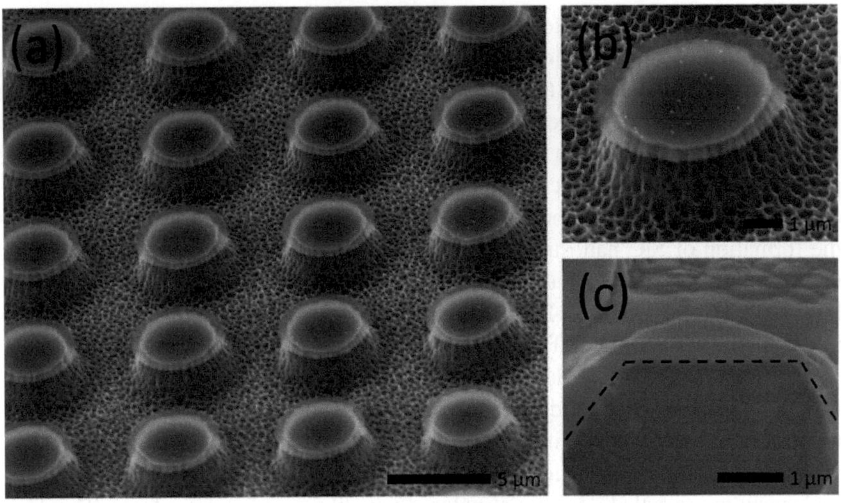

Fig. 3.9 SiC micropillar array with a porous surface produced by Pt-assisted chemical etching using a Pt dot pattern in a solution of 69.4 mM $K_2S_2O_8$ and 28.9 M HF under 254 nm UV exposure (11 mW cm^{-2}). **a** SEM image of an array of 5 μm pillars with 10 μm pitch. **b** An enlarged SEM image of a single micropillar. The Pt thin film remains on top of the pillars and appears transparent at the edges due to undercut suspension. **c** A FIB cross section of a micropillar showing that the Pt-covered surface. Reprinted with permission from [19]. Copyright 2021, John Wiley & Sons, Inc

Fig. 3.10 TEM image of porous silicon carbide. Reprinted with permission from [20]. Copyright 2003, Elsevier B.V

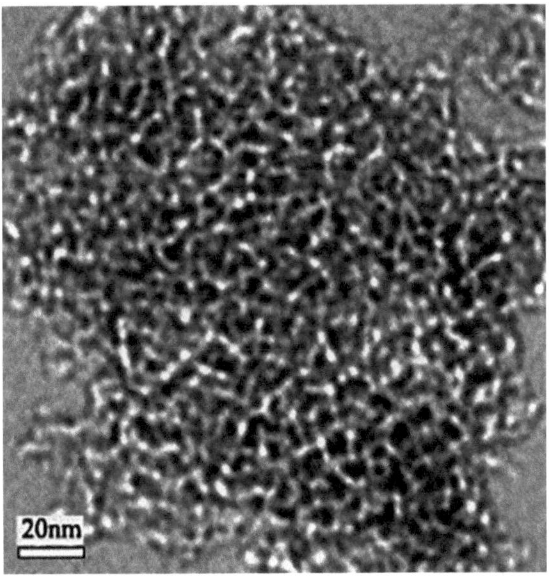

SiC(0001)Si. The etching rate of SiC(000$\bar{1}$)C exhibits an Arrhenius type temperature dependence with an activation energy of 15–20 kcal mol^{-1}. The activation energy and selectivity between the (000$\bar{1}$)C and the (0001)Si surfaces are similar to those for thermal oxidation, suggesting that the surface oxidation occurs during molten KOH etching of SiC and is the rate-limiting step for the etching. An electrolytic plasma-assisted chemical etching method without fluorine-containing reactants was developed to fabricate nanoporous structure on SiC wafer [23]. This method enables the formation of SiC nanostructures via a plasma-enveloped micro-tool traversing over the target material in an electrolyte bath. It produced a uniform nanoporous layer on a 4H–SiC wafer in KOH aqueous solution, with adjustable pore diameters in the range of 40–130 nm. Plasma diagnosis by optical emission spectrometry and surface microanalysis revealed that the nanoporous structure was produced by electrolytic plasma-assisted oxidation and subsequent thermochemical reduction of an oxide (Fig. 3.11a). The maximum etch rate and depth are 540 nm min^{-1} and 10 μm, respectively. The cross-sectional TEM image (Fig. 3.11b) of the nanoporous 4H–SiC specimen revealed dendrite-structured nanofibers and etching front. The TEM energy-dispersive X-ray spectroscopy (EDX) data and mapping (Fig. 3.11c) indicate that the nanofibers are mainly composed of Si and C. The higher proportion of oxygen in the nanofibers with respect to that in the intact substrate indicates apparent oxidation of the nanoporous layer. The effects of current density and KOH concentration on the etching of 4H–SiC were investigated [24]. The etching rate increases and the surface quality decreases with increasing current density and KOH concentration, leading to transition from the flat mode to the porous mode. An etching rate of 1 μm h^{-1} in the flat etching mode was realized. The sample surface was improved by sacrificial oxidation of SiC.

3.2 Luminescence from Nanoporous SiC

The luminescence properties of porous silicon carbide have attracted much interest since the report of strong luminescence from porous silicon at room temperature [25, 26]. Luminescence was observed from porous silicon carbide fabricated by electrochemical etching of (0001) *n*-type (nitrogen-doped) 6H–SiC wafers [27]. The pore size varies from tens of nanometers to sub-micrometers as shown in Fig. 3.12. The porous layer exhibited intense blue-green luminescence with a maximum at around 460 nm (2.7 eV) at room temperature (Fig. 3.13). The luminescence intensity is about 100 times stronger than that arising from donor–acceptor recombination in crystalline 6H–SiC. The peak energy is below the bandgap of crystalline SiC at room temperature, and therefore, this intense blue–green luminescence was suggested to arise from the surface region of porous SiC. Figure 3.14 shows the Raman spectra of the porous and crystalline materials. The porous sample shows a broadened 960 cm^{-1} LO-phonon mode, and the holding mode at below 800 cm^{-1} becomes stronger. The similar positions of the TO and LO phonons in both samples indicate that the porous

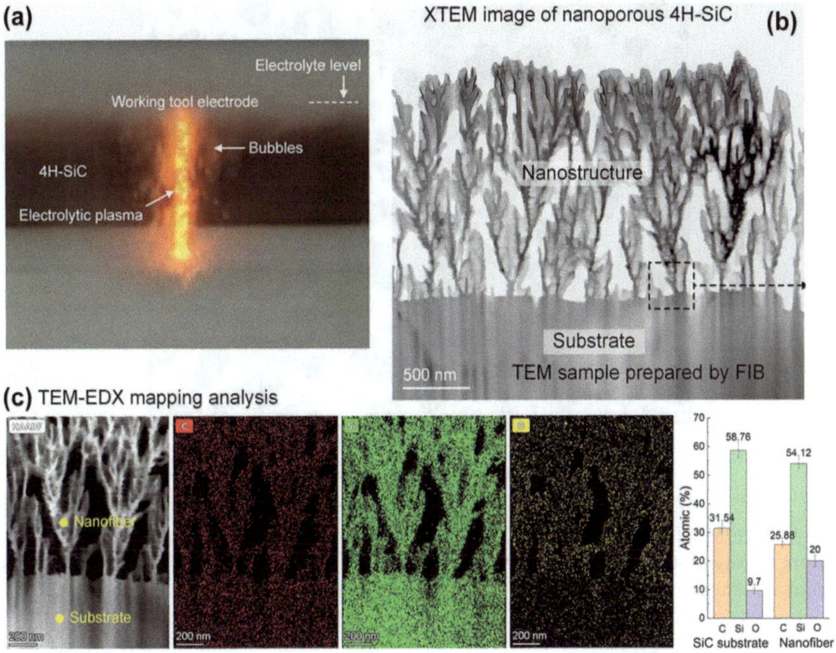

Fig. 3.11 **a** Typical electrolytic plasma during electrolytic plasma-assisted chemical etching process. **b** Cross-sectional TEM image of the cross section of nanoporous 4H–SiC showing the etching residue of dendritic nanofibers. **c** Cross-sectional EDX mapping of the nanofiber structure. Reprinted with permission from [23]. Copyright 2023, John Wiley & Sons

layer retains the crystalline structure of bulk material. The broadening and holding-mode enhancement observed from the porous material may arise from strain defects or phonon confinement. The luminescence decay reveals that the higher porosity (smaller microstructure size) leads to slower decay and a stronger emission. The lifetime varies from 100 picoseconds to nanoseconds, and it was suggested that the carriers in the core crystal were excited and the excited carriers were transferred from the core to the surface, consequently producing the luminescence. The authors further explored the blue light-emitting diode using porous 6H–SiC as a luminescent layer [28]. The diode structure is a Schottky-like junction between indium tin oxide and porous SiC. The junction showed a rectification behavior and blue electroluminescence under the forward bias. The electroluminescence has a broad spectrum from 350 to 780 nm with a peak at 470 nm, close to that of the PL. The electroluminescence spectrum is broader than the PL spectrum. Electroluminescence of porous SiC fabricated by electrochemical etching was also realized from a forward-biased 4H–SiC *p–i–n* junction [29]. A broad green PL centered at ~ 510 nm was supposed to stem from porous SiC. A lateral bipolar diffusion model was proposed, in which mobile carriers diffused laterally from the junction toward the porous SiC surface region driven by a lateral carrier concentration gradient.

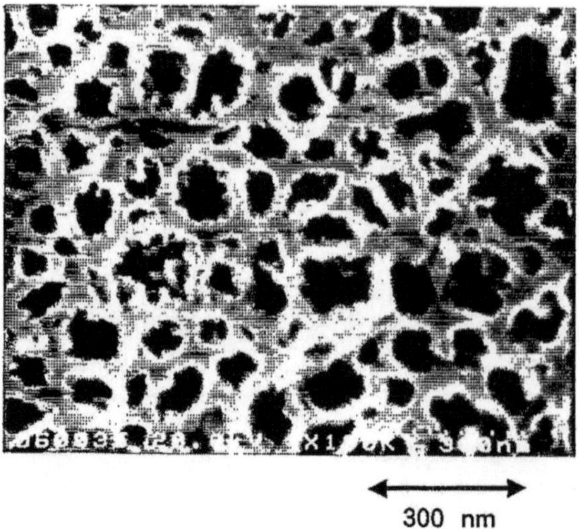

Fig. 3.12 SEM image of the porous 6H–SiC layer fabricated by electrochemical anodization at a current density of 40 mA cm^{-2}. Reprinted with permission from [27]. Copyright 1994, American Institute of Physics

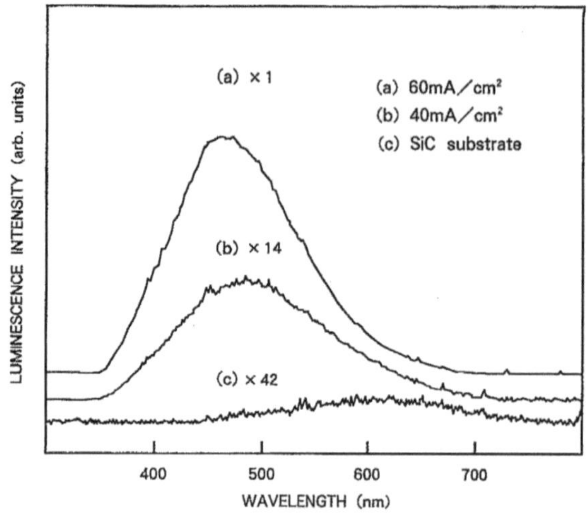

Fig. 3.13 Photoluminescence spectra obtained from porous SiC fabricated at anodization current densities of **a** 60 mA cm^{-2} and **b** 40 mA/cm^{-2}. The spectrum of the crystalline 6H–SiC is shown for comparison. Reprinted with permission from [27]. Copyright 1994, American Institute of Physics

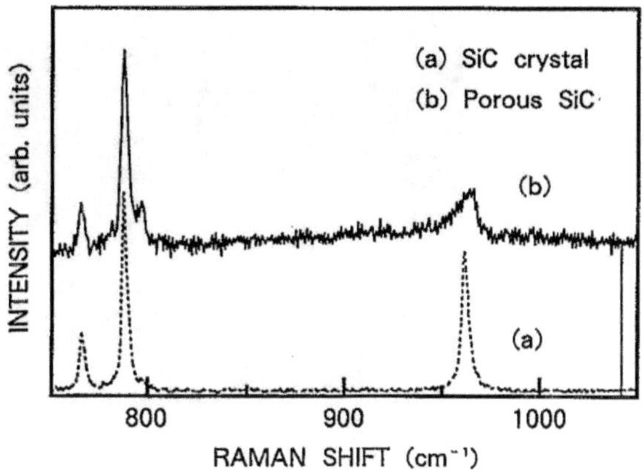

Fig. 3.14 Raman scattering spectra of porous SiC and crystalline SiC. Reprinted with permission from [27]. Copyright 1994, American Institute of Physics

Porous *p*-type 6H–SiC materials prepared by electrochemical anodization of *p*-type 6H–SiC wafers in HF have a different luminescence behavior [30]. Transmission electron microscopy revealed uniformly dispersed pores in the porous layer with an interpore spacing of l–10 nm (Fig. 3.15). The porous film exhibited as a single crystal. Cathodoluminescence comprises a broadband at about 500 nm (2.5 eV) from the sample anodized at 50 mA cm^{-2} (Fig. 3.16). A narrow and weaker peak arising at 3.7 eV (above bandgap) is likely related to the interband transition, but it is absent from the luminescence spectrum of the sample anodized at 5 mA cm^{-2}.

The porous silicon carbide materials obtained by anodization of (mostly *n*-type) 3C–, 6H–, and 4H–SiC crystals showed similar luminescence characteristics [31]. The porous material prepared without UV-light illumination has an average fiber size of 0.5–1 μm, whereas that obtained with the aid of UV light has a fiber size of smaller than 50 nm. Both materials have the dendritic structure and contain some much smaller fibers. The porous silicon carbide materials prepared from different polytypes yielded nearly the same luminescence peak at around 2.5 eV (Fig. 3.17). The close PL peak positions observed from the three different polytypes exclude deep defect centers in the bandgap as the origin of the observed PL, because the defects in the bulk materials generally show different emission photon energies due to the large difference in the bandgap energies of 3C, 6H, and 4H polytypes. For instance, the D_1 defect center has an energy comparable to that of the observed PL peak, but the D_1 PL peak position varies by 0.9 eV among these polytypes. Annealing in oxygen at 700 °C quenched the blue–green PL completely (Fig. 3.18), but the emission band was partially recovered by dipping the oxidized crystal in 40% HF, suggesting that PL is related to defect states in the etched surface. The luminescence peaks at 2.43, 2.22, 2.07, and 1.93 eV were identified in the PL spectra by using selective excitation of the PL by tuning excitation wavelength. The result shows that the change of the fiber size in the porous structure gave rise to no additional PL bands.

Photoluminescence from the porous 6H–SiC prepared by anodization of *n*-type 6H–SiC with a net doping n_D–n_A of 5 × 10^{18} cm^{-3} was studied [32]. The materials were electrochemically etched in HF (alcohol: 50 wt% HF = 1: 1) at a current

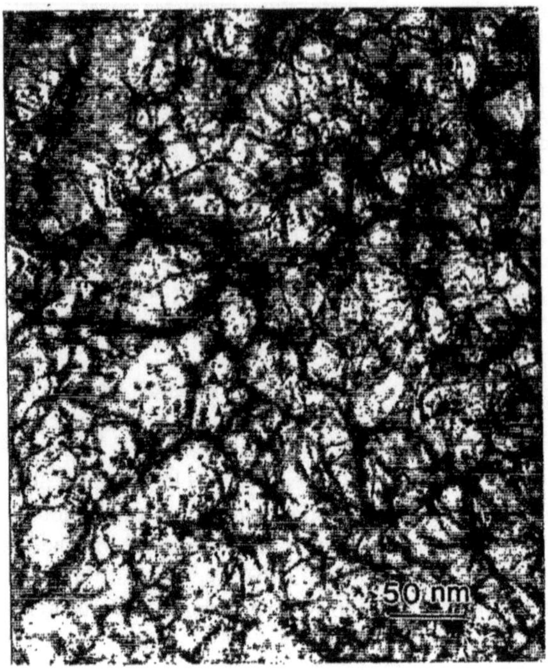

Fig. 3.15 TEM image of porous *p*-type 6H–SiC fabricated by electrochemical anodization at an anodization current density of 50 mA cm^{-2}. Reprinted with permission from [30]. Copyright 1994, American Institute of Physics

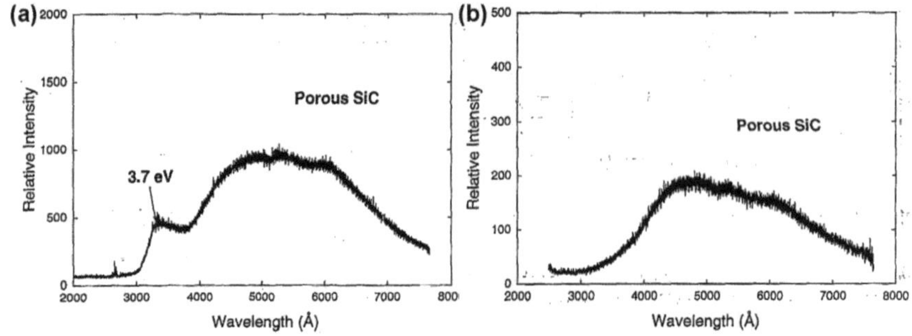

Fig. 3.16 Cathodoluminescence spectra of porous *p*-type 6H–SiC anodized at **a** 50 mA cm^{-2} and **b** 5 mA cm^{-2} at room temperature. Reprinted with permission from [30]. Copyright 1994, American Institute of Physics

density of 10–500 mA cm^{-2} under UV-light illumination. The size of the crystalline grains in the porous structure is typically 30 nm. The room-temperature luminescence spectrum of the porous material is similar to that of the bulk materials (Fig. 3.19), showing a broadband at around 2.65 eV. This luminescence was ascribed to the

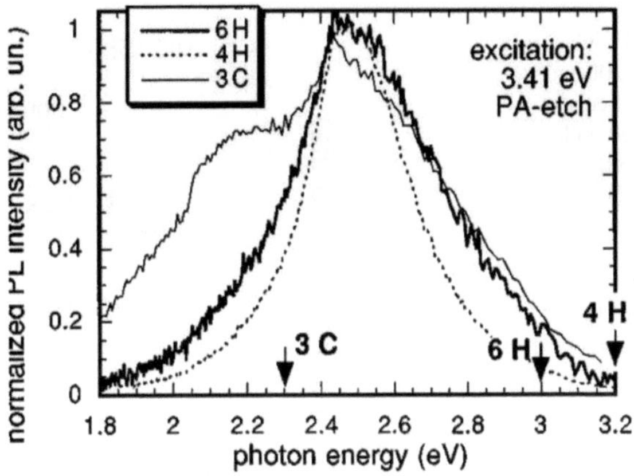

Fig. 3.17 Normalized photoluminescence spectra of the porous silicon carbide prepared by anodization of different polytypes of SiC crystals. The excitation photon energy is above the bandgaps of all the polytypes. Reprinted with permission from [31]. Copyright 1995, American Institute of Physics

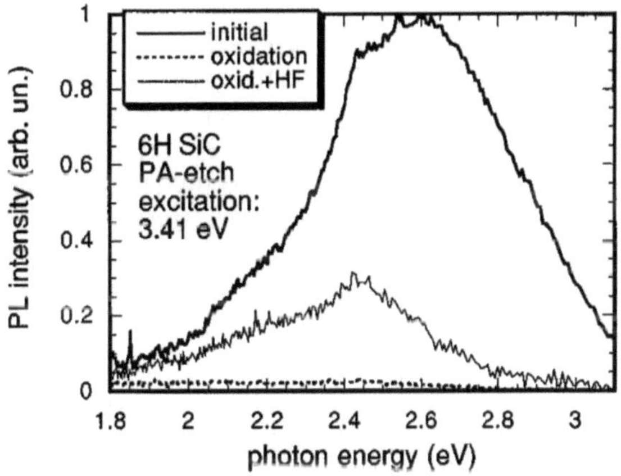

Fig. 3.18 Photoluminescence spectra of the as-prepared, oxidized, and oxidized and HF-dipped porous 6H–SiC. Reprinted with permission from [31]. Copyright 1995, American Institute of Physics

recombination associated with the donor–acceptor pairs, but it is stronger from the porous material by a factor of 100. With regard to porous silicon carbide, the intensity was further increased by a factor of two when the temperature dropped from 300 to 5 K. The large increase in PL intensity in porous material is due to the geometrical carrier confinement in the granular and porous material. But no evidence of quantum confinement was observed. Infrared transmission study (Fig. 3.20) indicates that the porous layer was predominantly terminated by CH_2 and CH_3. The strongest absorption in the 800–1000 cm^{-1} range was attributed to the Si–C TO and LO phonon

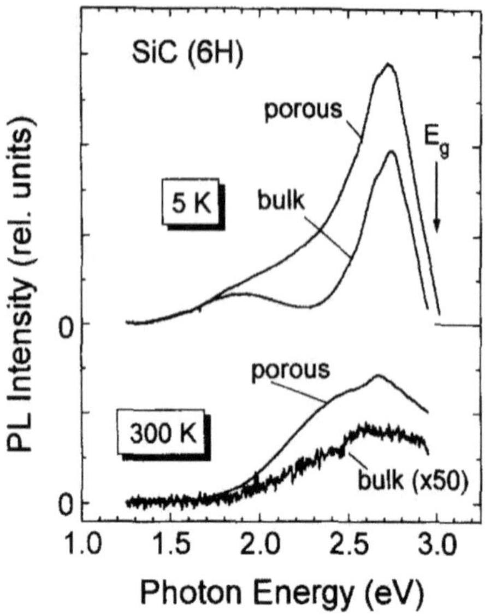

Fig. 3.19 PL spectra obtained from bulk and porous 6H–SiC at 300 and 5 K at an excitation energy of 3.5 eV. Reprinted with permission from [32]. Copyright 1995, Elsevier B.V

vibrations, but the Si–H signal expected at 2100–2200 cm^{-1} was not observed. This is because that the Si–H bond is readily oxidized, but the C–H bond has stronger resistance to oxidation, thus remaining stable under ambient conditions.

The reflectance of porous SiC (PSC) thin films on *p*-type 6H–SiC substrates was measured in the infrared Reststrahlen region by using FTIR reflectance spectroscopy and was compared to simulated spectra based on phenomenological and Bergman statistical effective-medium dielectric functions [33]. Figure 3.21a shows the IR spectra in the Reststrahlen region of thin (PSC1) and thick (PSC2) highly porous (porosity $f = 50.74$) *p*-PSC films on bulk 6H–SiC substrates, the corresponding bare bulk 6H–SiC substrate, and a thick, moderately porous ($f = 50.50$) *n*-PSC film on a bulk 4H–SiC substrate (PSC3). The IR reflectance features from 4H–SiC and 6H–SiC substrates are nearly identical. Figure 3.22b shows the reflectance of the PSC films at higher energies, with interference fringes within 1000–4000 cm^{-1}. The spectra of the *p*-PSC and *n*-PSC films are very different from that of bulk SiC (Fig. 3.21a). For porous SiC, the SiC Reststrahlen band splits into two bands: a broad one at lower frequency and a sharp higher-frequency one near the LO frequency. The low-frequency shoulder of the low-frequency band in porous SiC is similar to that of bulk, whereas the high-frequency shoulder of this band has a more gradual “descent”. This dip in the Reststrahlen band near 950–960 cm^{-1} appeared in all porous SiC samples. The evaluated phenomenological models include Bruggeman, cavity- and sphere-Maxwell–Garnett (C-MG and S-MG), Landau–Lifshitz/Looyenga (LLL), and Monecke models. Hybrid C-MG and LLL models were considered by mixing the statistical spectral density functions of the C-MG and LLL effective dielectric functions. The hybrid model agrees the best with experiments. The dip (or splitting)

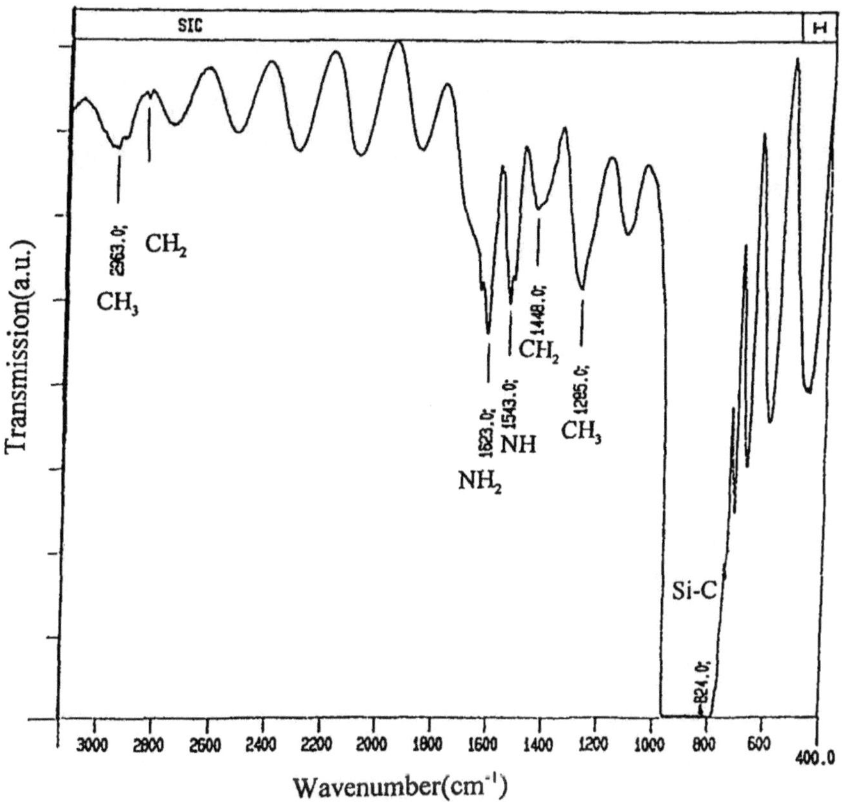

Fig. 3.20 Infrared transmission spectrum of the free standing porous 6H–SiC layer. Reprinted with permission from [32]. Copyright 1995, Elsevier B.V

in the porous SiC film Reststrahlen band was ascribed to surface optical phonon modes.

Blue green photoluminescence was observed from anodized *n*-type and *p*-type 6H–SiC [34]. The starting materials are 6H–SiC single crystals with $n_D–n_A = 2.5 \times 10^{17}$ cm^{-3} (*n*-type, N-doped) and $n_A–n_D = 8.3 \times 10^{17}$ cm^{-3} (*p*-type, Al-doped). The *p*-type samples were etched in darkness in a mixture of HF and ethanol (10–50 wt% HF: ethanol = 1: 1 in volume) at current densities of 20–60 mA cm^{-2}. The *n*-type SiC was etched under UV-light illumination at a constant anodization voltage of 3 V. SEM shows that the pore size and the pore spacing of the anodized *n*-type sample are on the order of 100 nm (Fig. 3.22). The anodized *p*-type sample shows very different dendritic structures with sizes as small as 10 nm. The *n*-type bulk wafer shows an interior defect-related luminescence band at 2.2 eV with a full width at half maximum of 0.5 eV (Fig. 3.23). In contrast, the porous sample shows a broader band at about 2.6 eV, whereas the initial luminescence at 2.2 eV was reduced substantially. The energy of this peak is below the bandgap of bulk 6H–SiC (2.86 eV), and there

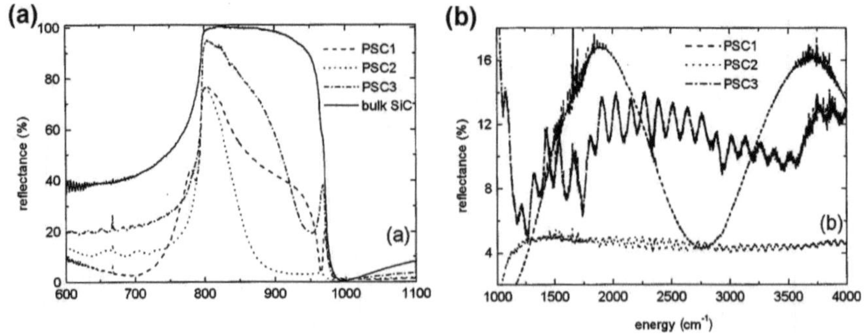

Fig. 3.21 Infrared reflectance, **a** in the Reststrahlen band from three porous SiC films, each on a SiC substrate, along with the IR reflectance from a bare 6H–SiC substrate, and **b** above the Reststrahlen band, showing fringes due to interference in the films. Reprinted with permission from [33]. Copyright 2000, American Physical Society

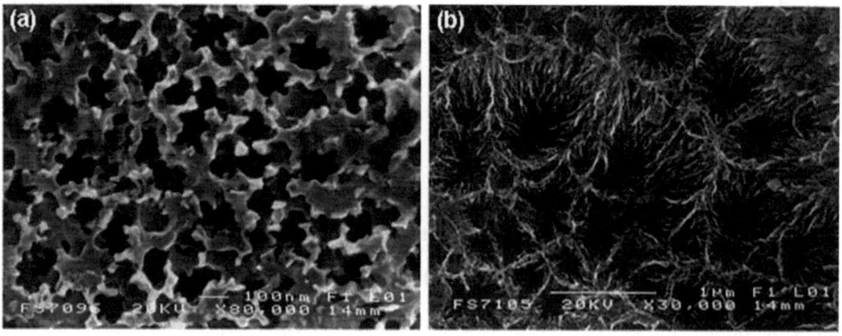

Fig. 3.22 SEM images of **a** porous *n*-type 6H–SiC layer and **b** supercritically dried porous *p*-type SiC layer. Reprinted with permission from [34]. Copyright 1997, Elsevier B.V

is no strong dependence on the illumination intensity and HF concentration. The PL spectrum of the porous *p*-type sample shows a strong peak at 2.5 eV, similar to the *n*-type sample. The time-dependent PL measurement shows that the blue emission band has a decay lifetime lying in the nanosecond and microsecond regime. It may arise from some surface defect, and no evidence of quantum size effect was observed. In another study of anodized porous *p*-type 6H–SiC, the transmission spectrum shows a very wide transmission edge which may be caused by surface states [35]. The porous sample exhibited room-temperature PL that is 20 times stronger than that of the bulk material. This PL is temperature independent.

The morphology and luminescence characteristics of porous *n*-type 6H–SiC have been investigated [36]. The Si-terminated *n*-type 6H–SiC(0001) surface was electrochemically etched by diluted HF (5%) at a current density of 10–80 mA cm^{-2} under UV-light irradiation. Most of the porous samples have irregular triangular etch pits, and the porous nanocrystallites have an average diameter of about 25 nm (Fig. 3.24).

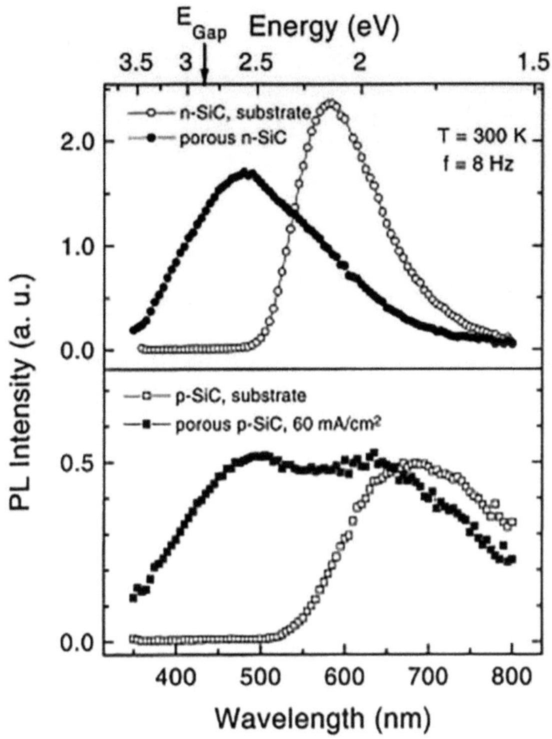

Fig. 3.23 Photoluminescence spectra of porous *n*-type and *p*-type SiC as well as initial bulk wafers. Reprinted with permission from [34]. Copyright 1997, Elsevier B.V

In some porous samples, the triangular etch pits are oriented regularly in planes parallel to the surface of the wafer. The cathodoluminescence spectra of the porous layers have peaks in the UV region (Fig. 3.25) with energies above the bandgap of bulk 6H–SiC. The photoluminescence also shifted to blue in the porous SiC, although with a smaller shift. The Raman spectroscopy indicates that the evolution of the blue-shifted cathodoluminescence is strongly correlated with the presence of a broadened low-frequency shoulder on the polar A_1 LO phonon mode of 6H–SiC. The shoulder can be quantitatively described when surface mode scattering is taken into account, implying that surface states induced by etching account for the observed blue-shifted luminescence. An electroless approach was utilized to produce porous silicon carbide [37]. Platinum was deposited on the *n*-type 4H– or 6H–SiC wafer prior to etching to serve as a catalyst for the reduction of the oxidant $K_2S_2O_8$. The oxidant in combination with UV-light illumination injected holes into the valence band, and the holes then induced oxidation of the wafer. The resulting silicon dioxide was dissolved in hydrofluoric acid, finally producing the porous structure.

The effects of current density on luminescence properties of anodized porous silicon carbide have been investigated [38]. The *n*-type (nitrogen-doped) 6H–SiC

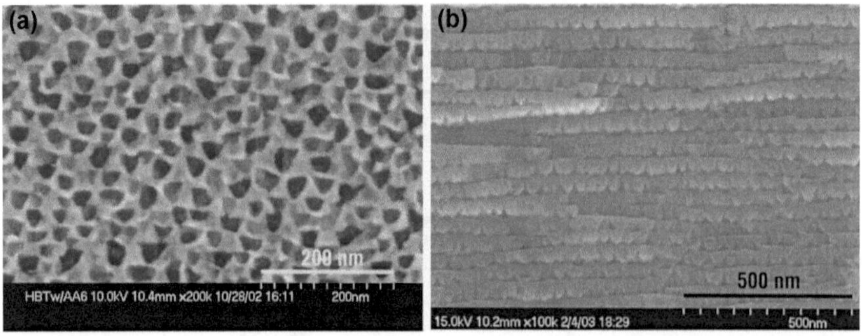

Fig. 3.24 Cross-sectional SEM images of two porous *n*-type 6H–SiC samples prepared by electrochemical etching at 35 mA cm^{-2}. One, **a** has small and triangular nanocrystallites and etching pits and the other, **b** has parallel etching pits. Reprinted with permission from [36]. Copyright 2004, American Institute of Physics

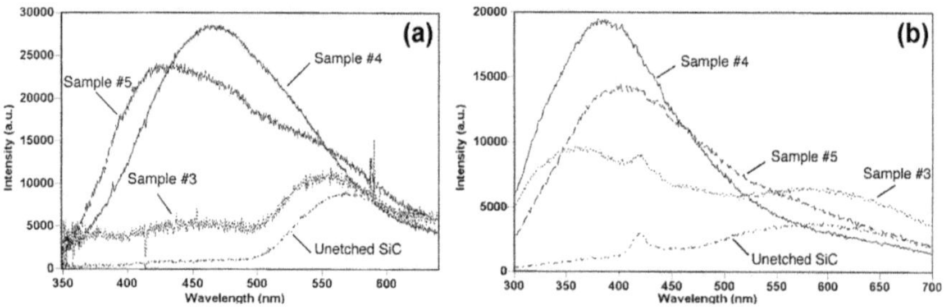

Fig. 3.25 Photoluminescence, **a** (excited at 325 nm) and cathodoluminescence, **b** spectra of anodized porous 6H–SiC samples and unetched 6H–SiC wafer. Reprinted with permission from [36]. Copyright 2004, American Institute of Physics

crystals were electrochemically etched in an electrolyte under UV-light irradiation. The electrolyte contains HF, ethylene glycol (ethane-1, 2-diol), and hydrogen peroxide (H_2O_2), with a volume ratio of 2: 1: 1. A pulsed current with a duration of 10 ms and interval of 4 ms was applied during etching, and the current densities are 10, 15, and 20 mA cm^{-2}. SEM (Fig. 3.26) shows that the morphology of the porous materials does not depend on the etching current density. The bulk 6H–SiC showed PL at about 410 nm (3.02 eV) at room temperature (Fig. 3.27). In contrast, the PL from the porous samples shifted to blue slightly and the shift increased with current density, thus showing sign of quantum confinement, but the full width at half maximum of the PL is rather narrow (13–22 nm) compared with that of conventionally observed PL in SiC at room temperature for unknown reasons. Raman scattering revealed an intensity enhancement for all the phonon modes with increasing etching current density. The A_1 LO phonon in the porous samples has a red shift compared to that of the bulk materials (Fig. 3.28), and the red shift is larger for bigger etching current density. This may be related to the tensile stress in the porous samples.

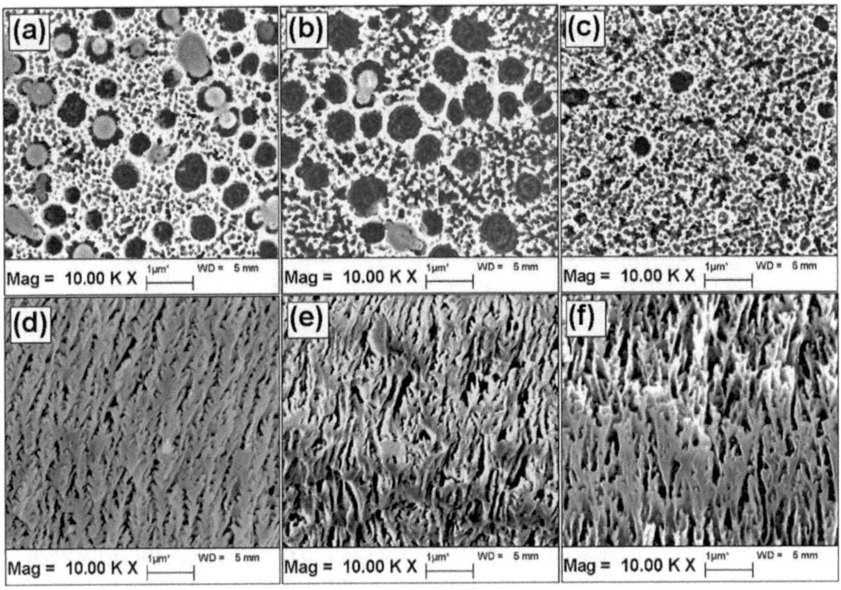

Fig. 3.26 Planar SEM micrographs of porous *n*-type 6H–SiC fabricated by UV-assisted electrochemical etching at current densities of **a** 10, **b** 15, and **c** 20 mA cm^{-2}. The cross-sectional images are shown in (**d**), (**e**), and **f**, respectively. Reprinted with permission from [38]. Copyright 2013, IOP Publishing

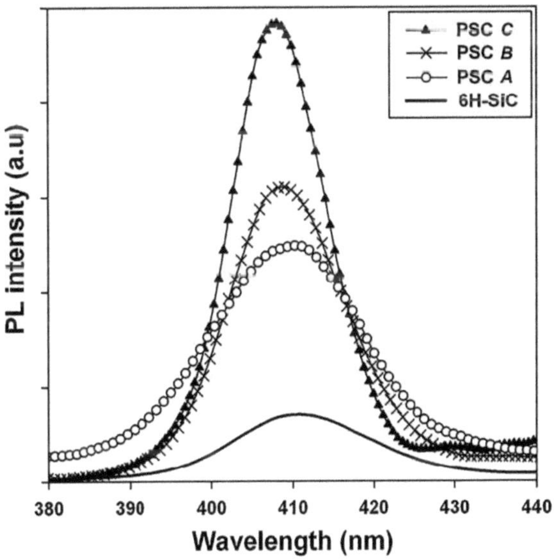

Fig. 3.27 Photoluminescence spectra of the porous 6H–SiC samples prepared at etching current densities of 10 mA cm^{-2} (PSC A), 15 mA cm^{-2} (PSC B), and 20 mA cm^{-2} (PSC C) as well as unetched 6H–SiC. Reprinted with permission from [38]. Copyright 2013, IOP Publishing

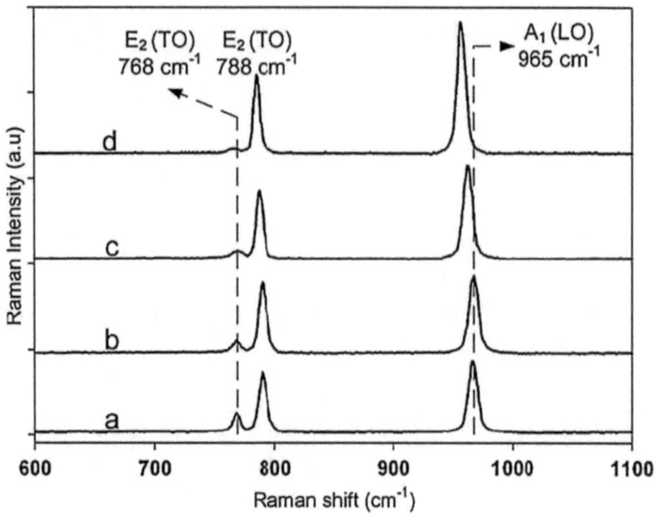

Fig. 3.28 Raman spectra of **a** unetched 6H–SiC and porous 6H–SiC prepared at etching current densities of **b** 10, **c** 15, and **d** 20 mA cm^{-2} in the 600–1100 cm^{-1} range. Reprinted with permission from [38]. Copyright 2013, IOP Publishing

Atomic layer deposition of Al_2O_3 films on the porous (B and N co-doped) 6H–SiC wafer fabricated by electrochemical etching generated significant PL (at ~ 520 nm) enhancement of the latter (up to 6.9 times) [39]. The optimized passivation conditions for porous SiC are as follows: 20 nm in thickness of Al_2O_3, deposition temperature of 160 °C, N_2 purge time of 20 s, and followed annealing at 350 °C for 5 min. Subsequent study reports complex luminescence from the Al_2O_3 film-covered porous 6H–SiC wafers, comprising the blue–green emission from the porous layer and yellow emission from the bulk SiC layer [40]. Two emission bands centered at 460 and 530 nm were observed from the porous layers fabricated from both commercial *n*-type and lab grown N-B co-doped 6H–SiC. It is well known that porous or nanostructured alumina can yield near-UV and visible luminescence; the role of the deposited alumina in the observed wideband luminescence of the Al_2O_3-porous 6H–SiC composite is unclear. Further study shows that the atomic-layer-deposited Al_2O_3 film can reach a passivation depth of 30 μm deep into the C-face porous SiC [41]. It was concluded that the pulsed-voltage etching is preferable for fabricating uniform porous structures compared with the constant-voltage etching.

3.3 Applications

Porous silicon carbide, like porous silicon, can be used in sensing applications. Because silicon carbide is chemically inert, sensors composed of porous SiC are well suited for applications in harsh environments. Preliminary studies indicate that porous SiC fabricated by electrochemical etching can be used to construct humidity sensors [42, 43]. The capacitance and resistance measured were used to derive the humidity in the environment. An NH_3 gas sensor based on porous silicon carbide was also produced [44], and preliminary data suggest that this sensor can detect changes in the ammonia concentration of about 1–2 ppm. A metal–semiconductor–metal ultraviolet photodetector was fabricated based on porous SiC [45]. SiC was deposited on porous silicon substrates via radio frequency magnetron sputtering. The as-deposited SiC was amorphous and converted into crystalline by annealing. To fabricate photodetectors, two inter-digitated Schottky contacts of Ni with four fingers for each electrode were deposited onto the porous substrates. The electrical characteristics of fabricated UV photodetectors on annealed substrates showed enhanced sensitivity and responsivity to ultraviolet radiation. Ordered nanoporous SiC can act as stable field emitter capable of electron emission at levels comparable to thermal sources [46]. The porous SiC structures were fabricated by electrochemical etching using two processing conditions [11]. The highly *n*-doped 6H–SiC wafers were electrochemically etched into a continuous and highly porous structure, which subsequently experienced ion etching to form a variety of arrays, while maintaining morphology and porosity. Therefore, the emitter is a monolithic structure. Stable emission of 6 A cm^{-2} was demonstrated in a macroscopic array. These emitters are sufficiently robust. The field emission properties are tunable through control of the nanoscale morphology and the macroscopic shape of the emitter array.

References

1. Brack K (1965) X-ray method for the determination of the polarity of SiC crystals. J Appl Phys 36:3560–3562
2. Shor JS, Zhang XG, Osgood RM (1992) Laser-assisted photoelectrochemical etching of n-type beta-SiC. J Electrochem Soc 139:1213–1216
3. Shor JS, Osgood RM, Kurtz AD (1992) Photoelectrochemical conductivity selective etch stops for SiC. Appl Phys Lett 60:1001–1003
4. Shor JS, Grimberg I, Weiss B-Z, Kurtz AD (1993) Direct observation of porous Sic formed by anodization in HF. Appl Phys Lett 62:2836–2838
5. Shor JS, Osgood RM (1993) Broad-area photoelectrochemical etching of n-Type beta-SiC. J Electrochem Soc 140:L123–L125
6. Shor JS, Kurtz AD (1994) Photoelectrochemical etching of 6H–SiC. J Electrochem Soc 141:778–781
7. Shi D, Chen Y, Li Z, Dong S, Li L, Hou M, Liu H, Zhao S, Chen X, Wong C-P, Zhao N (2022) Anisotropic charge transport enabling high-throughput and high-aspect-ratio wet etching of silicon carbide. Small Methods 6:2200329
8. Takazawa A, Tamura T, Yamada M (1993) Porous β-SiC fabrication by electrochemical anodization. Jpn J Appl Phys 32:3148–3149
9. Konstantinov AO, Harris CI, Janzén E (1994) Electrical properties and formation mechanism of porous silicon carbide. Appl Phys Lett 65:2699–2701
10. Shishkin Y, Choyke WJ, Devaty RP (2004) Photoelectrochemical etching of *n*-type 4H silicon carbide. J Appl Phys 96:2311–2322
11. Ke Y, Devaty RP, Choyke WJ (2008) Comparative columnar porous etching studies on n-type 6H SiC crystalline faces. Phys Stat Sol (B) 245:1396–1403

12. Zhao L, Chen S, Wang L, Gao F, Yao X, Yang W (2018) Large-scale fabrication of free-standing and transparent SiC nanohole array with tailored structures. Ceram Int 44:7280–7285
13. King SW, Nemanich RJ, Davis RF (1999) Wet chemical processing of (0001)Si 6H–SiC hydrophobic and hydrophilic surfaces. J Electrochem Soc 146:1910–1917
14. Dhar S, Seitz O, Halls MD, Choi S, Chabal YJ, Feldman LC (2009) Chemical properties of oxidized silicon carbide surfaces upon etching in hydrofluoric acid. J Am Chem Soc 131:16808–16813
15. Cambaz GZ, Yushin GN, Gogotsi Y, Lutsenko VG (2006) Anisotropic etching of SiC whiskers. Nano Lett 6:548–551
16. Sugita T, Hiramatsu K, Ikeda S, Matsumura M (2013) Fabrication of pores in a silicon carbide wafer by electrochemical etching with a glassy-carbon needle electrode. ACS Appl Mater Interfaces 5:2580–2584
17. Nakamura S, Kimoto T, Matsunami H, Tanaka S, Teraguchi N, Suzuki A (2000) Formation of periodic steps with a unit-cell height on 6H–SiC (0001) surface by HCl etching. Appl Phys Lett 76:3412–3414
18. Liu Y, Lin W, Lin Z, Xiu Y, Wong CP (2012) A combined etching process toward robust superhydrophobic SiC surfaces. Nanotechnology 23:255703
19. Michaels JA, Janavicius L, Wu X, Chan C, Huang H-C, Namiki S, Kim M, Sievers D, Li X (2021) Producing silicon carbide micro and nanostructures by plasma-free metal-assisted chemical etching. Adv Funct Mater 31:2103298
20. Jin G-Q, Guo X-Y (2003) Synthesis and characterization of mesoporous silicon carbide. Micropor Mesopor Mat 60:207–212
21. Sung I-K, Christian, Mitchell M, Kim D-P, Kenis PJA (2005) Tailored macroporous SiCN and SiC structures for high-temperature fuel reforming. Adv Funct Mater 15:1336–1342
22. Katsuno M, Ohtani N, Takahashi J, Yashiro H, Kanaya M (1999) Mechanism of molten KOH etching of SiC single crystals: comparative study with thermal oxidation. Jpn J Appl Phys 38:4661–4665
23. Zhan S, Liu B, Yu X, Chen X, Zeng G, Zhao Y (2023) Fabrication of homogeneous nanoporous structure on 4H-/6H–SiC wafer surface via efficient and eco-friendly electrolytic plasma-assisted chemical etching. Small 19:2205720
24. Yang S, Zhao S, Chen J, Li Y, Yan G, Guan M, Zhang Y, Sun G, Zeng Y, Liu X (2023) Electrochemical etching modes of 4H–SiC in KOH solutions. Semicond Sci Technol 38:055019
25. Lehmann V, Gösele U (1991) Porous silicon formation: a quantum wire effect. Appl Phys Lett 58:856–858
26. Canham LT (1990) Silicon quantum wire array fabrication by electrochemical and chemical dissolution of wafers. Appl Phys Lett 57:1046–1048
27. Matsumoto T, Takahashi J, Tamaki T, Futagi T, Mimura H, Kanemitsu Y (1994) Blue-green luminescence from porous silicon carbide. Appl Phys Lett 64:226–228
28. Mimura H, Matsumoto T, Kanemitsu Y (1994) Blue electroluminescence from porous silicon carbide. Appl Phys Lett 65:3350–3352
29. Bawa S, Zhang T, Dow L, Peter S, Kitai AH (2021) Porous SiC electroluminescence from p–i–n junction and a lateral carrier diffusion model. J Appl Phys 129:044501
30. Shor JS, Bemis L, Kuttz AD, Grimberg I, Weiss BZ, MacMillian MF, Choyke WJ (1994) Characterization of nanocrystallites in porous *p*-type 6H–SiC. J Appl Phys 76:4045–4049
31. Konstantinov AO, Henry A, Harris CI, Janzén E (1995) Photoluminescence studies of porous silicon carbide. Appl Phys Lett 66:2250–2252
32. Petrova-Koch V, Sreseli O, Polisski G, Kovalev D, Muschik T, Koch F (1995) Luminescence enhancement by electrochemical etching of SiC(6H). Thin Solid Films 255:107–110
33. Spanier JE, Herman IP (2000) Use of hybrid phenomenological and statistical effective-medium theories of dielectric functions to model the infrared reflectance of porous SiC films. Phys Rev B 61:10437–10450
34. Jessensky O, Müller F, Gösele U (1997) Microstructure and photoluminescence of electrochemically etched porous SiC. Thin Solid Films 297:224–228

35. Kim S, Spanier JE, Herman IP (2000) Optical transmission, photoluminescence, and Raman scattering of porous SiC prepared from *p*-type 6H SiC. Jpn J Appl Phys 39:5875–5878
36. Rittenhouse TL, Bohn PW, Hossain TK, Adesida I, Lindesay J, Marcus A (2004) Surface-state origin for the blueshifted emission in anodically etched porous silicon carbide. J Appl Phys 95:490–496
37. Rittenhouse TL, Bohn PW, Adesida I (2003) Structural and spectroscopic characterization of porous silicon carbide formed by Pt-assisted electroless chemical etching. Solid State Commun 126:245–250
38. Naderi N, Hashim MR, Saron KMA, Rouhi J (2013) Enhanced optical performance of electrochemically etched porous silicon carbide. Semicond Sci Technol 28:025011
39. Lu W, Iwasa Y, Ou Y, Jinno D, Kamiyama S, Petersena PM, Ou H (2017) Effective optimization of surface passivation on porous silicon carbide using atomic layer deposited Al_2O_3. RSC Adv 7:8090
40. Lu W, Tarekegne AT, Ou Y, Kamiyama S, Ou H (2019) Effective optimization of surface passivation on porous silicon carbide using atomic layer deposited Al_2O_3. Sci Rep 9:16333
41. Yanai K, Lu W, Yamane Y, Kodera K, Ou Y, Ou H, Kamiyama S, Takeuchi T, Iwaya M, Akasaki I (2022) Improved passivation depth of porous fluorescent 6H–SiC with Si/C faces using atomic layer deposition. Jpn J Appl Phys 61:035502
42. Connolly EJ, O'Halloran GM, Pham HTM, Sarro PM, French PJ (2002) Comparison of porous silicon, porous polysilicon and porous silicon carbide as materials for humidity sensing applications. Sensor Actuat A 99:25–30
43. Connolly EJ, Pham HTM, Groeneweg J, Sarro PM, French PJ (2004) Relative humidity sensors using porous SiC membranes and Al electrodes. Sensor Actuat B 100:216–220
44. Connolly EJ, Timmer B, Pham HTM, Groeneweg J, Sarro PM, Olthuis W, French PJ (2005) A porous SiC ammonia sensor. Sensor Actuat B 109:44–46
45. Naderi N, Hashim MR (2013) Porous-shaped silicon carbide ultraviolet photodetectors on porous silicon substrates. J Alloy Compd 552:356–362
46. Kang M-G, Lezec HJ, Sharifi F (2013) Stable field emission from nanoporous silicon carbide. Nanotechnology 24:065201

Chapter 4
SiC Nanoparticles and Quantum Dots

Silicon carbide nanoparticles generally refer to particles about 1–100 nm in size and are single crystals in most cases. Silicon carbide nanocrystals (NCs) smaller than around 10 nm are also called SiC quantum dots (QDs). In this size regime, on account of the strong spatial confinement of carriers, SiC nanocrystals exhibit obvious quantum size effects, especially the quantum confinement effect. As a result, the energy gap of the silicon carbide nanocrystals widens and their interband transition luminescence shifts toward the smaller wavelength region (blue shift) gradually as their sizes diminish. In addition, their luminescence quantum yields may be improved by several orders of magnitude relative to that of the bulk materials. Note that when sizes of the NCs are only a couple of nanometers or even below one nanometer, the energy levels deviate from the quasi-continuous ones exhibited in bulk crystals and become more and more discrete with decreasing size owing to strong quantum confinement effect, being like those of big molecules. In this case, terms "band" and "bandgap" are not so accurate again, but they are still used sometimes for brevity. Lowest unoccupied molecular orbital (LUMO) and highest occupied molecular orbital (HOMO) are better terms to substitute for "conduction band minimum" and "valence band maximum" for ultrasmall nanoparticles with rather discrete energy levels. Since the sizes of the nanoparticles are very small, their surfaces strongly affect their properties. Silicon carbide is a compound semiconductor and its surface has fruitful bonding structures involving Si, C, O, H, and other atoms, and so a variety of surface-defect-dominated color centers that accommodate localized orbitals can form, consequently leading to complex luminescence properties. In this chapter, the preparation methods, electronic structures, luminescence properties, and potential applications of silicon carbide nanoparticles are discussed.

J. Fan and P. K. Chu, *Silicon Carbide Nanostructures*, Engineering Materials and Processes, https://doi.org/10.1007/978-3-032-10986-6_4

4.1 Fabrication and Luminescence Properties

Generally, silicon carbide nanoparticles are synthesized by using vapor phase or solution-based chemical reactions. Chemical etching of large silicon carbide single crystals combined with ultrasonic dispersion is another important fabrication method. The luminescence properties of these synthesized SiC nanoparticles have been intensively studied. In general, recombination of delocalized conduction band electrons and valence band holes in the interior region is an important luminescence mechanism of SiC nanocrystals, whereas recombination of an electron–hole pair, with either or both carriers (electron and hole) located at a surface defect, is another important luminescence mechanism. The former mechanism shows the size dependence of the luminescence peak wavelength, and the luminescence obeys the quantum confinement effect. In contrast, the defects associated luminescence is expected to show a relatively constant peak wavelength. These two types of luminescence interweave and compete with each other, bring complex and fruitful luminescence behaviors of small SiC nanocrystals.

4.1.1 Chemical Vapor Synthesis

In 1983, a radiofrequency plasma tube was designed to overcome the problem of overheating in the conventional plasma systems [1]. Using this system, ultrafine and ultrapure silicon carbide powders can be synthesized by radial injection of silane, methane, and hydrogen into the tail flame of the plasma. The composition of the silicon carbide powder varies over a wide range from excess silicon to excess carbon, although β-SiC is always the dominant constituent in the product, as confirmed by X-ray diffraction. Transmission electron microscopy shows that the synthesized SiC particles are 10 to 20 nm in size (Fig. 4.1) and XRD reveals a domain size of 7.5 nm and crystal structure of β-SiC. Hypersonic plasma particle deposition has been employed to synthesize silicon carbide coatings on molybdenum substrates [2]. The $SiCl_4$ and CH_4 vapor as precursors in deposition was injected into the Ar–H_2 plasma. SEM, TEM, and XRD indicate that the silicon carbide coatings were partially consolidated, showing layered morphology, with all the layers containing nanostructured SiC grains of around 30 nm (Fig. 4.2). Aerosol probe measurement also demonstrates that the small particle residence time (about 50 μs) minimizes particle agglomeration. XRD and FTIR show that the silicon carbide coatings are amorphous (Fig. 4.2) and hydrogenated. XPS indicates that oxidation of the deposit is limited to the surface (Fig. 4.3).

β-SiC powders containing nanocrystallites smaller than 10 nm were synthesized by thermal decomposition of silicon organic precursors at reduced pressure [3]. The optimal conditions for synthesizing small grains of low agglomeration and high crystallinity are: tetramethylsilane [$Si(CH_3)_4$] precursor, total pressure below 1000 kPa, precursor partial pressure below 660 Pa, and reaction temperature above 1400 K.

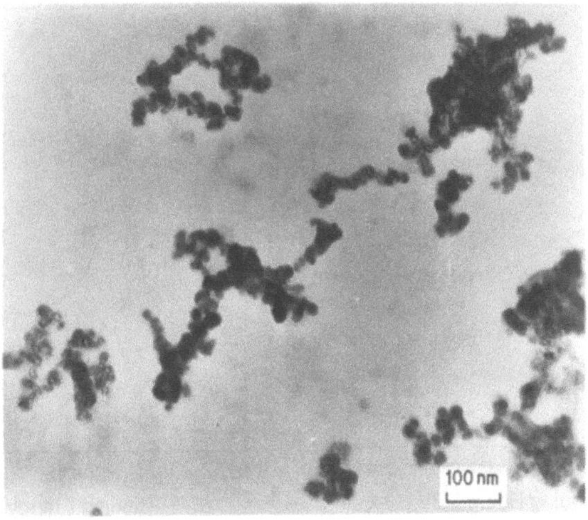

Fig. 4.1 TEM micrograph of the silicon carbide powder. Reprinted with permission from [1]. Copyright 1983, Springer Nature Limited

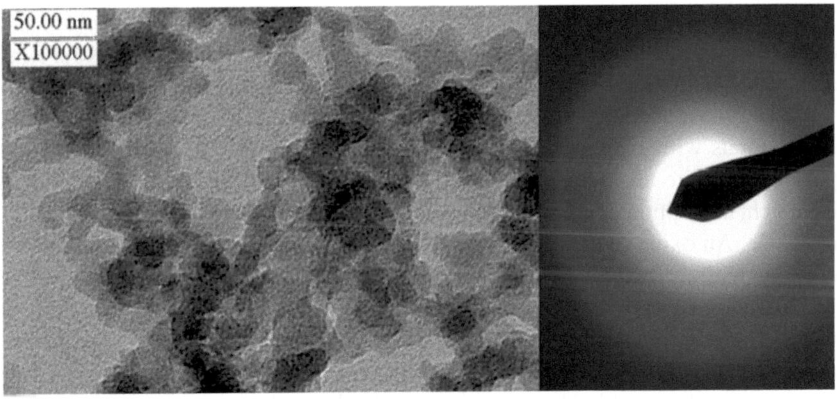

Fig. 4.2 TEM image of the silicon carbide particles collected by hypersonic impacting. Reprinted with permission from. Reprinted with permissions from [2]. Copyright 1998, Elsevier B.V

Decomposition of tetramethylsilane at lower temperatures and lower pressures can also produce SiC nanoparticles [4]. By employing a microwave plasma reactor in the synthesis, the acquired particles are amorphous, with sizes of 4–6 nm. The ratio of carbon and silicon can be adjusted by addition of hydrogen. Size-controllable synthesis of free-standing SiC NCs was realized through a plasma-based process by employing tetramethylsilane as the precursor [5]. The NCs have ultrasmall sizes down to 1.5 nm (Fig. 4.4). The PL quantum yields (excitation: 280 nm) for the NCs deposited on quartz substrates are 13.2%, 8.4%, and 2.6% for the NCs with a mean diameter of 1.5, 3.7, and 5.3 nm, respectively. The highest quantum yield

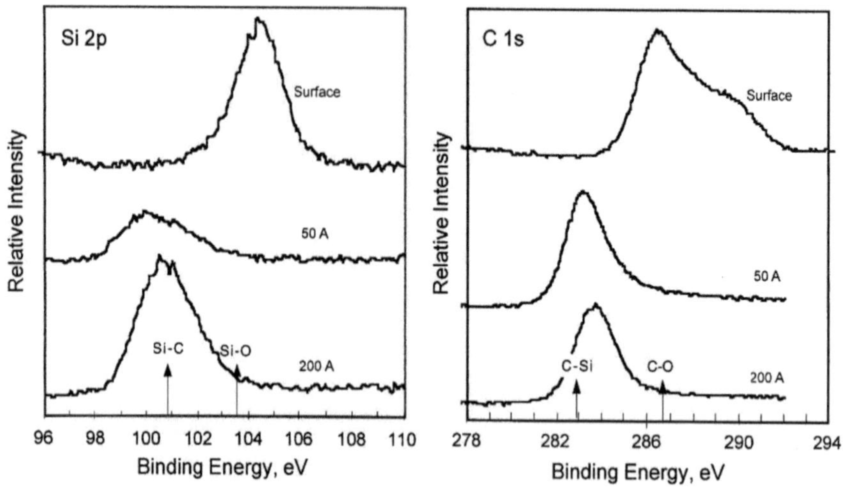

Fig. 4.3 XPS spectra of the Si 2p and C 1 s peaks as a function of depth. The proportion of oxygen-bonded silicon and carbon atoms decreases with increasing depth. Reprinted with permission from [2]. Copyright 1998, Elsevier B.V

of the smallest QDs was ascribed to quantum confinement in ultrasmall NCs. FTIR and XPS spectroscopy suggests that these SiC NCs possess low levels of oxygen-related surface functional groups such as C=O, O–Si, and OH. In another study, SiC nanocrystals with sizes of 2–4 nm were synthesized in low-pressure non-thermal plasma from the alkylsilane precursor tetramethylsilane [6]. These 3C–SiC NCs have light absorption that follows the quantum confinement effect, but without luminescence. An oxide shell was observed from the surface of the nanocrystalline SiC core. A fluidized bed CVD was developed for the synthesis of monodisperse SiC nanoparticles using a single precursor of hexamethyldisilane [7]. The synthesized nanoparticles have average sizes ranging from 10 to 200 nm via tuning the temperature and the gas ratio. A hierarchical structure with 5–8 nm SiC nanoparticles embedded in the graphite matrix was synthesized after heat treatment of the SiC–C nanospheres. Decomposition of hexamethyldisilane in an atmospheric non-thermal arc plasma may also yield ultrafine amorphous SiC nanoparticles based on magnetically stabilized gliding arc discharge [8]. The product is a mixture of amorphous SiC nanoparticles and free carbon. The average particle size is 9 nm. These nanoparticles exhibited blue–green luminescence. The low input energy favors synthesis of amorphous SiC nanoparticles, and the high input energy led to generation of β-SiC. The addition of hydrogen in plasma gas can reduce the particle size, improve crystallinity, and reduce carbon and oxygen proportions [9].

Laser pyrolysis is a popular method to prepare silicon carbide nanoparticles. Pulsed CO_2 laser-induced decomposition of a gaseous mixture of SiH_4 and C_2H_2 with varying concentrations of the hydrocarbon was used to produce silicon and silicon carbide nanoparticles [10]. Silicon carbide nanoclusters with diameters between 3 and 3.5 nm were deposited on various substrates, but significantly larger nanoparticles

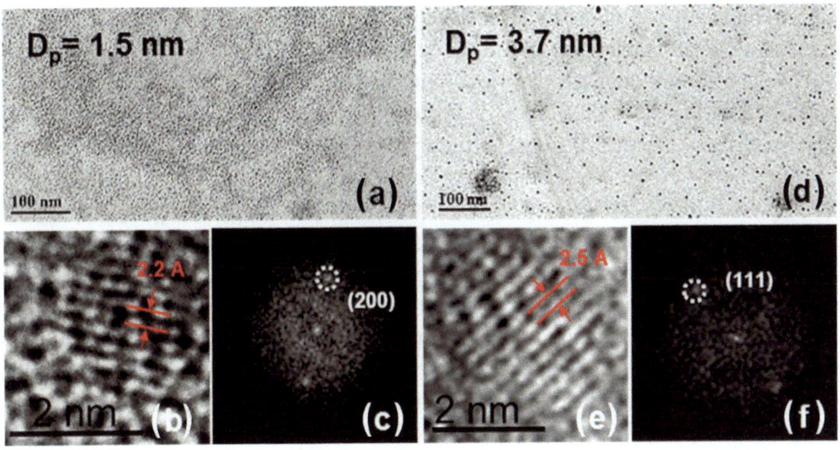

Fig. 4.4 TEM images and corresponding fast Fourier-transform (FFT) images for SiC NCs synthesized at three different precursor flow conditions: (**a–c**) 0.4 sccm, (**d–f**) 2.4 sccm. Reprinted with permission from [5]. Copyright 2016, Royal Society of Chemistry

were collected from the exhaust line of the reactor. FTIR and XRD reveal that the concentration of SiC in the nanoparticles increases with increasing acetylene flow rate. Besides SiC nanoparticles, there is a substantial portion of free crystalline silicon in the product when a stoichiometric SiH_4/C_2H_2 mixture was used in the synthesis. Pure SiC nanoparticles can only be synthesized with excess hydrocarbon, and XRD reveals that the SiC nanoparticles comprise crystalline β-SiC.

The properties of the SiC nanocrystallites synthesized by laser pyrolysis have been determined. The nonlinear optical property of the composite materials containing SiC nanocrystallites embedded in the photopolymer olygoetheracryalte was investigated [11]. The maximum photo-induced second harmonic generation for 0.95 SiC hexagonality is 10.1 pm V^{-1} in the χ_{222} tensor component. The main contribution is from the guest–host interface. The hexagonal SiC components play a key role in the photo-induced nonlinear optical effect. The electrical and dielectric properties of the SiC nanopowders were modified drastically by annealing at 1500 °C [12]. The materials exhibited a quasi-insulating characteristic and a large effective dielectric constant below this temperature. On the contrary, significant conductivity and a smaller dielectric constant were observed from the materials annealed at a higher temperature. The interface effect was suggested to play a key role in the observed conductivity and dielectric characteristic. The electro-optic phenomenon in electrically poled films based on large-sized SiC nanocrystallites as chromophores and polymethylmethacrylate as the host matrix was investigated [13]. For SiC nanocrystallite concentrations between 1 and 3 wt%, the largest effective linear electro-optic coefficient was measured to be 26 pm V^{-1} at 0.633 μm. Therefore, polarization at the SiC–polymer boundary plays a key role in the electro-optic phenomenon. Electron paramagnetic resonance was conducted on the core/shell nanocomposites, that is, the SiC nanoparticles encapsulated by conducting polyaniline [14]. Temperature-dependent EPR measurement reveals paramagnetic susceptibilities with Curie–Weiss-like features and thermally activated spins in the nanocomposites.

The photoluminescence properties of SiC nanoparticles were investigated by using Raman scattering [15]. The nanoparticles were synthesized by using CO_2 laser pyrolysis of a gaseous mixture of silane and acetylene. TEM reveals that SiC particles have a homogeneous composition with a narrow size range of around 10 nm (Fig. 4.5). The high-resolution TEM images reveal that each particle is composed of several crystallites with an average diameter of 2.5 nm, and these crystallites are embedded in an amorphous SiC matrix. Nuclear magnetic resonance indicates the coexistence of amorphous, hexagonal, and cubic SiC phases in the SiC nanoparticles. The Raman spectra obtained from the as-prepared sample SiC229 (Fig. 4.6) and oxidized ones are dominated by broad PL bands. Besides the Si–C and C vibrational signals, the Raman spectrum is dominated by a broad band within 514–800 nm that is attributed to PL. The sample oxidized at the highest temperature exhibited the most intense PL. The infrared transmission spectra (Fig. 4.7) revealed the existence of Si–Si and Si–O bonds on the particle surfaces. Surprisingly, two samples with similar particle sizes showed different PL response, and the sample with a larger proportion of amorphous layer exhibited stronger PL. In addition, after annealing in argon at a temperature low enough to preserve the average particle size but high enough to improve the crystalline structure, drastically diminished PL was observed. These observations suggest that the visible PL originates from the amorphous portion of the SiC nanoparticles and heterogeneous composition of the surface.

Raman scattering spectra were acquired from laser synthesized SiC nanoparticles with different sizes and structures [16], revealing the enhanced vibrational density of states. The numerically simulated spectrum profile shows agreement with the

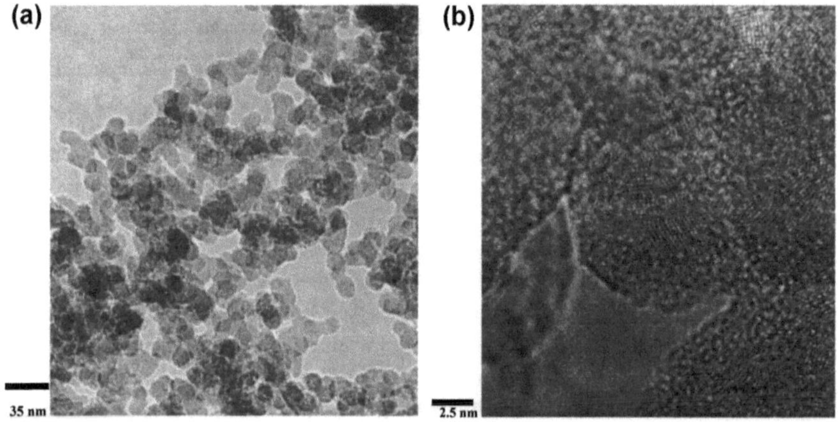

Fig. 4.5 TEM images of the SiC229 sample: low-resolution image (**a**) revealing the particle diameter of about 10 nm and high-resolution image (**b**) showing the amorphous background with a nanocrystallite diameter of about 2–3 nm. Reprinted with permission from [15]. Copyright 2002, American Physical Society

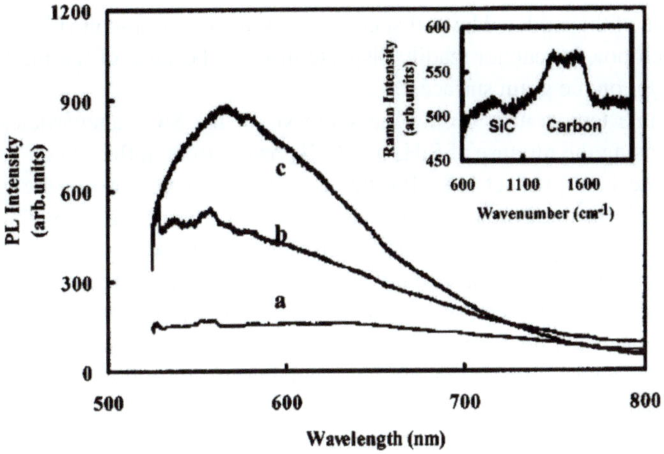

Fig. 4.6 Broad photoluminescence bands in the Raman spectra of SiC229 samples: (**a**) as-prepared sample and (**b, c**) samples heated in oxygen at 440 °C (**b**) and 970 °C (**c**). The intensity is normalized by weighting factors of 1/3 (**a**), 1 (**b**), and 1/10 (**c**). The inset shows the Raman spectrum of the as-prepared SiC229 sample exhibiting vibrational bands of C and SiC. Reprinted with permission from [15]. Copyright 2002, American Physical Society

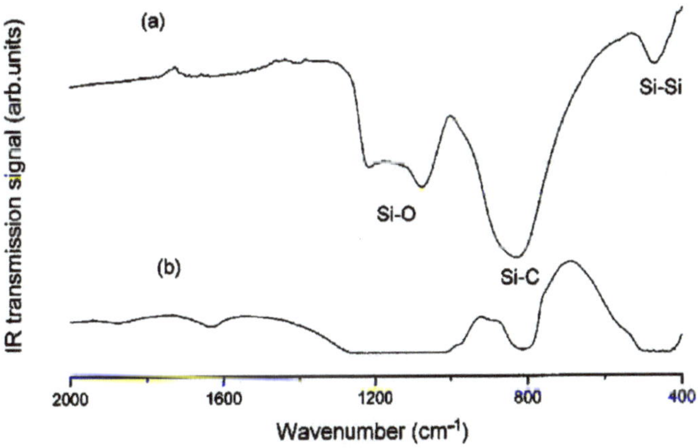

Fig. 4.7 Infrared transmission spectra of the SiC229 sample annealed for one hour in an oxygen atmosphere at 440 °C (curve a) and 970 °C (curve b). Reprinted with permission from [15]. Copyright 2002, American Physical Society

measured spectra for amorphous clusters. The major contribution to the vibrational density of states is the amorphous region in the SiC nanoparticles, and it depends on the nanoparticle size. Photon correlation spectroscopy revealed the dispersion behavior of these SiC nanoparticles suspended in water [17]. The SiC suspension consisting of stoichiometric or silicon-rich nanopowders was readily dispersed in water and stable with time. 95% of the nanoparticles were isolated, indicating the crucial role of the oxidized layer that covers the grain of the silicon-rich sample. The carbon-rich powder cannot readily disperse in water because of the inert graphitic carbon layer on the grain surface.

The flame temperature during the synthesis of the SiC nanoparticles by laser pyrolysis using the mixture of SiH_4 and C_2H_2 has a strong influence on the composition of the final product [18]. For the same C/Si atomic ratio in the gas phase, the C/Si ratio in the synthesized powder increased from 0.7 at 875 °C to 1.02 at 1100 °C, suggesting the effect of C_2H_2 dissociation on particle growth. Enhanced crystallization was observed with increasing reaction temperature by X-ray diffraction; the transmission electron microscopy disclosed the presence of 10 nm grains for all laser powers (or flame temperature). These grains are mostly amorphous if the synthesis temperature is low, but they contain some crystallites with a mean size of 2 nm, and the quantity increases with temperature as well.

A multiscale hybrid numerical simulation was performed to study the conditions necessary to obtain gas phase deposition of stoichiometric SiC quantum dots on the Si(100) substrate at the initial stage of growth [19]. The result shows that equal Si and k_{st} atom deposition fluxes result in nonstoichiometric nanodots owing to different surface fluxes of Si and C adatoms to the QDs. At this stage, the surface fluxes of Si and C adatoms to SiC nanodots can be controlled by manipulating the Si/C atom influx ratio and the Si(100) surface temperature. Effective strategies to maintain a stoichiometric (Si: $C = 1$: 1) elemental ratio during the initial stages of deposition of SiC/Si(100) QDs nuclei in a neutral/ionized gas-based process were proposed.

4.1.2 Electrochemical/Chemical Etching and Quantum Confinement

4.1.2.1 3C–SiC Nanocrystals and Quantum Confinement

An experimental protocol to prepare colloidal Si nanoparticles has been developed [20]. The *n*- or *p*-type Si crystals were electrochemically etched in an aqueous mixture of HF and 95% ethanol (1: 1 by volume) to generate porous Si that was then ultrasonically dispersed in water or other solvent to form a suspension with fine Si particles. Transmission electron microscopy revealed that these suspended Si particles are crystalline, with sizes ranging from many micrometers to a few nanometers. The Si nanoparticles exhibited luminescence bearing the quantum confinement effect. A similar procedure was employed to fabricate cubic SiC nanocrystals [21]. The polycrystalline 3C–SiC wafer was electrochemically etched in a mixture of HF and ethanol (HF: $C_2H_5OH = 2$: 1 by volume) under UV light illumination. The etching current densities are separately 100, 70, 60, 40, and 20 mA/cm^2. Subsequent sonication in water yielded a solution with suspended SiC nanoparticles, which are nearly spherical and have diameters of 1–7 nm (Fig. 4.8). The high-resolution TEM images disclose that these particles are crystalline, retaining the crystal structure of bulk

3C–SiC. The colloidal 3C–SiC nanocrystals exhibited strong photoluminescence (Fig. 4.9) visible to the naked eyes (Fig. 4.10). The PL spectrum is smooth and shows only one peak, implying a dominant luminescence mechanism. The PL peak shifts to red with decreasing excitation photon energy. Regardless of the etching conditions, the PL peak for each SiC nanoparticle solution finally reaches a threshold of around 550 nm (2.25 eV), which is close to the indirect bandgap of bulk 3C–SiC at room temperature (2.2 eV); no PL was observed if excited by light with a photon energy below the threshold [22]. These PL characteristics suggest the quantum confinement effect and corresponding PL of interband transition in these SiC nanocrystals.

The monotonic PL red shift observed from colloidal 3C–SiC nanocrystals with increasing excitation wavelength can be explained by quantum confinement by taking

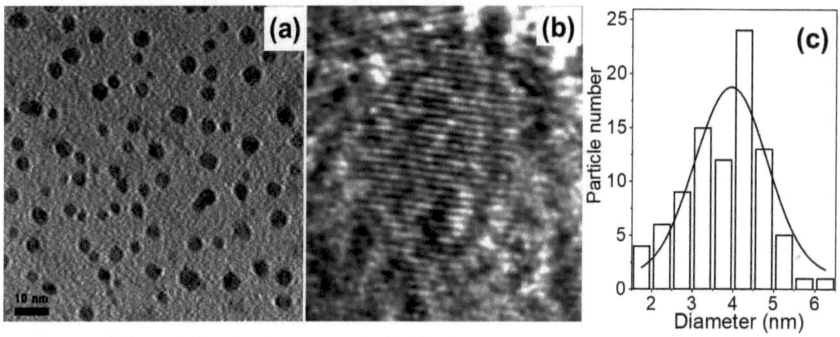

Fig. 4.8 **a** TEM image of the 3C–SiC nanocrystals. **b** HRTEM image of a typical particle. The lattice fringes are assigned to the (111) planes of 3C–SiC. **c** Particle size distribution. The curve is the Gaussian fit showing the most probable size of 3.9 nm. Reprinted with permission from [21]. Copyright 2005, American Physical Society

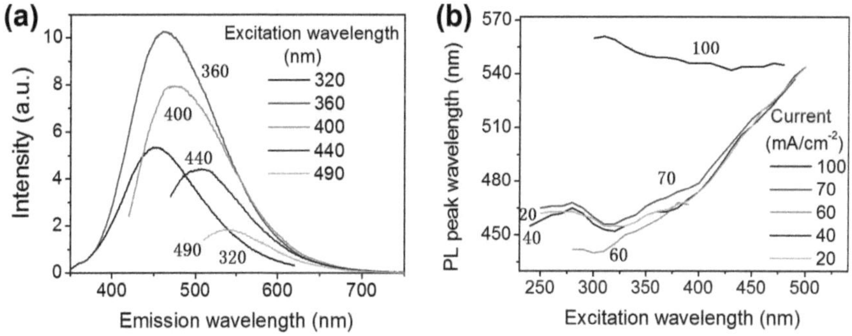

Fig. 4.9 **a** PL spectra of the 3C–SiC nanocrystal aqueous solution obtained at an etching current density of 40 mA cm^{-2}, excited with different wavelengths of light. **b** PL peak wavelength versus excitation wavelength, as determined from five 3C–SiC nanocrystal aqueous solutions fabricated at different etching current densities. Reprinted with permission from [21]. Copyright 2005, American Physical Society

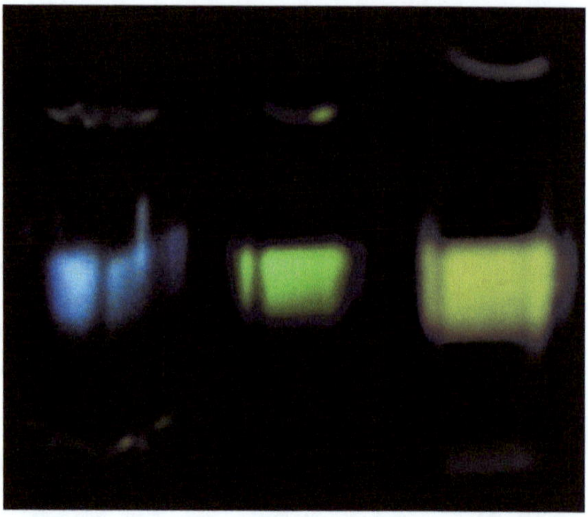

Fig. 4.10 Luminescence image of the 3C–SiC nanocrystal aqueous solution excited by 320, 400, and 450 nm light (from left to right), respectively. Reprinted with permission from [21]. Copyright 2005, American Physical Society

into account the broad size distribution of the nanocrystals. Figure 4.11 depicts the schematic of the proposed mechanism of PL peak red shift with increasing excitation wavelength, which applies to the general case of an ensemble of semiconductor QDs with significant particle size dispersion [23]. According to the quantum confinement effect, the smaller the nanocrystal, the wider is the energy gap and the bluer of the photons emitted by the interband transition of electrons; as a result, in a SiC colloid solution, the total PL spectrum is a superposition of all the emission bands from individual SiC nanocrystals of different sizes in the solution. When the excitation photon energy decreases gradually (excitation wavelength increases), the threshold size (smallest allowable size) of excitable nanocrystals (shaded area in Fig. 4.11) increases accordingly, and thus more smaller nanocrystals (blank area in Fig. 4.11) cannot be excited, and this leads to gradual red shift of the peak of the superposed PL spectrum. The 3C–SiC nanocrystals suspended in different solvents show similar luminescence behavior (Fig. 4.12), with weak solvent-dependence, implying more evidences of the quantum confinement effect.

Luminescence from Si nanocrystals either in porous Si or as separated Si nanocrystals exhibits the quantum confinement effect [24, 25]. However, no sound evidence of quantum confinement effect was observed during early studies involving porous SiC and various types of embedded SiC nanocrystals [26]. There are three reasons for this. First of all, although porous SiC has many nanoparticles and nanowires on the surface, these nanostructures are not separated but rather connected to the SiC substrate to form a single crystal and hence, electrons and holes undergoing recombination are not confined to the individual nanoparticle/nanowire. Secondly, the nanoparticles and nanowires on the porous SiC surface generally have dimensions of over 10 nm and in this size regime the quantum confinement effect becomes rather weak. Thirdly, although some embedded SiC nanoparticles may be as small as several nanometers, surface defects may highly quench the quantum confinement luminescence by offering nonradiative recombination pathways or dominate in

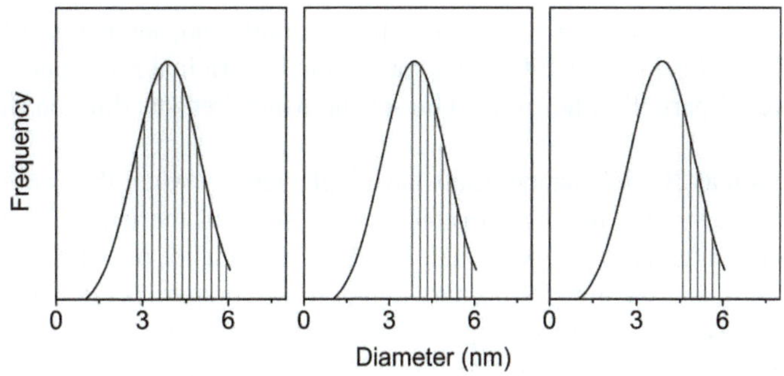

Fig. 4.11 Schematic showing the mechanism of the PL peak red shift with increasing excitation wavelength. The shaded area represents the size range of the excitable particles for a specific excitation wavelength. The excitation wavelength increases from left to right. Reprinted with permission from [23]. Copyright 2006, American Institute of Physics

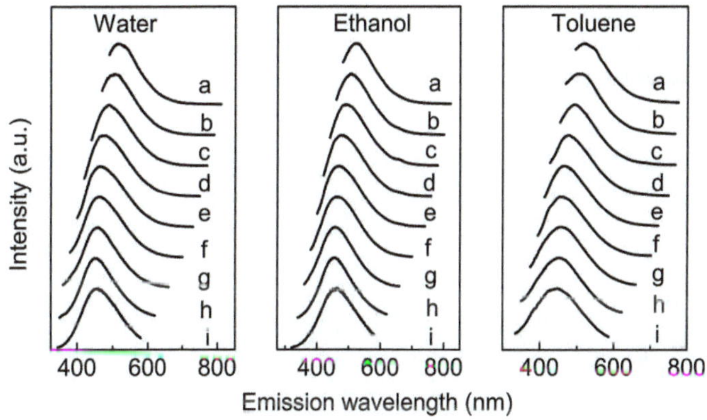

Fig. 4.12 **a–i** Normalized PL spectra of 3C–SiC nanocrystals in different solvents excited by 460, 440, 420, 400, 380, 360, 340, 320, and 300 nm light (from a to i), respectively. Reprinted with permission from [23]. Copyright 2006, American Institute of Physics

luminescence similar to the role played by Si=O in porous Si. In contrast, the separated ultrasmall colloidal SiC nanocrystals are more favorable structures to show the quantum confinement effect. SEM reveals that the main structure (within the SEM resolution) on the surface of porous SiC consists of necklace like wires with typical diameters around tens of nanometers (Fig. 4.13) [27]. Therefore, the ultrasonic treatment plays a key role in producing SiC nanocrystals only a couple of nanometers in size. The porous SiC (Fig. 4.13) shows a different reflectance spectrum compared to the pristine SiC (Fig. 4.14). The infrared reflectance spectrum of bulk polycrystalline 3C–SiC comprises a reststrahl band in the range of 750–1000 cm^{-1}, which is the typical spectrum of bulk 3C–SiC. The rise in the reststrahl peak occurs near the bulk TO phonon frequency at ~ 796 cm^{-1} and the sharp decrease occurs near the bulk LO phonon frequency at ~ 972 cm^{-1}. The as-anodized porous materials show significantly reduced intensity, especially the reststrahl band, and it is ascribed to the rough surface of the porous SiC. After sonication, the original reflectance spectrum was recovered partially due to breaking of the nanowires standing on the porous surface.

The colloidal 3C–SiC nanocrystals have high stability when they are dispersed in water. After storage in air for over 7 months, some of the larger 3C–SiC crystallites would aggregate and precipitate, whereas the smaller ones remained stably dispersed in water, retaining the primary luminescence properties [28]. However, the PL intensity decreased and the PL peak became narrower (Fig. 4.15). In practice, to one's surprise, our continuously long-term observation shows that smaller and stable SiC nanocrystals continue to be well dispersed and luminesce for over 10 years. This virtue renders the benign 3C–SiC nanocrystals superior micro-emitters, especially for biological and medical applications.

The UV–visible light absorption and carrier recombination properties of colloidal 3C–SiC QDs have been investigated [29]. The 3C–SiC QDs with an average diameter of 4 nm have an indirect-like energy gap (Fig. 4.16), and the main portion of the band-edge absorption spectrum obeys the following equation which describes absorption of indirect bandgap semiconductors:

Fig. 4.13 **a** and **b** SEM micrographs of porous polycrystalline 3C–SiC. **c** and **d** TEM images showing several residual 3C–SiC nanowires after ultrasonic treatment, along with 3C–SiC nanocrystals (small dots). Reprinted with permission from [27]. Copyright 2009, American Institute of Physics

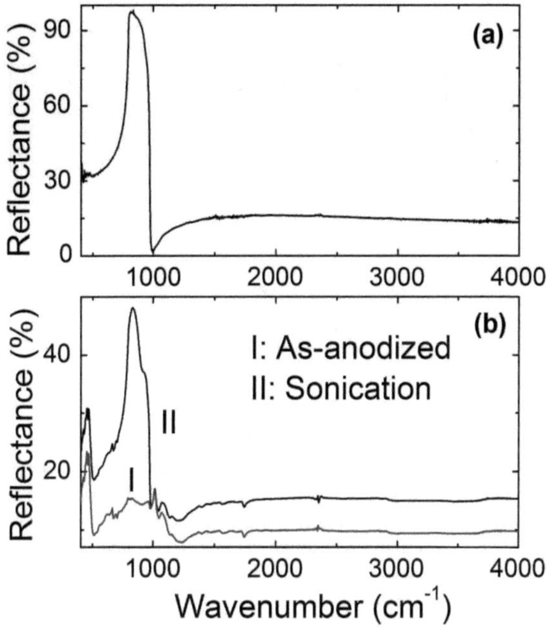

Fig. 4.14 Infrared reflectance spectra of **a** bulk polycrystalline 3C–SiC and **b** as-prepared (curve I) and sonicated (curve II) porous polycrystalline 3C–SiC. Reprinted with permission from [27]. Copyright 2009, American Institute of Physics

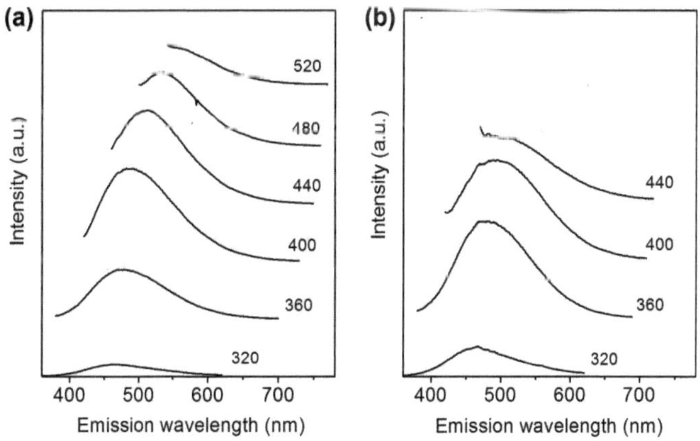

Fig. 4.15 PL spectra of **a** as-prepared and **b** stored (over 7 months) SiC nanocrystals in water measured under different excitation wavelengths. Reprinted with permission from [28]. Copyright 2006, Elsevier B.V

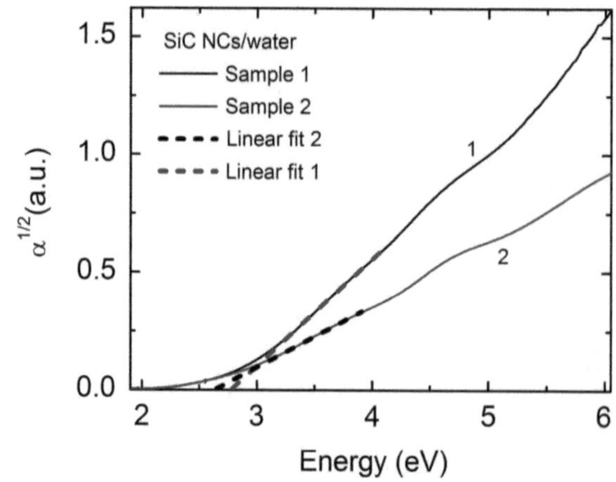

Fig. 4.16 UV–visible absorption spectra of the colloidal 3C–SiC nanocrystals in the aqueous solution for two different samples. The square root of the absorbance is plotted against photon energy. The broken lines are the linear fits of the major portion of the spectra, and the extrapolations yield an x-intercept of 2.78/2.63 eV for the respective sample. Reprinted with permission from [29]. Copyright 2010, American Institute of Physics

$$\alpha = A(h\nu - E_g)^2, \tag{4.1}$$

where α is the absorption coefficient, v is the photon frequency, h is Planck's constant, E_g is the energy gap of the crystallite, and A is a constant. The energy gap obtained by extrapolation of the fitted curve is significantly larger than the bandgap of bulk 3C–SiC, indicating strong quantum confinement. In contrast, ultrasmall SiC QDs 1 nm in size exhibited discrete and sharp absorption peaks (Fig. 4.17). Unlike the quasi-continuous distribution of energy levels in bulk silicon carbide, the ultrasmall SiC QDs possess discrete energy levels, and this feature lies between those of single crystals and molecules. The discrete peaks also imply that these SiC QDs have a relatively narrow size distribution such that the discrete absorption features are not smeared. The time-resolved PL spectra of the colloidal SiC nanocrystals (Fig. 4.18) can be fitted by the following triple-exponential function:

$$I(t) = A_1 e^{-t/\tau_1} + A_2 e^{-t/\tau_2} + A_3 e^{-t/\tau_3}, \tag{4.2}$$

where τ_i and A_i ($I = 1, 2, 3$) represent the lifetime and weight factor of each component. The three components have lifetimes on the order of nanoseconds to tens of nanoseconds and show dependence on the particle size. The fractional intensity (R_i) (ratio of number of photons emitted via the ith channel to total number of emitted photons) of each component can be evaluated by using the following equation:

$$R_i = \frac{\int_0^\infty (A_i e^{-t/\tau_i})\mathrm{d}t}{\sum_{j=1}^{3} \int_0^\infty (A_j e^{-t/\tau_j})\mathrm{d}t} = \frac{A_i \tau_i}{\sum_{j=1}^{3} (A_j \tau_j)} \quad (i = 1, 2, 3). \tag{4.3}$$

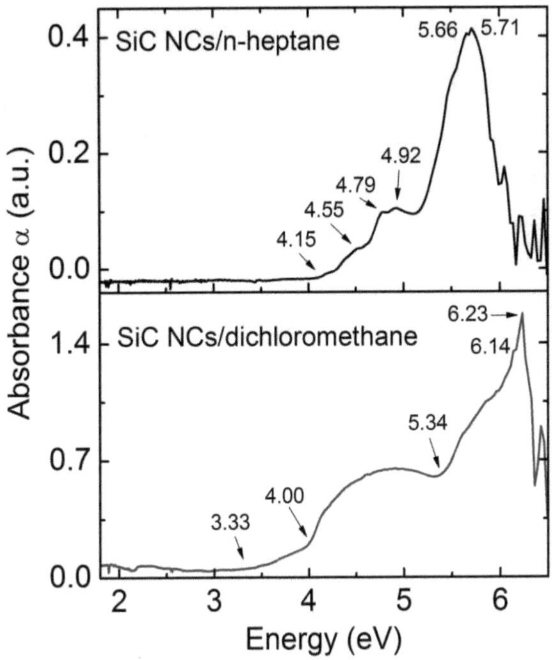

Fig. 4.17 UV–visible absorption spectra of the colloidal 3C–SiC nanocrystals in *n*-heptane and dichloromethane. The feigned peaks beyond 5.71 eV (upper curve) and 6.23 eV (lower curve) represent experimental errors. Reprinted with permission from [29]. Copyright 2010, American Institute of Physics

Besides electrochemical etching, direct chemical etching can produce porous SiC as well as SiC nanocrystals after ultrasonic treatment. The 3C–SiC powder with a micrometer size was etched in a mixture of nitric acid and hydrofluoric acid (65% HNO_3: 40% HF = 1: 3 in volume) at 100 °C [30]. Nitric acid oxidizes the surface of the SiC crystal into SiO_2 and then hydrofluoric acid etches SiO_2, leaving the porous SiC on the surface. Subsequent sonication in ethanol would break the porous film, generating SiC nanocrystals dispersed in ethanol. TEM reveals that the nanocrystals are spherical, with an average size of 6.5 nm (Fig. 4.19). The PL peak from the 3C–SiC nanoparticles in ethanol shifted from 420 to 512 nm when the excitation wavelength increased.

The SiC nanoclusters with an average diameter of only 7 Å (Fig. 4.20a) were fabricated by hydrothermally diminishing the sizes of the SiC microcrystals under high pressure and high temperature [31]. The SiC microcrystallites were first etched in a HF–HNO_3 (volume ratio of 3: 1) mixed solution at 100 °C for 1 h. The resultant porous microcrystallites underwent ultrasonic treatment in water for 20 min followed by centrifugation to produce SiC nanoparticles dispersed in an aqueous solution. Subsequently, this solution was hydrothermally treated in a Teflon-lined autoclave at 180 °C for 10 h such that the sizes of the nanoparticles were further reduced to

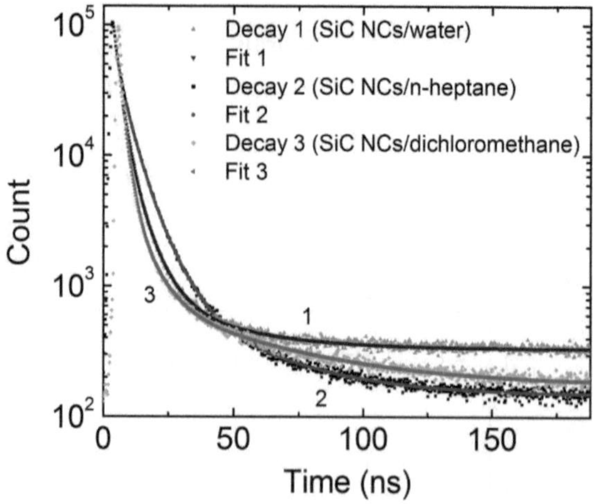

Fig. 4.18 Time-resolved photoluminescence spectra of the 3C–SiC nanocrystals in water, *n*-heptane, and dichloromethane solvents excited by 280 nm light and monitored at 450, 420, and 410 nm, respectively. The solid lines are triple-exponential fits with lifetimes and fractional intensities of [2.6 ns (0.34), 6.7 ns (0.51), 28.9 ns (0.15)], [3.5 ns (0.22), 7.7 ns (0.62), 31.8 ns (0.16)], and [2.2 ns (0.30), 7.3 ns (0.36), 44.7 ns (0.34)], respectively. Reprinted with permission from [29]. Copyright 2010, American Institute of Physics

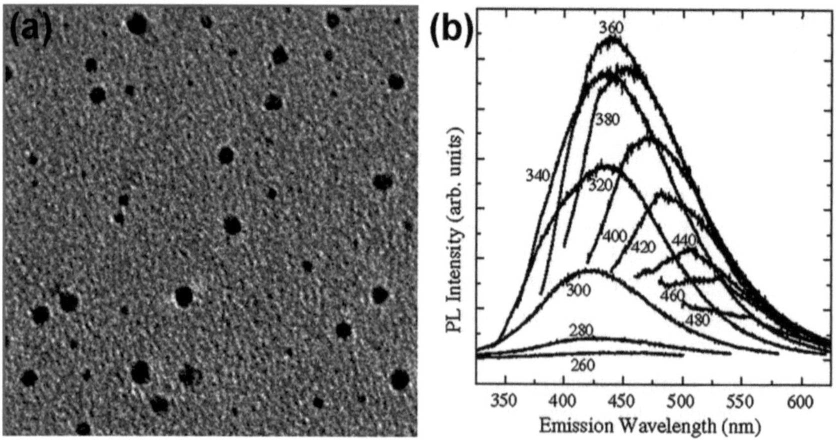

Fig. 4.19 TEM image (**a**) of the SiC nanoparticles produced by chemical etching followed by sonication. The PL spectra (**b**) of the SiC nanoparticles in ethanol under different wavelengths of excitation. Reprinted with permission from [30]. Copyright 2007, Elsevier

yield suspended SiC nanoclusters. The results reveal that the SiC nanoclusters have an indirect-like energy gap of 5.1 eV. Unlike the larger SiC NCs, the luminescence of the SiC nanoclusters seems to be dominated by two types of oxygen-related surface defects, and the maximum of their PL/PL excitation (PLE) spectrum lies at

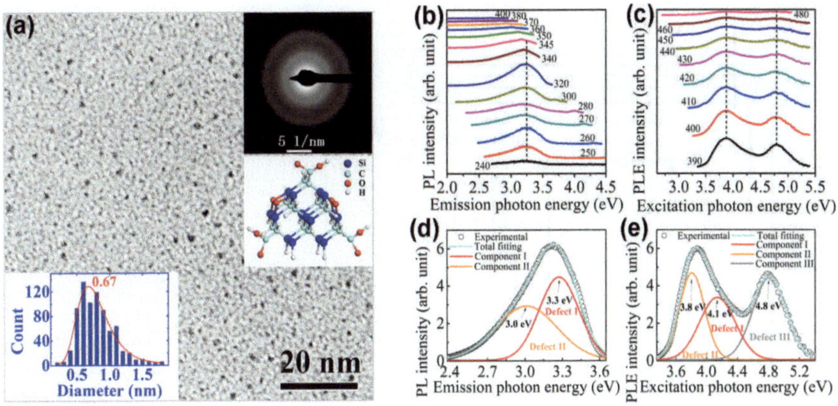

Fig. 4.20 **a** TEM image of SiC nanoclusters. Insets show selected-area electron diffraction pattern, particle size distribution, and schematic crystal structure, respectively. **b** PL spectra and **c** PL excitation spectra of SiC nanoclusters in water. The excitation/emission wavelength for each spectrum is indicated (in units of nm). Gaussian fit of **d** PL spectrum (excitation: 320 nm) and **e** PL excitation spectrum (emission: 390 nm), revealing contributions of different surface defects (I and II refer to oxygen-related surface traps, III might arise from Si–Si surface trap). Reprinted with permission from [31]. Copyright 2016, American Institute of Physics

4.1/3.3 and 3.8/3.0 eV, respectively (Figs. 4.20b–e and 4.21a), as revealed by the Gaussian-function-decomposed PL/PLE spectra. The decomposition takes the form

$$I(\nu) = \sum_{i=1}^{n} I_{0i} \exp\{-[4(\nu - \nu_{0i})^2/(\Delta\nu_{0i})^2] \ln 2\}, \quad (4.4)$$

where ν is the photon frequency, the ith component has a maximal light intensity I_{0i} at ν_{0i}, and $\Delta\nu_{0i}$ is its full width at half maximum. It is apparent that the PL/PLE spectrum is well decomposed. Considering the significant dispersive particle size distribution of such SiC nanoparticles or nanoclusters, it is less likely that their observed light emissions arise from interband transitions, although not fully excluded. By using the experimental light absorption data for several SiC QD or nanocluster ensembles of distinct average sizes, we conclude that the optical gap of the SiC nanoparticles with reference to bulk bandgap is approximately inversely proportional to the diameter (Fig. 4.21b), following

$$E_g(d) = 2.2 + 1.95/d^{0.97} \quad (d \text{ in nm}). \quad (4.5)$$

Figure 4.21b shows also the literature values calculated using the DFT [32] or time-dependent DFT (TDDFT) method [33], and each set of data is fit with $\Delta E_g(d) \propto d^{-1.17\ (\text{or}\ 1.33)}$. Apparently, the DFT and TDDFT calculations overestimate the increase rate of energy gap with decreasing particle size. For the particles with sizes below 1.5 nm, the DFT result matches the experimental result better;

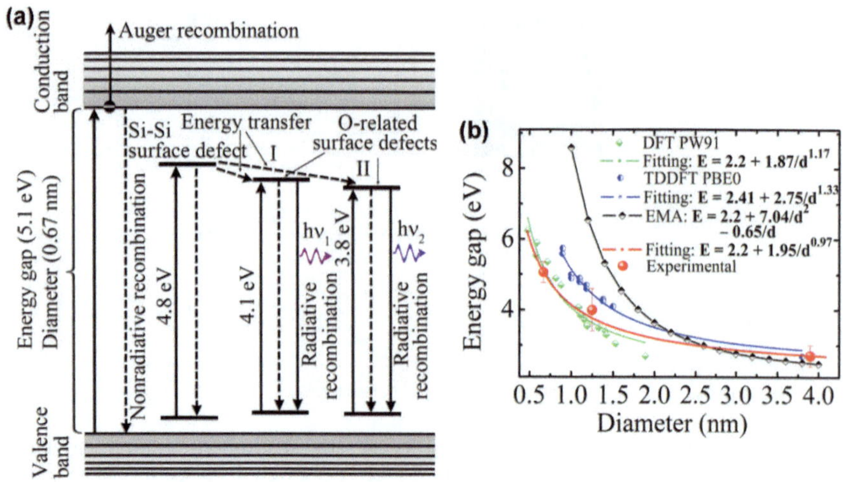

Fig. 4.21 **a** Schematic energy-level diagram of a SiC nanocluster showing energy levels of surface defects and correlated light absorption and emission processes. **b** Comparison of measured and calculated energy gaps as a function of diameter for the SiC nanoparticle. Green and blue signs correspond to calculated values by DFT and time-dependent DFT [32, 33]. Effective mass approximation (EMA) result is also displayed for comparison. Reprinted with permission from [31]. Copyright 2016, American Institute of Physics

whereas the TDDFT result matches better for the particles above 2.0 nm,. The difference between the experimental and calculated results as well as the difference between two calculated results owes to two reasons. The calculations adopted hydrogen passivation of the SiC QD, but the surfaces of the actual SiC nanoparticles under experimental investigation are highly oxidized. The surface chemistry usually has a significant influence on the energy gap of the nanoparticles. Secondly, different exchange–correlation functionals were adopted in different first-principles calculations. Figure 4.21b also depicts the energy gap–particle size relationship calculated using the effective mass approximation formula [34]:

$$E_{\mathrm{g}}(r) = E_{\mathrm{g}}^{\mathrm{bulk}} + \frac{\hbar^2\pi^2}{2r^2}\left(\frac{1}{m_e^*} + \frac{1}{m_h^*}\right) - \frac{1.8e^2}{4\pi\varepsilon_0\varepsilon_r r}, \tag{4.6}$$

where r is the particle radius, m_e^* (m_h^*) is the effective mass of the electron (hole). For 3C–SiC, $1/m_e^* + 1/m_h^* = 1/(0.214m_0)$ and the relative dielectric constant $\varepsilon_r =$ 7.95. It is apparent that the effective mass approximation highly overestimates the increase rate of the energy gap with decreasing size, and the calculated energy gaps for the particles below ~ 2.0 nm are fully unreliable. Another study involves 3C–SiC QDs fabricated by a similar hydrothermally chemical etching method, whose average diameters are 2.02, 1.75, 1.18, and 1.15 nm for etching duration times of 30, 1, 1.5, and 2 h, respectively [35]. The electrochemical gap of colloidal 3C–SiC QDs was determined by cyclic voltammetry, which showed a similar trend of the energy gaps of the QDs. The colloidal SiC QDs exhibited PL at ~ 483 nm for the sample etched for 2 h, compared to 501 nm for the sample etched for 1 h. The sample etched for 2 h showed a big Stokes shift of ~ 277 nm.

4.1.2.2 Surface Structures of 3C–SiC Nanocrystals

The surface structures of 3C–SiC nanocrystals are important because they determine the surface defect states and affect the luminescence properties. 3C–SiC nanocrystals prepared by electrochemical etching and ultrasonic treatment exhibited simple infrared absorption features [36]. Figure 4.22a shows the infrared transmission spectra of 3C–SiC crystallites with different sizes. The strongest light absorption at around 738 cm^{-1} is assigned to the TO phonon mode of SiC. The smaller wavenumber with respect to that of bulk SiC (796 cm^{-1}) is due to the phonon confinement effect, and the large spectral width is attributed to the dispersed particle size. The constant and second-strongest peak at 482 cm^{-1} may be ascribed to the Si–Si bonds caused by surface reconstruction of the SiC nanocrystals, although Si–O bonds have absorption peaks at lower wavenumbers. The absorption peaks at 1084 and 1124 cm^{-1} are assigned to Si–O–Si, C–O–C, or Si–O–C, with 1084 cm^{-1} likely due to Si–O–Si and 1124 cm^{-1} from C–O–C. The weak absorption peaks at 1628 and 3150–3700 cm^{-1} originate from residual water molecules. The Si–H signal expected at 2100–2250 cm^{-1} was not detected. This is different from as-prepared Si nanocrystals whose surface is mainly terminated by hydrogen atoms but unstable and oxidization occurs readily after storage in air for a few minutes. This oxidation changes the luminescence mechanism from quantum confinement to surface states (Si=O) related [37]. The polar Si–O–Si surface explains the hydrophilicity of the SiC nanocrystals since the hydrogen bonds can form readily between the bridged oxygen atom and surrounding water molecule. Concerning the 2-day-old SiC nanocrystal solution, the absorption peak at 482 cm^{-1} diminished while those at 1084 and 1124 cm^{-1} became more intense due to oxidation of Si–Si into Si–O–Si. The 2-month-old sample exhibited overall spectral features (Fig. 4.22b) similar to those of the fresh sample, confirming high surface stability. After ultrasonic treatment, the peak at 482 cm^{-1} was dramatically enhanced, but the 1084 and 1124 cm^{-1} peaks diminished. This is ascribed to the breakage of some bridged bonds and recovery of Si–Si bonds. For sample B which contains many micrometer-sized crystallites, the absorption peak of the LO phonon at 948 cm^{-1} arose (curve C in Fig. 4.22a). This peak shifts to smaller wavenumbers relative to the bulk value (972 cm^{-1}) due to the phonon confinement effect and other surface effects. The peak at 482 cm^{-1} still existed, showing occurrence of the Si–Si surface reconstruction. The absorption peaks at 1084 and 1124 cm^{-1} were not observed due to the reduced surface-atom proportion in the micrometer-sized crystallites relative to the nanocrystals. The TO and LO phonon modes of the larger pristine SiC powders were observed at 800 and 962 cm^{-1}, respectively, closer to the bulk values. As expected, the surface vibrational modes are absent. The absence of the LO phonon mode in the SiC QDs indicates that this mode is inactive on this scale, possibly caused by the breakdown of the spatial displacement symmetry. The 3C–SiC nanocrystals in ethanol showed similar infrared absorption to that of the nanocrystals suspended in water (Fig. 4.23a), except that the signals of the bridged bonds are now weaker. The absorption peak at 1743 cm^{-1} arises from the C=O-related bonds. The SiC nanocrystals in dichloromethane showed very weak infrared absorption (Fig. 4.23b), and this is consistent with the low solubility of SiC particles in this solvent of small polarity. No noticeable signals were detected from SiC nanocrystals suspended in the nonpolar solvent *n*-heptane, in which only the smallest SiC particles can be dissolved.

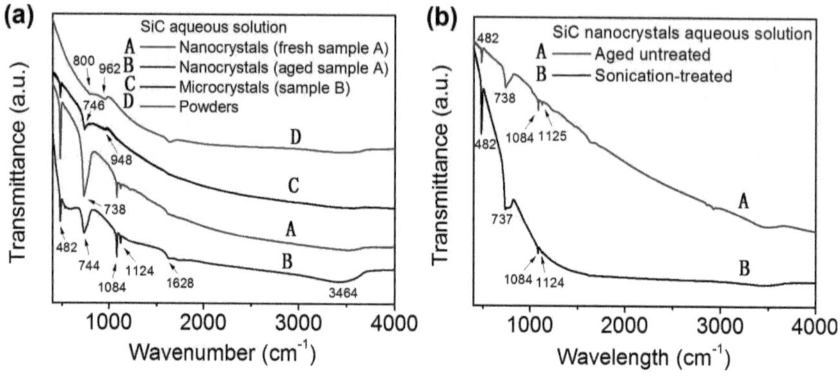

Fig. 4.22 **a** Infrared transmission spectra of the 3C–SiC crystallites embedded in KBr pellets obtained from the SiC aqueous solutions containing quantum dots with an average size of 4 nm (sample A), crystallites with sizes ranging from tens of nanometers to many micrometers (sample B), and powders large enough to be visible to the naked eyes, respectively. All the spectra were acquired right after evaporation of water from the SiC/KBr pellet. **b** Infrared transmission spectra of sample A after storage in air for 2 months with and without sonication before the measurement. Reprinted with permission from [36]. Copyright 2011, Elsevier B.V

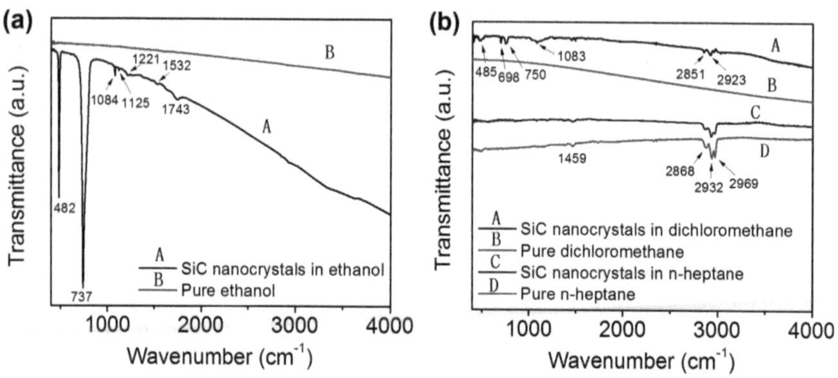

Fig. 4.23 Infrared transmission spectra of the 3C–SiC nanocrystals in KBr pellets obtained from **a** ethanol and **b** dichloromethane or *n*-heptane solutions, respectively. The spectra of the solvents are also shown. All the spectra were measured right after evaporation of the solvent from the SiC/KBr pellet. Reprinted with permission from [36]. Copyright 2011, Elsevier B.V

The surface structure of 3C–SiC nanocrystals with an average size of 5 nm was investigated by using attenuated total internal reflection infrared spectroscopy [38, 39] after evaporation of the solvent. The infrared spectrum (Fig. 4.24) comprises the Si–C band at 800 cm^{-1}. The absorption peak at 1595 cm^{-1} from the ethanol and *n*-butanol samples arises from surface –OH bending and C=O stretching vibration. The signals of Si–O–Si and C–O–C vibrations are present between 1000 and 1200 cm^{-1} and the bands at 1260 and 1750 cm^{-1} belong to symmetric and asymmetric COO– stretching vibration. The band at 1350 cm^{-1} is attributed to C–H bending vibration, but the signal of Si–H vibration was not observed. The Si–O–Si bonds are more stable than S–H bonds on the surfaces of the SiC nanocrystals, as reported previously [36]. The Raman spectra of the SiC nanocrystals and microcrystals are compared (Fig. 4.25) [38]. The strong TO phonon band at 792 cm^{-1} and LO phonon band at 961 cm^{-1} were observed for the microcrystals. The strong TO phonon vibration lies at nearly the same position for the nanocrystals, but the LO phonon signal is quite weak. This observation is consistent with the infrared absorption result, further indicating that the LO phonon mode is inactive in 3C–SiC nanocrystals [36]. A similar phenomenon has been observed [40], as shown in Fig. 4.26. The Raman spectra of both pure and polymer-encapsulated 3C–SiC nanocrystals show strong TO phonon signals at 788.2 and 790.4 cm^{-1}, respectively. In contrast, the signal of the LO phonon vibration is very weak. However, for larger particles embedded in the glycerol matrix, a strong signal of the LO phonon mode at 972.6 cm^{-1} was observed, and in this case the peaks of the TO and LO phonon vibrations are close to that of the bulk materials, suggesting that these particles are large.

The pH may influence the luminescence properties of silicon carbide colloids [41]. The pH of the aqueous solution of the 3C–SiC nanocrystals was adjusted by

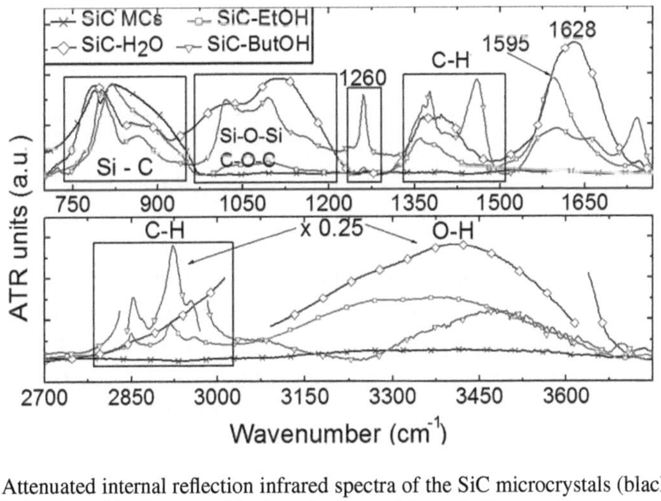

Fig. 4.24 Attenuated internal reflection infrared spectra of the SiC microcrystals (black) and SiC nanocrystals in ethanol (red), water (blue), and *n*-butanol (green). The spectra were recorded on a multi bounce germanium crystal after evaporation of the solvent. Reprinted with permission from [38]. Copyright 2011, American Institute of Physics

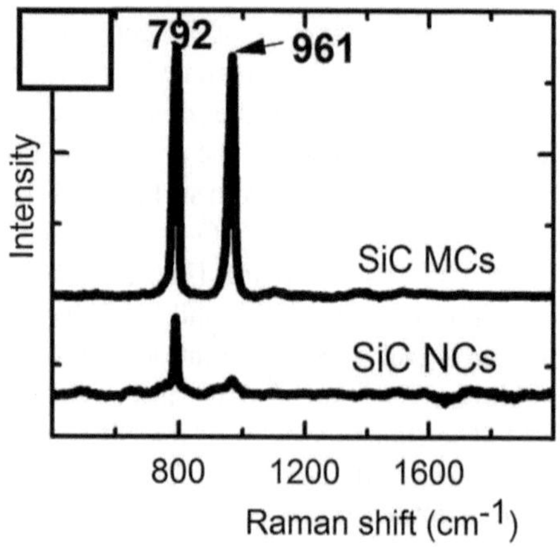

Fig. 4.25 Raman spectra of the SiC microcrystals and nanocrystals. Reprinted with permission from [38]. Copyright 2011, American Institute of Physics

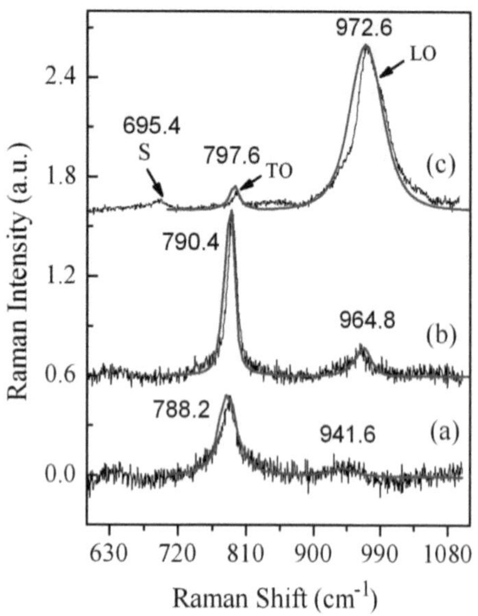

Fig. 4.26 Raman spectra of the **a** assembled 3C–SiC nanocrystals film and **b**, **c** 3C–SiC nanocrystals embedded in the glycerol matrix. **b** and **c** correspond to regions containing smaller and larger nanocrystals, respectively. Reprinted with permission from [40]. Copyright 2010, Optica Publishing Group

addition of HCl or NaOH. Besides the blue emission band that obeys the quantum confinement effect, another emission band at 510 nm was supposed to exist for excitation wavelengths larger than 350 nm and this band appeared from particles suspended in an acidic medium but not a basic one (Fig. 4.27). Based on infrared absorption spectroscopy, XPS, and theoretical calculations, the 510 nm emission may arise from structures involving attachment of the H^+ and OH^- to the Si dimers on the surface of the SiC nanocrystal. Surface surfactants were found to influence the luminescence properties of the SiC colloids [42], and in some studies the carboxyl groups were proposed to exist on the surfaces of SiC nanocrystals [42, 43]. If this is true, the 510 nm luminescence peak might be related to the carboxyl groups.

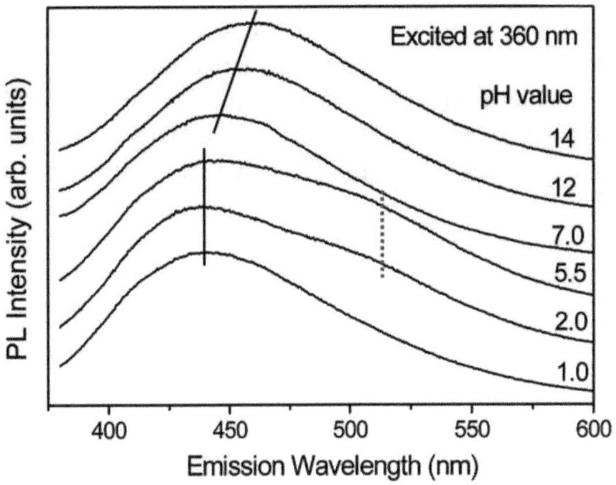

Fig. 4.27 PL spectra of the 3C–SiC nanocrystal aqueous solutions with different pH values as indicated. Reprinted with permission from [41]. Copyright 2009, American Chemical Society

The colloidal SiC QDs fabricated by either electrochemical etching of bulk SiC single crystals or chemical etching of micrometer-sized SiC powders followed by ultrasonic treatment contain fruitful surface functional groups, particularly for the latter type of SiC QDs. Some surface functional groups can generate localized surface states to give rise to significant light emissions. The origin of a blue defect emission band that often dominates the photoluminescence of colloidal SiC QDs was clarified by studying the surface dependence of PL of the SiC QDs fabricated by chemical etching of SiC powders [44]. The introduction of KOH caused nearly complete quenching of the blue emission band peaked at around 438 nm, whose peak wavelength kept constant when the excitation wavelength increased (Fig. 4.28), strongly suggesting luminescence origin of defects. Meanwhile, the previously hidden 470-nm band became remarkable, whose shifting peak with increasing excitation wavelength suggests significant contribution from the quantum confinement luminescence. The introduction of KOH also caused significant weakening of the previously strongest 326-nm PL excitation band, and the initial hidden 365-nm peak shifted to 375 nm, becoming most apparent. These phenomena suggest that the light absorptions at 326 and 365–375 nm correspond to the 438-nm constant blue emission band and the wavelength-tunable emission band (at over 470 nm), respectively. This implies that KOH can largely reduce the surface defects associated with the 438-nm emission band. After further HCl treatment of the KOH-treated colloidal SiC QDs, both the PL and PLE spectra resemble that of the pristine SiC QDs prior to the KOH treatment, suggesting partial recovery of the 438-nm luminescent surface color centers. The UV–Vis absorption and FTIR spectra further showed similar reversible variations during the alkali and subsequent acid surface treatments (Fig. 4.29), particularly disappearance and re-arising of the infrared absorption signals of the COO groups. Such information points out that the surface states related to C=O (in COO) functional groups are responsible for the peak-fixed 438-nm emission band. There is another UV absorption peak in the PLE spectrum, whose correlated excited carriers contribute to both C=O-related luminescence and quantum confinement luminescence of the SiC

QDs through electron transfer. It may arise from surface Si–Si bonds, and less likely belongs to the light absorption across the direct energy gap.

A later following work gives more supportive evidences about the surface origin of the blue emission band in the colloidal SiC QDs fabricated from chemical etching of micrometer-sized 3C–SiC powder in a HF: HNO_3 solution [45]. The sample with broad size distribution was centrifuged using a 30 kDa macrosep filter. The remaining solution was washed 10 times to remove most small particles, the obtained filtrate exhibited a blue peak similar to that of the as-prepared sample, but the red shoulder disappeared, and the peak wavelength did not shift with increasing excitation wavelength (Fig. 4.30). Hydroxyl- and silicon oxide-terminated SiC NCs can be obtained by further surface BH_4 or Zn/H^+ reduction of the colloidal SiC NCs [46]. Elimination of both carboxyl and hydroxyl groups by reduction was realized, causing dramatic blue shift of the PL. Combining the PL spectral result with the attenuated total reflectance infrared spectra and ab initio calculations, a common emitter observed

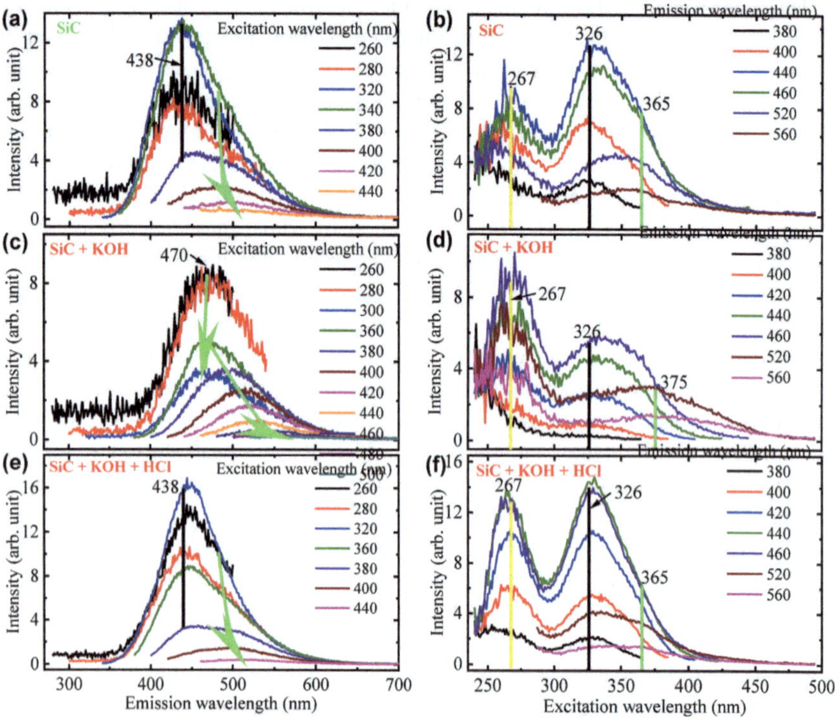

Fig. 4.28 PL and PL excitation spectra of (**a**) and **b** pristine SiC QD aqueous solution (pH = 1), **c** and **d** SiC QD solution + KOH (pH = 14), and **e** and **f** SiC QD solution + KOH + HCl (pH = 1). Black and green lines indicate peak positions of C=O surface state luminescence and intrinsic quantum confinement-dominated luminescence in the QDs. Yellow lines indicate absorption peaks probably by Si–Si surface defects. Reprinted with permission from [44]. Copyright 2015, American Institute of Physics

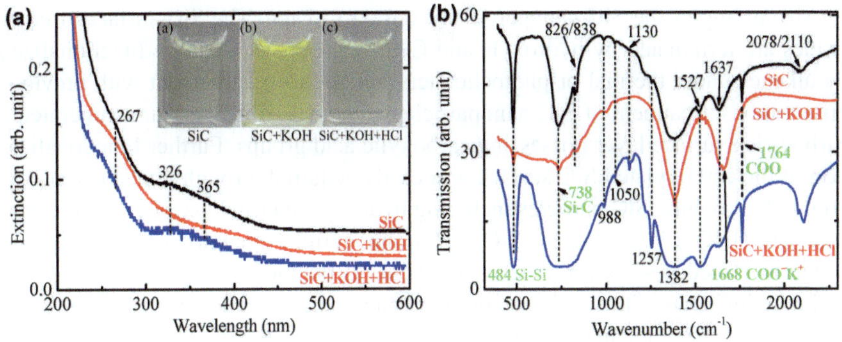

Fig. 4.29 **a** UV–Vis absorption spectra and **b** FTIR spectra of the pristine SiC QDs, SiC QDs + KOH, SiC QDs + KOH + HCl. The inset of **a** shows the daylight photos of the samples. Reprinted with permission from [44]. Copyright 2015, American Institute of Physics

in all the samples exhibiting light emissions at around 410–420 nm was supposed to be associated with SiO_x defect-related defects on the SiC NC surfaces.

Chemical transformations of surfaces of commercially available 3C–SiC nanoparticles containing 97% of 3C–SiC and having sizes of 10–20 nm were studied (Fig. 4.31) [47] The FTIR, XPS, and temperature-programmed desorption mass spectrometry show that thermal oxidation of these SiC nanoparticles led to formation of a hydroxylated SiO_2 surface layer, with C_3Si–H and CH_x groups at the SiO_2/SiC interface. Oxidation plus subsequent oxide dissolution by HF or KOH allowed

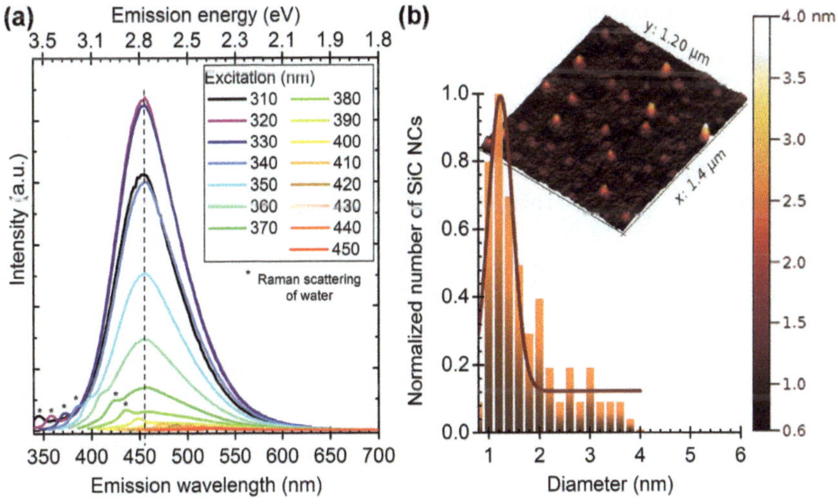

Fig. 4.30 **a** PL spectra of colloidal SiC NCs at different excitation wavelengths after SiC NCs were filtered through a 30 kDa centrifuge filter. **b** AFM image and size distribution, with an average size of 1.5 nm. Reprinted with permission from [45]. Copyright 2015, Royal Society of Chemistry

for size tuning of the SiC nanoparticles from 9 to 17 nm. The SiC surfaces free of oxides are terminated by hydroxyls and C_3Si-H groups and can be functionalized by alkenes under thermal or photochemical initiation. In agreement with previous observations, treatment of SiC nanoparticles with a HF/HNO_3 mixture generated a carbon-rich surface layer possessing carboxylic acid groups. Further functionalization of the carboxylated SiC surface was easily acquired through amide chemistry. Highly hydrophilic SiC particles including hydroxylated (SiC_KOH), carboxylated (SiC_HF/HNO_3), or aminated (SiC_NH_2) ones formed aqueous solutions that are stable in closed vessels for at least one year. In practice, as aforementioned, the colloidal SiC QDs with sizes of a couple of nanometers can be ultra-stable in water [28] for over 10 years, maintaining the light emissions upon excitation, as observed for our SiC QDs fabricated by either chemical or electrochemical etching methods.

Various applications of silicon carbide rely on its reliable surface, dry or wet oxidation at 850–1250 °C can yield an insulating oxide layer on the surface of bulk SiC. The research shows that the SiC QDs with sizes in the strong quantum confinement regime, that is, only a couple of nanometers, can be naturally oxidized at a much lower temperature of 220 °C to form core/shell heteroepitaxial SiC/SiO_2 QDs with well crystallized silica nanoshells (Fig. 4.32) [48]. The surface silica layer can not only enhance the radiative recombination rate of carriers in the core SiC QD via generating a carrier potential barrier but also suppress the nonradiative recombination by reducing the surface dangling bonds. As a result, the PL quantum yield (QY) is 2.8 times higher in the SiC/SiO_2 QDs than in the pristine SiC QDs. The radiative (τ_r) and nonradiative (τ_{nr}) transition lifetimes can be calculated by using the relations of

$$QY = (1/\tau_r)/(1/\tau_r + 1/\tau_{nr}), \tag{4.7}$$

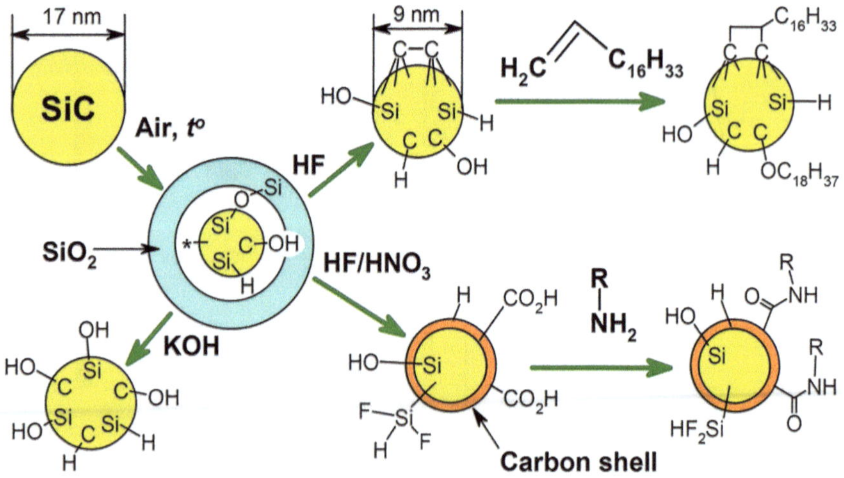

Fig. 4.31 Illustration of possible ways of surface treatments of colloidal SiC nanoparticles with initial sizes of 10–20 nm. Reprinted with permission from [47]. Copyright 2017, American Chemical Society

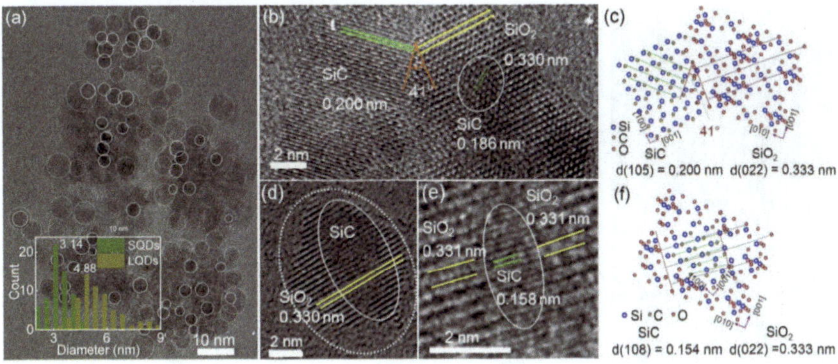

Fig. 4.32 **a** High-resolution TEM image of SiC–SiO$_2$ QDs. Inset shows particle size distribution ("SQDs"/"LQDs" refers to smaller/larger sized QDs). **b**, **d**, and **e** Magnified TEM images of the SiC-SiO$_2$ heteronanocrystal core/shell QDs and **c** and **f** corresponding schematic crystal structures. Reprinted with permission from [48]. Copyright 2022, American Institute of Physics

$$1/\tau = 1/\tau_r + 1/\tau_{nr}, \tag{4.8}$$

where τ is the measured overall PL lifetime. The result shows that the radiative transition is enhanced whereas the nonradiative transition is suppressed in the SiC/SiO$_2$ QDs compared with the pristine SiC QDs.

The SiC/SiO$_2$ QDs are very stable in air. They showed lower toxicity (particularly at higher QD concentrations) and higher brightness than the bare SiC QDs when they were taken by the A375 human cells (Fig. 4.33), hence they are more favorable for biological applications. The SiC/SiO$_2$ QDs are also promising building blocks for nanoscale electronic and photonic devices.

Little has been known about the role of Si=O functional groups in luminescence of the colloidal SiC QDs. In contrast, the wide investigations have confirmed that the Si=O surface-state orange luminescence (with a peak at ~ 590 nm, 2.1 eV) is the dominant surface-state luminescence of both Si QDs and porous Si fabricated by chemical or electrochemical etching [37]. Yellow defect luminescence peaked at around 565 nm has been observed from the colloidal SiC QDs [49], which decays much slowly (lifetime > 70 ns) compared with the previously observed blue–green fluorescence (< 15 ns). The PL spectrum of the as-prepared SiC QDs in deionized water comprises a strong blue band at around 429 nm arising from the C=O-related surface states (Fig. 4.34a). The corresponding excitation band lies at ~ 327 nm, as exhibited by the PLE spectrum. The introduction of Ag NPs in the SiC QD solution caused quenching of the strong blue band to expose the previously hidden weaker emission bands. This quenching of luminescence occurs efficiently when the overlap of the emission band of the donor and the absorption band of the acceptor is large and their distance is small (< 10 nm). The Ag NPs has a localized surface plasmon resonance peak at ~ 400 nm, so they can efficiently quench the blue emission of the SiC QDs. Figures 4.34c, d show the PL and PLE spectra of the aqueous

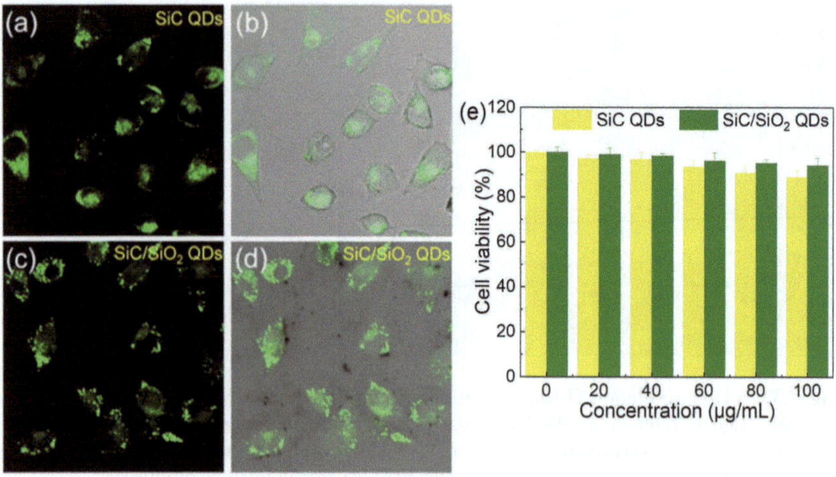

Fig. 4.33 **a–d** Dark and bright field fluorescence images of the A375 human malignant melanoma cells after uptake of the SiC or SiC/SiO_2 QDs. All the photos were taken under 405 nm excitation and each one has a width of 159 μm. **e** Cytotoxicity evaluation of the SiC and SiC/SiO_2 QDs. Reprinted with permission from [48]. Copyright 2022, American Institute of Physics

solution containing the mixed SiC QDs and Ag NPs. Now a green emission band at around 517 nm dominates, which did not shift with increasing excitation wavelength, suggesting its origin of surface states. This 517-nm emission band resembles the previously reported green defect band [41]. Apparently, the green band correlates with the strongest excitation band at around 452 nm in the PLE spectrum. There is another UV emission band lying at around 340 nm, which is ascribed to the quantum confinement luminescence of the hexagonal SiC QDs. Both the PL and PLE spectra change substantially after surface treatment (by hydrogen fluoride at 25 °C for 1 h) of the colloidal SiC QDs, as shown in Fig. 4.34e, f. In the presence of the Ag NPs in the surface-treated SiC QD solution, the green emission band vanished, confirming its origin of QD surface. In contrast, the UV band of the hexagonal SiC QDs remained. Meanwhile, a remarkable yellow emission band centered at ~ 565 nm became predominant. The blue and green bands are all much stronger than the UV band, whereas the yellow band is only a little stronger than the UV band; hence this yellow band may have existed prior to the surface treatment, but it was covered by the blue and green bands. The yellow band did not shift with increasing excitation wavelength, implying its origin of defects. The PL excitation spectrum comprises a remarkable excitation band at ~ 503 nm that correlates with the yellow emission band. The intensity of the yellow band increases by a factor of two after 15 days of storage of the treated SiC QD solution in air, implying its correlation with surface oxide groups of the SiC QDs. The 518-nm excitation band also became stronger after the storage, supporting its correspondence to the yellow emission band.

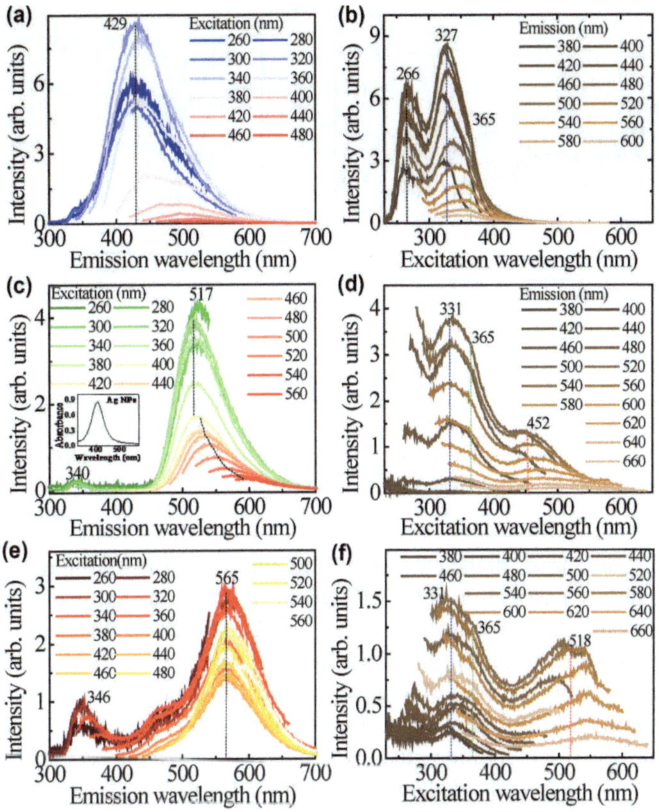

Fig. 4.34 **a** PL and **b** PLE spectra of SiC QD aqueous solution. **c** PL and **d** PLE spectra of SiC QD solution + Ag NPs. Inset shows UV–Vis absorption spectrum of Ag NPs. **e** PL and **f** PLE spectra of pristine and aged (for 15 d) acid-treated SiC QDs + Ag NPs. Reprinted with permission from [49]. Copyright 2021, American Chemical Society

Theoretical calculations concerning electronic and vibration properties of SiC QDs were performed [49]. Three types of SiC QDs were modeled: purely H-terminated SiC QD ($C_{168}Si_{216}H_{248}$, 1.75 nm; H–SiC QD), 4 Si=O groups-terminated SiC QD ($C_{168}Si_{216}H_{240}O_4$, 1.75 nm; 4O–SiC QD), and 16 Si=O groups-terminated SiC QD ($C_{168}Si_{216}H_{216}O_{16}$, 1.71 nm; 16O–SiC QD) (other surface dangling bonds are terminated by H atoms). The infrared spectra (Fig. 4.35) were calculated based on the semi-empirical (AM1) Hartree–Fock model. As far as the absorption peaks and their intensities are concerned, the calculated infrared spectrum matches well the experimental data. The major TO phonon peak arises at 721 cm^{-1}, in agreement with the measured 744 cm^{-1} considering the stronger phonon confinement effect of the calculated SiC QDs with smaller sizes. The split TO phonon peaks lie at 574, 609, 630, 665, 777, and 840 cm^{-1}, in agreement with the measured ones at 576, 613, 634, 667, 777, and 839 cm^{-1}. The LO phonon mode that splits into a twin peak at 959/980 cm^{-1} agrees with the measured one at 955/982 cm^{-1}. An intrinsic absorption peak

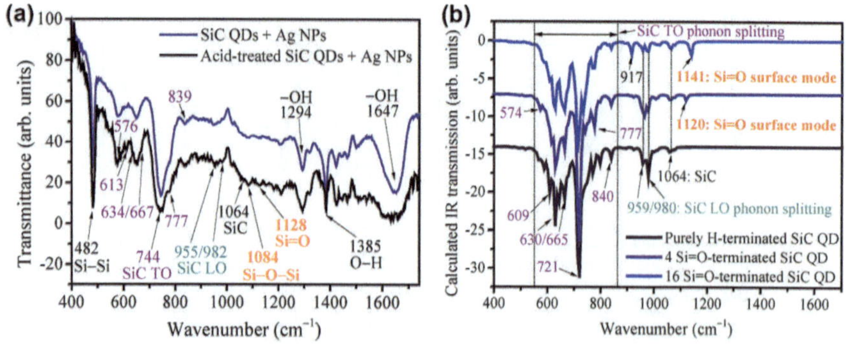

Fig. 4.35 **a** FTIR spectra of SiC QDs + Ag NPs and acid surface-treated SiC QDs + Ag NPs. **b** Calculated infrared transmission spectra of purely H-terminated SiC QD ($C_{168}Si_{216}H_{248}$), 4 Si=O groups-terminated SiC QD ($C_{168}Si_{216}H_{240}O_4$), and 16 Si=O groups-terminated SiC QD ($C_{168}Si_{216}H_{216}O_{16}$) (other dangling bonds are terminated by H atoms). Reprinted with permission from [49]. Copyright 2021, American Chemical Society

at 1064 cm^{-1} also arises in the measured FTIR spectrum. The most remarkable variation after surface passivation of the SiC QD by Si=O lies in arising of a remarkable peak at 1120 cm^{-1}, which shifts to 1141 cm^{-1} for a higher Si=O coverage, confirming its assignment to the surface Si=O bonds. This peak agrees with the measured absorption peak at 1128 cm^{-1}; indeed, this absorption peak became stronger after storage of the SiC QDs in air for 15 days, supporting its assignment to the Si=O bonds. Therefore, the surface Si=O vibration mode is infrared active and may act as a fingerprint for identifying Si=O bonds on the surface of the SiC QD.

Figure 4.36a shows the calculated highest occupied molecule orbital (HOMO) and lowest unoccupied molecule orbital (LUMO) of the H–SiC QD, 4O-SiC QD, and 16O-SiC QD by using TDDFT employing HSE06 hybrid exchange–correlation functional and the 6-31G(d) basis set [49]. As aforementioned, the hybrid exchange–correlation functional is more accurate than the PBE functional in predicting energy gaps. Both HOMO and LUMO of the H–SiC QD are delocalized, mainly distributed in the core region. The HOMO of the 4O-SiC QD becomes localized at Si=O, and the LUMO is hybridized, connecting the Si=O bonds on the opposite surfaces of the QD. With increasing Si=O coverage, both HOMO and LUMO become highly localized at Si=O groups. Figure 4.36b shows the energy gap versus Si=O coverage (in number) calculated using different combinations of exchange–correlation functional/basis set. The more accurate HSE06/6-31G(d) approach yielded an energy gap of 2.71 eV for the H–SiC QD; for the 4O-SiC QD and 16O-SiC QD, the energy gaps between the surface-localized LUMO and HOMO decreases to 2.50 and 1.76 eV, respectively, and the peak energy (2.39 eV) of the measured PL excitation spectrum of the SiC QDs lie between them, supporting its origin of Si=O surface states. The calculated energy gap for a specific size of SiC QD is dependent on the adopted exchange–correlation functional and basis set, but each method predicted substantial narrowing of the LUMO–HOMO energy gap with increasing Si=O coverage of the SiC QD. This

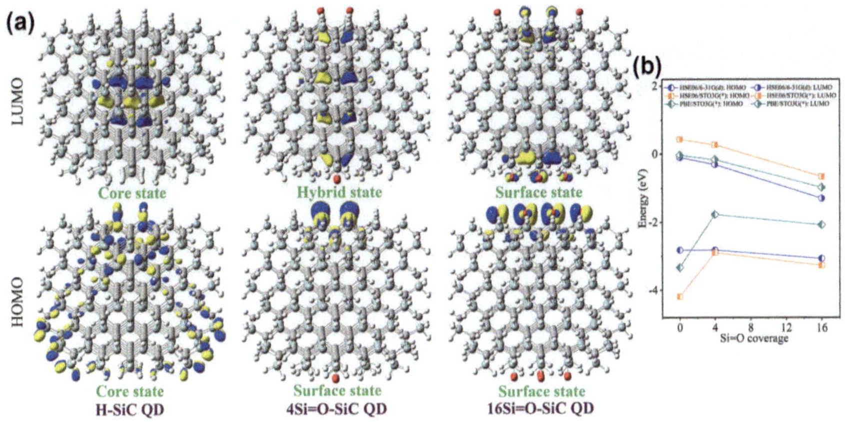

Fig. 4.36 **a** TDDFT [HSE06/6-31G(d)] calculated HOMO and LUMO of H–SiC QD, 4 Si=O–SiC QD, and 16 Si=O–SiC QD. **b** Energy gap versus Si=O coverage (in number) of the SiC QD calculated using three different combinations of exchange–correlation functional/basis set. Reprinted with permission from [49]. Copyright 2021, American Chemical Society

supports the assignment of the yellow emission band to the Si=O-related surface states.

4.1.2.3 Luminescent Films Composed of 3C–SiC Nanocrystals

Colloidal 3C–SiC nanocrystals exhibit stable and strong yellow–green to blue luminescence, but it is uncertain whether these luminescence properties can be retained if these nanocrystals are incorporated into solid films. If so, SiC nanocrystals will have useful applications in solid-state lighting and displays. 3C–SiC nanocrystals 1–6 nm in size were embedded in a polystyrene film obtained through casting the toluene solution dissolved with 3C–SiC nanoparticles and polystyrene [50]. The SiC nanoparticle/polystyrene composite film exhibited blue photoluminescence under 325 nm excitation (Fig. 4.37). The 3C–SiC nanocrystal/SiO_2 core/shell nanostructures embedded in the Si matrix have been prepared and they exhibited robust blue luminescence (Fig. 4.38) [51]. The 3C–SiC nanocrystals were introduced into the pores of the naturally oxidized porous Si, followed by annealing in argon to form the 3C–SiC nanocrystal/SiO_2 core/shell nanostructures in Si. Luminescence from the 3C–SiC nanocrystal/SiO_2 nanocomposites is very stable at high temperature.

The 3C–SiC nanocrystals embedded in a glycerol film showed stable and tunable PL in the 360–540 nm range (Fig. 4.39) [52]. XPS indicates that the silicon terminated nanocrystal surface was bonded to the glycerol molecules. Calculations based on the density functional theory suggest that the glycerol molecules bonded to the nanocrystals produced dramatic change in the SiC nanocrystal surface structure, leading to a quasi-continuous electronic band associated with the tunable emission

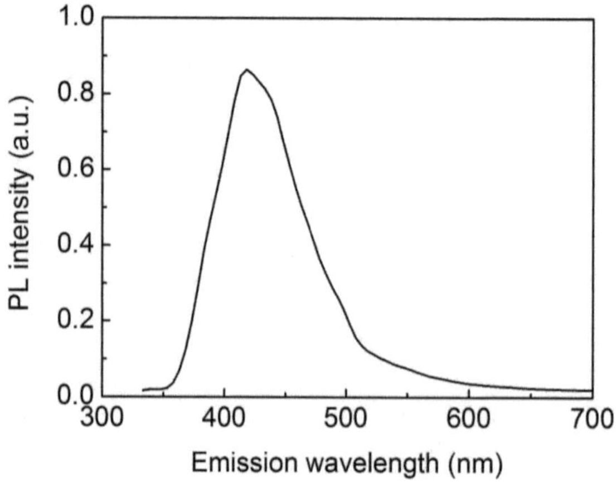

Fig. 4.37 PL spectrum of the 3C–SiC nanocrystallites embedded in a polystyrene film excited by 325 nm light. Reprinted with permission from [50]. Copyright 2005, American Institute of Physics

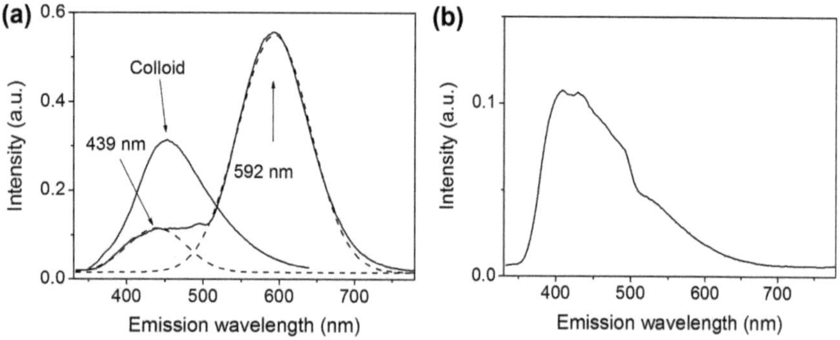

Fig. 4.38 **a** PL spectra of the 3C–SiC nanocrystals embedded in the oxidized porous Si. Dashed lines are double-Gaussian fit of the measured spectrum. "Colloid" refers to the original 3C–SiC nanocrystal aqueous solution. **b** PL spectrum of the 3C–SiC nanocrystal/SiO_2 core/shell nanostructures embedded in the Si matrix. Reprinted with permission from [51]. Copyright 2006, Springer-Verlag GmbH Germany, part of Springer Nature

wavelengths. Luminescence of as-prepared and glycerol-treated 3C–SiC NC solutions was compared [53]. The PL, blinking, and time-resolved PL indicate that surface passivation by glycerol enhanced the PL and improved the PL stability. The PL decay lifetimes kept the same irrespective of the emission wavelength, implying that the broad emission band correlates with the size distribution of the 3C–SiC NCs. The surface passivation did not affect the PL lifetimes.

The 3C–SiC nanocrystal–sodium dodecyl sulfonate (SDS) inorganic–organic interlinked solid network showed intense photoluminescence [54]. Figure 4.40 shows the high-resolution TEM image of the 3C–SiC nanocrystals embedded in the SDS film. There are many 2–4 nm SiC particles, which have good crystallinity, with

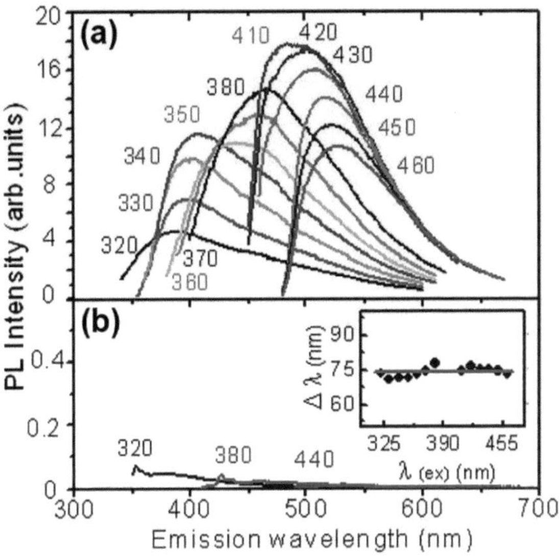

Fig. 4.39 PL spectra of the (**a**) 3C–SiC nanocrystals/glycerol solid film on a Si wafer under different excitations and **b** pure glycerol excited at three wavelengths. The inset shows the difference between emission and excitation wavelengths as a function of excitation wavelength. Reprinted with permission from [52]. Copyright 2010, American Chemical Society

random crystallographic orientations. PL from this SiC nanocrystal–SDS film shifted from 417 to 491 nm with increasing excitation wavelength (Fig. 4.41), obeying quantum confinement effect. The light emission is very strong and visible to the naked eyes (Fig. 4.42). The time-resolved PL follows a triple-decay law, with an average lifetime of 13.65 ns.

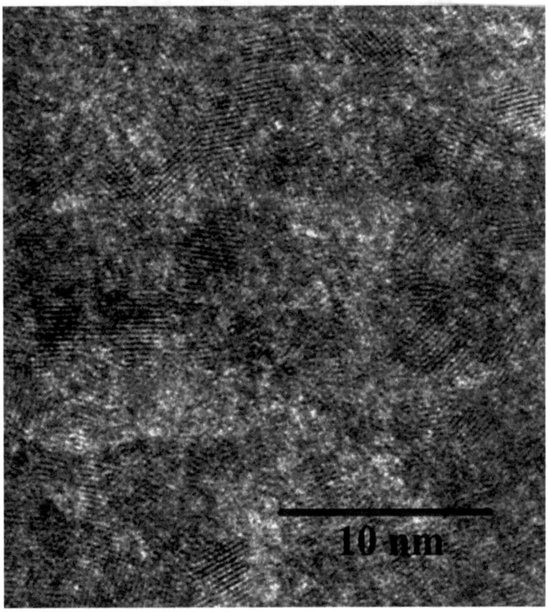

Fig. 4.40 HRTEM image of the 3C–SiC nanocrystal–sodium dodecyl sulfonate composite network film. Reprinted with permission from [54]. Copyright 2012, Royal Society of Chemistry

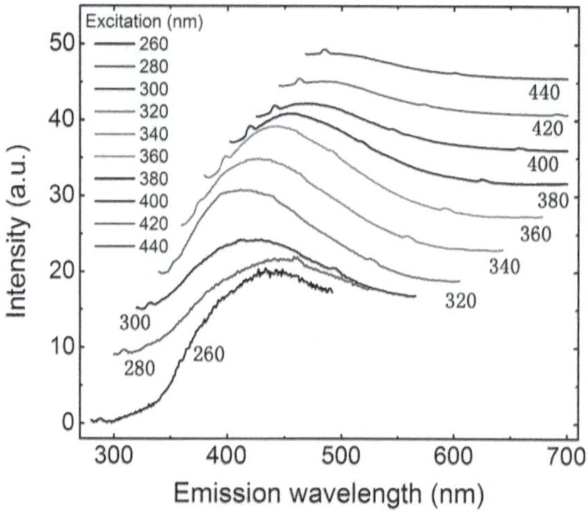

Fig. 4.41 PL spectra of the 3C–SiC nanocrystal–sodium dodecyl sulfonate composite network film under different excitations. Reprinted with permission from [54]. Copyright 2012, Royal Society of Chemistry

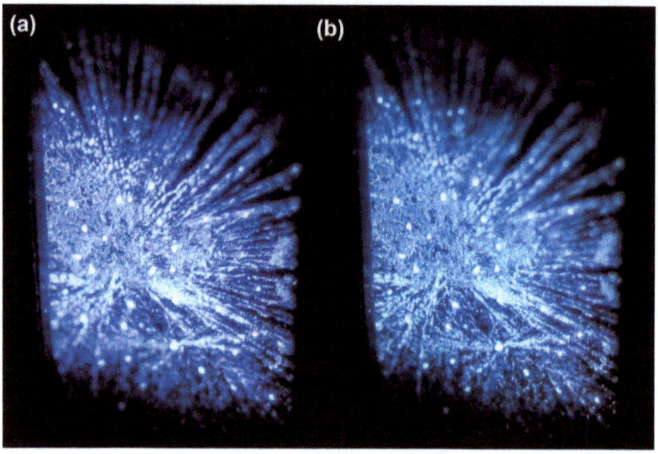

Fig. 4.42 Luminescence photos of the 3C–SiC nanocrystal–sodium dodecyl sulfonate composite network film on a Si wafer excited by **a** 320 nm and **b** 360 nm lines of the Xe lamp. Reprinted with permission from [54]. Copyright 2012, Royal Society of Chemistry

Some other 3C–SiC nanocrystals incorporated luminescent solid films have been fabricated and investigated, for instance, 3C–SiC nanocrystal/silica nanocomposites [55] and 3C–SiC nanocrystals embedded in the binary poly(allylamine hydrochloride)–sodium poly(styrene sulfonate) polyelectrolytes solid matrix [56]. The SiC QDs embedded in the binary poly(allylamine hydrochloride)-sodium poly(styrene sulfonate) polyelectrolytes solid matrix exhibited apparent photon reabsorption phenomenon. The PL peak exhibited red shift with increasing concentration of the embedded QDs. The analysis shows that this shift arises from reabsorption of PL of smaller SiC QDs by larger ones in the matrix. Deposition of Ag nanoparticles on the surface of the solid matrix can dramatically reduce light reabsorption owing to surface plasmon-induced concentration of incident light. Meanwhile, the PL intensity was enhanced by a factor of 1.4.

In fact, the photon reabsorption among the different luminescent nanoparticles of an ensemble of colloidal nanoparticles can substantially influence the profile and peak position of the instrument-detected nominal "PL spectrum," sometimes rather astonishing and unexpected. Figure 4.43 compares the PL spectra of the freshly prepared carbon QD solution and those of its 16-times-diluted solution [57]. The pristine QD solution exhibited strongest light emission at ~ 515 nm. However, after dilution, the strongest emission band shifted to ~ 424 nm, revealing that there was strong photon reabsorption among the QDs of the pristine solution and this yielded a detected fictitious red emission band rather than the true blue emission band. Figure 4.44 shows the variations of the wavelength and intensity of the strongest emission in nine samples vs. particle concentration. For comparison, the data for another two groups of QD aqueous solutions are also presented, whose original solutions before dilution have pH of 7 and 3, respectively. The values of the PL intensities are normalized in such a way that all the resulting values correspond to the emissions of a nearly identical collection of CQDs. The result indicates that with gradual dilution of the QD solution, the PL of a fixed collection of QDs shifts monotonically to blue and meanwhile the emission becomes more and more strong, as a result of increasing mitigation of the photon reabsorption. These results reveal that the phenomenon of photon reabsorption is very important in explaining the luminescence properties of the colloidal nanoparticles and should be seriously taken into account during the optical spectroscopy of colloidal nanoparticles; otherwise, distinct or even wrong conclusions may be drawn if different concentrations of nanoparticles have been utilized in studying their luminescence properties. A general quantitative model describing the influence of the photon reabsorption among the colloidal QDs on the detected nominal PL spectrum has been constructed, and the theoretical result agrees well with the experimental result by taking the ensemble of colloidal zinc oxide QDs as a prototype system [58].

Synthesis of 3C–SiC QDs/cetyltrimethylammonium bromide (CTAB) multilayered vesicle core/shell nanostructures has been reported [59]. The Coulomb attraction force between the negatively charged surface of the QD and positively charged ionized CTAB molecules caused formation of the core/shell nanostructures. The zeta potential distribution of the 3C–SiC nanocrystals suspended in ethanol (pH = 3) is shown in Fig. 4.45. Most particles show a positive zeta potential (net charge) probably because the negatively charged oxygen atom of the Si–O–Si group on the surface attracts positive ions from the surrounding solvent, resulting in an overall positive charge on most particles.

3C–SiC nanocrystals can be directly incorporated into a solid film without using encapsulation materials [60]. Figure 4.46a shows the PL spectra of the self-assembled film of the 3C–SiC nanocrystals. The spectrum is smooth, with a full width at

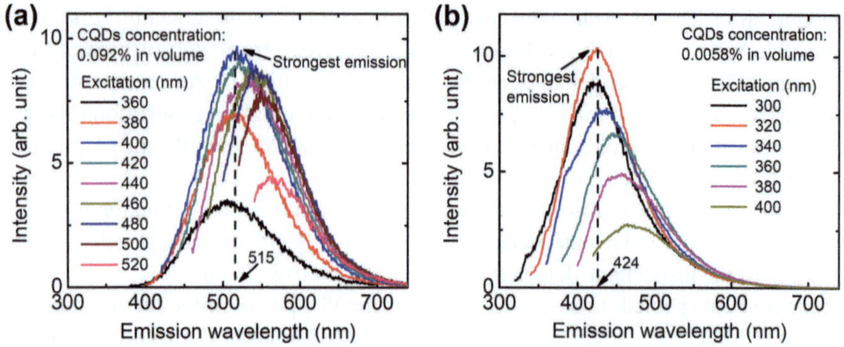

Fig. 4.43 PL spectra of the carbon QD aqueous solution under different wavelengths of excitation. **a** Freshly prepared QD solution (pH = 13) and **b** its diluted solution. These two solutions contain an identical collection of CQDs. Reprinted with permission from [57]. Copyright 2014, American Institute of Physics

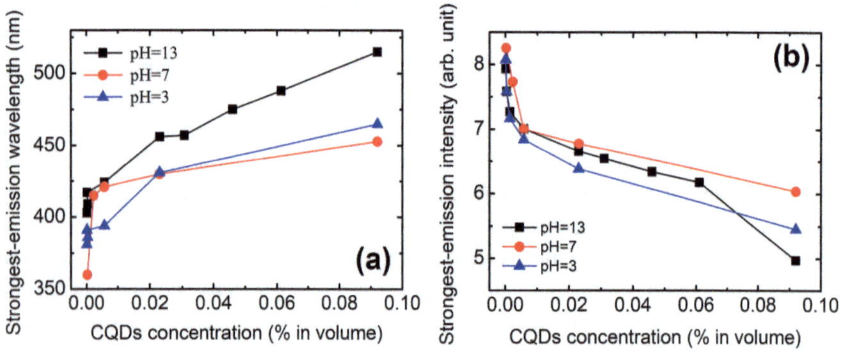

Fig. 4.44 Plot of **a** wavelength and **b** intensity of the strongest emission of the carbon QD aqueous solutions vs. concentration. Each group of nine data on the same curve are measured values for the nine solutions which were obtained through dilution of the same original QD solution (with pH = 13, 7, or 3). The intensities are such normalized that they correspond to the emissions of an identical collection of QDs. Reprinted with permission from [57]. Copyright 2014, American Institute of Physics

half maximum of about 110 nm. The PL shifted to red with increasing excitation wavelengths (Fig. 4.46b). These spectral features imply the quantum confinement effect.

4.1.2.4 More Polytypes of SiC Nanocrystals

There have been many reports on the luminescence properties of cubic silicon carbide nanocrystals. In contrast, relatively little is known about the properties of hexagonal silicon carbide nanocrystals. Unlike cubic silicon carbide, no definite evidence of

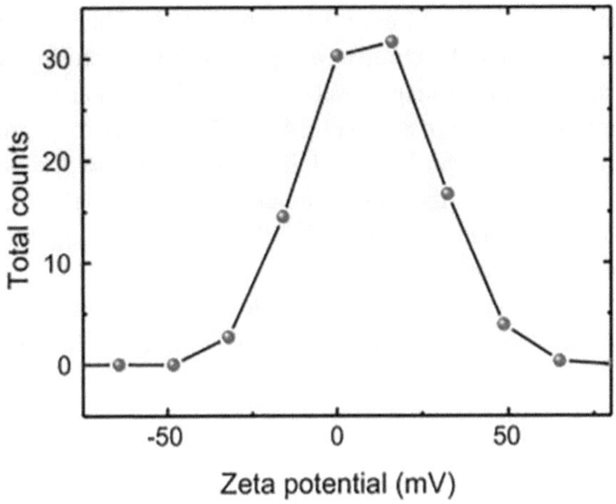

Fig. 4.45 Zeta potential distribution of the 3C–SiC quantum dots suspended in ethanol showing an average value of 9.62 mV. Reprinted with permission from [59]. Copyright 2013, Elsevier B.V

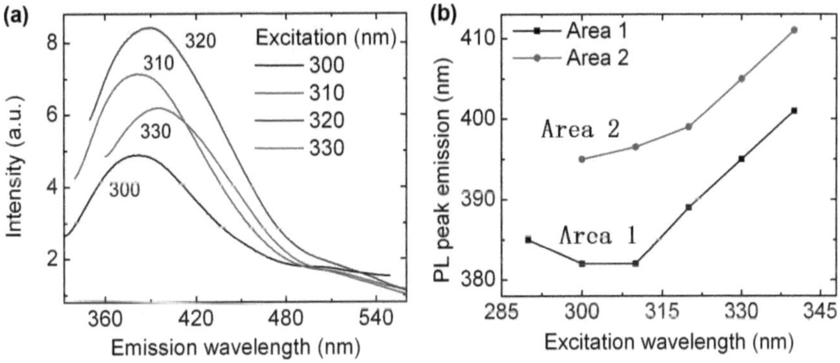

Fig. 4.46 **a** PL spectra of the close-packed 3C–SiC nanocrystal film (area 1) on a glass slide under different excitations. **b** PL peak wavelength as a function of excitation wavelength (area 1). The values of another area (area 2) on the same film are also shown. Reprinted with permission from [60]. Copyright 2011, American Institute of Physics

quantum confinement effect had previously been observed from hexagonal silicon carbide.

The free porous 6H–SiC layer was prepared by electrochemical etching of *n*-type 6H–SiC. Mechanical grinding of the porous free layer can produce porous SiC nanopowders [61, 62]. The samples exhibited similar photoluminescence features, with a strong band at around 2.65 eV (Fig. 4.47). The emission was proposed to arise from recombination of the N–Al donor–acceptor electronic states as well as surface states. The diluted nanopowders suspended in water yielded a PL peak at around 3.74 eV, which was suggested to arise from the quantum-confined carriers considering its above-bandgap energy. The 6H–SiC nanocrystals dispersed in water and hydrofluoric acid showed different PL features [63]. These nanoparticles were fabricated by electrochemical etching of *n*-type bulk 6H–SiC followed by ultrasonic treatment. Transmission electron microscopy reveals that the particles are crystalline, with diameters ranging from 1 to 8 nm (Fig. 4.48). The nanoparticles in water showed a broad emission band at 2.78 eV (Fig. 4.49), which is below the bandgap of bulk 6H–SiC. The PL spectrum of the nanoparticles suspended in HF/ethanol showed an above-bandgap emission band at around 3.5 eV, which was suggested to arise from the quantum confinement effect.

How would different polytypes of nanoscale SiC behave in luminescence properties when their sizes diminish to a couple of nanometers? The comparative study

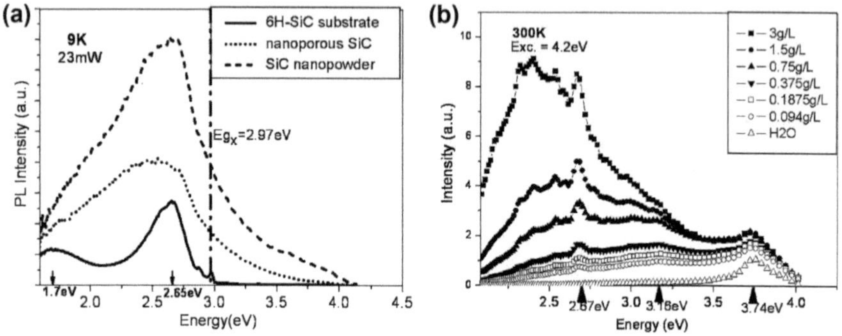

Fig. 4.47 PL spectra **a** of 6H–SiC wafer, porous SiC, and SiC nanopowder. Concentration dependent normalized (relative to water Raman peak) PL spectra, **b** of the nanopowders suspended in water. Reprinted with permission from [61]. Copyright 2007, American Institute of Physics

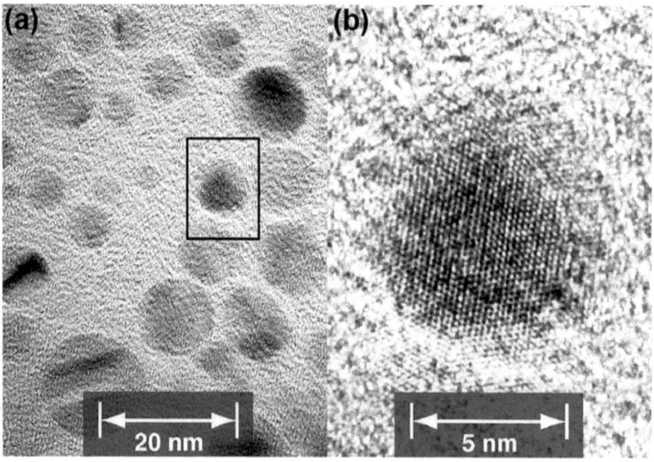

Fig. 4.48 **a** TEM image of 6H–SiC nanoparticles. **b** High-resolution image of a typical nanoparticle showing lattice fringes. Reprinted with permission from [63]. Copyright 2008, American Institute of Physics

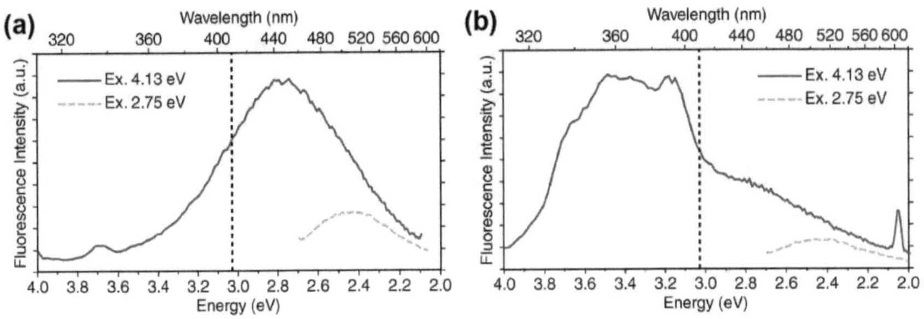

Fig. 4.49 PL spectra of 6H–SiC nanocrystals dispersed in **a** water and **b** HF/ethanol. Reprinted with permission from [63]. Copyright 2008, American Institute of Physics

of photoluminescence from the SiC QDs fabricated through diminishing the sizes of bulk 3C–, 6H–, and 4H–SiC single crystals via electrochemical etching followed by ultrasonic dispersion in a liquid highly improves our understanding of the characteristics of SiC QDs [64]. Particularly, a dominant blue defect emission band was observed, and the SiC QDs exhibited ability of phase transition among different polytypes with the aid of sonication. These particles are 1–8 nm in size. Figure 4.50a–c show their PL spectra under different wavelengths of excitation. Three ensembles of QDs fabricated from different polytypes of bulk SiC crystals show similar spectral features. All the PL spectra are smooth and intense. The FWHM values are between 100 and 130 nm. The large linewidth is attributed to the broad size distribution of the nanoparticles as well as overlapping of different origins of emission bands. The PL peak wavelength varies similarly with increasing excitation wavelength for the three samples (Fig. 4.50d). There is a nearly constant emission band at around 452 nm for excitation wavelengths smaller than 320 nm (Zone I), implying that this blue emission band belongs to the defect-related luminescence similar to that observed from porous 6H–SiC [65]. The PL peak wavelength shifted to red nearly linearly with increasing excitation wavelengths over 400 nm before reaching a maximum at around 565 nm (2.2 eV), which is very close to the bandgap of bulk 3C–SiC. These features are characteristic of interband recombination luminescence from 3C–SiC nanocrystals following the quantum confinement effect. The 6H sample shows flatter PL peaks than the 3C and 4H samples probably due to the different size distributions.

The PL excitation measurement reveals more similarity among the three ensembles of SiC QDs (Fig. 4.51). The PLE spectra of each ensemble show three distinct peaks, implying that there are three types of light absorption processes. The PLE spectra monitored at 580 nm for all the samples show an identical peak at about 365 nm (Fig. 4.51d), indicating their same origin. Furthermore, the three PLE bands vanish at about 563 nm, further confirming that the redder part of the PL in each ensemble corresponds to interband recombination luminescence of the 3C–SiC QDs. Hence the wide excitation band at around 365 nm is associated with light absorption by the cubic SiC QDs. However, this PLE band can neither be fitted with a single Gaussian function nor a Lorentzian function. Instead, it is the superimposition of a broad Gaussian band that shifts with monitored emission wavelength and a constant

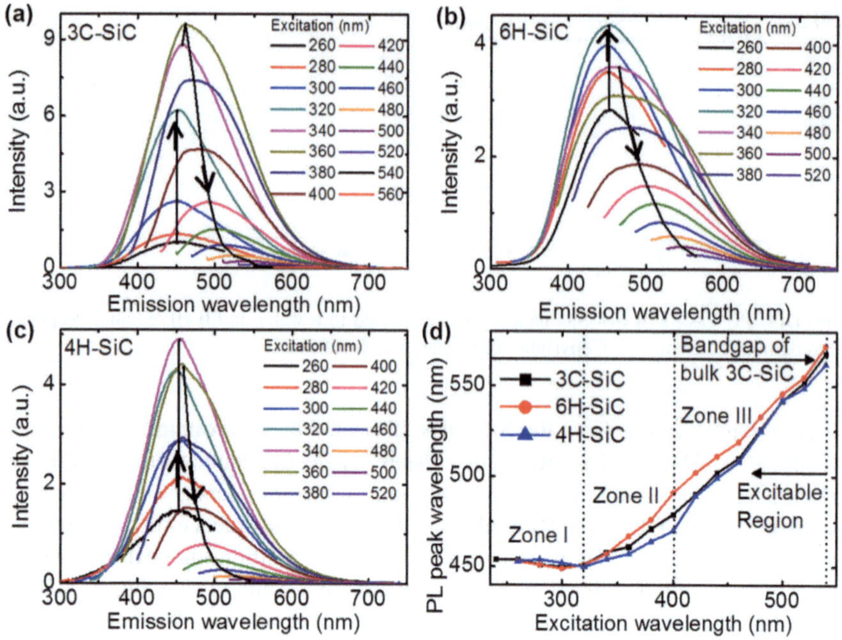

Fig. 4.50 **a–c** PL spectra of three aqueous solutions of SiC quantum dots obtained by electrochemical etching of bulk 3C–, 6H–, and 4H–SiC, respectively. Black lines mark the variation trend of the PL peak and the arrows indicate the direction of increasing excitation wavelength. **d** PL peak wavelength versus excitation wavelength. Reprinted with permission from [64]. Copyright 2012, American Institute of Physics

narrow Gaussian band peaked at ~ 365 nm, the former is ascribed to the interband light absorption of cubic SiC QDs, whereas the latter could be attributed to some surface defect (such as C–O–C related).

Figure 4.52 shows the time-resolved PL spectra of the three ensembles of SiC QDs stemming from different polytypes of bulk SiC crystals. Each spectrum can be fitted by a triple-exponential function. The lifetime and fractional intensities of the three components extracted from the fits are summarized in Table 4.1. Each component in one ensemble has a counterpart from any of the other two ensembles, concluded based on the closeness in the lifetimes and factional intensities. The faster decay component with a lifetime of around 4 ns dominates the blue-band emission of each sample with a fractional intensity of over 60%, suggesting that it is associated with the blue emission band. The slower decay component has a lifetime of over 10 ns and fractional intensity of around 20%. This may be related to the tunable quantum confinement luminescence from the 3C–SiC QDs.

These results indicate that the three ensembles of SiC QDs have the same luminescence mechanisms: one defect associated blue luminescence and one tunable quantum confinement-dominated luminescence. A polytypic transformation model was proposed to explain this strange behavior (Fig. 4.53). Based on this model, some

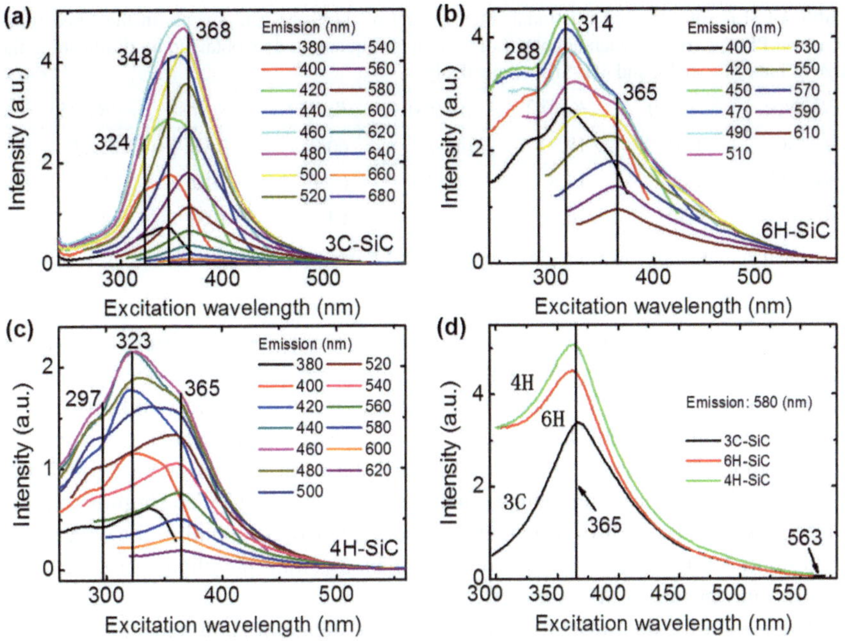

Fig. 4.51 **a–c** PL excitation spectra recorded at different emission wavelengths. The black lines mark the peak positions. **d** PL excitation spectra for the 580 nm emission. Reprinted with permission from [64]. Copyright 2012, American Institute of Physics

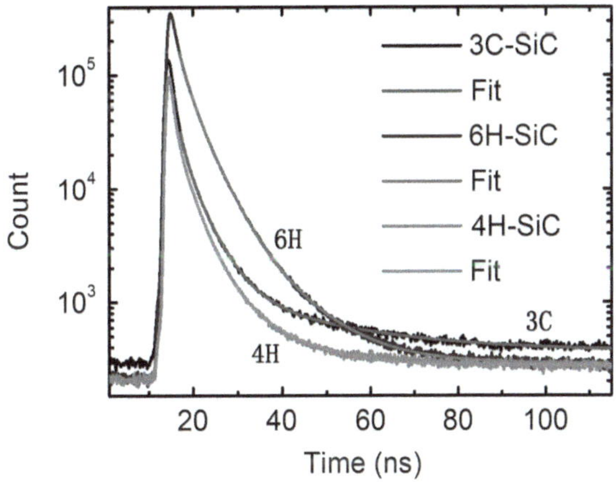

Fig. 4.52 Time-resolved PL spectra excited by 280 nm light and monitored at 420 nm (3C) and 450 nm (6H, 4H). Each spectrum can be fitted by a triple-exponential function. The component lifetimes and fractional intensities are summarized in Table 4.1. Reprinted with permission from [64]. Copyright 2012, American Institute of Physics

hexagonal QDs can be transformed into cubic ones (and vice versa) by applying ultrasonic waves during the preparation of the QDs from the etched porous SiC if the energy offered by the ultrasonic wave is big enough to overcome the potential energy barrier of the polytypic transformation. In practice, as aforementioned, many theoretical calculations have indicated that bulk 3C-, 6H-, and 4H–SiC have very close

Table 4.1 Lifetime (τ) and fractional intensity (R) of the three components in the fitted time-resolved PL spectrum of each aqueous solution of SiC quantum dots obtained by diminishing the size of bulk 3C–, 6H–, and 4H–SiC, respectively

Polytype	τ_1 (ns)	R_1 (%)	τ_2 (ns)	R_2 (%)	τ_3 (ns)	R_3 (%)
3C–SiC	1.63	18.7	4.67	60.0	24.8	21.3
6H–SiC	1.66	14.8	4.28	67.1	10.9	18.1
4H–SiC	0.89	14.1	3.38	64.4	10.6	21.5

cohesive energies [66]. Thus, the difference in cohesive energy of different polytypes of SiC QDs may also be small, making the ultrasonic wave-driven polytypic transformation occur readily. This accounts for the PL and PLE spectral similarity among the three ensembles of SiC QDs with distinct bulk SiC origins. A subsequent experimental study offers more evidences of the α to β phase transformation for the SiC QDs suspended in solutions at ambient temperature and pressure [67]. The high-resolution TEM as well as light absorption and emission spectroscopy confirm the occurrence of this phase transformation. The rare phenomenon of multiple-phonon-assisted light absorption was observed in these SiC QDs.

The reason for the difficulty in observing the quantum confinement effect from hexagonal silicon carbide compared to cubic silicon carbide has been explored [64]. A semiconductor QD exhibits remarkable quantum confinement effect only when its size is close to or smaller than its bulk exciton Bohr diameter [68]. The Bohr radius of a Mott–Wannier exciton in silicon carbide can be evaluated using the following equation [69]:

$$R = \frac{4\pi\varepsilon_0\varepsilon_r\hbar^2}{\mu e^2} = \frac{0.053\varepsilon_r}{\mu/m_0}(\text{nm}), \quad (4.9)$$

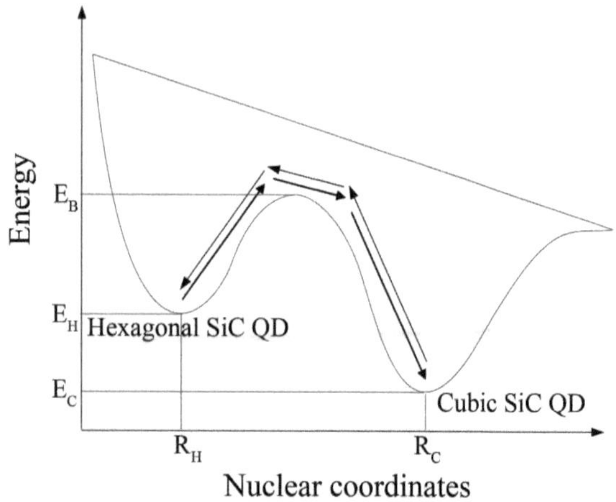

Fig. 4.53 Schematic illustration of hexagonal–cubic polytypic transformation of the SiC quantum dot on a potential energy vs. nuclear coordinate curve. Reprinted with permission from [64]. Copyright 2012, American Institute of Physics

where ε_0 and ε_r are the dielectric constant in vacuum and relative permittivity, e is the electron charge, m_0 is the electron mass, $\hbar$ is the reduced Planck's constant, and $\frac{1}{\mu} = \frac{1}{m_e^*} + \frac{1}{m_h^*}$ is the reduced effective mass of the electron–hole pair, with m_e^* and m_h^* effective masses of electron and hole at the conduction band minimum and valence band maximum. $m_{\mathrm{e,h}}^* = \hbar^2/(\partial^2 E_{\mathrm{c,v}}(\boldsymbol{k})/\partial k^2)$, where $E_{\mathrm{c,v}}$ is the electron energy at the conduction band minimum or the valence band maximum. The effective masses of electrons and holes of silicon carbide are taken from Ref. [22]. By using Eq. (4.5), the exciton Bohr radii were evaluated to be 2.0, 0.7, and 1.2 nm for 3C–, 6H–, and 4H–SiC, respectively. The most probable diameter of the SiC QDs obtained by electrochemical etching is generally around 3.9 nm which is close to the exciton Bohr diameter of 3C–SiC (4.0 nm) but larger than those of 4H–SiC (2.4 nm) and 6H–SiC (1.4 nm). Therefore, even smaller sizes of SiC QDs are necessary to observe the quantum confinement effect in hexagonal SiC QDs than in cubic SiC QDs. Note that one can also calculate the binding energy (ionization energy) of the ground-state Mott–Wannier exciton by using

$$E_{\mathrm{b}} = \mu e^4/\left[2(4\pi\hbar\varepsilon_0\varepsilon_r)^2\right]. \tag{4.10}$$

By quenching the strong blue (C=O related) defect emission of the SiC QDs through coupling them with Ag nanoparticles (NPs), one indeed can observe ultraviolet quantum confinement luminescence of colloidal hexagonal SiC QDs [70]. The experimental result in combination with the theoretical calculation reveals that there is strong resonance coupling between the photon-excited electron–hole pair in the SiC QD and the localized surface plasmon in the Ag NP. This causes efficient nonradiative resonance energy transfer from the excited SiC QD to the Ag NP, leading to quenching of the intense blue surface-defect luminescence of the SiC QDs. As a result, the previously hidden near-UV emission band was observed (Fig. 4.54). The study shows that this near-UV emission stems from interband transition of the carriers in the 6H–SiC QDs and it exhibited remarkable quantum confinement effect.

In practice, the high degree of spectral overlap between the blue emission band of the SiC QDs and the absorption spectrum of the Ag NPs (with a localized surface plasmon resonance peak at ~ 400 nm, inset of Fig. 4.54a) renders the resonance energy transfer between the SiC QDs and the nearby Ag NPs very efficient. This resonance energy transfer is an electrodynamic phenomenon, where two very close oscillating dipoles interact with each other via the near field, through which the electromagnetic energy is transferred nonradiatively from the excited donor (SiC QD) to the ground-state acceptor (Ag NP). This energy transfer occurs very efficiently if there is significant overlap between the emission spectrum of the donor and the absorption spectrum of the acceptor and their distance is smaller than ~ 10 nm. This additional nonradiative relaxation channel results in luminescence quenching of the donor. The recombination of the electron–hole pair in the SiC QD and the collective oscillation of the free electrons in the Ag NP can be taken as oscillating dipoles to the lowest order approximation from the classical viewpoint. The rate of energy transfer between a donor and an acceptor at a distance of r follows [71]

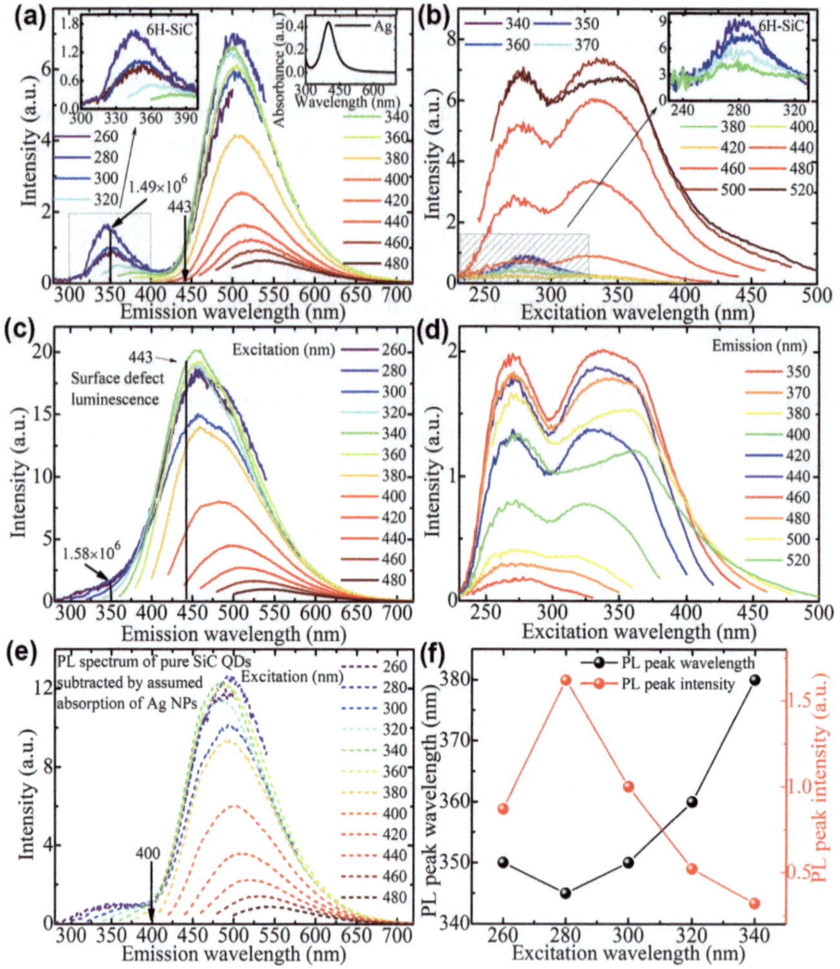

Fig. 4.54 **a** PL and **b** PLE spectra of SiC QDs and Ag NPs in ethanol. Insets show magnified images of shadowed regions as well as UV–Vis absorption spectrum of Ag NPs. **c** PL and **d** PLE spectra of SiC QDs in ethanol. **e** PL spectra of SiC QDs [in (**c**)] subtracted by calculated light absorption of Ag NPs. **f** PL peak wavelength and intensity vs. excitation wavelength for UV emission band in inset of (**a**). Reprinted with permission from [70]. Copyright 2017, American Institute of Physics

$$R_{DA} = 3/(64\pi^5 r^6 \tau_{sp}) \times \int_0^\infty (c/n\nu)^4 g_D(\nu)\sigma_A(\nu)d\nu, \quad (4.11)$$

where τ_{sp} is the spontaneous emission lifetime of the donor, n is the refractive index of the liquid surrounding the SiC QDs ($n = 1.36$ for ethanol), $g_D(\nu)$ and $\sigma_A(\nu)$ are the lineshape function of the donor and the absorption cross section of the acceptor, respectively. The absorbance of the Ag NP solution follows

$$A_{Ag} = \log[I_0/I(l)] = \log[I_0/(I_0 e^{-\alpha l})] = \alpha l/\ln 10 \approx \sigma N_{Ag} l/\ln 10, \qquad (4.12)$$

where α is the absorption coefficient, σ is the absorption cross section of a single Ag NP and N_{Ag} is the concentration of the Ag NPs, I_0 is the intensity of light incident on the front surface of the square cuvette, and l (1 cm) is the width of the cuvette along the light path. One gets $\alpha = A_{Ag} \ln 10/l$ and $\sigma \approx (A_{Ag} \ln 10)/(N_{Ag} l)$. Substituting the latter into Eq. (4.11) and substituting the normalized spectrum of the blue emission band $f_{SiC}(\nu)/\int_0^\infty f_{SiC}(\nu')d\nu'$ (obtained from double-Gaussian function decomposition of the total PL spectrum of the pure SiC QD solution) for $g_D(\nu)$, one obtains [70]

$$R_{DA} = \frac{3}{64\pi^5 r^6 \tau_{sp\_SiC}} \int_0^\infty \left(\frac{c}{n\nu}\right)^4 \frac{f_{SiC}(\nu)}{\int_0^\infty f_{SiC}(\nu')d\nu'} \frac{A_{Ag}(\nu)\ln 10}{N_{Ag} l} d\nu \approx \frac{0.436}{\tau_{sp\_SiC}}, \qquad (4.13)$$

where λ_0 is the wavelength in vacuum, $r = 4.75$ nm, and $N_{Ag} = 1.4 \times 10^{16}$ mL^{-1}. The estimated energy transfer rate is comparable to the spontaneous emission rate $1/\tau_{sp\_SiC}$ of the blue emission band of the SiC QDs and this explains the efficient quenching of the blue emission band of the SiC QDs.

Note that the absorption of the Ag NPs might also weaken the detected nominal "luminescence spectrum" of the SiC QDs. To clarify the role of such absorption in the detected weakened blue band, we consider the effect of photon absorption by the Ag NPs. The light intensity at the exit surface of the cuvette if there are no Ag NPs in the solution (Fig. 4.54c) was used for the calculation. Assume uniform suspension of the SiC QDs and Ag NPs in the ethanol solution, and then the light intensity at the exit surface of the cuvette can be evaluated by taking into account the absorption of the Ag NPs:

$$I = \int_0^l dI \times e^{-\alpha(l-z)} = \int_0^l \frac{I_{orig}}{l} dz \times e^{-\alpha(l-z)} = \frac{I_{orig}}{A_{Ag}\ln 10}(1 - 10^{-A_{Ag}}), \qquad (4.14)$$

where I_{orig} is the light intensity at the exit surface of the cuvette in the absence of Ag NPs in the solution (Fig. 4.54c). The calculated PL spectra are shown in Fig. 4.54e. It is apparent that the absorption of the Ag NPs merely slightly alters the profile of the original PL spectrum of the pure SiC QDs in the near-UV to blue region. Therefore, the absorption of the Ag NPs cannot account for the complete vanishing of the blue emission band and dominance of the near-UV band. The absorption effect of the Ag NPs is not only weak but also nonselective, that is, it is independent of the origin of the photons to be absorbed (arising from blue defect luminescence or quantum confinement luminescence). By contrast, the PL quenching resulting from resonance energy transfer between the SiC QDs and the Ag NPs is emission-band-selective, the coupling between the electron–hole pair associated with the blue emission of the SiC QD and the localized surface plasmon of the Ag NP is very efficient because of their high degree of spectral overlap, hence it causes selective and nearly complete quenching of the blue emission band.

4.1.3 Solution-Based Chemical Synthesis

Compared to other group IV nanoparticles such as Si and Ge nanoparticles [72], fewer chemical synthesis methods have been developed to produce SiC nanoparticles. One reason is that silicon carbide crystals have very high melting points resulting from the strong Si–C bonds and consequently, chemical synthesis of silicon carbide crystals typically needs a very high temperature at ambient pressure.

A one-step reaction involving a single precursor without catalytic assistance can produce SiC with a large surface area under autogenic pressure at an elevated temperature [73]. In this synthesis, dissociation of triethylsilane at 800 °C yielded the silicon carbide–carbon nanocomposite and at 1000 °C produced SiC according to the following reactions:

$$C_6H_{16}Si \xrightarrow{800^\circ C} SiC_{(s)} + 2C_{(s)} + nH_{2(g)} + hydrocarbons_{(g)}, \tag{4.15}$$

$$C_6H_{16}Si \xrightarrow{1000^\circ C} SiC_{(s)} + nH_{2(g)} + hydrocarbons_{(g)}, \tag{4.16}$$

Transmission electron microscopy reveals that the synthesized particles have diameters of 6–12 nm, and high-resolution TEM and electron diffraction (Fig. 4.55) indicate that they are crystalline 3C–SiC. Nanocasting and carbothermal reduction also produced 8–12 nm size 3C–SiC nanocrystals with large surface areas [74].

Thermal reduction of phenylsiloxane polymers yielded size controlled, oxide embedded, and free-standing SiC nanocrystals [75]. Compositional tuning of the polymers was realized by varying the relative amounts of phenyl trichlorosilane ($C_6H_5SiCl_3$) and silicon tetrachloride ($SiCl_4$) during hydrolysis and condensation (Fig. 4.56). Thermal reduction of the condensed copolymers produced the oxide embedded SiC nanocrystals. X-ray diffraction (Fig. 4.57a) indicates that the average particle size depends on the relative $C_6H_5SiCl_3$ concentration in the initial precursor mixture. In the XRD spectra, the broad reflections at 36°, 60°, and 72° are assigned to the (111), (220), and (311) crystal planes of 3C–SiC, respectively. The peaks broadened as the polymer phenyl concentration was reduced (from samples A3 to C3). The magnified spectra of (220) reflection (Fig. 4.57b) show this trend more clearly. The SiC particle size was calculated to be 9, 7, and 5 nm for samples A3, B3, and C3, respectively, using the XRD peak width based on Scherrer equation. The phenylsiloxane polymer composition controls the particle size. TEM provides similar results. Free-standing SiC nanocrystals were prepared by oxidation of carbon and chemical etching of the SiO_2 matrix. The high-resolution TEM micrographs (Fig. 4.58) disclose good crystallinity and the 3C polytype. The FTIR spectrum

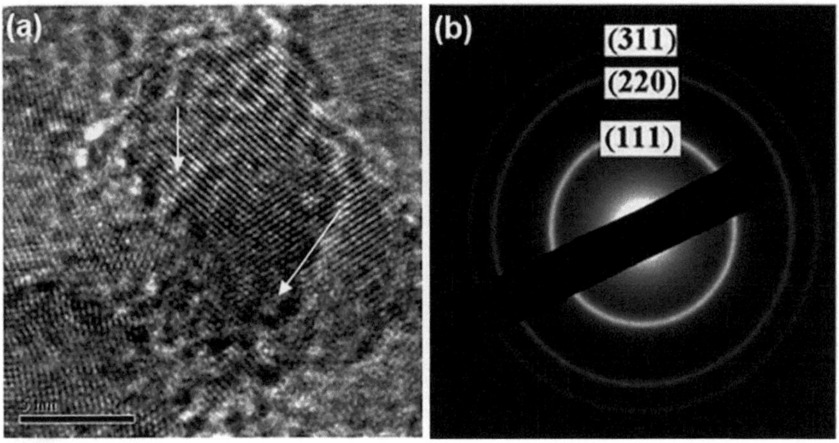

Fig. 4.55 **a** HRTEM image of the SiC sample. **b** Selected-area electron diffraction pattern of the SiC sample. Reprinted with permission from [73]. Copyright 2005, American Chemical Society

obtained from the liberated SiC nanocrystals shows a broad dominant peak at 835 cm^{-1} ascribed to optical phonon vibrations of SiC, a broad peak at 3400 cm^{-1} arising from O–H stretching vibration, and another peak at about 1100 cm^{-1} stemming from Si–O bonds. These free-standing SiC particles exhibited no luminescence due to the complex surface chemistry and relatively big crystal sizes. Thermal processing of the sol–gel derived polymers $[(HSiO_{1.5})_{0.95}(RSiO_{1.5})_{0.05}]_n$ ($R = C_8H_{17}, C_{10}H_{21}$) can also give rise to SiC nanocrystals [76].

4.1.4 Laser Ablation Method

Femtosecond laser has long been explored as an efficient way of producing SiC nanoparticles on the surfaces of SiC films or single crystals. An array of highly orientated SiC nanoparticles on the surface of 3C–SiC thin films was produced without inducing radiation defects in deeper layers of SiC film through defect activation and consequent Coulomb explosion by the femtosecond laser irradiation at a fluence well below the single shot thermal modification threshold [77]. The nanoparticles are approximately spherical and have uniform sizes of 100–120 nm. The conversion of 3C–SiC layer into nanoparticles occurred to a depth of ~ 350 nm. AES reveals that the chemical composition of the produced nanoparticles is identical to that of the original thin film. XRD indicates that the crystal structure of the irradiated region still remained the initial cubic structure, although lattice rearrangement due to lattice heating was observed. The nanoparticles aligned along a preferred lattice orientation <100> . Nanoripples with periods of 150 and 80 nm can form on the surfaces of 6H–SiC crystals irradiated by the *p*-polarized 800 nm and the *s*-polarized 400 nm femtosecond lasers, respectively [78]. Simultaneous irradiation of two collinear laser

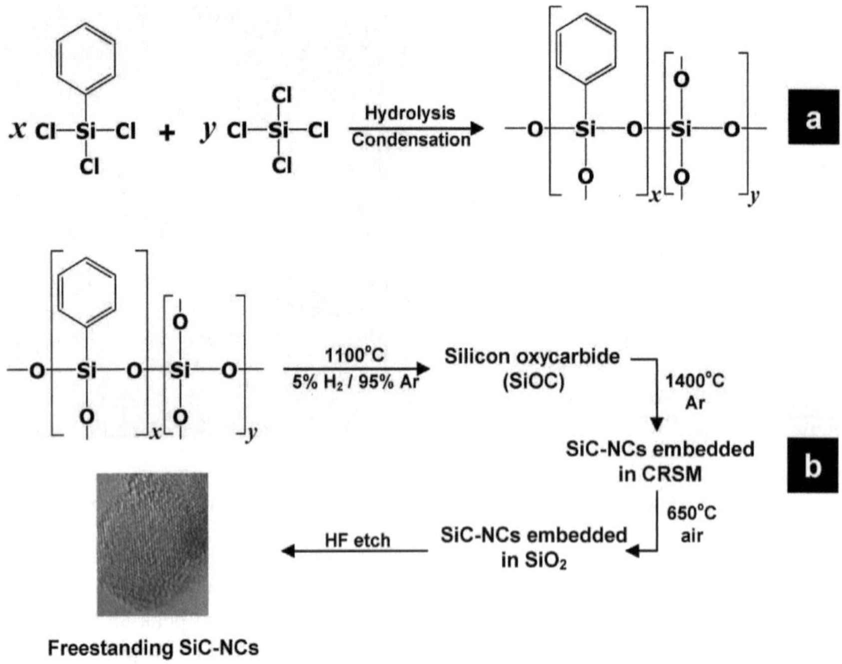

Fig. 4.56 **a** Hydrolysis and condensation of various mixtures of $C_6H_5SiCl_3$ and $SiCl_4$ yielding phenylsiloxane polymers $[(C_6H_5SiO_{1.5})_x(SiO_2)_y]_n$. **b** Thermal reduction of $[(C_6H_5SiO_{1.5})_x(SiO_2)_y]_n$ polymers yielding carbon-rich silicon matrix-embedded SiC nanocrystals. Subsequent oxidation of matrix carbon and chemical etching of the SiO_2 matrix yielded free-standing size-controlled SiC nanocrystals. Reprinted with permission from [75]. Copyright 2009, American Chemical Society

beams on the sample surface produced nanoparticles on the whole ablation area, which aligned into parallel lines (Fig. 4.59). The second harmonic generation in the sample surface excited by the incident lasers was supposed to play an important role in the formation of these SiC nanostructures.

Laser ablation of silicon wafers immersed in ethanol followed by etching was used to fabricate β-SiC QDs [79]. Under laser irradiation, the active carbon atoms dissociated from ethanol reacted with active silicon atoms from the Si wafer, producing SiC and Si nanoparticles. Subsequent rinsing in an aqueous HF–H_2O_2 solution dissolved the Si particles, leaving only SiC nanoparticles. The use of a mixture of ethanol and toluene (ethanol: toluene = 7: 1 by volume) leads to production of a higher proportion of β-SiC nanoparticles. TEM shows that the sample contains β-SiC nanocrystals a few nanometers in size (Fig. 4.60) and they produced strong and stable violet emission (Fig. 4.61a), with the peak wavelength blue shifting significantly relative to the bulk 3C–SiC bandgap, suggesting quantum confinement effect. The ring-like 3C–SiC nanostructures (Fig. 4.61b, c) have also been observed from the samples prepared using a similar method [80, 81]. About fifteen percent of the nanoparticles in the pure SiC sample have ring structures.

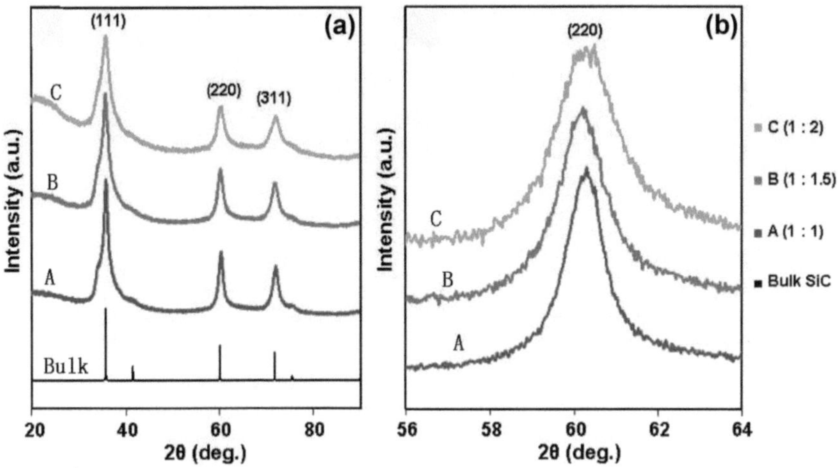

Fig. 4.57 XRD spectra of oxide embedded SiC nanocrystals indicating size tunability related to phenylsiloxane composition (the $C_6H_5SiO_{1.5}/SiO_2$ ratio). **a** XRD patterns of carbon-rich silica matrix-embedded SiC nanocrystals A3, B3, and C3. They are characteristic of 3C–SiC. **b** Enlarged XRD patterns of the (220) faces of the embedded SiC nanocrystals showing clear crystal size control. Reprinted with permission from [75]. Copyright 2009, American Chemical Society

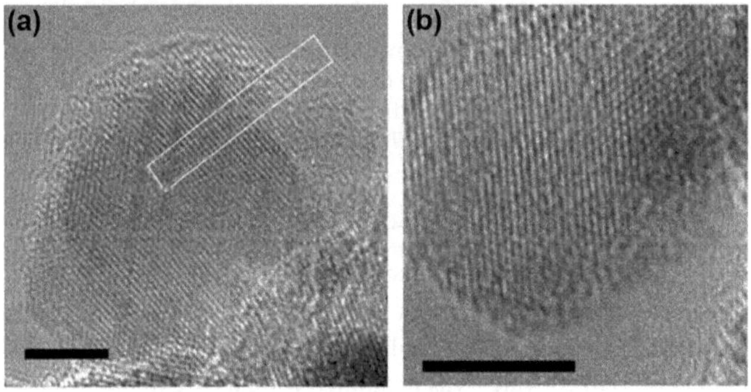

Fig. 4.58 High-resolution TEM images of free-standing SiC nanocrystals from samples (**a**) A5 and (**b**) C5, respectively. The scale bars represent 5 nm. Reprinted with permission from [75]. Copyright 2009, American Chemical Society

Direct laser ablation of polycrystalline 3C–SiC in water can produce luminescent SiC nanocrystals [82]. Figure 4.62 shows the TEM image of the synthesized SiC nanocrystallites less than 10 nm in size. Most of the nanocrystals are smaller than 5 nm and they have a zeta potential of – 49 mV. Based on the infrared spectroscopy, it was proposed that surface charges related to carboxylate anions account for the long-term stability of the colloidal solutions. These SiC particles showed photoluminescence at about 2.75 eV, which is independent of the laser power. Another study shows that ethanol rather than water as the collecting medium is also favorable for production

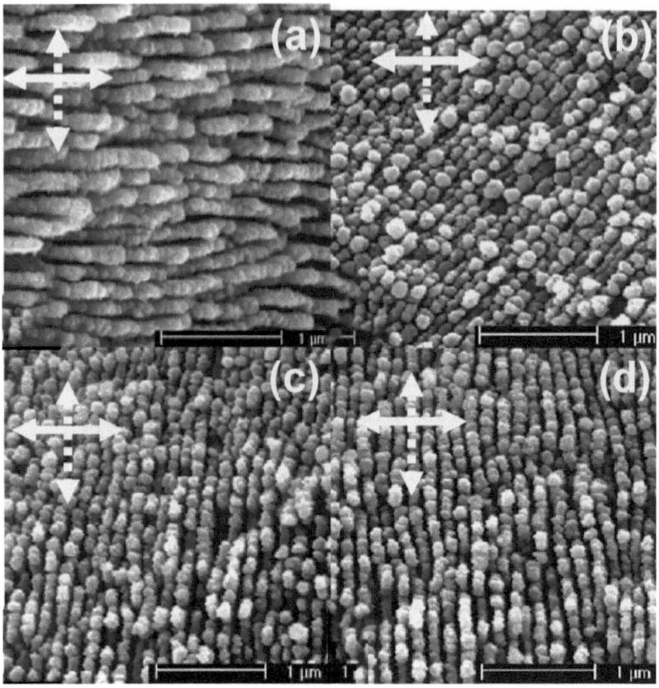

Fig. 4.59 The rotation of the nanostructures with the pulse energies of 400 and 800 nm lasers. **a** 800 nm (18 μJ), **b** 400 nm (0.1 μJ) + 800 nm (1.9 μJ), **c** 400 nm (1.1 μJ) + 800 nm (2.1 μJ), and **d** 400 nm (1.8 μJ) + 800 nm (2.2 μJ). Solid and dashed arrows show 800 and 400 nm polarization, respectively. Reprinted with permission from [78]. Copyright 2006, American Institute of Physics

of SiC NCs by ablation of pelletized SiC microparticles with a continuous scanning mode laser [83]. The SiC NPs exhibited excellent colloidal stability for extended periods of time. 3C–SiC QDs with near uniform sizes of around 2 nm can also be produced by laser ablation of 6H–SiC polycrystalline target [84]. The polycrystalline 6H–SiC target in deionized water was irradiated for 2 h by 248 nm laser with a 10 ns pulse duration (320 mJ per pulse) and a 10 Hz repetition rate. The high temperature and pressure during the growth of QDs caused phase transformation from 6H–SiC to 3C–SiC.

4.1.5 Other Methods

The particle-size dependence of optical phonon frequencies and damping of two types of SiC small particles were investigated by using Raman scattering and infrared absorption [85]. The particles were fabricated by pyrolysis of polycarbosilane or a reaction between silica and carbon. The spectra can be explained qualitatively by

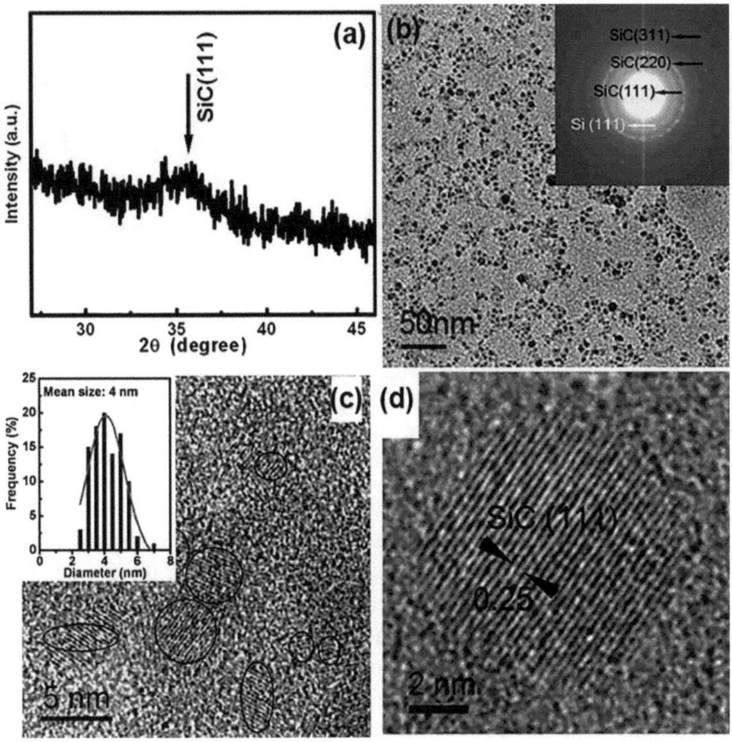

Fig. 4.60 **a** XRD spectrum, **b** TEM image, and **c**, **d** high-resolution TEM images of as-prepared sample of laser-ablated Si wafer in a mixed solution (ethanol: toluene = 7: 1 by volume). The inset in (**b**) shows the selected-area electron diffraction pattern. The inset in (**c**) shows the size distribution of the nanoparticles. Reprinted with permission from [79]. Copyright from 2009, Royal Society of Chemistry

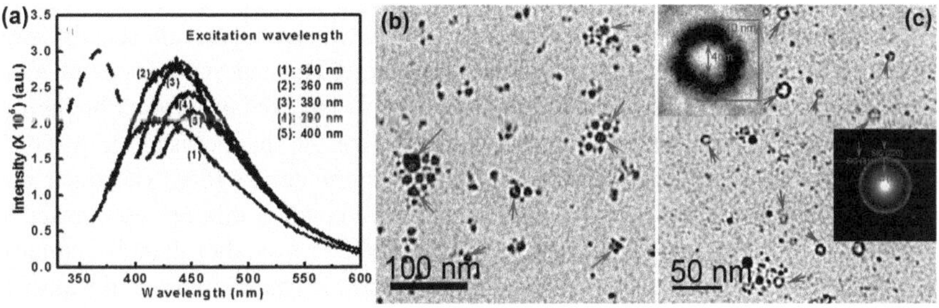

Fig. 4.61 **a** PL spectra of the synthesized β-SiC nanoparticles excited by 340–400 nm light. The dashed line is the PL excitation spectrum at 430 nm [79]. **b** and **c** TEM images of SiC nanorings and smaller Si particles. The left inset shows the high-resolution image of a nanoring and right inset shows the selected-area electron diffraction pattern of the nanorings corresponding to 3C–SiC. Reprinted with permission from [80]. Copyright 2011, American Chemical Society

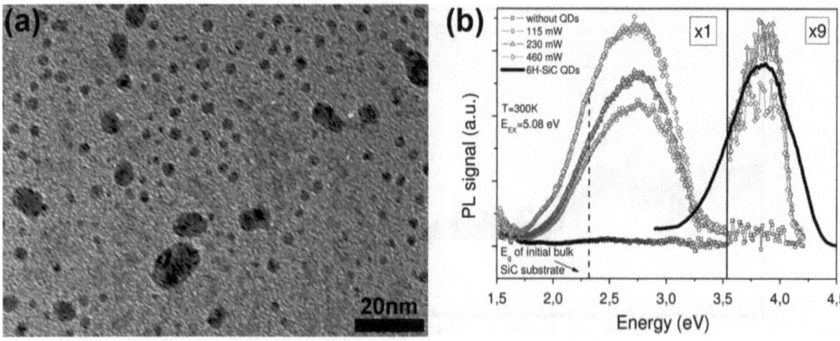

Fig. 4.62 **a** TEM image of the SiC nanocrystals obtained by laser ablation. **b** PL spectra of the SiC nanocrystals in water prepared by using different powers of laser. PL spectra of pure water and 6H–SiC nanocrystals reported in Ref. 62 are shown for comparison. The excitation photon energy is 5.08 eV. Reprinted with permission from [82]. Copyright 2011, John Wiley & Sons, Inc

a model of core plasmon and carrier-free layer on the particle surface for particles larger than 30 nm, on the basis of Mie's scattering equation or the average dielectric function of Maxwell–Garnet. The particle surface layers have thicknesses between 5 and 40 nm. For particles smaller than 200 nm, the TO-phonon frequency (ω_T), LO-phonon frequency (ω_L), and corresponding damping depend on the particle size instead of crystallite size. For smaller particles, ω_T and ω_L-ω_T become smaller and phonon damping becomes larger. The absorption spectrum of the particles a few nanometers in size exhibited both crystal-like component and broad symmetric line attributed to the 1 nm thick amorphous surface layer.

High-energy ball milling of mixed silicon and graphite powders at room temperature produced nanoscale crystalline SiC powders [86]. The SiC structure depends on the milling speed, charge ratio, and milling time. A similar method was used to synthesize β-SiC nanoparticles [87]. The starting graphite and silicon mixture reacted completely to form β-SiC nanoparticles with an average grain size of 8 nm. The grain size decreased gradually with milling time. The nanoparticles exhibited a broad photoluminescence band at 387 nm under 325 nm excitation. A carbothermal reduction method has been developed to synthesize SiC nanocrystalline powders using a solution of sugar in silica sol as the starting materials [88]. The sugar was first converted to carbon particles, and the silica and carbon mixture was heated to 1550–1800 °C in an argon atmosphere. The particles are smaller than 500 nm and the crystallites are smaller than 100 nm. Silicon carbide nanoparticles and nanorods were prepared by the solid-state metathesis reaction of silica, carbon, and magnesium at 600 °C [89]. Electron microscopy indicates that the SiO_2 particles with diameters of 30.6, 66.8, and 213.8 nm yielded SiC particles with diameters of 51.3, 92.8, and 278.3 nm, respectively (Fig. 4.63). SiC rods and fibers were fabricated by similar reactions between SiO_2 rods/fibers and C and Mg powders (Fig. 4.64). Commercial glass wool was also used to synthesize long SiC fibers by using this method, and X-ray diffraction confirms that all the products are crystalline β-SiC. Rice husks can also yield 3C–SiC nanoparticles after magnesiothermic reduction at a relative low temperature of 600 °C [90]. The SiC nanoparticles inherited the morphology of biogenetic nano-SiO_2 in rice husks with sizes of 20–30 nm. SiO_2 was first reduced to Mg_2Si in the rapid exothermic process, the intermediate Mg_2Si further reacted with residual SiO_2 and C to produce SiC. These SiC nanoparticles showed electromagnetic wave absorption, with a minimal reflection loss of – 5.88 dB and a reflection loss bandwidth (< – 5 dB) of 1.78 GHz. Sintering nanoparticles under high pressure and high temperature can yield nanocrystalline 3C–SiC bulks [91]. These 3C–SiC bulks were densely sintered, with average grain sizes down to 10 nm, depending on sintering conditions. The Vickers hardness increases with decreasing grain size, following the Hall–Petch relation. The bulk sintered under 25 GPa and at 1400 °C has an average grain size of 10 nm and an asymptotic Vickers hardness of 41.5 GPa.

3C–SiC nanoparticles can be synthesized by reaction of 1,2-dimenthoxyethane $CH_3OCH_2CH_2OCH_3$, $SiCl_4$, and Mg in an autoclave at 600 °C [92]. The as-synthesized products are composed of 3C–SiC nanoparticles encapsulated in branched carbon nanotubes with wavelike walls. These nanoparticles possess 3C–SiC cores with diameters of ~ 20–40 nm and carbon shells with thicknesses of ~ 3–5 nm. Both TO (Γ) phonons and LO (Γ) phonons show red shift of Raman lines by 6 cm^{-1} with respect to those of bulk 3C–SiC.

Fig. 4.63 SEM images of SiC nanocrystals with diameters of about, **a** 51.3 nm, **b** 92.8 nm, **c** 278.3 nm; **d** HRTEM image of a SiC nanocrystal; **e** Fourier-transform image of (**d**) indicating the lattice spacing; and **f** selected-area electron diffraction pattern of the SiC nanocrystals. Reprinted with permission from [89]. Copyright 2013, Royal Society of Chemistry

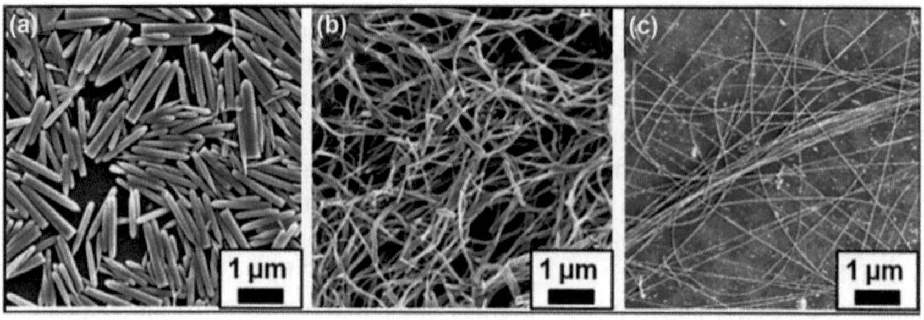

Fig. 4.64 SEM images of SiC **a** rods (length: 1.5 μm, diameter: 0.20 μm), **b** fibers (length > 10 μm, diameter: 0.18 μm), and **c** fibers (length > 25 μm, diameter: 0.050 μm). Reprinted with permission from [89]. Copyright 2013, Royal Society of Chemistry

4.1.6 Silicon Carbide Nanospheres

3C–SiC cages, cubes, and nanoparticles were synthesized by vapor–liquid reactions employing methylchlorosilanes and Na as the starting materials [93]. This route is a simplified solvent-free Yajima process, and the precursors are polycarbosilane and NaCl. 3C–SiC with different morphologies were produced with different precursors. Figure 4.65 shows the electron microscopy images of two types of SiC structures: Ia and Ib. Ia contains hollow cubic cages with an edge length of 60–400 nm and edge thickness of 40 nm, whereas Ib contains cubes with an edge length of 1–2 μm. Energy-dispersive X-ray spectroscopy reveals the presence of Si, C, and some O in both structures, and TEM indicates that Ia has empty inner space. These hollow cages show a diffused ring electron diffraction pattern indicative of a polycrystalline structure. Ib is composed of hollow polycrystalline 3C–SiC cubes with a wall thickness of 20 nm.

The reaction between $SiCl_4$ (or Si powders), C_6Cl_6, and sodium at 600 °C produced SiC nanospheres [94]. The TEM image and electron diffraction pattern of the product are shown in Fig. 4.66. The SiC hollow nanospheres are 50–100 nm in diameter and 10 nm in shell thickness. The three polycrystalline rings in the electron diffraction pattern correspond to the (111), (220), and (311) planes of 3C–SiC. There are some nanowires and nanoparticles in the product as well, and the SiC hollow nanospheres account for three quarters of the materials. The reaction processes leading to the SiC hollow nanospheres are as follows:

$$\begin{aligned} nC_6Cl_6 + 6nNa &\rightarrow 6nC \quad (\text{cluster}) + 6nNaCl, \\ &\rightarrow C \quad (\text{sphere aggregate}) \end{aligned} \tag{4.17}$$

$$SiCl_4 + 4Na \rightarrow Si + 4NaCl, \tag{4.18}$$

$$Si + C \rightarrow SiC. \tag{4.19}$$

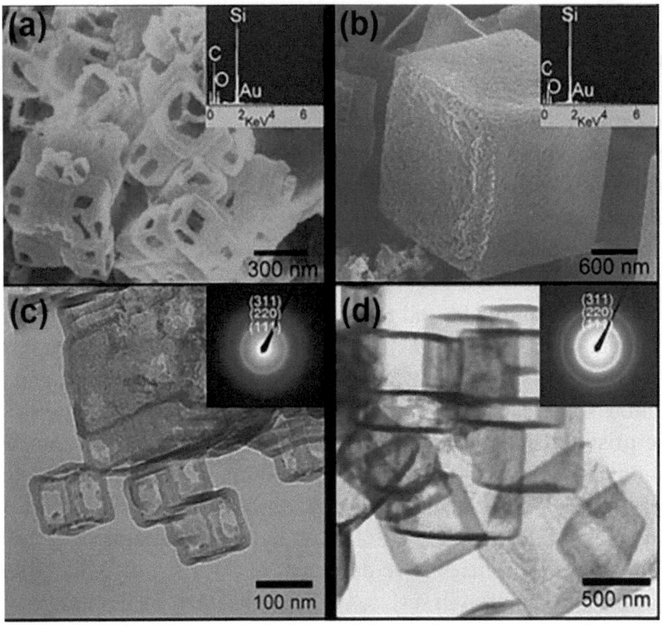

Fig. 4.65 Images of the SiC cubes prepared from $SiMe_2Cl_2$ and Na. SEM images and energy-dispersive X-ray spectroscopy spectra (insets) of two types of structures, **a** Ia and **b** Ib. Transmission electron microscopy images and electron diffraction patterns (insets) of **c** Ia and **d** Ib. Reprinted with permission from [93]. Copyright 2005, John Wiley & Sons, Inc

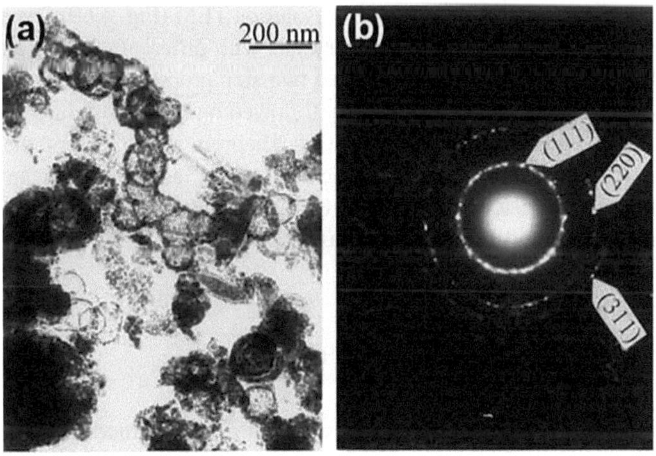

Fig. 4.66 **a** TEM image and **b** electron diffraction pattern of hollow SiC nanospheres. Reprinted with permission from [94]. Copyright 2003, Elsevier B.V

The three-dimensional hollow SiC sphere assembly was fabricated by embedding polymethylsilane and polysilazane into sacrificial three-dimensional ordered macroporous carbon templates followed by ashing in air after pyrolysis in nitrogen [95]. Figure 4.67 shows the electron microscopy images of the product. The SiC spheres expose partly the empty core through small holes in the shells. The morphology of SiC spheres depends on the sacrificial carbon templates and polymer concentration. The hollow SiC spheres have an outer diameter of 135–896 nm and a shell thickness of 14–79 nm. A vapor–solid reaction between solid-carbon spheres and Si vapor was employed to synthesize SiC hollow spheres [96]. The sphere size is tunable from microscale to nanoscale, and the shell thickness is also tunable depending on the experimental parameters. The vapor–solid reaction between carbon nanoparticles and silicon monoxide vapor produced from a mixture of silicon and silica also yielded hollow spherical SiC NCs [97]. The size of the hollow spheres depends on that of the pristine carbon nanoparticles. The thickness can be controlled by the siliconization degree and carbon nanoparticle multilayer structures. 3C–SiC hollow nanospheres were prepared with 80% yield through the reaction between $SiCl_4$, CBr_3H, and Na–K alloy at 130 °C followed by $HClO_4$ treatment at 180 °C [98]. These hollow SiC nanospheres have diameters of 80–120 nm and an average shell thickness of 15 nm. Their rough surfaces indicate that they are composed of nanoparticles. The reaction between waste plastics and Si powders at 350–500 °C produced 3C–SiC nanomaterials including nanospheres and coexisting amorphous graphite particles [99]. The molten salt synthesis technique was used to produce SiC shells on carbon black particle templates [100]. Oxidation in air at 600 °C removed the residual carbon cores to obtain the hollow SiC spheres (Fig. 4.68). The SiC shell has an average thickness of 37 nm, and high-resolution TEM (Fig. 4.69) reveals that the SiC shell is composed of 3C–SiC nanocrystals with different orientations. Hollow SiC nanoparticles can be synthesized via a two-step process involving non-thermal plasma synthesis of silicon nanoparticles, followed by their in-flight carbonization, also initiated by non-thermal plasma [101]. Theoretical analysis of the diffusion kinetics indicates that interaction with the ionized gas induces significant nanoparticle heating, allowing for fast transport of carbon into the silicon particle and for subsequent nucleation of the 3C–SiC phase.

4.2 Surface Chemical Functionalization

There have been relatively few reports concerning surface modification of silicon carbide nanoparticles. Preparation of aqueous α-SiC suspensions with the acidic form of polyethylene imine has been reported [102]. The SiC powder has an average size of 150 nm. Adsorption of the polyelectrolyte depends on the polyelectrolyte type and charge density on the SiC powder surface in water. The efficiency of the dispersant was determined by the electrostatic interaction between the positive site of the acidic form of polyethylene imine and oxidized surface of silicon carbide in the acidic media. Good dispersion and stability were achieved by electrostatic and

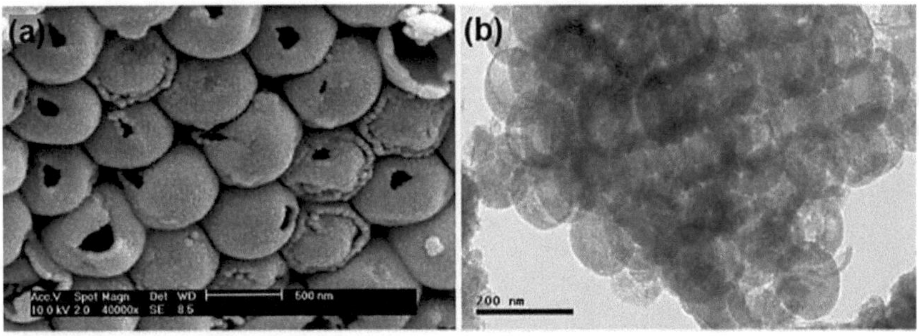

Fig. 4.67 **a** SEM and **b** TEM images of 500 nm and 135 nm hollow SiC sphere assemblies, respectively. Reprinted with permission from [95]. Copyright 2004, Royal Society of Chemistry

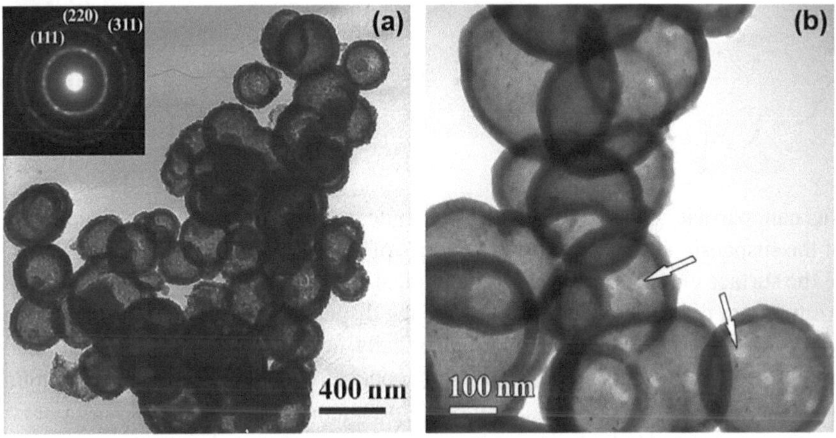

Fig. 4.68 **a** TEM image of hollow SiC spheres after removal of carbon core followed by NaOH treatment. The inset is the selected-area electron diffraction pattern of the SiC shell. **b** HRTEM image of hollow SiC spheres. Reprinted with permission from [100]. Copyright 2012, Elsevier B.V

steric stabilization. The stability of the SiC nanoparticle suspension was investigated [103]. The sample consists of 3C–SiC monocrystalline spherical particles with an average diameter of 40 nm and are covered by an amorphous carbon surface layer. Free carbon was eliminated by heating the sample at 650 °C in air, leaving an excess of amorphous SiO_2 on the particle surface, and the silica coating was stripped by dipping in HF. Different techniques have been used to characterize the hydrophilic/hydrophobic characteristics of passively oxidized SiC grains [104]. The SiC grains are less soluble in water than silica, and their surface properties do not change with the exposure time to water. The SiC surface appeared to be composed of hydrophobic and hydrophilic parts, and the hydrophilic component constituted 22–42%. The effects of ionization of polyethylenimine on dispersion of the SiC nanoparticles in water have been investigated by using sedimentation and rheological measurements [105]. Polyethylenimine with a lower degree of ionization is beneficial to dispersion of the

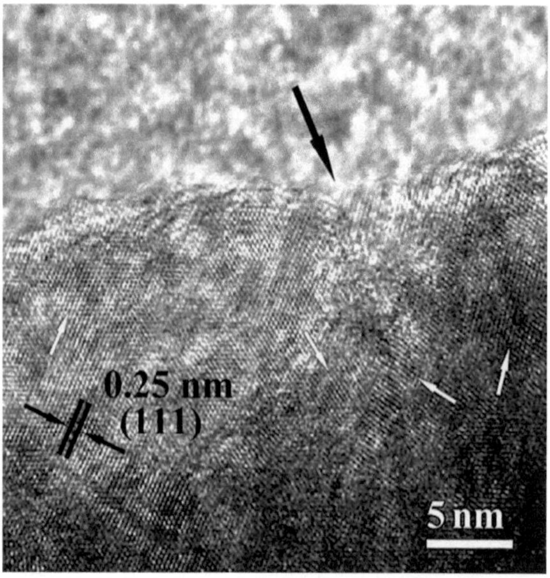

Fig. 4.69 HRTEM image of hollow SiC spheres showing lattice fringes of SiC shell. Reprinted with permission from [100]. Copyright 2012, Elsevier B.V

SiC nanoparticles, whereas that with a higher degree of ionization caused flocculation of the suspension. The dispersion properties of the suspension were studied in terms of the surface charge, particle size, rheological characteristics, and adsorption [106]. Ammonium polycarboxylate was used as the dispersant to stabilize the suspension. The powder surface charge varied dramatically in the presence of the ammonium polycarboxylate dispersant. The dispersant concentration that leads to the maximum degree of dispersion is 2.4 mg g^{-1} of SiC (corresponding to the adsorbed amount of 1.10 mg g^{-1}) at a pH of 7.5. This is lower than complete monolayer coverage (corresponding to adsorbed amount of 1.75 mg g^{-1}) of the particle surface by the dispersant. Surface functionalization with organic molecules can improve the dispersion of SiC nanoparticles in nonaqueous media. Grafting of polyacetal onto the SiC nanoparticle surface was investigated [107]. Coating of the SiC nanoparticle surface caused almost complete decomposition of the secondary structure of the agglomerate and formation of primary structures. The sedimentation experiment revealed that the coated SiC nanoparticles are very stable in butanone. Surface modification of the SiC nanoparticles by three types of azo radical initiators was investigated [108]. The radicals generated from the azo initiators reacted with the unsaturated hydrocarbons on the surfaces of the SiC nanoparticles. As a result, the hydrophobic SiC surface became hydrophilic, resulting in improved dispersion stability of the SiC nanoparticles in the aqueous solution. The stability of the SiC nanoparticles in aqueous solutions at various pH values can be controlled by the structure of the azo radical initiators. Owing to the surface carboxyl groups, the stability of the SiC nanoparticles grafted by 2,2'-azobisisobutyronitrile was improved at a pH over 5. The SiC nanoparticles modified with 2,2′-azobis(2-methylpropionamidine)dihydrochloride were stabilized at a pH below 3 due to the formation of amine groups, and SiC nanoparticles modified by

2,2′-azobis[N-(2-carboxyethyl)-2-methylpropionamidine)*n*-hydrate were stabilized at a pH of 3 or 11 because this azo radical initiator consists of both amine and carboxyl groups.

Surface modification of SiC nanoparticles by electroless nickel plating was investigated [109]. XRD, TEM, and XPS show that a nanoscale agglomerate nickel layer formed on the surfaces of the SiC nanoparticles (Fig. 4.70). The real part of the dielectric permittivity and dielectric loss of the nickel-coated SiC nanoparticles were improved significantly compared to the pristine material. The enhanced dielectric properties stem from the increased electrical conductivity rendered by the nickel-coated SiC nanoparticles. The nickel layer as a thin film of contaminant and corresponding skin effect also contributed to the enhancement.

Polyvinyl alcohol (PVA) grafted SiC (PVA–g-SiC)/PVA nanocomposite was synthesized by incorporating PVA grafted SiC NCs inside PVA matrix [110]. Refractive index of PVA increased from 1.5 to 2.0 for PVA-g-SiC/PVA nanocomposite film containing 0.015 wt% of PVA–g-SiC NCs at 550 nm. Microhardness measurements performed at an applied load of 9.8 mN showed an increase in the Knoop microhardness number of PVA containing 0.015 wt% PVA–g-SiC NCs. A thermaloxidation chemical-etching method was developed for deaggregating and producing dispersed SiC nanoparticles [111]. A polydopamine coating with controllable thickness was used for decorating gold nanoparticles on the surface. A polyglycerol coating gave excellent dispersity to SiC nanoparticles. A single-pot method was developed to produce mono/multifunctional polyglycerol-modified SiC nanoparticles, through which CD44 proteins on cell surfaces were selectively labeled through biotin-mediated immunostaining.

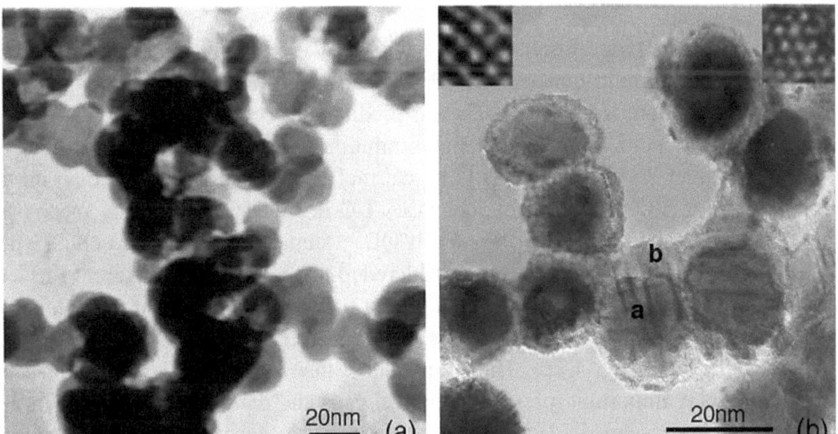

Fig. 4.70 HRTEM images of **a** original SiC nanoparticles and **b** nickel-coated SiC nanoparticles [109]. Reprinted with permission from. Copyright 2006, Elsevier B.V

4.3 Theoretical Perspectives

In comparison with experimental investigations, there are fewer theoretical studies on the structural and electronic properties of silicon carbide nanoparticles. The major reason lies in the difficulty in fulfilling the requirements for computing the properties of nanocrystals with the sizes of experimental interest of generally over 1 nm. The large polytype number, diverse surface reconstruction, and complex surface chemical functional groups of SiC nanocrystals render theoretical assessments even more challenging, because in general the surface chemistry of SiC nanocrystals cannot be fully determined experimentally. Therefore, it is hard to make a meaningful comparison between the computed and measured results. However, with continuous improvement in the computation power, advances do have been made in theoretical investigations of SiC nanocrystals, especially by using first-principles method, and the results offer insights into the structural and electronic properties of SiC nanocrystals. One key characteristic of a semiconductor quantum dot is the electronic band structure and associated energy gap. The energy gap is the energy difference between the highest occupied molecular orbital and lowest unoccupied molecular orbital of the quantum dot. It is well known that the density functional theory employing the LDA or GGA exchange–correlation functional generally underestimates the energy gaps of semiconductors by around 1 eV. However, previous studies of the size dependence of energy gaps of semiconductors such as silicon indicate that the HOMO–LUMO energy gap obtained by the DFT calculations and the gap obtained by the more accurate Green's function *GW* and quantum Monte Carlo calculations show a similar size dependence [112, 113]. Hence, the calculated energy gaps of SiC QDs are expected to impart at least qualitatively information about the size and polytype dependence of the energy gap.

Four silicon carbide clusters SiC_3 and Si_2C_2 were studied by using ab initio calculations using multi-configurational self-consistent-field wavefunctions [114]. The global minimum isomer for SiC_3 was predicted to be a C-infinity linear triplet with a terminal silicon atom. The global minimum for Si_2C_2 is a rhombic structure. The linear triplet Si–C–C–Si is just 1.0 kcal mol^{-1} bigger in energy. The ab initio computation of SiC QDs with diameters of 1–3 nm was performed by means of the DFT–GGA approach [115]. For each QD, a cubic β-SiC core structure with six different surface geometries was considered together with three different surface terminations (C, Si, or Si-rich) and two different surface structures, that is, the ideally terminated or reconstructed (100) facets. Hydrogen termination of the surface was taken into account (Fig. 4.71). The results show that the surface composition and termination play important roles in the optical gaps and thermodynamic stability of these nanoparticles (Fig. 4.72). Irrespective of size, hydrogenated QDs with the unreconstructed C-terminated surface and (2×1)-reconstructed Si-terminated surface are the most stable, and they can have a gap difference as large as 1–1.5 eV for the same diameter. In the calculated phase diagram displayed in Fig. 4.72 (right panel), the horizontal dotted line indicates the value of hydrogen chemical potential μ_H, above which bulk β-SiC spontaneously dissociates into CH_4 and SiH_4. The two thick vertical gray bars represent the size and surface dependent transition region where the value of $\delta\mu$ is so high (low) that pure Si (C) QDs are more stable than SiC QDs. The region highlighted in red corresponds to the stable region of the SiC QDs, where the (1 × 1) C-terminated and (2 × 1) Si-terminated structures are the most stable. The computation indicates that the energy gap of β-SiC QDs can be engineered as a function of size and surface composition to obtain absorption and emission from UV to green.

The dependence of the energy gap and stability on the polytype in silicon carbide nanoclusters was investigated using the DFT–GGA calculations [32]. Spherical SiC nanoclusters with diameters up to 2 nm from bulk 2H, 3C, and 4H polytype crystals were considered. Prior to the calculation, the dangling bonds of the surface atoms in the clusters were terminated by hydrogen atoms. Figure 4.73 shows the atomic arrangements in two SiC nanoclusters of different sizes obtained from different polytypes. Clusters 0.70 nm in size have similar structures independent of the parent polytypes. Polytypism of the parent crystals is completely lost in these clusters due to the limited number of bilayers. The 1.11 nm clusters exhibit dramatic structural difference among polytypes. Different from the bulk characteristics, the binding energy difference between different polytypes of clusters with diameters between 0.5 and 2 nm is negligible. This implies that the problem associated with the synthesis of bulk SiC polytypes may disappear for small clusters. For clusters smaller than 1 nm, the variation in the energy gaps with size in three polytypes shows indistinguishable trends (Fig. 4.74). In contrast, the energy gaps of clusters larger than 1 nm exhibit distinct size dependence for different polytypes.

The stability of fullerene-like SiC clusters was investigated by using the DFT calculations [116]. The carbon fullerene-based SiC clusters stemming from the C_{20}

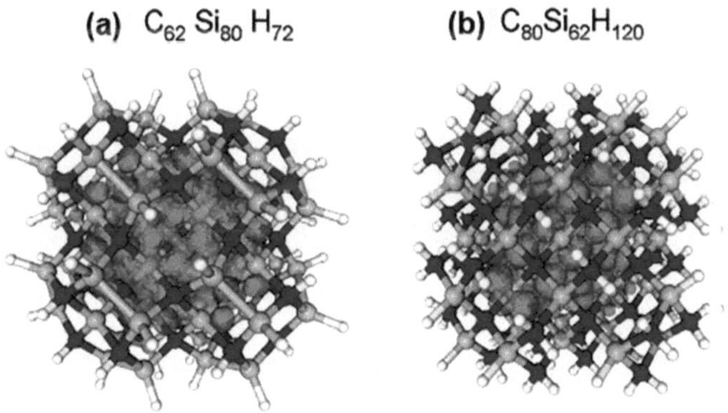

Fig. 4.71 Structure model of two β-SiC quantum dots with the same shape and radius (1.13 nm) but with different surface termination and reconstruction: **a** $C_{62}Si_{80}H_{72}$ with a (2 × 1) Si-terminated surface; **b** $C_{80}Si_{62}H_{120}$ with a (1 × 1) C-terminated surface. Black (gray) spheres represent C (Si) atoms and white spheres represent H atoms. The gray isosurface contains 95% of the spatial distribution probability of the LUMO state of each quantum dot. Reprinted with permission from [115]. Copyright 2004, American Chemical Society

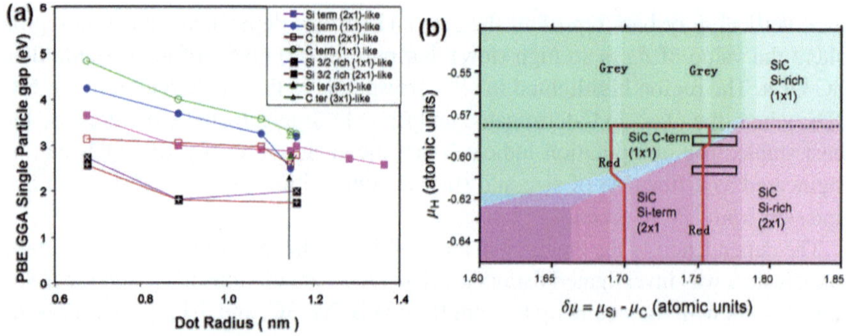

Fig. 4.72 (left panel) Computed single-particle electronic energy gap of β-SiC quantum dots as a function of size for different surface structures. The arrow indicates the particle sizes shown in Fig. 4.69. (right panel) Relative stability of SiC dots with different surface structures as a function of the difference between the Si and C chemical potentials ($\delta\mu = \mu_{Si} - \mu_C$) and the hydrogen chemical potential (μ_H). The horizontal dotted line indicates the value of μ_H, above which SiC dissociates into silane and methane. The vertical thick gray lines denote transition regions where SiC dots are not energetically favored compared to pure Si (left) or pure C dots (right), depending on size and surface structure. SiC quantum dots are stable within the area enclosed in red. The horizontal rectangles denote examples of values of $\delta\mu$ and μ_H that can be reached experimentally. Reprinted with permission from [115]. Copyright 2004, American Chemical Society

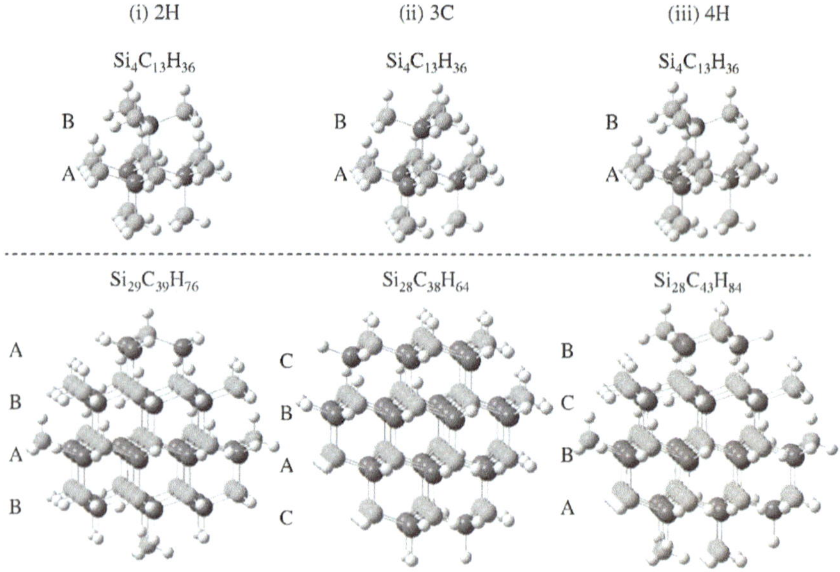

Fig. 4.73 Atomic arrangements of 2H-, 3C-, and 4H–SiC nanoclusters with two sizes. Clusters of 0.70 nm (upper panel) exhibit nearly identical structures regardless of the polytype, whereas for 1.11 nm clusters, dramatic structural difference can be observed among different polytypes. Reprinted with permission from [32]. Copyright 2007, American Institute of Physics

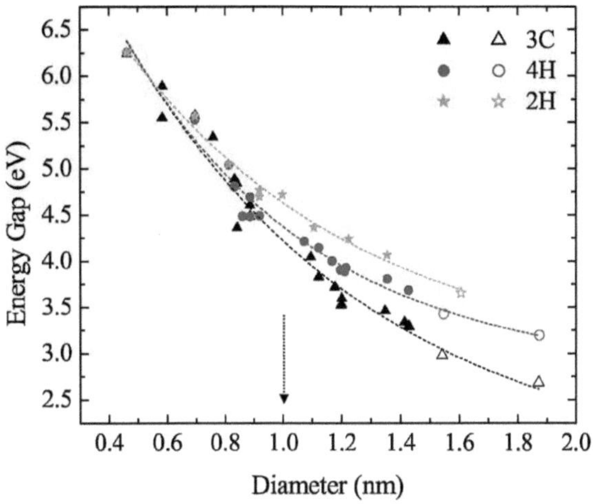

Fig. 4.74 Calculated HOMO–LUMO gap as a function of size for 3C–, 4H–, and 2H–SiC nanoclusters. The solid data points are the energy gaps of the relaxed clusters. Open data points correspond to the gaps of larger geometry-unrelaxed clusters. The energy gap difference between relaxed and unrelaxed clusters is expected to be within 0.15 eV. Dashed lines serve as visual guide. Reprinted with permission from [32]. Copyright 2007, American Institute of Physics

fullerene structure are distorted, with open structures. The Si atoms tend to be clustered even if there are only three Si substitutions. Cage derived Si–C clusters are more energetically favorable than the bowl derived ones. The HOMO–LUMO gaps of both structures decrease as the number of Si atoms in the cluster increases. The binding energy declines linearly with increasing number of Si atoms, suggesting overall reduced stability. Therefore, it was concluded that there is little chance that SiC fullerene can form from carbon-based fullerenes.

The electronic structure and absorption spectra of hydrogenated cubic silicon carbide nanocrystals were investigated by using first-principles calculations [33]. The crystal structure was constructed by cutting approximately spherical clusters out of the perfect cubic SiC crystal in such a way that each surface atom has at most two dangling bonds. The diameter of the resulting clusters excluding hydrogen atoms is between 0.9 and 3.2 nm, and the surface dangling bonds were saturated by hydrogen atoms (Fig. 4.75). The reconstructed surface can substantially change the absorption onset and spectrum shape at higher energies (Fig. 4.76a). The optical gap of small SiC nanocrystals is reduced by surface reconstruction, and different surface reconstructions give rise to different energy gaps for the same particle size (Fig. 4.76b). The calculated absorption energy gap by means of extrapolation agrees well with the measured value [117]. Absorption by 0.9–1.4 nm oxygenated silicon carbide nanoparticles was studied by using TDDFT calculations [118]. The result shows that Si–O and C–O single bonds result in relatively large energy gaps in the ultraviolet region, whereas Si=O and C=O double bonds dramatically reduce the energy gap in the blue and red regions, respectively.

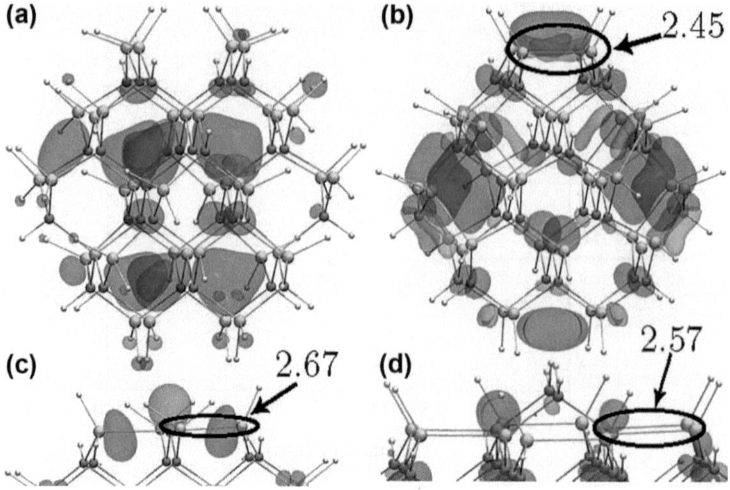

Fig. 4.75 **a** Unreconstructed $d = 1.0$ nm SiC nanocrystal (NC), **b** Si 2 × 1-type reconstruction in $d = 1.0$ nm SiC NC, **c** 3 × 1-type reconstructions in $d = 1.5$ nm SiC NC, and **d** step-like reconstruction in $d = 2.1$ nm SiC NC. The bond lengths of the long Si–Si bonds are denoted in angstrom. Small (big) gray (white) and smallest white balls represent C (Si) and H atoms, respectively. The lobes are the isosurfaces of the LUMO states. Reprinted with permission from [33]. Copyright 2010, American Institute of Physics

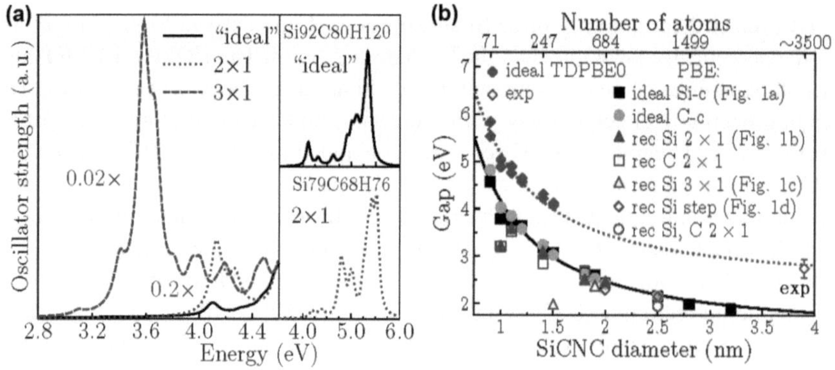

Fig. 4.76 **a** Calculated absorption spectrum of the ensemble of SiC nanocrystals with diameters between 1.0 and 1.5 nm. Solid curve: ideal nonreconstructed NCs; dotted curve with 0.2 scaling: 2 × 1-type reconstructed NCs; dashed curve with 0.02 scaling: 3 × 1-type reconstructed NCs. Right part shows absorption spectrum of individual NCs: (top) an ideal $d = 1.5$ nm and (bottom) a 2 × 1 reconstructed $d = 1.4$ nm NC. **b** Calculated energy gaps of SiC NCs with different methods. The curves are the fits. The experimental data are from Ref. 117. "Ideal" ("rec") refers to nonreconstructed (reconstructed). Si–c (C–c) indicates $Si_xC_yH_z$ ($Si_yC_xH_z$) NCs. The sign after "rec" indicates the type of reconstruction, for example, "rec Si 2 × 1" means 2 × 1 reconstructed surface with long Si–Si bonds [33]. Reprinted with permission from. Copyright 2010, American Institute of Physics

SiC QDs terminated by C–H or Si–H bonds were studied by using DFT calculations [119]. The size of the SiC QDs was chosen to be 0.94–2.49 nm. The calculation shows that the quantum confinement effect is present (absent) in the highest occupied (lowest unoccupied) molecular orbital of the SiC QDs regardless of surface termination (Fig. 4.77). The electron affinity of the SiC QD is negative for C–H termination and positive for Si–H termination. The influence of surface passivation (–H, –OH, $–NH_2$) on the electronic structures of the stoichiometric Si_nC_n (n = 10–68) QDs was investigated by using DFT calculations [120]. The HOMO–LUMO gap of SiC nanoparticles depends on the surface chemistry, and the –H passivated QDs have the largest energy, followed by the –OH and $–NH_2$ passivated QDs. The band energy diagram of SiC NCs spanning the regime from strong to weak quantum confinement was analyzed [121]. The absolute positions of the HOMO and LUMO show size dependence. The HOMO level follows the expected behavior for quantum-confined electronic states, the LUMO energy shifts below the bulk conduction band minimum as a result of the interplay between quantum confinement and formation of surface states due to partial and site-selective oxygen passivation.

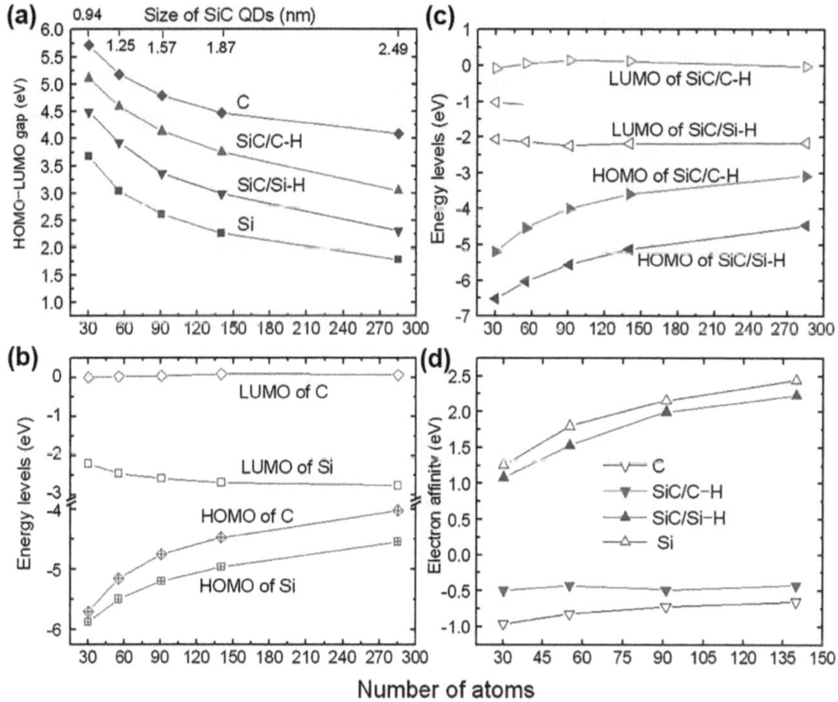

Fig. 4.77 Size dependence of **a** the HOMO–LUMO gaps of C, Si, and SiC quantum dots (QDs). **b** the HOMO and LUMO levels of C and Si QDs, **c** HOMO and LUMO energy levels of SiC QDs terminated by C–H or Si–H bonds and **d** the electron affinities of the C, Si, and SiC QDs. Reprinted with permission from [119]. Copyright 2012, Royal Society of Chemistry

Molecular dynamics simulations can give additional and distinct insights into the properties of the SiC nanoparticles. Large-scale molecular dynamics simulations were conducted to study defect generation in nanocrystalline SiC with average grain sizes of 5–21 nm [122]. Primary knock-on atom kinetic energies of 10 keV were simulated. Cascade structures in nanocrystalline SiC with a grain size below 12 nm are different from those produced in single-crystalline SiC. The local stresses near the grain boundaries highly affect the behavior of the knock-on atom and secondary recoil atoms. The grain boundaries serve as sinks for deposition of kinetic energy. The knock-on atom and secondary recoil atoms favorably deposit energy along the grain boundaries for grains smaller than 12 nm, leading to atomic displacements within the grain boundaries. For larger grain sizes, most defects arise in the grains. The defect production in gains increases with grain size. The most common defects generated in nanocrystalline SiC are antisites, there are also vacancies and interstitials. In contrast, the dominate defects in a single-crystalline SiC are Frenkel pairs. Defect production efficiency increases with grain size, with a typical value of 0.18 for small grains and 0.5 for larger grains. Cascade simulations in single crystal and nanocrystalline SiC were conducted in order to determine the role of grain boundaries and grain size on defect production during primary radiation damage [123]. Cascades were performed with 4 and 10 keV silicon as the primary knock-on atom. Total defect production increases with decreasing grain size, due to increased production in grain boundaries and changing grain boundary volume fraction. The presence of grain boundaries does not affect the total normalized in-grain defect production significantly. Defect production in the single grain containing the knock-on atom increases for smaller grain sizes. For smaller grain sizes, the defect production decreases with increasing distance from the grain boundary, whereas for larger grain sizes, the effect of the presence of the grain boundaries on defect generation is negligible. Under the C displacement condition, molecular dynamics simulation was used to examine the influence of grain boundaries on the amorphization of nanocrystalline silicon carbide by point defect accumulation [124]. The interstitials are preferentially absorbed and accumulated at grain boundaries that provide the sinks for defect annihilation at low doses and driving force to initiate amorphization in the nanocrystalline SiC at higher doses. The major surviving defects are C interstitials. The concentration of defect clusters increases with increasing dose, and they primarily distribute along the grain boundaries. These small clusters subsequently coalesce and form amorphous domains at the grain boundaries during accumulation of carbon defects. At a dose of 0.55 displacements per atom, the pair correlation function lacks long range order, indicating complete amorphization of the nanocrystalline SiC. Effects of uniaxial compressive and tensile stress loading on twin-free and nanotwinned nanocrystalline 3C–SiC ceramic were achieved by using large-scale molecular dynamics simulations [125]. The simulation indicates a strong and unique tension–compression asymmetry in strength of nanocrystalline ceramics. Primarily intergranular crack propagation and fracture are along the grain boundary transpires. A high density of nanoscale twins in 3C–SiC NCs can give rise to alternation of the crack path from intergranular to intragranular type induced by shear, which brings about the glide of Shockley partials along the coherent twin boundary and subsequent formation of coherent twin boundary steps and twin plane migration.

Facet formation in spherical and cubic SiC nanoparticles and in SiC nanowires was studied using molecular dynamics simulations reaching microseconds [126]. The equilibrium shapes are identical, irrespective of the initial shape of SiC nanoparticles or nanowires. For spherical and cubic nanoparticles, (110) facets form within 10 ns by lateral liquid-like atomic diffusion. Faceting in SiC nanowires also involves normal diffusional motion with a higher energy barrier and much longer faceting times. Molecular dynamics simulations reveal a transition from strengthening to softening in the shear deformation of nanocrystalline SiC ceramics as a function of grain size [127]. The simulation shows a well-defined maximum in the shear strength for grain sizes of 6.2–7.7 nm. Further decrease in grain size causes diminishing strength, following an inverse Hall–Petch behavior. With decreasing grain size, the increasing grain boundary regions result in homogenization of shear stresses across the microstructure, which allows for lower local shear stress levels at higher macroscopically applied stresses. This accounts for the Hall–Petch behavior. For grain sizes below 6.2 nm, the compliant nature of the disordered grain boundary regions dominates the mechanical response, reducing the shear strength and triggering conversion into the inverse Hall–Petch behavior. First-principles molecular dynamics simulation and classical force-field approach were used to investigate the mechanical properties of 1–2 nm Si and SiC nanoparticles [128]. The simulations reveal that the strength continues to increase down to such sizes and that the theoretical bulk strength can be reached or even exceeded in some cases. Most nanoparticles yield amorphization at strains greater than 20%. Homogeneous formation of a dislocation loop embryo for the <111> compression of SiC nanoparticles was also confirmed.

Large-scale molecular dynamics simulations with a reactive force-field elucidate the complex roles of atomic oxygen reservoir conditions and grain size on oxidation kinetics and the nature of oxides generated in monocrystalline and nanocrystalline 3C–SiC at 1100–2000 K [129]. The result reveals that nanocrystalline SiC exhibits two distinct oxidation kinetics with a transition point at 1500 K due to surface melting. The introduction of a grain boundary network produces twice decrease in oxidation activation energies compared to monocrystalline SiC below 1500 K. Above 1500 K, the activation energies increase dramatically due to formation of a liquid Si phase at the SiC/Si oxide interface. Incoherent grain boundaries in the crystalline SiC promote the stability of the interfacial liquid phase. Excited state molecular dynamics was used to study how oxygen impurities arise in the plasma synthesis of SiC NCs [130]. The PL from various oxygen impurities was investigated using first-principles modeling and density matrix dissipative dynamics, governed by the Redfield equation of motion. The intrinsic model shows the highest PL quantum yield of 55.37%. The surface oxygen impurities $SiOCH_3$ and SiOH show PL quantum yields of 41.49% and 40.15%.

4.4 Preliminary Applications

Silicon carbide is a unique wide bandgap semiconductor with excellent properties such as high hardness, low density, high thermal stability, good radiation resistance, and biological compatibility. Moreover, silicon carbide nanocrystals have tunable energy gaps, showing adjustable and robust luminescence spanning the visible and near UV regions. Therefore, SiC nanoparticles have immense application potential.

4.4.1 Enhancement of Mechanical Properties

SiC nanoparticles are widely used to improve the mechanical properties of metals and ceramics due to their superior mechanical characteristics. Nanocomposites consisting of a nanocrystalline Ni matrix with a grain size of 15–20 nm were reinforced with up to 3 wt% submicrometer SiC particles [131]. The microhardness values are about two times larger than those of polycrystalline Ni–SiC composites. Further study was conducted on nanocomposites containing up to 10.5 vol% SiC particulates (average size of 400 nm) [132]. These nanocomposites possess improved mechanical properties including hardness, yield, and tensile stress than conventional Ni–SiC composites with micrometer grain sizes. The materials have tensile strength up to four times that of conventional polycrystalline Ni and twice that of conventional polycrystalline Ni–SiC with a comparable SiC content. The nanocomposites with a SiC content of less than 2 vol% have higher tensile and yield strength as well as increased tensile ductility than pure nanocrystalline nickel with a comparable grain size. However, an excessively large SiC content reduces the strength and ductility due to possible particle clustering.

SiC nanoparticles affect the mechanical properties of Ni-Co/SiC nanocomposite coatings prepared by electrodeposition [133]. The SiC nanoparticles have a mean diameter of 50 nm, and the microhardness and wear and corrosion resistance of the nanocomposite coatings increase with nano-SiC concentration (Fig. 4.78). The co-deposited SiC nanoparticles are uniformly distributed in the Ni–Co matrix and contribute to the enhanced microhardness and wear resistance. For the Ni/SiC nanocomposite coatings containing SiC nano-particulates with a mean size of 50 nm, the microhardness and wear and corrosion resistance also increase with SiC nanoparticle concentration [134].

The fracture behavior of Al_2O_3 containing 5 vol% 15 nm SiC particles was investigated [135]. The strength increased from 560 MPa for Al_2O_3 to 760 MPa for the composite materials. The average strength of the composites increased to about 1000 MPa after annealing at 1300 °C for 2 h. The observed strengthening and toughening were believed to be due to machining induced compressive surface stress. In another study involving alumina containing 17 vol% silicon carbide, the minimum creep rate of the nanocomposite is three orders of magnitude smaller and the creep life is 10 times longer than those of pure alumina [136]. The bending

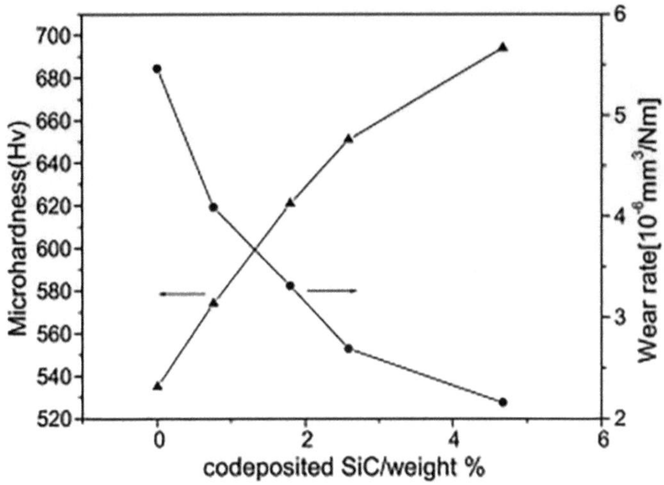

Fig. 4.78 Microhardness and wear rate of the Ni-Co/SiC nanocomposite coatings versus weight percentage of SiC particulates in the nanocomposite coating. Reprinted with permission from [133]. Copyright 2006, Elsevier B.V

strength of the Al_2O_3/5 vol% SiC densified by spark plasma sintering at 1450 °C is as high as 1000 MPa [137]. The SiC particulates are 70 nm in size and the electron microscopy indicates that the SiC particles are mainly concentrated in the Al_2O_3 grains. SiC nanoparticles can also enhance the mechanical properties of aluminum. The microstructures and mechanical properties of aluminum-based nanocomposites containing SiC nanoparticles smaller than 30 nm were determined [138]. The yield strength of the A356 alloy was improved by more than 50% by addition of 2.0 wt% of SC nanoparticles.

The Si_3N_4 matrix dispersed with less than 10 vol% SiC nanoparticles showed the same order of resistivity and dielectric properties as the pure materials, but improved mechanical properties [139]. The strength of the Si_3N_4/SiC nanocomposites at 1400 °C is 70–100% of its room-temperature strength [140]. The creep and oxidation resistance are very high and comparable to or better than those of the best Si_3N_4-based materials previously reported.

Mg was reinforced with SiC nanoparticles with an average diameter of 30 nm [141]. In the case of mixed powders (Fig. 4.79), the nanoparticles decorated the elongated coarsen grained Mg grain boundaries after extrusion; for the milled powders, a submicrometer grained Mg structure in which the SiC nanoparticles decorated the shear band in the heavily deformed Mg formed and at least partly dispersed in the submicrometer matrix grains after extrusion. The milled composite has the largest flow stress and smallest creep rates in comparison with other composites and pure magnesium. Another study shows that the microhardness of SiC nanoparticle-reinforced magnesium composites was improved with increasing fraction of SiC nanoparticles, and the microhardness of AZ91D/5 wt% SiC increased by 75% compared to AZ91D [142].

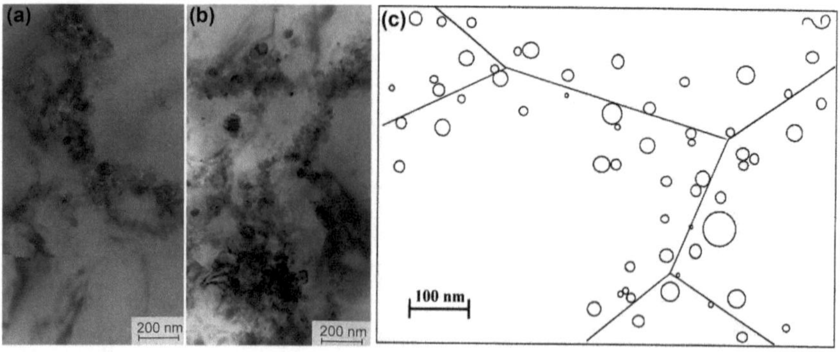

Fig. 4.79 Transmission electron micrographs of Mg/3 vol% SiC nanoparticles: powder only mixed, **a** and also milled for 8 h before extrusion (**b**). **c** Schematic of the milled composite after extrusion. The SiC nanoparticles decorated the grain boundaries of the metal matrix and formed together with the matrix along the grain boundaries a nanostructured linked network. Reprinted with permission from [141]. Copyright 2001, Elsevier B.V

SiC nanoparticles can improve the properties of epoxies [143]. The nanoparticles are superior to micrometer size particles in enhancing the thermal and mechanical properties of the SiC/epoxy composite. Nanoparticle infusion improved the thermal stability of the composite by enhancing cross-linking in the polymer. The SiC nanoparticles acted as fillers in the voids of the composites to boost the mechanical properties. Carbon nanotube (80 nm) and SiC nanoparticle composites have superior mechanical properties [144]. The three-point bending strength and fracture toughness were improved by about 10% relative to monolithic SiC ceramics fabricated by a similar process due to strengthening and toughening caused by carbon nanotubes in the matrix.

4.4.2 Superconductivity Enhancement

SiC nanoparticles have been demonstrated to be able to significantly enhance the properties of MgB_2 superconductors. Doping of MgB_2 with (10–100 nm) SiC nanoparticles and the improved flux pinning in $MgB_{2-x}(SiC)_{x/2}$ with $x = 0, 0.2$, and 0.3 and 10 wt% nano-SiC-doped MgB_2 samples were studied [145]. Co-substitution of B by Si and C counterbalanced the single element doping effect, reducing T_c (critical temperature) by only 1.5 K. Moreover, intra-grain pinning centers effective at high fields and high temperature were generated, and J_c (critical current density) (Fig. 4.80) and H_{irr} (irreversibility field) were significantly enhanced. The critical current density of the 10 wt% doped sample increased by a factor of 42 at 20 K and 5 T compared to the undoped sample and it reached 2.4×10^5 A cm^{-2} at 20 K and 2 T, which is comparable to the J_c values of the best Ag/Bi-2223 tapes. The critical current density at 20 K and 4 T is twice that of the best MgB_2 thin films and an order of magnitude larger than that of the best Fe/MgB_2 tapes.

The effects of the incorporation of 30 nm SiO_2 and SiC on the microstructures and superconducting properties of in situ processed MgB_2/Fe tapes were studied [146]. Mg or MgH_2 powders are the Mg source, and the use of MgH_2 instead of Mg powders improved the correlation between the MgB_2 grains and improved J_C. SiC doping effectively enhanced J_C with regard to the starting powder mixtures of MgH_2 + B and Mg + B. In contrast, SiO_2 is only effective for the Mg + B powder mixture. The highest J_C value of 6500 A cm^{-2} at 4.2 K and 12 T was observed from the 10 at% SiC-doped tape. The SiC doping improved the irreversibility field from 17 to 23 T and pinning characteristics. However, excessive SiC doping degraded J_C in the low field region. Further measurements on the 10 at% SiC-doped MgB_2 indicate that the upper critical field H_{c2} was systematically enhanced by SiC doping and by lowering reaction temperature [147]. H_{c2} (10 K) exceeded 33 T and the extrapolated zero temperature value exceeded 40 T. Only 8–17% of the MgB_2 cross section carried current, as revealed by the Rowell analysis. A higher reaction temperature improved the connectivity but degraded H_{c2} and flux pinning.

The mechanism pertaining to the enhancement of the electromagnetic properties of MgB_2 by doping with SiC nanoparticles was studied [148]. The mechanism is based on the sintering temperature effect on the lattice parameters, carbon content, and electromagnetic properties. The optimal doping effect was attained when carbon substitution and MgB_2 formation took place at the same time at a low temperature. Carbon substitution led to the enhancement in both H_{c2} and flux pinning. Carbon substitution for boron caused disorder in lattice sites and increase in the resistivity and consequent enhancement in H_{c2}. Carbon substitution combined with low temperature

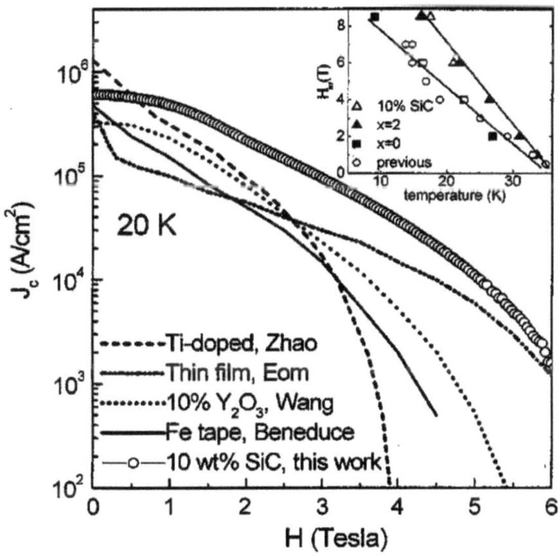

Fig. 4.80 Comparison of magnetic $J_c(H)$ at 20 K for the 10 wt% SiC-doped MgB_2 sample and samples from other works: Ti doped, Y_2O_3 doped, thin film with strong pinning, and Fe/MgB_2 tape. The inset shows the temperature dependence of the irreversibility field for SiC-doped MgB_2 with different SiC content (triangles and squares) and for previously prepared undoped MgB_2 (round symbols). Reprinted with permission from [145]. Copyright 2002, American Institute of Physics

processing led to reduced grain size, fluctuation in T_c, extra defects, and embedded inclusions which enhanced flux pinning.

4.4.3 Surface Plasmon Polariton and Luminescence Enhancement

Luminescence from silicon carbide nanoparticles can be enhanced by surface plasmon polaritons of nearby metals. Photoluminescence enhancement was observed from the SiC nanocrystals/PSS-PAH bilayers/Ag nanoparticles sandwiched structures [149], where PSS refers to poly(styrene sulfonic acid) sodium salt and PAH refers to poly(allylamine hydrochloride). The PSS–PAH polyelectrolyte bilayers served as the spacing dielectric. The SiC nanocrystals are about 3.8 nm in size and the silver crystallites have an irregular shape and an average diameter of 47 nm. The enhancement depends on the bilayer number. The integrated PL intensity was improved by up to 17 times (Fig. 4.81), and the luminescence from the sandwiched structures is bright and visible to the naked eyes. Large fluorescence enhancement was observed from 1 to 6 nm diameter SiC nanocrystals dispersed in a sodium dodecyl sulfonate dielectric film coated with Ag nanoparticles [150]. The enhancement in the integrated fluorescence intensity is 176 folds under 360 nm excitation (Fig. 4.82) and the PL peak intensity showed a 53.3-fold enhancement. Finite element simulation reveals that the absorption enhancement caused by light concentration at near field of the Ag nanoparticles contributes only slightly to the overall PL enhancement. The Ag nanoparticles exhibited a resonance extinction peak at about 420 nm (Fig. 4.82d), which is very close to the PL peak (432 nm) of the SiC nanoparticles. It was concluded that the strong resonant coupling between the excited SiC nanocrystals and the localized surface plasmons of the Ag nanoparticles plays a dominant role in the strong fluorescence enhancement.

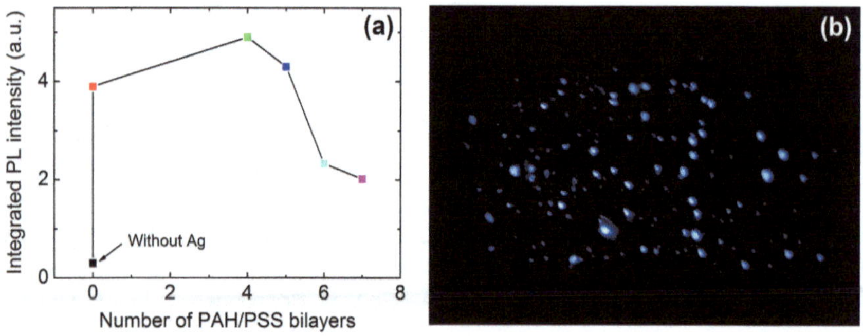

Fig. 4.81 **a** Integrated intensity of the photoluminescence spectrum versus number of bilayers for the SiC nanocrystals/PSS–PAH bilayers/Ag nanoparticles sandwiched structures. **b** Luminescence graph of the sandwiched structure with four PSS–PAH bilayers. Reprinted with permissions from [149]. Copyright 2012, Elsevier B.V

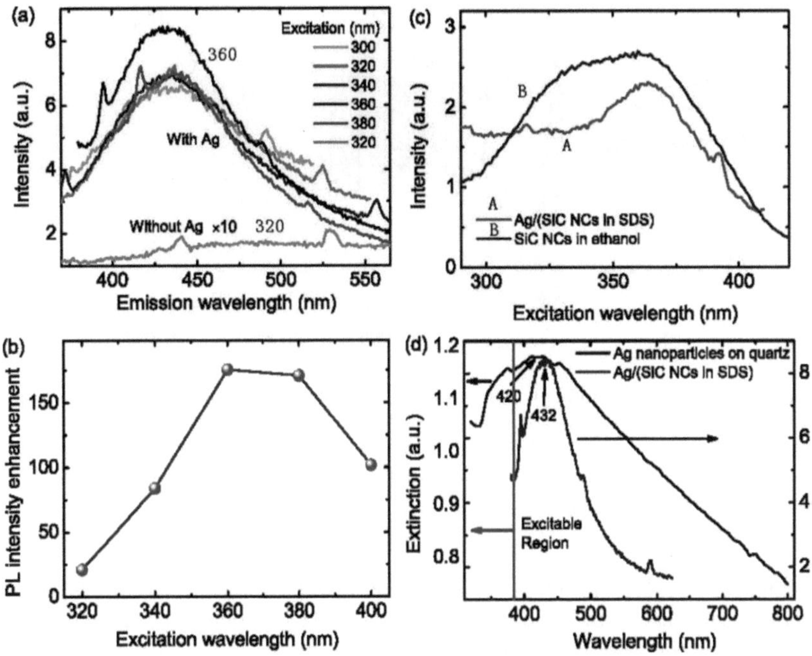

Fig. 4.82 **a** PL spectra of the Ag nanoparticles/(SiC nanocrystals (NCs) in sodium dodecyl sulfonate (SDS) dielectric)/Si wafer under different excitations. The spectrum before Ag-deposition is also shown. **b** Ratio of the integrated PL intensity between samples with and without Ag-deposition. **c** PL excitation spectra of the Ag/(SiC NCs in SDS dielectric)/Si wafer and the SiC NCs in ethanol for emission wavelengths of 430 and 440 nm, respectively. **d** Comparison of the UV–Vis extinction spectrum (resonance peak: 420 nm) of the Ag nanoparticles on a quartz wafer and the PL spectrum of the Ag/(SiC NCs in SDS dielectric)/Si wafer (peak: 432 nm) excited at 360 nm. Reprinted with permission from [150]. Copyright 2013, IOP Publishing

Gold colloids encapsulated by silica shells were mixed with SiC nanoparticles to improve photoluminescence [151, 152]. The enhancement depends on the silica shell thickness, and an enhancement factor of 12.5 was achieved from gold colloids 20 nm in diameter with a 25 nm shell. The $Au/SiO_2/SiC$ composite exhibited blue shifted (from 582 to 523 nm) photoluminescence compared to pure SiC sample. Photoluminescence enhancement was observed from SiC QDs deposited on the nano-Ag/SiN_X substrate [153]. The maximum enhancement factor is 15. The finite-difference time domain simulation shows that the enhancement arises from the near-field coupling between the SiC nanoparticles and multipolar localized plasmons of the silver nanoparticles. The nonlinear response from the SiC nanoparticles was observed from the same nano-Ag/SiN_X substrate [154]. The 72 fold enhancement in the second harmonic generation was observed, which was caused by localization of the SiC nanoparticles in the strong field regions in the case of spectral overlap

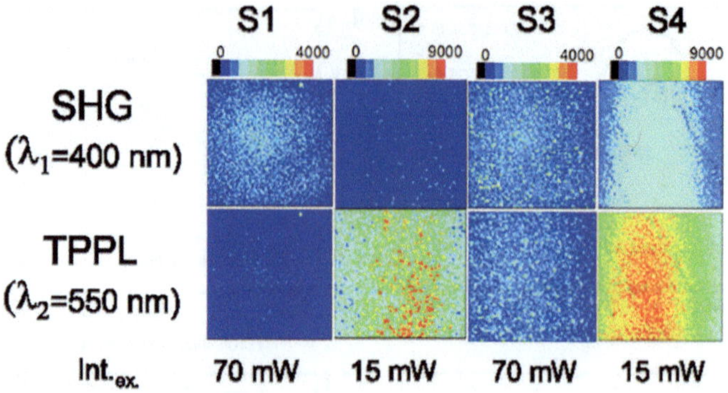

Fig. 4.83 Fluorescence microscopy images of the second-harmonic generation (400 nm, upper four) and two-photon-excited photoluminescence (550 nm, lower four) excited at 800 nm. The samples are SiN_X (S1), nano-Ag/SiN_X (S2), SiC nanoparticles/SiN_X (S3), and SiC nanoparticles/nano-Ag/SiN_X (S4). The excitation laser power is indicated. Reprinted with permission from [154]. Copyright 2013, IOP Publishing

between the multi-polar plasmon modes and excitation and second-harmonic generation lines. The maximum 287 fold enhancement in the two-photon-excited luminescence from the SiC nanoparticles was observed from this nano-Ag/SiN_X substrate and the fluorescence images are shown in Fig. 4.83.

4.4.4 Electroluminescence Toward LED

Silicon carbide is a favorable semiconductor to fabricate light-emitting diodes (LEDs) because blue to near-UV luminescence can be realized due to the wide bandgap. Some SiC-based LED products are available on the market although the low luminescence efficiency of bulk silicon carbide is the major limitation. The quantum yields of porous silicon carbide and silicon carbide nanocrystals can be several orders of magnitude higher than that of bulk SiC crystals. For example, the quantum yield of 3C–SiC QDs suspended in water can reach ~ 17% [116], which is comparable to that of some direct-bandgap semiconductors. Therefore, the SiC nanocrystals are promising materials in blue and near-UV LEDs.

A blue LED using porous SiC as a luminescent layer has been produced [155]. The ITO/porous SiC junction showed a rectification behavior and blue electroluminescence under forward biasing. The electroluminescence has a very broad spectrum from 350 to 780 nm, with a peak at 470 nm (Fig. 4.84), which is similar to the photoluminescence spectrum. The blue electroluminescence at low current level was suggested to arise from recombination of the carriers injected from both the SiC substrate and ITO electrode. At a high current level, luminescence was limited by the holes injected from the ITO electrode. However, the quantum efficiency of the porous SiC LED is rather low ($< 10^{-5}$), possibly ascribed to the limited light-emitting area of porous SiC.

Electroluminescence was observed from an Al/polymer-encapsulated 3C–SiC QD/ITO/glass structure [156]. The primary poly(acrylic acid)-encapsulated 3C–SiC QDs smaller than 7 nm in size produced photoluminescence with peak wavelengths between 400 and 540 nm. The device showed an electroluminescence spectrum (Fig. 4.85), with the peak wavelength varying between 490 and 460 nm when the applied voltage increased from 5 to 10 V. The luminescence may stem from carrier recombination in the embedded QDs. The device has a low turn-on voltage of 4–5 V. No electroluminescence was observed at voltages lower than 4 V or higher than 11 V, and no quantum efficiency value was measured.

Si QDs and SiC QDs co-doped Si-rich Si_xC_{1-x} *p–i–n* junction LEDs have been prepared [157]. When the *i*-SiC thickness varied from 25 to 100 nm, the turn-on voltage changed from 6.1 to 10.8 V. The maximum electroluminescence power of

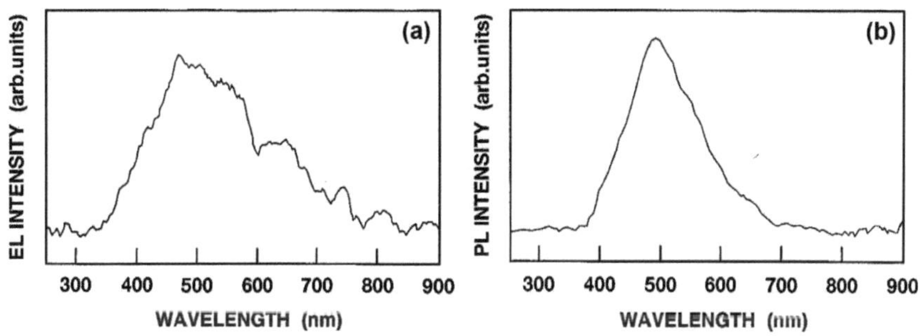

Fig. 4.84 **a** Electroluminescence spectrum from the ITO/porous SiC junction. **b** PL spectrum of a similar porous SiC excited by 325 nm [155]. Reprinted with permission from. Copyright 1994, American Institute of Physics

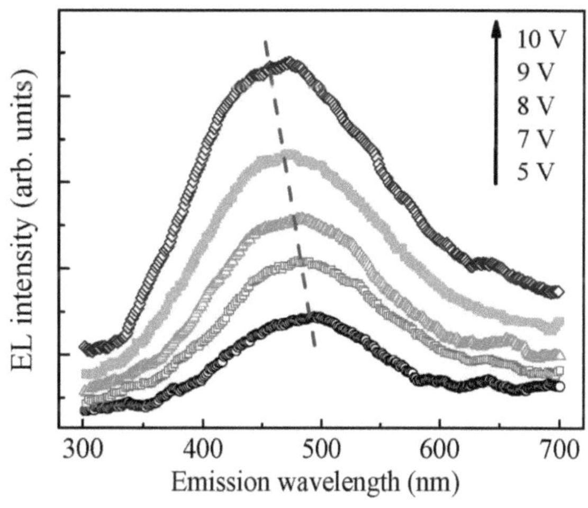

Fig. 4.85 Electroluminescence spectra of the Al/ polymer-encapsulated 3C–SiC quantum dot/ITO/ glass. The applied voltages are indicated. Reprinted with permission from [156]. Copyright 2012, American Institute of Physics

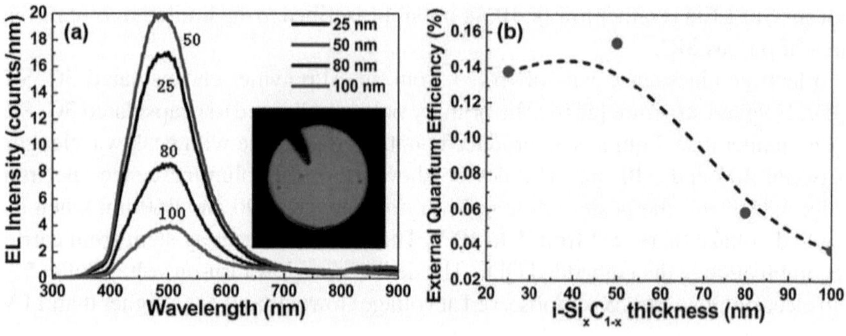

Fig. 4.86 **a** Electroluminescence spectra of Si-rich Si_xC_{1-x} *p–i–n* junction LEDs with different *i*-Si_xC_{1-x} thicknesses. The inset is the electroluminescence emission pattern. **b** External quantum efficiency as a function of *i*-Si_xC_{1-x} thickness [157]. Reprinted with permission from. Copyright 2014, IEEE

the LED with 50 nm thick *i*-SiC was observed to be 136 nW. The LED showed green electroluminescence at 500 nm (Fig. 4.86a). The external quantum efficiency of the LED varied with the *i*-SiC thickness, reaching 0.158% for a thickness of 50 nm (Fig. 4.86b). The carrier injection efficiency was enhanced by 46% via increasing the doping concentration to 10^{16} cm^{-3}, resulting in nearly one order of magnitude improvement in the external quantum efficiency.

Quasi-white electroluminescence in SiC QDs-based light-emitting devices was observed [158]. The SiC QDs were fabricated by electrochemical etching of bulk 6H–SiC single crystals. Three types (A, B, and C) of LEDs were designed. Device A has three layers: indium tin oxide (ITO), 120-nm-thick SiC QDs, and Al film (150 nm). Device B comprise more layers: ITO, 100-nm-thick SiC QDs, 1,3,5-tris(1-phenyl-1H-benzimidazol-2-yl) benzene (TPBI, 12 nm), and LiF/Al (2/150 nm). Device C has a more complex structure: ITO, SiC QDs embedded in poly allylamine hydrochloride-sodium (PAH:SiC QDs, 100 nm), TPBI (12 nm), and LiF/Al (2/150 nm). Figure 4.87 shows their electroluminescence spectra measured under different biases. The electroluminescence spectra are wide, crossing the visible region of 350–900 nm, with FWHM values of ~ 340–360 nm. The similar spectral profiles of the three devices indicate their identical origin, from recombination of the injected carriers in the SiC QD layer. The electroluminescence spectrum is much wider than the PL spectrum. For electroluminescence, the electrons and holes are injected into the QD layer from two opposite electrodes, so they can occupy much more empty energy levels in the conduction and valence bands of the QDs. This accounts for a wider electroluminescence spectrum than the PL spectrum. Light emissions from ample surface defects in the SiC QD film contribute also to the wide electroluminescence spectrum. Although device A contains only SiC QDs besides the ITO and Al electrodes, it also exhibited full-visible-region electroluminescence; this strongly supports that the SiC QDs are responsible for the quasi-white electroluminescence. Analysis of the current density vs. bias curves and spatial distribution of energy levels

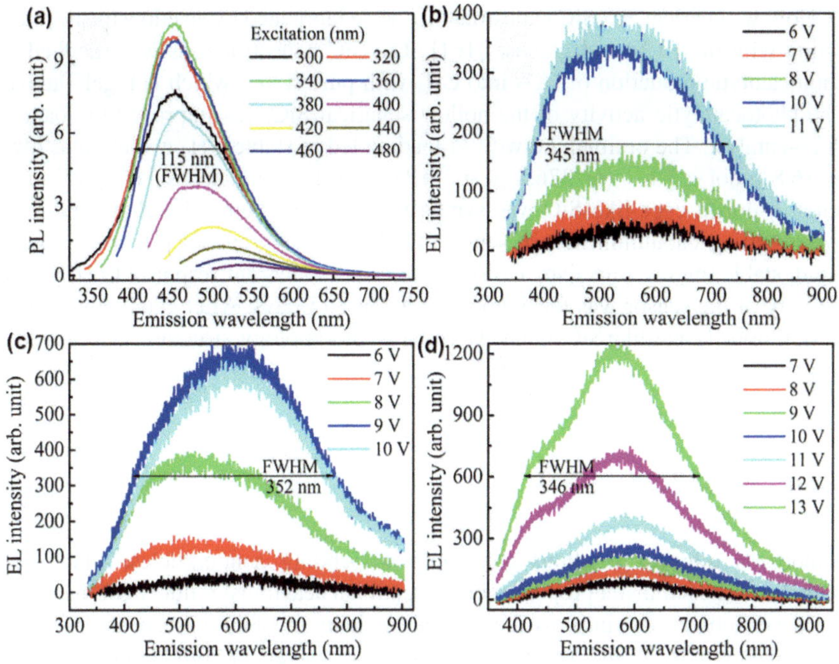

Fig. 4.87 **a** PL spectra of SiC QDs in ethanol. Electroluminescence spectra of **b** device A, **c** device B, and **d** device C. Reprinted with permission from [158]. Copyright 2018, John Wiley & Sons, Inc

shows that the carrier transport is dominated by quantum tunneling at low biases and direct injection at high biases, respectively.

4.4.5 Catalysis

Electrochemical reduction of water to hydrogen in the proximity of the 3C–SiC nanocrystals film supported by glassy carbon substrate was investigated [159]. The 3C–SiC nanocrystals showed superior electrocatalytic activity in the hydrogen evolution reaction due to the large reduction in the activation barrier on the nanocrystal surface enabling efficient dissociation of H_2O molecules. These nanocrystals with surface autocatalytic effects can be used to split water with a high efficiency. Theoretical studies indicate that the hydrogen evolution reaction on SiC electrodes happens at pre-existing hydrogenated Si–H surface sites by a Heyrovsky–Volmer mechanism [160]. The Heyrovsky reaction takes place where an electron from the substrate reacts with a hydronium adsorbed at a Si–H site, yielding a H_2 molecule and Si dangling bond. The following Volmer reaction regenerates the Si–H and the rise in the conduction band due to quantum confinement accounts for the electrochemical activity of small SiC nanocrystals.

Hollow spherical 3C–SiC nanostructures were fabricated by an ethylsilicate interaction with the P123 and glucose [161]. The SiC nanostructures were applied to photocatalytic reduction of CO_2 into CH_4 with pure water, which is highly active. The photocatalytic activity of the hollow spherical SiC was improved by loading Pt co-catalyst. The optimal 2.0 wt% Pt loading led to stable CH_4 evolution as high as 16.8 μmol g^{-1} h^{-1} (or 376.4 μl g^{-1} h^{-1}) with the simulated solar light irradiation. F-doped cage-like SiC was synthesized by wet chemical etching and used for designing cataluminescence sensors [162]. The as-prepared products showed rapid, stable, highly selective, and sensitive cataluminescent response to H_2S. The stability of the sensor was fairly good for at least 15 days. After cataluminescence tests, F-doped cage-like SiC retained its pristine morphology, structure, and chemical composition.

4.4.6 *Nanofluids*

Nanofluids refer to solutions containing suspending nanoparticles, which can improve the heat transfer properties of conventional heat transfer fluids. SiC nanoparticles were observed to improve the thermal conductivity of liquid suspensions [163]. Spherical/cylindrical SiC particles with average diameters of 26/600 nm were separately dispersed in distilled water and ethylene glycol. The thermal conductivity of the SiC suspensions was significantly enhanced relative to the base liquids. At low concentrations, the thermal conductivity enhancement ratio increased roughly linearly with the volume fraction of the particles. The thermal conductivity of the SiC suspension is proportional to that of the base liquid and the morphology of the suspended particles affected the thermal conductivity enhancement ratio.

Thermal conductivity and mechanical effects of SiC nanoparticles dispersed in water were investigated [164]. The SiC particles have an average size of 170 nm and a polydispersity of ~ 30%, as determined from small-angle XRD and dynamic light scattering techniques. Room-temperature viscosity of the nanofluids ranges from 2 to 3 cP for nominal nanoparticle loadings of 4–7 vol%. Viscosity of the nanofluids did not remarkably change with test temperature up to 85 °C. Optical microscopy showed no agglomeration of the nanoparticles in the diluted nanofluid. Enhancement in thermal conductivity is about 28% over that of water at 7 vol% particle loadings under ambient conditions (Fig. 4.88).

Heat transfer experiments were performed with a water-based nanofluid containing 170-nm SiC nanoparticles at a 3.7% volume concentration [165]. Heat transfer coefficient increased 50–60% above the base fluid water when compared on the basis of constant Reynolds number. This enhancement is 14–32% bigger than that predicted by a standard single-phase turbulent heat transfer correlation, pointing to heat transfer mechanisms that involve particle interactions. The experiment shows that the particle size is a very powerful parameter in nanofluid property management [166]. Both thermal conductivity and viscosity of nanofluids favor the use of larger

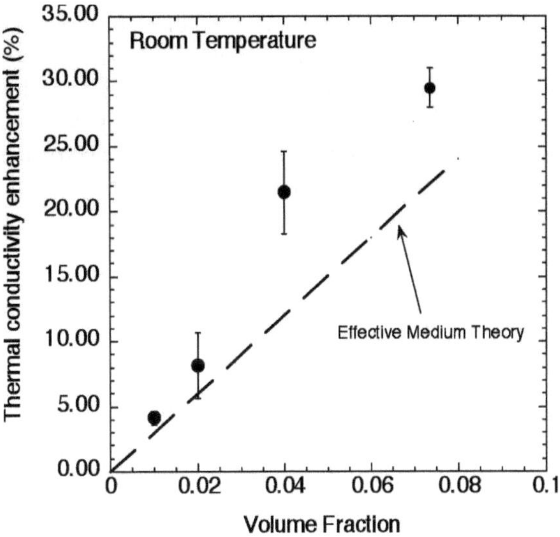

Fig. 4.88 Plot of the thermal conductivity enhancement as a function of SiC nanoparticle loading at room temperature. Reprinted with permission from [164]. Copyright 2009, American Institute of Physics

particles over small ones. But the usage of larger particles may diminish the stability of suspensions and bring erosion damage to the equipment.

The thermophysical properties of SiC/deionized water nanofluids were investigated [167]. For the SiC dispersion stability, the iso-electric point was identified as pH ~ 6 and the pH control apart from the point was successful to keep the stability. The maximum relative viscosity for the 3 vol% SiC/deionized water nanofluids compared with deionized water is 102%. The maximum relative thermal conductivity for the 3 vol% SiC/deionized water nanofluids compared with deionized water is 7.2%. The overall effectiveness ($C_\mu/4C_k$) of the current SiC nanofluids is over ~ 3.3, which is desirable below ~ 1. The rheological behavior of ethylene glycol-based nanofluids containing spherical-shaped SiC nanoparticles was studied using a HAAICE MARS III rheometer at different volume fractions and temperatures [168]. The nanofluids exhibited an interesting theological behavior of the viscosity curves. The theoretical analysis reveals that the theological behavior of the studied nanofluids strongly depends on shear rates and particle concentration. The viscosity of the nanofluids increases with volume fraction but decreases with temperature. The stability, optical properties, and thermal conductivity of SiC/saline water-based nanofluids were investigated [169]. The SiC nanofluids yielded increase in thermal conductivity with respect to base fluids by over 6% at 0.4 vol% SiC nanofluids, and showed good capacity of solar absorption (< 1% luminousness at 0.4 vol% SiC nanofluids). Increase in salt concentration showed negative influence on the stability and thermal conductivity of the nanofluids. 0.4 vol% SiC nanofluid prepared with natural seawater had a sufficient stable period. The thermal conductivity of seawater-based nanofluid increased by 5.2% with respect to that of the base fluid. The thermophysical properties of nanofluid composed of water and SiC nanoparticles for use in the cooling of a PV/T system was studied [170]. The addition of 3 wt% of the nanoparticles into water led to increase in fluid density by 0.0082% and increase in viscosity by 1.8%. Thermal conductivity increased by 8.2% for the tested temperature of 25–60 °C. After 6 months, the thermal conductivity decreased by 0.003 W m^{-1} K^{-1}. The use of 3 wt% SiC nanofluid increased the electrical efficiency by 24.1% compared to the PV system alone, and the thermal efficiency increased by 100.19% compared to the use of water for cooling.

4.4.7 Other Applications

Bi_2Te_3 thermoelectric materials incorporated with *n*-type 100 nm diameter SiC nanoparticles have been prepared [171] and showed improved Seebeck coefficient, decreased electrical conductivity, and reduced thermal conductivity compared to the Bi_2Te_3 matrix at 323–523 K. The maximum dimensionless figure of merit was improved from 0.99 for Bi_2Te_3 to 1.04 for 0.1 vol% SiC dispersed Bi_2Te_3 at 423 K. The mechanical properties including Vickers hardness, Young's modulus, and fracture toughness were improved after addition of SiC nanoparticles.

SiC nanoparticle doping can change the conductivity of polypyrrole/SiC nanocomposites [172]. The polypyrrole in the polymer nanocomposites showed improved thermal stability and tolerated a higher decomposition temperature. The electron transport in the polymer nanocomposites follows a quasi 3-d variable range hopping conduction mechanism. The polypyrrole/SiC nanocomposites have higher conductivity than the pure polypyrrole.

Nitrogen-doped SiC nanoparticles with a mean size of 5.96 nm showed favorable microwave absorption when they were mixed with paraffin [173]. The complex permittivity of the composites was changed by the SiC particles. The real and imaginary parts of the complex permittivity and dielectric dissipation factor of the composites increased with the volume filling factor of the nitrogen-doped SiC particles. The single layer composites with 7 wt% doped SiC particles and a thickness of 2.96 mm showed a reflection loss smaller than – 10 dB (90% absorption) at 9.8–15.8 GHz and a minimum value of – 63.41 dB at 12.17 GHz. A synergistic control of morphology and particle size of SiC/C composites by varying the calcination temperature and pH environment can bring efficient microwave absorption [174]. The hollow mesoporous SiC@C microsphere (HPSiC@C) calcined at 1350 °C showed the best microwave absorption performance for an amount of ammonia of 3 mL. The effective absorption bandwidth is 5.61 GHz at 2.1 mm and the maximum reflection loss is – 45.61 dB at 13.2 GHz.

Photobleaching of the luminescent surface defects of the colloidal 3C–SiC QDs obtained by chemical etching of micrometer-sized SiC powder can apply to sensing of UV irradiation [175]. The combined experimental and DFT calculation results indicate that the violet emission at ~ 392 nm of the SiC QDs arises from carboxylic acid groups-related surface defects. Upon exposure of the colloidal SiC QDs to UV irradiation, the 392 nm emission of the surface defects showed photobleaching, but it is recoverable, whereas the 438 nm emission of surface defects showed irreversible photobleaching. The UV irradiation caused ionization of the luminescent carboxy groups on the SiC QD surface, either into non-luminescent COO^- plus H^+, or into CO_2 plus H^+, and the former process is reversible, whereas the latter is irreversible. In contrast, the UV irradiation caused conversion of the 438-nm luminescent COOR groups on the SiC QD surface into CO_2 plus R–H, which is irreversible. The ~ 392-nm emission became stronger with increase in H^+ concentration. The photobleaching property of the SiC QDs was utilized to detect the solar UV irradiation with high accuracy.

References

1. Hollabaugh CM, Hull DE, Newkirk LR, Petrovic JJ (1983) R.F.-plasma system for the production of ultrafine, ultrapure silicon carbide powder. J Mater Sci 18:3190–3194
2. Rao NP, Tymiak N, Blum J, Neuman A, Lee HJ, Girshick SL, McMurry PH, Heberlein J (1998) Hypersonic plasma particle deposition of nanostructured silicon and silicon carbide. J Aerosol Sci 29:707–720
3. Klein S, Winterer M, Hahn H (1998) Reduced-pressure chemical vapor synthesis of nanocrystalline silicon carbide powders. Chem Vap Deposition 4:143–149
4. Lin H, Gerbec JA, Sushchikh M, McFarland EW (2008) Synthesis of amorphous silicon carbide nanoparticles in a low temperature low pressure plasma reactor. Nanotechnology 19:325601
5. Askari S, Ul Haq A, Macias-Montero M, Levchenko I, Yu F, Zhou W, Ostrikov K, Maguire P, Svrcekg V, Mariotti D (2016) Ultra-small photoluminescent silicon-carbide nanocrystals by atmospheric-pressure plasmas. Nanoscale 8:17141–17149
6. Petersen RJ, Thomas SA, Anderson KJ, Pringle TA, May S, Hobbie EK (2022) Silicon-carbide nanocrystals from nonthermal plasma: surface chemistry and quantum confinement. J Phys Chem C 126:12935–12943
7. Liu R, Liu M, Chang J (2017) Large-scale synthesis of monodisperse SiC nanoparticles with adjustable size, stoichiometric ratio and properties by fluidized bed chemical vapor deposition. J Nanopart Res 19:26
8. Zhou J, Wang C, Song M, Chen X, Xia W (2021) Simple synthesis of ultrafine amorphous silicon carbide nanoparticles by atmospheric plasmas. Mater Lett 299:130072
9. Wang C, Zhou J, Song M, Lu Z, Chen X, Zheng Y, Xia W (2021) Synthesis of ultrafine silicon carbide nanoparticles using nonthermal arc plasma at atmospheric pressure. J Am Ceram Soc 104:3883–3894
10. Huisken F, Kohn B, Alexandrescu R, Cojocaru S, Crunteanu A, Ledoux G, Reynaud C (1999) Silicon carbide nanoparticles produced by CO_2 laser pyrolysis of SiH_4/C_2H_2 gas mixtures in a flow reactor. J Nanopart Res 1:293–303
11. Kityk IV, Makowska-Janusik M, Kassiba A, Plucinski KJ (2000) SiC nanocrystals embedded in oligoetheracrylate photopolymer matrices; new promising nonlinear optical materials. Opt Mater 13:449–453
12. Kassiba A, Tabellout M, Charpentier S, Herlin N, Emery JR (2000) Conduction and dielectric behaviour of SiC nano-sized materials. Solid State Commun 115:389–393
13. Bouclé J, Kassiba A, Makowska-Janusik M, Sanetra J, Herlin-Boime N, Bulou A, Kodjikian S (2005) Electro-optic phenomena in guest–host films of PMMA and SiC nanocrystals. Opt Commun 246:415–420
14. Kassiba A, Bednarski W, Pud A, Errien N, Makowska-Janusik M, Laskowski L, Tabellout M, Kodjikian S, Fatyeyeva K, Ogurtsov N, Noskov Y (2007) Hybrid core-shell nanocomposites

based on silicon carbide nanoparticles functionalized by conducting polyaniline: electron paramagnetic resonance investigations. J Phys Chem C 111:11544–11551
15. Kassiba A, Makowska-Janusik M, Bouclé J, Bardeau JF, Bulou A, Herlin-Boime N (2002) Photoluminescence features on the Raman spectra of quasistoichiometric SiC nanoparticles: experimental and numerical simulations. Phys Rev B 66:155317
16. Makowska-Janusik M, Kassiba A, Bouclé J, Bardeau JF, Kodjikian S, Désert A (2005) Vibrational density of states in silicon carbide nanoparticles: experiments and numerical simulations. J Phys Condens Matter 17:5101–5110
17. Bouclé J, Herlin-Boime N, Kassiba A (2005) Influence of silicon and carbon excesses on the aqueous dispersion of SiC nanocrystals for optical application. J Nanopart Res 7:275–285
18. Herlin-Boime N, Vicens J, Dufour C, Ténégal F, Reynaud C, Rizk R (2004) Flame temperature effect on the structure of SiC nanoparticles grown by laser pyrolysis. J Nanopart Res 6:63–70
19. Rider AE, Levchenko I, Ostrikov K (2007) Surface fluxes of Si and C adatoms at initial growth stages of SiC quantum dots. J Appl Phys 101:044306
20. Heinrich JL, Curtis CL, Credo GM, Kavanagh KL, Sailor MJ (1992) Luminescent colloidal silicon suspensions from porous silicon. Science 255:66–68
21. Wu XL, Fan JY, Qiu T, Yang X, Siu GG, Chu PK (2005) Experimental evidence for the quantum confinement effect in 3C–SiC nanocrystallites. Phys Rev Lett 94:026102
22. Madelung O (2004) Semiconductors: data handbook, 3rd edn. Springer, Berlin
23. Fan JY, Wu XL, Li HX, Liu HW, Siu GG, Chu PK (2006) Luminescence from colloidal 3C–SiC nanocrystals in different solvents. Appl Phys Lett 88:041909
24. Lehmann V, Gösele U (1991) Porous silicon formation: a quantum wire effect. Appl Phys Lett 58:856–858
25. Canham LT (1990) Silicon quantum wire array fabrication by electrochemical and chemical dissolution of wafers. Appl Phys Lett 57:1046–1048
26. Fan JY, Wu XL, Chu PK (2006) Low-dimensional SiC nanostructures: fabrication, luminescence, and electrical properties. Prog Mater Sci 51:983–1031
27. Fan JY, Li HX, Cui WN (2009) Microstructure and infrared spectral properties of porous polycrystalline and nanocrystalline cubic silicon carbide. Appl Phys Lett 95:021906
28. Fan JY, Wu XL, Zhao PQ, Chu PK (2006) Stability of luminescent 3C–SiC nanocrystallites in aqueous solution. Phys Lett A 360:336–338
29. Fan JY, Li HX, Cui WN, Dai DJ, Chu PK (2010) Excitation and recombination photodynamics in colloidal cubic SiC nanocrystals. Appl Phys Lett 97:191911
30. Zhu J, Liu Z, Wu XL, Xu LL, Zhang WC, Chu PK (2007) Luminescent small-diameter 3C–SiC nanocrystals fabricated via a simple chemical etching method. Nanotechnology 18:365603
31. Guo X, Chen X, Fan B, Zhang Y, Fan J (2016) Photon absorption and emission properties of 7A° SiC nanoclusters: electronic gap, surface state, and quantum size effect. Appl Phys Lett 109:013104
32. Peng XH, Nayak SK, Alizadeh A, Varanasi KK, Bhate N, Rowland LB, Kumar SK (2007) First-principles study of the effects of polytype and size on energy gaps in SiC nanoclusters. J Appl Phys 102:024304
33. Vörös M, Deák P, Frauenheim T, Gali A (2010) The absorption spectrum of hydrogenated silicon carbide nanocrystals from *ab initio* calculations. Appl Phys Lett 96:051909
34. Brus L (1986) Electronic wave functions in semiconductor clusters: experiment and theory. J Phys Chem 90:2555
35. Olaoye AO, Sani M, Ooi MDJ, Hussin MH, Pakhuruddin MZ, Hashim MR, Rashid M (2024) Determination of quantum size effect of colloidal SiC quantum dots by cyclic voltammetry. Emergent Mater. 7:1417–1428
36. Fan JY, Li HX, Zhang N, Lu RF (2011) Identification of the reconstruction and bonding structure of SiC nanocrystal surface by infrared spectroscopy. Appl Surf Sci 258:627–630
37. Wolkin MV, Jorne J, Fauchet PM, Allan G, Delerue C (1999) Electronic states and luminescence in porous silicon quantum dots: the role of oxygen. Phys Rev Lett 82:197–200
38. Beke D, Szekrényes Z, Balogh I, Veres M, Fazakas É, Varga LK, Kamarás K, Czigány Z, Gali A (2011) Characterization of luminescent silicon carbide nanocrystals prepared by reactive bonding and subsequent wet chemical etching. Appl Phys Lett 99:213108

39. Beke D, Szekrényes Z, Balogh I, Czigány Z, Kamarás K, Gali A (2013) Preparation of small silicon carbide quantum dots by wet chemical etching. J Mater Res 28:44–49
40. Liu LZ, Wang J, Wu XL, Li TH, Chu PK (2010) Longitudinal optical phonon–plasmon coupling in luminescent 3C–SiC nanocrystal films. Opt Lett 35:4024–4026
41. Wu XL, Xiong SJ, Zhu J, Wang J, Shen JC, Chu PK (2009) Identification of surface structures on 3C–SiC nanocrystals with hydrogen and hydroxyl bonding by photoluminescence. Nano Lett 9:4053–4060
42. Zakharko Y, Botsoa J, Alekseev S, Lysenko V, Bluet J-M, Marty O, Skryshevsky VA, Guillot G (2010) Influence of the interfacial chemical environment on the luminescence of 3C–SiC nanoparticles. J Appl Phys 107:013503
43. Li Y, Chen C, Li J-T, Yang Y, Lin Z-M (2011) Surface charges and optical characteristic of colloidal cubic SiC nanocrystals. Nanoscale Res Lett 6:454
44. Dai D, Guo X, Fan J (2015) Identification of luminescent surface defect in SiC quantum dots. Appl Phys Lett 106:053115
45. Beke D, Szekrényes Z, Czigány Z, Kamarás K, Gali Á (2015) Dominant luminescence is not due to quantum confinement in molecular-sized silicon carbide nanocrystals. Nanoscale 7:10982–10988
46. Beke D, Jánosi TZ, Somogyi B, Major DÁ, Szekrényes Z, Erostyák J, Kamarás K, Gali A (2016) Identification of luminescence centers in molecular-sized silicon carbide nanocrystals. J Phys Chem C 120:685–691
47. Alekseev S, Shamatulskaya E, Volvach M, Gryn S, Korytko D, Bezverkhyy I, Iablokov V, Lysenko V (2017) Size and surface chemistry tuning of silicon carbide nanoparticles. Langmuir 33:13561–13571
48. Li Y, Liu X, Liang T, Liu W, Fan J (2022) Native surface oxidation yields SiC–SiO_2 core–shell quantum dots with improved quantum efficiency. J Chem Phys 156:094705
49. Li Y, Liu X, Liang T, Fan J (2021) Core and surface electronic states and phonon modes in SiC quantum dots studied by optical spectroscopy and hybrid TDDFT. J Phys Chem C 125:7259–7266
50. Fan JY, Wu XL, Kong F, Qiu T, Huang GS (2005) Luminescent silicon carbide nanocrystallites in 3C–SiC/polystyrene films. Appl Phys Lett 86:171903
51. Fan JY, Wu XL, Li HX, Liu HW, Huang GS, Siu GG, Chu PK (2006) Si-based solid blue emitters from 3C–SiC nanocrystals. Appl Phys A 82:485–487
52. Wang J, Xiong SJ, Wu XL, Li TH, Chu PK (2010) Glycerol-bonded 3C–SiC nanocrystal solid films exhibiting broad and stable violet to blue-green emission. Nano Lett 10:1466–1471
53. Gan Z, Wu X, Xu H, Zhang N, Nie S, Fu Y (2016) Electron transition pathways of photoluminescence from 3C–SiC nanocrystals unraveled by steady-state, blinking and time-resolved photoluminescence measurements. J Phys D: Appl Phys 49:275107
54. Dai D, Zhang N, Zhang W, Fan J (2012) Highly bright tunable blue-violet photoluminescence in SiC nanocrystal-sodium dodecyl sulfonate crosslinked network. Nanoscale 4:3044–3046
55. Dai D, Fan J, Zhang N (2011) Synthesis and luminescence properties of silica-coated cubic silicon carbide nanocrystal composites. Micro Nano Lett. 6:878–880
56. Zhang N, Dai D, Zhang W, Fan J (2012) Photoluminescence and light reabsorption in SiC quantum dots embedded in binary-polyelectrolyte solid matrix. J Appl Phys 112:094315
57. Zhang W, Dai D, Chen X, Guo X, Fan J (2014) Red shift in the photoluminescence of colloidal carbon quantum dots induced by photon reabsorption. Appl Phys Lett 104:091902
58. Fan B, Guo X, Zhang Y, Fan J (2017) Analytical model of photon reabsorption in ZnO quantum dots with size and concentration dependent dual-color photoluminescence. J Appl Phys 121:054309
59. Dai D, Guo X, Fan J (2013) Synthesis and photoluminescence of semiconductor quantum dots/cetyltrimethylammonium bromide vesicle core/shell nanostructures. Appl Surf Sci 276:359–362
60. Fan JY, Li HX, Wang QJ, Dai DJ, Chu PK (2011) UV-blue photoluminescence from close-packed SiC nanocrystal film. Appl Phys Lett 98:081913

61. Botsoa J, Bluet JM, Lysenko V, Marty O, Barbier D, Guillot G (2007) Photoluminescence of 6H–SiC nanostructures fabricated by electrochemical etching. J Appl Phys 102:083526
62. Botsoa J, Bluet JM, Lysenko V, Sfaxi L, Zakharko Y, Marty O, Guillot G (2009) Luminescence mechanisms in 6H–SiC nanocrystals. Phys Rev B 80:155317
63. Rossi AM, Murphy TE, Reipa V (2008) Ultraviolet photoluminescence from 6H silicon carbide nanoparticles. Appl Phys Lett 92:253112
64. Fan J, Li H, Wang J, Xiao M (2012) Fabrication and photoluminescence of SiC quantum dots stemming from 3C, 6H, and 4H polytypes of bulk SiC. Appl Phys Lett 101:131906
65. Matsumoto T, Takahashi J, Tamaki T, Futagi T, Mimura H, Kanemitsu Y (1994) Blue-green luminescence from porous silicon carbide. Appl Phys Lett 64:226–228
66. Käckell P, Wenzien B, Bechstedt F (1994) Influence of atomic relaxations on the structural properties of SiC polytypes from ab initio calculations. Phys Rev B 50:17037–17046
67. Guo X, Dai D, Fan B, Fan J (2014) Experimental evidence of $\alpha \rightarrow \beta$ phase transformation in SiC quantum dots and their size-dependent luminescence. Appl Phys Lett 105:193110
68. Yoffe AD (2001) Semiconductor quantum dots and related systems: electronic, optical, luminescence and related properties of low dimensional systems. Adv Phys 50:1–208
69. Wolfe JP (1982) Thermodynamics of excitons in semiconductors. Phys Today 35:46–54
70. Guo X, Zhang Y, Fan B, Fan J (2017) Quantum confinement effect in 6H–SiC quantum dots observed via plasmon–exciton coupling-induced defect-luminescence quenching. Appl Phys Lett 110:123104
71. Svelto O, Hanna DC (1998) Principles of lasers, 4th edn. Springer, New York
72. Fan J, Chu PK (2010) Group IV nanoparticles: synthesis, properties, and biological applications. Small 6:2080–2098
73. Pol VG, Pol SV, Gedanken A (2005) Novel synthesis of high surface area silicon carbide by RAPET (reactions under autogenic pressure at elevated temperature) of organosilanes. Chem Mater 17:1797–1802
74. Lu A-H, Schmidt W, Kiefer W, Schüth F (2005) High surface area mesoporous SiC synthesized via nanocasting and carbothermal reduction process. J Mater Sci 40:5091–5093
75. Henderson EJ, Veinot JGC (2009) From phenylsiloxane polymer composition to size-controlled silicon carbide nanocrystals. J Am Chem Soc 131:809–815
76. Dasog M, Rachinsky C, Veinot JGC (2011) From Si and C encapsulated SiO_2 to SiC: exploring the influence of sol–gel polymer substitution on thermally induced nanocrystal formation. J Mater Chem 21:12422–12427
77. Dong Y, Molian P (2004) Coulomb explosion-induced formation of highly oriented nanoparticles on thin films of 3C–SiC by the femtosecond pulsed laser. Appl Phys Lett 84:10–12
78. Jia TQ, Zhao FL, Huang M, Chen HX, Qiu JR, Li RX, Xu ZZ, Kuroda H (2006) Alignment of nanoparticles formed on the surface of 6H–SiC crystals irradiated by two collinear femtosecond laser beams. Appl Phys Lett 88:111117
79. Yang S, Cai W, Zeng H, Xu X (2009) Ultra-fine β-SiC quantum dots fabricated by laser ablation in reactive liquid at room temperature and their violet emission. J Mater Chem 19:7119–7123
80. Yang S, Cai W, Zhang H, Zeng H, Lei Y (2011) A general strategy for fabricating unique carbide nanostructures with excitation wavelength-dependent light emissions. J Phys Chem C 115:7279–7284
81. Yang S, Kiraly B, Wang WY, Shang S, Cao B, Zeng H, Zhao Y, Li W, Liu Z-K, Cai W, Huang TJ (2012) Fabrication and characterization of beaded SiC quantum rings with anomalous red spectral shift. Adv Mater 24:5598–5603
82. Zakharko Y, Rioux D, Patskovsky S, Lysenko V, Marty O, Bluet JM, Meunier M (2011) Direct synthesis of luminescent SiC quantum dots in water by laser ablation. Phys Status Solidi-R 5:292–294
83. Ortega-Liébana MC, Hueso JL, Arenal R, Lahoz R, de la Fuente GF, Santamaría J (2015) Continuous-mode laser ablation at the solid–liquid interface of pelletized low-cost materials for the production of luminescent silicon carbide nanocrystals. J Phys Chem C 119:2158–2165
84. Zhu J, Hu S, Xia W-W, Li TH, Fan L, Chen H-T (2014) Photoluminescence of ~2 nm 3C–SiC quantum dots fabricated from polycrystalline 6H–SiC target by pulsed laser ablation. Mater Lett 132:210–213

85. Sasaki Y, Nishina Y, Sato M, Qkamura K (1989) Optical-phonon states of SiC small particles studied by Raman scattering and infrared absorption. Phys Rev B 40:1762–1772
86. Yang Z-G, Shaw LL (1996) Synthesis of nanocrystalline SiC at ambient temperature through high energy reaction milling. Nanostruct Mater 7:873–886
87. Zhang GQ, Wei GD, Zheng KZ, Li L, Xu DP, Wang DY, Xue YF, Su WH (2010) The synthesis of beta-SiC nanoparticles by high-energy mechanical ball milling and their photoluminescence properties. J Nanosci Nanotechnol 10:1951–1955
88. Martin H-P, Ecke R, Miiller E (1998) Synthesis of nanocrystalline silicon carbide powder by carbothermal reduction. J Eur Ceram Soc 18:1737-1742
89. Dasog M, Smith LF, Purkait TK, Veinot JGC (2013) Low temperature synthesis of silicon carbide nanomaterials using a solid-state method. Chem Commun 49:7004–7006
90. Su JJ, Gao B, Chen ZD, Fu JJ, An WL, Peng X, Zhang XM, Wang L, Huo KF, Chu PK (2016) Large-scale synthesis and mechanism of β-SiC nanoparticles from rice husks by low-temperature magnesiothermic reduction. ACS Sustainable Chem Eng 4:6600–6607
91. Sun RX, Wei XD, Hu WT, Ying P, Wu YJ, Wang LY, Chen S, Zhang X, Ma MD, Yu DL, Wang L, Gao GY, Xu B, Tian YJ (2022) Nanocrystalline cubic silicon carbide: a route to superhardness. Small 18:2201212
92. Xi G, Yu S, Zhang R, Zhang M, Ma D, Qian Y (2005) Crystalline silicon carbide nanoparticles encapsulated in branched wavelike carbon nanotubes: synthesis and optical properties. J Phys Chem B 109:13200–13204
93. Wang CH, Chang YH, Yen MY, Peng CW, Lee CY, Chiu HT (2005) Synthesis of silicon carbide nanostructures via a simplified Yajima process-reaction at the vapor-liquid interface. Adv Mater 17:419–422
94. Shen G, Chen D, Tang K, Qian Y, Zhang S (2003) Silicon carbide hollow nanospheres, nanowires and coaxial nanowires. Chem Phys Lett 375:177–184
95. Wang H, Yu JS, Li XD, Kim DP (2004) Inorganic polymer-derived hollow SiC and filled SiCN sphere assemblies from a 3DOM carbon template. Chem Commun 2352–2353
96. Zhang Y, Shi EW, Chen ZZ, Li XB, Xiao B (2006) Large-scale fabrication of silicon carbide hollow spheres. J Mater Chem 16:4141–4145
97. Liu Z, Ci L, Jin-Phillipp NY, Ru1hle M (2007) Vapor-solid reaction for silicon carbide hollow spherical nanocrystals. J Phys Chem C 111:12517 12521 (2007)
98. Li P, Xu L, Qian Y (2008) Selective synthesis of 3C–SiC hollow nanospheres and nanowires. Cryst Growth Des 8:2431–2436
99. Ju Z, Xu L, Pang Q, Xing Z, Ma X, Qian Y (2009) The synthesis of nanostructured SiC from waste plastics and silicon powder. Nanotechnology 20:355604
100. Ye J, Zhang S, Lee WE (2012) Novel low temperature synthesis and characterisation of hollow silicon carbide spheres. Micropor Mesopor Mat 152:25–30
101. Coleman D, Lopez T, Yasar-Inceoglu O, Mangolini L (2015) Hollow silicon carbide nanoparticles from a non-thermal plasma process. J Appl Phys 117:193301
102. Baklouti S, Pagnoux C, Chartier T, Baumard JF (1997) Processing of aqueous α-Al_2O_3, α-SiO_2 and α-SiC suspensions with polyelectrolytes. J Eur Ceram Soc 17:1387–1392
103. Tartaj P, Reece M, Moya JS (1998) Electrokinetic behavior and stability of silicon carbide nanoparticulate dispersions. J Am Ceram Soc 81:389–394
104. Médout-Marère V, Ghzaoui AEl, Charnay C, Douillard JM, Chauveteau G, Partyka S (2000) Surface heterogeneity of passively oxidized silicon carbide particles: hydrophobic-hydrophilic partition. J Colloid Interf Sci 223:205–214
105. Zhu X, Tang F, Suzuki TS, Sakka Y (2003) Role of the initial degree of ionization of polyethylenimine in the dispersion of silicon carbide nanoparticles. J Am Ceram Soc 86:189–191
106. Singh BP, Jena J, Besra L, Bhattacharjee S (2007) Dispersion of nano-silicon carbide (SiC) powder in aqueous suspensions. J Nanopart Res 9:797–806
107. Che J, Wang X, Xiao Y, Wu X, Zhou L, Yuan W (2007) Effect of inorganic-organic composite coating on the dispersion of silicon carbide nanoparticles in non-aqueous medium. Nanotechnology 18:135706

108. Iijima M, Kamiya H (2008) Surface modification of silicon carbide nanoparticles by azo radical initiators. J Phys Chem C 112:11786–11790
109. Zou G, Cao M, Lin H, Jin H, Kang Y, Chen Y (2006) Nickel layer deposition on SiC nanoparticles by simple electroless plating and its dielectric behaviors. Powder Technol 168:84–88
110. Saini I, Sharma A, Dhiman R, Aggarwal S, Ram S, Sharma PK (2017) Grafted SiC nanocrystals: for enhanced optical, electrical and mechanical properties of polyvinyl alcohol. J Alloy Compd 714:172–180
111. Sotoma S, Abe H, Miyanoiri Y, Ohshima T, Harada Y (2023) Highly dispersed 3C silicon carbide nanoparticles with a polydopamine/polyglycerol shell for versatile functionalization. ACS Appl Mater Interfaces 15:21413–21424
112. Zhao X, Wei CM, Yang L, Chou MY (2004) Quantum confinement and electronic properties of silicon nanowires. Phys Rev Lett 92:236805
113. Puzder A, Williamson AJ, Grossman JC, Galli G (2002) Surface chemistry of silicon nanoclusters. Phys Rev Lett 88:097401
114. Rintelman JM, Gordon MS (2001) Structure and energetics of the silicon carbide clusters SiC_3 and Si_2C_2. J Chem Phys 115:1795–1803
115. Reboredo FA, Pizzagalli L, Galli G (2004) Computational engineering of the stability and optical gaps of SiC quantum dots. Nano Lett 4:801–804
116. Huda MN, Ray AK (2008) Evolution of SiC nanocluster from carbon fullerene: a density functional theoretic study. Chem Phys Lett 457:124–129
117. Fan J, Li H, Jiang J, So LKY, Lam YW, Chu PK (2008) 3C–SiC nanocrystals as fluorescent biological labels. Small 4:1058–1062
118. Vörös M, Deák P, Frauenheim T, Gali A (2010) The absorption of oxygenated silicon carbide nanoparticles. J Chem Phys 133:064705
119. Zhang Z, Dai Y, Yu L, Guo M, Huang B, Whangbo M-H (2012) The surface termination effect on the quantum confinement and electron affinities of 3C–SiC quantum dots: a first-principles study. Nanoscale 4:1592–1597
120. Saha S, Sarkar P (2012) Tuning the HOMO-LUMO gap of SiC quantum dots by surface functionalization. Chem Phys Lett 536:118–122
121. Haq AU, Buerkle M, Askari S, Rocks C, Ni C, Švrček V, Maguire P, Irvine JT, Mariotti D (2020) Controlling the energy-level alignment of silicon carbide nanocrystals by combining surface chemistry with quantum confinement. J Phys Chem Lett 11:1721–1728
122. Gao F, Chen D, Hu WY, Weber WJ (2010) Energy dissipation and defect generation in nanocrystalline silicon carbide. Phys Rev B 81:184101
123. Swaminathan N, Kamenski PJ, Morgan D, Szlufarska I (2010) Effects of grain size and grain boundaries on defect production in nanocrystalline 3C–SiC. Acta Mater 58:2843–2853
124. Chen D, Gao F, Liu B (2015) Grain boundary resistance to amorphization of nanocrystalline silicon carbide. Sci Rep 5:16602
125. Chavoshi SZ, Xu SZ (2018) Tension-compression asymmetry in plasticity of nanotwinned 3C–SiC nanocrystals. J Appl Phys 124:095103
126. Sveinsson HA, Hafreager A, Kalia RK, Nakano A, Vashishta P, Malthe-Sorenssen A (2020) Direct atomic simulations of facet formation and equilibrium shapes of SiC nanoparticles. Cryst Growth Des 20:2147–2152
127. Chavoshi SZ, Branicio PS, An Q (2021) Transition between Hall-Petch and inverse Hall-Petch behavior in nanocrystalline silicon carbide. Phys Rev Mater 5:073606
128. Pizzagalli L, Godet J (2023) Ultrahigh strength and pasticity mechanisms of Si and SiC nanoparticles revealed by first-principles molecular dynamics. Phys Rev Lett 131:236201
129. Capaldi L, Sansoz F (2023) High-temperature active oxidation of nanocrystalline silicon-carbide: a reactive force-field molecular dynamics study. Acta Mater 258:119229
130. Granlie JD, Hobbie EK, Kilin DS (2023) Formation and luminescence of single oxygen impurities on the surface of SiC nanocrystals. J Phys Chem Lett 14:6202–6208
131. Zimmerman AF, Clark DG, Aust KT, Erb U (2002) Pulse electrodeposition of Ni-SiC nanocomposite. Mater Lett 52:85–90

132. Zimmerman AF, Palumbo G, Aust KT, Erb U (2002) Mechanical properties of nickel silicon carbide nanocomposites. Mat Sci Eng A 328:137–146
133. Shi L, Sun C, Gao P, Zhou F, Liu W (2006) Mechanical properties and wear and corrosion resistance of electrodeposited Ni-Co/SiC nanocomposite coating. Appl Surf Sci 252:3591–3599
134. Vaezi MR, Sadrnezhaad SK, Nikzad L (2008) Electrodeposition of Ni-SiC nano-composite coatings and evaluation of wear and corrosion resistance and electroplating characteristics. Colloid Surface A 315:176–182
135. Zhao J, Stearns LC, Harmer MP, Chan HM, Miller GA (1993) Mechanical behavior of alumina-silicon carbide "nanocomposites." J Am Ceram Soc 76:503–510
136. Ohj T, Nakahira A, Hirano T, Niihara K (1994) Tensile creep behavior of alumina/silicon carbide nanocomposite. J Am Ceram Soc 77:3259–3262
137. Gao L, Wang HZ, Hong JS, Miyamoto H, Miyamoto K, Nishikawa Y, Torre SDDL (1999) Mechanical properties and microstructure of nano-SiC-Al_2O_3 composites densified by spark plasma sintering. J Eur Ceram Soc 19:609–613
138. Yang Y, Lan J, Li X (2004) Study on bulk aluminum matrix nano-composite fabricated by ultrasonic dispersion of nano-sized SiC particles in molten aluminum alloy. Mat Sci Eng A 380:378–383
139. Sawaguchi A, Toda K, Niihara K (1991) Mechanical and electrical properties of silicon nitride-silicon carbide nanocomposite material. J Am Ceram Soc 74:1142–1144
140. Rendtel A, Hübner H, Herrmann M, Schubert C (1998) Silicon nitride/silicon carbide nanocomposite materials: II, hot strength, creep, and oxidation resistance. J Am Ceram Soc 81:1109–1120
141. Ferkel H, Mordike BL (2001) Magnesium strengthened by SiC nanoparticles. Mat Sci Eng A 298:193–199
142. Lan J, Yang Y, Li X (2004) Microstructure and microhardness of SiC nanoparticles reinforced magnesium composites fabricated by ultrasonic method. Mat Sci Eng A 386:284–290
143. Chisholm N, Mahfuz H, Rangari VK, Ashfaq A, Jeelani S (2005) Fabrication and mechanical characterization of carbon/SiC-epoxy nanocomposites. Compos Struct 67:115–124
144. Ma RZ, Wu J, Wei BQ, Liang J, Wu DH (1998) Processing and properties of carbon nanotubes-nano-SiC ceramic. J Mater Sci 33:5243–5246
145. Dou SX, Soltanian S, Horvat J, Wang XL, Zhou SH, Ionescu M, Liu HK, Munroe P, Tomsic M (2002) Enhancement of the critical current density and flux pinning of MgB_2 superconductor by nanoparticle SiC doping. Appl Phys Lett 81:3419–3421
146. Matsumoto A, Kumakura H, Kitaguchi H, Hatakeyama H (2003) Effect of SiO_2 and SiC doping on the powder-in-tube processed MgB_2 tapes. Supercond Sci Technol 16:926–930
147. Matsumoto A, Kumakura H, Kitaguchi H, Senkowicz BJ, Jewell MC, Hellstrom EE, Zhu Y, Voyles PM, Larbalestier DC (2006) Evaluation of connectivity, flux pinning, and upper critical field contributions to the critical current density of bulk pure and SiC-alloyed MgB_2. Appl Phys Lett 89:132508
148. Dou SX, Shcherbakova O, Yeoh WK, Kim JH, Soltanian S, Wang XL, Senatore C, Flukiger R, Dhalle M, Husnjak O, Babic E (2007) Mechanism of enhancement in electromagnetic properties of MgB_2 by nano SiC doping. Phys Rev Lett 98:097002
149. Zhang N, Dai DJ, Fan JY (2012) Plasmon-assisted photoluminescence enhancement of SiC nanocrystals by proximal silver nanoparticles. Appl Surf Sci 258:10140–10143
150. Dai D, Dong Z, Fan J (2013) Giant photoluminescence enhancement in SiC nanocrystals by resonant semiconductor exciton-metal surface plasmon coupling. Nanotechnology 24:025201
151. Sui N, Monnier V, Zakharko Y, Chevolot Y, Alekseev S, Bluet J-M, Lysenko V, Souteyrand E (2012) Plasmon-controlled narrower and blue-shifted fluorescence emission in (Au@SiO_2)SiC nanohybrids. J Nanopart Res 14:1004
152. Sui N, Monnier V, Zakharko Y, Chevolot Y, Alekseev S, Bluet J-M, Lysenko V, Souteyrand E (2013) Fluorescent (Au@SiO_2)SiC nanohybrids: influence of gold nanoparticle diameter and SiC nanoparticle surface density. Plasmonics 8:85–92

153. Zakharko Y, Serdiuk T, Nychyporuk T, Géloën A, Lemiti M, Lysenko V (2012) Plasmon-enhanced photoluminescence of SiC quantum dots for cell imaging applications. Plasmonics 7:725–732
154. Zakharko Y, Nychyporuk T, Bonacina L, Lemiti M, Lysenko V (2013) Plasmon-enhanced nonlinear optical properties of SiC nanoparticles. Nanotechnology 24:055703
155. Mimura H, Matsumoto T, Kanemitsu Y (1994) Blue electroluminescence from porous silicon carbide. Appl Phys Lett 65:3350–3352
156. Xiao B, Wu XL, Xu W, Chu PK (2012) Tunable electroluminescence from polymer-passivated 3C–SiC quantum dot thin films. Appl Phys Lett 101:123110
157. Tai H-Y, Cheng C-H, Lin G-R (2014) Blue-green light emission from Si and SiC quantum dots co-doped Si-rich SiC *p-i-n* junction diode. IEEE J Sel Top Quant Electron 20:8200507
158. Chen XF, Guo XX, Wu WH, Fan JY (2018) Quasi-white light-emitting devices based on SiC quantum dots. Phys Status Solidi RRL 12:1800171
159. He C, Wu X, Shen J, Chu PK (2012) High-efficiency electrochemical hydrogen evolution based on surface autocatalytic effect of ultrathin 3C–SiC nanocrystals. Nano Lett 12:1545–1548
160. Shen X, Pantelides ST (2013) Atomic-scale mechanism of efficient hydrogen evolution at SiC nanocrystal electrodes. J Phys Chem Lett 4:100–104
161. Wang Y, Zhang LN, Zhang XY, Zhang ZZ, Tong YC, Li FY, Wu JCS, Wang XX (2017) Openmouthed β-SiC hollow-sphere with highly photocatalytic activity for reduction of CO_2 with H_2O. Appl Catal B Environ 206:158–167
162. Wu LQ, Zhang LC, Sun MX, Liu R, Yu LZ, Lv Y (2017) Metal-free cataluminescence gas sensor for hydrogen sulfide based on its catalytic oxidation on silicon carbide nanocages. Anal Chem 89:13666–13672
163. Xie H, Wang J, Xi T, Liu Y (2002) Thermal conductivity of suspensions containing nanosized SiC particles. Int J Thermophys 23:571–580
164. Singh D, Timofeeva E, Yu W, Routbort J, France D, Smith D, Lopez-Cepero JM (2009) An investigation of silicon carbide-water nanofluid for heat transfer applications. J Appl Phys 105:064306
165. Yu WH, France DM, Smith DS, Singh D, Timofeeva EV, Routbort JL (2009) Heat transfer to a silicon carbide/water nanofluid. Int J Heat Mass Tran 52:3606–3612
166. Timofeeva EV, Smith DS, Yu W, France DM, Singh D, Routbort JL (2010) Particle size and interfacial effects on thermo-physical and heat transfer characteristics of water-based α-SiC nanofluids. Nanotechnology 21:215703
167. Lee SW, Park SD, Kang S, Bang IC, Kim JH (2011) Investigation of viscosity and thermal conductivity of SiC nanofluids for heat transfer applications. Int J Heat Mass Tran 54:433–438
168. Li XK, Zou CJ, Wang TY, Lei XY (2015) Rheological behavior of ethylene glycol-based SiC nanofluids. Int J Heat Mass Tran 84:925–930
169. Chen WJ, Zou CJ, Li XK, Li L (2017) Experimental investigation of SiC nanofluids for solar distillation system: stability, optical properties and thermal conductivity with saline water based fluid. Int J Heat Mass Tran 107:264–270
170. Al-Waeli AHA, Sopian K, Chaichan MT, Kazem HA, Hasan HA, Al-Shamani AN (2017) An experimental investigation of SiC nanofluid as a base-fluid for a photovoltaic thermal PV/T system. Energy Convers Manage 142:547–558
171. Zhao L-D, Zhang B-P, Li J-F, Zhou M, Liu W-S, Liu J (2008) Thermoelectric and mechanical properties of nano-SiC-dispersed Bi_2Te_3 fabricated by mechanical alloying and spark plasma sintering. J Alloys Compd 455:259–264
172. Mavinakuli P, Wei S, Wang Q, Karki AB, Dhage S, Wang Z, Young DP, Guo Z (2010) Polypyrrole/silicon carbide nanocomposites with tunable electrical conductivity. J Phys Chem C 114:3874–3882

173. Zhao D-L, Luo F, Zhou W-C (2010) Microwave absorbing property and complex permittivity of nano SiC particles doped with nitrogen. J Alloys Compd 490:190–194
174. Liu J, Tao JQ, Gao LL, He XX, Wei B, Gu YS, Yao ZJ, Zhou JT (2022) Morphology-size synergy strategy of SiC@C nanoparticles towards lightweight and efficient microwave absorption. Chem Eng J 433:134484
175. Dai DJ, Zhang CY, Wang L, Wang Y, Zhang BH, Deng KM, Lu W, Fan JY (2019) Reversible/irreversible photobleaching of fluorescent surface defects of SiC quantum dots: mechanism and sensing of solar UV irradiation. Adv Mater Interfaces 6:1900272

Chapter 5
SiC Nanowires

There have been a large amount of studies concerning the fabrication and properties of silicon carbide nanowires. In fact, much more knowledge has been learnt about SiC nanowires than other low-dimensional SiC nanostructures. A lot of methods have been developed to synthesize SiC nanowires and some can yield large-scale and high-quality SiC nanowires and nanowire arrays. Filed electron emission is the most intensively investigated characteristic of SiC nanowires and there are also some studies involving the luminescence properties of SiC nanowires. SiC nanorods, nanobelts, and various other forms of one-dimensional nanostructures have also been synthesized and studied.

5.1 Synthesis and Optical Properties

5.1.1 Vapor–Solid Growth

SiC nanowires were synthesized using the vapor–solid reactions during the earliest years of investigations. In 1995, investigators began to use carbon nanotubes as templates for the space-limited synthesis of various carbide nanorods [1]. The SiC nanorods with diameters similar to or much smaller than that of carbon nanotubes were fabricated through reactions between the carbon nanotubes and SiO or Si–I_2. The TEM images show that the SiC nanorods synthesized by the reaction of carbon nanotubes with Si and I_2 at 1200 °C have lengths on the order of 1 μm and diameters of 2–20 nm, similar to the diameter of the carbon nanotube reactant. X-ray diffraction shows the zinc blende β-SiC structure of the product. Transmission electron microscopy reveals that the nanorods possess a high density of planar defects. The defects in <111> oriented rods are rotational twin stacking faults. All the nanorods produced at 1300–1400 °C using SiO as the gaseous reactant have rod axis along the <111> direction, but the nanorods prepared at 1100–1200 °C using Si and I_2 show

J. Fan and P. K. Chu, *Silicon Carbide Nanostructures*, Engineering Materials and Processes, https://doi.org/10.1007/978-3-032-10986-6_5

varied rod directions, some along the <100> direction, although in these rods small defect regions with a <111> direction are also present. This method has been developed into a two-step reaction scheme for the synthesis of SiC nanorods at 1400 °C [4]. SiO vapor produced via silicon reduction of silica reacted with carbon nanotubes, yielding SiC nanorods. Transmission electron microscopy reveals that the nanorods are single crystalline β-SiC with diameters ranging from 3 to 40 nm. The infrared absorption spectrum comprises a strong Si–C stretching vibration peak at 814 cm^{-1} and another weak Si–O stretching vibration peak at about 1100 cm^{-1}. This indicates the nanorods mainly consist of SiC, but with minor amount of silicon oxide. The Raman spectrum shows a sharp peak at 784 cm^{-1} attributed to TO phonon mode of SiC. The nanorods have photoluminescence peaked at around 430 nm under 260 nm excitation at room temperature (Fig. 5.1). A two-step growth model for the SiC nanorods was proposed as:

$$SiO_2(s) + Si(s) \rightarrow 2SiO(v), \tag{5.1}$$

$$SiO(v) + 2C(s) \rightarrow SiC(s) + CO(v). \tag{5.2}$$

High-resolution TEM observation reveals that the β-SiC nanorods synthesized by the carbon nanotube confined reaction possess many layer faults [5]. In addition, there is a thick amorphous layer on the nanorod surface. The SiC nanorods with uniform diameters can be prepared by suppressing the epitaxial growth on the surface of the initially formed SiC nanorods by effectively carrying away the CO gas by-product from the reaction area [6]. The diameter distribution is consistent with that of the starting carbon nanotubes. The TEM image (Fig. 5.2) of the nanorods produced at 1400 °C shows that they have uniform diameters ranging from 20 to 25 nm, slightly larger than that of starting carbon nanotubes. The shape of the nanorods can be controlled by thermally adjusting the carbon nanotube shape during the reaction process. Silicon carbide (SiC) nanorods were synthesized by reacting multi-walled

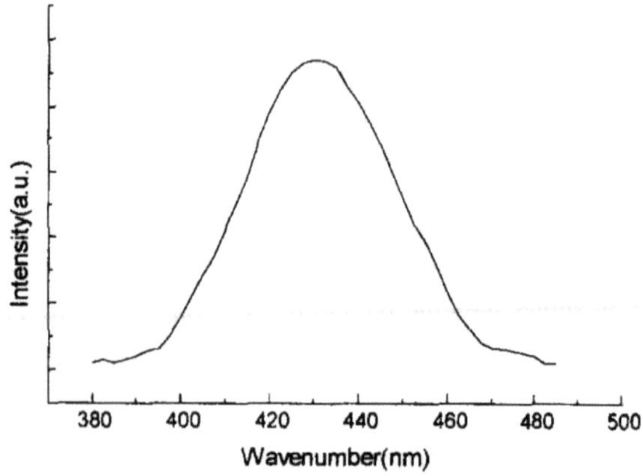

Fig. 5.1 Photoluminescence spectrum of SiC nanorods. Reprinted with permission from [4]. Copyright 1995, Springer Nature Limited

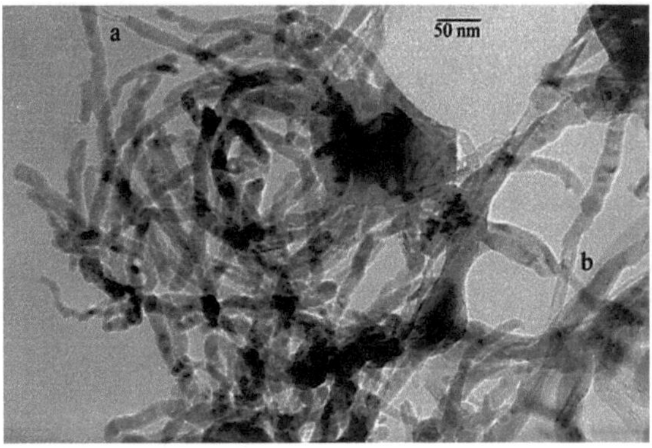

Fig. 5.2 TEM image of SiC nanorods synthesized at 1400 °C. Reprinted with permission from [6]. Copyright 2000, Elsevier B.V

carbon nanotubes with Si particles in a NaCl–NaF binary salt at 1100–1200 °C in Ar for 4 h [7]. At 1100 °C, 3C–SiC is detectable by XRD, but unreacted Si and CNTs are dominant; at 1200 °C, only SiC peaks are detectable. The resultant SiC nanorods mostly follow the morphologies of the as-received carbon nanotubes. Most SiC nanorods have complex stacking sequences, containing various forms of stacking faults, and only some are fault-free.

Raman scattering was conducted on the amorphous silica-sheathed SiC nanorods with diameters of 3–30 nm and lengths of over 1 μm produced by carbon nanotube confinement reaction [8], and most nanorods contain high-density defects including rotational twins and stacking faults. The Raman spectrum (Fig. 5.3a) comprises two strong peaks at 791 and 924 cm^{-1}, corresponding to the TO and LO phonon modes of β-SiC, respectively. They shift to small wavenumbers with respect to bulk values (at the Γ-point of the Brillouin zone) of β-SiC, which are separately 796 and 972 cm^{-1} for TO and LO phonon modes [9]. Calculation shows that the phonon confinement contributes only a small portion to the large shift of the LO mode frequency and that the Raman spectra can be considered as the Fröhlich-enhanced effective density of states of optical phonons (Fig. 5.3b), which becomes crucial because of the broken translation symmetry caused by the interior structure defects in the nanorods. The weak peak at 864 cm^{-1} which has no counterpart in bulk SiC was assigned to the interface-like mode. The calculation and the experiment agree well.

SiC nanowires can also be synthesized by employing other solid carbon sources other than carbon nanotubes. Carbothermal reduction of SiO_2 (or SiO/Si) powder by carbon black powder produced β-SiC whiskers [10]. The mixed powders were heated at 1300–1550 °C in hydrogen atmosphere followed by oxidation at 700 °C in air to eliminate excess carbon. The product contains β-SiC whiskers and spherical particles. The β-SiC whiskers formed by the solid–gas (carbon–SiO gas) reaction and have high density of stacking faults, whereas the solid–solid (carbon–Si) reaction yielded β-SiC spherical particles with low density of stacking faults. The sizes of the whiskers and particles decreased with decreasing reaction temperature and time. The stacking fault density increased with heating rate due to increased reaction rate. Three kinds of 3C–SiC nanostructures were obtained by carbothermal reduction

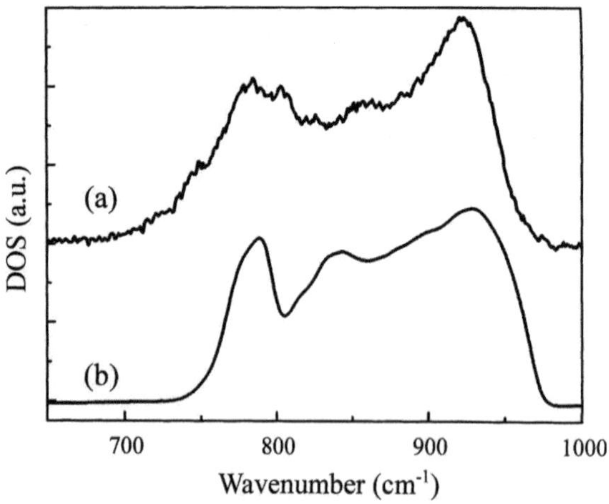

Fig. 5.3 **a** Reduced Stokes Raman spectrum of SiC nanorods. **b** Calculated effective density of optical phonon states for a rectangular SiC nanorod with a cross-section of 24 × 24 mono-layers by employing a Gaussian broadening factor of 5 cm^{-1}. Reprinted with permission from [8]. Copyright 1999, Elsevier B.V

of the SiO_2 xerogels under different conditions [11]. The powders obtained from carbothermal reduction at 1650 °C for 2.5 h in static argon atmosphere are 3C–SiC nanoparticles with diameters ranging from 30 to 100 nm. The products synthesized by carbothermal reduction at 1650 °C for 1.5 h followed by heating at 1800 °C for 30 min in flowing argon atmosphere are nanowires, which consist of crystalline 3C–SiC cores with diameters of 10–25 nm and amorphous SiO_2 shells with outer diameters of 20–70 nm. The products formed by carbothermal reduction at 1650 for 2.5 h in flowing argon atmosphere are bare SiC nanowires with diameters of 15–30 nm.

The solid-reaction methods have been developed to synthesize SiC nanowires [12]. These methods involve heating silica gel or fumed silica with activated carbon in a reduction atmosphere. One involves heating silica gel with activated carbon at 1360 °C in H_2 or NH_3, and the activated carbon was prepared by thermal decomposition of polyethylene glycol at 700 °C in an argon atmosphere. The SEM images (Fig. 5.4) show that the nanowires have diameters of around 350 nm and lengths up to several tens of micrometers. The proposed reactions are as follows:

$$SiO_2 + C \rightarrow SiO + CO, \tag{5.3}$$

$$SiO + 2C \rightarrow SiC + CO. \tag{5.4}$$

The NH_3 and H_2 facilitate the conversion of SiO_2 into SiO, which reacts with carbon to yield SiC. The nanowires have wide photoluminescence band centered at about 400 nm (Fig. 5.5). The infrared absorption spectrum comprises a strong peak at 810 cm^{-1} and this value is larger than the TO phonon mode value of bulk β-SiC. The Raman spectrum comprises a broad band at around 780 cm^{-1} corresponding to the TO phonon mode of β-SiC, but it shifts to lower wavenumbers by 16 cm^{-1} compared with the bulk value caused by phonon confinement and stacking faults.

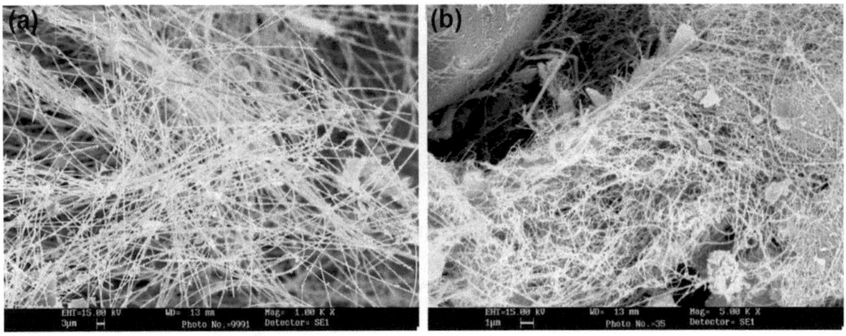

Fig. 5.4 SEM images of β-SiC nanowires synthesized by heating the gel containing activated carbon and silica at 1360 °C in NH_3 for **a** 4 h and **b** 7 h. The scale bar is 3 μm for (**a**) and 1 μm for (**b**). Reprinted with permission from [12]. Copyright 2002, Royal Society of Chemistry

The LO phonon expected to arise at 972 cm^{-1} is absent in the spectrum and this can be explained by the Raman selection rule for the given experimental geometry of Raman scattering measurement.

Reaction between SiO and amorphous activated carbon at 1380 °C produced SiC nanorods with diameters of 20–100 nm and lengths of 10–100 μm [13]. Transmission electron microscopy and energy dispersive spectroscopy reveal that the SiC nanorods grew on either a chain or from facets of SiC nanoparticles. Most of them are straight and orientated along the <111> direction. Branched structures are also present (Fig. 5.6), and on a typical tip of the straight SiC nanorod there is a thin amorphous layer with thickness of 1–3 nm, which consists of Si, C, and a small quantity of O. Single-crystalline SiC–SiO_2 core shell nanowires have been grown using electrospun polyacrylonitrile nanofibers as templates and silica and graphite

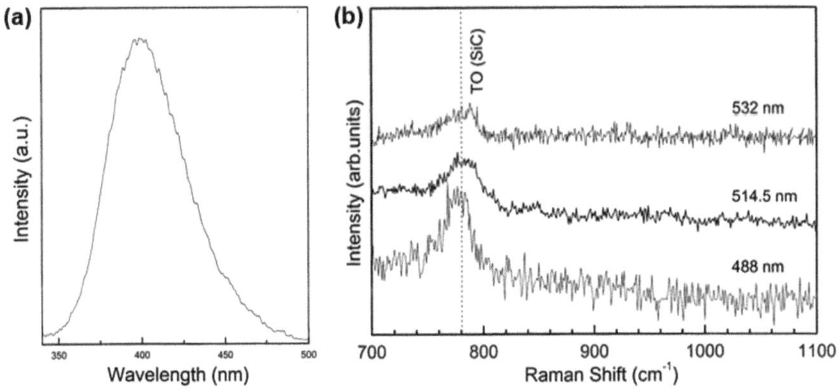

Fig. 5.5 Photoluminescence (**a**) and Raman (**b**) spectra of SiC nanowires synthesized by heating the gel containing activated carbon and silica at 1360 °C in NH_3 for 7 h. Reprinted with permission from [12]. Copyright 2002, Royal Society of Chemistry

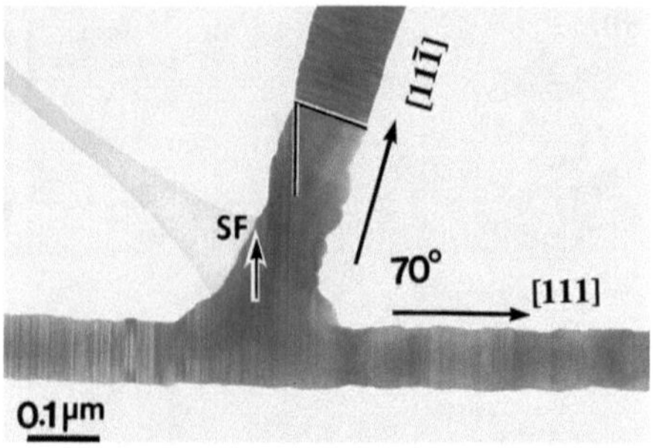

Fig. 5.6 TEM image of a branched SiC nanorod. "SF" denotes stacking fault [13]. Reprinted with permission from. Copyright 2002, Springer Science + Business Media, LLC, part of Springer Nature

powder as source materials [14]. The produced SiC NWs have uniform diameters, well-ordered structures, and very low concentrations of stacking faults. There are also some bent or helical SiC NWs. Thermal evaporation of SiO powder and activated carbon fiber can also give rise to 3C–SiC nanowires coated with SiO_2 [15]. Raman scattering characterization shows that the TO and LO phonon modes of the SiC nanowires have larger red shifts compared to those of bulk 3C–SiC. The electrical measurement from a single 3C–SiC nanowire-based field effect transistor shows that 3C–SiC nanowires are characteristic of *n*-channel behavior.

Arc-discharge method was used to synthesize β-SiC SiC nanowires with preferential <111> growth direction using SiO_2 powder as filler in a graphite anode [16]. The nanowires have diameters of 3–15 nm and lengths up to several micrometers. The growth of the nanowires was suggested to follow a catalyst-free vapor–solid mechanism. The SiC nanowires have photoluminescence with a single wide emission band centered at 472 nm (Fig. 5.7). The photoluminescence excitation spectrum comprises a single peak at 305 nm.

Thermal evaporation of ZnS and solid carbon on silicon wafer at 1100 °C yielded aligned SiC nanowire arrays [17]. The nanowires have uniform sizes, with a most probable diameter of 8 nm and a smallest diameter of 3 nm. They usually grew along the <111> direction. They possess a thin oxide shell and contain few stacking faults.

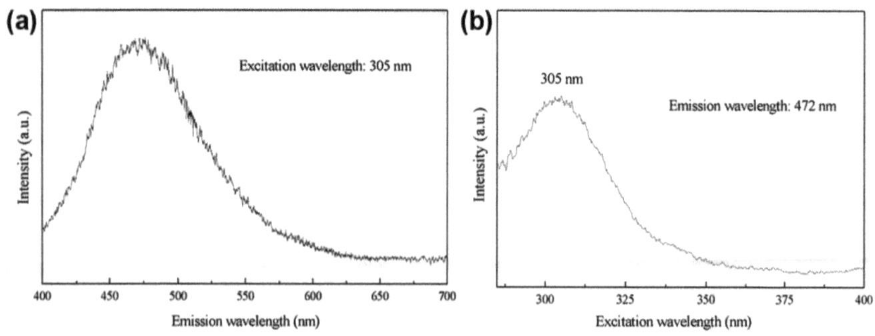

Fig. 5.7 **a** Photoluminescence and (**b**) photoluminescence excitation spectra of the SiC nanowires. Reprinted with permission from [16]. Copyright 2007, American Chemical Society

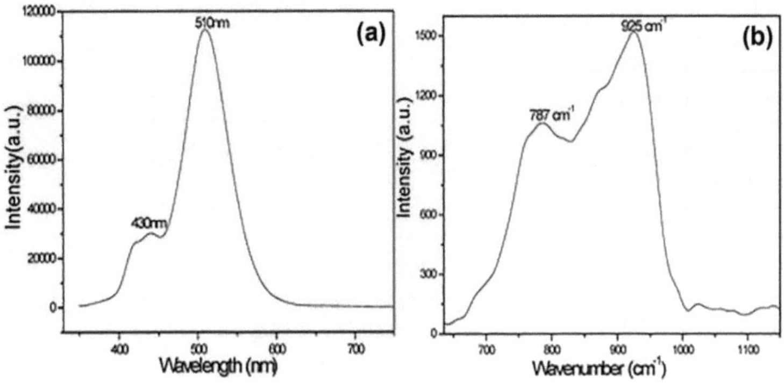

Fig. 5.8 **a** Photoluminescence (under 325 nm excitation) and (**b**) Raman spectra of the oriented SiC nanowires. Reprinted with permission from [17]. Copyright 2007, American Chemical Society

The PL spectrum (Fig. 5.8a) consists of a usual 430 nm emission band and another more intense emission band at around 510 nm. The Raman spectrum (Fig. 5.8b) comprises two strong peaks at 787 and 925 cm^{-1}, corresponding to the TO and LO phonon modes of cubic SiC, respectively. They all shift to lower wavenumbers relative to bulk material. In particular, the LO phonon line has a large shift of 47 cm^{-1}. SiC–SiO_2 core–shell nanowires were synthesized by thermal evaporation of the graphite powder on a Si substrate [18]. These SiC nanowires grew along the <111> direction, and they have diameters of 50–100 nm and lengths of several hundred micrometers. Each nanowire consists of a single-crystalline SiC core and a thin amorphous SiO_2 shell. 3C–SiC nanowires were synthesized on the surface of graphite by partially reacting with silicon powders in NaF–NaCl-based salt at 1150–1400 °C in argon [19]. The formation of SiC nanowires start at about 1200 °C, and the amount of SiC nanowires in the resultant powder increased with temperature. The nanowires have diameters of 10–50 nm and various lengths; their preferred growth direction is <111>.

Using CH_3SiCl_3 and H as precursors, large quantities of 3C–SiC nanowires have been fabricated by using a CVD process without employing catalysts [20]. The as-grown SiC nanowires have diameters of about 70 nm. When the deposition temperature increased or H_2: MTS molar ratio decreased, the SiC crystal size increased, and the products changed from nanowires to grains. 3C–SiC coaxial nanocables with amorphous wrapping layers were also acquired by oxidation of the SiC nanowires. Thermal decomposition of commercial silicone oil (dimethyl polysiloxane) at a relatively low temperature (800 °C) can yield 3C–SiC nanorods with a high yield (75%) [21]. No catalyst, template, or solvent was employed in this synthesis. The reversible hydrogen take-up and release (1.5 wt%) by the high surface area (563 m^2 g^{-1}) SiC nanorods was observed at room temperature.

Raman scattering from a single 3C–SiC nanowire has been investigated by using surface-enhanced Raman scattering approach [22]. The 3C–SiC nanowires were synthesized by thermal heating of the mixed silicon, silicon oxide, and carbon

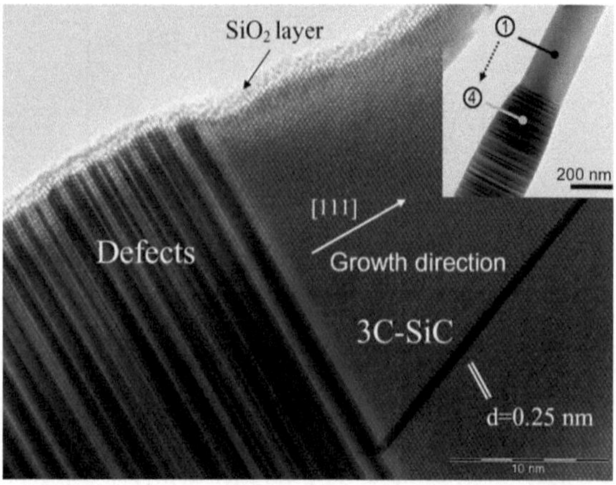

Fig. 5.9 TEM image of a typical SiC nanowire exhibiting an area of random network of stacking faults (dark part) and a pure 3C–SiC section (bright part). Inset shows HRTEM image of the interface region between two crystallographic zones. Reprinted with permission from [22]. Copyright 2007, John Wiley & Sons, Inc

powders at 1400 °C under an argon flow. The Raman scattering was enhanced by a structure based on a tungsten tip covered by a thin gold layer to which the SiC nanowire was attached. Figure 5.9 shows the TEM image of a typical SiC nanowire with a diameter of 100 nm. It contains a homogenous region (bright) corresponding to 3C–SiC that grows in the <111> orientation and a second region (dark) exhibiting a random network of stacking faults. The cross-sectional dimension of the 3C–SiC part is uniform, whereas that of the dark zone varies due to the strain relaxation caused by the structural defects. Figure 5.10a shows the enhanced Raman spectrum of an individual SiC nanowire with 40 nm in diameter. The peak centered at 797 cm^{-1} is attributed to a TO mode and the two shoulders centered at 925 and 955 cm^{-1} are attributed to the LO modes. The peaks at 842 and 896 cm^{-1} were assigned to Fröhlich transitions enhanced by structural defects. The absence of the conventional LO phonon line at around 970 cm^{-1} suggests that this nanowire is not a pure 3C–SiC crystal and contains many stacking faults. The high-resolution TEM image of a nanowire with comparable size is shown in Fig. 5.10b. Note that the whole nanowire is composed of stacking faults and this is different from the case of the nanowires with greater diameters (Fig. 5.9) where the pure 3C crystalline parts are present. The free carrier concentration was calculated to be 10^{18}–10^{19} cm^{-3} by using the Fano interference features of sharp phonon lines.

5.1.2 *Vapor–Liquid–Solid Growth*

The vapor–liquid–solid (VLS) growth mechanism is a common nanowire growth mechanism, which dominates the crystal growth process of various types of nanowires. In 1964, Wagner and Ellis proposed the concept of VLS growth during studying the crystal growth of Si whiskers [23]. The Si whiskers were grown on a Si wafer (Fig. 5.11a), on which the gold particle was placed prior to the nanowire

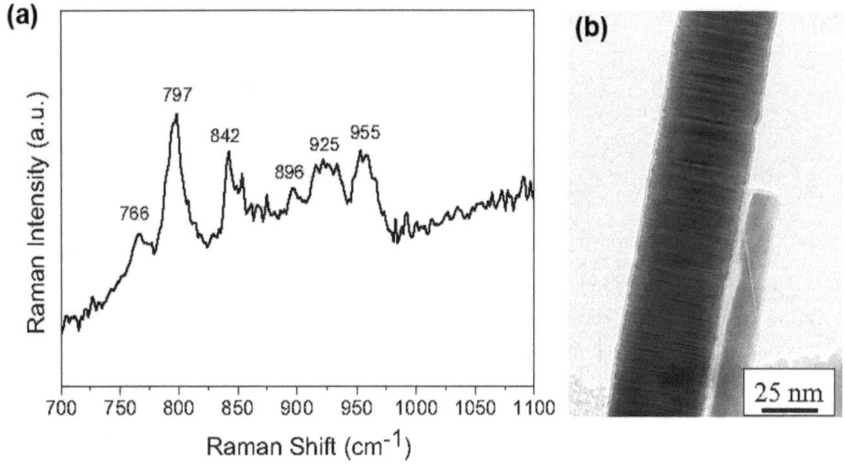

Fig. 5.10 **a** Raman spectrum of a single SiC nanowire with diameter of 40 nm. **b** TEM image of a typical nanowire of the same diameter. The sequence of stacking faults is clearly seen. Reprinted with permission from [22]. Copyright 2007, John Wiley & Sons, Inc

growth. At 950 °C, the gold melts into a droplet of Au–Si alloy. The hydrogen reduction of $SiCl_4$ gives rise to free Si atoms which prefer to dissolve in the alloy droplet to form a supersaturated solution, where the Si whisker starts to grow downward by precipitation of Si atoms. During this process, the Au–Si droplet stays on the tip of the growing Si whisker until Au is consumed or the growth condition is changed. A wide range of cross-sectional dimensions of the Si whiskers can be acquired through this VLS method (Fig. 5.11b).

The SiC nanowhiskers were prepared via arc-discharge between two graphite electrodes, and the anode was filled with a mixed silicon/graphite powder containing Fe catalyst [24]. Transmission electron microscopy reveals two types of structures in the product, rod-like structures and nanosized particles (Fig. 5.12). The rods are about 10 nm in diameter and larger than 100 nm in length, the particles have diameters in the range of 10–20 nm. The rods were only observed when the iron catalyst was used in the arc-discharge reaction. The nanowhiskers have lattice fringes either parallel to or perpendicular to the rod axis, as shown in Fig. 5.13. The rods are surrounded by an amorphous layer. The lattice fringe spacings are 0.243 ± 0.010 nm, suggesting that the rods are single-crystalline 3C–SiC with a preferred <111 > growth orientation. In the electron energy loss spectrum of an individual rod, the Si–L, C–K, and Si–K edges are present and there are no signals of Fe, O, and N. A two-stage vapor–liquid–solid growth mechanism was proposed for the whisker growth. At 1220 °C, a Si/Fe liquid alloy acts as the site at which the whisker begins to grow, at elevated temperatures above 1410 °C (i.e., the melting point of Si), the whisker can grow without the catalyst through solution and precipitation of carbon in the liquid Si tip. For the typical temperature of 2000 °C at the cathode, the remaining iron catalyst will evaporate, which accounts for the absence of Fe in the final whiskers. An arc-discharge caused

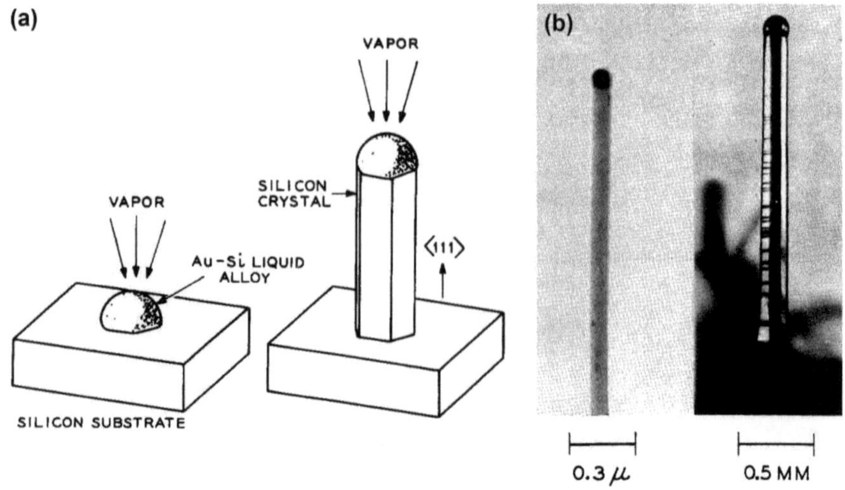

Fig. 5.11 **a** Growth of a Si crystal by vapor–liquid–solid mechanism. (left) Initial state with liquid droplet on the Si substrate. (right) Growing crystal with liquid droplet at the tip. **b** TEM images of the Si wires showing solidified Au–Si alloy at the tip, left is a 100 nm diameter whisker and right is a twelve-sided 200 μm diameter needle. Reprinted with permission from [23]. Copyright 1964, American Institute of Physics

reaction of the SiC rod containing iron impurity as the anode with the quartz tube produced SiC nanorods coated with amorphous SiO_2 [25]. The prepared nanorods have a 3C–SiC crystal core with a uniform diameter of 5–20 nm and an amorphous SiO_2 shell tens of nanometers in thickness, and their lengths range from several hundred nanometers to several micrometers.

Pyrolysis of polysilazane at 1700 °C under catalysis of $FeCl_2$ and under nitrogen gas flow yielded β-SiC nanorods [26]. The synthesized nanorods have diameters of 80–200 nm and lengths of about 4 μm. They preferentially grew along the <111>

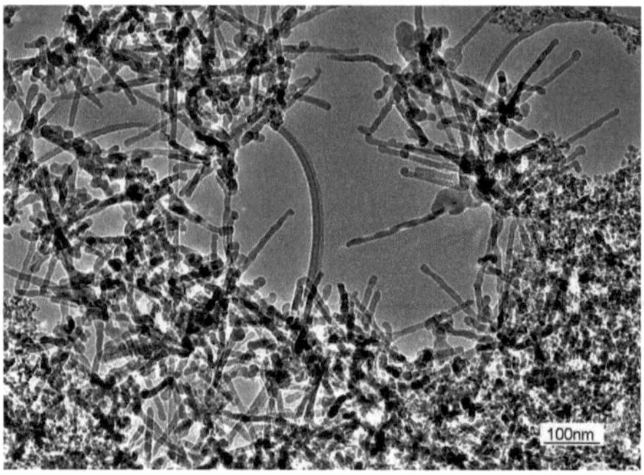

Fig. 5.12 TEM micrograph of SiC whiskers formed by arc-discharge evaporation of a Si/Fe-filled graphite electrode. They are about 10 nm in diameter and more than 100 nm in length. Some small particles can be seen at the bottom. Reprinted with permission from [24]. Copyright 2000, John Wiley & Sons, Inc

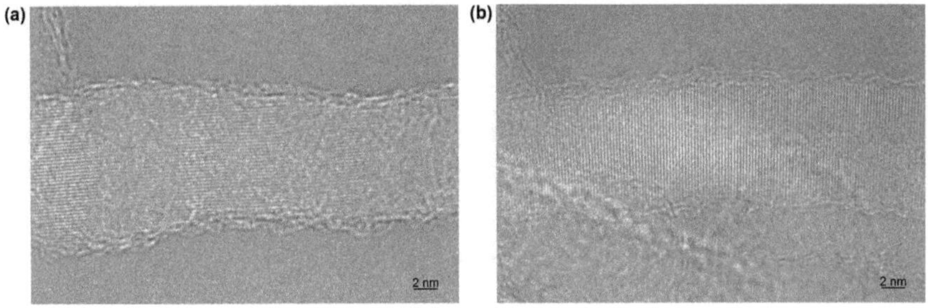

Fig. 5.13 HRTEM images of two SiC nanowhiskers. The lattice fringe of the whisker is oriented parallel (**a**) or perpendicular (**b**) to its axis. The fringe spacing is 0.243 ± 0.010 nm. There is an outer layer of disordered graphite for each whisker. Reprinted with permission from [24]. Copyright 2000, John Wiley & Sons, Inc

direction. A solid–liquid–solid mechanism was proposed to explain the growth of the SiC nanorods. Pyrolysis of the polymer generated an amorphous SiCN phase, which then reacted with Fe to form the liquid Si–Fe–C alloy. However, SiC nanorods did not form at this stage because the Si_3N_4 phase formed first, which is more stable at temperatures up to 1500 °C in the presence of N_2. When the temperature was improved to 1500 °C at which SiC phase is thermodynamically stable, the SiC nanorods began to grow by dissolving of Si_3N_4 and free carbon in the Si–Fe–C liquid alloy. These nanorods showed PL with two strong bands (Fig. 5.14), one is a strong sharp UV line at 378 nm (3.3 eV) and another is a broad green emission band centered at around 510 nm [27]. Stacking faults could be the origin of the UV emission. The time-resolved PL spectrum of the nanorods can be fit by a double-exponential function, giving rise to two time constants of 1.0 and 6.5 ns. The morphology of the SiC nanorods can be controlled by changing the source species pressure during the vapor–liquid–solid growth of the nanowires by pyrolysis of polysilazane [28]. Nanowires with Eiffel-tower shape, spindle shape, and modulated diameters and periods have been synthesized using this technique. The precisely controlled growth of SiC nanowires with a conical shape can be realized by tailoring the cooling rate [29]. The SiC nanowires with needle-like, pyramid-like, and ball-like tips were synthesized by adjusting the cooling rate at the respective temperature. The diverse morphologies of the synthesized SiC nanowires result from different degrees of supersaturation and different alloy droplet sizes caused by different cooling rates. The SiC nanorods with periodically tuned cross-sectional dimensions can also be grown by using alternatively interchanged cooling and heating processes (Fig. 5.15). The periodicity of the obtained nanorods shows consistence with the prescribed pyrolysis scheme.

Carbothermal reduction of sol–gel derived silica by activated carbon, whose mixture was embedded with Fe nanoparticles, produced bulk quantities of 3C–SiC nanowires [30]. The nanowires consist of 10–30 nm diameter single crystalline cores wrapped with amorphous SiO_2 layers and have lengths up to several tens of micrometers. The formation of the 3C–SiC nanowires involves a vapor–liquid–solid

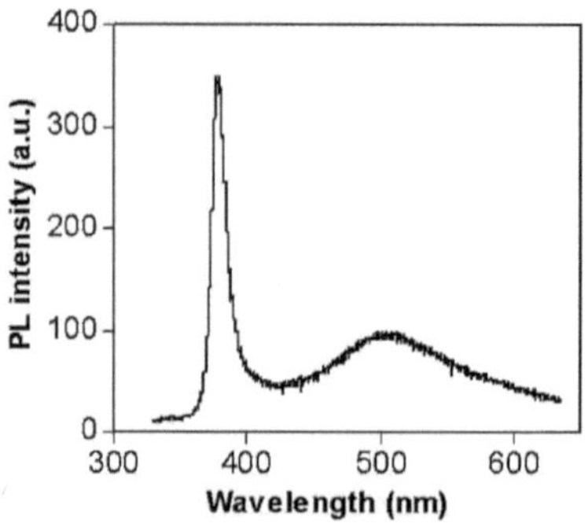

Fig. 5.14 Photoluminescence spectrum of the SiC nanorods under 325 nm excitation. Reprinted with permission from [27]. Copyright 2006, American Institute of Physics

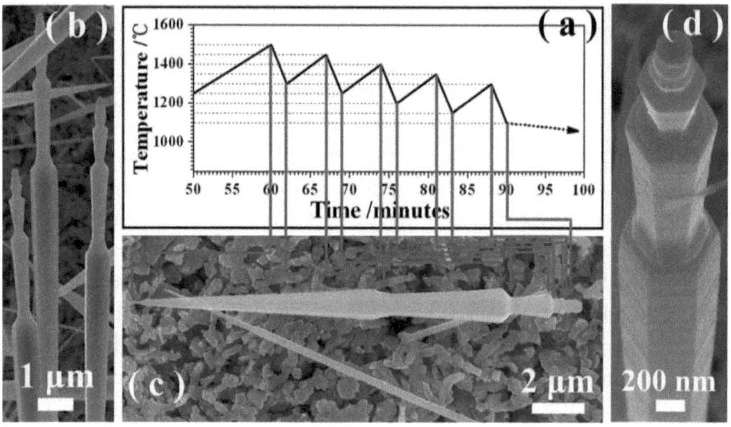

Fig. 5.15 **a** Pyrolysis scheme with an alternatively repeated cooling and heating process (cooling rate: 100 °C min^{-1}, heating rate: 30 °C min^{-1}). **b–d** SiC nanowires with periodic structures. Reprinted with permission from [29]. Copyright 2012, Royal Society of Chemistry

growth mechanism. Two broad photoluminescence peaks at around 340 and 440 nm were observed at room temperature. Note that most of the methods developed for synthesizing SiC nanowires that were called "carbothermal reduction" methods by the investigators actually occupy the vapor–liquid–solid growth mechanism. The carbothermal reduction method usually involves various types of solid silicon and carbon source materials, such as carbonaceous silicon xerogel [31], SiO_2 powders and carbon black [32], amorphous SiO_2 powder and natural nanoporous bamboo [33], as both the renewable carbon source and the template for the formation of catalyst particles.

The SiC nanowires can exhibit different morphologies at different growth stages [34]. The nanowires were synthesized by thermal evaporation and reaction of the mixed SiO_2 and C powders under catalysis of iron at 1500 °C. The nanowires grew by

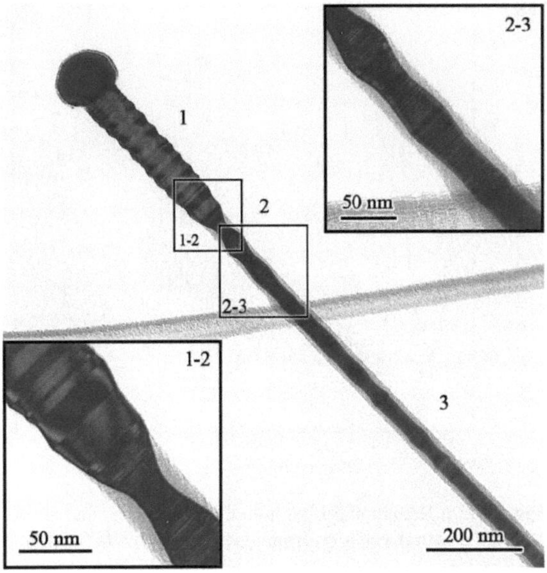

Fig. 5.16 TEM image of a SiC nanowire showing three distinct growth stages. Reprinted with permission from [34]. Copyright 2007, American Institute of Physics

a vapor–liquid–solid process and they showed different types of structures during the successive growth stages (Fig. 5.16). In the first stage, the SiC nanowires exhibited as periodically twinned crystals. In the second stage, they took the form of core–shell structures. The core is crystalline and its cross-sectional area varies with position along the nanowire growth direction. The shell is amorphous. In the third stage, SiC nanowires still exhibited core–shell structures, but the core diameter became uniform.

SiC nanowires can be synthesized by reaction of gas and solid source materials at elevated temperatures. A reduction-carburization method was employed to synthesize β-SiC nanowires at 700 °C by taking Si powder and tetrachloride (CCl_4) as source materials and Na as the reductant [35]. The nanowires have diameters of 15–20 nm and lengths of 5–10 μm. They grew by the vapor–liquid–solid mechanism. The Raman spectrum (Fig. 5.17a) comprises a strong β-SiC TO phonon mode at 776 cm^{-1} and a very weak LO phonon mode at 954 cm^{-1}. Both Raman lines shift to lower wavenumbers by 18–20 cm^{-1} with respect to bulk values due to the small nanowire diameter and the stacking faults. The PL spectrum (Fig. 5.17b) comprises a single peak at around 448 nm.

The synthesized SiC nanowires usually belong to 3C polytype, but few SiC nanowires show photoluminescence characteristics that follow the quantum confinement effect in 3C–SiC. The observed PL spectra in the SiC nanowires generally have significant blue shift relative to the bandgap of bulk 3C–SiC (~2.2 eV at room temperature). At first glance, this behavior seems to be an indication of quantum confinement in that the quantum confinement can result in a larger energy gap in the SiC nanowire than in bulk material. However, most of the synthesized SiC nanowires

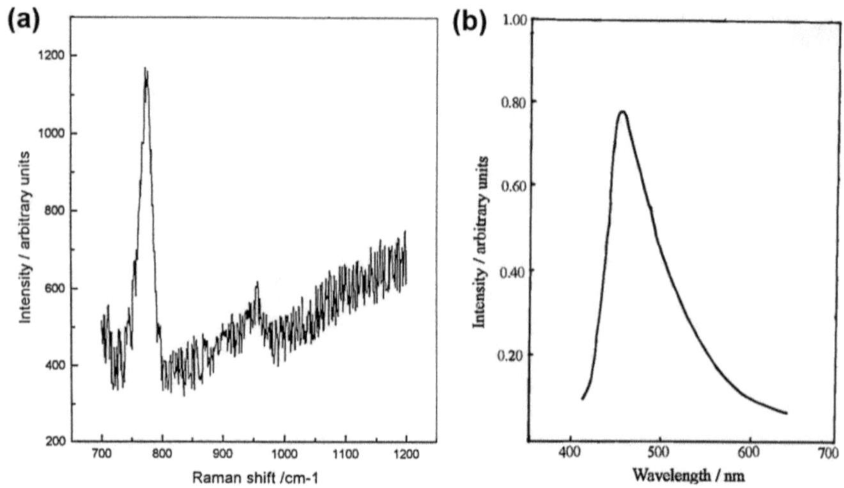

Fig. 5.17 **a** Raman and **b** photoluminescence (excited at 385 nm) spectra of the SiC nanowires. Reprinted with permission from [35]. Copyright 2000, American Chemical Society

have diameters over 10 nm, and the quantum confinement effect in this size regime is rather weak, so their energy gaps are expected to be close to the bulk bandgap. Therefore, the usually observed PL in the SiC nanowires with significant blue shift can hard be explained by the quantum confinement effect. Instead, they may originate from some defects including surface defects, interior defects, and defects in the outer silicon oxide shell.

Occasionally, the synthesized SiC nanowires have photoluminescence with peaks close to the bulk bandgap [36]. In that study the 3C–SiC nanowhiskers with <111> axial orientation were synthesized on a Fe-deposited Si substrate in a microwave plasma CVD system in hydrogen and methane atmosphere. The SiC whiskers have diameters between 5 and 38.5 nm with a most probable diameter of 15.4 nm. Figure 5.18 shows the photoluminescence spectra of the nanowhiskers synthesized at different temperatures. The dominant emission band has a peak between 500 and 600 nm, close to the bulk bandgap value of 3C–SiC. There is another weak emission band centered at around 800 nm. The PL intensity increased with the substrate temperature and the Fe film thickness due to the increase in nanowire quantity as demonstrated by the SEM observation.

Direct or indirect CVD has been utilized to synthesize SiC nanowires that usually grow by the vapor–liquid–solid mechanism. A one-step procedure has been developed to synthesize β-SiC nanorods from the solid carbon and silicon sources on a Si substrate by hot filament CVD at 1000 °C with catalysis of impurity metal particles [37]. The atomic hydrogen etches the solid source materials, giving rise to gaseous hydrocarbon radicals (CH_x), hydrosilicon radicals (SiH_x), or organosilicon molecules. These vapor molecules dissociate into Si and C atoms and dissolve in the liquid metal catalyst, leading to growth of SiC nanowires upon supersaturation. The

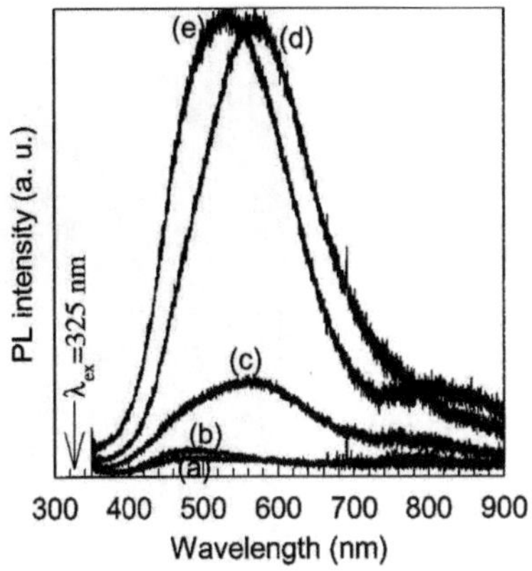

Fig. 5.18 Photoluminescence spectra of several SiC nanowhisker samples synthesized at (a) 800 °C, (b) 850 °C, (c) 900 °C, (d) 950 °C, and (e) 1000 °C, respectively, with the same thickness of Fe film on the Si substrate. The excitation wavelength is 325 nm. Reprinted with permission from [36]. Copyright 2002, American Institute of Physics

nanorods contain a crystalline β-SiC core and an amorphous silicon oxide shell and grew along the <100> direction (Fig. 5.19). The nanowire diameters are between 10–30 nm and the lengths are less than 1 μm. SiC nanowires can also be synthesized at lower temperatures but under higher pressures. A plate made of a C–Si–SiO_2 powder mixture can also be used as carbon and silicon sources for the CVD growth of the SiC nanowires [38]. Hydrogen acts both as a reactant and as a mass transporting medium. 3C–SiC nanorods were synthesized through a one-step reaction at 400 °C under high pressure [39]. The reaction was carried out in an autoclave using $SiCl_4$ and CCl_4 as reactants and metal Na as coreductant. The nanorods have diameters of 10–40 nm and lengths up to several micrometers. The pressure in the autoclave resulting from vaporization of $SiCl_4$ and CCl_4 reaches about 9 MPa and as a result, the crystalline SiC can form at relatively low temperatures. This reaction is described as:

$$SiCl_4 + CCl_4 + 8Na \rightarrow SiC + 8NaCl. \tag{5.5}$$

The reaction between $SiCl_4$ (or Si powders), C_6Cl_6 (or carbon powder), and sodium can produce various β-SiC nanostructures at different temperatures, including hollow nanospheres, nanowires, and coaxial nanowires [40].

Silicon substrate can accommodate growth of highly aligned 3C–SiC nanowires. Aligned SiC nanowires were synthesized through a reaction of the Si substrates and a methane–hydrogen mixture at 1100 °C [41]. The catalysts are Ga metal, GaN powder, and Fe nanoparticles deposited on the Si substrate. The nanowires have an average diameter of 40 nm and an average length of 500 μm, they grew along the <111> direction. Highly aligned SiC nanowires have been synthesized by using CVD by taking methyltrichlorosilane (CH_3SiCl_3) as the source precursor under hydrogen

Fig. 5.19 HRTEM image of a β-SiC nanorod. Reprinted with permission from [37]. Copyright 1999, American Institute of Physics

flow at 950 °C [42]. Ni-deposited Si wafer was taken as the substrate. The nanowires have diameters smaller than 50 nm and lengths up to several micrometers (Fig. 5.20). X-ray diffraction and transmission electron microscopy reveal that the nanowires are single crystalline 3C–SiC, with a preferential growth direction of <111> . They exhibited photoluminescence with a wide emission band centered at around 420 nm under 325-nm excitation.

SiC powder or bulk material can also be used to synthesize SiC nanowires by employing metal-catalyzed vapor–liquid–solid process. Laser ablation of a SiC target at 900 °C yielded SiC nanowires with an average diameter of 80 nm and lengths of tens of micrometers [43]. High-resolution TEM observation reveals that the nanowires contain a high density of stacking faults and are wrapped by a silicon oxide layer with thicknesses of ~ 17 nm. The grown mechanism is a typical vapor–liquid–solid mechanism, in which nano-sized iron particles act as catalyst at the tip of each SiC nanowire. The Raman scattering shows that the TO and LO phonon modes of the SiC

Fig. 5.20 SEM image of vertically aligned SiC nanowires grown on a Si substrate at 1100 °C under hydrogen flow. The lower inset is the TEM image of a SiC nanowire. Reprinted with permission from [42]. Copyright 2004, Elsevier B.V

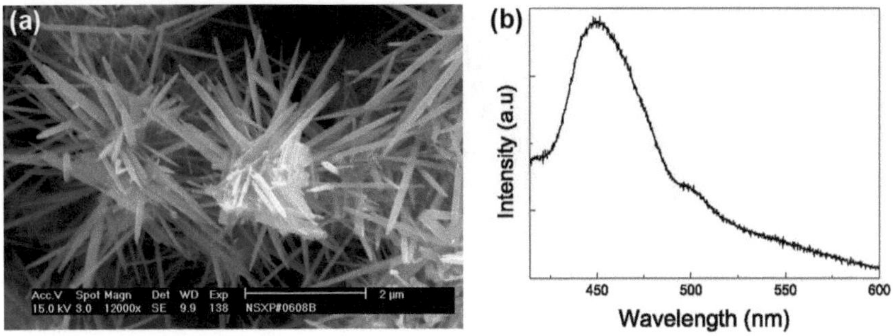

Fig. 5.21 **a** SEM micrograph of bunches of SiC nanowires. **b** PL spectrum of the SiC nanowires under 400 nm excitation at room temperature. Reprinted with permission from [46]. Copyright 2003, Elsevier B.V

nanowires shift to lower wavenumbers by 12 and 34 cm^{-1} relative to bulk values. The peak shift, band broadening, and asymmetry of the Raman band were explained by the size confinement effect in both radial and growth directions. The needle-shaped 3C–SiC nanowires were synthesized by thermal evaporation of SiC powder with iron or Al as a catalyst in flowing argon atmosphere at 1700 °C [44–46]. These bunched SiC nanowires have diameters of 20–50 nm and lengths of 1–2 μm (Fig. 5.21) [46]. They showed photoluminescence with a broad band and a dominant peak centered at 450 nm under 400-nm excitation. The full width at half maximum of the dominant peak is 60 nm. There is another weak emission peak at around 500 nm.

The SiC nanowires can be grown by a catalyst-assisted sublimation method [47]. It involves microwave heating-assisted physical vapor transport from a source 4H–SiC wafer to a nearby substrate 4H–SiC wafer in nitrogen atmosphere at 1650–1750 °C. The 4H–SiC substrate was coated with a 5-nm thickness metal (Fe, Ni, Pd, Pt) catalyst film prior to the growth. The nanowires grew by the vapor–liquid–solid mechanism. They are 15–300 nm in diameter and 10–30 μm in length (Fig. 5.22a–c). Electron backscatter diffraction and selected-area electron diffraction confirm the 3C–SiC structure of the nanowires. The nanowire caps were identified as Fe_2Si, Ni_3Si, Pd_2Si, and PtSi phases for the nanowires grown with Fe, Ni, Pd, and Pt catalysts, respectively. The nanowires grew along the <112> orientation, different from the commonly observed <111> direction. The micro-Raman spectrum (Fig. 5.22d) from single nanowires indicates regions with varying compressive strain in the nanowires and shows modes not arising from the Brillouin zone center, possibly related to defects.

Combustion synthesis of a mixture containing Si-containing compounds (Si elemental, FeSi, $CaSi_2$) and halocarbons [poly(tetrafluoroethylene), $C_{12}F_{10}$,C_2Cl_6, C_6Cl_6] in a calorimetric bomb produced cubic phase silicon carbide nanofibers [48]. The combustion process produced a huge amount of heat, resulting in high temperatures, under this condition the reactants decomposed to form gaseous radical species for the synthesis. The synthesized SiC nanofibers are 20–100 nm in diameter. Raman scattering shows that the product is composed of mostly cubic polytype of SiC. The

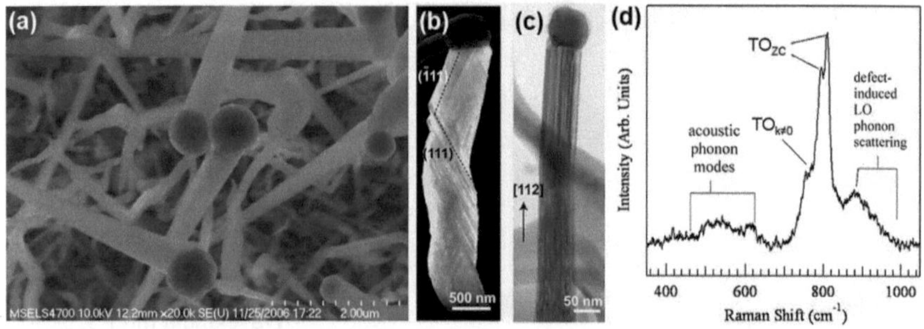

Fig. 5.22 **a** Cone-shaped 3C–SiC nanowires with metal-Si compound tips. **b** and **c** Diffraction contrast TEM images of two types of 3C–SiC nanowires with metal-Si compound tips. **b** Twin-like defects were observed on different sets of {111} planes. The twin structure was confirmed by selected-area electron diffraction pattern. **c** Planar defects parallel to {111} planes along the wire axis. **d** Micro-Raman spectrum from an isolated SiC nanowire. TO_{ZC} refers to TO phonon at the Brillouin zone center. $TO_{k\neq0}$ refers to TO phonon from non-zone center ($k \neq 0$) points, whose scattering is forbidden in a perfect single crystal but becomes active in the presence of defects which destroy translation symmetry. Reprinted with permission from [47]. Copyright 2007, American Chemical Society

combustion product contains also some uniform carbon nanoparticles with diameters of 20–30 nm. Cathodoluminescence in these nanowires at temperatures of 77–300 K has been investigated [49]. As shown in Fig. 5.23a, the bulk 3C–SiC has a strong emission band at 2.38 eV (at 77 K), which may be the exciton recombination band. There is another weaker shoulder at around 1.9 eV. The 2.38 eV band vanishes in the spectrum of the nanowires and the luminescence is dominated by a single strong band centered at 1.9 eV. This emission band diminished with increasing temperature (Fig. 5.23b). The inset shows the triple-Gaussian decomposition of the spectrum measured at 290 K. The three components are located at around 1.98, 2.30, and 2.77 eV, respectively. The 1.9-eV emission was proposed to arise from some defects in the nanowires since the photon energy is smaller than the bandgap of bulk 3C–SiC.

With the extensive studies of SiC nanowires, more and more fabrication methods have been developed. To date, large scale of SiC nanowires can be readily synthesized by using many methods. For example, large areas of centimeter-long SiC nanowires were synthesized by pyrolysis of a polymer precursor (polysilacarbosilane) with ferrocene as the catalyst by CVD at 1300 °C [50]. These nanowires have diameters of 100–200 nm and lengths of several centimeters (Fig. 5.24), they are single-crystal β-SiC growing along the <111> direction by the vapor–liquid–solid mechanism. X-ray diffraction pattern of the nanowires consists of several strong and sharp peaks, indicating good crystallinity of the nanowires, whose positions (35.7°, 41.5°, 60.1°, 71.8°, and 75.5°) and intensities agree well with those of β-SiC (111), (200), (220), (311), and (222) planes, respectively. This confirms that β-SiC is the main crystalline phase of the nanowires. The Raman spectrum comprises two sharp peaks at 791.7 and 966.9 cm^{-1}, corresponding to the TO and LO phonon modes at the Γ point of β-SiC, but with 4.3 and 5.1 cm^{-1} shifts to low wavenumbers, respectively.

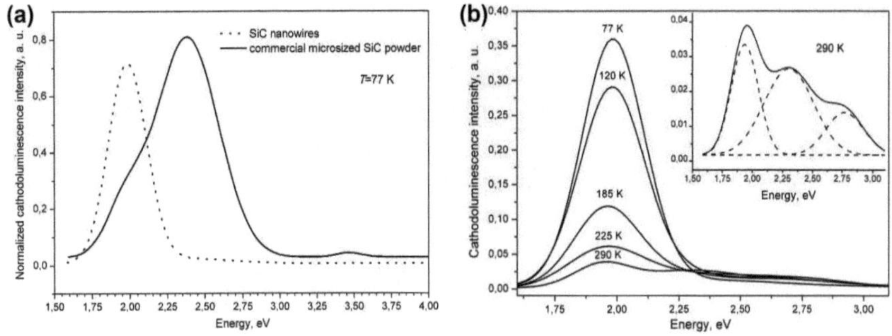

Fig. 5.23 **a** Normalized cathodoluminescence spectra of bulk 3C–SiC and SiC nanowires. **b** Cathodoluminescence spectra of SiC nanowires at different temperatures. The inset shows the triple-Gaussian fit of the spectrum measured at 290 K. Reprinted with permission from [49]. Copyright 2009, John Wiley & Sons, Inc

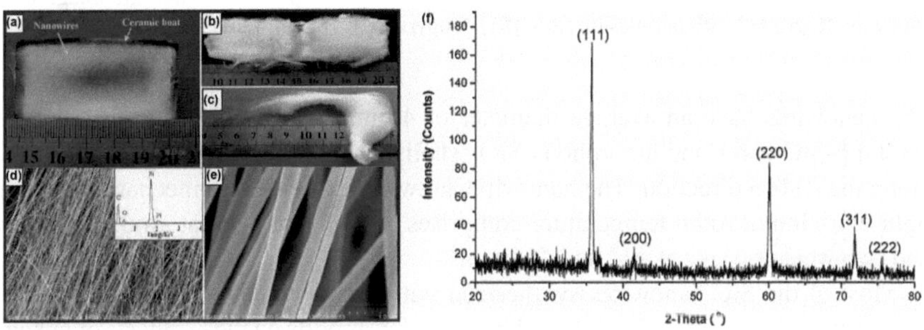

Fig. 5.24 **a** SiC nanowires with lengths of about 1 cm grown vertically on the ceramic wall. **b** Cotton-like products covering the outside back of the ceramic boat. **c** A bundle of fibers with lengths up to several centimeters. The ruler unit is centimeter. **d** and **e** SEM images of the SiC nanowires. The inset of **d** shows the energy-dispersive X-ray spectrum. **f** XRD pattern of the SiC nanowires. The peak positions and intensities agree well with those of β-SiC. Reprinted with permission from [50]. Copyright 2009, American Chemical Society

The SiOC nanocomposites were adopted to synthesize large-scale SiC nanowires [51]. The mixture of silica sol and sucrose powder was consecutively heated at 90 °C in air and 850 °C in argon atmosphere to give rise to a black powder of SiOC nanocomposites. These nanocomposites were heated up to 1500 °C in argon atmosphere to produce a white cotton-like product (Fig. 5.25) that is 12 cm in diameter and 1–2 cm in thickness. Scanning electron microscopy reveals that the cotton-like sample contains a large amount of smooth and straight nanowires with uniform diameters of about 100 nm. High-resolution TEM observation reveals that the nanowires are β-SiC single crystals which preferentially grew along the <111> direction.

Large-scale centimeter-long single-crystal β-SiC nanowires were prepared using CH_4 as the carbon source and SiO or the mixture of Si and SiO_2 as the silicon source by using catalyst-free CVD under super-atmospheric pressure at 1350–1600 °C [52].

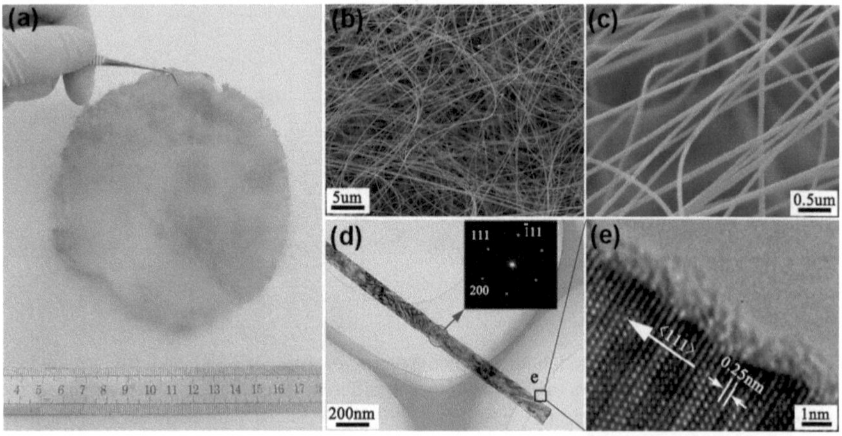

Fig. 5.25 **a** Cotton-like bundles of SiC nanowires. **b** and **c** SEM images of the nanowires. **d** TEM image of one nanowire and corresponding electron diffraction pattern. **e** HRTEM image of a nanowire. Reprinted with permission from [51]. Copyright 2010, IOP Publishing

The nanowires have an average diameter of 40 nm (Fig. 5.26). They have single crystal β-SiC cores and amorphous SiO_2 shells of 1–30 nm in thickness and grew along the <111> direction. The nanowires grew by a vapor–solid mechanism. Their light emission at room temperature comprises a dominant peak at around 460 nm and two weaker peaks at 442 and 418 nm.

Most of the SiC nanowires synthesized with various methods have cubic zinc-blende crystal structures. There have been very few reports on the synthesis of hexagonal polytypes of SiC nanowires. 2H–SiC nanowhiskers were synthesized by reaction of SiO with methane at 1300 °C [53]. The whiskers have diameters below 200 nm and lengths over 2 μm (Fig. 5.27a and b). Electron diffraction and high-resolution TEM observation indicate that the nanowhisker is 2H polytype except for a 3C–SiC tip. These nanowhiskers grew along the [0001] orientation and they contain many stacking faults and micro-twins. Their Raman spectrum consists of a strong peak at 796.6 cm^{-1} and a weak peak at 763 cm^{-1} (Fig. 5.27c). Since generally there is only one TO phonon peak at 791 cm^{-1} for 3C–SiC whereas there are two TO phonon peaks at 764 and 799 cm^{-1} for 2H–SiC, thus the observed Raman signals were assigned to 2H–SiC. No catalyst appeared on the tip of the nanowhisker and an oxide-assisted growth mechanism was thus proposed, with the reaction:

$$SiO(g) + CH_4(g) \rightarrow SiC(s) + SiO_2(s) + 2H_2(g). \quad (5.6)$$

There is a report on the synthesis of 2H–SiC nanoflakes [54]. These nanoflakes were synthesized through a one-step solvothermal route involving reaction of $SiCl_4$ and CaC_2 in an autoclave at 180 °C. The reaction was suggested to be:

$$2CaC_2 + SiCl_4 \rightarrow SiC + 2CaCl_2 + 3C. \quad (5.7)$$

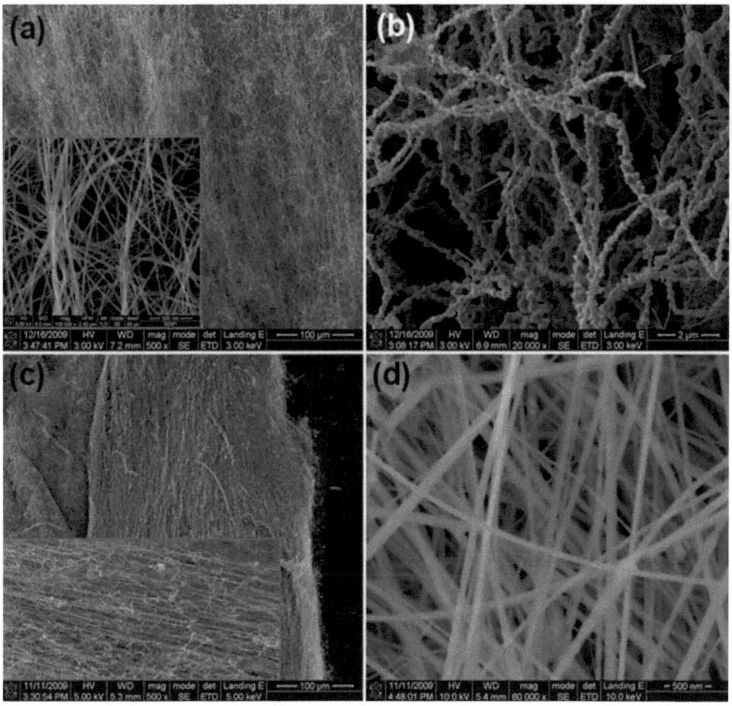

Fig. 5.26 **a** and **b** SEM images of SiC nanowires. **c** and **d** SEM images of the SiC nanowires collected from another area in the ceramic boat. Reprinted with permission from [52]. Copyright 2011, Royal Society of Chemistry

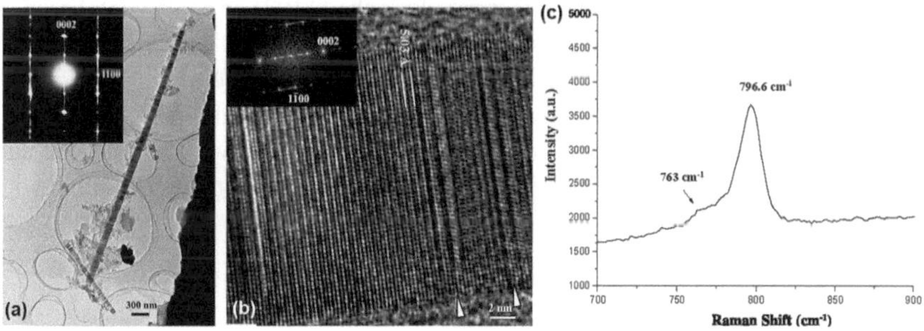

Fig. 5.27 **a** TEM image of a typical SiC nanowhisker. The inset shows the electron diffraction pattern indicating the 2H–SiC polytype. **b** HRTEM image of a 2H–SiC nanowhisker (inset is the electron diffraction pattern). **c** Raman spectrum of the SiC nanowhiskers. Reprinted with permission from [53]. Copyright 2003, Elsevier B.V

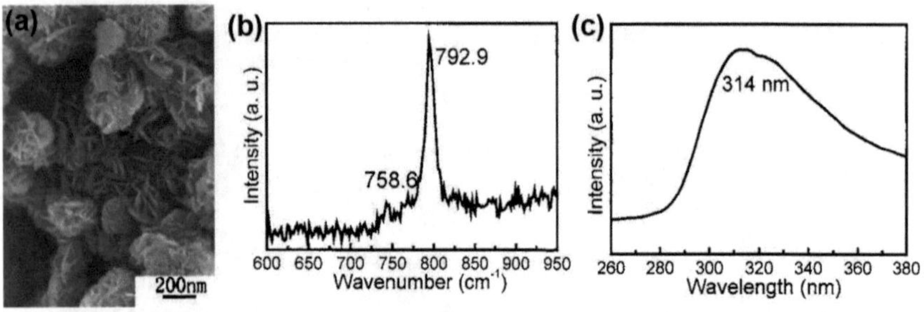

Fig. 5.28 SEM image (**a**), Raman spectrum (**b**), and photoluminescence spectrum (**c**) (under 325 nm excitation) of the 2H–SiC nanoflakes. Reprinted with permission from [54]. Copyright 2006, American Institute of Physics

The X-ray diffraction as well as infrared and Raman spectral analysis confirms that the nanoflakes have a crystalline structure of 2H–SiC. They are 200–500 nm in diameter and 15 nm in thickness (Fig. 5.28) and grew along the [001] direction. The Raman spectrum consists of a strong peak at 792.9 cm^{-1} and a weak shoulder at 758.6 cm^{-1}, characteristic of 2H–SiC, Their infrared spectrum comprises a strong and wide peak centered at 820 cm^{-1} that was assigned to the optical phonon mode of SiC. The nanoflakes exhibited rare photoluminescence centered at 314 nm, possibly arising from the defects in the surface silicon oxide layer.

The catalyst-assisted pyrolysis of polyaluminasilazane (or polysilazane) precursor at 1350–1450 °C produced pure or Al-doped 6H-SiC nanowires [55–57]. The catalysts are Fe, $FeCl_2$, or $Fe(NO_3)_3$ powder. Using this vapor–liquid–solid growth approach, the highly oriented SiC nanowire arrays were grown on the 6H-SiC single crystal substrate [56]. The nanowires grew along the [$\bar{1}$102] or equivalent direction on 6H-SiC (0001) substrates and along the [10$\bar{1}$0] or equivalent direction on 6H-SiC (10$\bar{1}$0) and (11$\bar{2}$0) substrates (Fig. 5.29). This method can yield highly aligned SiC nanowire arrays. This pyrolysis method also yielded 6H-SiC nanowires with a triangular prism shape [57]. The nanowires have an average edge width of 700 nm and lengths up to tens of micrometers (Fig. 5.30). The wires are single-crystalline with 1 at% Al dopant and grew along the <102> direction.

5.2 Field Electron Emission

The field electron emission properties of silicon carbide nanowires have been extensively studied. In 1928, Fowler and Nordheim calculated the electron emission of cold metal in intensive electric fields using the potential barrier tunneling theory of quantum mechanics [59]. They derived an equation that has been widely used in the study of field electron emissions of SiC nanowires:

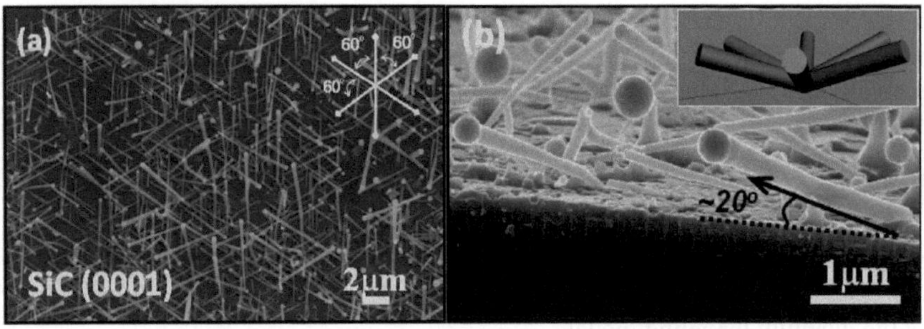

Fig. 5.29 **a** Top-view SEM image of SiC nanowires grown on a 6H-SiC (0001) substrate showing the preferred orientation with six growth directions. **b** Side-view SEM image suggesting the oblique growth of the nanowires on the substrate with an angle of 20° as illustrated by the model in the inset. Reprinted with permission from [56]. Copyright 2010, American Chemical Society

Fig. 5.30 **a** SEM image of the 6H-SiC nanowires with prism-like shape. **b–d** Magnification images of the nanowires. Reprinted with permission from [57]. Copyright 2012, Royal Society of Chemistry

$$J \propto A \frac{\mu^{\frac{1}{2}}}{(\chi + \mu)\chi^{\frac{1}{2}}} E^2 e^{-\frac{B\chi^{\frac{3}{2}}}{E}}, \tag{5.8}$$

in which J is the electron current density at the emitter, E is the local electric field at the emitter which could be different from the applied field due to the field enhancement effect of the emitter, A and B are constants, χ is the thermionic work function, and μ is the usual parameter of the electron distribution in the Fermi–Dirac statistics that is equivalent to the thermodynamic partial potential of an electron. χ and μ

are dependent on the material and geometric characteristics of the emitter. When researchers study the field electron emission properties of the SiC nanowires, they generally use a relationship derived from Eq. (5.8):

$$\ln \frac{J}{E^2} = C - D\frac{1}{E}, \tag{5.9}$$

where C and D are two constants. Therefore, the ln (J/E^2) versus $1/E$ curve is subjected to exhibit as a straight line if the considered electron emission is governed by the quantum tunneling model.

The field emission properties of the SiC nanowires grown on a Si substrate have been investigated [60]. The nanowires were synthesized by using hot-filament-assisted CVD with solid silicon and carbon sources. Scanning electron microscopy reveals that the SiC nanowires are randomly aligned and have diameters of around 20 nm. The electron field emission of the SiC nanowires was measured in a vacuum chamber with a pressure of 5×10^{-8} Torr (Fig. 5.31). Figure 5.32a shows the emission current–voltage (I–V) curve of the SiC nanowire film at an anode–sample separation of 20–120 μm. These I–V curves can be roughly fit by an exponential function. The estimated turn-on field for a current density of 0.01 mA cm^{-2} is 20 V μm^{-1} and comparable to that of many other field emitters such as diamond films and carbon nanotubes. The lowest emission current density for actual applications of flat-panel displays is about 10 mA cm^{-2}. The threshold field to give rise to this current density is 30 V μm^{-1} for the SiC nanowire film. Figure 5.32b displays the Fowler–Nordheim plot of ln(J/E^2) versus $1/E$ for the SiC nanowire film at different anode–sample separations. These roughly straight lines indicate that the observed currents do originate from the quantum process of electron barrier-tunneling. The turn-on field was reduced to 13–17 V μm^{-1} in the β-SiC nanowires with diameters of 5–20 nm and lengths of 1 μm [61] and to 9 V μm^{-1} in those with diameters of 20–70 nm and lengths of 1 μm [62]. A lower threshold field of 6.2 V μm^{-1} was observed from the SiC nanowires fabricated using detonation soot powders and Si wafers [63]. The nanowires consist of single-crystalline 3C–SiC cores with diameters of 30–100 nm and lengths of 0.5–1.5 μm, which are wrapped by very thin amorphous oxide layers.

Highly aligned SiC nanowire array possesses superior field emission performance [64]. The nanowire array was fabricated by reaction of SiO with freestanding highly aligned carbon nanotube templates at 1400 °C. The SEM images (Fig. 5.33) show that the nanowires have similar diameters (10–40 nm), spacing values (about 100 nm), and lengths (up to 2 mm) to those of the aligned carbon nanotubes. The top end of the nanowire array was covered with a layer of the iron/silica nanoparticles resulting from those particles present on the top end of the initial nanotube array. The bottom end of the SiC nanowire array (Fig. 5.33b) is composed of highly oriented and well-separated SiC nanowire tips with a very high density of 10^9–10^{10} cm^{-2}. The electron diffraction pattern reveals that the nanowires are β-SiC crystals growing along the <111> direction. Figure 5.34 shows the field emission current density versus applied electric field curve of the nanowire array. The turn-on field corresponding to an emission current density 10 μA cm^{-2} is 0.7–1.5 V μm^{-1}, the threshold field corresponding

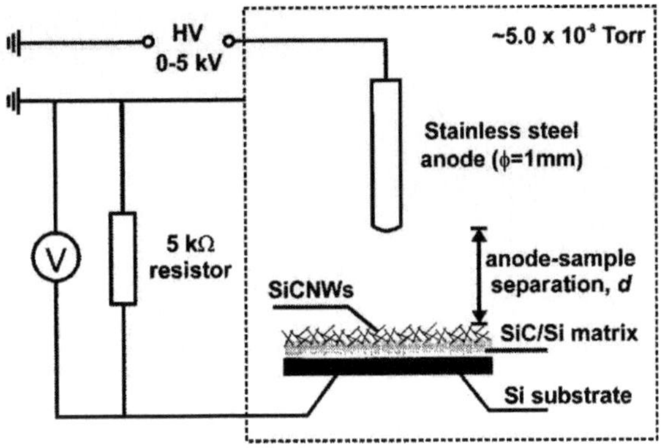

Fig. 5.31 Schematic diagram showing the vacuum chamber setup for measurement of field electron emission of the SiC nanowires. Reprinted with permission from [60]. Copyright 1999, American Institute of Physics

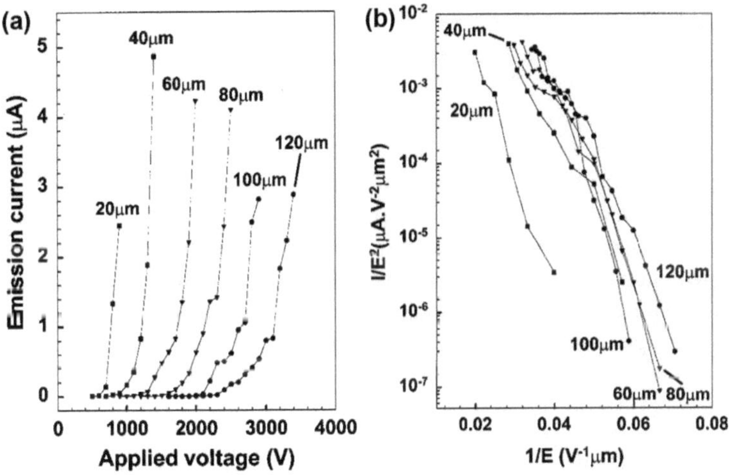

Fig. 5.32 **a** Emission current versus applied voltage for the SiC nanowire film at different anode–sample separations. **b** Fowler–Nordheim plots of the SiC nanowire film at different anode–sample separations. Reprinted with permission from [60]. Copyright 1999, American Institute of Physics

to an emission current density 10 mA cm^{-2} is 2.5–3.5 V μm^{-1}, which are among the lowest (turn-on or threshold) electric fields for all reported field-emitting materials, indicating the promise of the oriented SiC nanowires for practical applications in vacuum microelectronic devices. The nearly linear relationship between $\ln(J/E^2)$ and $1/E$ (inset of Fig. 5.34) clearly demonstrates the field electron emission origin of the measured currents. The field electron emission is very stable, the current fluctuation is only ± 3% during 24 h of continuous operation at 5 mA/cm^2, and there is no significant change in the morphology of the nanowire array during its 24 h continuous

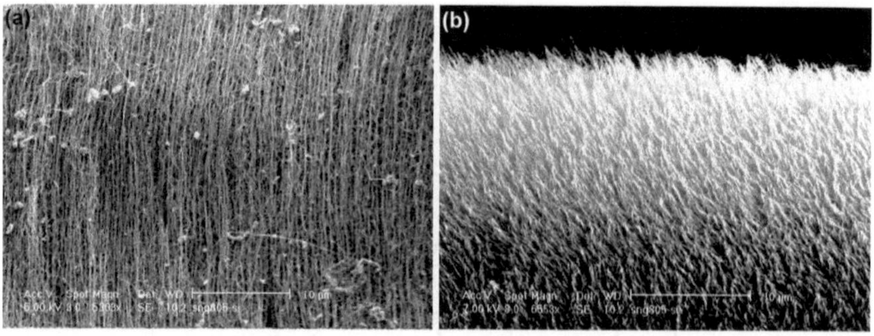

Fig. 5.33 **a** SEM image of oriented SiC nanowire array. The wires are highly aligned and have uniform diameters and interspacing values. There are some Si-containing particles on the array surface formed during reaction. **b** Bottom structure of the SiC nanowire array showing high density, highly oriented, and well-spaced nanowire tips. Reprinted with permission from [64]. Copyright 2000, John Wiley & Sons, Inc

emission. Highly oriented SiC porous nanowire arrays on a Si substrate obtained via in situ carbonizing aligned Si nanowire arrays on the Si substrate exhibited also excellent field emission [65]. The achieved SiC NW arrays inherited the diameters and lengths of the mother Si nanowire arrays. They possess a large field emission current density at a very low electric field. The recorded field emission current density maintained at 0.57 mA cm^{-2} for 1200 min at an applied voltage of 3000 V. No obvious degradation of current density was observed over this period. The SiC nanowire density also affects the field emission performance, as demonstrated by a study of the 3C–SiC nanowires oriented along the <111> direction [66]. The nanowires grew on a Si substrate by thermal evaporation of WO_3 and graphite powder under NiO catalysis at high temperature. The NiO catalyst concentration tuned the density of the nanowires. The field emission efficiency depends strongly on the density of the SiC nanowires. The lowest turn-on field of 1.8 V μm^{-1} and highest field enhancement factor of 5.9×10^3 were observed for the medium-density SiC nanowire sample.

Bunches of needle-shaped silicon carbide nanowires exhibited good field electron emission performance [44]. The needle-shaped nanowires were synthesized by thermal evaporation of SiC powders with ion catalysis. They have diameters of 20–50 nm and lengths of 1–2 μm (Fig. 5.35a). The needle shape of the nanowires renders them ideal field electron emitters. The turn-on field and threshold field which give rise to an emission current density of 10 μA cm^{-2} and 10 mA cm^{-2} are 5 and 8.5 V μm^{-1}, respectively (Fig. 5.35b). The stable-emission current density of 30.8 mA cm^{-2} was acquired at an electric field of 9.6 V μm^{-1}, and the maximum emission current density is 83 mA cm^{-2}. The good linearity of the curve in the Fowler–Nordheim plot (Fig. 5.35c) demonstrates the field emission nature of the nanowires. The current fluctuation is within ± 4% during 100 h continuous emission at a current of 75 μA that corresponds to a current density of 31 mA cm^{-2}. The morphology of the sample has no obvious change after the operation.

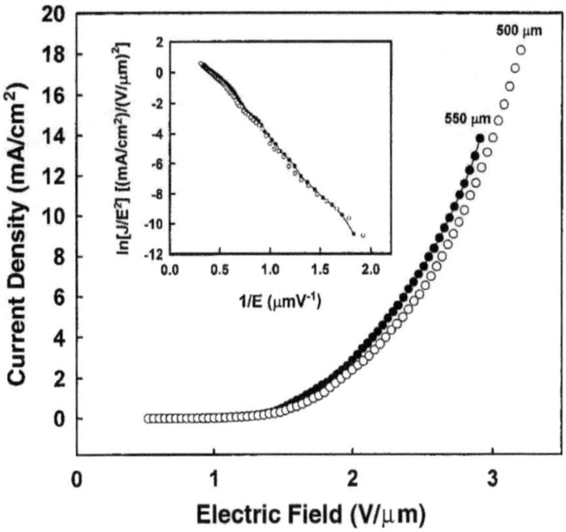

Fig. 5.34 Plot of emission current density versus applied electric field for oriented SiC nanowire film with an emitting area of 3.65 mm^2. The average turn-on field and threshold field are 0.9 and 2.7 V mm^{-1}, respectively. The inset is the Fowler–Nordheim plot whose linearity indicates the field emission nature. Reprinted with permission from [64]. Copyright 2000, John Wiley & Sons, Inc

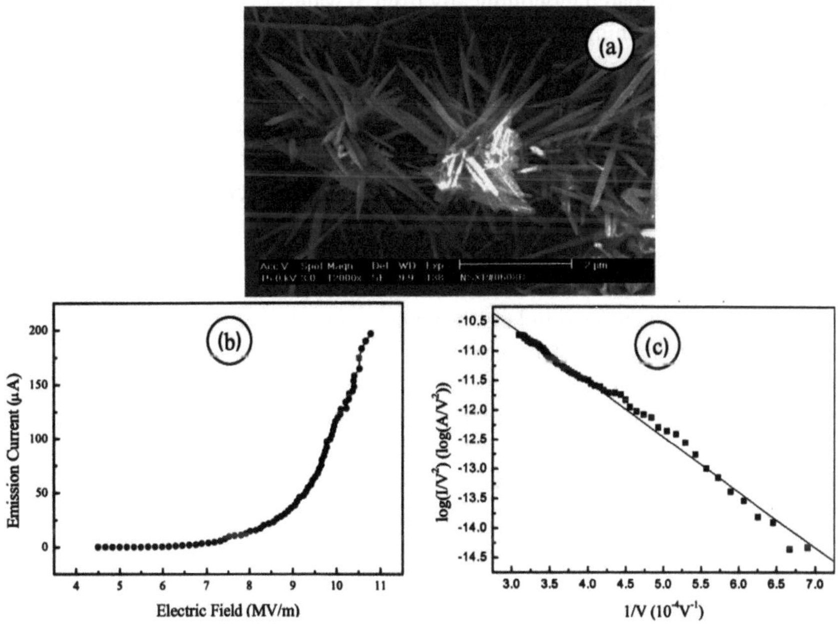

Fig. 5.35 **a** SEM image of bunches of needle-shaped SiC nanowires produced by heating SiC particles in flowing argon atmosphere at 1700 °C. **b** Field electron emission current versus applied electric field for the nanowires. **c** Corresponding Fowler–Nordheim plot. Reprinted with permission from [44]. Copyright 2002, American Institute of Physics

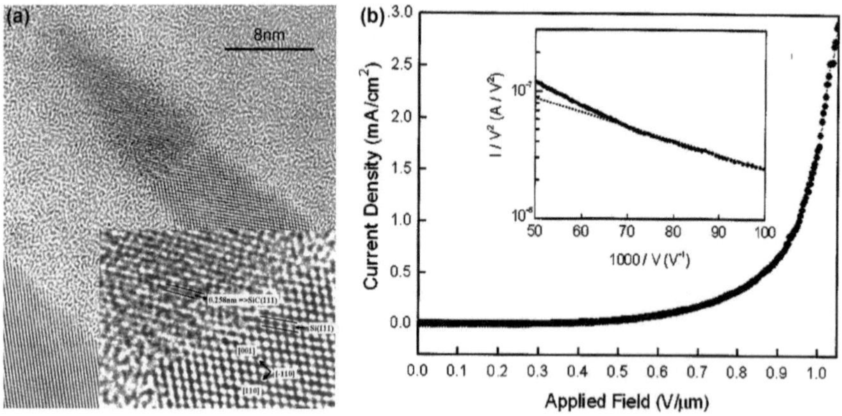

Fig. 5.36 **a** Cross-section TEM micrograph of a SiC-capped Si nanotip fabricated by dry etching of *p*-type Si (100) substrate. The inset shows magnified lattice image at the interface between Si and SiC. **b** Field emission data of the SiC-capped Si nanotips. The inset shows the Fowler–Nordheim plot and the dashed line represents a linear relationship for an ideal field emitter. Reprinted with permission from [67]. Copyright 2003, American Institute of Physics

The SiC–Si composite nanotips showed good field emission properties [67]. The Si nanotips were prepared monolithically from Si wafers by using electron cyclotron resonance plasma etching technique at 200 °C. They have diameters of only about 1 nm and the nanotip density ranges from 10^9 to 3×10^{11} cm^{-2}. High-resolution TEM (Fig. 5.36a) and Auger electron spectroscopy analysis demonstrate that the nanotips are composed of monolithic silicon and nanometer-sized SiC cap at the top. The nanotips have a very low turn-on field of 0.35 V μm^{-1} which gives rise to a 10 μA cm^{-2} emission current density and have a high field emission current density 3.0 mA cm^{-2} at an applied field as low as 1.0 V μm^{-1} (Fig. 5.36b).

The study shows that the SiC nanowires with sharp ends have better field emission performance in comparison with usual SiC nanowires [68]. The aligned β-SiC nanowires were synthesized in large scale using magnesium-catalyzed coreduction of $SiCl_4$ and 2-ethoxyethanol in an autoclave at 600 °C. The β-SiC rod-like or needle-like nanostructures can be separately produced by employing different ratio of reactants. The infrared spectrum of the nanowires shows a strong absorption peak at around 814 cm^{-1} attributed to the optical phonon vibration of β-SiC. The Raman spectrum comprises two sharp peaks at 795 and 973 cm^{-1}, which are very close to the TO and LO phonon wavenumbers of bulk β-SiC, respectively. The SiC nanowires synthesized using 1 mL of $SiCl_4$ and 0.21 mL of 2-ethoxyethanol are needle-like with sharp tips (Fig. 5.37a) and thus are favorable for field emission applications. Figure 5.37b shows the emission current density versus applied electric field curve for the needle-like nanowires, and from it the turn-on field was derived as 1.6 V μm^{-1}. In comparison, the turn-on fields for two nanorod samples acquired by taking different reactant ratios are 8.4 and 12 V μm^{-1}, respectively. It is clear that the SiC nanowires with sharp tips have better field emission performance. Another study of the SiC

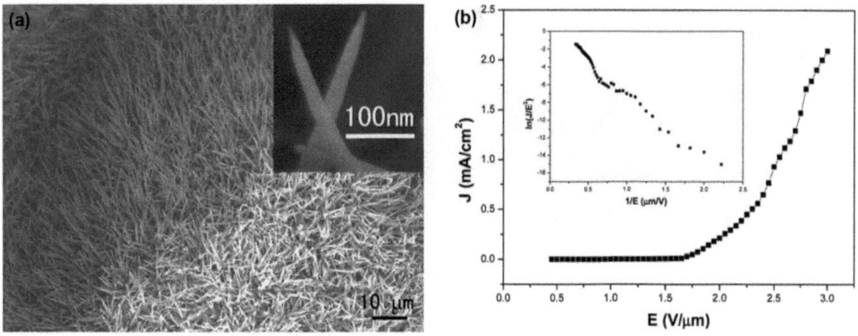

Fig. 5.37 **a** SEM image and **b** field emission current density versus electric field curve of the SiC nanowires synthesized by magnesium-catalyzed co-reduction of a mixture of 1.4 mL $SiCl_4$ and 0.3 mL 2-ethoxyethanol in an autoclave at 600 °C. The inset in right panel shows the Fowler–Nordheim plot. Reprinted with permission from [68]. Copyright 2006, American Chemical Society

nanowires grown on the surface of the bulk SiC ceramic substrate reveals that the morphology of the nanowire film and the diameters of the nanowires are sensitive to the thickness of the catalyst film, and both of them strongly affect the field emission performance [69]. The SiC nanowires of small diameters have very low turn-on and threshold fields for electron emission. The occurrence of an insulator-to-metal-like transition in a field emitting nanowire was proposed to explain the field emission behaviors.

The SiC nanowires with sharp ends can be produced by pyrolysis of polymer precursors. The 3C–SiC nanowire clusters were synthesized via pyrolysis of polysilazane with catalysis of $Al(NO_3)_3$ at 1550 °C in an argon atmosphere [70]. The nanowires have a tapered and bamboo-like structure with tiny tips that are several to tens of nanometers in diameter (Fig. 5.38a–c). X-ray diffraction and Al-element imaging (Fig. 5.38d, e) confirm the Al doping in the nanowires. The nanowires possess an extremely low field emission turn-on field of 0.55–1.54 V μm^{-1} ascribed to the special tapered and bamboo-like structure of the nanowires and to Al doping. A prototype of the diode-type low-voltage field emission display device employing the β-SiC nanowire cathode (Fig. 5.39) has been demonstrated [71]. The nanowires have diameters of 10–50 nm and lengths of several tens of micrometers. This field emission display device (flat vacuum lamps) has a low threshold electric field (2 V μm^{-1}), a high-emission current density, and long-term stable performance.

BN coatings can affect antioxidation ability and field emission properties of SiC nanowires [58]. Under the oxidizing condition, SiC nanowires without BN coatings were cracked into nanoparticles or almost fully converted into SiO_2 nanowires at above 800 K, depending on the crystallization degree of the SiC nanowires. In contrast, at a temperature higher than 1273 K, the BN-coated SiC nanowires still exhibited strong oxidation resistance due to the excellent chemical stability of BN at high temperatures. The BN coating also reduced the turn-on field of SiC nanowires from larger than 10 V μm^{-1} to lower than 6 V μm^{-1}.

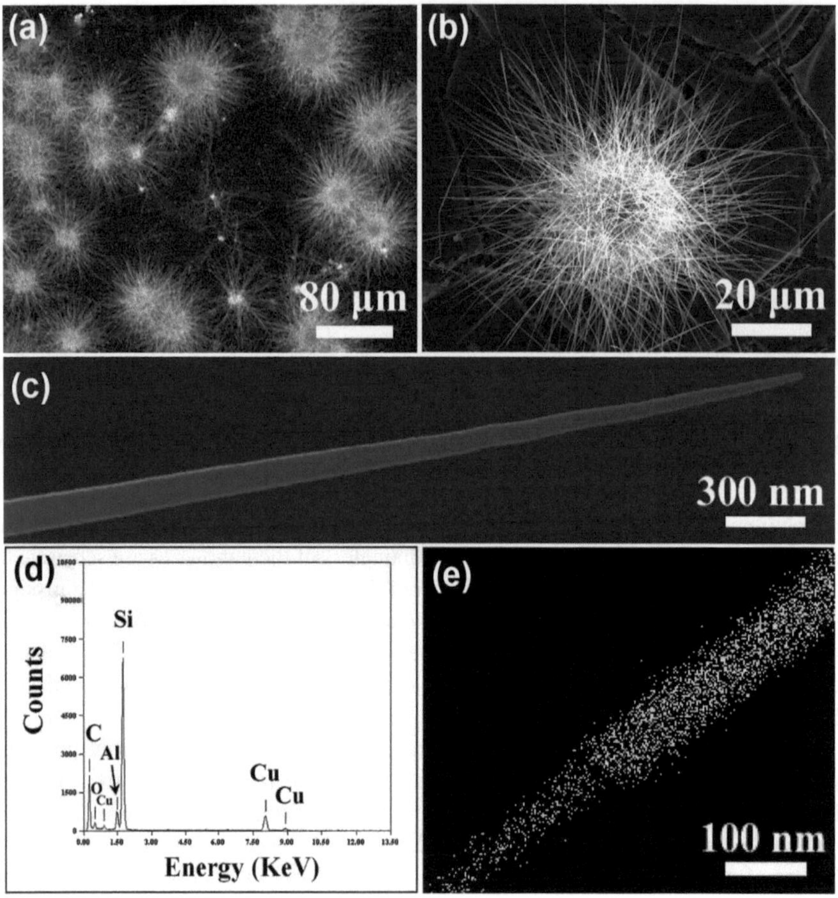

Fig. 5.38 **a** and **b** TEM images of tapered SiC nanowires. **c** Magnification image of a typical tapered SiC nanowire. **d** Energy-dispersive X-ray spectrum of the nanowires. **e** Al-element mapping within one nanowire. Reprinted with permission from [70]. Copyright 2010, American Chemical Society

Porous SiC can also show strong field electron emissions. The performance of the field emission of porous SiC can be significantly improved when regular porous structures were designed [72]. The porous structure was conventionally fabricated by electrochemical etching of bulk 6H-SiC. The morphology of the porous SiC depends on the electrochemical conditions used in the synthesis. The porous nanostructures with different morphologies have different field emission characteristics because different porous structures have different local field enhancement. The mesa, pillar, and fin arrays of porous SiC can be separately produced by reactive ion etching, photolithography, or focused ion beam treatment of the original porous structures. The pillar array (Fig. 5.40) has the best field emission performance with a turn-on field (for an emission current of 10 μA) as low as 4.4 V μm^{-1} and the stable emission current density can reach as high as 6 A cm^{-2}.

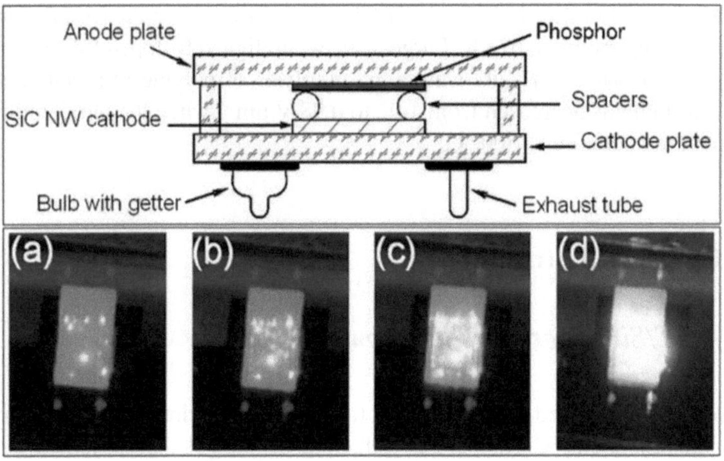

Fig. 5.39 Scheme of field electron emission display device employing 5 × 5 mm^2 SiC nanowire-based cathode. Field emission (phosphor screen) images were recorded at increasing electric field of **a** 2.3 V μm^{-1}, **b** 2.7 V μm^{-1}, **c** 3.0 V μm^{-1}, and **d** 3.3 V μm^{-1}. Reprinted with permission from [71]. Copyright 2008, IOP Publishing

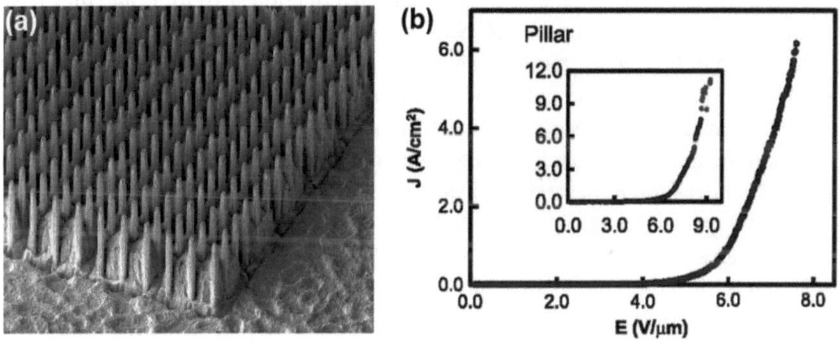

Fig. 5.40 **a** SEM image of porous SiC pillar arrays fabricated by focused ion beam. Scale bar is 50 μm. **b** Field emission current intensity versus electric field for the pillar arrays. The turn-on field for a current of 10 μA is 4.4 V μm^{-1}. The inset is the *J*–*E* curve extended to the failure point of the pillar array showing the emission instability (red) at the material limit. Reprinted with permission from [72]. Copyright 2013, IOP Publishing

Flexible N-doped *n*-type SiC nanoneedles grew on carbon fabric substrates via pyrolysis of polymeric precursor [73]. The turn-on field and threshold field decreased from 1.37 to 0.67 V μm^{-1} and 1.87 to 0.96 V μm^{-1} when the temperature rose from room temperature to 400 °C, respectively. The current emission stability values of the *n*-type SiC nanoneedles at room temperature and 200 °C are ~ 7.7% and 14.1%, respectively. The P-doped SiC nanoparticles with numerous sharp edges and corners were grown on the carbon fabric substrates via pyrolysis of polymeric precursors [74].

They exhibited a low threshold field of 1.03–0.73 V μm^{-1} when the temperature rose from room temperature to 673 K. Large-scale well-aligned B-doped SiC nanoneedle arrays were grown on 6H-SiC wafers via pyrolysis of polymeric precursors [75]. Their turn-on fields decreased from 1.92 to 0.98 V μm^{-1} when the temperature rose from room temperature to 500 °C.

5.3 Core–Shell Structures

5.3.1 *SiC/SiO$_2$ Core/Shell Nanowires*

Most of the synthesized core/shell-structured SiC nanowires possess amorphous silicon dioxide shells. Carbothermal reduction of sol–gel derived silica xerogels containing carbon nanoparticles can yield pure or core/shell structured β-SiC nanorods [11, 76, 77]. The β-SiC nanorods encapsulated by amorphous SiO_2 were achieved by consecutive reduction at 1650 and 1800 °C. These nanorods are up to 20 μm long, the nanorod core has a diameter ranging from 10 to 30 nm (Fig. 5.41a) and the amorphous SiO_2 shell has a thickness of 10–35 nm. The β-SiC nanorods without amorphous layers were obtained by carbothermal reduction at 1650 °C and they have the same diameters to that of the SiC/SiO_2 nanorod cores. The high-resolution TEM image (Fig. 5.41b) reveals that the nanorod core is crystalline β-SiC growing along the <111> direction and the outer shell is amorphous. Electron diffraction (inset of Fig. 5.41), energy-dispersive X-ray spectroscopy, and XRD confirm that the core is β-SiC and the amorphous shell is SiO_2, which formed by the reaction of SiO vapor and O_2. Using a similar method, the β-SiC nanowires can be synthesized via vapor–liquid–solid growth taking activated carbon and Fe nanoparticle-embedded sol–gel derived silica as starting materials at 1400 °C [30]. The nanowires are composed of 10–30 nm β-SiC cores and amorphous SiO_2 shells. The XRD spectrum (Fig. 5.42a) shows three sharp lines that are assigned to the (111), (220), and (311) lattice planes of β-SiC. The weak peak at $2\theta = 33.5°$ which is close to β-SiC (111) peak is commonly assigned to stacking faults in the nanowires. The background halo pattern is characteristic of that of amorphous materials. The SiC nanowires showed photoluminescence (Fig. 5.42b) with two distinct bands centered at around 340 and 440 nm, respectively. The ultraviolet band results from the defects in the amorphous SiO_2 shells and the blue band is associated with the SiC cores of the SiC/SiO_2 core/shell nanowires.

The side-by-side biaxially structured SiC–SiO_x ($x = 1$–2) nanowires have been synthesized [78]. The mixture of amorphous SiO and carbon/graphite was heated at 1500 °C to generate the nanowires. Figure 5.43 (left panel) shows the TEM image of the nanowires situated on a carbon film. The nanowires are uniform with diameters of 50–80 nm and lengths as large as 100 μm. Three types of nanowires are present: pure SiO_x nanowires, coaxially SiO_x sheathed β-SiC nanowires (50%), and biaxial β-SiC–SiO_x nanowires (30%). The coaxial β-SiC–SiO_x nanowires grew along the β-SiC <111> direction with a high density of twins and stacking faults, whose interfaces

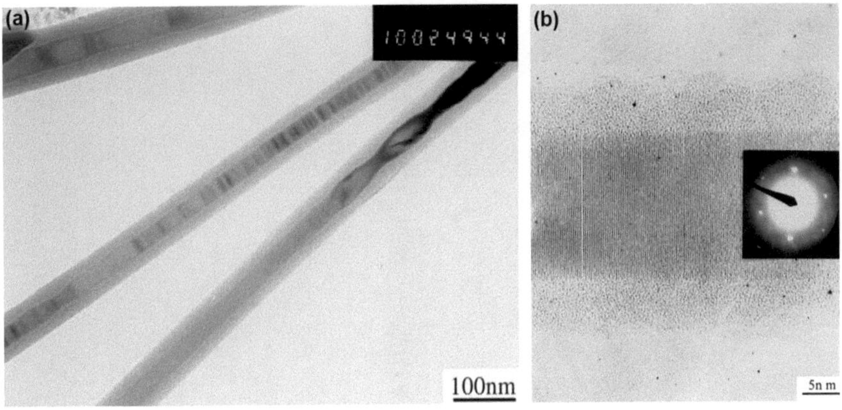

Fig. 5.41 **a** TEM image of the SiC nanorods produced by consecutive carbothermal reductions at 1650 °C and 1800 °C. **b** HRTEM image of a typical nanorod. The lattice fringes correspond to (111) planes of β-SiC. The inset is the selected-area electron diffraction pattern. Reprinted with permission from [76]. Copyright 1998, Springer Nature Limited

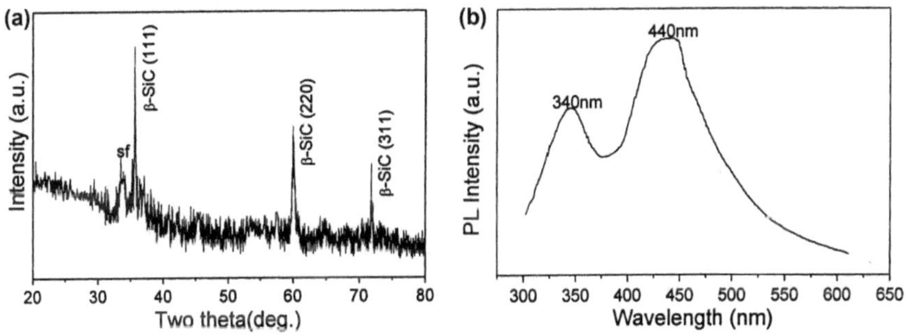

Fig. 5.42 **a** XRD spectrum and **b** PL spectrum under 260 nm excitation of the β-SiC nanowires. Reprinted with permission from [30]. Copyright 2000, Elsevier B.V

are perpendicular to the growth direction. The cross-sectional TEM images (Fig. 5.43, left panel) reveal the geometrical shapes of the coaxial and biaxial nanowires. High-resolution TEM images (Fig. 5.43, right panel) reveal that the biaxial SiC–SiO_x nanowire is composed of side-by-side silica and β-SiC sub-nanowires. There is also a thin layer of silica on the surface of the β-SiC sub-nanowire and it possesses a high density of stacking faults and a few twins. The presence of the planar defects leads to the [311] growth direction for the biaxial nanowires, whereas the biaxial nearly defect-free nanowires possess the <211> growth direction. The Young's modulus of the biaxial nanowires is between 50–70 GPa.

Two-dimensional nanoscale SiC networks can be prepared by heating SiC–Fe–Co mixture at 1500 °C in CO atmosphere [80]. High-resolution TEM measurement reveals that most nanowires have a crystalline β-SiC core with lengths up to 100 μm

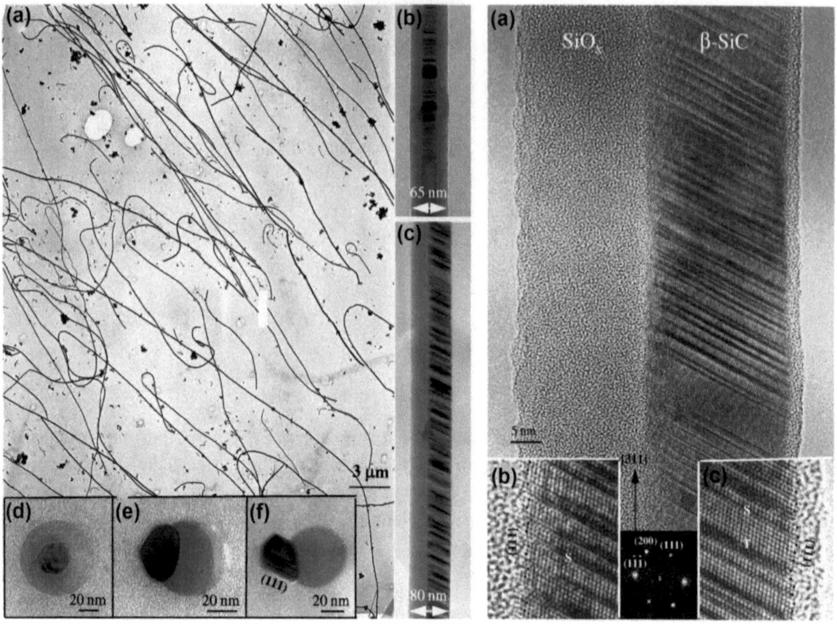

Fig. 5.43 (left panel) (**a**) TEM image of the nanowires showing large lengths and uniform diameters. **b** and **c** Side views of coaxially and biaxial structure SiC–SiO_x nanowires. **d–f** Cross-sectional TEM images of coaxial and biaxial structure SiC–SiO_x nanowires. (right panel) (**a**) HRTEM image of a biaxial structure β-SiC–SiO_x nanowire. The bottom Fourier transform image from a defect-free area reveals the [311] growth direction. Magnification images of SiC–SiO_x interface regions on the left (**b**) and right (**c**) sides of the SiC sub-nanowire showing the {111} facets at the interface. "S" and "T" stand for stacking fault and twin, respectively. Reprinted with permission from [78]. Copyright 2000, American Institute of Physics

and an amorphous SiO_x ($x = 1$–2) shell. The binary Fe–Co catalyst accounts for the SiC network growth. An arc-discharge approach has been developed to synthesize the SiC nanorods sheathed in amorphous SiO_2 by taking the SiC rod containing iron impurity as the anode [25]. The crystalline β-SiC cores have uniform diameters of 5–20 nm and lengths ranging from hundreds of nanometers to several micrometers, the amorphous silica shells have thicknesses of tens of nanometers. Nanofiber templates were used to synthesize high-quality SiC nanowires without assistance of any catalyst. The β-SiC nanowires have been synthesized by heating the template of electrospun polyacrylonitrile nanofibers and a mixture of silica and graphite powders (to give rise to SiO and CO vapors) at 1600 °C [81]. The as-prepared nanowires have SiC/graphite core/shell structures, and chemical etching in aqua regia removed the graphite shell. The resultant nanowires (Fig. 5.44) with diameters of 20–40 nm show amorphous SiO_2 sheaths (~ 5 nm in thickness) as determined by electron energy loss spectroscopy, and the shells formed during etching since the nanowires without etching contained no SiO_2. The SiC/SiO_2 nanowires are uniform in the axial direction and they have very few stacking faults, as confirmed by the sharp and distinct lines in the XRD spectrum.

The SiC–SiO_2 core–shell nanowires have been prepared by heating WO_3/C powder-covered Si substrate with NiO catalysis at 1000–1100 °C [82]. Transmission electron microscopy reveals that the SiC cores have diameters of 15–25 nm and the SiO_2 shells have an average thickness of 20 nm. The thickness of the SiO_2 shell can be reduced by hydrofluoric acid etching (Fig. 5.45a), and the nanowires with complete removal of the SiO_2 shells possess ~ 5-nm thickness carbon shells. The SiC–SiO_2 nanowires have field electron emission with performance depending on the shell thickness, and the nanowires sheathed by 10-nm thickness SiO_2 layers showed the best field emission characters including the lowest turn-on field (Fig. 5.45b).

High-purity SiC/SiO_x core/shell nanowires can be produced by arc-discharge in deionized water between two graphite electrodes, with the anode filled with a mixture

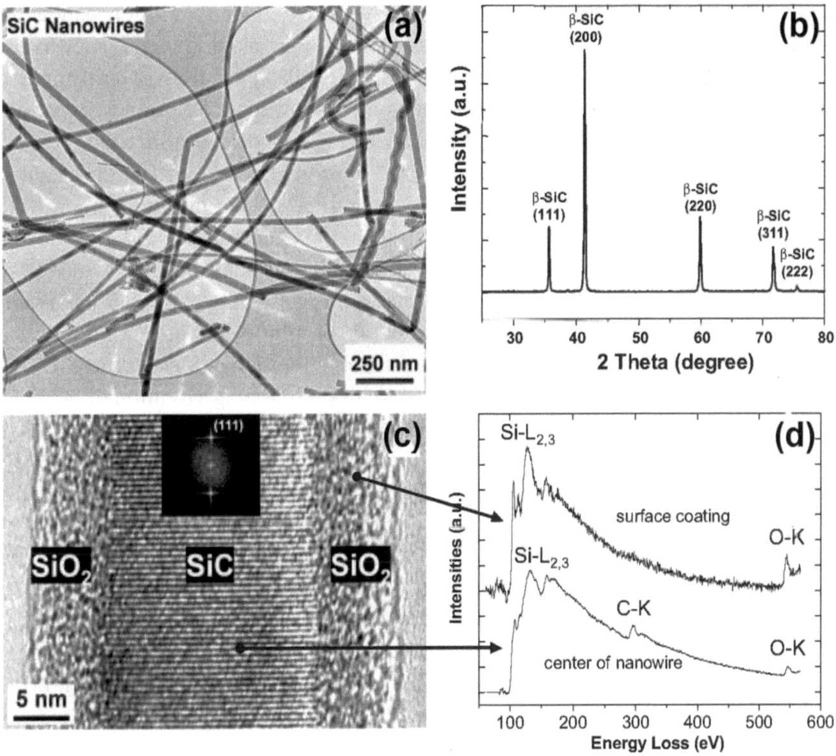

Fig. 5.44 **a** TEM image of SiC nanowires after surface etching with an average diameter of 30 nm. **b** XRD pattern of the nanowires. **c** HRTEM image showing the crystalline SiC core sheathed by an amorphous shell. The fast Fourier transform image in the inset indicates the growth direction of β-SiC (111). **d** Electron energy loss spectroscopy conducted on specified points revealing the silica structure of the shell. Reprinted with permission from [81]. Copyright 2005, John Wiley & Sons, Inc

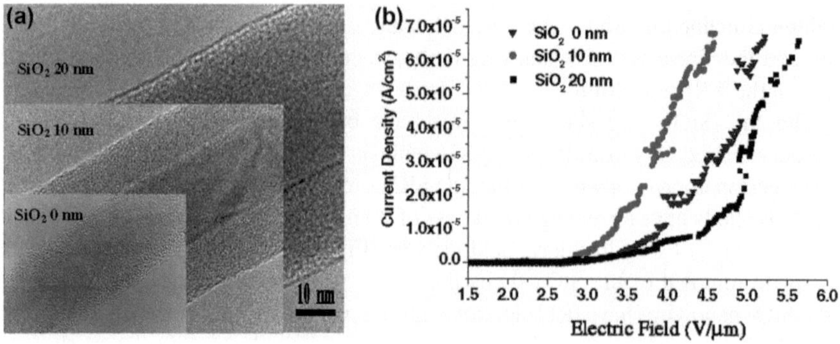

Fig. 5.45 **a** TEM image of β-SiC–SiO_2 core–shell nanowires for various SiO_2 layer thicknesses controlled by HF etching. **b** Field emission current density versus applied electric field curves for the nanowires. Reprinted with permission from [82]. Copyright 2005, IOP Publishing

of graphite, Si and Fe powders as reactants [83]. The obtained SiC/SiO_x nanowires consist of uniform cubic β-SiC cores and amorphous silicon oxide shells. They have a very small average core diameter of 5 nm and their lengths range from several hundred nanometers to several micrometers. The nanowires showed photoluminescence with double emission bands (Fig. 5.46). The two bands are located at around 317 and 368 nm, respectively for a sample with an average core diameter of 10 nm. This first band kept nearly constant whereas the second shifted to 393 nm in a sample with an average core diameter of 20 nm. After anneal in vacuum at 950 °C, the 317 nm emission band diminished, whereas the other emission band became stronger. A study shows that the FTIR and Raman spectra of the SiC/SiO_2 core/shell nanowires are also special [84]. There are two infrared absorption peaks at around 479 and 1100 cm^{-1}, which were assigned to Si–O stretching vibration of amorphous SiO_2; an absorption peak at ~ 788 cm^{-1} was assigned to Si–C stretching vibration of 3C–SiC. Their Raman spectrum comprises a peak at ~ 785 cm^{-1} belonging to the TO phonon at the Γ point of 3C–SiC and another peak at 916 cm^{-1} ascribed to amorphous SiO_2. The TO phonon line shifts toward lower wavenumbers by 11 cm^{-1} compared to that of bulk SiC.

The influence of the amorphous SiO_2 shell on the cathodoluminescence of the 3C–SiC/SiO_2 core/shell nanowires have been investigated [85]. The nanowires were prepared by reaction between carbon monoxide and native oxide on (001) Si substrates at 1050–1100 °C catalyzed by nickel nitrate. The main luminescence peak lies at around 2.41 eV. The emission intensity decreased by HF etching, showing the enhancement effect of the SiO_2 shell on the luminescence, possibly arising from carrier diffusion from the SiO_2 shell to the SiC core. The cathodoluminescence in two samples of 3C–SiC/SiO_2 nanowires with different shell thicknesses has been comparatively investigated [86]. Two samples (Fig. 5.47a) have similar nanowire core diameter distributions between 15–30 nm but have distinct core-to-shell ratios of 1: 1 and 1: 5, respectively. The cathodoluminescence measured at 10 and 300 K (Fig. 5.47b) showed two emission peaks for both samples, at around 2.5 and 4.0 eV.

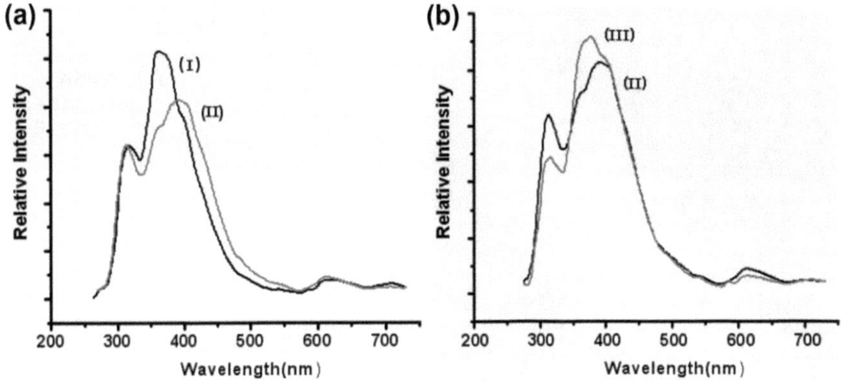

Fig. 5.46 **a** PL spectra of the SiC/SiO_x core/shell nanowires with an average core diameter of 10 nm (I) and 20 nm (II), respectively. **b** PL spectra of the SiC/SiO_x core/shell nanowires with an average core diameter of 20 nm before (II) and after (III) annealing at 950 °C in vacuum. Reprinted with permission from [83]. Copyright 2005, IOP Publishing

The emission band at around 4.0 eV is stronger in the sample with a thicker silica shell, whereas the emission band at 2.5 eV is stronger in the sample with a thinner shell. This observation suggests that the 4.0-eV emission arises from the silica shell and the 2.5-eV emission stems from the core SiC (possibly from surface defects).

Centimeter-long β-SiC/SiO_2 core/shell nanowires have been synthesized by using a catalyst-free CVD at 1250–1300 °C in argon atmosphere [87]. The reactants were divided into two groups (Si and SiO_2 powder, WO_3, and graphite powder) and were placed at two sites in the furnace, whereas the product was collected on the graphite paper placed at a third site. The nanowires have diameters of 50–100 nm and lengths up to 10 mm (Fig. 5.48). They have photoluminescence with two distinct peaks centered at 289 and 394 nm which may arise from the surface silica shell. Ni-assisted CVD approach can also yield ultralong SiC/SiO_2 nanowires [88]. The nanowires consist of single-crystalline cubic SiC nanowire cores and uniform amorphous SiO_2 shells, which entangle with each other and construct a macroscopic thin film with a length of ~ 17 cm. The yield of the as-synthesized SiC/SiO_2 nanowires reaches gram scale (2.164 g) under the assistance of the ingenious homemade reactor.

Very long SiC-based coaxial nanocables were synthesized by employing SiO and polypropylene as solid precursors of Si and C [89]. This technique is capable of changing the chemical composition of the nanocable outer layer from silica to carbon and boron nitride by varying the proportion or the chemical composition of the solid reactants. The nanocables consist of SiC cores of 30 nm in diameter and lengths up to several hundred nanometers. The thickness of the coating is 2–10 nm. A vapor–solid growth mechanism was proposed. In practice, the shell thicknesses of the SiC/SiO_2 core–shell nanowires are also tunable through appropriate time of treatment in NaOH solution [90]. Pure SiC nanowires were also acquired by alkaline cleaning in 1 mol L^{-1} NaOH solution at 70 °C for 40 min. PL measurements show that the SiC nanowires coated with an optimum SiO_2 thickness (3.03 nm) have better

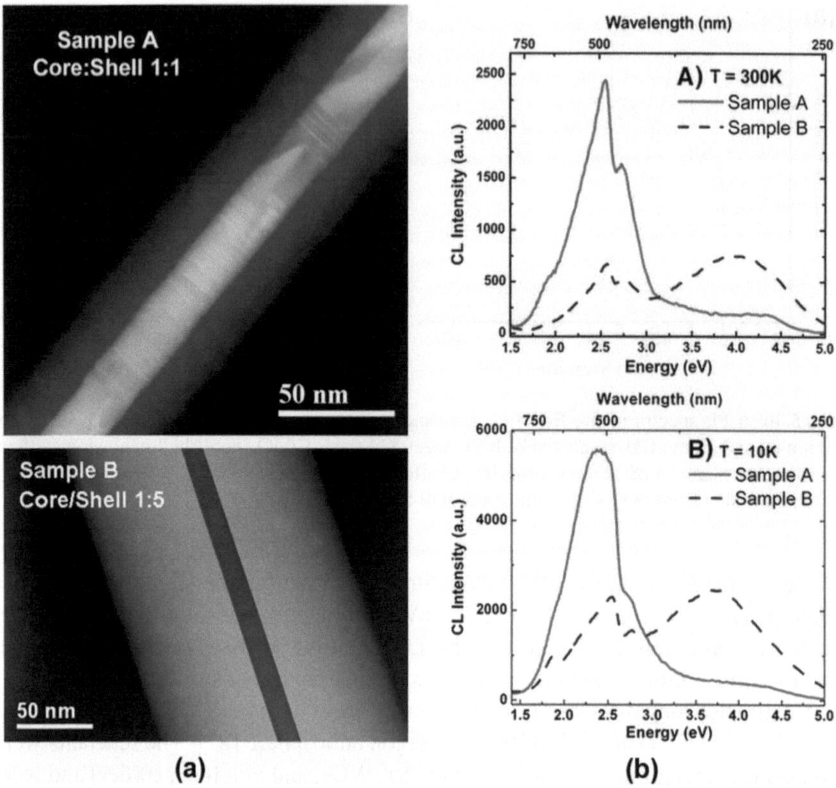

Fig. 5.47 **a** False-color TEM images of two typical SiC/SiO_2 core/shell nanowires from two samples. **b** Cathodoluminescence spectra at room and low temperatures for two samples. Reprinted with permission from [86]. Copyright 2012, Elsevier B.V

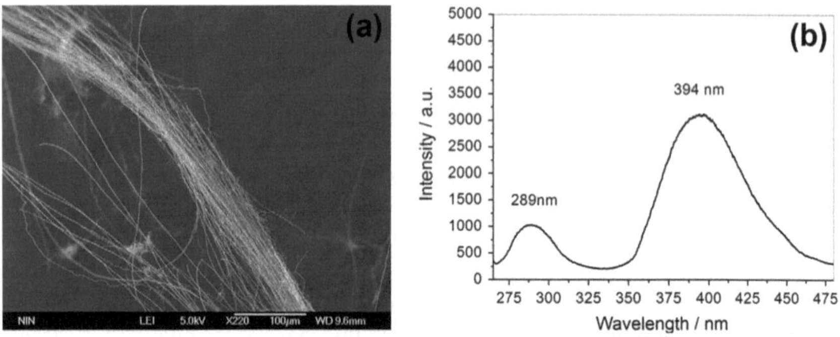

Fig. 5.48 **a** SEM image of ultralong SiC nanowires. **b** PL spectrum of the SiC nanowires under 260 nm excitation at room temperature. Reprinted with permission from [87]. Copyright 2011, Elsevier B.V

luminescence properties than either the bare SiC nanowires or SiC nanowires with thicker coatings of SiO_2.

5.3.2 Carbon-Coated SiC Nanowires

Although most of the reported SiC nanowires possess silica shells, but researchers have observed that some SiC nanowires obtained under specific conditions possess carbon shells or silica–carbon mixed shells. A thermal evaporation method has been developed to synthesize the SiC–SiO_2-carbon nanotube coaxial nanocables [91]. The nanocables were synthesized by heating a mixture of SiO and Fe powders under flowing CH_4 gas at 1550 °C. These coaxial nanocables consist of a single-crystalline β-SiC cores, amorphous SiO_2 intermediate layers, and graphitic carbon outer sheaths (Fig. 5.49a). The nanocables are 20–50 nm in diameter and the nanocable cores are 10–30 nm in diameter. The nanochains of SiC–carbon nanotube heterojunctions (Fig. 5.49b) were acquired by annealing the as-prepared nanocables at 1600 °C in vacuum. The study shows that a SiC nanowire can be transformed into a carbon nanotube or a SiC–C core–shell nanowire owing to Joule heating by applying a current on the order of 10 mA [92].

SiC nanowhiskers can also be transformed into different carbon nanostructures through high-temperature extraction of Si atoms by treatment in Cl_2 or HCl or by vacuum decomposition [93]. High-resolution TEM and Raman analysis show that chlorination of SiC nanowhiskers generated amorphous carbon-sheathed SiC whiskers and finally amorphous nanoporous carbon (Fig. 5.50). High-temperature treatment of SiC in HCl yielded fullerene-like structures, whereas high-temperature vacuum decomposition yielded graphite. SiC–C coaxial nanocables can be synthesized via a solvothermal process using $SiCl_4$, C_6Cl_6 and Na as starting materials at 250 °C under a pressure of 2.2 MPa [94]. The diameters of the nanocables are around

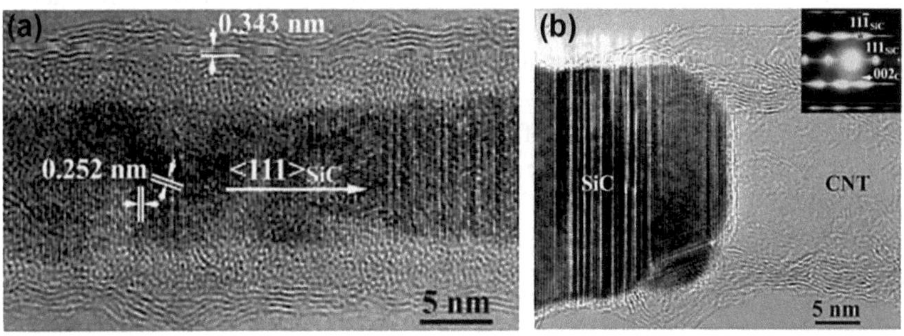

Fig. 5.49 **a** HRTEM image of coaxial nanocable showing single-crystalline β-SiC core, amorphous SiO_2 intermediate layer, and outer graphitic C sheath. **b** HRTEM image of SiC nanorod–carbon nanotube heterojunction. The inset shows the selected-area electron diffraction pattern. Reprinted with permission from [91]. Copyright 2004, John Wiley & Sons, Inc.

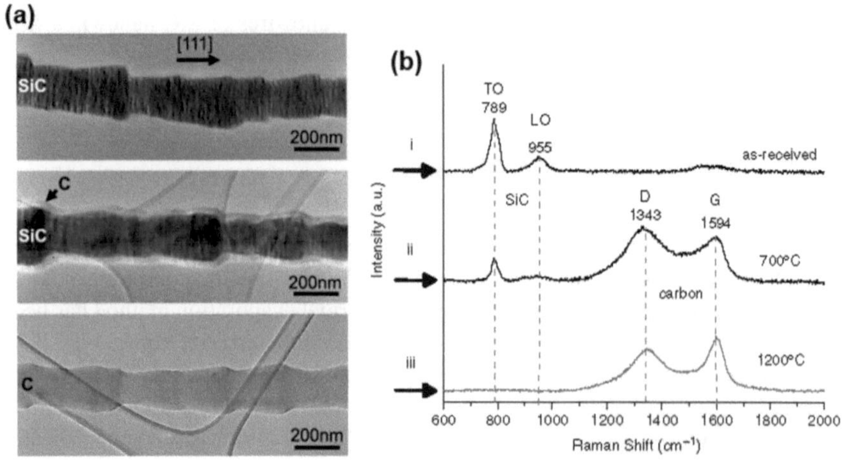

Fig. 5.50 **a** TEM images and **b** Raman spectra of SiC nanowhiskers measured at different stages of transformation to carbon: (i) initial: as-prepared SiC nanowhisker, (ii) intermediate: carbon coating on SiC nanowhisker, and (iii) final: complete transformation into amorphous carbon. The nanowhiskers were treated in Cl_2 at 700 °C (ii) and 1200 °C (iii). Reprinted with permission from [93]. Copyright 2006, John Wiley & Sons, Inc

50 nm and the lengths are up to one micrometer. The nanocable contains a β-SiC shell coated by an amorphous carbon shell. The field electron emission has been observed from the carbon-coated β-SiC nanowires grown on the Si substrate [95]. The nanowires have diameters between 20 and 50 nm and carbon shell thicknesses of 2–3 nm. The turn-on field corresponding to an emission current density of 10 μA cm^{-2} is about 4.2 V mm^{-1}. 3C–SiC nanowires wrapped by carbon nanosheets were synthesized by using CVD using methyltrichlorosilane as the precursor [96]. The core has a diameter of about 30 nm. The carbon nanosheet is polycrystalline graphite and it has a thickness of about several nanometers. Carbon-coated SiC nanowires were fabricated by epitaxial growth on SiC [97]. The epitaxial carbon layers (amorphous carbon or graphite layer) grew epitaxially on the surfaces of the SiC nanowires in a high-vacuum and high-temperature environment, leading to carbon-coated pearl-chain-shaped SiC nanowires. Some 3C–SiC particles orderly grew along the surfaces of the SiC nanowires. The growth of carbon shell was ascribed to thermal evaporation of Si or SiO_2 on the surfaces of the SiC nanowire.

5.3.3 BN-Coated SiC Nanowires

SiC/SiO_2/(BN + C) core/shell/shell nanocables with diameters of a few tens of nanometers and lengths up to 50 μm were synthesized by means of reactive laser ablation [98]. Transmission electron microscopy and analysis of elemental profile across a nanocable help determine the nanocable structures. The result shows that the nanocable consists of a β-phase silicon carbide core, an amorphous silicon oxide intermediate layer, and a graphitic outer shell made of boron nitride and carbon layers separated in the radial direction (Fig. 5.51). In the laser ablation, a mixed and compressed powder of BN, C, SiO, and Li_3N was taken as the starting material. The growth of nanocables comprises two steps. First, the SiC–SiO_2 nanowire formed through a reaction C (solid or vapor) + SiO (gas) → SiC (solid) + SiO_2 (solid). The second step involves coating of the BCN sheath and phase separation of BN and C into nanodomains. The graphitic shell can only form in the presence of Li_3N in the starting material, otherwise, only the SiC–SiO_2 wires without any graphitic sheaths formed because C reacted readily with SiO to form SiC/SiO_2. The existence of lithium can advance the formation of graphite-like BN.

The carbon nanotube template method has been employed to synthesize the silicon carbide nanowires sheathed with boron nitride nanotubes [99]. A mixture of silica and silicon powders reacted with nearby B_2O_3 powder covered with carbon nanotubes at 1753 K in a flowing nitrogen atmosphere to give rise to the SiC/BN nanowires. Through the carbon nanotube–substitution reaction, carbon nanotubes reacted with boron oxide vapor in the presence of nitrogen gas to yield BN nanotubes. The reaction of SiO vapor within carbon nanotubes with the inner carbon layers or volatile

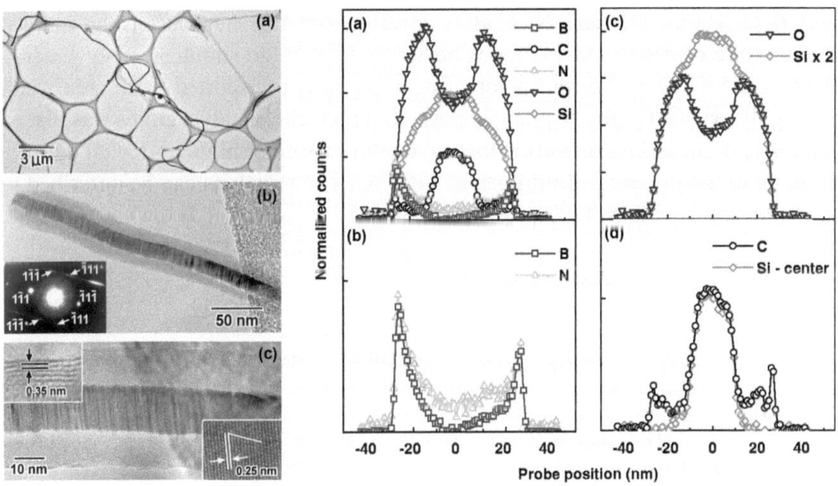

Fig. 5.51 (left panel) **a** TEM image of nanocables. **b** Magnified image showing a crystalline core and an amorphous layer. Selected-area electron diffraction pattern (inset) indicates that the core is β-SiC with the <110> axis parallel to the electron beam. The asterisks represent twinning planes. **c** High-resolution image of the nanocable. Top inset shows lattice fringes of the outer C–BN layer and bottom inset shows lattice fringes of the β-SiC core. (right panel) Elemental profiles across a nanocable. **a** Profiles of all elements in the nanocable. **b** B and N profiles showing the tubular feature of the outer sheath and unity atomic ratio. **c** Si and O profiles, in which the Si concentration is doubled; the Si:O ratio in the intermediate layer is 1:2. **d** C profile and reduced profile of Si in the center core, acquired by subtracting the contribution of Si comprised in the SiO_2 layer from the total Si concentration. The Si:C ratio in the wire core is unity. Reprinted with permission from [98]. Copyright 1998, American Association for the Advancement of Science

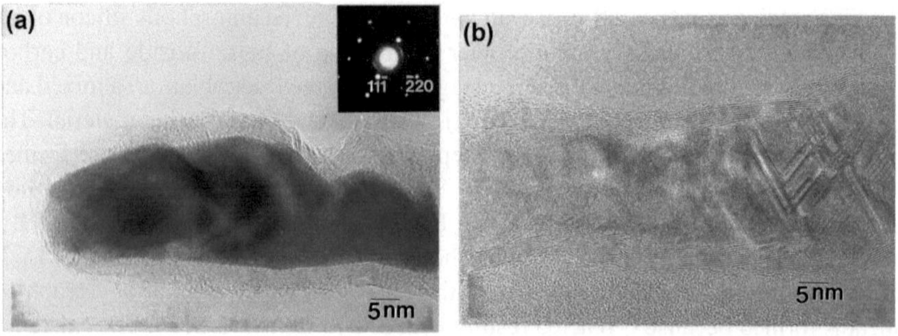

Fig. 5.52 High-resolution TEM images of SiC-filled BN nanotubes. **a** Tip of filled BN nanotube. The inset shows the selected-area electron diffraction pattern which corresponds to β-SiC and the electron beam is along the <112> axis. **b** SiC-filled BN nanotube with high density of planar defects. Reprinted with permission from [99]. Copyright 1999, American Institute of Physics

carbon mono-oxide in the interior produced SiC nanowires. The obtained sheathed nanowires are as long as the starting carbon nanotubes. High-resolution TEM images (Fig. 5.52) reveal the β-SiC core structure and good crystallinity of the outer BN shell.

Uniform BN-coated SiC nanowires were synthesized via the vapor–liquid–solid growth [3, 100]. Nanoscale Ni–C alloy covering graphite sheet acts as the catalyst. The mixture of boron and silica was heated at 1500 °C to simultaneously generate B_2O_2 and SiO gas-phase precursors. The vapor was transported to an area where the highly oriented pyrolitic graphite covered with nickel catalyst grains was placed. Silicon carbide and boron nitride formed on the surface of highly oriented pyrolitic graphite in the presence of nitrogen at 1400 °C. The reactions can be described by the following equations [100]:

$$2B(s) + 2SiO_2(s) \rightarrow B_2O_2(g) + 2SiO(g), \tag{5.10}$$

$$SiO(g) + 2C(s) \rightarrow 2SiC(s) + CO(g), \tag{5.11}$$

$$B_2O_2(g) + 2C(s) + N_2(g) \rightarrow 2BN(s) + 2CO(g). \tag{5.12}$$

The nanowire core with a diameter ranging from several nanometers to 80 nm is β-SiC growing along the <111> direction. It contains a great amount of stacking faults and microtwins, as shown in Fig. 5.53a. The nanowire surface is sheathed with well-crystallized BN with a thickness of 2–4 nm. The bicrystalline SiC nanowires with BN coating are also present and they have a [311] growth axis. The effects of BN coating on antioxidation ability and field emission properties of the SiC nanowires have been studied [58]. Upon oxidation, the SiC nanowires without BN coating cracked into nanoparticles or almost fully converted into SiO_2 nanowires at temperatures above 800 K, depending on crystallization degree of the SiC nanowires. After BN coating, the antioxidation ability of the SiC nanowires was improved due to excellent chemical

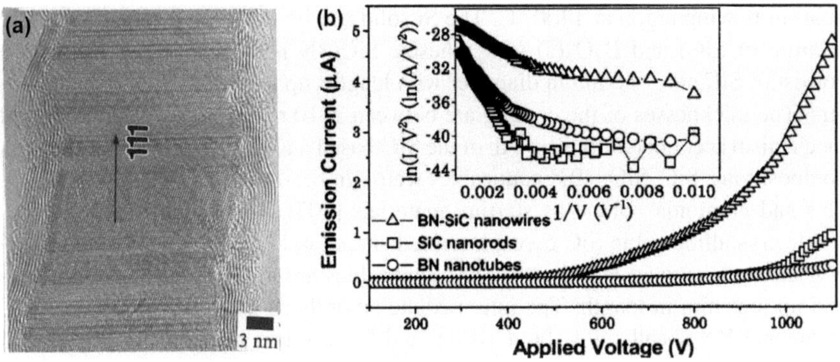

Fig. 5.53 **a** HRTEM image of the BN-coated SiC nanowire. The β-SiC core grew along the <111> direction and has high-density planar defects [3]. **b** Field electron emission current versus applied electric voltage for single-crystal SiC nanowires, BN nanotubes, and BN-coated SiC nanowires, respectively. The inset shows the Fowler–Nordheim plot. Reprinted with permission from [58]. Copyright 2003, American Institute of Physics

stability of BN at high temperatures. The BN-coated SiC nanowires maintained high resistance to oxidation even at temperatures higher than 1273 K. The field electron emission performance of the SiC nanowires was also improved after BN sheath. The turn-on field, which is the field to give rise to a current density 10 μA cm^{-2}, was reduced from above 10 V μm^{-1} for the SiC nanowires without coating to below 6 V μm^{-1} for the BN-coated SiC nanowires (Fig. 5.53b). The Fowler–Nordheim plots for the three investigated samples show an approximately linear relationship between $\ln(I/V^2)$ and 1/V for applied voltages over 300 V, suggesting that a quantum tunneling process is responsible for the observed electron emissions.

BN nanotube-coated β-SiC nanowires, with 10–15 nm spacing between the SiC core and the BN nanotube sheath, were synthesized using CVD [101]. Specified post-treatment caused consecutive formations of BN–SiO_2–SiC nanocables, BN nanotube-encapsulated SiC nanowires with reduced diameters, and SiC or C–SiC nanocables encapsulated by C–BN composite nanotubes. A mixture of Si and In_2O_3 powders was heated in a small BN crucible in flowing CH_4 gas at 1600 °C to yield the composite nanowires. The decomposition of In_2O_3 during heating produced the oxygen gas source. The electric current versus voltage measurement shows that the BN nanotube sheath provides excellent electrical insulation for the encapsulated semiconducting nanowires. β-SiC nanowires can be grown by thermal treatment of Si particles in a graphite crucible in nitrogen atmosphere at 1200 °C [102]. Using a similar method, treatment of a mixture of the boron nitride and Si powders under argon atmosphere produced hexagonal-BN-coated β-SiC nanowires. A technique has been developed to synthesize very long SiC-based coaxial nanocables [89] with varied chemical composition of the nanocable outer layers, from silica to carbon and to boron nitride, depending on the composition of the reactants. The SiC nanowires with silica or carbon sheath were prepared by thermal reaction of polypropylene (in an alumina boat) and an equimolar mixture of Si(s) and SiO_2(s) (in another alumina boat) in flowing argon at 1400 °C. The Si solid can be substituted by an equimolar mixture of Si(s) and B_2O_3(s) to synthesize SiC/BN nanocables. The nanocables consist of SiC cores 30 nm in diameter with lengths up to several hundred nanometers. The thicknesses of the coating are between 2–10 nm. The vapor–solid growth mechanism accounts for the growth of the SiC-based nanocables of various chemical compositions. SiC–SiO_2–BN nanocables were synthesized using SiC–SiO_2 nanocables and ammonia borane as starting materials [103]. The nanocables consist of single crystalline cubic SiC cores, intermediate amorphous SiO_2 layers, and single crystalline hexagonal BN sheaths. The nanocables are around 100 nm in diameter and up to 1 mm in length. The intermediate amorphous SiO_2 layers and the outer hexagonal BN sheaths are about 10 nm and 5 nm thick, respectively. Increase in the amount of ammonia borane caused the transformation of SiC–SiO_2 nanocables into BN nanotubes. A strategy to produce GaN/3C–SiC heterostructure nanowires was developed [104]. 3C–SiC nanowires (diameter: $\sim$ 100 nm) possessing periodic stacking fault sections synthesized by CVD served as cores of heterostructures. Hexagonal wurtzite-type GaN shell grew on the surface of the 3C–SiC wire core. The stacking faults in 3C–SiC nanowires play a key role in the epitaxial growth of GaN because they are areas with higher surface energies. The disordered atoms in the SiC defect area promoted lattice-matching with the GaN phase, leading to nucleation.

5.4 Special Structures

Besides pure SiC nanowires and core/shell structured SiC nanowires, various more complex one-dimensional SiC nanostructures have been synthesized and investigated, such as the bamboo- or bead-like SiC nanowires, helical SiC nanowires, and interconnected SiC nanowire networks, etc. Silicon carbide can exist as rather fruitful morphologies of one-dimensional nanostructures.

5.4.1 *Bamboo-Like Structures*

Bamboo-like SiC nanofibers have been synthesized by carbothermal reduction of carbonaceous silicon xerogel containing lanthanum nitrate [31]. The xerogel was heated to 1300 °C in flowing argon gas to synthesize the product, which was then heated in air at 700 °C to eliminate residual carbon and treated by nitric acid and hydrofluoric acid to remove unreacted silica and other impurities. The product contains bamboo-like β-SiC nanofibers with diameters of 40–100 nm and lengths of tens of micrometers to hundreds of micrometers. The nanofibers contain planar stacking faults with the plane direction perpendicular to the fiber axis. A modified method produced the periodically twinned SiC nanowires by carbothermal reduction of a carbonaceous silica xerogel prepared from tetraethoxysilane and biphenyl with iron nitrate as an additive [105]. The twinned β-SiC nanowires have hexagonal cross

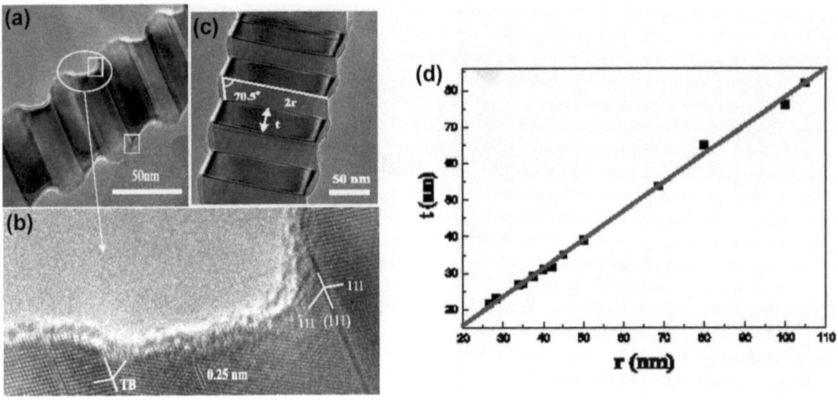

Fig. 5.54 **a** TEM image of a bamboo-like SiC nanowire. **b** HRTEM image of one part of the nanowire. **c** and **d** depict the linear relationship between the twin segment thickness (t) and the nanowire radius (r). Reprinted with permission from [105]. Copyright 2008, IOP Publishing

sections, diameters of 50–300 nm, and lengths of tens of micrometers to hundreds of micrometers. They have periodically twinned segments with uniform thicknesses along the entire nanowire axis (Fig. 5.54). The formation of the periodic twin structure ensures the minimum surface energy and strain energy for the synthesized nanowire. The thickness of the twinned segment shows a linear dependence on its diameter.

Single-crystalline bamboo-like β-SiC nanowires with hexagonal cross sections were produced by thermal evaporation of mixed SiO + C + GaN powders at 1350 °C in an argon atmosphere [106]. The nanowires grew along the <111> direction by a vapor–liquid–solid mechanism. The nanowire comprises a hexagonal stem decorated by larger-diameter knots along the whole nanowire axis (Fig. 5.55a–f). The electron diffraction patterns from the stem and knot areas have the same orientation, indicating that the whole bamboo-like β-SiC nanowire is a single crystal growing along the <111> direction. The electron diffraction pattern of the stem consists of only bright spots, whereas the pattern of the knot comprises both bright spots and streaks, indicating that the defects exist only in the knot areas. High-resolution TEM images of both stem and knot parts of the β-SiC nanowire reveal the lattice spacing of 0.25 nm, corresponding to the {111} planes of β-SiC. The stem shows a uniform structure, whereas the knot possesses stacking faults. Figure 5.55g shows the plot of emission current density versus applied electric field at an anode–sample distance of 100 μm. The turn-on field which corresponds to the generation of a current density 10 μA cm^{-2} is 10.1 V $μm^{-1}$. The Fowler–Nordheim plot shows a linear curve for applied voltages over 1/0.0025 V.

Twinned zigzag 3C–SiC nanoneedles were synthesized by the reaction between silicon vapor and solid multiwalled carbon nanotubes at 1500 °C [2]. The Si atom in the vapor phase reacted with carbon to form SiC nuclei in the active part of carbon nanotubes, which then grew along the <111> direction to form SiC nanoneedles. The SiC nanoneedle has a zigzag morphology (Fig. 5.56) and it is composed of (111)

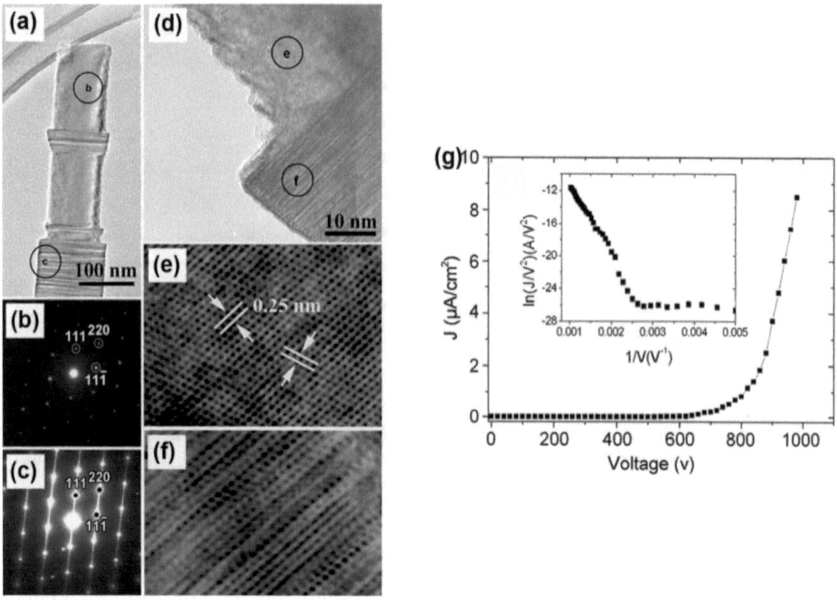

Fig. 5.55 **a** TEM image of a single bamboo-like β-SiC nanowire. Selected-area electron diffraction patterns taken from the stem (**b**) and knot (**c**) regions of the bamboo-like β-SiC nanowire. **d** TEM image of the stem/knot interface. HRTEM images of the stem (**e**) and knot (**f**). **g** Field emission current density versus applied voltage curve for the bamboo-like β-SiC nanowires. The inset shows the Fowler–Nordheim plot. Reprinted with permission from [106]. Copyright 2006, IOP Publishing

$[11\bar{2}]$ periodic twined segments. The zigzag morphology was observed from the $[01\bar{1}]$ zone axis. The formation of the twin structures gave rise to SiC nanoneedles with lower cohesive energies than SiC nanoneedles without twin structures, whereas the formation of the periodic $(11\bar{1})$ facets on the nanoneedle surface may ensure the lowest surface energy. The room temperature PL spectrum (excitation: 325 nm) of the SiC nanoneedles comprises a dominant peak at 403 nm (3.12 eV) and was ascribed to the nanometer-sized diameters of the SiC nanoneedles. Another study reports the dominant emission band at around 470 nm (2.64 eV) from the weakly bamboo-like SiC nanowires [107].

Single-crystalline bamboo-like β-SiC nanowires were synthesized by using catalyst-free vapor deposition using silicon and graphite carbon as starting materials [108]. The morphology of the SiC nanowires varied from cylinder, hexagonal prism, to bamboo shapes as the reaction temperature increased from 1470°, 1550°, to 1630 °C. Figure 5.57 shows the electron microscopy images of the bamboo-like SiC nanowires. The nanowire comprises many decorated knots and truncated hexagonal pyramids. It possesses high-density stacking faults, especially at the tip. The electron diffraction pattern contains spots belonging to β-SiC {220} planes and streaks attributed to the stacking faults. The nanowires grew by a vapor–solid mechanism.

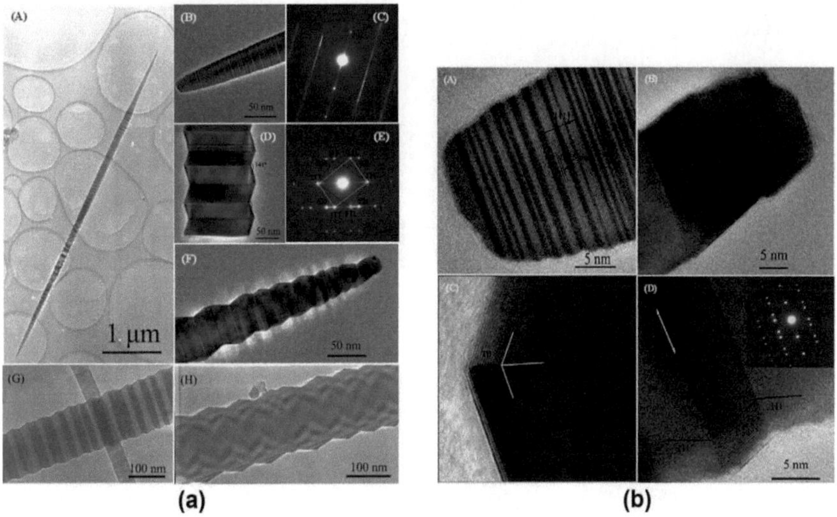

Fig. 5.56 **a** A: TEM image of the SiC nanoneedle with upper flat surface and lower zigzag surface. B: Magnification image of flat surface. C: Electron diffraction pattern of the flat part showing streaks of $(\bar{2}0\bar{2})$ planes. D: Magnification image of zigzag surface. E: Electron diffraction pattern of the zigzag part showing well-defined (111) twinning relationship. F: Magnification image of zigzag surface. G and H: TEM images of another nanoneedle. **b** HRTEM image of a SiC nanoneedle: A: flat part, B: tip of the zigzag part, C and D: neighboring twin segments of the zigzag part from the $[11\bar{1}]$ direction. Reprinted with permission from [2]. Copyright 2007, American Chemical Society

5.4.2 Beaded Structures

There exist two kinds of beaded SiC nanowire chains, divided according to whether the outside beads attached to the core nanowires belong to SiC or SiO_2. The necklace-like β-SiC nanowires have been synthesized by using catalyst-free CVD using SiO_2 and graphite powder as starting materials [109]. The SiC nanowires have diameters of 50–100 nm and lengths up to several tens of micrometers (Fig. 5.58). The bead diameters of the necklace-like nanowires range from 0.5 to 1.0 μm. The nanowires may grow by a vapor–solid mechanism since there was no metallic droplet in the product.

The SiC/SiO_2 chain-like nanostructures have been synthesized via catalyst-free chemical vapor reaction using a mixture of Si–SiO_2 powder and CH_4 as source materials at 1200 °C [110]. The main body of the chainlike nanostructure is a single crystalline β-SiC nanowire with a diameter of 20–30 nm and a length up to several tens of micrometers (Fig. 5.59a). The β-SiC nanowire preferentially grew along the <111> direction and it contains a high density of stacking faults and twin defects. The core nanowire was wrapped by a series of 80–100 nm diameter amorphous SiO_2 spheres. The SiC/SiO_2 nanochains showed photoluminescence at room temperature, with a narrow band centered at 410 nm and a wide band centered at 490 nm (Fig. 5.59b).

Fig. 5.57 **a** and **b** SEM images of bamboo-like SiC nanowires. **c** TEM image of a SiC nanowire and corresponding electron diffraction pattern (inset). **d** TEM image of a nanowire with consecutive truncated cones at the tip. **e** HRTEM image of the junction area. Reprinted with permission from [108]. Copyright 2008, IOP Publishing

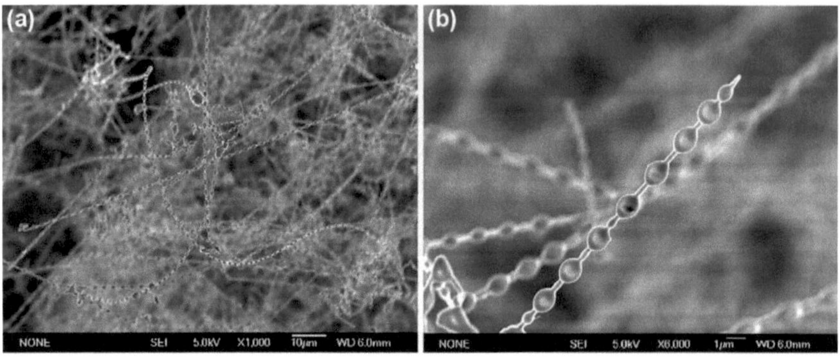

Fig. 5.58 **a** and **b** SEM images of necklace-like SiC nanowires. Reprinted with permission from [109]. Copyright 2006, Elsevier B.V

1D SiC/SiO_2 nanochains and 2D SiC/SiO_2 X-junction and Y-junction nanochains were synthesized using a microwave method [111]. The nanochains consist of 3C–SiC stems with diameters of 20–80 nm and outside attached periodic SiO_2 beads with diameters of 100–400 nm. The presence of SiO_2 beads enhanced the luminescence of the SiC stems. A two-step growth mechanism of the nanochains was proposed. SiC/SiO_x core–shell nanochain heterojunctions were synthesized on a Si substrate

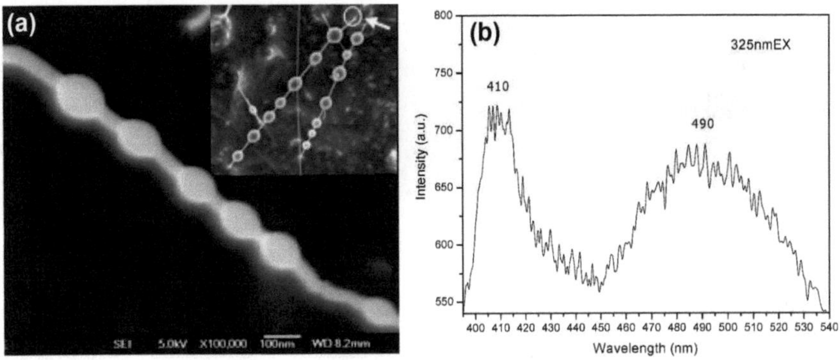

Fig. 5.59 **a** SEM image of the chain-like β-SiC nanowire. The scale bar is 100 nm. The inset shows the SEM image of two nanowires on the graphite substrate. **b** Photoluminescence spectrum of the chain-like β-SiC nanowires under 325 nm excitation. Reprinted with permission from [110]. Copyright 2009, American Chemical Society

via a thermal evaporation method at 1500 °C without using catalysts, template, or flowing gases [112]. The cross-sectional elemental mapping reveals a core–shell structure that consists of a SiC core and a SiO_x shell. The SiC core is about 50 nm in diameter. The formation of the SiO_x beads correlates with the Rayleigh instability and poor wettability between SiC and SiO_x as well as the existence of a high density of stacking faults within the SiC nanowires. The nanochains showed photoluminescence peaked at 407 nm. Necklace-like SiC/SiO_x heterojunctions were synthesized by using a vapor reaction technique at 1200 °C, using the Si powder as raw material, without using catalysts or flowing gases [113]. The heterojunctions have lengths up to millimeters. They are composed of SiC/SiO_2 core/shell chains with diameters of 0.3–1 μm and amorphous SiO_2 beads with diameters of 2–5 μm. The TEM image of a typical heterojunction shows that the stem of the heterojunction comprises a SiC core of 350 nm in diameter and an amorphous silica shell of 2 nm in thickness (Fig. 5.60a–c). The SiC core exhibits a lattice spacing of 0.25 nm, matching that of the (111) plane of 3C–SiC, indicating that the core nanowire prefers to grow along the <111> orientation. The SAED pattern confirms that the bead outside the stem is amorphous SiO_2, consistent with the XRD result (Fig. 5.60d). The FTIR spectrum comprises two strong absorption peaks at 794 and 950 cm^{-1} that are assigned to the TO and LO modes of Si–C stretching vibration (Fig. 5.60e). The absorption peak at ~ 436 cm^{-1} corresponds to the Si–O–Si stretching vibration of amorphous SiO_2. The heterojunctions may form via the vapor–solid growth mechanism. The formation of SiO_2 beads was attributed to Rayleigh instability and surface tension. The heterojunctions showed photoluminescence, and the PL and PL excitation spectra have single peaks at 407 and 330 nm, respectively (Fig. 5.60f). The luminescence was attributed to defects in the SiO_2 beads or the interface between the SiO_2 shell and SiC core, such as oxygen vacancy.

Silicon carbide beaded nanochains were prepared from carbothermal reduction of carbonaceous silica xerogel using cetyltrimethylammonium bromide and lanthanum

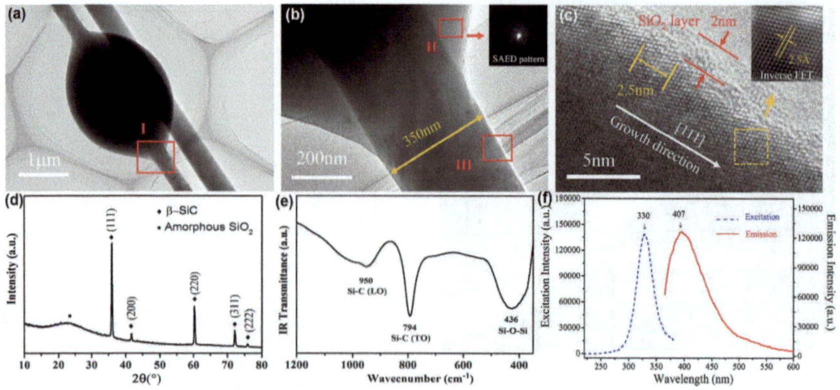

Fig. 5.60 **a** TEM image of a necklace-like structure. **b** Magnified TEM image of marked region in (**a**). Inset shows SAED pattern of marked region. **c** High-resolution TEM image of marked region III in (**b**). Inset shows inverse FFT image. **d** XRD pattern, (**e**) FTIR absorbance spectrum, and (**f**) PL and PL excitation spectra of as-prepared samples. Reprinted with permission from [113]. Copyright 2017, Elsevier B.V

nitrate as additives [114]. The nanochain consists of a stem with a diameter of about 50 nm and uniform beads with diameters of 100–200 nm (Fig. 5.61). The proportion of the nanochains in the sample is larger than 60%. X-ray diffraction pattern (Fig. 5.61e) reveals that the sample is pure cubic SiC and the peak on the left shoulder of the (111) peak is ascribed to the stacking faults in the nanowires. Most of the nanochains have nearly uniform periodicity, but others have varying periodicity. The dimensions of the beads and stems are approximately homogeneous in one nanochain but they vary among different nanochains. Many nanochains have amorphous thin shells. The nanochains preferentially grew along the <111> orientation. The beads of the nanowire comprise a higher concentration of stacking faults as compared with the stem area of the nanowire.

Long 3C–SiC microwhiskers with a necklace-like morphology have been synthesized through carbothermal reduction without using catalysts [115]. The wood flour/silica/phenolic composite was chosen as silicon and carbon sources. The microwhiskers grew along the <111> direction, with lengths up to hundreds of micrometers. The whiskers consist of 3C–SiC strings with diameters of 1–2 μm and periodic 3C–SiC beads with diameters of 3–5 μm. Rather homogenous β-SiC/SiO_2 chain-like nanowires can be synthesized by chemical vapor reaction using Si–SiO_2 powder and C_3H_6 gas as source materials with assistance of anodic aluminum oxide templates [116]. The SEM image (Fig. 5.62) shows that the synthesized nanostructures have uniform structures. Each nanowire comprises many spheres with sizes of 25–30 nm that are connected together by a long core wire with a diameter of 15–20 nm. No catalyst particles were found to attach to the resultant nanochains, so they may grow via a vapor–solid mechanism. High-resolution TEM images reveal that the SiC nanochains contain a high density of stacking faults and the nanochain surface was wrapped by a thin amorphous silica layer with a thickness of 2–5 nm. The

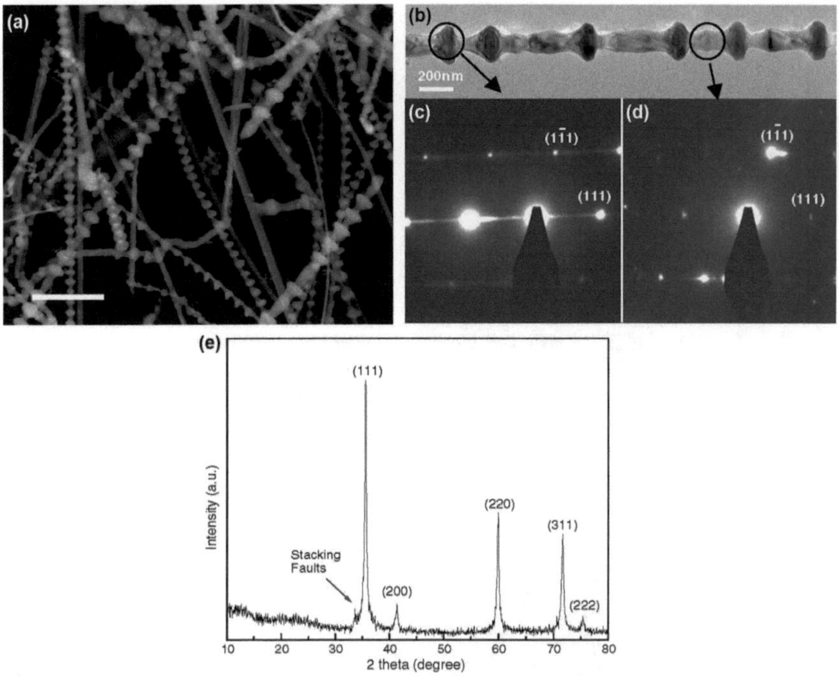

Fig. 5.61 **a** SEM and **b** TEM images of beaded SiC nanochains. The electron diffraction patterns **c**, **d** indicate the <111> growth direction and the streaks are assigned to the stacking faults. **e** XRD pattern of the nanochains. Reprinted with permission from [114]. Copyright 2006, IOP Publishing

TEM observation also indicates that the SiC core grew along the <111> direction. The nanochains showed photoluminescence with a single emission band centered at around 408 nm under 270 nm excitation.

5.4.3 Other Periodic Structures

Hierarchical single-crystalline β-SiC nanoarchitectures were synthesized using a catalyst-assisted CVD method [117]. The nanostructures were prepared by heating the mixed SiO, Ga_2O_3, and graphite powder at 1350 °C in argon atmosphere. Electron microscopy and X-ray diffraction reveal that the nanostructures are single-crystalline β-SiC. Each of them is composed of a core stem growing along the <111> orientation and a series of platelets growing perpendicular to the stem (Fig. 5.63). The core stems have diameters of 10–20 nm and the branched nanoplatelets have thicknesses of 5–10 nm. The electron diffraction pattern contains both bright spots and streaks and the streaks are assigned to the stacking faults in the nanostructures. The energy-dispersive X-ray spectroscopy reveals strong signals of Si and C from both the stems and the platelets and confirms their SiC structure. The hierarchical structures grew

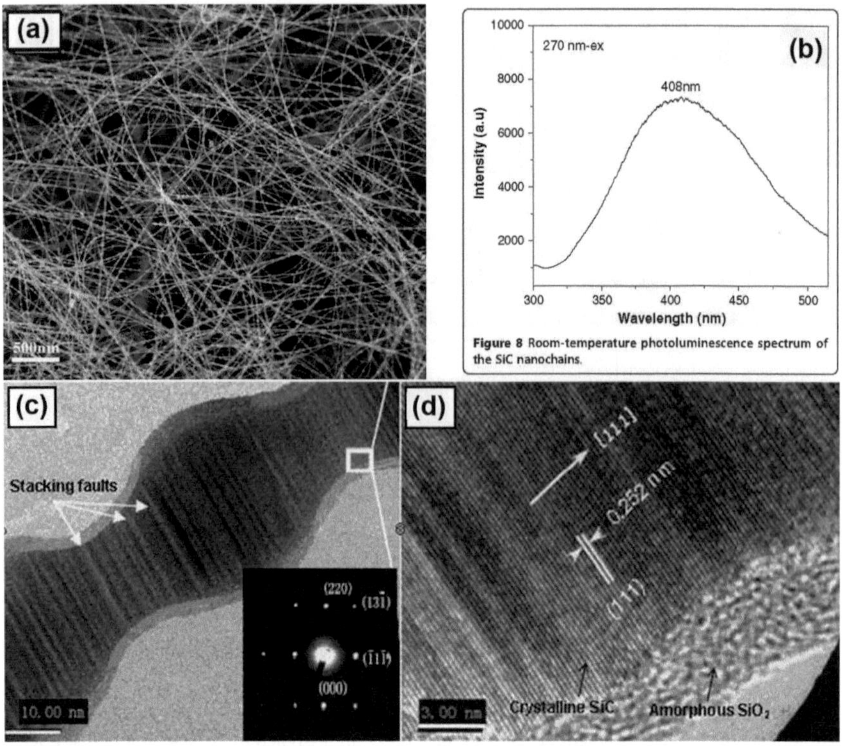

Fig. 5.62 **a** SEM image of the β-SiC/SiO_2 chain-like nanowires. **b** PL spectrum of the SiC nanochains under 270 nm excitation. **c** and **d** TEM image of a SiC nanochain. The inset is the electron diffraction pattern. (lower right) HRTEM image of the SiC nanochain. Reprinted with permission from [116]. Copyright 2011, Springer Nature Limited

by a two-step process. They showed field electron emission with a turn-on field of 12 V μm^{-1}.

β-SiC nanowires with branched fins were prepared by CVD using Si and SiO_2 powders and CH_4 as starting materials [118]. The fin diameter is 100–120 nm and the core stem diameter is 60–70 nm. Compressibility of periodically twinned β-SiC nanowires was studied using in situ high-pressure X-ray diffraction [119]. The twinned nanowires were synthesized by carbothermal reduction of a carbonaceous silicon xerogel using tetraethoxysilane and biphenyl as source materials. They are 50–300 nm in diameter and tens to hundreds of micrometers in length (Fig. 5.64). They have periodically twinned segments, hexagonal cross sections, and zigzag outer surfaces. They grew along the <111> direction. The cubic crystal structure was maintained under pressures up to 37.7 GPa. The bulk modulus of the nanowires is 316 GPa, 20–40% greater than that of SiC with other morphologies, showing that twinning improves the mechanical properties of the SiC nanostructures. 1D periodic 3C–SiC nanostructures with different diameters and morphologies were synthesized by sol–gel carbothermal reduction method [120]. Smooth nanowires,

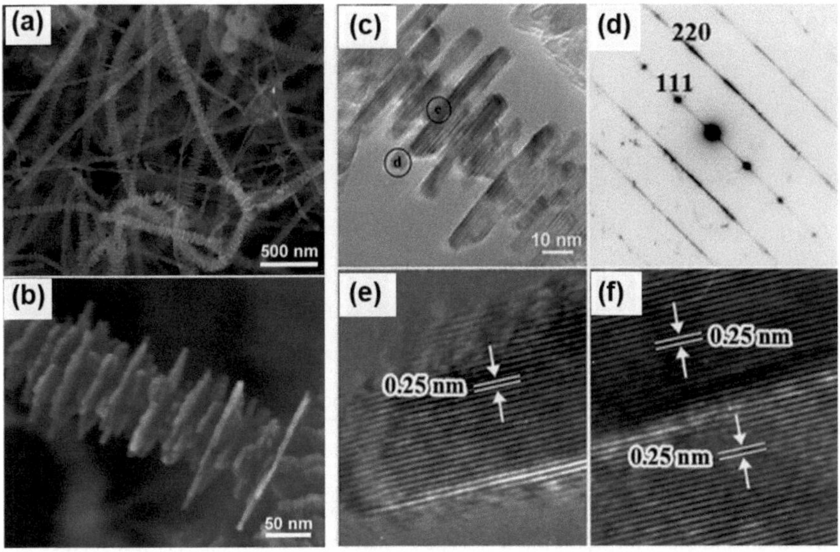

Fig. 5.63 **a** and **b** SEM images of one-dimensional hierarchical β-SiC nanostructures. Each hierarchical structure contains numerous thin nanoplatelets parallel to each other and perpendicular to the core stem. **c** TEM and **e, f** HRTEM images and **d** electron diffraction pattern of the nanostructures. Reprinted with permission from [117]. Copyright 2007, American Chemical Society

hierarchical nanodishes, and short nanorods formed at a reaction temperature of 1500, 1550, and 1600 °C, respectively. The vapor–solid epitaxial mechanism was proposed. Hierarchical and twinned SiC nanowires were synthesized by carbothermal reduction of the xerogel via a vapor–liquid–solid process [121]. The dominant polytype of the nanowires is 3C, but some take the 2H polytype. As a doping material, the SiC nanowires can be doped into the matrix of the nanocrystalline magnesium, where they showed good interlocking with the Mg matrix due to their special hierarchical structure. There are well-bonded interfaces between the doped SiC nanowires and the Mg matrix.

5.4.4 Connected Structures

There are many reports on synthesis of the SiC nanowires with interconnected structures. The isolated and branched β-SiC whiskers were synthesized from SiO_2 and carbon black powders via carbothermal reduction at 1420 °C [122]. The reactant mixture that contains synthesized whiskers was subjected to additional heating at 700 °C in air to eliminate excess carbon via oxidation. The nanowires formed via the gas–solid reaction between SiO and carbon black. Growth planes and growth directions of the whisker depend on the surrounding growth conditions including growth rate, content of stacking faults, and supply of carbon and SiO source. The

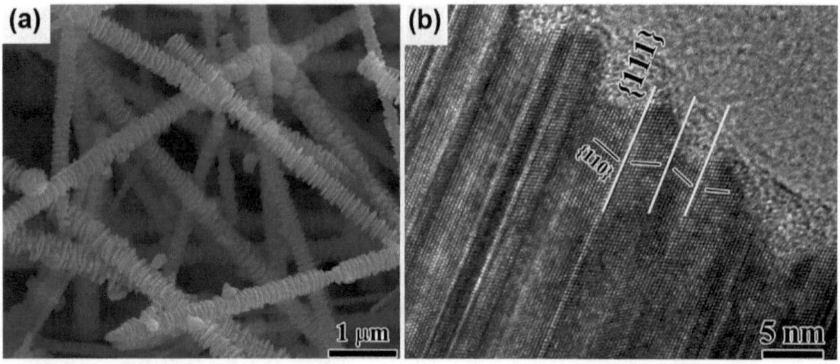

Fig. 5.64 **a** SEM image of periodically twinned SiC nanowires. **b** HRTEM image showing the twinned structure. Reprinted with permission from [119]. Copyright 2010, Elsevier B.V

whiskers were classified into three types (Fig. 5.65) in terms of morphology, growth direction, and stacking-fault planes. Type A whisker has a relatively flat surface and the stacking-fault planes are perpendicular to the growth direction. Type B whisker has a rough surface and the stacking fault planes are inclined at an angle of 35° to the growth direction. Type C has a rough sawtooth surface and the stacking faults exist concurrently in three different {111} planes. The high-resolution TEM images of the whiskers (Fig. 5.66) show that type A whisker contains many stacking faults perpendicular to the growth direction including many twin faults. The surface energy of the {111} planes in β-SiC is much smaller than those of other crystal planes, hence β-SiC whisker grew easily in the <111> direction and thus stacking faults can readily be inserted in the {111} planes. The 3C–SiC with a stacking fault has a lower energy than that without stacking faults and this explains the frequent occurrence of stacking faults in the SiC whiskers. In type B whiskers, the stacking faults that are inclined to the growth direction are present only in a branched whisker with a small thickness, and these whiskers mainly consist of twin faults. The electron diffraction patterns in type C whiskers contain three different directions of streaks that are perpendicular to respective set of stacking fault layers. The defected and branched whiskers have separation angles 125°, 70°, and 109°, respectively. These whiskers separately consist of mixtures of type A and type B, type A, or parallel growth by two pairs of type A and type B whiskers. The difference in growth speed between different types of whiskers caused the whisker deflection.

The β-SiC nanowire networks were synthesized by using chemical vapor reaction using a mixture of milled Si and SiC powder and C_3H_6 as starting materials [123], and the single-crystal β-SiC nanowires with diameters of 20–70 nm grew along the <111> direction. The SiC/SiO_2 core/shell nanowire networks were synthesized by high-temperature evaporation reaction using mixed Si, SiO_2, and graphite powders as source materials at 1500 °C in argon atmosphere [124]. The SEM observation (Fig. 5.67a) indicates that the product is an interconnected network of wires with diameters of about 180 nm. The TEM examination (Fig. 5.67b, c) reveals that the

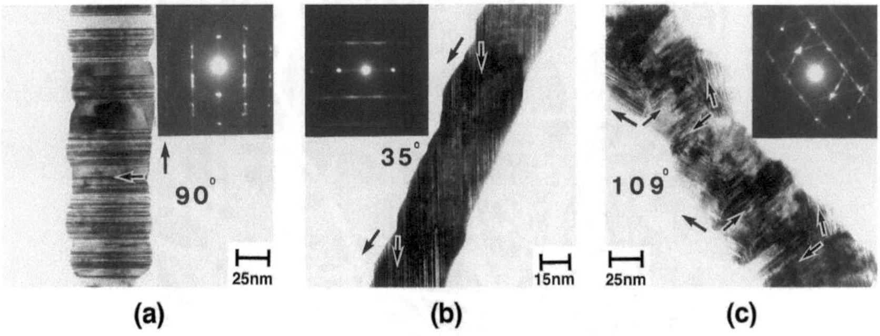

Fig. 5.65 TEM micrographs of three types of β-SiC whiskers: **a**) type A, **b** type B, and **c** type C. They are aligned with the electron beam parallel to the <110> zone axis. The insets are the corresponding electron diffraction patterns. Reprinted with permission from [122]. Copyright 2000, John Wiley & Sons, Inc

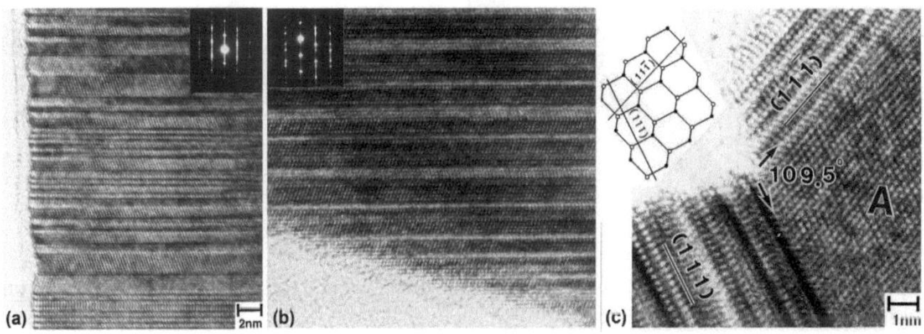

Fig. 5.66 HRTEM images of three types of β-SiC whiskers: **a** type A, **b** type B, and **c** type C. The insets show electron diffraction patterns or lattice configuration schematic. Reprinted with permission from [122]. Copyright 2000, John Wiley & Sons, Inc

connected wires are composed of core nanowires about 20 nm in diameter and outer sheaths of amorphous silica.

Biaxial β-SiC nanowires were fabricated by using high-temperature chemical vapor reaction using multiwalled carbon nanotubes and Si powder as starting materials at 1473–1573 K in vacuum [125]. Transmission electron microscopy reveals Y-shaped structure of the SiC bi-nanowires (Fig. 5.68). The individual nanowires grew along the <111> direction by a vapor–liquid–solid mechanism before they merged and co-grew to form a bi-nanowire. The bi-nanowire grew along the <110> , <113>, or <112> orientation. The SiC nanowires synthesized by employing polymer matrices [126] or by usual carbothermal reduction [127] can also exhibit interconnected structures.

Three-dimensional flower-like SiC nanostructures were synthesized via thermal reaction of gallium nitride powder, silicon wafer, and methane gas at 1100 °C [128]. Scanning electron microscopy reveals flower-like structures with diameters of 1–2 μm and lengths of 3–5 μm that grew on and are normal to the Si wafer (Fig. 5.69a).

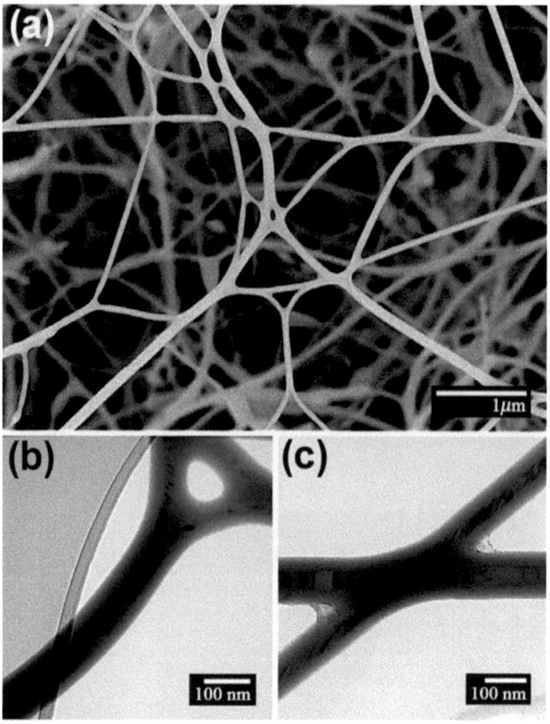

Fig. 5.67 **a** SEM and **b**, **c** TEM images of SiC nanowire networks. Reprinted with permission from [124]. Copyright 2007, IOP Publishing

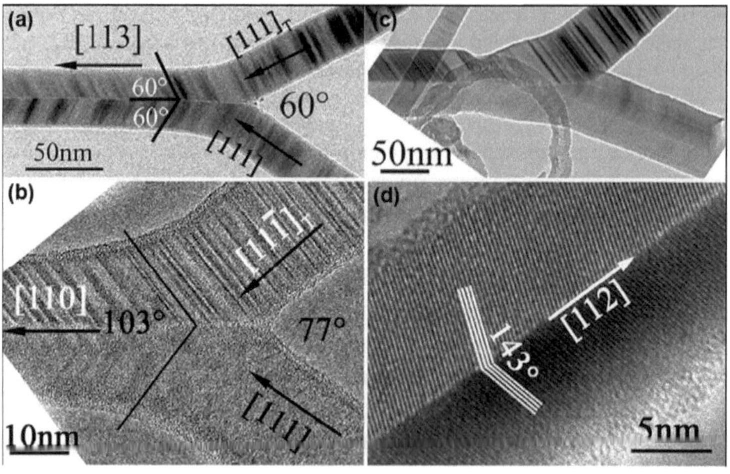

Fig. 5.68 TEM images of SiC bi-nanowires that grew along, **a** [108] axis, **b** [106] axis, and **d** [107] axis as well as **c** a SiC bi-nanowire growing from three nanowires. Reprinted with permission from [125]. Copyright 2010, IOP Publishing

Fig. 5.69 **a** SEM image of SiC nanowire flowers grown on the Si substrate. **b** Cross-sectional view showing highly aligned and closely packed nanowires. Reprinted with permission from [128]. Copyright 2004, IOP Publishing

A typical flower consists of a stem of interwoven nanowires, a head comprising a bundle of nanowires with diameters of 100–200 nm, and a single catalyst particle attached to the top end of the flower. The existence of the catalyst on the top of the product suggests a vapor–liquid–solid growth mechanism. Note that the thermal decomposition of gallium nitride gave rise to gallium which served as the catalyst. The cross-sectional view through the layer (Fig. 5.69b) reveals that each flower has a hollow stem with a diameter of 400–800 nm which is predominantly normal to the Si substrate. Each flower has a catalyst particle on its tip.

SiC tetrapods have also been synthesized. The adamantane embedded in a methyltrimethoxysilane was spin-coated onto a SiO_2 on Si wafer and the coated wafer was treated in microwaves at 850 °C in methane and hydrogen plasma to produce the SiC tetrapods [129]. Figure 5.70a shows the SEM image of the tetrapods. The tetrapod legs are 40 ± 10 nm in diameter and 90 ± 20 nm in length, and the legs have hexagonal cross sections. The TEM observation (Fig. 5.70b) reveals that the legs of the tetrapods are 4H–SiC crystals. Figure 5.70c shows the photoluminescence spectra of individual tetrapods at room temperature. The sharp luminescence line varies among different tetrapods in the 550–800 nm range and it has a very small full width at half maximum of about 5 nm. This variation may be due to the difference in the tetrapod structure. The luminescence lifetime of the tetrapods varies between 1.4–6.3 ns. Only the broad luminescence band (without sharp lines) was observed in the measured regions with high density of tetrapods.

5.4.5 *Helical Structures*

Helical crystalline β-SiC nanowires sheathed with amorphous silica (SiC/SiO_2) were synthesized by CVD at 1100 °C [130]. The nanowires formed on a Si substrate through the reaction $2SiO + C \rightarrow SiC + SiO_2$, decomposition of CH_4 with Fe

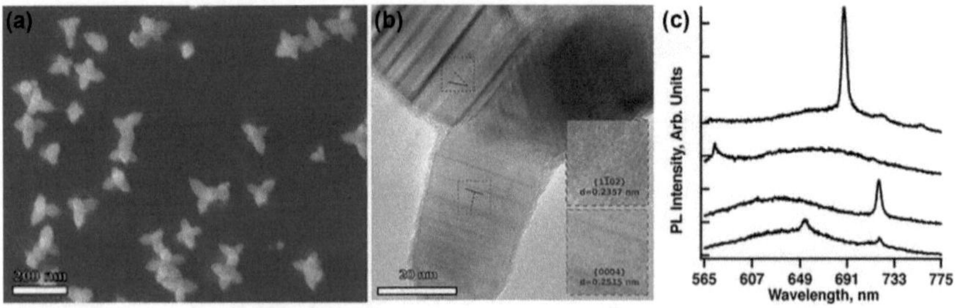

Fig. 5.70 **a** SEM image of the SiC tetrapods grown on the substrate. **b** HRTEM image of a tetrapod showing the lattice spacings for two legs with different orientations. The insets indicate the lattice spacings of 4H–SiC. **c** PL spectra of several single tetrapods excited at 532 nm. Reprinted with permission from [129]. Copyright 2013, American Chemical Society

catalysis provided the carbon source and the Si substrate surface offered the SiO source. The SiC core has a diameter of 10–40 nm and a helical periodicity of 40–80 nm; it is encapsulated by a uniform layer of amorphous silica with a thickness of 30–60 nm (Fig. 5.71). The growth of the helical nanowires may be driven by screw dislocations.

Amorphous SiC nanosprings and biphase (crystalline core/amorphous shell) helical nanowires were synthesized by using CVD [131]. Both structures grew by the vapor–liquid–solid mechanism, as demonstrated by the NiSi catalyst cap on the nanowire tip (Fig. 5.72a). The formation of the amorphous SiC nanosprings (Fig. 5.72b) was explained by the contact angle anisotropy model primarily proposed to explain the formation of amorphous boron carbide nanosprings. The formation of the biphase helical nanowires was explained by a modified contact angle anisotropy model, and there exists temperature gradient in the catalyst in this model.

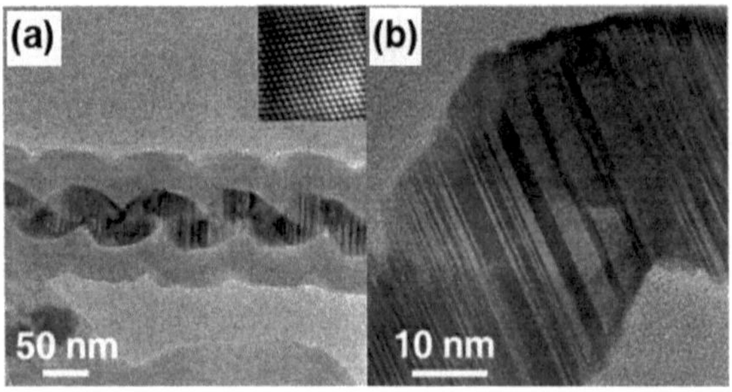

Fig. 5.71 **a** TEM image of a helical SiC nanowire. The inset shows the fast Fourier transform filtered lattice image of one part of the helical SiC nanowire in the right panel, showing that the SiC core is highly crystalline. **b** HRTEM image of a portion of a helical SiC nanowire. Reprinted with permission from [130]. Copyright 2002, American Chemical Society

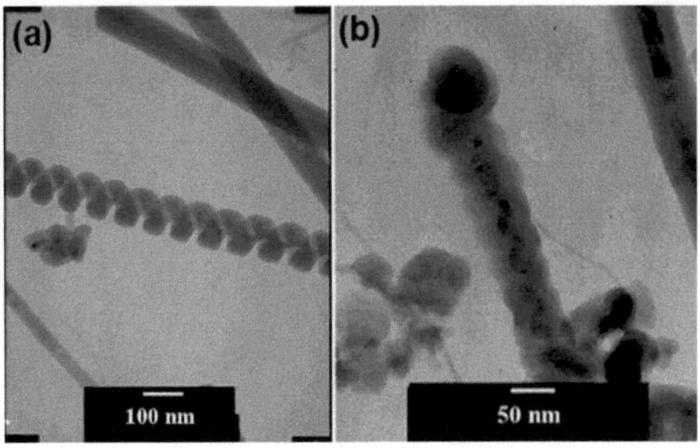

Fig. 5.72 **a** TEM image of amorphous SiC nanospring. **b** TEM image of biphase helical SiC nanowire with NiSi catalyst on its tip. Reprinted with permission from [131]. Copyright 2003, American Chemical Society

5.4.6 Nanobelts

Nanobelt is a thin, narrow, and long belt-like nanostructure. In comparison with SiC nanowires, there have been very few reports on the SiC nanobelts. A lithium-assisted synthetic route has been developed for the growth of 3C–SiC nanobelts at a mild temperature [132]. The nanobelts were prepared by reaction of CH_3CH_2OH, $SiCl_4$, and Li in an autoclave at 600 °C. They were characterized by various methods (Fig. 5.73). The X-ray diffraction pattern shows strong and sharp lines that belong to 3C–SiC, indicating good crystallization of the nanobelts. The sharp line on the shoulder of the <111> peak arises from the stacking faults in the nanobelts. Electron microscopy confirms the 3C–SiC structure of the nanobelts and shows that the nanobelts are 50–200 nm wide, 20–60 nm thick, and up to tens of micrometers long. The Raman spectrum of the nanobelts comprises two strong peaks at 788.5 and 960.5 cm^{-1}, characteristic of the TO and LO phonons of 3C–SiC, respectively. These two peaks shift to lower wavenumbers by 7–12 cm^{-1} relative to that of bulk 3C–SiC caused by phonon confinement and stacking faults. The nanobelts showed photoluminescence with one dominant emission band centered at around 402 nm under 229-nm excitation and there is an additional weak band at around 415 nm. Although there is metal lithium in the starting materials, but the SEM and TEM analysis indicates that the nanobelt growth was not governed by the vapor–liquid–solid mechanism and instead, the lithium acted as a reactant and its reaction with ethanol generated CLi_x gas, which reacted with $SiCl_4$ to give rise to the SiC nanobelts.

SiC nanobelts were synthesized by reaction of Si and carbon nanotubes without catalysts [133]. The nanobelts have a bicrystalline structure, which has growth directions, i.e., <121> and <111>, splitting along the twin boundary that exists at the central line. The nanobelts have widths of 100–200 nm and thicknesses of 10–20 nm. Their lengths are up to several tens of micrometers. The bicrystalline SiC nanobelts grew with the vapor–solid mechanism. The PL spectrum of the bicrystalline SiC nanobelts at room temperature comprises a strong and sharp emission peak centered at 418 nm and a weak broad band on the red side. Microwave-assisted reaction can produce SiC nanobelts. The 3C–SiC nanobelts with lengths of many micrometers and an average thickness of 4.4 nm were synthesized by microwave-assisted carbothermal

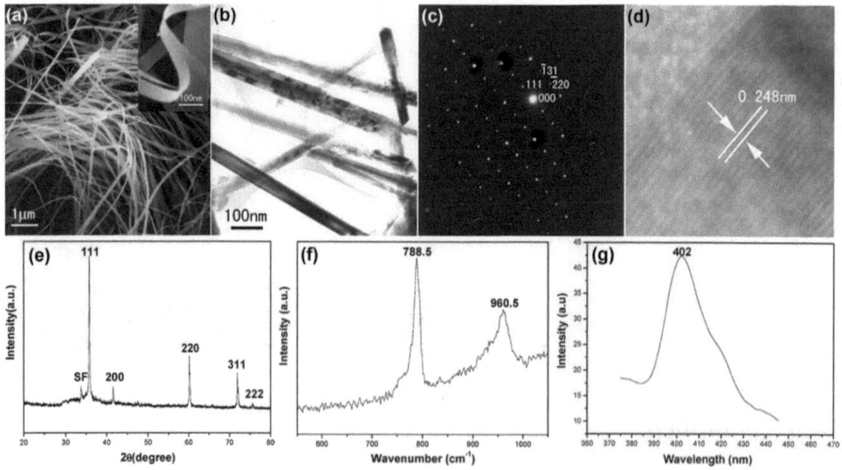

Fig. 5.73 **a** SEM and **b** TEM images of 3C–SiC nanobelts. **c** Electron diffraction pattern and **d** HRTEM image of a 3C–SiC nanobelt. **e** X-ray diffraction pattern, **f** Raman spectrum, and **g** PL spectrum (excited at 229 nm at room temperature) of the 3C–SiC nanobelts. Reprinted with permission from [132]. Copyright 2004, American Chemical Society

reaction between Si, SiO_2, and charcoal powders [134]. The SEM image (Fig. 5.74) shows that the nanobelts are 20–200 nm wide and up to many micrometers long. The nanobelts showed field electron emission with a low turn-on field of 3.2 V μm^{-1}. The photoluminescence of the nanobelts showed a relatively narrow band centered at around 391 nm under 325-nm excitation.

The SiC nanoribbons have been synthesized by direct thermal reaction of silicon and carbon black powders at 1250–1500 °C in argon atmosphere [135]. The nanoribbons are tens to hundreds of micrometers long, a few micrometers wide, and tens of nanometers thick (Fig. 5.75). The TEM, energy dispersive X-ray spectroscopy, X-ray diffraction, Raman scattering, and X-ray photoelectron spectroscopy reveal that the nanoribbons are wurtzite 2H–SiC and they grew by a vapor–solid mechanism with the growth direction of $[10\bar{1}0]$. Thermal treatment of wheat husks at over 1450 °C

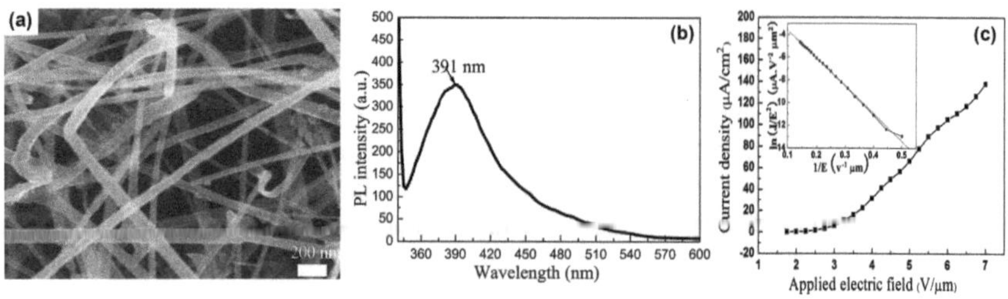

Fig. 5.74 **a** SEM image, **b** PL spectrum (under 325 nm excitation), and **c** field emission current density versus applied electric field curve for 3C–SiC nanobelts. Reprinted with permission from [134]. Copyright 2008, Elsevier B.V

Fig. 5.75 **a** and **b** SEM and **c**, **d** TEM images of the SiC nanobelts. The electron diffraction pattern is also shown. Reprinted with permission from [135]. Copyright 2010, Springer Nature Limited

can produce SiC nanobelts, along with nanotubes, nanowires, and nanorods [136]. The samples consist of 2H and 3C polytypes of SiC.

5.4.7 Legs on Fibers

Silicon carbide nanowires can grow on some nanofiber templates, giving rise to hierarchical nanostructures. The needle-shaped 3C–SiC nanowire arrays were grown from the polyacrylonitrile carbon fiber substrate using thermal evaporation of silicon on the carbon fibers in argon atmosphere at 1600 °C [137]. The polyacrylonitrile carbon fiber is composed of crystallized graphite and amorphous carbon. The SEM observation reveals that many bundles of needle-shaped nanowires grew on the outer surface of each carbon fiber (Fig. 5.76) and they have an average diameter of 100 nm and lengths of 5–10 μm. X-ray diffraction reveals that the nanoneedles are 3C–SiC. The growth of these SiC nanoneedles could be governed by the vapor–solid mechanism because no metal catalyst was found in the reactants and in the products. The SiC nanoneedles showed photoluminescence centered at around 468 nm under 380 nm excitation at room temperature.

Fig. 5.76 SEM images of needle-shaped SiC nanowires grown on polyacrylonitrile carbon fibers. **a** is higher magnification and **b** is lower magnification images. Reprinted with permission from [137]. Copyright 2008, Elsevier B.V

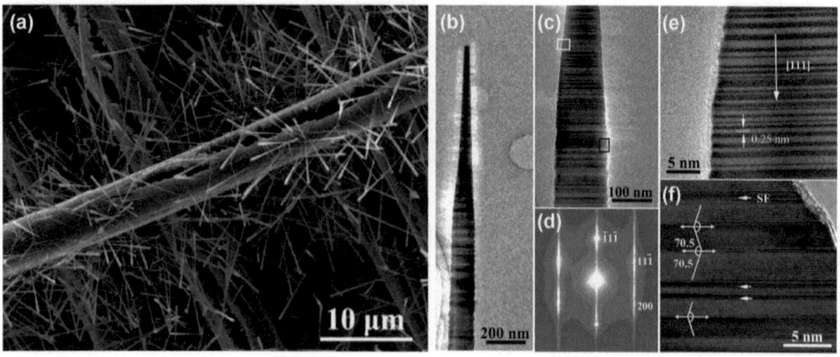

Fig. 5.77 **a** SEM image of tapered SiC nanowires on the carbon fabric. (right panel) (**b**) and **c** TEM images of a typical SiC nanowire showing tapered and zigzag facet morphologies. **d** Electron diffraction pattern of tapered SiC nanowires taken along [011] zone axis of cubic SiC. The streaks arise from stacking faults. **e** and **f** HRTEM images of the marked regions in (**c**). Reprinted with permission from [138]. Copyright 2012, American Chemical Society

Tapered SiC nanowires grew on the surface of a flexible carbon fabric by using CVD [138]. The thermal reaction of Si powder with iron nitrate-coated carbon fabric at 1500 °C in flowing argon gas yielded the SiC nanowires which are single-crystalline β-SiC growing along the <111> direction. The nanowire has zigzag facets on the outer surface caused by the quasi-periodic arrangement of twinning structures along the wire axis (Fig. 5.77). This twining structure ensures a lowest surface energy during nanowire growth. The structural characterizations and thermodynamics analysis demonstrate that the tapered SiC nanowires grew through the Fe-assisted vapor–liquid–solid process. The nanowires showed field electron emission with a very low turn-on field of 1.2 V μm^{-1}.

The β-SiC nanorods on electrospun nanofibers were synthesized by carbothermal reduction of Tb-doped SiO_2 nanofibers with graphite at 1250 °C [139, 140]. The nanorods are 100–300 nm in diameter and 2–3 μm in length (Fig. 5.78) and they grew by the vapor–liquid–solid mechanism with Tb serving as the catalyst. Transmission electron microscopy reveals that the SiC nanorods grew preferentially along the <111> direction. The nanorods exhibited field electron emission with a low turn-on field of 8 V μm^{-1} [140]. Carbon fibers or cloths prove to be very favorable substrates for accommodating the deposited SiC nanowires [141–143]. In addition, SiC backbone nanowires can also act as substrates for epitaxial growth of high-density SiC dendrites [144].

5.4.8 Nanowire Arrays

Highly oriented SiC nanowire arrays were synthesized by chemical vapor reaction of the mixed Si and SiO_2 powders on a piece of carbon cloth with assistance of

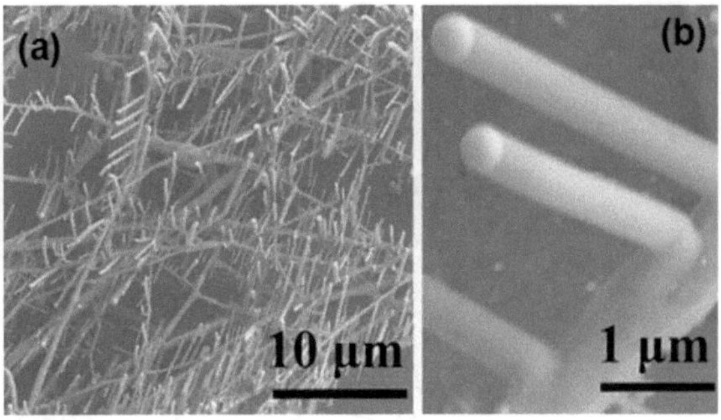

Fig. 5.78 SEM images of the SiC nanorods growing on the silica fibers. Reprinted with permission from [139]. Copyright 2009, Springer Nature Limited

an ordered nanoporous anodic aluminum oxide template under C_3H_6 gas flow at 1230 °C [145]. The nanowires are single-crystalline β-SiC with diameters of 30–60 nm and lengths of about 8 μm. They are uniformly oriented along the direction of the pores of the anodic aluminum oxide template (Fig. 5.79a). The nanowires grew along the <111> direction and they contain a high density of planar defects. The Raman spectrum comprises two strong and wide peaks at around 794 and 966 cm^{-1} belonging to TO and LO phonon modes of β-SiC. The photoluminescence spectrum (Fig. 5.79b) has a single strong peak centered at around 385 nm and this position is far bluer in comparison with most of the reported PL peaks in the SiC nanowires.

Hexagonal phase SiC nanowire arrays can grow vertically on single-crystalline SiC substrates [146]. The Raman spectroscopy and atomic resolution microscopy confirm that the nanowires display a polytypic distribution of predominantly 2H and 4H (Fig. 5.80). The polytypic distribution of hexagonal phase nanowires was predicted in the regime of high supersaturation. Disorder-induced phonon density of states can be reduced while maintaining nanowire morphology through a post-growth anneal.

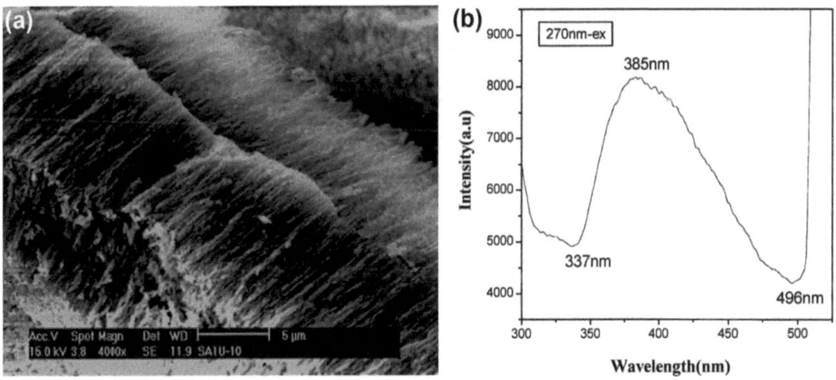

Fig. 5.79 **a** SEM image of oriented SiC nanowire arrays. **b** Photoluminescence spectrum of oriented SiC nanowires under 270 nm excitation. Reprinted with permission from [145]. Copyright 2006, American Chemical Society

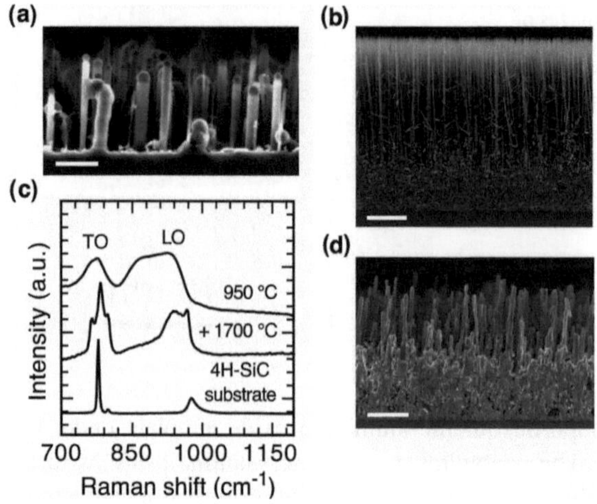

Fig. 5.80 **a** Cross-sectional SEM image of SiC nanowires growing on a high-quality 4H–SiC (0001) wafer for 2 min. Scale bar is 200 nm. **b** Cross-sectional SEM image of nanowires that grew for 64 min. Scale bar is 2 μm. **c** Raman spectra highlighting crystallinity repair of SiC nanowires with 3 min post-growth anneal at 1700 °C under inert argon gas. **d** Cross-sectional SEM image of SiC nanowires heat-treated to 1700 °C. Scale bar is 2 μm. Reprinted with permission from [146]. Copyright 2016, American Chemical Society

Large-scale highly oriented 4H–SiC gourd-shaped nanowire arrays were fabricated via electrochemical anodic oxidation of 4H–SiC bulk crystals assisted by pulsed voltage [147]. The formation of the gourd-shaped nanowires was ascribed to the applied pulsed voltage. Their field emission had a low turn-on field of 0.95 V μm^{-1}. A hybrid anodic and metal-assisted chemical etching method was proposed to fabricate SiC nanowires based on the wet etching approach [148]. The metal component acted as a catalyst to produce hole carriers and introduced band bending in SiC to accumulate sufficient holes for etching. By tuning the etching electrical bias and time, various nanostructures can be obtained, and the diameters of the obtained pores and nanowires range from tens to hundreds of nanometers.

As a wide-bandgap semiconductor, SiC has excellent mechanical and electrical properties and is a crucial template for epitaxially growing other semiconductors. The monolithic 6H-SiC nanowire arrays prove to be favorable substrates for epitaxially grown lead halide perovskites with superior photophysical properties [149]. The 6H-SiC nanowire arrays were fabricated by using the electrochemical anodization of an *n*-type single crystalline 6H-SiC wafer. They have diameters of around 100–400 nm (Fig. 5.81a–c). The small lattice mismatch (0.8%) between SiC and $CsPbBr_3$ ensures perfect heteroepitaxial growth of the $CsPbBr_3$ QDs and nanosheets over the SiC nanowire arrays (Fig. 5.81d–f). The heteroepitaxial perovskites showed intense multiband blue luminescence stemming from the strongly quantum-confined excitons with a twice prolonged lifetime compared with free nanocrystals. Figure 5.82 shows the 2D contour PL spectra of the as-prepared and aged (in air, for 4 days) samples. There are four discrete narrow (FWHM: 12–27 nm) emission bands centered at 462, 479, 490, and 516 nm. They are characteristic of the luminescence of the $CsPbBr_3$ nanocrystals, and the discrete peaks suggest there are discrete-sized nanocrystal ensembles owing to the hierarchical structure of the monolithic SiC nanowire arrays.

The deep layers of the forest-like 4H- and 6H-SiC nanowire (NW) arrays (intrinsically bound as a single crystal) may lie in microcavities that formed naturally underneath the outmost surface of the etched porous SiC wafer. As a result, the

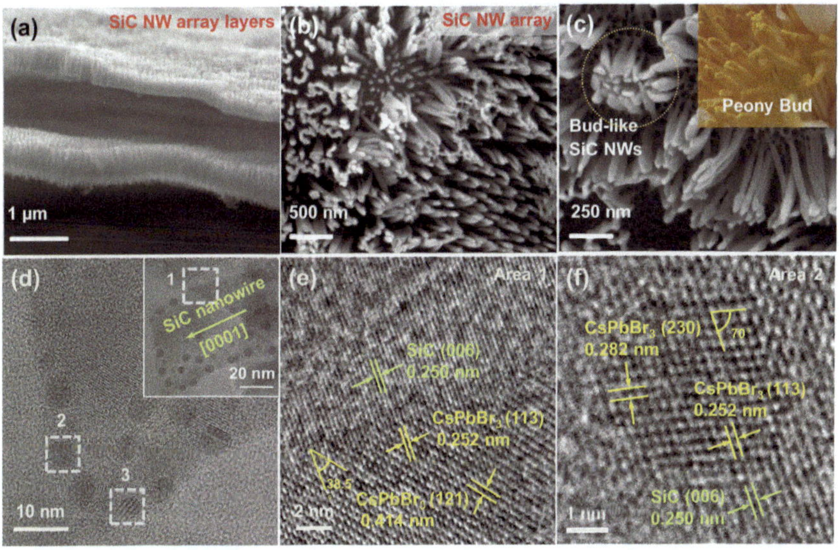

Fig. 5.81 **a** Lateral view and **b**, **c** top-view SEM images of 6H-SiC NW arrays after peeling off the surface layer. **d** High-resolution TEM image of perovskite–SiC heteronanocrystals and **e**, **f** magnified images. Reprinted with permission from [149]. Copyright 2023, IOP Publishing

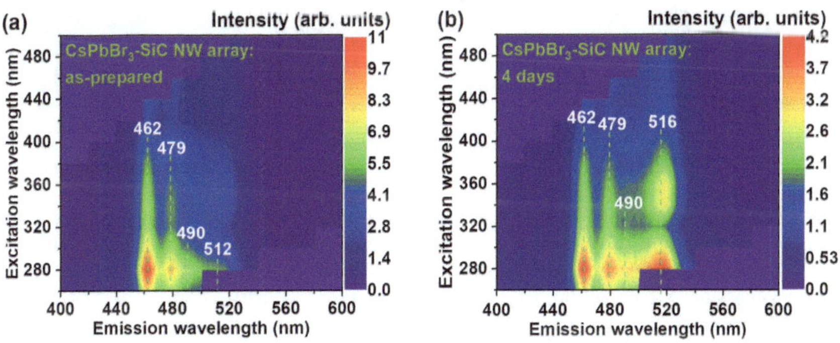

Fig. 5.82 2D contour PL spectra of **a** as-prepared and **b** air-stored heteroepitaxial perovskite–SiC NW arrays. Reprinted with permission from [149]. Copyright 2023, IOP Publishing

nanowire arrays can exhibit interesting broadband cavity-mode light emissions spanning the red–green–blue spectral region [150]. The naturally formed Fabry–Pérot cavity microcavity (Fig. 5.83a, b) encapsulating the SiC nanowires yielded discrete and multimodal emission lines (Fig. 5.83c). Time-resolved PL spectrum of the 4H–SiC NWs (emission: 576 nm, excitation: 400 nm) (Fig. 5.83d) satisfies well the stretched-exponential decay law

$$I(t) = I_0 \exp\left[-\left(\frac{t}{\tau}\right)^{1/h}\right] + C, \tag{5.13}$$

where I_0 is the initial PL intensity, τ is a constant measuring the relaxation lifetime, h measures the distribution range of the lifetimes, and C represents the background signal. The fitting gives τ and$h$ for two decay PL spectra. The average lifetime follows $\tau_{\text{ave}} = h \times \tau \times \Gamma(h)$ (Γ: gamma function). The femtosecond laser can penetrate into the deep layer of the etched SiC wafer, reach and excite the interior NWs within the cavity, leading to cavity-mode PL decay with a very short average lifetime ($\tau_{\text{cav\_ave}}$) of 200 ps (Fig. 5.83d). In contrast, the faint light coming from the NanoLED can only excite the surface-exposed NWs, leading to free-space mode PL

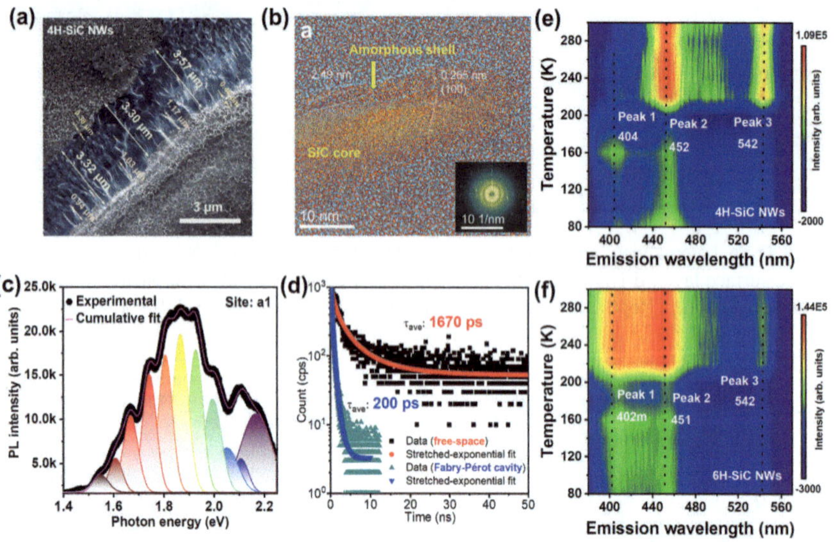

Fig. 5.83 **a** Side-view SEM image of 4H–SiC forest-like nanowire array with a cavity structure. **b** High-resolution TEM image of a 4H–SiC nanowire. **c** Cavity-mode PL spectra of 4H–SiC nanowire array under 532 nm laser (202.8 μW) excitation. **d** Time-resolved PL spectra (excitation: 400 nm, emission: 576 nm) of the 4H–SiC nanowire array in the Fabry–Pérot cavity excited under intense femtosecond laser and that of the exposed free-space 4H–SiC nanowire array excited under weak NanoLED light. **e** and **f** 2D contour temperature-dependent PL spectra (excitation: 325 nm) of 4H–SiC nanowires and 6H-SiC nanowires. Reprinted with permissions from [150]. Copyright 2025, SPIE

decay with a much longer lifetime (τ_{free_ave}) of 1670 ps. Quantum electrodynamics predicts that the optical microcavity can highly reduce or prolong the spontaneous emission lifetime of the embedded emitter in it if the light emission is in resonance or detuning with the cavity mode, as a result of highly improved or reduced photon density of states. The improvement in near-resonance photon density of states and improvement in spontaneous emission rate are called Purcell effect. The linewidth of the Fabry–Pérot cavity resonance mode (Fig. 5.83c) of the 4H–SiC NWs follows [151, 152]

$$\frac{c}{2\sum_i n_i l_i}\frac{1-(R_1R_2)^{1/2}}{\pi(R_1R_2)^{1/4}}, \tag{5.14}$$

where c is the light speed in vacuum, and R_1 and R_2 are the reflectivity of two cavity surfaces. For simplicity, assume an air–SiC interface on both upper and lower surfaces. The calculation gives a linewidth of 0.034 eV for the 2.17 eV peak. The spontaneous emission rate in the case of weak light–matter coupling is proportional to the local photon density of states (Fermi's golden rule). Consequently, the Purcell factor, that is, the ratio between the radiative transition rate of the cavity mode and that of the free-space mode, follows [153]

$$F_p = \frac{1/\tau_{cav\_ave}}{1/\tau_{free\_ave}} = \frac{3\lambda_c^3}{4\pi^2 n^3}\frac{Q}{V_m}, \tag{5.15}$$

where λ_c is the free-space wavelength, n is the refractive index of the medium surrounding the emitter, the cavity quality factor $Q \approx \upsilon_c/\Delta\upsilon_c = 63.8$, as obtained using the linewidth 0.034 eV for the 2.17 eV peak, and the mode volume $V_m = \pi d^2 H/4$, with H the cavity length and d the cavity-mode beam diameter. Assume that the mode-limited spontaneous emission from the single cylinder-like SiC NW has a mode cross section close to its diameter of ~ 42.9 nm. The refractive index of the SiC NWs (2.67 at 571 nm) is significantly greater than unity of air and the waveguide effect renders the light field more confined in the NWs, leading to a reduction in the effective mode volume. The Purcell factor is calculated to be 8.6 using Eq. (5.15), in agreement with $F_p = \left(1/\tau_{cav\_ave}\right)/\left(1/\tau_{free\_ave}\right) = 8.4$, as obtained from the measured lifetimes with assumption of a constant quantum yield (it may bring error but not qualitatively affect the conclusion). The big Purcell factor indicates significant increase in the spontaneous emission rate of the SiC NWs in the cavity compared to that in free space.

The low-temperature photoluminescence feature and work function show significant dependence on the nanowire polytype. To one's surprise, the photoluminescence exhibited peculiar staircase-function enhancement when the temperature increased and reached around 200 K, and after that, the luminescence intensity kept stable at least up to room temperature (Fig. 5.83e, f). It shows the contrary trend to the usual PL quenching behavior with elevating temperature:

$$I(T) = \frac{I_0}{1 + C * \exp[-(E_A/k_B T)]}, \tag{5.16}$$

where I_0 and E_A are the saturation intensity and activation energy of thermal quenching, and C is a constant. This peculiar abrupt luminescence enhancement might be ascribed to suppression of the nonradiative transition channels at elevated temperatures. The electronic structures of the single-root 4H- and 6H-SiC nanowire forests were further investigated. The work function of the sample was calculated from the contact potential difference (CPD) measured using the Kelvin probe force microscope (KPFM). CPD is defined as the potential difference between the sample and the AFM tip: $V_{CPD} = (\phi_{tip} - \phi_{sample})/e$, where ϕ_{sample} and ϕ_{tip} are work functions of the sample and the tip, and e is the elementary charge. Figure 5.84c shows the statistical CPD value distributions of the 4H and 6HSiC nanowires. The CPD values are approximately uniform across the whole measured region. The average CPD of the 4H–SiC nanowires and 6H-SiC nanowires are 595.2 and 487.8 mV, respectively, indicating a smaller work function and a higher Fermi energy of the former. The ultraviolet photoelectron spectroscopy (UPS) (He I energy: 21.22 eV) of the 4H- and 6H-SiC NWs provides more information (Fig. 5.84a, b). The linear fit of the higher energy tails of the UPS spectra gives work functions of 3.78 and 4.37 eV for the 4H- and 6H-SiC nanowires, respectively. The smaller work function of the 4H–SiC nanowires is consistent with the KPFM result. The linear fit of the lower energy tails gives valence band maximum (VBM) energies of – 2.46 and – 2.52 eV (relative to Fermi energy) for the 4H- and 6H-SiC NWs, respectively. The SiC nanowires without apparent quantum confinement effect share the same bandgap with the SiC bulk crystal. The bandgap of 4H–SiC is ~ 3.28 eV and that of 6H-SiC is ~ 3.08 eV. The energy band diagrams for the 4H- and 6H-SiC nanowires are illustrated on the basis of these results (Fig. 5.84d). Apparently, these nanowires are *n*-type semiconductors and this feature inherits from that of the mother bulk crystals. The 4H–SiC nanowires have an affinity energy of 2.96 eV, being small than that of the 6H-SiC NWs (3.81 eV).

Electron beam lithography can create more ordered SiC nanorod arrays, although the size of a single nanorod is usually much bigger than that obtained by using the chemical or electrochemical etching methods. The spectrum of near-field thermal radiation from arrays of 6H-SiC nanopillars was measured using an internal-reflection-element-based spectroscopy technique [154]. The bottom and top diameters of the nanopillars are 700 and 500 nm in sample A, and 650 nm and 500 nm in sample B, respectively (Fig. 5.85). The interpillar spacings (bottom-to-bottom gap between two adjacent nanopillars) are 300 and 350 nm for samples A and B, respectively. The heights of the nanopillars in samples A and B are 1 and 1.6 μm, respectively. The height and width of the conical trench around each nanopillar is 50 nm in dample A. Figure 5.86 shows the spectra at 150 °C for sample A and a flat 6H-SiC sample. The spectrum of the flat SiC has a peak at 935 cm^{-1} corresponding to thermal emission of propagating surface phonon polaritons in the Reststrahlen band of 6H-SiC (796–965 cm^{-1}). The near-field spectrum of sample A differs substantially from that of the flat surface. It comprises five peaks at 777, 843, 876, 910,

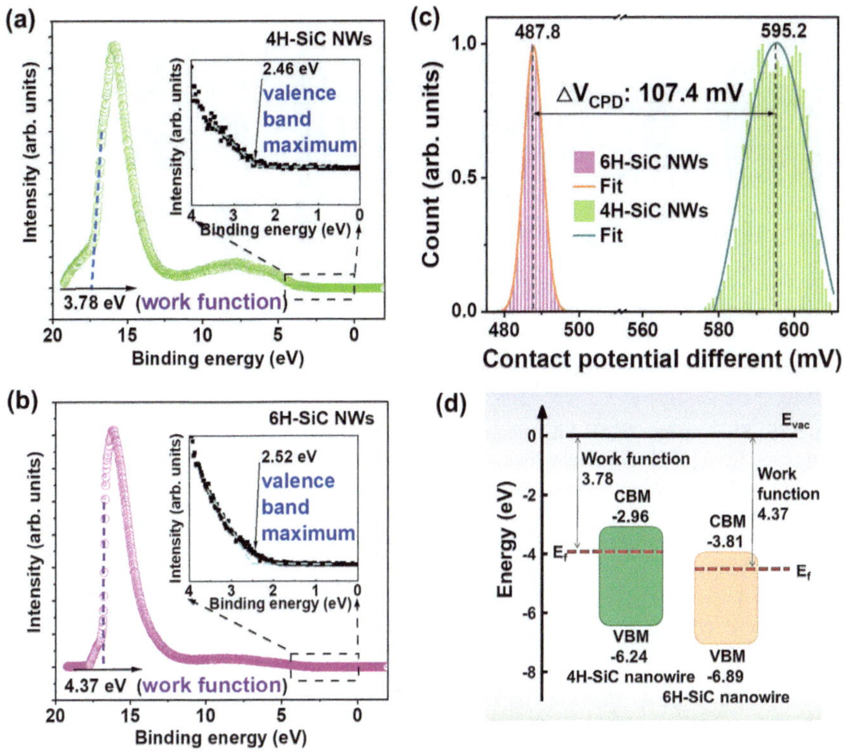

Fig. 5.84 **a** and **b** UPS spectra of 4H- and 6H-SiC nanowires. **c** Kelvin probe force microscope contact potential difference value statistical diagrams of 4H–SiC nanowires and 6H-SiC nanowires. **d** Corresponding energy band diagrams with denoted UPS-derived work function as well as valence band maximum (VBM) and conduction band minimum (CBM). Reprinted with permissions from [150]. Copyright 2025, SPIE

and 933 cm^{-1}. The simulated spectrum comprises five peaks at 794, 832, 899, 919, and 946 cm^{-1}. These peaks arise from thermal emission of the transverse dipole, quadrupole, and octupole as well as longitudinal monopole from 6H–SiC nanopillars at a near-field distance from the array. The spectral locations of the longitudinal monopole and transverse dipole are affected by the near-field coupling between neighboring nanopillars and coupling between the nanopillars and the substrate. The spectrum of near-field thermal radiation from the 6H-SiC nanopillar arrays is tunable by varying the sizes of the nanopillars.

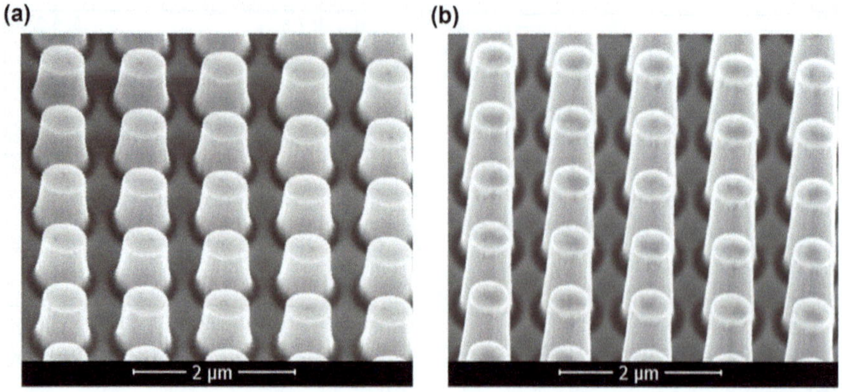

Fig. 5.85 SEM images of 6H-SiC nanopillar arrays: **a** sample A and **b** sample B. Reprinted with permission from [154]. Copyright 2023, American Chemical Society

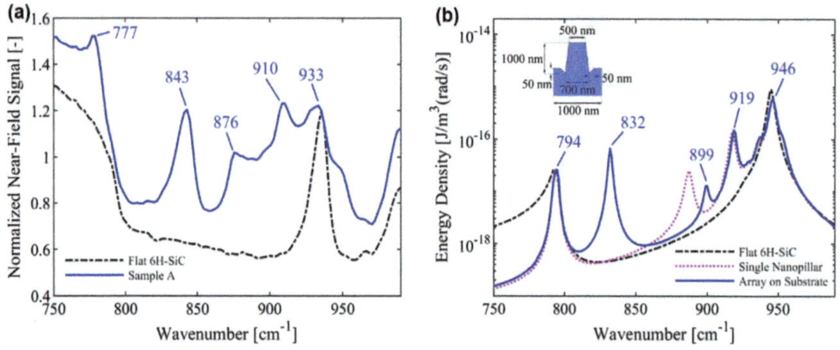

Fig. 5.86 **a** Measured near-field thermal radiation spectra for sample A and a flat 6H-SiC sample at 150 °C and a distance of 100 nm above the samples. The near-field spectra are normalized by the far-field thermal emission for a carbon nanotube at the same temperature. **b** Theoretical near-field energy density for sample A, a flat 6H-SiC sample, and a single free-standing nanopillar at 150 °C and a distance of 100 nm. The inset shows the dimensions of the modeled unit cell of sample A. Reprinted with permission from [154]. Copyright 2023, American Chemical Society

5.5 Some Characteristics

Besides the commonly concerned field electron emission and luminescence properties, some other properties of the SiC nanowires have also been investigated. The electrical transport as well as optical properties of single crystalline 3C–SiC nanowires have been investigated [155]. The nanowires were fabricated by using CVD using Si/SiO_2 wafer and methyltrichlorosilane (CH_3SiCl_3) as source materials under Ni catalysis in flowing hydrogen at 950 °C. These nanowires have diameters smaller than 100 nm and lengths of several micrometers. Transmission electron microscopy reveals that the nanowires grew along the <111> direction. They exhibited blue

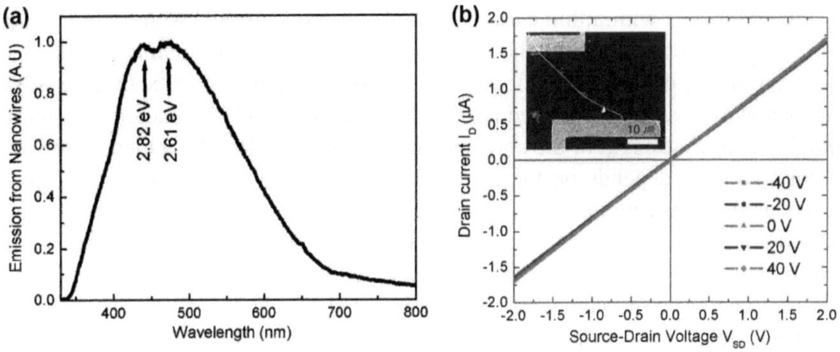

Fig. 5.87 **a** PL spectrum of the SiC nanowires. **b** *I–V* curve for the SiC nanowire connected by Ti/Au electrodes after annealing at 700 °C for different back gating from – 40 to 40 V. The inset shows the SEM image of the two-terminal SiC nanowire device. Reprinted with permission from [155]. Copyright 2004, American Institute of Physics

photoluminescence centered at around 450 nm (Fig. 5.87a) at room temperature. The *I–V* measurement of the nanowires (Fig. 5.87b) indicates that they have a low resistivity of 2.2×10^{-2} Ω cm for 0 V gate voltage at room temperature. The estimated electron mobility of the nanowires is 15 $cm^2 V^{-1} s^{-1}$. A *p*-type SiC nanowire field-effect transistor using an individual Al-doped 3C–SiC nanowire was realized [156]. The Raman spectroscopy indicates that the linewidth and peak intensity of the LO phonon bands are sensitive to temperature variations.

Like bulk SiC, the SiC nanowires have good mechanical and thermal properties. The unusually large strain plasticity of ceramic SiC nanowires at near room temperature has been observed by using in situ transmission electron microscopy [157]. The continuous plasticity of the SiC nanowires was accompanied by a process of increased dislocation density at an early stage, followed by an obvious lattice distortion, and finally reached complete amorphization at the most strained region of the nanowire. Superplasticity of single-crystal β-SiC <111> nanowires for over 200% elongation was observed by using in situ axial-tensile scanning electron microscopy [158]. The axial localized plasticity and super-plasticity only result from the 3C segments in the SiC nanowires, through dislocation generation, propagation, and amorphization in contrast to the highly defected structural segments that show elastic-deformation only due to the lack of slip systems. The measurement of the high mechanical *Q* factor for singly clamped SiC nanowires by field emission in ultrahigh vacuum reveals a room temperature *Q* factor of 159,000 after high temperature in situ cleaning [159, 160]. A laser Doppler vibrometer determined the Young's modulus of Fe-catalyzed 3C–SiC nanowires at 300–575 K [161]. The nanowires grew along the <111> direction and exhibited different cross-sectional geometries of circle, rectangle, hexagon, ellipse, trapezoid, and triangle. When the effective diameters of the nanowires decreased from 200 to 55 nm, their room-temperature Young's modulus decreased from ~ 550 GPa (bulk value) to ~ 460 GPa, and their temperature coefficient of Young's modulus varied from -47.4 ± 1.7 ppm K^{-1} (bulk value) to -78.1 ± 5.9 ppm K^{-1}.

SiC nanowires with a thin carbon coating can reinforce SiC matrix composites with a high efficiency [162]. The composites were fabricated in situ by a process based on chemical vapor infiltration. When ~ 6 vol% randomly oriented single-crystal SiC nanowires were incorporated into the SiC matrices, both the fracture toughness and the flexural strength of the composites were doubled. Reinforcement efficiency of the nanowire depends on the amount of C deposited on the nanowires which serves as the nanowire/matrix interfacial layer. The SiC nanowires with diameters of 100 nm exhibited whisker strength of about 24 GPa, showing their promise for reinforcing the performance of bulk ceramic composites. Quantitative mechanical characterization of SiC nanowires was implemented through in situ tensile tests inside SEM using a microelectromechanical system [163]. The nanowires synthesized with a vapor–liquid–solid process grew along the <111> orientation. They have a pure 3C structure, 3C structure with an inclined stacking fault, or highly defective structure. The SiC NWs deformed elastically until brittle fracture. The fracture origin was identified using the 3C structures with inclined stacking faults. Their fracture strength increased when the nanowire diameter decreased from 45 to 17 nm and it approached the theoretical strength of 3C–SiC. The size effect on fracture strength was ascribed to the size-dependent defect density rather than the surface effect which dominates for single crystalline nanowires. Loop bending in air caused the <111>-oriented 3C–SiC nanowires with stacking faults elastically deformed until brittle breaking [164]. The measured fracture strain lies in 4–7%, about two orders of that for its bulk counterpart. TEM analysis reveals that electron irradiation caused amorphization in the strained SiC nanowire.

In situ electrical measurement by TEM of an individual <111> -oriented 3C–SiC SiC nanowire subjected to tensile strain reveals fracture strain approaching 10% [165]. The conductance increased monotonously with tensile strain. The calculated piezoresistance coefficient $- 1.15 \times 10^{-11}$ Pa^{-1} of this SiC nanowire is close to that of bulk material. The piezoresistance behaviors of *n*-type 3C–SiC nanowires showed favorable performances [166]. The very low current in pA levels can be fed back responding to the applied compressive forces in nN levels. The transverse piezoresistance coefficient of the nanowire lies in $0.75–7.7 \times 10^{-11}$ Pa^{-1} for an applied force of 25.59–153.56 nN. B-doped *p*-type 3C–SiC nanowires synthesized by catalyst-assisted pyrolysis of polysilazane showed giant piezoresistance effect [167]. The resistance of these nanowires increased with increasing compressed stress at the same bias, showing negative piezoresistance behaviors. The negative piezoresistance coefficient fell in the range of $- 8.83$ to $- 103.42 \times 10^{-11}$ Pa^{-1} when the applied loading force increased from 51.7 to 181.0 nN. The giant gauge factor is up to – 620.5. Highly sensitive pressure sensors were implemented using piezoresistive 3C–SiC nanowires [168]. Experimental results show threefold enhancement in sensitivity compared to that using conventional structures.

The measured thermal conductivities of the single and double β-SiC nanowires are 82 and 73 W mK^{-1}, respectively [169] and these values are smaller than that of bulk material. The time-domain thermoreflectance measurements probed the thermal conductivity of SiC nanowires as a function of their length and temperature, and revealed that quasi-ballistic heat conduction may persist on the microscale at operating temperatures of power devices [170]. The deviation of the thermal conductivity from the diffusive limit in nanowires shorter than a few micrometers indicates the transition into a quasi-ballistic thermal conduction regime. Although the deviation is larger at lower temperatures, the effect persisted even at over room temperature. The Monte Carlo simulations of phonon transport support the experimental results and suggest how phonons with long mean free paths carry substantial heat, leading to quasi-ballistic conduction. Thermal transport measurements of bent SiC nanowires at 20–300 K reveal reduction in thermal conductivity compared to their straight counterparts [171]. The relative change is up to 55% at 20 K and decreases with temperature owing to the inhomogeneous strain-induced phonon scattering.

The dielectric properties of the SiC nanowires depend strongly on their chemical compositions, especially on the carbon content [172]. The complex permittivity calculated from the measured reflectivity decreased with increasing frequency and the nanowires with higher carbon content showed higher complex permittivity. 3C–SiC nanowires with high stacking faults density showed higher permittivity in 2–18 GHz [173]. The stacking faults of 2H polytype formed in the 3C–SiC matrix, leading to formation of the type-II 3C–SiC/2H–SiC heterostructures in that both the valence-band maximum and conduction-band minimum of 3C–SiC are lower than that of 2H–SiC. The charge separation at the type-II heterointerface yields plenty of interface dipoles, leading to big dipole polarization loss. This may account for the high dielectric permittivity of these SiC nanowires.

The Raman spectra of SiC nanowires are usually distinct from that of SiC powders of various grain sizes [174]. The temperature and pressure dependence of the Raman spectra of the SiC powders is similar to that of bulk crystals, but the SiC nanowires show distinct dependence. SiC nanowires with two different mean diameters, 20 nm (NW-A) and 200 nm (NW-B), and lengths approaching several hundred micrometers were obtained by reaction between multiwalled carbon nanotubes and silicon in the liquid phase at 1450 °C (NW-B) and in the solid phase at 1250 °C (NW-A). In both samples the concentrations of defects, especially stacking faults, are very high, as revealed by the frequency separation between the TO peak and its lower frequency shoulder, large bandwidths of phonon peaks, and enhanced intensity of the defect-sensitive band at ~ 860 cm^{-1}. The concentration of defects varied within the sample from one site to another in NW-A. With decrease of the grain size, the TO phonon shifts to higher frequency, whereas the LO phonon to lower frequency. The hydrostatic pressure dependence of the TO and LO phonons frequencies of bulk SiC was investigated [24, 25]. For nanosized SiC powders and nanowires at room temperature and pressures up to 50 kbar (5 GPa), the TO and LO phonon frequencies shift to higher values and their peaks broaden with increasing pressure. Raman intensity enhancement of the LO phonon was observed when the excitation laser wavelength decreased from 780 to 514 nm.

5.6 Theoretical Studies

The mechanical and electronic properties of the silicon carbide nanowires have been widely investigated theoretically using molecular dynamics simulations and first-principles calculations. The external loading effect on the elastic properties of the 3C–SiC nanowires with different diameters and structural changes was studied using molecular dynamics simulations [175]. The elastic modulus and limit of structural integrity are highly independent of the wire diameter under uniaxial compressive and tensile loading (Fig. 5.88). In contrast, the elastic response and structural deformation mechanism near the elastic limit depend strongly on the nanowire diameter under nonaxial torsional and bending strains. The critical angle of torsion is approximately inversely proportional to the wire radius, and the torsional rigidity increases with the wire radius following a power law. The bending loading shows similar effects. The nonlinear effects play an important role in defining the external bending and torsional strain effect of the nanowires with diameters smaller than 1 nm. Some mechanical properties of the <111> -oriented SiC nanowires are comparable to or even better than that of carbon nanotubes of similar diameters.

The size dependence of surface stress, surface stiffness, and Young's modulus of a prism-cross-section nanowire was theoretically investigated [176]. The calculation shows that the edge phase induces the size dependence of the nominal specific surface energy, surface stress, and surface stiffness, whereas the surface phase and the edge phase cause the size dependence of the nominal Young's modulus (Fig. 5.89). The edge and surface effects become more important as the cross-sectional area of a nanowire becomes smaller. Molecular dynamics simulation conducted on hexagonal prism <111> β-SiC nanowires confirms the soundness of the theoretical approach. The mechanical properties of the <111>-oriented β-SiC nanowires under tension, compression, torsion, and combined loading were investigated using molecular dynamics simulations employing the Tersoff bond-order potential [177]. The collapse of the SiC nanowires is dependent on the size and the temperature under axial tensile

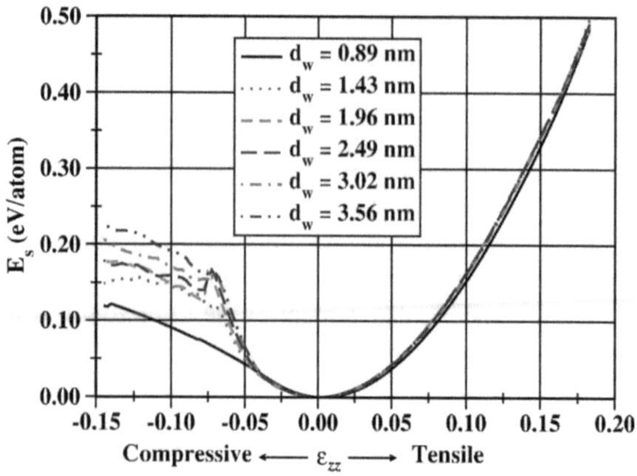

Fig. 5.88 Normalized (per atom) elastic energy as a function of uniaxial compressive strain and tensile strain for six SiC nanowires with different diameters. Reprinted with permission from [175]. Copyright 2006, American Physical Society

strain. Under axial compressive strain, the collapse of the SiC nanowires by yielding or column buckling mode is dependent on the length and diameter of the nanowires. Under torsion strain, the SiC nanowires collapse through a crystal to amorphous phase transformation in several atomic layers. Molecular dynamics simulations with the Tersoff potential reveal the response of twinned SiC nanowires under tensile and compressive strain [178]. Twin stacking faults can improve the critical strain of the twinned nanowires, and their critical strains are bigger than those of perfect nanowires with the same diameters. Under axial tensile strain, the bonds of the nanowires are stretched just before failure. The failure behavior depends on the twin segment thickness and the diameter of the nanowire. Under axial compressive strain, the collapse of twinned SiC nanowire has two failure modes, depending on its length and diameter: shell buckling for short nanowires and columnar buckling for longer ones. The molecular dynamic simulation reveals the influence of microstructure on the brittleness and plasticity of <111>-oriented SiC nanowires [179]. The result shows that the plastic deformation is mainly induced by the anti-parallel sliding of 3C grains along an intergranular amorphous film parallel to the (111) plane and inclined at an angle of 19.47° relative to the nanowire axis. Using first-principles calculations, the mechanical and electronic properties of hydrogen-passivated 3C–, 2H–, 4H–, and 6H–SiC nanowires were calculated [180]. The bandgaps of the nanowires are bigger than corresponding bulk values and decrease with increasing diameter. All nanowires exhibit direct bandgaps, distinct from the indirect bandgaps of bulk SiC. The bandgap dependence on the strain is different for each polytype of nanowire. The bandgaps increase (decrease) with tensile (compressive) strain for 3C–SiC nanowires. The influence of strain on the bandgap is more pronounced in thicker wires for 4H- and 6H–SiC nanowires. Combined DFT and molecular dynamics simulations reveal the atomistic origin of diameter-dependent extreme mechanical behavior of <111> 3C–SiC nanowires [181]. Heterogeneity in atomic stress and variations in diameter-dependent potential-energy density strongly affect extreme mechanical properties in the nanowires. The heterogeneity in stress evolves from the nonuniform bond lengths mediated by low coordinated surface atoms and it penetrates the entire cross section in thinner nanowires and constitutes the atomistic basis for their large reduction in fracture strain, toughness, and strength. Stress heterogeneity is much higher in ultrathin nanowires, whereas its intensity drops and saturates rapidly in larger ones. Stiffness, toughness, strength, and fracture strain of the nanowires increase nonlinearly with increasing diameter.

The geometrical and electronic structures of 3C–SiC nanowires growing along the <100> axis were investigated using DFT calculations [182]. The hydrogen passivated SiC nanowires show the quantum confinement effect in both C- and Si-rich wires, exhibiting as bandgap broadening with decreasing diameter. For the pure SiC nanowires, the dangling bonds strongly reconstruct and the nanowire exhibits metallic property due to the formation of surface states, and the conduction is nearly totally sustained by the outer layer of the nanowire. Figure 5.90 shows the calculated distribution of the density of states and the electronic band structure of the C-coated SiC nanowire.

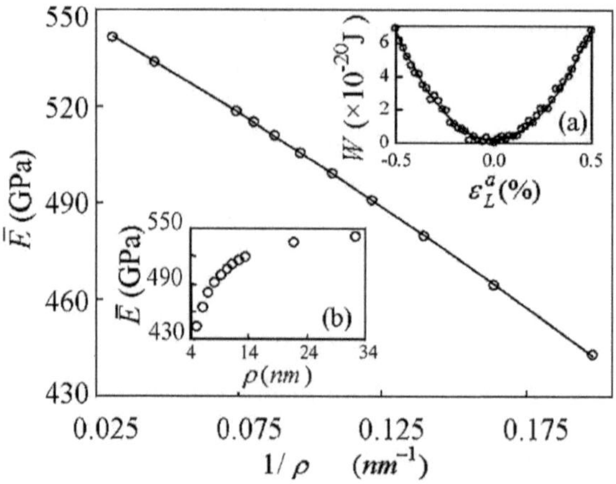

Fig. 5.89 Nominal Young's modulus as a function of the inverse of the cross-sectional perimeter for a <111> β-SiC nanowire. The solid line is the model fit. The inset *a* shows the work *W* done on a nanowire (6 layers) versus the applied strain. The inset *b* shows the nominal Young's modulus of the nanowires versus the cross-section perimeter. Reprinted with permission from [176]. Copyright 2008, American Institute of Physics

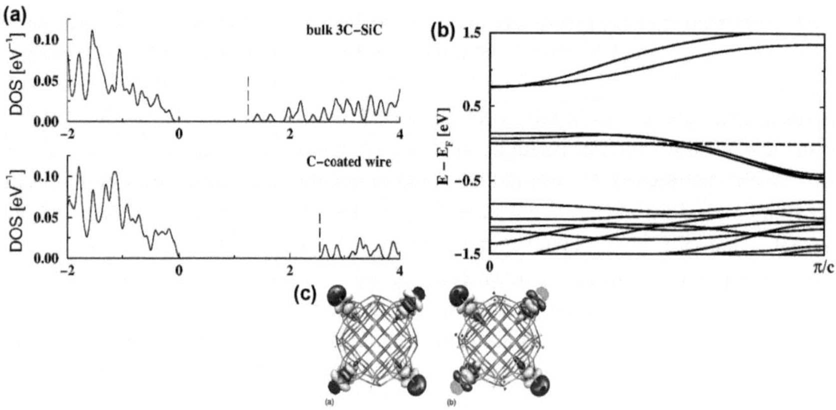

Fig. 5.90 **a** Electronic density of states of bulk 3C–SiC and carbon-coated SiC nanowire. The density of states are normalized to the number of states to allow for comparison between systems calculated with different simulation cells and aligned to the valence band maximum. The underestimation of the bandgap is not compensated for. **b** Electronic band structure of the C-coated SiC nanowire showing four reconstruction-induced surface states (degenerate two by two) that cross the Fermi level. **c** Wave function isosurfaces of two of the four surface states of the C-coated SiC nanowire. Reprinted with permission from [182]. Copyright 2005, American Physical Society

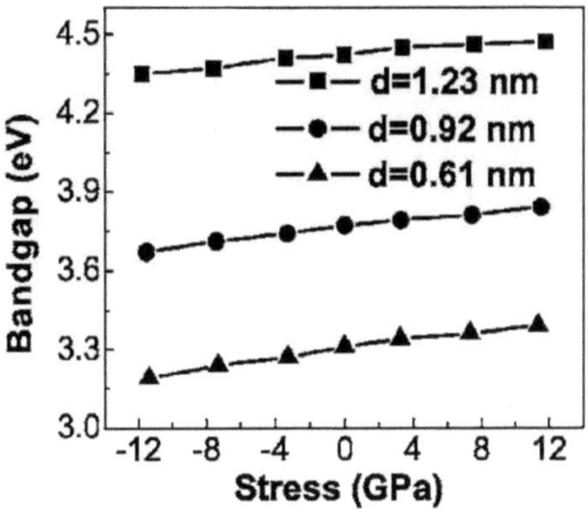

Fig. 5.91 Bandgap as a function of uniaxial stress for <111> β-SiC nanowires with different diameters. Reprinted with permission from [183]. Copyright 2006, American Institute of Physics

The first-principles calculation shows that the bandgap of β-SiC nanowire increases as the diameter decreases as a result of the quantum confinement effect and increases (decreases) slightly with increasing tensile (compressive) stress up to about 12 GPa (Fig. 5.91) [183]. The calculated Young's modulus and tensile strength of the SiC nanowire are about 620 and 52 GPa, respectively, in agreement with the experimental data.

The DFT calculation indicates that the pure SiC nanowires with wurtzite structure that grow along the [0001] direction are semiconductors [184], however, their bandgap decreases with decreasing diameter due to the presence of surface states. The hydrogen surface termination induces broadened bandgap that follows the quantum confinement effect. The energetics and electronic structures of 2H–SiC nanowires and nanotubes with radii ranging from 0.45 to 2.9 nm were investigated using DFT calculations [185]. The formation energy of the nanowire decreases with increasing wire radius, and the formation energy of the nanotube decreases with increasing wall thickness. The SiC nanotube with faceted single-crystalline wall is energetically more favorable than the cylindrical nanotube. The pure faceted SiC nanowire and nanotube possess indirect bandgaps related to surface states, with gap values smaller than that of bulk SiC. The hydrogen-passivated SiC nanowire and nanotube have direct bandgaps with values greater than that of bulk SiC. The DFT calculation performed on the hydrogenated 3C- and 2H–SiC nanowires shows that the <111>-orientated 3C–SiC nanowire is energetically more stable than other types of nanowires with similar sizes [186]. This result agrees with the experimental observation that most of the synthesized SiC nanowires are the 3C–SiC nanowires that grow along the <111> orientation. The calculation shows that all the nanowires have direct bandgaps except those grow along the <112> direction (Fig. 5.92, left panel). The direct bandgap character is maintained for 3C–SiC nanowires with diameters up to 2.8 nm. The bandgap of the nanowire increases with decreasing wire size (Fig. 5.92, right panel). The DFT calculation shows that both saturated and unsaturated twinned 3C–SiC nanowires that grow along the <111> direction exhibit semiconducting characteristics [187]. The saturated nanowires have direct bandgaps that increase with

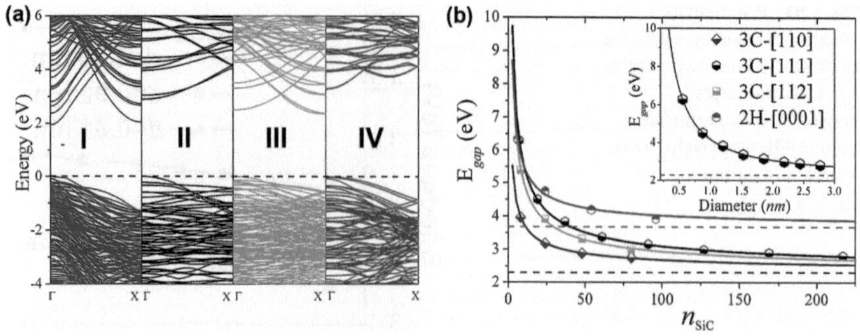

Fig. 5.92 **a** Electronic band structures of the SiC nanowires corresponding to (I) 3C <110>, (II) 3C <111>, (III) 3C <112>, and (IV) 2H [0001]. **b** Bandgap of SiC nanowires as a function of nanowire size (the number of SiC units per cell). The solid lines are the fitted results. The inset shows the bandgap of [2]-oriented 3C–SiC nanowire as a function of diameter. Reprinted with permission from [186]. Copyright 2009, American Chemical Society

decreasing diameter. The 2H stacking inside the 3C stacking shows no influence on the electronic properties of the nanowires.

A computational model for the axial conductivity of one-dimensional nanowires was proposed on the basis of semiconductor transport theory [188]. The conductivity, carrier concentration, and mobility of phosphorus doped SiC nanowires before and after passivation were numerically calculated. Hydrogen passivation highly improves the conductivity of P-SiC nanowires, and the conductivity increases by two orders of magnitude above room temperature. Hydrogen passivation saturates the surface dangling bonds, leading to vanishing of discrete impurity band of P-SiC nanowires. The surface dangling bonds lead to larger thermal instability of conductivity at room temperature, although this thermal instability diminishes rapidly with increasing temperature. DFT calculations reveal influence of hydrogen and hydroxyl passivation on the structure, electrical and optical properties of SiC nanowires [189]. Passivation improves the stability of SiC nanowire structure, and hydroxyl has better passivation effect than hydrogen. Both hydrogen and hydroxyl passivation increase the bandgap of SiC nanowires, and the varying trend of bandgap correlates with the polarity of the covalent bond formed by the passivation of surface atoms. Passivation enhances the stability of the optical properties of SiC nanowires, leading to narrowing of light absorption and photoconductivity spectra and shift of the response peak toward the deep ultraviolet region. The effects of hydrogen, fluorine, and chlorine atom passivation on the conductivity and dielectric properties of SiC nanowires were numerically calculated on the basis of transport and polarization relaxation theories [190]. Passivation reduces the dark conductivity of SiC nanowires and improves its ultraviolet photoconductivity. Photoconductivity of univalent (H) passivated SiC nanowires is better than that of seven-valent (Cl, F) passivated ones. The passivated SiC nanowires exhibit strong dielectric response in both deep ultraviolet and microwave regions. Hydrogen passivation generates the strongest dielectric response in deep ultraviolet, while fluorine passivation causes the strongest dielectric relaxation in the microwave band.

The DFT calculation shows that a uniform surface passivation with either hydrogen or with mixed hydrogen/hydroxyl groups for the surfaces of SiC nanowires can effectively remove the surface states from their band gaps [191]. Both types of coverage address the issue of charge-state instability of the near-surface V_{Si}^{-1} defect. The defect's high spin state ($S = 3/2$) remains stable over a large range of electronic chemical potential in an *n*-doped SiC nanowire. Surface passivation with fluorine should be avoided in that it cannot fully eliminate surface-related states from the valence-band edge. Hydrogen/hydroxyl passivation not only increases the photostability of V_{Si}^{-1} but also restores the frequency of its quantum emission.

Thermal conductivity of 3C and 2H polytypes of SiC nanowires was studied using nonequilibrium molecular dynamics simulations [192]. The thermal conductivity of the diameter modulated nanowires may be even smaller than that of the constant diameter nanowire owing to a significant thermal boundary resistance displayed by the constriction. The constriction resistance correlates with the confinement of low-frequency modes. Monte Carlo simulations indicate that the constriction resistance value may be explained by the specular reflections of such modes on the surface surrounding the constriction. Molecular dynamics simulations show that the thermal conductivity of period modulated antiphase superlattice nanowires can be reduced by up to a factor of two at room temperature compared to pristine nanowires [193]. New vibrational modes emerge on the interfaces due to formation of Si–Si and C–C bonds. The increased phonon interfacial scattering is the predominant factor hindering the thermal transport along the wires with periods > 6 nm. Phonon coherent transport also arises in the structures with periods < 6 nm, leading to a minimum thermal conductivity at the period of 6 nm.

5.7 Promising Applications

5.7.1 Electromagnetic Wave Absorption

There have been a number of investigations on the electromagnetic wave absorption of silicon carbide, mostly emphasizing the synthesis and technical details of the practical applications. This section gives a bird's eye view of this field without intending to cover all the advances in this field. As chemically inert and strongly microwave absorbing materials, sintered SiC cylinders can be applied as passive heating elements in microwave-assisted organic synthesis in nonpolar solvents [194]. They can absorb microwave energy and subsequently transfer the generated thermal energy via conduction phenomena to the reaction mixture. The use of passive heating elements allows for effective heating of otherwise microwave transparent or poorly absorbing solvents such as hexane, carbon tetrachloride, tetrahydrofuran, dioxane, or toluene to temperatures far over their boiling points (200–250 °C) under sealed

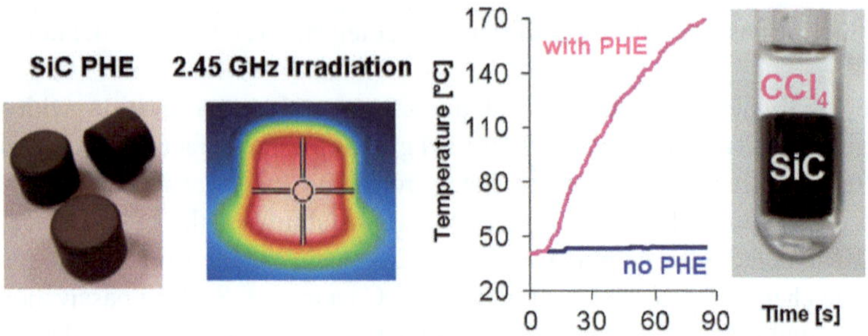

Fig. 5.93 SiC can act as passive heating element (PHE) in microwave-assisted organic synthesis. Reprinted with permission from [194]. Copyright 2006, American Chemical Society

vessel microwave conditions (Fig. 5.93). The SiC heating elements are thermally and chemically resistant to 1500 °C and compatible with nearly any solvent or reagent. Using reaction vials made of SiC ceramic in microwave reactors can mimic a conventionally heated autoclave experiment, while retaining the rapid heating and excellent process-control features inherent to microwave chemistry [195]. The high microwave absorbtivity, thermal conductivity, and thermal effusivity of SiC allow for rapid heat exchange with its surroundings. High chemical resistance of SiC vials allows for usage of corrosive reagents not suitable for processing in standard glass vials. High temperatures can be acquired rapidly by exposure to microwave irradiation. By employing the SiC vial, any effects of the electromagnetic field on the reaction mixture can be excluded.

The electromagnetic wave absorption properties of the 2 mm thick SiC nanowire-epoxy composite have been investigated in the range of 2–40 GHz [196]. The single-crystalline β-SiC nanowires grew along the <111> direction and they have diameters of 30–70 nm and lengths of 4–10 μm. The 35 wt% SiC nanowire composites showed remarkable electromagnetic wave absorptions of –31.7 and –9.8 dB at 8.3 and 27 GHz (Fig. 5.94), respectively. The composite containing 25 wt% SiC nanowire exhibited the minimum reflection loss of –32.4 dB at 31.1 GHz. The electromagnetic wave absorption may arise from the electric conductance loss caused by network-like SiC nanowires in the resin matrix and from the relaxation polarization loss at the interface of the SiC nanowires and the epoxy resin. The microwave absorption property of the β-SiC nanowires synthesized by reaction of multiwalled carbon nanotubes and Si powder in the presence of molten salts at 1250 °C has been investigated [197]. These nanowires grew along the <111> direction and they have diameters of 20–80 nm and lengths up to several tens of micrometers. A minimum reflection loss of 217.4 dB at 11.2 GHz was acquired for the silicone matrix incorporated by 30 wt% SiC nanowires.

Purely SiC nanowires are favorable microwave absorbing materials. 3C–SiC nanowires with stacking faults were fabricated, and the stacking fault concentration decreased significantly with increasing heating temperature [198]. The stacking fault planes were inclined at an angle of 35° to the growth direction at a heating temperature of 1400 °C, which were converted to be perpendicular to the growth direction at 1600 °C. The inclined stacking fault planes in the domain induced the

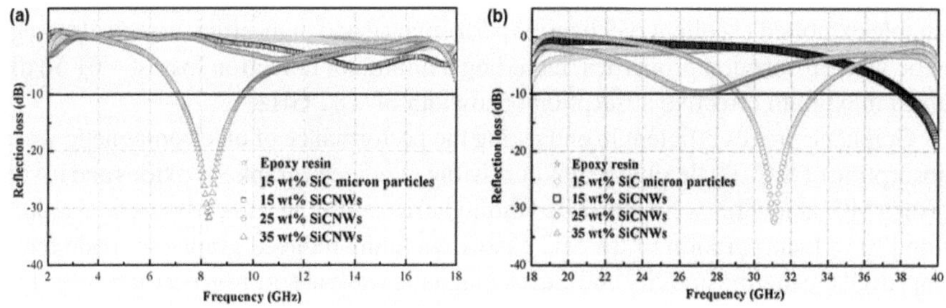

Fig. 5.94 Frequency dependences of reflection loss for epoxy resin, composite with 15 wt% SiC micrometer-sized particles, and composites with 15, 25, and 35 wt% SiC nanowires at 2–18 GHz (**a**) and 18–40 GHz (**b**), respectively. Reprinted with permissions from [196]. Copyright 2010, American Chemical Society

strongest dielectric resonance. The SiC NWs synthesized at 1400 °C possess the biggest stacking fault concentration and numerous dipoles, which caused extensive polarization and energy dissipation under an altering electromagnetic field. The lowest reflection loss of – 30 dB and an effective absorption band (reflection loss ≤ 10 dB) of 3.7 GHz were realized. The dielectric and microwave absorption properties of the SiC-nanowire-reinforced SiCf/SiC composites were investigated within 8.2–12.4 GHz at 25–600 °C [199]. The composites exhibited a minimum reflection loss of – 47.5 dB at 11.4 GHz and an effective bandwidth of 2.8 GHz at 600 °C. The excellent microwave absorption performance was ascribed to dielectric loss originating from the SiC nanowires with abundant stacking faults and heterostructure interfaces. Lightweight and strong SiC aerogels with in situ grown SiC nanowires were fabricated by direct graphitization and subsequent carbothermal reduction of eggplants [200]. The aerogels of SiC nanowires-decorated SiC frameworks showed excellent electromagnetic wave absorption, thermal insulation, and mechanical properties. Highly porous and interconnected SiC networks facilitated multiple reflection–absorption effects of infiltrated electromagnetic waves and extended the routes of heat transmission. SiC nanowires with numerous stacking faults and heterojunctions play a key role in dissipating electromagnetic energy, obstructing heat transfer, and reinforcing the SiC skeleton. The SiC aerogels with a 97.6% porosity showed a minimum reflection of – 43 dB and effective bandwidth of 4 GHz at 2 mm thickness. The SiC aerogels showed a very low thermal conductivity of 0.035 W m^{-1} K^{-1}. SiC nanowires/SiC whiskers foam showed efficient high-temperature electromagnetic wave absorption [201]. At room-temperature, the maximum electromagnetic wave effective absorption bandwidth and minimum reflection coefficient reached 4 GHz and –57 dB at room temperature and 3 GHz and – 15 dB at 600 °C, respectively. Even after oxidation of the foam, the values lie in the range of 2.7–3.9 GHz and from – 16 dB to – 64 dB at 1000–1500 °C. The foam showed flexural strength as high as 17.05 MPa. Multifunctional and monolithic SiC ceramic aerogel was fabricated using a freeze casting technique followed by carbonation and calcination [202]. The SiC nanofibers and nanowires inside the SiC aerogel-2 reinforced the stable porous structure. The sample exhibited excellent mechanical performance and high-efficiency electromagnetic wave absorption properties, including a minimum reflection loss of – 61.56 dB and a maximum effective absorption bandwidth of 9.82 GHz.

Graphene proves efficient in enhancing the performance of electromagnetic wave absorption of SiC. 3D flexible foams consisting of reduced graphene oxides and in situ grown SiC nanowires were fabricated through freeze-drying and carbothermal reduction [203]. Incorporation of the SiC nanowires in the reduced graphene oxide foam improved the thermostability and electromagnetic wave absorption performance. The foams were thermally stable beyond ~ 630 °C. The foams achieved effective absorption in the entire X-band (8.2–12.4 GHz) with a thin thickness of 3 mm. The SiC nanowires with abundant stacking faults, twinning interfaces, arid bridged junctions play an important role in enhancing the electromagnetic wave absorption performance. Light spongy bone-like aerogels consisting of graphene and SiC whiskers have unique hierarchically ordered structure and possess low density (72 mg cm^{-3}) and high microwave absorption performance [204]. Their minimum reflection loss reached –47.3 dB at 10.52 GHz with a thickness of 3 mm, and the effective bandwidth (reflection loss < – 10 dB) reached 4.7 GHz. The incorporation of the SiC nanowire and conductive polymer polypyrrole in graphene aerogel helps to realize an effective absorption bandwidth of 6.4 GHz in the frequency range of 8.2–14.6 GHz at a thickness of 2.32 mm [205]. The foam can acquire tunable effective absorption between the whole X-band and Ku-band by adjusting the loading of polypyrrole or SiC nanowires. The minimum reflection loss of – 60.05 dB at 17.12 GHz with a sample thickness of 2.00 mm and effective absorption bandwidth (< – 10 dB) of 6.72 GHz were realized using the SiC whiskers/reduced graphene oxide aerogels [79].

Carbon fibers or other carbon materials can also enhance the performance of electromagnetic wave absorption of SiC. A mixture of nanosize carbon black and nanosize silicon carbide improved the microwave absorption properties [206]. Addition of certain mass of SiC reduced the percolation threshold of the carbon black/epoxide resin composite and improved the microwave absorption of the composite. When 5 wt% carbon black was blended with 50 wt% SiC to fabricate a composite with a 2 mm thickness, the maximum reflection loss reached – 41 dB at 9 GHz, and the – 10 dB bandwidth reached 6 GHz. Effects of SiC on the mechanical and electromagnetic properties of the carbon fiber-reinforced SiC matrix composites were studied [207]. With increase of SiC content from 21.5 to 42.2 vol%, the total porosity decreased from 38.5 to 17.8 vol%, the flexural strength and fracture toughness increased from 38 ± 4 to 375 ± 10 MPa and from 6.2 ± 0.7 to 21 ± 0.3 MPa $m^{1/2}$. The electromagnetic interference shielding effectiveness decreased from 43 ± 1.4 to 31 ± 1.1 dB over 8.2–12.4 GHz with increasing SiC content. With increasing SiC content, the electrical conductivity diminished, leading to drop of the conductive loss and absorption loss of electromagnetic waves. The SiC nanowires@carbon fiber composites showed good electromagnetic wave absorption performance, with a maximum reflection loss of – 45.98 dB at 4.4 GHz [208]. After further coating of the conductive polymer polypyrrole, the nanocomposites showed superior electromagnetic wave absorption ability, with a maximum reflection loss of – 50.19 dB at

14.2 GHz and an effective bandwidth of 6.2 GHz. Many structures have been designed to improve the electromagnetic wave absorption performances, such as multiwalled carbon nanotube/SiC porous nanocomposites, polymer derived SiC composed of SiC nanocrystals and free carbon nanodomains, SiC nanowires/nanofibers carbon fibers, SiC nanowires-reinforced SiC/carbon foam composites [209–215], and all these composites exhibited excellent microwave absorption performances.

Oxides can highly enhance the wave absorption ability of SiC. NiO nanorings on the surfaces of SiC powders exhibited enhanced dielectric properties and excellent microwave absorption at 373–773 K due to the special ring-like morphology of the NiO nanocrystals [216]. The minimum reflection loss is – 46.9 dB at 673 K, more than three times bigger than that of pure SiC. The absorption bandwidths cover the whole X-band at 673–773 K when the reflection loss is lower than – 10 dB. The imaginary dielectric constant and loss tangent increased over four times and three times with increasing temperature [217]. Formation of shells of Fe_3O_4 nanoparticles on the surfaces of the SiC nanowires showed enhanced microwave absorption [218]. The monodisperse Fe_3O_4 nanoparticles have sizes of 10 nm. The coverage density of the nanoparticles is tunable by changing the weight ratio of the precursors. The Fe_3O_4 nanoparticles increased the magnetic loss and improved the impedance matching conditions compared to untreated SiC nanowires. The reflection loss can be as low as – 51 dB.

Various materials, such as SiOC, $ZrN_{0.4}B_{0.6}$, ZrO_2, Si_3N_4, have been explored for constructing composites with SiC to enhance the microwave absorption performances and proved very efficient [219–222]. Metal Fe and Al, and even polymer bound with SiC can bring favorable effects on the microwave absorption performances [223–225].

5.7.2 *Thermal Management and Fireproofing*

Free-standing SiC nanowire paper consisting of interweaved ultralong SiC nanowires synthesized by carbothermal reduction was fabricated through a vacuum filtration method [226]. The SiC nanowire has a width of ~ 100 nm and a length up to several hundred micrometers. The SiC nanowire paper showed high flexibility and strong mechanical stability. The paper exhibited excellent nonflammability in fire and remained well preserved without visible damage when it underwent heating in an electric oven at 1000 °C for 3 h. Ultralight and strong 3D SiC structures were fabricated by carbothermal reduction of SiO with a graphene foam [227]. The foams have an average height of 2 mm and densities of 9–17 mg cm^{-3}. AFM reveals an average flake thickness of 2–3 nm and an average lateral size of 2 μm. Compared with most existing lightweight foams, the 3D SiC form exhibited superior compression strength and significant recovery after compression strain of 70%. Highly porous 3D SiC NW aerogel assembled by interweaved 3C–SiC nanowires of 20–50 nm in diameter and tens to hundreds of micrometers in length were fabricated [228]. It possesses ultralow density (~ 5 mg cm^{-3}), large recoverable compression strain (> 70%) and fatigue resistance, refractory property, oxidation and high temperature resistance (Fig. 5.95) and thermal insulating property (0.026 W m^{-1} K^{-1} at room temperature in N_2). The SiC NW aerogel exhibited an adsorption selectivity of low-viscosity organic solvents with a high absorption capacity (130–237 g g^{-1}).

Vertically aligned and interconnected SiC nanowire networks can act as efficient fillers for polymer composites, leading to highly enhanced thermal conductivity [229]. The SiC NW networks were produced by freeze-casting nanowire aqueous suspensions followed by thermal sintering to consolidate the nanowire junctions, generating a hierarchical architecture with aligned honeycomb-like SiC NW layers. The composite achieved by infiltrating SiC NW networks with epoxy resin, at a SiC NW loading of 2.17 vol%, showed high through-plane thermal conductivity (1.67 W m^{-1} K^{-1}) compared to the pure matrix. A graphene hybrid paper was fabricated by intercalation of silicon source and in situ growth of SiC nanorods between graphene sheets through carbothermal reduction reaction [230]. Due to the formation of covalent C–Si bonding at the graphene–SiC interface, the hybrid paper possesses a superior through-plane thermal conductivity of 10.9 W m^{-1} K^{-1} and can reach 17.6 W m^{-1} K^{-1} under packaging conditions at 75 psi. The cooling efficiency of the

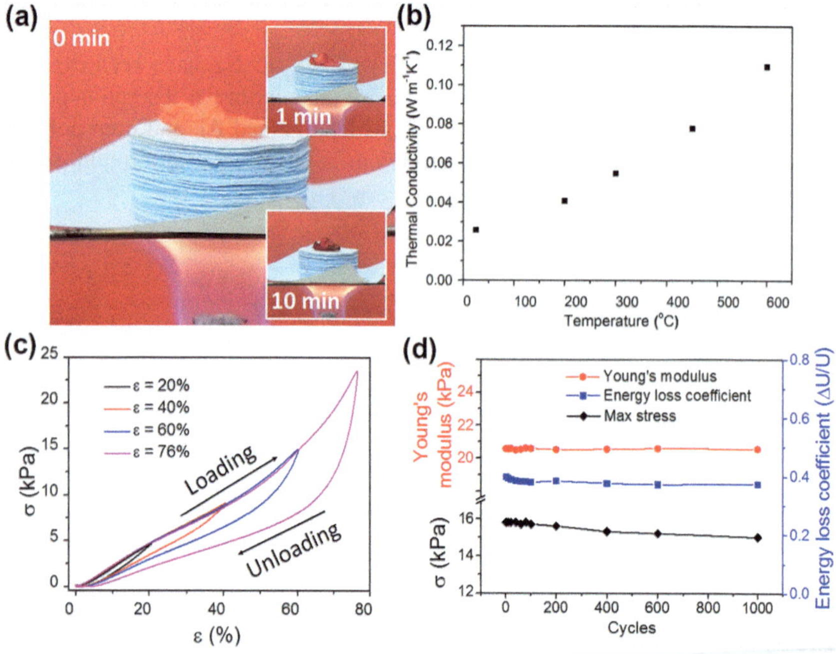

Fig. 5.95 Thermal insulation performance of the SiC NW aerogel. **a** Fresh petal protected by a piece of aerogel with a thickness of 10 mm from withering or carbonization under the heating of an alcohol lamp for 10 min. **b** Thermal conductivity of the SiC NW aerogel in N_2 at different temperatures. **c** One thousand loading–unloading fatigue cycles at set $\varepsilon = 60\%$. **d** History of the Young's modulus, maximum stress, and energy loss coefficient as a function of the compressive test cycles. Reprinted with permission from [228]. Copyright 2018, American Chemical Society

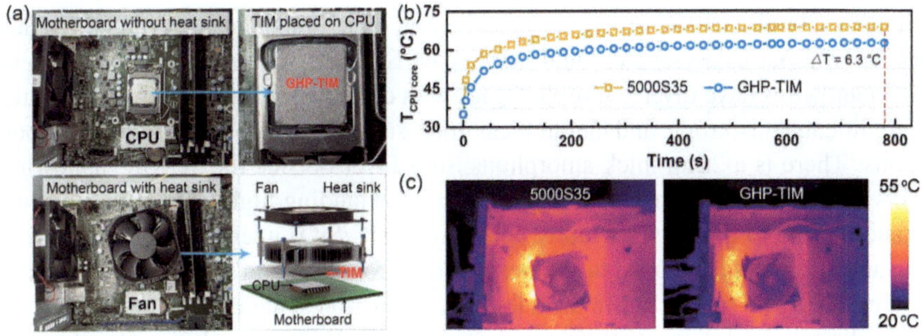

Fig. 5.96 **a** Experimental setup for comparing the cooling efficiency between graphene@SiC nanowire hybrid paper (GHP) and state-of-the-art thermal pad based on desktop computers. **b** CPU core temperature evolution as a function of running time. **c** Comparative IR images of the motherboard when 5000S535 and GHP were used as thermal interface materials (TIMs), respectively. Reprinted with permission from [230]. Copyright 2019, American Chemical Society

hybrid paper was significantly improved with respect to that of the state-of-the-art thermal pads (Fig. 5.96).

Combined CVD process and layer-by-layer self-assembly led to construction of hyperelastic SiC nanofibrous aerogels with 3D porous architecture [231]. The aerogels is capable of quickly recovering from large compressive deformation (50% strain), functioning in a wide range of temperatures – 196 to 1100 °C in air, maintaining high particle matter removal efficiency of > 99.96%, and rapidly absorbing various organic solvents and oils with high capacity and robust recoverability. An efficient method for large-scale production of resilient SiC nanowire aerogels with tunable densities and desired shapes was developed [232]. Their maximum working temperatures in Ar and air can reach 1400 and 1000 °C, respectively. They have outstanding flame-erosion resistance and low thermal conductivity (2S mW m^{-1} K^{-1}). Multifunctional SiC nanofiber aerogel spring was fabricated [233], which showed ultralow density (9 mg cm^{-3}), ultralow thermal conductivity (0.029 W m^{-1} K^{-1} at 20 V), excellent ablation and oxidation resistance, and a stable 3D structure composed of interlacing 3C–SiC nanofibers with diameters of 300–500 nm and lengths of tens to hundreds of micrometers. They have permanent deformation of only 1.3% at 20 °C after 1000 cycles. They exhibited robust hyperelasticity and cyclic fatigue resistance at both low (–196 °C) and high (700 °C) temperatures. 3D constructed resilient SiC nanowire aerogels possessing programmed geometries and engineered mechanical properties were prepared via additive manufacturing [234]. Their Young's moduli are tunable within 0.012–5.800 MPa. These lightweight and resilient SiC nanowire aerogels showed a low thermal conductivity of 0.046 W m^{-1} K^{-1}. An air suction effect induction strategy was proposed to fabricate a super thermally insulating SiC aerogel [235]. It used the air suction effect to regulate the directional flow of the SiO gas, which can induce directional growth and assembly of SiC nanowires to form a directional lamellar structure. The aerogels showed significant improvement in compression and elasticity performance. They have an ultralow thermal conductivity of 0.019 W m^{-1} K^{-1}. Carbothermal reduction of the agricultural residue loofah without catalysts can yield large-scale SiC nanowire sponges [236]. These SiC nanowires have diverse morphologies with diameters of 50–400 nm. They have a significant anisotropic and hierarchical microstructure with planar aligned bamboo joints. There is a 2-nm thick amorphous SiO_2 layer outside the 3C–SiC nanowire. This loofah-based SiC nanowire sponge has outstanding high-temperature performance owing to the stacking faults, SiO_2 layer, SiC/SiO_2 interface of the nanowire. An approach was developed to synthesize SiC aerogels within seconds and over liter scale [237]. The reason lies in the adopted combustion synthesis and fast expansion from the powder reactant to the aerogel product over 1000% in volume. The obtained SiC aerogels showed low thermal conductivity, high elasticity, and damage tolerance.

5.7.3 *Other Preliminary Applications*

There has been increasing interest on the applications of silicon carbide nanowires. The photocatalytic property of the SiC nanowires has been investigated by measuring the photodegradation rate of acetaldehyde catalyzed by SiC [238]. The β-SiC nanowires coated with amorphous SiO_2 were synthesized by using the thermal evaporation method. These nanowires exhibited excellent photocatalytic property, as demonstrated by the efficient decomposition of the acetaldehyde under UV light irradiation (Fig. 5.97). The photocatalytic reaction was measured by monitoring the amount of CO_2 gas produced in this process. The SiO_2-coated SiC nanowires showed better catalysis effect than the oxide-free SiC nanowires obtained by HF etching. The 3C–SiC nanowires can absorb visible light and exhibited excellent photocatalytic hydrogen evolution performance from pure water under visible light irradiation [239]. The hydrogen evolution rate can reach more than 60 μL g^{-1} h^{-1}. Water has been decomposed into H_2 and O_2. The oriented arrays of SiC nanowires on the Si substrate prepared by carbonizing the Si nanowire arrays with the graphite powder are porous and have high surface-to-volume ratios [240]. They were used as the photocatalyst for water splitting under UV irradiation, exhibiting an enhanced and stable photocatalytic performance. Hierarchical 3C–SiC nanowires can advance the photocatalytic degradation of methylene blue solution under visible light irradiation [241]. Methylene blue was degraded efficiently after 5 h irradiation. Platinum nanoparticle-decorated SiC nanowires can apply to photocatalytic water splitting reaction [242]. The average H_2 evolution rate reached 4572 μL g^{-1} h^{-1}. The platinum acts as an acceptor of photoelectrons to effectively restrain the electron–hole pair recombination of the SiC nanowires. B-doped 3C–SiC nanowires showed efficient activity as high as 108.4 μmol h^{-1} g^{-1} for H_2 production, about 20 times higher than that of 3C–SiC nanowhiskers [243]. Multiwalled carbon nanotube–SiC nanowire heterostructures were used for photocatalysis of the H_2 generation under visible light irradiation [244]. The H_2 evolution rate reached 108 μmol h^{-1} g^{-1}, 3.1 times higher than that of pure SiC without carbon nanotubes. Mesoporous SiC nanofibers with embedded graphitic carbon catalyzed the hydrogen evolution efficiently [245].

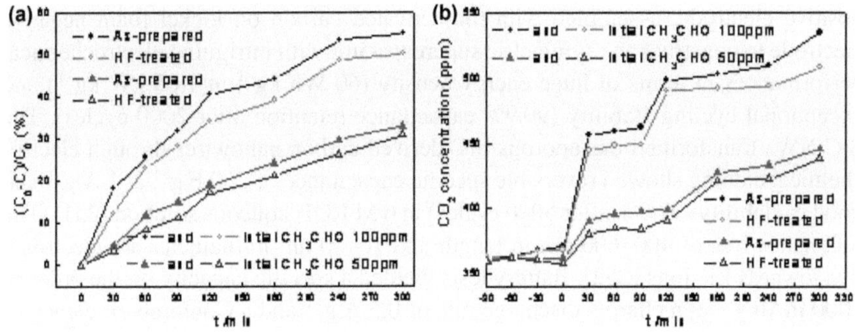

Fig. 5.97 **a** Photodegradation rate of acetaldehyde gas as a function of irradiation time of UV light and the degradation was catalyzed by the SiC nanowires. C_0 is initial concentration of gaseous acetaldehyde (CH_3CHO) and C is the CH_3CHO concentration at time t. **b** CO_2 concentration (generated by decomposition of acetaldehyde) as a function of irradiation (starting from the zero instant) time. Reprinted with permission from [238]. Copyright 2006, American Institute of Physics

The efficiency is 180.2 μmol h^{-1} g^{-1} under irradiation of simulated solar light and 31.0 μmol h^{-1} g^{-1} under visible light irradiation in high-pH solution. The composites of SiC nanowires on carbon fibers are promising support for catalysts used under harsh environments [246]. They exhibited excellent structural stability at high temperatures in air, in concentrated acidic/alkaline solutions after electrochemical stressing for 2000 cycles, and in oxygen evolution reaction after 10 h of continuous operation.

SiC nanowires are also promising robust electrode materials for supercapacitors [247]. The Si nanowires produced by low-temperature electrochemical etching corroded during charge/discharge cycling in the aqueous environment, but the corrosion was mitigated upon coating with a SiC passivation layer. The passivated NWs achieved capacitance values up to ~ 1.7 mF cm^{-2} projected area (comparable to carbon based micro-supercapacitor electrodes). They exhibited robust cycling stability, and maintained capacitive behavior over a wide range of charge/discharge rates. The SiC NWs growing on a SiC thin film coated with a thin Ni catalyst layer were tested [248]. A specific capacitance in the range of ~ 240 μF cm^{-2} was demonstrated. The SiC nanowires are exceptionally stable, with 95% capacitance retention after 2×10^5 charge/discharge cycles in an environmentally benign, aqueous electrolyte. The supercapacitor electrodes made of SiC nanowires growing on flexible carbon fabric showed excellent cycle stability [249]. A high areal capacitance of 23 mF cm^{-2} was achieved at a scan rate of 50 mV s^{-1} at room temperature. The capacitance increased with increasing temperature. No decrease in capacitance at room temperature was observed after 10^5 cycles. At 60 °C, 90% of the initial capacitance was retained after 10^5 cycles. Ultrathin $NiCo_2O_4$/NiO nanosheets on SiC nanowires were demonstrated to be stable hybrid electrodes for asymmetric supercapacitors [250]. The electrode delivered an ultrahigh specific capacitance of 1801 F g^{-1} at 1 mA cm^{-2} and a remarkable rate capability of 1499 F g^{-1} at 10 mA cm^{-2}. The functionalized SiC NWs@$NiCo_2O_4$/NiO nanosheets on carbon cloth act as the positive electrode, assembled with the activated carbon on nickel foam negative electrode to construct an asymmetric supercapacitor with intriguing electrochemical performances in terms of huge energy density (60 Wh kg^{-1} at 1.66 kW kg^{-1}) and exceptional cycling stability (90.9% capacitance retention after 2000 cycles). The SiC NWs transformed mesoporous SiC-derived carbon nanowires through electrochemical etching showed reversible specific capacitance of 260 F g^{-1} at 1 A g^{-1} and good cyclability (97.9% after 5000 cycles) in 6 M KOH aqueous solution [251]. The SiC nanowires of 400–600 nm in length and 10–40 nm in diameter are electroactive towards Li^+ ions [252]. Battery tests yielded a specific capacity on the order of 1000 mAh g^{-1} at a charge–discharge rate of 0.5 A g^{-1} and a Coulomb efficiency of more than 99% over hundreds of cycles. High-temperature supercapacitors based on SiC nanowires as the electrode materials were realized [253]. In 2.0 m KCl aqueous electrolyte, the electrode based on etched SiC nanowires delivered a specific capacitance of 23.6 mF cm^{-2} (29.5 F g^{-1}) at a current density of 0.2 mA cm^{-2} (0.25 A g^{-1}), about 3.3 times that of the counterpart without etching. The as-constructed ionic-liquid-based supercapacitors can undergo the operation temperatures up to 150 °C, with a capacitance retention of 80% for 10,000 cycles. Even under progressive variation at temperatures between 0 and 150 °C, the capacitance retentions maintained over 76% for 12,000 cycles.

Silicon carbide nanowires can be used to construct field-effect transistors toward integrated circuits and sensing applications [254]. A prototype of the field effect transistor based on the β-SiC nanowires has been demonstrated [255]. These nanowires were synthesized by thermal reaction between the activated carbon fiber and the SiO powder without using catalysts at 1450 °C. They have diameters of 10–25 nm and lengths up to 10 μm. The measured carrier mobility of the *n*-type SiC field effect transistor is 6.4 and 15.9 cm^2 V^{-1} s^{-1} for applied drain–source voltage of 0.01 and 0.05 V at room temperature (Fig. 5.98), respectively. The SiC field effect transistor has bigger drain current and higher carrier mobility at an elevated temperature than at room temperature.

The electrical properties of 3C–SiC nanowires contacted with two types of ohmic contacts (Ti/Au and Ni/Au) were studied [256]. The Ni/Au ohmic contact on SiC nanowires exhibited a lower contact resistance of 5.9×10^{-6} Ω cm^2 in comparison with that of the Ti/Au contact (2.6×10^{-4} Ω cm^2). The SiC nanowire field effect transistor with Ni/Au ohmic contact has much lower contact resistance and better performance relative to the transistor with Ti/Au contact. The back-gated field effect transistors based on 3C–SiC nanowires (Fig. 5.99) have been fabricated [257]. The device showed either ohmic contact or rectifying contact. The ohmic-contact transistor has a very weak gating effect and the device switching off is unachievable because of the high electron concentration along the nanowire. The Schottky contact barrier at source/drain regions is more favorable for improvement of the transistor performance by suppressing the off current. The Schottky contact barrier at high positive gate voltages (over 20 V) is more transparent, leading to I_{ON}/I_{OFF} ratio of 103 in contrast to the weak gating effect of the ohmic contact 3C–SiC nanowire transistor. On account of the practical application of the SiC nanowires in electronic devices, the doping level of the SiC nanowires must be controllable. There has been

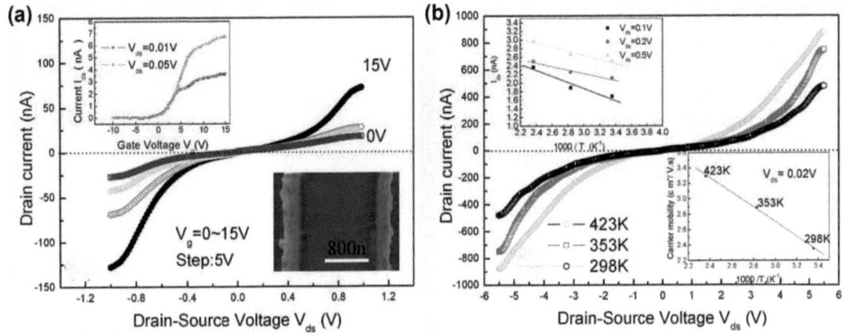

Fig. 5.98 Characteristics of the *n*-type SiC nanowire-based field effect transistor. **a** Drain current versus voltage at different gate voltages at room temperature. Lower inset shows the SiC nanowire bridging two electrodes. Upper inset shows drain current versus gate voltage at drain voltage 0.01 and 0.05 V, respectively. **b** Drain current as a function of drain–source voltage at gate voltage 5 V. Lower inset is the Arrhenius plot of mobility. Upper inset shows the temperature dependence of drain current of the SiC nanowire-based field effect transistor with Au contact electrodes at gate voltage 5 V. Reprinted with permission from [255]. Copyright 2006, IEEE

a report on fabrication of SiC NW Schottky diodes [258]. Each diode consists of one nanowire with a diameter of ~ 160 nm. The device could maintain similar values for ideality factor, barrier height, and effective Richardson constant upon proton irradiation with a fluence of 10^{16} ion cm^{-2} at 873 K, showing application potential for harsh environments.

The nitrogen doping of the 3C–SiC nanowires has been demonstrated through a reaction of catalyst-assisted pyrolysis of polymeric precursors with $Co(NO_3)_2$ as catalysts [259]. A relatively low partial pressure is crucial for the doping of nitrogen to the SiC nanowires. The concentration of the doped nitrogen decreased with increasing pyrolysis temperature during growth of the SiC nanowires, this character can be used for controlled doping of nitrogen in the SiC nanowires.

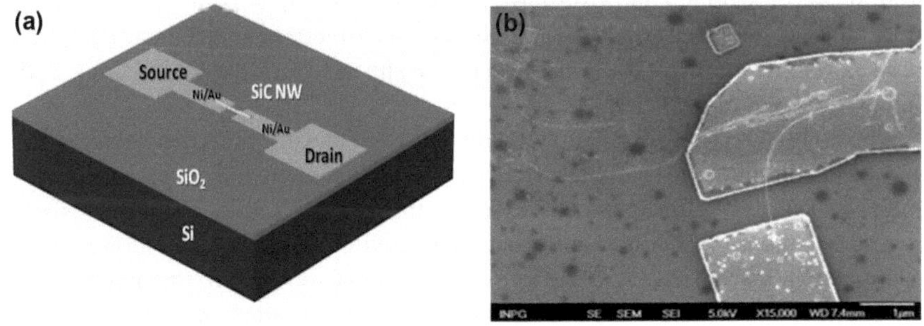

Fig. 5.99 **a** Schematic of SiC nanowire-based field effect transistor. **b** SEM image of the fabricated device showing a SiC nanowire between two metallic contacts. Reprinted with permission from [257]. Copyright 2011, IEEE

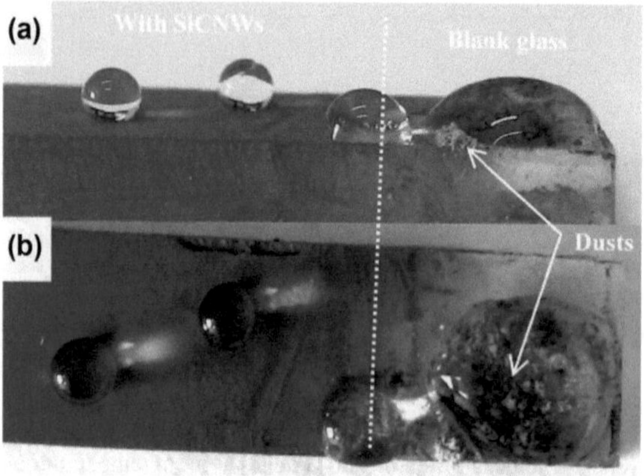

Fig. 5.100 **a** Cross-section view and **b** top view photos of self-cleaning test with dust on SiC nanowires-coated glass and control blank glass. Reprinted with permission from [260]. Copyright 2009, American Chemical Society

The self-cleaning glass can be fabricated by coating superhydrophobic SiC nanowires in tetraethyl orthosilicate (TEOS) solution [260]. The coated glass by 10-cycles coating of the SiC nanowires with a high roughness of 1928.9 nm showed a high water-contact angle up to 160° (Fig. 5.100) and a low sliding angle down to 5°. The coated sample showed high chemical stability even after immersion in water for 14 days.

Polyurethane-based nanocomposite film was fabricated by incorporating carbon-coated SiC nanowires into the polymer matrix [261]. Measurement of the electric field-induced strain indicates that incorporation of 0.5 wt% nanowires in the polymer increased the strain level by a factor of 1.7 at field strength 6.5 V μm^{-1}. The improvement of the electromechanical response is related to a stronger space charge effect in the nanocomposite than in the pure polymer. The electrostatically actuated contact-mode nanoelectromechanical switches based on very thin SiC nanowires have been demonstrated [262]. The SiC nanowires were lithographically patterned from a SiC layer with a thickness of 50 nm that was heteroepitaxially grown on a Si substrate (Fig. 5.101). Several types of in-plane electrostatic SiC nanowire switches were realized, and the nanowires have diameters as small as 20 nm and the lateral switching gaps are as narrow as 10 nm. These contact-mode nanomechanical switches showed very low turn-on voltages down to 1 V and very short switching times in the submicrosecond range.

The SiC–SiO_2 core–shell nanowire film has good anti-friction performance [263] and the maximum static frictional force between the nanowire film and a macroscopic solid surface is 5–12 times that between two macroscopic solids. The wear-resistant nanoscale silicon carbide tips have been prepared by annealing the surface of the nanoscale Si tip exposed to carbon ions [264]. The SiC-covered Si tip maintained its nanoscale sharpness and exhibited wear resistance that is orders of magnitude larger than that of conventional Si tips and at least 100-fold higher than that of SiO-doped diamond-like-carbon tips. The sharpness, high wear resistance, and high thermal stability of these tips render them favorable for scanning probe applications.

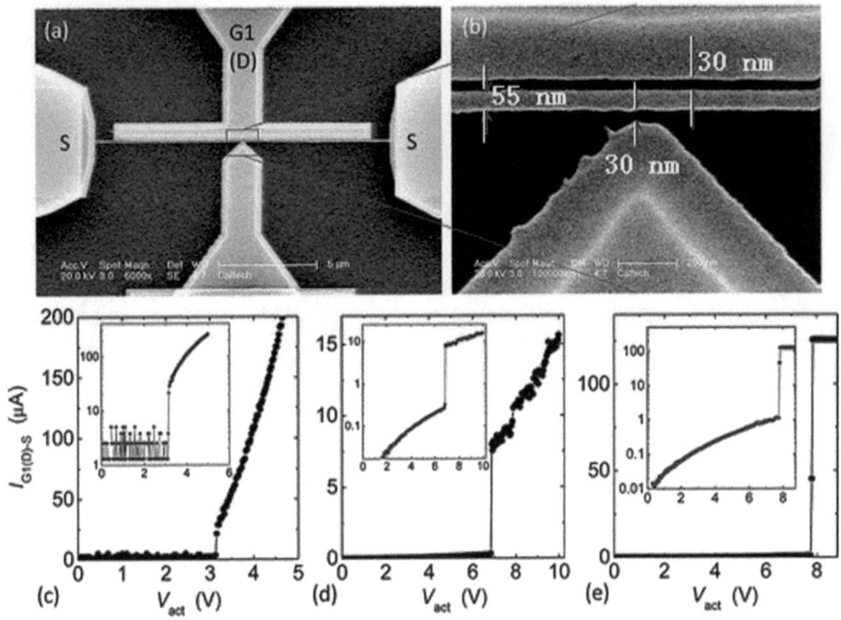

Fig. 5.101 Low-voltage nanoelectromechanical switches based on metallized SiC nanowires. **a** A 15 μm long and 55 nm wide SiC nanowire metallized with 30 nm Al on top. **b** Magnification micrograph showing the 55 nm wide nanowire and the 30 nm wide gap. **c** and **d** Nanoelectromechanical switching events from two 20 μm-long devices, with V_{on} of 3.1 and 7.0 V, respectively. **e** Switching event from a 15 μm-long device, with V_{on} of 7.8 V. The insets in (**c**–**e**) are the data shown with measured current in the logarithmic scale. Reprinted with permission from [262]. Copyright 2010, American Chemical Society

The SiC nanowires can serve as the host for accommodating metal catalysts. The Pd nanoparticles attached on the surfaces of the selectively-etched periodically twinned SiC nanowires [265] showed prohibited migration and coalescence. As a result, such attached Pd nanoparticles showed improved stability during their catalysis for combustion of methane. An oxidation protective Si–SiC coating with randomly oriented SiC nanowires was prepared on the SiC-coated carbon/carbon composites [266]. First, a porous network of SiC nanowires was synthesized by using CVD. This material experienced pack cementation to infiltrate the porous layer with a mixture of Si and SiC. The nanowires in the coating efficiently suppressed the cracking of the coating. The thermogravimetric analysis and thermal shock demonstrated that the coating has excellent oxidation protection for C/C composites between room temperature and 1500 °C. The coating can efficiently protect the C/C composites from oxidation at 900 °C for more than 313 h or at 1400 °C for more than 112 h.

A two-step method has been developed to synthesize the heterojunction of the ZnO nanorods on the SiC nanowires [267]. The SiC nanowires were prepared by heating the Si wafer and the WO_3/C mixed powders under catalysis of NiO in nitrogen

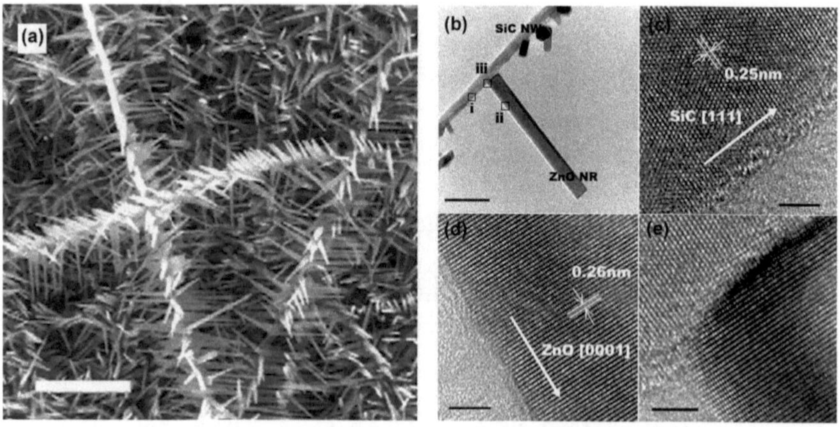

Fig. 5.102 **a** ZnO nanorods grown on β-SiC nanowire. (right panel) **b** TEM image of the ZnO nanorod/SiC nanowire heterojunction. **c**–**e** High-resolution micrographs of the marked regions in (**b**), corresponding to the SiC nanowire (i), ZnO nanorod (ii), and nanojunction (iii), respectively. The scale bar is 100 nm in (**b**) and 2.5 nm in (**c**–**e**). Reprinted with permission from [267]. Copyright 2005, IOP Publishing

atmosphere at 1050 °C. The diethylzinc was subsequently used as metal–organic source to grow ZnO nanorods on the as-synthesized SiC nanowires at 450 °C. High-resolution TEM images (Fig. 5.102) show that the heterojunction has an atomically abrupt interface and there is no interfacial layer between the ZnO nanorod and the SiC nanowire. The [0001] direction of the ZnO nanowire is perpendicular to the <111> direction of the SiC nanowire at the heterojunction. The optical transmission spectra reveal that the fabric composed of assembled 2H–SiC nanowires absorbed most incident 200–400 nm light, but is transparent to visible light [268]. A polydimethylsiloxane–SiC fabric–polydimethylsiloxane sandwich film device exhibited stable electrical output even when repeatedly stretched by up to 50%. The transparent and stretchable SiC fabric exhibited considerable photoelectric activity and exhibited rapid photoresponse (rise and decay time < 30 ms) to 340–400 nm light, covering most UV-A spectral region.

References

1. Dai H, Wong EW, Lu YZ, Fan S, Lieber CM (1995) Synthesis and characterization of carbide nanorods. Nature 375:769–772
2. Wu R, Pan Y, Yang G, Gao M, Wu L, Chen J, Zhai R, Lin J (2007) Twinned SiC zigzag nanoneedles. J Phys Chem C 111:6233–6237
3. Tang CC, Bando Y, Sato T, Kurashima K, Ding XX, Gan ZW, Qi SR (2002) SiC and its bicrystalline nanowires with uniform BN coatings. Appl Phys Lett 80:4641–4643
4. Han W, Fan S, Li Q, Liang W, Gu B, Yu D (1997) Continuous synthesis and characterization of silicon carbide nanorods. Chem Phys Lett 265:374–378

5. Zhu J, Fan S (1999) Nanostructure of GaN and SiC nanowires based on carbon nanotubes. J Mater Res 14:1175–1177
6. Tang CC, Fan SS, Dang HY, Zhao JH, Zhang C, Li P, Gu Q (2000) Growth of SiC nanorods prepared by carbon nanotubes-confined reaction. J Cryst Growth 210:595–599
7. Xie W, Möbus G, Zhang S (2011) Molten salt synthesis of silicon carbide nanorods using carbon nanotubes as templates. J Mater Chem 21:18325
8. Zhang S-L, Zhu B-F, Huang F, Yan Y, Shang E-Y, Fan S, Han W (1999) Effect of defects on optical phonon Raman spectra in SiC nanorods. Solid State Commun 111:647–651
9. Feldman DW, Parker JH, Choyke WJ, Patrick L (1968) Phonon dispersion curves by Raman scattering in SiC, polytypes 3C, 4H, 6H, 15R, and 21R. Phys Rev 173:787–793
10. Seo WS, Koumoto K (1996) Stacking faults in β-SiC formed during carbothermal reduction of SiO_2. J Am Ceram Soc 79:1777–1782
11. Meng GW, Cui Z, Zhang LD, Phillipp F (2000) Growth and characterization of nanostructured β-SiC via carbothermal reduction of SiO_2 xerogels containing carbon nanoparticles. J Cryst Growth 209:801–806
12. Gundiah G, Madhav GV, Govindaraj A, Seikh MM, Rao CNR (2002) Synthesis and characterization of silicon carbide, silicon oxynitride and silicon nitride nanowires. J Mater Chem 12:1606–1611
13. Gao YH, Bando Y, Kurashima K, Sato T (2002) SiC nanorods prepared from SiO and activated carbon. J Mater Sci 37:2023–2029
14. Ye HH, Titchenal N, Gogotsi Y (2005) SiC nanowires synthesized from electrospun nanofiber templates. Adv Mater 17:1531–1535
15. Zhou WM, Liu X, Zhang YF (2006) Simple approach to β-SiC nanowires: synthesis, optical, and electrical properties. Appl Phys Lett 89:223124
16. Chiu S-C, Huang C-W, Li Y-Y (2007) Synthesis of high-purity silicon carbide nanowires by a catalyst-free arc-discharge method. J Phys Chem C 111:10294–10297
17. Niu JJ, Wang JN (2007) A simple route to synthesize scales of aligned single-crystalline SiC nanowires arrays with very small diameter and optical properties. J Phys Chem B 111:4368–4373
18. Chen K, Huang ZH, Huang JT, Fang MH, Liu YG, Ji HP, Yin L (2013) Synthesis of SiC nanowires by thermal evaporation method without catalyst assistant. Ceram Int 39:1957–1962
19. Ding J, Deng CJ, Yuan WJ, Zhu HX, Zhang XJ (2014) Novel synthesis and characterization of silicon carbide nanowires on graphite flakes. Ceram Int 40:4001–4007
20. Fu QG, Li HJ, Shi XH, Li KZ, Wei J, Hu ZB (2006) Synthesis of silicon carbide nanowires by CVD without using a metallic catalyst. Mater Chem Phys 100:108–111
21. Pol VG, Pol SV, Gedanken A, Lim SH, Zhong Z, Lin J (2006) Thermal decomposition of commercial silicone oil to produce high yield high surface area SiC nanorods. J Phys Chem B 110:11237–11240
22. Bechelany M, Brioude A, Cornu D, Ferro G, Miele P (2007) A Raman spectroscopy study of individual SiC nanowires. Adv Funct Mater 17:939–943
23. Wagner RS, Ellis WC (1964) Vapor-liquid-solid mechanism of single crystal growth. Appl Phys Lett 4:89–90
24. Seeger T, Kohler-Redlich P, Rühle M (2000) Synthesis of nanometer-sized SiC whiskers in the arc-discharge. Adv Mater 12:279–282
25. Li YB, Xie SS, Zou XP, Tang DS, Liu ZQ, Zhou WY, Wang G (2001) Large-scale synthesis of β-SiC nanorods in the arc-discharge. J Cryst Growth 223:125–128
26. Yang W, Miao H, Xie Z, Zhang L, An L (2004) Synthesis of silicon carbide nanorods by catalyst-assisted pyrolysis of polymeric precursor. Chem Phys Lett 383:441–444
27. Zhang L, Yang W, Jin H, Zheng Z, Xie Z, Miao H, An L (2006) Ultraviolet photoluminescence from 3C–SiC nanorods. Appl Phys Lett 89:143101
28. Wang H, Xie Z, Yang W, Fang J, An L (2008) Morphology control in the vapor-liquid-solid growth of SiC nanowires. Cryst Growth Des 8:3893–3896
29. Feng W, Ma J, Yang W (2012) Precise control on the growth of SiC nanowires. CrystEngComm 14:1210–1212

30. Liang CH, Meng GW, Zhang LD, Wu YC, Cui Z (2000) Large-scale synthesis of β-SiC nanowires by using mesoporous silica embedded with Fe nanoparticles. Chem Phys Lett 329:323–328
31. Hao YJ, Jin GQ, Han XD, Guo XY (2006) Synthesis and characterization of bamboo-like SiC nanofibers. Mater Lett 60:1334–1337
32. Lee JS, Byeun YK, Lee SH, Choi SC (2008) In situ growth of SiC nanowires by carbothermal reduction using a mixture of low-purity SiO_2 and carbon. J Alloy Compd 456:257–263
33. Tao XY, Li YP, Du J, Xia Y, Yang YC, Huang H, Gan YP, Zhang WK, Li XD (2011) A generic bamboo-based carbothermal method for preparing carbide (SiC, B_4C, TiC, TaC, NbC, $Ti_xNb_{1-x}C$, and $Ta_xNb_{1-x}C$) nanowires. J Mater Chem 21:9095
34. Shim HW, Huang H (2007) Three-stage transition during silicon carbide nanowire growth. Appl Phys Lett 90:083106
35. Hu JQ, Lu QY, Tang KB, Deng B, Jiang RR, Qian YT, Yu WC, Zhou GE, Liu XM, Wu JX (2000) Synthesis and characterization of SiC nanowires through a reduction-carburization route. J Phys Chem B 104:5251–5254
36. Zhang Y, Nishitani-Gamo M, Xiao C, Ando T (2002) Synthesis of 3C–SiC nanowhiskers and emission of visible photoluminescence. J Appl Phys 91:6066–6070
37. Zhou XT, Wang N, Lai HL, Peng HY, Bello I, Wong NB, Lee CS, Lee ST (1999) β-SiC nanorods synthesized by hot filament chemical vapor deposition. Appl Phys Lett 74:3942–3944
38. Lai HL, Wong NB, Zhou XT, Peng HY, Au FC, Wang N, Bello I, Lee CS, Lee ST, Duan XF (2000) Straight β-SiC nanorods synthesized by using C-Si-SiO_2. Appl Phys Lett 76:294–296
39. Lu Q, Hu J, Tang K, Qian Y (1999) Growth of SiC nanorods at low temperature. Appl Phys Lett 75:507–509
40. Shen G, Chen D, Tang K, Qian Y, Zhang S (2003) Silicon carbide hollow nanospheres, nanowires and coaxial nanowires. Chem Phys Lett 375:177–184
41. Kim HY, Park J, Yang H (2006) Direct synthesis of aligned silicon carbide nanowires from the silicon substrates. Chem Commun 256–257
42. Choi H-J, Seong H-K, Lee J-C, Sung Y-M (2004) Growth and modulation of silicon carbide nanowires. J Cryst Growth 269:472–478
43. Shi W, Zheng Y, Peng H, Wang N, Lee CS, Lee S-T (2000) Laser ablation synthesis and optical characterization of silicon carbide nanowires. J Am Ceram Soc 83:3228–3230
44. Wu ZS, Deng SZ, Xu NS, Chen J, Zhou J, Chen J (2002) Needle-shaped silicon carbide nanowires: synthesis and field electron emission properties. Appl Phys Lett 80:3829–3831
45. Deng SZ, Wu ZS, Zhou J, Xu NS, Chen R, Chen J (2002) Synthesis of silicon carbide nanowires in a catalyst-assisted process. Chem Phys Lett 356:511–514
46. Feng DH, Jia TQ, Li XX, Xu ZZ, Chen J, Deng SZ, Wu ZS, Xu NS (2003) Catalytic synthesis and photoluminescence of needle-shaped 3C–SiC nanowires. Solid State Commun 128:295–297
47. Sundaresan SG, Davydov AV, Vaudin MD, Levin I, Maslar JE, Tian Y-L, Rao MV (2007) Growth of silicon carbide nanowires by a microwave heating-assisted physical vapor transport process using group VIII metal catalysts. Chem Mater 19:5531–5537
48. Huczko A, Bystrzejewski M, Lange H, Fabianowska A, Cudziło S, Panas A, Szala M (2005) Combustion synthesis as a novel method for production of 1-D SiC nanostructures. J Phys Chem B 109:16244–16251
49. Huczko A, Dabrowska A, Savchyn V, Popov AI, Karbovnyk I (2009) Silicon carbide nanowires: synthesis and cathodoluminescence. Phys Status Solidi B 246:2806–2808
50. Li G-Y, Li X-D, Chen Z-D, Wang J, Wang H, Che R-C (2009) Large areas of centimeters-long SiC nanowires synthesized by pyrolysis of a polymer precursor by a CVD route. J Phys Chem C 113:17655–17660
51. Zhang X, Huang X, Wen G, Geng X, Zhu J, Zhang T, Bai H (2010) Novel SiOC nanocomposites for high-yield preparation of ultra-large-scale SiC nanowires. Nanotechnology 21:385601

52. Lin L (2011) Synthesis and optical property of large-scale centimetres-long silicon carbide nanowires by catalyst-free CVD route under superatmospheric pressure conditions. Nanoscale 3:1582–1591
53. Yao Y, Lee ST, Li FH (2003) Direct synthesis of 2H–SiC nanowhiskers. Chem Phys Lett 381:628–633
54. Zou G, Dong C, Xiong K, Li H, Jiang C, Qian Y (2006) Low-temperature solvothermal route to 2H–SiC nanoflakes. Appl Phys Lett 88:071913
55. Gao F, Yang W, Wang H, Fan Y, Xie Z, An L (2008) Controlled Al-doped single-crystalline 6H-SiC nanowires. Cryst Growth Des 8:1461–1464
56. Wang H, Lin L, Yang W, Xie Z, An L (2010) Preferred orientation of SiC nanowires induced by substrates. J Phys Chem C 114:2591–2594
57. Gao F, Feng W, Wei G, Zheng J, Wang M, Yang W (2012) Triangular prism-shaped p-type 6H-SiC nanowires. CrystEngComm 14:488–491
58. Tang CC, Bando Y (2003) Effect of BN coatings on oxidation resistance and field emission of SiC nanowires. Appl Phys Lett 83:659–661
59. Fowler RH, Nordheim LW (1928) Electron emission in intense electric fields. Proc R Soc London Ser A 119:173–181
60. Wong KW, Zhou XT, Au FCK, Lai HL, Lee CS, Lee ST (1999) Field-emission characteristics of SiC nanowires prepared by chemical-vapor deposition. Appl Phys Lett 75:2918–2920
61. Zhou XT, Lai HL, Peng HY, Au FCK, Liao LS, Wang N, Bello I, Lee CS, Lee ST (2000) Thin β-SiC nanorods and their field emission properties. Chem Phys Lett 318:58–62
62. Lai HL, Wong NB, Zhou XT, Peng HY, Au FCK, Wang N, Bello I, Lee CS, Lee ST, Duan XF (2000) Straight β-SiC nanorods synthesized by using C-Si-SiO_2. Appl Phys Lett 76:294–296
63. Yang GZ, Cui H, Sun Y, Gong L, Chen J, Jiang D, Wang CX (2009) Simple catalyst-free method to the synthesis of β-SiC nanowires and their field emission properties. J Phys Chem C 113:15969–15973
64. Pan Z, Lai H-L, Au FCK, Duan X, Zhou W, Shi W, Wang N, Lee C-S, Wong N-B, Lee S-T, Xie S (2000) Oriented silicon carbide nanowires: synthesis and field emission properties. Adv Mater 12:1186–1190
65. Yang YJ, Meng GW, Liu XY, Zhang LD, Hu Z, He CY, Hu YM (2008) Aligned SiC porous nanowire arrays with excellent field emission properties converted from Si nanowires on silicon wafer. J Phys Chem C 112:20126–20130
66. Senthil K, Yong KJ (2008) Enhanced field emission from density-controlled SiC nanowires. Mater Chem Phys 112:88–93
67. Lo HC, Das D, Hwang JS, Chen KH (2003) SiC-capped nanotip arrays for field emission with ultralow turn-on field. Appl Phys Lett 83:1420–1422
68. Xi G, Liu Y, Liu X, Wang X, Qian Y (2006) Mg-catalyzed autoclave synthesis of aligned silicon carbide nanostructures. J Phys Chem B 110:14172–14178
69. Deng SZ, Li ZB, Wang WL, Xu NS, Zhou J, Zheng XG, Xu HT, Chen J, She JC (2006) Field emission study of SiC nanowires/nanorods directly grown on SiC ceramic substrate. Appl Phys Lett 89:023118
70. Zhang X, Chen Y, Xie Z, Yang W (2010) Shape and doping enhanced field emission properties of quasialigned 3C–SiC nanowires. J Phys Chem C 114:8251–8255
71. Kim D-W, Choi Y-J, Choi KJ, Park J-G, Park J-H, Pimenov SM, Frolov VD, Abanshin NP, Gorfinkel BI, Rossukanyi NM, Rukovishnikov AI (2008) Stable field emission performance of SiC-nanowire-based cathodes. Nanotechnology 19:225706
72. Kang M-G, Lezec HJ, Sharifi F (2013) Stable field emission from nanoporous silicon carbide. Nanotechnology 24:065201
73. Chen SL, Ying PZ, Wang L, Wei GD, Yang WY (2014) Temperature-dependent field emission of flexible n-type silicon carbide nanoneedle emitters. Appl Phys Lett 105:133106
74. Chen SL, Shang MH, Gao FM, Wang L, Ying PZ, Yang WY, Fang XS (2016) Extremely stable current emission of P-doped SiC flexible field emitters. Adv Sci 3:1500256
75. Wang L, Wei GD, Gao FM, Li CM, Yang WY (2015) High-temperature stable field emission of B-doped SiC nanoneedle arrays. Nanoscale 7:7585

76. Meng GW, Zhang LD, Mo CM, Zhang SY, Qin Y, Feng SP, Li HJ (1998) Preparation of β-SiC nanorods with and without amorphous SiO_2 wrapping layers. J Mater Res 13:2533–2538
77. Meng GW, Zhang LD, Qin Y, Mo CM, Phillipp F (1999) Synthesis of β-SiC nanowires with SiO_2 wrappers. Nanostructured Mater 12:1003–1006
78. Wang ZL, Dai ZR, Gao RP, Bai ZG, Gole JL (2000) Side-by-side silicon carbide-silica biaxial nanowires: synthesis, structure, and mechanical properties. Appl Phys Lett 77:3349–3351
79. Chen JP, Jia H, Liu Z, Kong QQ, Hou ZH, Xie LJ, Sun GH, Zhang SC, Chen CM (2020) Construction of C-Si heterojunction interface in SiC whisker/reduced graphene oxide aerogels for improving microwave absorption. Carbon 164:59–68
80. Zhu YQ, Hu WB, Hsu WK, Terrones M, Grobert N, Hare JP, Kroto HW, Walton DRM, Terrones H (1999) SiC-SiO_x heterojunctions in nanowires. J Mater Chem 9:3173–3178
81. Ye H, Titchenal N, Gogotsi Y, Ko F (2005) SiC nanowires synthesized from electrospun nanofiber templates. Adv Mater 17:1531–1535
82. Ryu Y, Tak Y, Yong K (2005) Direct growth of core-shell SiC-SiO_2 nanowires and field emission characteristics. Nanotechnology 16:S370–S374
83. Liu X-M, Yao K-F (2005) Large-scale synthesis and photoluminescence properties of SiC/SiO_x nanocables. Nanotechnology 16:2932–2935
84. Meng A, Li ZJ, Zhang JL, Gao L, Li HJ (2007) Synthesis and Raman scattering of β-SiC/SiO_2 core-shell nanowires. J Cryst Growth 308:263–268
85. Fabbri F, Rossi F, Attolini G, Salviati G, Iannotta S, Aversa L, Verucchi R, Nardi M, Fukata N, Dierre B, Sekiguchi T (2010) Enhancement of the core near-band-edge emission induced by an amorphous shell in coaxial one-dimensional nanostructure: the case of SiC/SiO_2 core/shell self-organized nanowires. Nanotechnology 21:345702
86. Fabbri F, Rossi F, Attolini G, Salviati G, Dierre B, Sekiguchi T, Fukata N (2012) Luminescence properties of SiC/SiO_2 core-shell nanowires with different radial structure. Mater Lett 71:137–140
87. Wei J, Li K, Chen J, Yuan H (2011) Synthesis of centimeter-scale ultra-long SiC nanowires by simple catalyst-free chemical vapor deposition. J Cryst Growth 335:160–164
88. Zhang M, Zhao J, Li ZJ, Ding SQ, Wang YQ, Qiu GH, Meng A, Li QD (2018) Ultra-long SiC/SiO_2 nanowires: simple gram-scale production and their effective blue-violet photoluminescence and microwave absorption properties. ACS Sustainable Chem. Eng. 6:3596–3603
89. Bechelany M, Brioude A, Stadelmann P, Ferro G, Cornu D, Miele P (2007) Very long SiC-based coaxial nanocables with tunable chemical composition. Adv Funct Mater 17:3251–3257
90. Li ZJ, Zhao J, Zhang M, Xia JY, Meng A (2014) SiC nanowires with thickness-controlled SiO_2 shells: fabrication, mechanism, reaction kinetics and photoluminescence properties. Nano Res 7:462–472
91. Li Y, Bando Y, Golberg D (2004) SiC-SiO_2-C coaxial nanocables and chains of carbon nanotube-SiC heterojunctions. Adv Mater 16:93–96
92. Kohno H, Mori Y, Ichikawa S, Ohno Y, Yonenaga I, Takeda S (2009) Transformation of a SiC nanowire into a carbon nanotube. Nanoscale 1:344–346
93. Cambaz ZG, Yushin GN, Gogotsi Y, Vyshnyakova KL, Pereselentseva LN (2006) Formation of carbide-derived carbon on β-silicon carbide whiskers. J Am Ceram Soc 89:509–514
94. Wang X-Y, Zhai H-Z, Cao C-B, Cai H-N, Wang Y, Chan HLW (2009) One-step synthesis of orientation accumulation SiC-C coaxial nanocables at low temperature. J Mater Chem 19:2958–2962
95. Ryu YH, Park BT, Song YH, Yong K (2004) Carbon-coated SiC nanowires: direct synthesis from Si and field emission characteristics. J Cryst Growth 271:99–104
96. Qiang XF, Li HJ, Zhang YL, Wang ZZ, Ba ZX (2016) Synthesis and toughening effect of SiC nanowires wrapped by carbon nanosheet on C/C composites. J Alloy Compd 676:245–250
97. Cheng WQ, Xiong YL, Ding LJ, Li HL, Yang JH, Zhu MM, Chen JJ, Zhang GJ (2022) Graphene-coated pearl-chain-shaped SiC nanowires. J Cryst Growth 599:126903
98. Zhang Y, Suenaga K, Colliex C, Iijima S (1998) Coaxial nanocable: silicon carbide and silicon oxide sheathed with boron nitride and carbon. Science 281:973–975

99. Han W, Redlich P, Ernst F, Rühle M (1999) Synthesizing boron nitride nanotubes filled with SiC nanowires by using carbon nanotubes as templates. Appl Phys Lett 75:1875–1877
100. Tang C, Bando Y, Sato T, Kurashima K (2002) Uniform boron nitride coatings on silicon carbide nanowires. Adv Mater 14:1046–1049
101. Li Y, Dorozhkin PS, Bando Y, Golberg D (2005) Controllable modification of SiC nanowires encapsulated in BN nanotubes. Adv Mater 17:545–549
102. Saulig-Wenger K, Cornu D, Chassagneux F, Ferro G, Epicier T, Miele P (2002) Direct synthesis of β-SiC and h-BN coated β-SiC nanowires. Solid State Commun 124:157–161
103. Zhong B, Song L, Huang X, Zhang X, Wen G, Zhou Y (2011) Novel coaxial SiC-SiO_2-BN nanocable: large-scale synthesis, formation mechanism and photoluminescence property. J Mater Chem 21:14432–14440
104. Liu BD, Yang B, Yuan F, Liu QY, Shi D, Jiang CH, Zhang JS, Staedler T, Jiang X (2015) Defect-induced nucleation and epitaxy: A new strategy toward the rational synthesis of WZ-GaN/3C–SiC core-shell heterostructures. Nano Lett 15:7837–7846
105. Wang D-H, Xu D, Wang Q, Hao Y-J, Jin G-Q, Guo X-Y, Tu KN (2008) Periodically twinned SiC nanowires. Nanotechnology 19:215602
106. Shen G, Bando Y, Ye C, Liu B, Golberg D (2006) Synthesis, characterization and field-emission properties of bamboo-like β-SiC nanowires. Nanotechnology 17:3468–3472
107. Chen JJ, Tang WH, Xin LP, Shi QA (2011) Band gap characterization and photoluminescence properties of SiC nanowires. Appl Phys A 102:213–217
108. Wu R, Li B, Gao M, Chen J, Zhu Q, Pan Y (2008) Tuning the morphologies of SiC nanowires via the control of growth temperature, and their photoluminescence properties. Nanotechnology 19:335602
109. Wei J, Li K-Z, Li H-J, Fu Q-G, Zhang L (2006) Growth and morphology of one-dimensional SiC nanostructures without catalyst assistant. Mater Chem Phys 95:140–144
110. Li Z, Gao W, Meng A, Geng Z, Gao L (2009) Large-scale synthesis and Raman and photoluminescence properties of single crystalline β-SiC nanowires periodically wrapped by amorphous SiO_2 nanospheres 2. J Phys Chem C 113:91–96
111. Wei GD, Qin WP, Zheng KZ, Zhang DS, Sun JB, Lin JJ, Kim RJ, Wang GF, Zhu PF, Wang LL (2009) Synthesis and properties of SiC/SiO_2 nanochain heterojunctions by microwave method. Cryst Growth Des 9:1431–1435
112. Liu HT, Huang ZH, Huang JT, Fang MH, Liu YG, Wu XW (2014) Thermal evaporation synthesis of SiC/SiO_x nanochain heterojunctions and their photoluminescence properties. J. Mater. Chem. C 2:7761
113. Li W, Jia QL, Liu XH, Zhang J (2017) Large scale synthesis and photoluminescence properties of necklace-like SiC/SiO_x heterojunctions via a molten salt mediated vapor reaction technique. Ceram Int 43:2950–2955
114. Hao Y-J, Wagner JB, Su DS, Jin G-Q, Guo X-Y (2006) Beaded silicon carbide nanochains via carbothermal reduction of carbonaceous silica xerogel. Nanotechnology 17:2870–2874
115. Li Z, Shi TJ, Tan DX (2010) Long β-Silicon carbide necklace-like whiskers prepared by carbothermal reduction of wood flour/silica/phenolic composite. J Am Ceram Soc 93:3499–3503
116. Meng A, Zhang M, Gao W, Sun S, Li Z (2011) Large-scale synthesis of β-SiC nanochains and their Raman/photoluminescence properties. Nanoscale Res Lett 6:34
117. Shen G, Bando Y, Golberg D (2007) Self-assembled hierarchical single-crystalline β-SiC nanoarchitectures. Cryst Growth Des 7:35–38
118. Guo JZ, Zuo Y, Li ZJ, Gao WD, Zhang JL (2007) Preparation of SiC nanowires with fins by chemical vapor deposition. Physica E 39:262–266
119. Lin ZJ, Wang L, Zhang J, Guo X-Y, Yang W, Mao H-K, Zhao Y (2010) Nanoscale twinning-induced elastic strengthening in silicon carbide nanowires. Scripta Mater 63:981–984
120. Xin L, Shi Q, Chen J, Tang W, Wang N, Liu Y, Lin Y (2012) Morphological evolution of one-dimensional SiC nanomaterials controlled by sol-gel carbothermal reduction. Mater Charact 55–61

121. Pozuelo M, Kao WH, Yang J-M (2013) High-resolution TEM characterization of SiC nanowires as reinforcements in a nanocrystalline Mg-matrix. Mater Charact 77:81–88
122. Seo W-S, Koumoto K, Aria S (2000) Morphology and stacking faults of β-silicon carbide whisker synthesized by carbothermal reduction. J Am Ceram Soc 83:2584–2592
123. Li HJ, Li ZJ, Meng AL, Li KZ, Zhang XN (2003) SiC nanowire networks. J Alloy Compd 352:279–282
124. Shim HW, Huang H (2007) Nanowebs and nanocables of silicon carbide. Nanotechnology 18:335607
125. Wu C, Liao X, Chen J (2010) The formation of symmetric SiC bi-nanowires with a Y-shaped junction. Nanotechnology 21:405303
126. Lu P, Huang Q, Mukherjee A, Hsieh YL (2011) Effects of polymer matrices to the formation of silicon carbide (SiC) nanoporous fibers and nanowires under carbothermal reduction. J Mater Chem 21:1005–1012
127. Qi XX, Liang J, Yu CY, Ma SF, Liu XG, Xu BS (2014) Facile synthesis of interconnected SiC nanowire networks on silicon substrate. Mater Lett 116:68–70
128. Ho GW, Wong ASW, Kang D-J, Welland ME (2004) Three-dimensional crystalline SiC nanowire flowers. Nanotechnology 15:996–999
129. Magyar AP, Aharonovich I, Baram M, Hu EL (2013) Photoluminescent SiC tetrapods. Nano Lett 13:1210–1215
130. Zhang H-F, Wang C-M, Wang L-S (2002) Helical crystalline SiC/SiO_2 core-shell nanowires. Nano Lett 2:941–944
131. Zhang D, Alkhateeb A, Han H, Mahmood H, McIlroy DN (2003) Silicon carbide nanosprings. Nano Lett 3:983–987
132. Xi G, Peng Y, Wan S, Li T, Yu W, Qian Y (2004) Lithium-assisted synthesis and characterization of crystalline 3C–SiC nanobelts. J Phys Chem B 108:20102–20104
133. Wu RB, Wu LL, Yang GY, Pan Y, Chen JJ, Zhai R, Lin J (2007) Fabrication and photoluminescence of bicrystalline SiC nanobelts. J Phys D: Appl Phys 40:3697–3701
134. Wei G, Qin W, Kim R, Sun J, Zhu P, Wang G, Wang L, Zhang D, Zheng K (2008) Quantum confinement effect and field emission characteristics of ultrathin 3C–SiC nanobelts. Chem Phys Lett 461:242–245
135. Zhang H, Ding W, He K, Li M (2010) Synthesis and characterization of crystalline silicon carbide nanoribbons. Nanoscale Res Lett 5:1264–1271
136. Qadri SB, Rath BB, Gorzkowski EP, Feng J, Qadri SN, Caldwell JD (2015) Nanotubes, nanobelts, nanowires, and nanorods of silicon carbide from the wheat husks. J Appl Phys 118:104904
137. Chen J, Wu R, Yang G, Pan Y, Lin J, Wu L, Zhai R (2008) Synthesis and photoluminescence of needle-shaped 3C–SiC nanowires on the substrate of PAN carbon fiber. J Alloy Compd 456:320–323
138. Wu R, Zhou K, Wei J, Huang Y, Su F, Chen J, Wang L (2012) Growth of tapered SiC nanowires on flexible carbon fabric: toward field emission applications. J Phys Chem C 116:12940–12945
139. Zhou J-Y, Chen Z-Y, Zhou M, Gao X-P, Xie E-Q (2009) SiC nanorods grown on electrospun nanofibers using Tb as catalyst: fabrication, characterization, and photoluminescence properties. Nanoscale Res Lett 4:814–819
140. Zhou J-Y, Chen Z-Y, Xu X-B, Zhou M, Ma Z-W, Zhao J-G, Li R-S, Xie E-Q (2010) Terbium-catalyzed selective area growth of SiC nanorods: synthesis, optimal growth, and field emission properties. J Am Ceram Soc 93:488–493
141. Wu RB, Zhou K, Qian XK, Wei J, Tao Y, Sow CH, Wang LY, Huang YZ (2013) Well-aligned SiC nanoneedle arrays for excellent field emitters. Mater Lett 91:220–223
142. Qiang XF, Li HJ, Zhang YL, Tian S, Wei JF (2013) Synthesis and Raman scattering of SiC nanowires decorated with SiC polycrystalline nanoparticles. Mater Lett 107:315–317
143. Liu HM, Li KZ, Chen H, Liu B, Yin XM (2022) Facile growth of oriented SiC nanowires arrays on carbon fiber cloth via CVD. Ceram Int 48:34543–34549
144. Liu JH, Yang Y, Song K, Li WJ, Wang HL, Vityuk N, Klymenko N, Krupska TV, Turov VV, Zhao XS, Yang WY, Chen SL, Li XX (2023) Three-dimensional branched SiC nanowire

field emitters with single-crystal integrated structures and increased emission sites: ultralow turn-on field and high stability. J Mater Chem C 11:16473
145. Li Z, Zhang J, Meng A, Guo J (2006) Large-area highly-oriented SiC nanowire arrays: synthesis, Raman, and photoluminescence properties. J Phys Chem B 110:22382–22386
146. Luna LE, Ophus C, Johanson J, Maboudian R, Carraro C (2016) Demonstration of hexagonal phase silicon carbide nanowire arrays with vertical alignment. Cryst Growth Des 16:2887–2892
147. Chen CM, Chen SL, Shang MH, Gao FM, Yang ZB, Liu Q, He ZY, Yang WY (2016) Fabrication of highly oriented 4H–SiC gourd-shaped nanowire arrays and their field emission properties. J Mater Chem C 4:5195–5201
148. Chen Y, Zhang C, Li LY, Zhou S, Chen X, Gao J, Zhao N, Wong CP (2019) Hybrid anodic and metal-assisted chemical etching method enabling fabrication of silicon carbide nanowires. Small 15:1803898
149. Ji X, Liang TY, Miao RA, Zhu JH, Xing JL, Wang ZQ, Fan JY (2023) Co-wide-bandgap conformally heteroepitaxial perovskite-monolithic SiC nanowire array with quantum-confined blue luminescence. Appl Phys Express 16:065004
150. Sun X, Ling Q, Miao R, Wu H, Fan J (2025) Purcell-enhanced picosecond emission in semiconducting 4H/6H-SiC monocrystalline nanowire forest microcavity. Adv Photon Nexus 4:056006
151. Svelto O, Hanna DC (1998) Principles of lasers, 4th edn. Springer, New York
152. Liang T, Liu W, Liu X, Li Y, Fan J (2021) Fabry-Perot mode-limited high-Purcell-enhanced spontaneous emission from in situ laser-induced $CsPbBr_3$ quantum dots in CsP_b2B_r5 microcavities. Nano Lett 22:355
153. Fox M (2006) Quantum optics: an introduction. Oxford University Press, Oxford
154. Zare S, Pouria R, Chow PK, Tiwald T, Tripp CP, Edalatpour S (2023) Probing near-field thermal emission of localized surface phonons from silicon carbide nanopillars. ACS Photonics 10:401–411
155. Seong H-K, Choi H-J, Lee S-K, Lee J-I, Choi D-J (2004) Optical and electrical transport properties in silicon carbide nanowires. Appl Phys Lett 85:1256–1258
156. Chen YQ, Zhang XN, Zhao Q, He L, Huang CK, Xie ZP (2011) P-type 3C–SiC nanowires and their optical and electrical transport properties. Chem Commun 47:6398–6400
157. Han XD, Zhang YF, Zheng K, Zhang XN, Zhang Z, Hao YJ, Guo XY, Yuan J, Wang ZL (2007) Low-temperature in situ large strain plasticity of ceramic SiC nanowires and its atomic-scale mechanism. Nano Lett 7:452–457
158. Zhang Y, Han X, Zheng K, Zhang Z, Zhang X, Fu J, Ji Y, Hao Y, Guo X, Wang Z (2007) Direct observation of super-plasticity of beta-SiC nanowires at low temperature. Adv Funct Mater 17:3435–3440
159. Perisanu S, Vincent P, Ayari A, Choueib M, Purcell ST, Bechelany M, Cornu D (2007) High Q factor for mechanical resonances of batch-fabricated SiC nanowires. Appl Phys Lett 90:043113
160. Perisanu S, Gouttenoire V, Vincent P, Ayari A, Choueib M, Bechelany M, Cornu D, Purcell ST (2008) Mechanical properties of SiC nanowires determined by scanning electron and field emission microscopies. Phys Rev B 77:165434
161. Yibibulla T, Jiang YJ, Wang SL, Huang H (2021) Size- and temperature-dependent Young's modulus of SiC nanowires determined by a laser-Doppler vibration measurement. Appl Phys Lett 118:043103
162. Yang W, Araki H, Tang C, Thaveethavorn S, Kohyama A, Suzuki H, Noda T (2005) Single-crystal SiC nanowires with a thin carbon coating for stronger and tougher ceramic composites. Adv Mater 17:1519–1523
163. Cheng GM, Chang TH, Qin QQ, Huang HC, Zhu Y (2014) Mechanical properties of silicon carbide nanowires: effect of size-dependent defect density. Nano Lett 14:754–758
164. Wang SL, Wu YQ, Lin LW, He YH, Huang H (2015) Fracture strain of SiC nanowires and direct evidence of electron-beam induced amorphisation in the strained nanowires. Small 11:1672–1676

165. Shao R, Zheng K, Zhang Y, Li Y, Zhang Z, Han X (2012) Piezoresistance behaviors of ultra-strained SiC nanowires. Appl Phys Lett 101:233109
166. Bi JH, Wei GD, Wang L (2013) Highly sensitive piezoresistance behaviors of n-type 3C–SiC nanowires. J Mater Chem C 1:4514–4517
167. Li XX, Chen SL, Ying PZ, Gao FM, Liu Q, Shang MH, Yang WY (2016) A giant negative piezoresistance effect in 3C–SiC nanowires with B dopants. J Mater Chem C 4:6466–6472
168. Phan HP, Dowling KM, Nguyen TK, Dinh T, Senesky DG, Namazu T, Dao DV, Nguyen NT (2018) Highly sensitive pressure sensors employing 3C–SiC nanowires fabricated on a free standing structure. Mater Design 156:16–21
169. Lee KM, Choi TY, Lee SK, Poulikakos D (2010) Focused ion beam-assisted manipulation of single and double β-SiC nanowires and their thermal conductivity measurements by the four-point-probe 3-ω method. Nanotechnology 21:125301
170. Anufriev R, Wu YH, Volz S, Nomura M (2024) Quasi-ballistic thermal transport in silicon carbide nanowires. Appl Phys Lett 124:022202
171. Chen WK, Wu XS, Deng SC, Yang N, Xu XF (2024) Strain effect on thermal conductivity of 3C–SiC nanowire. Appl Phys Lett 124:242201
172. Jänis A, Yao Y, Klement U (2011) Dielectric properties of SiC nanowires with different chemical compositions. IEEE T Nanotechnol 10:751–756
173. Kuang JL, Cao WB (2013) Stacking faults induced high dielectric permittivity of SiC wires. Appl Phys Lett 103:112906
174. Wieligor M, Wang YJ, Zerda TW (2005) Raman spectra of silicon carbide small particles and nanowires. J Phys Condens Matter 17:2387–2395
175. Makeev MA, Srivastava D, Menon M (2006) Silicon carbide nanowires under external loads: an atomistic simulation study. Phys Rev B 74:165303
176. Zhang T-Y, Luo M, Chan WK (2008) Size-dependent surface stress, surface stiffness, and Young's modulus of hexagonal prism [111] β-SiC nanowires. J Appl Phys 103:104308
177. Wang Z, Zu X, Gao F, Weber WJ (2008) Atomistic simulations of the mechanical properties of silicon carbide nanowires. Phys Rev B 77:224113
178. Wang ZG, Li JB, Gao F, Weber WJ (2010) Tensile and compressive mechanical behavior of twinned silicon carbide nanowires. Acta Mater 58:1963–1971
179. Wang J, Lu C, Wang Q, Xiao P, Ke F, Bai Y, Shen Y, Liao X, Gao H (2012) Influence of microstructures on mechanical behaviours of SiC nanowires: a molecular dynamics study. Nanotechnology 23:025703
180. Oliveira JB, Morbec JM, Miwa RH (2017) Mechanical and electronic properties of SiC nanowires: An ab initio study. J Appl Phys 121:104302
181. Elahi F, Ma L, Hossain ZM (2018) Heterogeneity governs diameter-dependent toughness and strength in SiC nanowires. Phys Rev B 98:174111
182. Rurali R (2005) Electronic and structural properties of silicon carbide nanowires. Phys Rev B 71:205405
183. Yan B, Zhou G, Duan W, Wu J, Gu B-L (2006) Uniaxial-stress effects on electronic properties of silicon carbide nanowires. Appl Phys Lett 89:023104
184. Durandurdu M (2006) Electronic and mechanical properties of wurtzite type SiC nanowires. Phys Stat Sol (B) 243:R37–R39
185. Wang Z, Zhao M, He T, Zhang X, Xi Z, Yan S, Liu X, Xia Y (2009) First-principles study of faceted single-crystalline silicon carbide nanowires and nanotubes. J Phys Chem C 113:856–861
186. Wang Z, Zhao M, He T, Zhang H, Zhang X, Xi Z, Yan S, Liu X, Xia Y (2009) Orientation-dependent stability and quantum-confinement effects of silicon carbide nanowires. J Phys Chem C 113:12731–12735
187. Wang Z, Wang S, Zhang C, Li J (2011) First principles study of the electronic properties of twinned SiC nanowires. J Nanopart Res 13:185–191
188. Li YJ, Li SL, Gong P, Li YL, Fang XY, Jia YH, Cao MS (2018) Effect of surface dangling bonds on transport properties of phosphorous doped SiC nanowires. Physica E 104:247–253

189. Jia YH, Gong P, Li SL, Ma WD, Fang XY, Yang YY, Cao MS (2020) Effects of hydroxyl groups and hydrogen passivation on the structure, electrical and optical properties of silicon carbide nanowires. Phys Lett A 384:126106
190. Ma Y, Yan H, Yu XX, Gong P, Li YL, Ma WD, Fang XY (2024) Effects of different atomic passivation on conductive and dielectric properties of silicon carbide nanowires. J Appl Phys 135:054101
191. Ngomsi CAS, Joshi T, Dev P (2024) Optimum surface-passivation schemes for near-surface spin defects in silicon carbide. Phys Rev Mater 8:056202
192. Termentzidis K, Barreteau T, Ni YX, Merabia S, Zianni X, Chalopin Y, Chantrenne P, Volz S (2013) Modulated SiC nanowires: molecular dynamics study of their thermal properties. Phys Rev B 87:125410
193. Xiong SY, Latour B, Ni YX, Volz S, Chalopin Y (2015) Efficient phonon blocking in SiC antiphase superlattice nanowires. Phys Rev B 91:224307
194. Kremsner JM, Kappe CO (2006) Silicon carbide passive heating elements in microwave-assisted organic synthesis. J Org Chem 71:4651–4658
195. Obermayer D, Gutmann B, Kappe CO (2009) Microwave chemistry in silicon carbide reaction vials: separating thermal from nonthermal effects. Angew Chem Int Ed 48:8321–8324
196. Chiu S-C, Yu H-C, Li Y-Y (2010) High electromagnetic wave absorption performance of silicon carbide nanowires in the Gigahertz range. J Phys Chem C 114:1947–1952
197. Wu R, Zhou K, Yang Z, Qian X, Wei J, Liu L, Huang Y, Kong L, Wang L (2013) Molten-salt-mediated synthesis of SiC nanowires for microwave absorption applications. CrystEngComm 15:570–576
198. Zhang HY, Xu YJ, Zhou JG, Jiao JF, Chen YJ, Wang H, Liu CY, Jiang ZH, Wang ZJ (2015) Stacking fault and unoccupied densities of state dependence of electromagnetic wave absorption in SiC nanowires. J Mater Chem C 3:4416
199. Han T, Luo RY, Cui GY, Wang LY (2019) Effect of SiC nanowires on the high-temperature microwave absorption properties of SiCf/SiC composites. J Eur Ceram Soc 39:1743–1756
200. Liang CY, Wang ZJ (2019) Eggplant-derived SiC aerogels with high-performance electromagnetic wave absorption and thermal insulation properties. Chem Eng J 373:598–605
201. Lv XY, Ye F, Cheng LF, Zhang LT (2022) 3D printing "wire-on-sphere" hierarchical SiC nanowires/SiC whiskers foam for efficient high-temperature electromagnetic wave absorption. J Mater Sci Technol 109:94–104
202. Song LM, Wu CW, Zhi Q, Zhang F, Song BZ, Guan L, Chen YQ, Wang HL, Zhang R, Fan BB (2023) Multifunctional SiC aerogel reinforced with nanofibers and nanowires for high-efficiency electromagnetic wave absorption. Chem Eng J 467:143518
203. Han MK, Yin XW, Hou ZX, Song CQ, Li XL, Zhang LT, Cheng LF (2017) Flexible and thermostable graphene/SiC nanowire foam composites with tunable electromagnetic wave absorption properties. ACS Appl Mater Interfaces 9:11803–11810
204. Jiang Y, Chen Y, Liu YJ, Sui GX (2018) Lightweight spongy bone-like graphene@SiC aerogel composites for high-performance microwave absorption. Chem Eng J 337:522–531
205. Cheng YH, Hu P, Zhou SB, Yan LW, Sun BQ, Zhang XH, Han WB (2018) Achieving tunability of effective electromagnetic wave absorption between the whole X-band and Ku-band via adjusting PPy loading in SiC nanowires/graphene hybrid foam. Carbon 132:430–443
206. Liu XX, Zhang ZY, Wu YP (2011) Absorption properties of carbon black/silicon carbide microwave absorbers. Compos Part B-Eng 42:326–329
207. Chen LQ, Yin XW, Fan XM, Chen M, Ma XK, Cheng LF, Zhang LT (2015) Mechanical and electromagnetic shielding properties of carbon fiber reinforced silicon carbide matrix composites. Carbon 95:10–19
208. Yan LW, Hong CQ, Sun BQ, Zhao GD, Cheng YH, Dong S, Zhang DY, Zhang XH (2017) In situ growth of core-sheath heterostructural SiC nanowire arrays on carbon fibers and enhanced electromagnetic wave absorption performance. ACS Appl Mater Interfaces 9:6320–6331
209. Zhu HL, Bai YJ, Liu R, Lun N, Qi YX, Han FD, Bi JQ (2011) In situ synthesis of one-dimensional MWCNT/SiC porous nanocomposites with excellent microwave absorption properties. J Mater Chem 21:13581

210. Li Q, Yin XW, Duan WY, Kong L, Hao BL, Ye F (2013) Electrical, dielectric and microwave-absorption properties of polymer derived SiC ceramics in X band. J Alloy Compd 565:66–72
211. Wu RB, Yang ZH, Fu MS, Zhou K (2016) In-situ growth of SiC nanowire arrays on carbon fibers and their microwave absorption properties. J Alloy Compd 687:833–838
212. Zhou W, Long L, Xiao P, Li Y, Luo H, Hu WD, Yin RM (2017) Silicon carbide nano-fibers in-situ grown on carbon fibers for enhanced microwave absorption properties. Ceram Int 43:5628–5634
213. Li BB, Mao BX, Wang XB, He T, Huang HQ (2020) Novel, hierarchical SiC nanowire-reinforced SiC/carbon foam composites: lightweight, ultrathin, and highly efficient microwave absorbers. J Alloy Compd 829:154609
214. Yu ZJ, Lv X, Mao KW, Yang YJ, Liu AH (2020) Role of in-situ formed free carbon on electromagnetic absorption properties of polymer-derived SiC ceramics. J Adv Ceram 9:617–628
215. Du B, Zhang DY, Qian JJ, Cai M, He C, Zhou P, Shui AZ (2021) Multifunctional carbon nanofiber-SiC nanowire aerogel films with superior microwave absorbing performance. Adv Compos Hybrid Mater 4:1281–1291
216. Yang HJ, Cao MS, Li Y, Shi HL, Hou ZL, Fang XY, Jin HB, Wang WZ, Yuan J (2014) Enhanced dielectric properties and excellent microwave absorption of SiC powders driven with NiO nanorings. J Mater Chem A 2:16397–16402
217. Yang HJ, Cao WQ, Zhang DQ, Su TJ, Shi HL, Wang WZ, Yuan J, Cao MS (2015) NiO hierarchical nanorings on SiC: enhancing relaxation to tune microwave absorption at elevated temperature. ACS Appl Mater Interfaces 7:7073–7077
218. Liang CY, Liu CY, Wang H, Wu LN, Jiang ZH, Xu YJ, Shen BZ, Wang ZJ (2014) SiC-Fe_3O_4 dielectric-magnetic hybrid nanowires: controllable fabrication, characterization and electromagnetic wave absorption. J Mater Chem A 2:16397
219. Duan WY, Yin XW, Ye F, Li Q, Han MK, Liu XF, Cai YZ (2016) Synthesis and EMW absorbing properties of nano SiC modified PDC–SiOC. J Mater Chem C 4:5962–5969
220. Jian X, Tian W, Li JY, Deng LJ, Zhou ZW, Zhang L, Lu HP, Yin LJ, Mahmood N (2019) High-temperature oxidation-resistant $ZrN_{0.4}B_{0.6}$/SiC nanohybrid for enhanced microwave absorption. ACS Appl Mater Interfaces 11:15869–15880
221. Li X, Li MH, Lu XK, Zhu WJ, Xu HL, Xue JM, Ye F, Liu YS, Fan XM, Cheng LF (2021) A sheath-core shaped ZrO_2-SiC/SiO_2 fiber felt with continuously distributed SiC for broad-band electromagnetic absorption. Chem Eng J 419:129414 (2021)
222. Cai ZX, Su L, Wang HJ, Niu M, Tao LT, Lu D, Xu L, Li MZ, Gao HF (2021) Alternating multilayered Si_3N_4/SiC aerogels for broadband and high-temperature electromagnetic wave absorption up to 1000 °C. ACS Appl Mater Interfaces 13:16704–16712
223. Kuang JL, Jiang P, Liu WX, Cao WB (2015) Synergistic effect of Fe-doping and stacking faults on the dielectric permittivity and microwave absorption properties of SiC whiskers. Appl Phys Lett 106:212903
224. Kuang JL, Jiang P, Ran FY, Cao WB (2016) Conductivity-dependent dielectric properties and microwave absorption of Al-doped SiC whiskers. J Alloy Compd 687:227–231
225. Zhang BX, Wu C, Ye KC, Sun CY, Wang ZJ (2023) Dual-induced SiC/SME multifunction composite for high-Efficiency broadband electromagnetic wave absorption. Carbon 213:118253
226. Chen JJ, Liao X, Wang MM, Liu ZX, Zhang JD, Ding LJ, Gao L, Li Y (2015) Highly flexible, nonflammable and free-standing SiC nanowire paper. Nanoscale 7:6374–6379
227. Chabi S, Rocha VG, García-Tuñón E, Ferraro C, Saiz E, Xia YD, Zhu YQ (2016) Ultralight, strong, three-dimensional SiC structures. ACS Nano 10:1871–1876
228. Su L, Wang HJ, Niu M, Fan XY, Ma MB, Shi ZQ, Guo SW (2018) Ultralight, recoverable, and high-temperature-resistant SiC nanowire aerogel. ACS Nano 12:3103–3111
229. Yao YM, Zhu XD, Zeng XL, Sun R, Xu JB, Wong CP (2018) Vertically aligned and interconnected SiC nanowire networks leading to significantly enhanced thermal conductivity of polymer composites. ACS Appl Mater Interfaces 10:9669–9678

230. Dai W, Lv L, Lu JB, Hou H, Yan QW, Alam FE, Li YF, Zeng XL, Yu JH, Wei QP, Xu XF, Wu JB, Jiang N, Du SY, Sun R, Xu JB, Wong CP, Lin CT (2019) A Paper-like inorganic thermal interface material composed of hierarchically structured graphene/silicon carbide nanorods. ACS Nano 13:1547–1554
231. Ren B, Liu JJ, Rong YD, Wang L, Lu YJ, Xi XQ, Yang JL (2019) Nanofibrous aerogel bulk assembled by cross-linked SiC/SiO_x core-shell nanofibers with multifunctionality and temperature-invariant hyperelasticity. ACS Nano 13:11603–11612
232. Lu D, Su L, Wang HJ, Niu M, Xu L, Ma MB, Gao HF, Cai ZX, Fan XY (2019) Scalable fabrication of resilient SiC nanowires aerogels with exceptional high-temperature stability. ACS Appl Mater Interfaces 11:45338–45344
233. Song LM, Fan BB, Chen YQ, Gao QC, Li Z, Wang HL, Zhang XY, Guan L, Li HX, Zhang R (2022) Ultralight and hyperelastic SiC nanofiber aerogel spring for personal thermal energy regulation. J. Adv. Ceram. 11:1235–1248
234. Guo PF, Su L, Peng K, Lu D, Xu L, Li MZ, Wang HJ (2022) Additive manufacturing of resilient SiC nanowire aerogels. ACS Nano 16:6625–6633
235. Yan MY, Zhang HP, Fu YY, Pan YL, Lun ZY, Zhang ZX, He P, Cheng XD (2022) Implementing an air suction effect induction strategy to create super thermally insulating and superelastic SiC aerogels. Small 18:2201039
236. Li HL, Dong Q, Zheng XP, Chen JJ, Lou YW, Yang JH, Zhu MM, Lv LM, Zhu HL, Yang XY (2023) Large-scale fabrication of high-performing single-crystal SiC nanowire sponges using natural loofah. ACS Sustainable Chem Eng 11:2554–2563
237. Han LJ, Chen SL, Li HH, Dong YH, Wang CA, Li JT (2024) Rapid and inexpensive synthesis of liter-scale SiC aerogels. Nat Commun 15:6959
238. Zhou W, Yan L, Wang Y, Zhang Y (2006) SiC nanowires: a photocatalytic nanomaterial. Appl Phys Lett 89:013105
239. Hao JY, Wang YY, Tong XL, Jin GQ, Guo XY (2012) Photocatalytic hydrogen production over modified SiC nanowires under visible light irradiation. Int J Hydrogen Energy 37:15038–15044
240. Liu HL, She GW, Mu LX, Shi WS (2012) Porous SiC nanowire arrays as stable photocatalyst for water splitting under UV irradiation. Mater Res Bull 47:917–920
241. Zhang JD, Chen JJ, Xin LP, Wang MM (2014) Hierarchical 3C–SiC nanowires as stable photocatalyst for organic dye degradation under visible light irradiation. Mater Sci Eng B 179:6–11
242. Wang MM, Chen JJ, Liao X, Liu ZX, Zhang JD, Gao L, Li Y (2014) Highly efficient photocatalytic hydrogen production of platinum nanoparticle-decorated SiC nanowires under simulated sunlight irradiation. Int J Hydrogen Energy 39:14581–14587
243. Yang T, Chang XW, Chen JH, Chou KC, Hou XM (2015) B-doped 3C–SiC nanowires with a finned microstructure for efficient visible light-driven photocatalytic hydrogen production. Nanoscale 7:8955
244. Zhou XF, Li X, Gao QZ, Yuan JL, Wen JQ, Fang YP, Liu W, Zhang SS, Liu YJ (2015) Metal-free carbon nanotube-SiC nanowire heterostructures with enhanced photocatalytic H_2 evolution under visible light irradiation. Catal Sci Technol 5:2798–2806
245. Wang B, Wang YD, Lei YP, Wu N, Gou YZ, Han C, Xie S, Fang D (2016) Mesoporous silicon carbide nanofibers with in situ embedded carbon for co-catalyst free photocatalytic hydrogen production. Nano Res 9:886–898
246. Zhang H, Ouyang Q, Yu LL, Hu R, Wan J, Song B, Huang Q, Yao YG (2023) Transient and in situ growth of nanostructured SiC on carbon fibers toward highly durable catalysis. Adv Funct Mater 33:2301375
247. Alper JP, Vincent M, Carraro C, Maboudian R (2012) Silicon carbide coated silicon nanowires as robust electrode material for aqueous micro-supercapacitor. Appl Phys Lett 100:163901
248. Alper JP, Kim MS, Vincent M, Hsia B, Radmilovic V, Carraro C, Maboudian R (2013) Silicon carbide nanowires as highly robust electrodes for micro-supercapacitors. J Power Sources 230:298–302

249. Gu L, Wang YW, Fang YJ, Lu R, Sha J (2013) Performance characteristics of supercapacitor electrodes made of silicon carbide nanowires grown on carbon fabric. J Power Sources 243:648–653
250. Zhao J, Li ZJ, Zhang M, Meng A, Li QD (2016) Direct growth of ultrathin $NiCo_2O_4$/NiO nanosheets on SiC nanowires as a free-standing advanced electrode for high-performance asymmetric supercapacitors. ACS Sustainable Chem Eng. 4:3598–3608
251. Zou XL, Ji L, Hsu HY, Zheng K, Pang ZY, Lu XG (2018) Designed synthesis of SiC nanowire-derived carbon with dual-scale nanostructures for supercapacitor applications. J Mater Chem A 6:12724
252. Vishnu DSM, Sure J, Kim HK, Kumar RV, Schwandt C (2020) Solid state electrochemically synthesised β-SiC nanowires as the anode material in lithium ion batteries. Energy Storage Mater 26:234–241
253. Li XX, Li WJ, Liu Q, Chen SL, Wang L, Gao FM, Shao G, Tian Y, Lin ZF, Yang WY (2021) Robust hgh-temperature supercapacitors based on SiC nanowires. Adv Funct Mater 31:2008901
254. Zekentes K, Choi J, Stambouli V, Bano E, Karker O, Rogdakis K (2022) Progress in SiC nanowire field-effect-transistors for integrated circuits and sensing applications. Microelectron Eng 255:111704
255. Zhou WM, Fang F, Hou ZY, Yan LJ, Zhang YF (2006) Field-effect transistor based on β-SiC nanowire. IEEE Electr Device L 27:463–465
256. Jang C-O, Kim T-H, Lee S-Y, Kim D-J, Lee S-K (2008) Low-resistance ohmic contacts to SiC nanowires and their applications to field-effect transistors. Nanotechnology 19:345203
257. Rogdakis K, Bano E, Montes L, Bechelany M, Cornu D, Zekentes K (2011) Rectifying source and drain contacts for effective carrier transport modulation of extremely doped SiC nanowire FETs. IEEE T Nanotechnol 10:980–984
258. Chen KY, Tripathy PK, Mondal K, Zhang HL, Couet A, Andrews JB (2023) Solution processed Schottky diodes enabled by silicon carbide nanowires for harsh environment applications. Nano Lett 23:2816–2821
259. He Z, Wang L, Gao F, Wei G, Zheng J, Cheng X, Tang B, Yang W (2013) Synthesis of n-type SiC nanowires with tailored doping levels. CrystEngComm 15:2354–2358
260. Niu JJ, Wang JN (2009) A novel self-cleaning coating with silicon carbide nanowires. J Phys Chem B 113:2909–2912
261. Guiffard B, Guyomar D, Seveyrat L, Chowanek Y, Bechelany M, Cornu D, Miele P (2009) Enhanced electroactive properties of polyurethane films loaded with carbon-coated SiC nanowires. J Phys D Appl Phys 42:055503
262. Feng XL, Matheny MH, Zorman CA, Mehregany M, Roukes ML (2010) Low voltage nanoelectromechanical switches based on silicon carbide nanowires. Nano Lett 10:2891–2896
263. Shim HW, Kuppers JD, Huang H (2009) Strong friction of silicon carbide nanowire films. Nanotechnology 20:025704
264. Lantz MA, Gotsmann B, Jaroenapibal P, Jacobs TDB, O'Connor SD, Sridharan K, Carpick RW (2012) Wear-resistant nanoscale silicon carbide tips for scanning probe applications. Adv Funct Mater 22:1639–1645
265. Guo X-N, Shang R-J, Wang D-H, Jin G-Q, Guo X-Y, Tu KN (2010) Avoiding loss of catalytic activity of Pd nanoparticles partially embedded in nanoditches in SiC nanowires. Nanoscale Res Lett 5:332–337
266. Chu YH, Li HJ, Fu QG, Wang HP, Hou XH, Zou X, Shang GN (2012) Oxidation protection of C/C composites with a multilayer coating of SiC and Si plus SiC plus SiC nanowires. Carbon 50:1280–1288
267. Tak YJ, Ryu YH, Yong K (2005) Atomically abrupt heteronanojunction of ZnO nanorods on SiC nanowires prepared by a two-step process. Nanotechnology 16:1712–1716
268. Sun B, Sun Y, Wang CX (2018) Flexible transparent and free-standing SiC nanowires fabric: stretchable UV absorber and fast-response UV-A detector. Small 14:1703391

Chapter 6
SiC Nanotubes and Two-Dimensional Monolayers

In most cases, the silicon carbide nanotubes have been fabricated with assistance of some templates. Unlike carbon nanotubes, the silicon carbide nanotubes contain two types of elements, silicon and carbon; therefore, the structure of the SiC nanotube is basically different from that of the carbon nanotube. In fact, most of the currently reported SiC nanotubes are not standard sp^2-bonded nanotubes and instead, the walls of these nanotubes have sp^3-bonded bulk-SiC-like crystal structures. Some are only the rolled-up nanostructures of bulk SiC, and others are composed of polycrystalline SiC particles. Several factors account for the difficulty in synthesizing actual SiC nanotubes or two-dimensional honeycomb-structured monolayers. Firstly, the Si atoms on the surface of the SiC nanotubes or monolayers are readily oxidized even at room temperature and this will destroy the honeycomb structure. Secondly, the theoretical study shows that the SiC nanowires may be more stable than the SiC nanotubes of the same diameter in the common experimentally approachable (a few nanometers or larger) size regime. Thirdly, unlike graphene which can be readily prepared by exfoliation of layer-structured graphite, there is no graphite-like sp^2-structured SiC crystal in nature from which the actual SiC monolayers or rolled nanotubes could be readily extracted. Therefore, the research field of SiC nanotubes and monolayers is somewhat strange in that the experimentalists often (not always) synthesize bulk-like sp^3-structured SiC nanotubes or ultrathin SiC layers, whereas the theorists generally investigate the structures and properties of actual sp^2-structured SiC nanotubes or monolayers. However, there have been some breakthroughs in the synthesis of sp^2-structured SiC in recent years, although still in infancy.

J. Fan and P. K. Chu, *Silicon Carbide Nanostructures*, Engineering Materials and Processes, https://doi.org/10.1007/978-3-032-10986-6_6

6.1 Fabrication Using Templates

In general, silicon carbide nanotubes are fabricated by using some templates, especially carbon nanotubes, where the silicon source material reacts with the carbon nanotube to yield SiC nanotube that maintains the morphology of the mother nanotube. Silicon carbide nanotube can also be synthesized as the shell of some core material, and subsequent removal of the core nanowire by acid etching gives rise to relatively pure SiC nanotube.

6.1.1 Synthesis Using Carbon Nanotubes/Nanofibers

Shape memory synthesis is one of the earliest methods used to synthesize SiC nanotubes [1, 2]. The SiO vapor generated by heating a Si/SiO_2 mixture at 1200–1350 °C served as the silicon source material, the subsequent gas–solid reaction between SiO and a shaped carbon template at a lower temperature of 1000–1250 °C yielded β-SiC material that has the same shape as the initial carbon material. The gas–solid reaction between SiO vapor and carbon nanotubes produced SiC nanotubes with internal diameters of 30–100 nm, whereas the reaction between SiO and carbon nanofibers produced SiC nanotubes with internal diameters of 10–80 nm (Fig. 6.1). X-ray diffraction confirms that the product is pure β-SiC with relatively good crystallization. The lattice fringes are parallel to the axis of the tube (Fig. 6.1) and have an interlayer distance of 0.36 nm, which is consistent with the spacing of two C planes in β-SiC (0.359 nm). The SiC nanotubes can act as favorable catalyst support in the reaction of low-temperature selective oxidation of H_2S into sulfur [3]. The SiC nanotubes improved both desulfurization activity of the catalyst and the solid sulfur storage capacity.

Silicon carbide nanowires and nanotubes have been prepared by thermal reaction between SiO and multiwalled carbon nanotubes [4]. Both β-SiC nanowires and biaxial SiC–SiO_x nanowires (Fig. 6.2a) were observed, but the most interesting structure found is the multiwalled SiC nanotube (Fig. 6.2b), which has an interlayer spacing of 3.5–4.5 Å. These nanotubes were produced by reaction of Si (via disproportionation of SiO) and the templates of the multiwalled carbon nanotubes at 935 °C. The elemental mapping confirms that there are silicon and carbon elements in the outer layer of the nanotube. The EELS measurement of a single multiwalled SiC nanotube indicates that the Si $L_{3,2}$-edge shifts to a higher energy at 101.2 eV from 99.3 eV of pure Si and the C K-edge shifts to a lower energy at 282.5 eV from 284 eV of pure carbon, showing strong chemical bonding between Si and C. The EELS spectra of the SiC nanotubes and the β-SiC nanowires (Fig. 6.3a) are quite similar except for the pre-edge peaks, further confirming the β-SiC structure of the nanotubes. The multiwalled SiC nanotubes will gradually transform to β-SiC crystals after continuous electron beam annealing under TEM (Fig. 6.3b), and the ultimately formed crystalline structure possesses a lattice spacing of 2.5 Å, corresponding to that

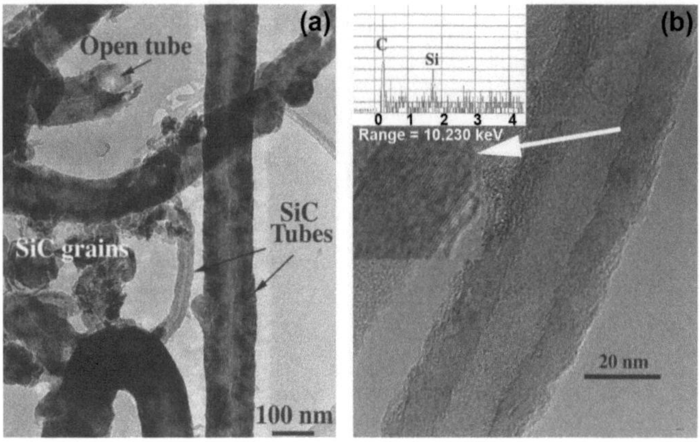

Fig. 6.1 **a** TEM and **b** high-resolution TEM images of the SiC nanotubes obtained by shape memory synthesis from carbon nanofibers. Reprinted with permission from [1]. Copyright 2001, Elsevier B.V.

of the {111} planes of β-SiC. This further confirms that the as-synthesized nanotube is indeed a SiC nanotube.

Single-phase SiC nanotubes have been synthesized by reaction of carbon nanotubes with Si powder at 1200 °C for 100 h [5, 6]. Transmission electron microscopy reveals the existence of both single-phase SiC nanotubes and C–SiC coaxial nanotubes, the latter are carbon nanotubes sheathed by SiC layers. The single-phase SiC nanotubes are polycrystalline and comprise grains with sizes of around 20 nm. The ratio of the number of single-phase SiC nanotubes (Fig. 6.4)

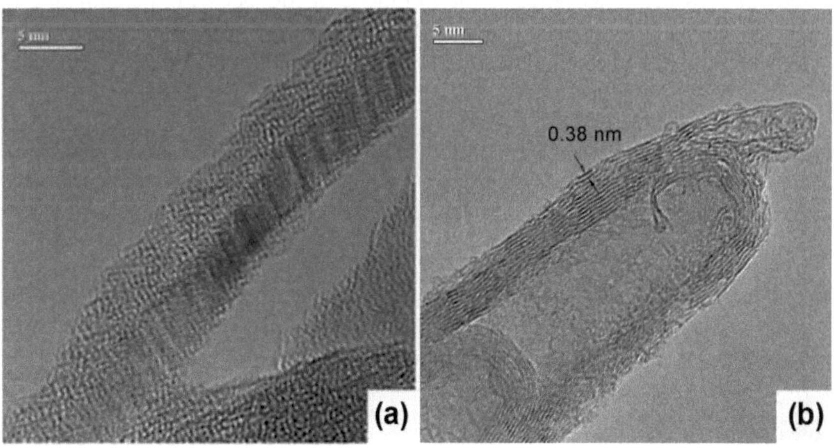

Fig. 6.2 **a** HRTEM image of the biaxial SiC–amorphous SiO_x nanowire obtained by reaction between Si and carbon nanotubes. **b** HRTEM image of a SiC nanotube with 3.8 Å interlayer spacing. Reprinted with permission from [4]. Copyright 2002, American Chemical Society

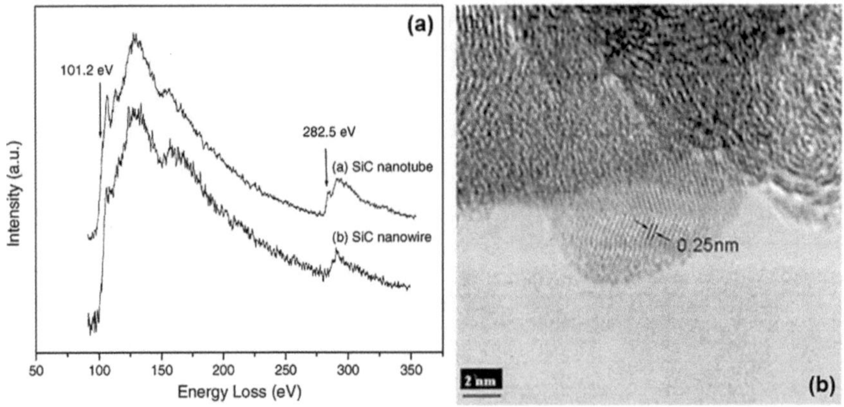

Fig. 6.3 **a** Electron energy loss spectra of a single SiC nanotube and a single SiC nanowire obtained from the same sample. **b** HRTEM image of the SiC nanotube after electron beam annealing for 5 min. Reprinted with permission from [4]. Copyright 2002, American Chemical Society

to that of the C–SiC nanotubes increased after heating the sample at 600 °C in air, resulting from the removal of the remaining carbon. The energy-dispersive X-ray spectroscopy analysis reveals that the atomic ratio of Si to C in single-phase SiC nanotubes is almost unity, and they contain nearly stoichiometric SiC grains.

The SiC nanotubes have been synthesized by high-temperature reaction between multiwalled carbon nanotubes and Si powder in flowing ammonia at 1350 °C [7]. The SEM and TEM observations (Fig. 6.5a) reveal that the SiC nanotubes are carbon-filled and open-ended; they have lengths between 0.7–1 μm and diameters between 45–70 nm. Raman spectrum (Fig. 6.5b) of the nanotubes comprises a strong peak at 900 cm^{-1}, which is assigned to the LO mode of SiC but with a significant red shift relative to bulk value due to the phonon confinement effect. This Raman peak is asymmetric and has a broad tail toward lower frequencies, which may be caused by

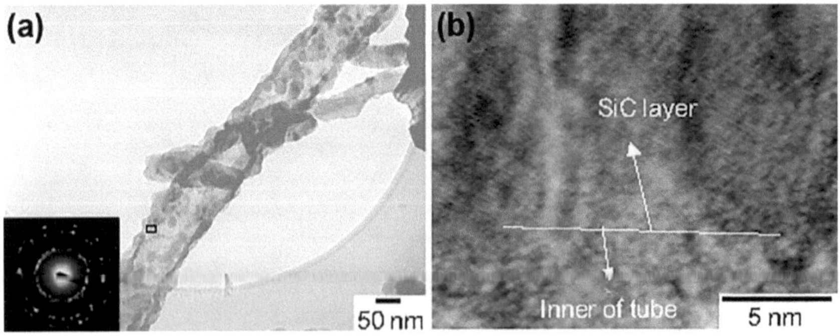

Fig. 6.4 **a** TEM and **b** HRTEM images of single-phase silicon carbide nanotube. Reprinted with permission from [5]. Copyright 2005, John Wiley & Sons, Inc.

mixing of hexagonal and cubic polytypes in the nanotubes. The electron diffraction analysis shows that cubic polytype SiC dominates over other hexagonal polytypes SiC in the sample. The infrared absorption spectrum (Fig. 6.5c) comprises a sharp band at around 790 cm^{-1} (TO phonon) and a shoulder at 950 cm^{-1} (LO phonon), and both shifts to lower frequencies relative to bulk values due to spatial confinement. Another study shows that during the reaction of the multiwalled carbon nanotubes and Si powder, besides SiC nanotubes, there could be also SiC nanowires and SiC/SiO_2 core–shell nanocables as products [8]. The SiC nanowires are ~ 100 nm in diameter and several tens of micrometers in length, and the SiC nanocables have SiC cores with diameters of 20–30 nm and amorphous SiO_2 shells with thicknesses of ~ 20 nm. The nanotubes have narrow hollow channels and have outer diameters of ~ 20 nm. Microwave heating of blended silicon dioxides and multiwalled carbon nanotubes can also produce SiC nanotubes [9]. High-yield conversion of SiO_2 and carbon nanotubes to SiC nanotubes was achieved for a SiO_2 and carbon nanotube molar ratio of 1:3.

Photoluminescence properties of SiC nanotubes synthesized by reaction of multiwalled carbon nanotubes and Si powder were examined [10]. Heat treatment in air eliminated remaining carbon atoms in the nanotubes and unreacted carbon nanotubes. PL spectra of SiC nanotubes with and without alkali and acid treatments were measured at room temperature. The PL spectrum of the SiC nanotubes that were heat-treated at 700 °C comprises a dominant peak at ~ 2.5 eV. The heat-treated nanotubes at 800 and 900 °C exhibited no PL owing to the formation of the SiO_2 layers on the surfaces of the SiC nanotubes. After treatment by NaOH followed by HCl solution to remove the SiO_2 layers after the heat treatment, the stronger PL band emerged. This suggests that the PL of the SiC nanotubes is closely related to their surface defects. In situ TEM was used to observe the microstructural changes of these SiC nanotubes under ion irradiation [11]. The nanotubes showed better resistance against amorphization by irradiation compared to bulk SiC, having a higher

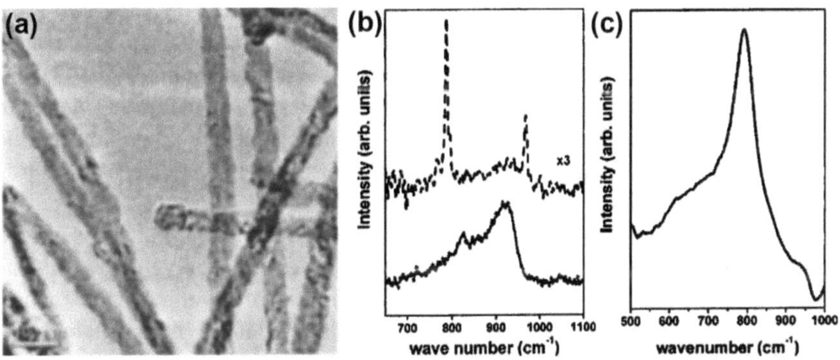

Fig. 6.5 **a** TEM image of the SiC nanotubes. **b** Raman spectra of the as-produced SiC nanotubes (lower) and reference sample (upper). **c** Infrared absorption spectrum of the SiC nanotubes. Reprinted with permission from [7]. Copyright 2005, American Institute of Physics

critical irradiation dose at room temperature. Both the inner and outer diameters of the nanotubes increased during irradiation till complete amorphization, and then they decreased. The lattice plane spacing of SiC crystals increased with increasing ion fluence due to increasing disorder. The ion irradiation at 700 °C changed very little the inner and outer diameters; the lattice plane spacing decreased slightly.

6.1.2 Formation on Core Nanowires

The SiC/ZnS nanocables and SiC nanotubes and other structures have been synthesized using ZnS nanowires, nanoribbons, and sheets as templates by a two-stage thermal reaction [12]. During this synthesis, a mixture of SiO and ZnS powders were first heated at 1150 °C in flowing argon and then heated at 1400 °C in flowing CH_4 gas. The obtained SiC/ZnS nanocables comprise ZnS cores with diameters of 50–130 nm and polycrystalline SiC shells with thicknesses of 15–50 nm. The removal of the ZnS cores from the nanocables by HCl etching produced open-ended, straight, and uniform SiC nanotubes (Fig. 6.6a–f). The high-resolution TEM observation of a SiC tube wall reveals the presence of numerous nanocrystals with sizes of 4–10 nm, which exhibits (111) lattice fringes of β-SiC. The electron diffraction pattern shows continuous polycrystalline rings corresponding to (111), (220), and (311) planes of β-SiC, confirming that the SiC nanotube is a polycrystal rather than a single crystal. The energy-dispersive X-ray spectrum reveals the chemical composition of SiC in the nanotube wall. The SiC nanotubes showed room-temperature photoluminescence with a strong symmetric emission band centered at 517 nm. Large and irregular thin SiC tubes or bags were prepared by HCl etching of the as-prepared ZnS sheets encapsulated by SiC layers (Fig. 6.6g–i). The electron diffraction confirms that these structures are also polycrystalline β-SiC.

Various hollow SiC nanostructures can be synthesized by employing ZnO nanostructure templates [13]. The SiC shell layers were deposited on one-, two-, and three-dimensional complex-shaped ZnO nanostructures via chemical vapor deposition using SiH_4 and CH_4 gases as starting materials at a temperature of 250–300 °C. Fruitful SiC-shelled nanostructures were obtained after removal of the ZnO templates by HCl etching (Fig. 6.7). The synthesized SiC nanoshells are amorphous due to the low reaction temperature employed, as demonstrated by the halo-like electron diffraction pattern (inset of Fig. 6.7).

Silicon carbide nanotubes have been fabricated by magnetron sputtering at room temperature using SiC target as source material and electrospun polyvinylpyrrolidone nanofibers as templates [14]. The polyvinylpyrrolidone core was eliminated by anneal at 500 °C and further anneal at 1250 °C yielded well-crystallized SiC nanotube. SEM and TEM measurements (Fig. 6.8a, b) reveal that the nanotubes have an average diameter of 150 nm and wall thicknesses of about 40 nm. The energy-dispersive X-ray spectroscopy shows that the nanotubes comprise Si, C, and O elements. The infrared absorption (Fig. 6.8c, d), X-ray diffraction, and electron diffraction measurements indicate that the annealed nanotubes are polycrystalline SiC crystals coated with SiO_2

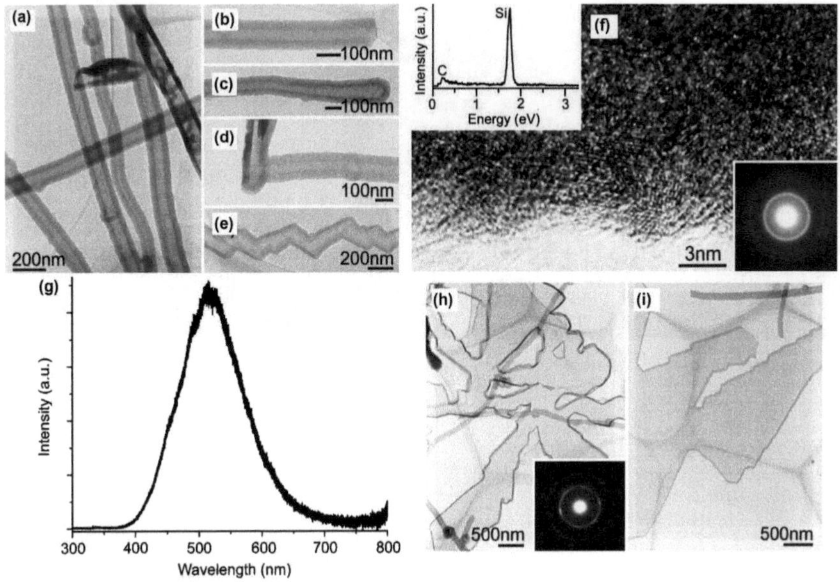

Fig. 6.6 **a–e** TEM images of the SiC nanotubes. **f** HRTEM image of a SiC nanotube wall. Left inset is the energy-dispersive X-ray spectrum. Right inset is the electron diffraction pattern showing (111), (220), and (311) plane reflections of β-SiC. **g** Room-temperature PL spectrum of the SiC nanotubes. **h, i** TEM images of large SiC thin tubes or bags with irregular shapes. Reprinted with permission from [12]. Copyright 2004, American Institute of Physics

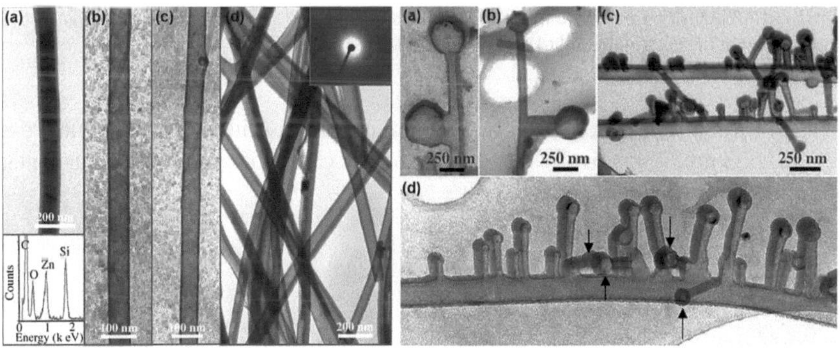

Fig. 6.7 (Left panel) **a** TEM image of ZnO–SiC core–shell nanobelt. The inset is the energy-dispersive X-ray spectrum showing elements of C, Si, O, and Zn. **b–d** TEM images of the SiC nanotubes after removal of ZnO cores. The inset is an electron diffraction pattern. (Right panel) TEM images of the SiC hollow nanojunctions. Reprinted with permission from [13]. Copyright 2006, John Wiley & Sons, Inc.

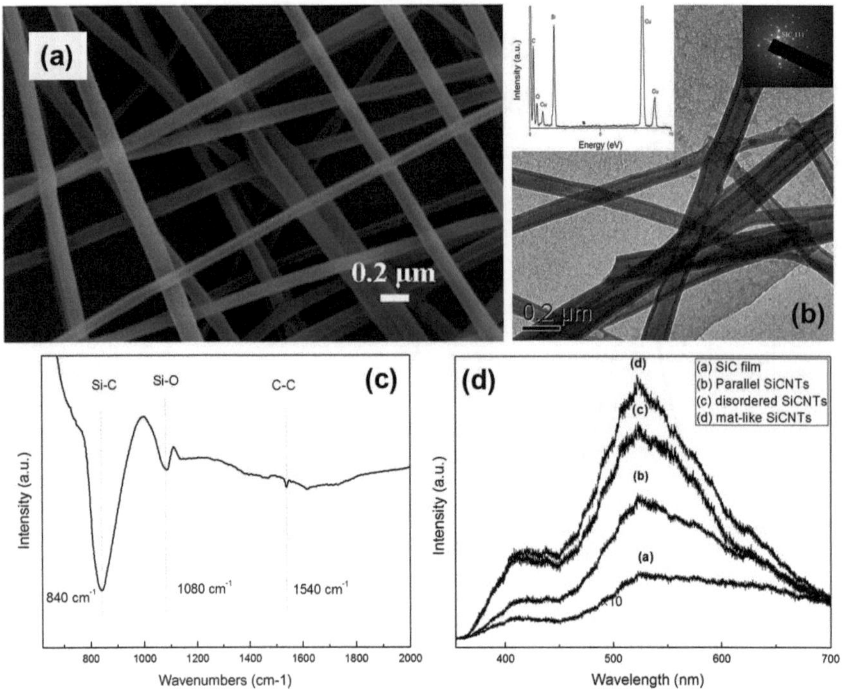

Fig. 6.8 **a** SEM and **b** TEM images of the SiC nanotubes. Left inset is the energy-dispersive X-ray spectrum showing C, Si, and O elements. Right inset is the electron diffraction pattern, in accordance with β-SiC (111) lattice planes. **c** Infrared transmission and **d** PL spectrum of the SiC nanotubes. Reprinted with permission from [14]. Copyright 2009, Elsevier B.V.

and free carbon. The nanotubes have photoluminescence with two peaks centered at 420 and 523 nm, respectively. The SiC nanotubes can also be synthesized using Ge or Si nanowires as templates [15, 16].

6.1.3 Other Templates

The alumina template has been used to synthesize well-aligned tubular SiC arrays by thermal decomposition of polymethylsilane at 1250 °C in argon atmosphere [17]. The template was removed by HF etching. The wall morphology (Fig. 6.9) can be tuned by changing the number of infiltration steps; the initial infiltration yielded tubular SiC arrays with wall thicknesses of 35 nm and diameters of 200 nm, whereas the repeated infiltration yielded SiC nanorods.

Tubular and bamboo-like SiC nanostructures can be synthesized by using CVD or pyrolysis inside porous alumina (thickness: 60 μm, pore size: 250 nm) using carbosilane as precursors in flowing nitrogen gas at 1000 °C [18]. The alumina

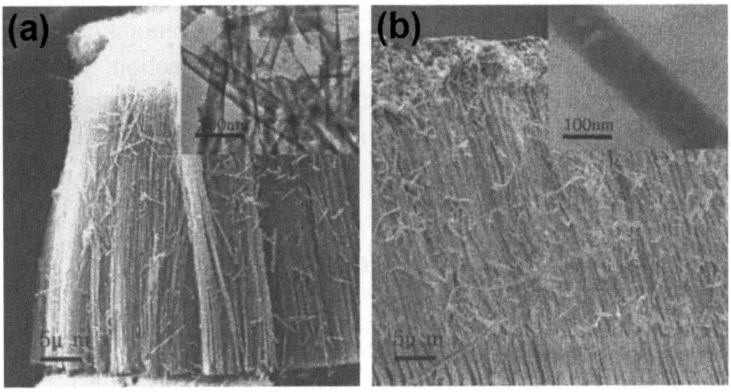

Fig. 6.9 SEM images of 100 nm diameter tubular SiC **a** nanotubes and **b** nanorods obtained from polymethylsilane. The insets show TEM images. Reprinted with permission from [17]. Copyright 2005, American Institute of Physics

template was removed via HF etching. A post-treatment anneal at 1600 °C in argon altered the morphology of the product. The as-prepared SiC nanotubes by CVD are amorphous and transformed into polycrystalline β-SiC nanotubes after annealing at 1600 °C (Fig. 6.10a, b). The SiC nanotubes have outer diameters of 200–300 nm and wall thicknesses of 20–40 nm and lengths up to the thickness of the original alumina template. Pyrolysis of the polymer gave rise to amorphous bamboo-like SiC nanofibers (Fig. 6.10c, d), and they converted into a collection of single-crystal SiC nanofibers and other small particles after anneal at 1600 °C.

Mesoporous silica has been employed to synthesize high-surface-area SiC materials, including whiskers and nanotubes [19]. The SiC materials were prepared by carbothermal reduction of mesoporous silica/sucrose composites. The SiC whiskers growing in the [111] direction were achieved at a reaction temperature of 1250 or 1300 °C for 14 h. The whiskers have diameters of 50–90 nm and lengths larger than 20 μm. The reaction with a prolonged period of 20 h at 1250 or 1300 °C yielded SiC nanotubes with diameters of 60–100 nm and lengths larger than 10 μm. The C/β-SiC/Si hybrid microtubes have been fabricated by releasing pre-stressed C/Si

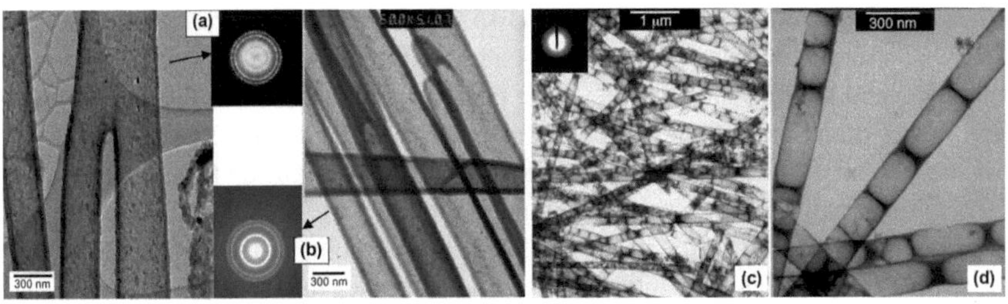

Fig. 6.10 TEM images and selected-area electron diffraction patterns for the **a**, **b** CVD-derived SiC nanotubes after anneal in argon at 1600 °C and **c**, **d** bamboo-like SiC nanofibers obtained from pyrolysis of polymer. Reprinted with permission from [18]. Copyright 2005, Elsevier B.V.

bilayer structures followed by annealing [20]. The β-SiC nanoparticles with sizes of 3–5 nm formed at the C/Si interface via a solid phase reaction. The Raman spectrum of the hybrid microtubes comprises two peaks at 776–783 and 921–941 cm^{-1}, corresponding to the TO and LO phonon modes of β-SiC nanoparticles.

6.2 Fabrication by Other Methods

The SiC nanotubes with very small diameters have been synthesized under supercritically hydrothermal conditions without the use of metal catalysts [21]. The nanotubes were prepared by hydrothermal reaction of 50 nm diameter SiC powder and 30 nm diameter SiO_2 powder in distilled water under stirring in a sealed reaction kettle at a temperature of 470 °C and a pressure of 8.0 MPa. The product comprises crystalline β-SiC nanotubes and some carbon grains. The nanotubes (Fig. 6.11a, b) are multi-walled, with very thin hollow cored, and their outer diameters are smaller than 10 nm and their lengths are many micrometers. The lattice fringes have a spacing of 0.25 nm, in consistent with the spacing of β-SiC (111) planes. The nanotubes have a thin amorphous silica sheath. X-ray diffraction and the electron diffraction confirm the single-crystalline β-SiC phase of these nanotubes. A small amount of SiC nanotubes with interlayer distances ranging from 0.26 to 0.30 nm are also present, suggesting that these nanotubes are carbon-deficient SiC nanotubes. The Raman spectrum of the nanotubes (Fig. 6.11c) comprises two peaks at 775.7 and 997.7 cm^{-1}, which are assigned to the TO and LO phonon modes of β-SiC nanostructures; however, the LO mode shows a significant blue shift rather than generally observed red shift relative to bulk value. The signals at 1329.8 and 1568.8 cm^{-1} are graphite E_{2g} and A_{1g} modes, indicating that the graphitic carbon coexists in the SiC nanotube sample. The infrared spectrum of the sample (Fig. 6.11d) comprises a strong absorption band at 791 cm^{-1}, corresponding to the TO phonon mode of β-SiC; the absorption bands at 472 and 1107 cm^{-1} are attributed to silica, including silica shells of SiC nanotubes and unreacted SiO_2 in the sample.

CVD approach can also yield SiC nanotubes by using methyltrichlorosilane as the gaseous source material and ferrocene and thiophene as catalysts [22]. The SiC tubes with 2080 nm in outer diameter and 1535 nm in inner diameter were observed. The wall structure similar to that of carbon nanotubes was not observed. The tubular single-crystal β-SiC nanostructures have been prepared by simple thermal evaporation of C_{60} on a Si substrate without using templates and catalysts [23]. The tubular SiC nanostructures have diameters of 40–100 nm and lengths up to several micrometers. TEM analysis shows that the tubular nanostructure is the β-SiC single crystal growing along the [111] direction (Fig. 6.12). The tubular β-SiC showed good field electron emission character with a turn-on field smaller than 5 V μm^{-1}.

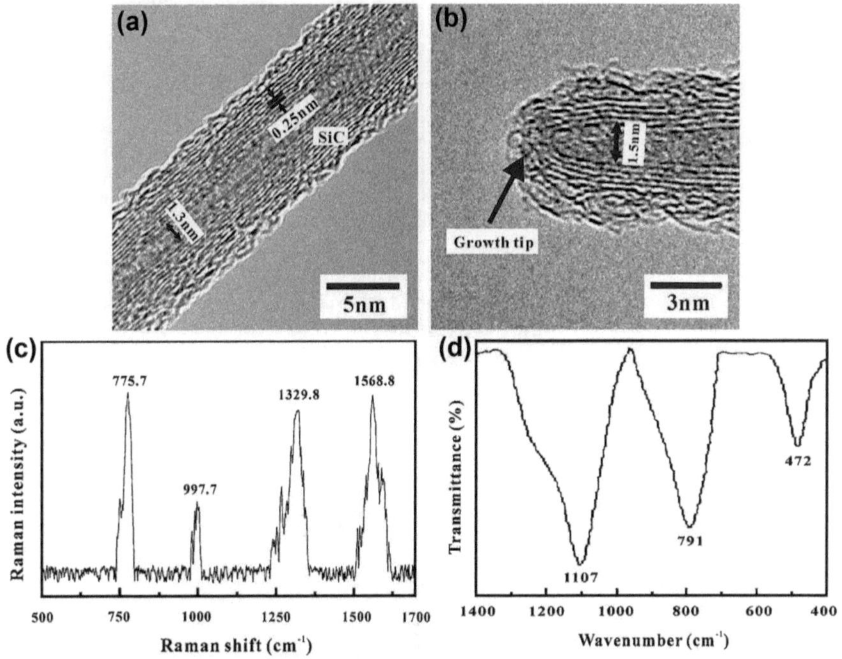

Fig. 6.11 HRTEM images of SiC nanotubes showing multiwalled crystalline SiC layer. **a** Tubular body with interlayer spacing of 0.25 nm corresponding to β-SiC {111} planes. **b** Nanotube tip with semicircular structure and diameter of 1.5 nm in the inner pore. **c** Raman spectrum of the SiC nanotubes. The peaks at 775.7 and 997.7 cm^{-1} belong to TO and LO phonon modes of β-SiC and peaks at 1329.8 and 1568.8 cm^{-1} correspond to graphite E_{2g} and A_{1g} modes. **d** Infrared spectrum of the SiC nanotubes. Reprinted with permission from [21]. Copyright 2006, American Institute of Physics

6.3 Theoretical Studies

True single-walled silicon carbide nanotube is the rolled two-dimensional honeycomb-structured monolayer, similar to that of the single-walled carbon nanotube, but a single type of carbon atoms is substituted by two types of carbon and silicon atoms. The single-walled carbon nanotube is acquired by cutting a two-dimensional graphene sheet and rolling it up along one of its Bravais lattice vectors ***C*** joining two equivalent sites on the original graphene lattice to form a specific type of nanotube. The circumferential vector is

$$\boldsymbol{C} = n\boldsymbol{a}_1 + m\boldsymbol{a}_2 \quad (n \geq m), \tag{6.1}$$

where $\boldsymbol{a}_1$ and $\boldsymbol{a}_2$ are the primitive cell lattice vectors of graphene, and $m = 0$ for all zigzag tubes, $n = m$ for all armchair tubes, otherwise chiral ones.

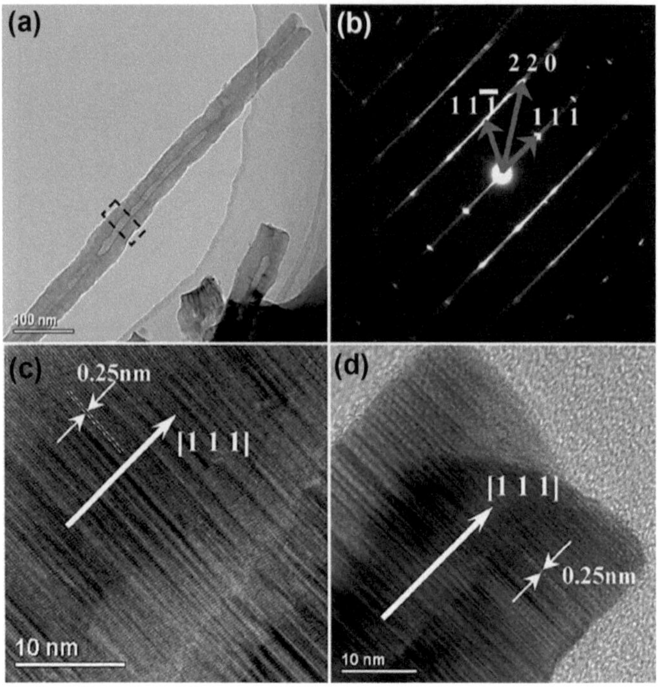

Fig. 6.12 **a** TEM image of tubular SiC nanostructure. **b** Corresponding selected-area electron diffraction pattern. HRTEM images taken from (**c**) the marked area in (**a**) and from (**d**) the top of an open end. Reprinted with permission from [23]. Copyright 2009, Royal Society of Chemistry

6.3.1 Structural and Electronic Properties

The DFT calculation predicts the existence of graphitic and tubular form of silicon carbide [24]. Figure 6.13a, b shows the atomic geometries and corresponding electronic band structures of SiC nanotubes with helicities of (5,5) and (9,0). The band structure possesses an indirect gap for the (5,5) nanotube and a direct gap for the (9,0) nanotube. The optical transition between the highest occupied band and lowest unoccupied band is allowed in the (9,0) nanotube, and thus there should be strong light absorption and emission. The direct bandgap of the (n,0) nanotube (n = 6–8) decreases with decreasing number n. The elastic energy of the SiC nanotube increases with decreasing diameter (Fig. 6.13c) and is smaller than that of the carbon nanotube of the same diameter.

In comparison with carbon nanotubes, silicon carbide nanotubes may have more types because there are two types of elements in the SiC nanotubes. The structure and stability of various SiC nanotube structures have been investigated using the generalized tight-binding molecular-dynamics scheme and ab initio method [25]. The system size is limited to a maximum of 72 atoms due to the computational demand of the ab initio method. Hydrogen atoms are attached to the edge atoms to passivate all

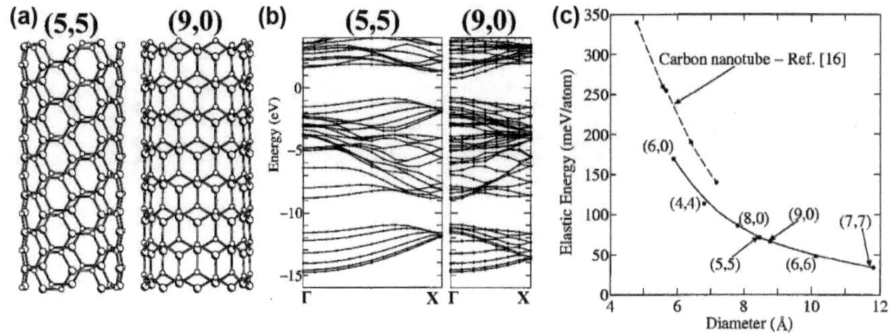

Fig. 6.13 **a** Atomic geometries and **b** corresponding electronic band structures of armchair-type (5,5) and zigzag-type (9,0) nanotubes. The hatched and open circles denote Si and C atoms, respectively. **c** Strain energy of SiC nanotubes versus diameter. The helical indices of SiC nanotubes are denoted. The solid and dotted lines are guides to the eye. Reprinted with permission from [24]. Copyright 2002, American Institute of Physics

dangling bonds. The calculation demonstrates two highly stable distinct geometries for each chirality (zigzag or armchair) of SiC nanotube, as shown in Fig. 6.14. Both nanotubes are constructed by rolling up the corresponding graphene-like two-dimensional sheet. Both sheets contain a Si to C ratio of 1:1. Type 1 hexagonal network consists of alternating Si and C atoms, and each Si atom has only C atoms as nearest neighbors and vice versa. Type 2 structure is obtained when each Si atom has two C atoms and one Si atom as nearest neighbors and vice versa. The total energy calculation using ab initio method demonstrates that type 1 structure is more stable than type 2 structure by 0.43 eV per SiC atom pair. Among the type 1 SiC nanotubes, the armchair (6,6) nanotube is slightly more stable than the zigzag (12,0) nanotube by 0.05 eV per Si–C atomic pair. On the contrary, among type 2 SiC nanotubes, the zigzag (12,0) nanotube is more stable than the armchair (6,6) nanotube by 0.45 eV per Si–C atomic pair. However, the energy differences between all structures considered are expected to diminish with increasing diameter of the nanotube. The Si–C bond length is 1.87 Å, irrespective of the nanotube type, and this value is slightly smaller than that of 3C-SiC structure, 1.89 Å. The calculation of the density of states (Fig. 6.15a) shows that these two types of SiC nanotube have quite different bandgap values. Type 1 SiC nanotubes have large bandgaps, 3.23 and 3.53 eV for zigzag and armchair chiralities, respectively. These values are similar to the bandgaps of bulk SiC (3–3.3 eV). In contrast, the bandgaps for type 2 nanotubes are much smaller, 0.7 and 0.5 eV for zigzag and armchair chirality, respectively. These values are very close to those of carbon nanotubes. This bandgap difference in two types of SiC nanotubes results in their different behaviors in the *I*–*V* characteristics (Fig. 6.15b); the current value is much higher for type 2 SiC nanotubes at the same voltage.

The geometrics, strain energies, and electronic structures of SiC nanotubes with different diameters and chirality have been investigated using density functional calculations [26]. In the considered nanotubes, the Si and C atoms are alternatively

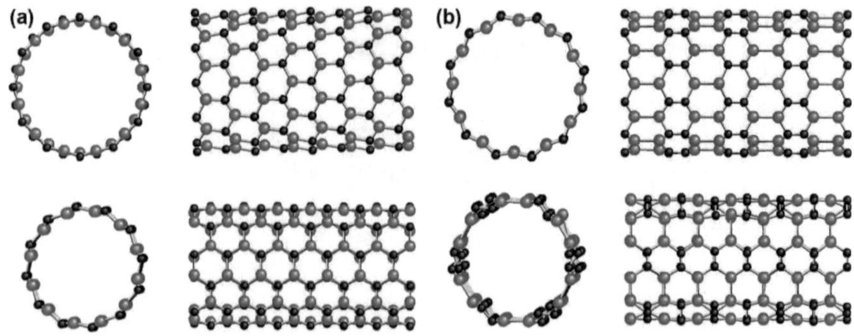

Fig. 6.14 **a** Type 1 and **b** type 2 single-walled SiC nanotubes of zigzag (top) and armchair (bottom) chirality obtained by rolling up corresponding graphene-like sheet and performing generalized tight-binding molecular-dynamics relaxation. Gray (black) balls denote Si (C) atoms. Si and C atoms lie in different rolled planes. Reprinted with permission from [25]. Copyright 2004, American Physical Society

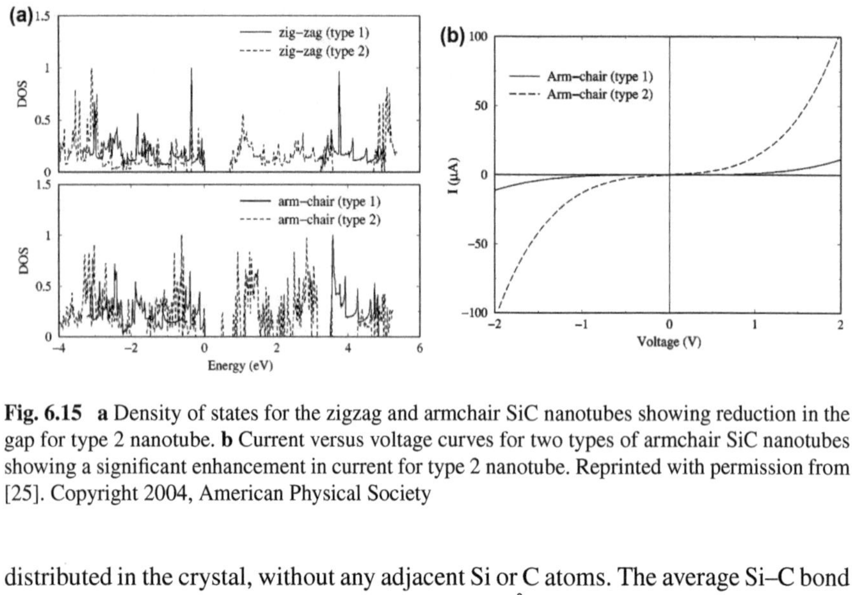

Fig. 6.15 **a** Density of states for the zigzag and armchair SiC nanotubes showing reduction in the gap for type 2 nanotube. **b** Current versus voltage curves for two types of armchair SiC nanotubes showing a significant enhancement in current for type 2 nanotube. Reprinted with permission from [25]. Copyright 2004, American Physical Society

distributed in the crystal, without any adjacent Si or C atoms. The average Si–C bond length in equilibrium configuration is about 1.80 Å. The strain energy in (5,5) SiC nanotubes (relative to 3C-SiC) is as high as 0.686 eV per atom. The strain energy decreases with increasing nanotube diameter (Fig. 6.16a). All the considered SiC nanotubes are semiconductors, and the bandgap increases with increasing diameter (Fig. 6.16b). The armchair and chiral SiC nanotubes have indirect bandgaps, similar to 3C-SiC, whereas the zigzag SiC nanotubes have direct bandgaps at the Γ point (Fig. 6.17). The highest occupied valence band and the lowest unoccupied conduction band states highly localize to C and Si atoms, respectively. When an H atom is chemically adsorbed above a C atom at the exterior of the tube, the hybridization of the C atom will change from sp^2 to sp^3, breaking the nearby p–p bonds and generating a local state close to the lowest unoccupied conduction band. This local state is half-occupied and therefore serves as a donor state, and thus this hydrogen-decorated SiC nanotube shows the character of n-type semiconductors. By contrast, an H atom chemically adsorbed on a Si atom results in a local state with the energy close to the highest occupied valence band. This half-occupied state acts as an acceptor state, and this hydrogen-decorated SiC nanotube shows the character of p-type semiconductors. Further calculation [27] shows that hydrogenation of all the Si atoms renders the (8,0) and (6,6) SiC nanotubes metallic with very high density of states at the Fermi level. Hydrogenation of all the C atoms results in increase of the bandgap of the (8,0) SiC nanotube to 2.6 eV and decrease of the bandgap of the (6,6) SiC nanotube to 1.47 eV. Hydrogenation of all the Si and C atoms results in a highly increased bandgap of the SiC nanotube. Besides hydrogen atoms, other chemical groups adsorbed on the SiC nanotubes may also strongly alter their electronic structures [28]. The calculation shows that the XH_3 radical ($X = \{C, Si\}$) can be chemically adsorbed on Si sites or C sites on the tube wall, with the adsorption energies ranging from -2.01 to -2.90 eV for the (5,5) SiC nanotubes. The electronic structure of the XH_3-decorated SiC nanotube exhibits characteristics of n-type and p-type semiconductors in the cases of XH_3 adsorption on a C atom and a Si atom, respectively.

The structural, elastic, and electronic properties of single-walled graphitic SiC, BN, and BeO nanotubes in zigzag and armchair configurations have been investigated using density functional theory employing self-interaction-corrected pseudopotentials [29]. The structural and elastic properties of all zigzag and armchair nanotubes considered are largely independent of the nanotube helicity. In contrast, the electronic structures of SiC and BN nanotubes are sensitively dependent on nanotube helicity (Fig. 6.18a). The bandgaps of zigzag and armchair BeO nanotubes vary peculiarly for small diameters; no bandgap breakdown occurs, and the bandgaps have a minimum for the zigzag BeO nanotubes. These phenomena are caused by

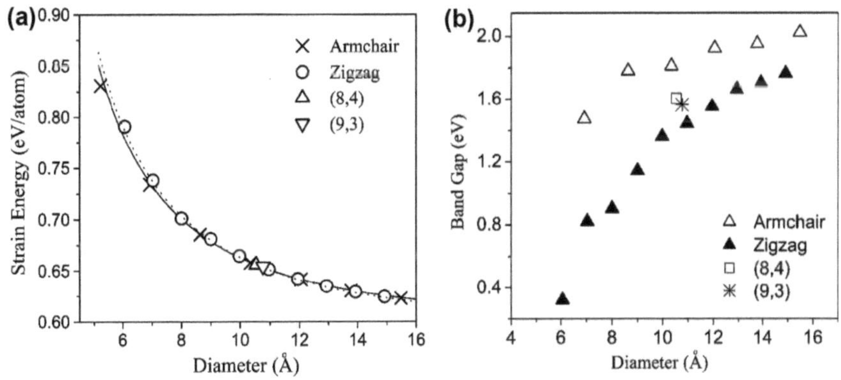

Fig. 6.16 **a** Strain energy in SiC nanotubes relative to 3C-SiC as a function of diameter. The solid and dotted lines are the fit. **b** Bandgap of armchair, zigzag, and chiral SiC nanotubes as a function of diameter. Reprinted with permission from [26]. Copyright 2005, American Physical Society

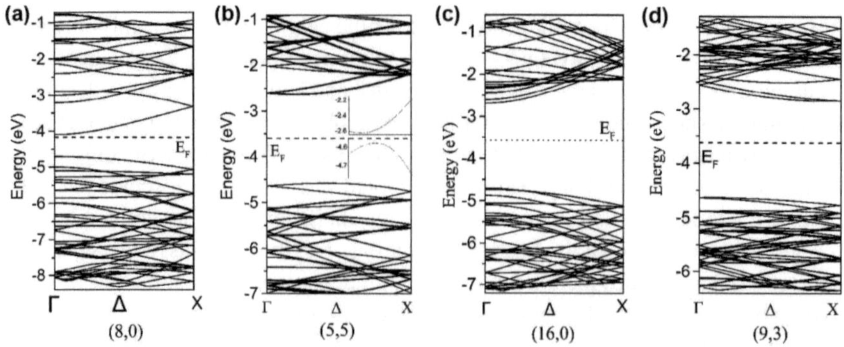

Fig. 6.17 Band structures along the nanotube axis of **a** zigzag (8,0), **b** armchair (5,5), **c** zigzag (16,0), and **d** chiral (9,3) SiC nanotubes. The Fermi levels E_F of these nanotubes are denoted by dashed lines. Reprinted with permission from [26]. Copyright 2005, American Physical Society

the qualitative difference in the nature of the lower conduction band states in SiC and BN nanotubes (Fig. 6.18b) as compared to BeO nanotubes and by the increasing ionicity from SiC, BN, to BeO.

The geometric and electronic structures of three types of armchair SiC nanotubes have been investigated using hybrid density functional theory and the finite cluster approximation [30]. Figure 6.19a–c displays the atomic arrangements of the three types of nanotubes. The dangling bonds are terminated by hydrogen atoms. The cohesive energy saturates at 4.63 eV for type 1 nanotube and at around 4.44 eV for type 2 and 3 nanotubes. The bandgaps of type 1 nanotubes are bigger than the bulk 3C-SiC bandgap and range from 2.78 to 2.91 eV (Fig. 6.19d). In contrast, type 2 and type 3 nanotubes have much lower bandgaps. The bandgap of the newly

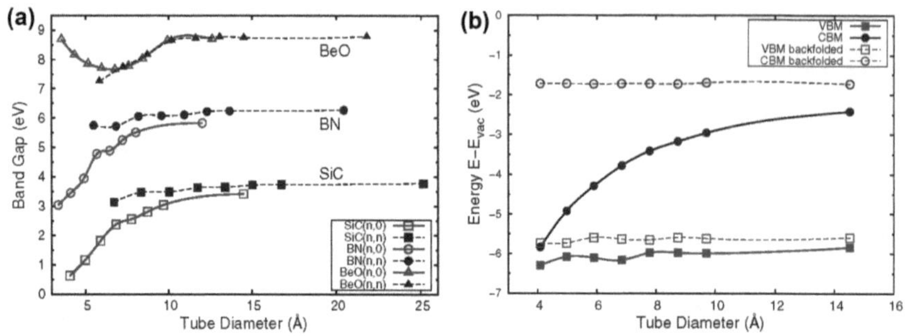

Fig. 6.18 **a** Fundamental gap of graphitic SiC, BN, and BeO (*n*,0) zigzag and (*n*,*n*) armchair nanotubes as a function of tube diameter. **b** Valence band maximum and conduction band minimum energies in zigzag SiC nanotubes as a function of tube diameter. Full symbols result from nanotube calculations, and open symbols follow from backfolding the bands of the graphitic SiC sheet. The solid and dashed lines are guides to the eye. Reprinted with permission from [29]. Copyright 2007, American Physical Society

proposed type 3 SiC nanotube decreases monotonically with increasing tube diameter, from 1.22 to 0.79 eV. This behavior is different from that of type 1 and type 2 nanotubes. All three types of tubes are nonmagnetic. The properties of single-walled hydrogen-terminated zigzag SiC nanotubes have been investigated similarly [31]. Figure 6.20a–c displays the atomic arrangements of the three types of nanotubes. Type 1 structure is more stable than type 2 structure. The cohesive energy per atom of type 3 nanotubes lies between those of type 1 and type 2. The bandgap of the type 3 nanotube decreases monotonically with increasing diameter, whereas those of type 1 and type 2 nanotubes oscillate with increasing diameter (Fig. 6.20d). The structural and electronic properties of chiral SiC nanotubes have also been investigated using hybrid density functional theory [32]. All the SiC nanotubes considered are semiconductors, and their bandgaps range from 0.20 to 2.9 eV. Like other types of SiC nanotubes, the structural stability of the chiral nanotubes increases with diameter, and it is independent of the chiral angle.

The optimized structures of graphitic-like sheets (single-layered and multilayered) as well as tubular structures (both single-wall and double-wall) of SiC including their electronic structures were investigated by using DFT–GGA calculations [33]. SiC graphitic-like structures form sp^2-like bonding, although the bulk phases of SiC exhibit sp^3 bonding. The interplanar spacing in a multilayer graphitic sheet depends on how the atoms are arranged in the adjacent SiC bilayers (such as Si–C or C–C interplanar order). The interplanar spacing is smaller for the Si–C ordering (~ 3.7 Å) than for the C–C ordering (~ 4.5 Å) (Fig. 6.21). The electrostatic interaction between the bilayers is responsible for the smaller interplanar spacing for the Si–C

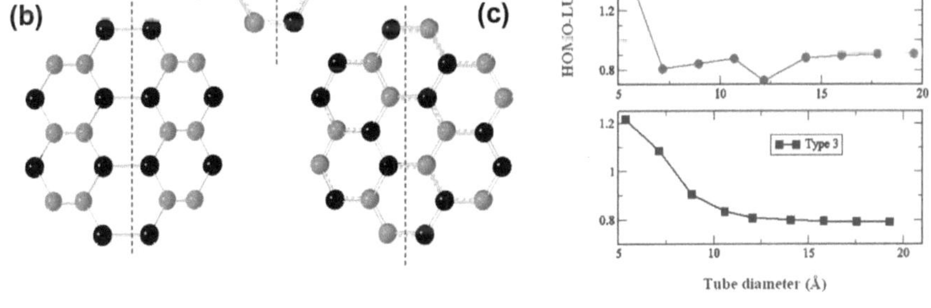

Fig. 6.19 Atomic arrangements of **a** type 1, **b** type 2, and **c** type 3 armchair SiC nanotubes. The carbon (silicon) atoms are gray (black). The dashed lines represent the orientation of tube axis. **d** Highest occupied molecular orbital–lowest unoccupied molecular orbital gap versus tube diameter for three types of nanotubes. Reprinted with permission from [30]. Copyright 2008, American Physical Society

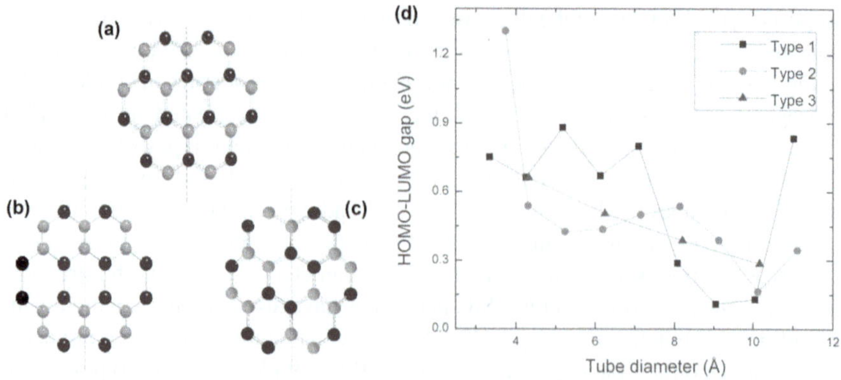

Fig. 6.20 Atomic arrangements of **a** type 1, **b** type 2, and **c** type 3 zigzag SiC nanotubes. The carbon (silicon) atoms are gray (black). The dashed lines represent the orientation of tube axis. **d** Highest occupied molecular orbital–lowest unoccupied molecular orbital gap versus tube diameter for three types of nanotubes. Reprinted with permission from [31]. Copyright 2007, IOP Publishing

ordering. These findings are helpful toward understanding the usual experimental observations of two different values for intershell distances (~ 3.8 or 4.8 Å) in the SiC-based multiwall nanotubes.

Fullerene hemisphere-capped finite SiC nanotubes were studied by using DFT calculations based on the hybrid functional B3LYP and an all electron 3-21G* basis set [34]. Capping of a SiC nanotube changes the cohesive energy, HOMO–LUMO gap, and other electronic and geometric properties of a SiC nanotube. The carbon-capped SiC nanotubes are more energetically preferable compared to silicon-capped tubes. The hydrogen-terminated SiC nanotube (5,5) possessing five unit cells has

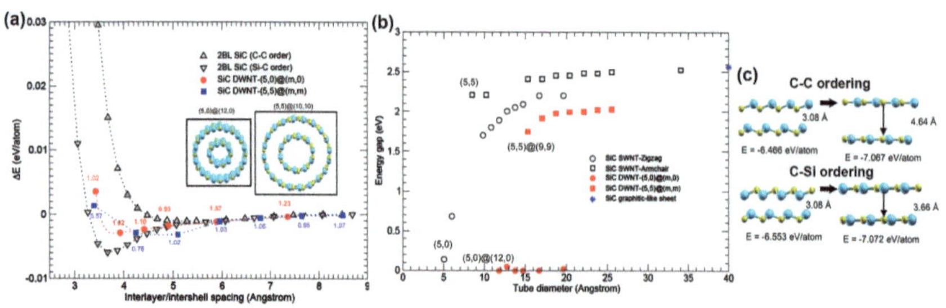

Fig. 6.21 **a** Relative energy per atom corresponding to two-bilayered (2BL) graphitic-like SiC structures and SiC double-wall nanotubes (DWNTs) are shown as a function of interlayer spacing and intershell spacing, respectively. The equilibrium interlayer spacing for 2BL SiC graphitic structures occurs at 3.67 and 4.82 Å, respectively, corresponding to Si–C and C–C ordering between the bilayers. **b** Energy gap as a function of tube diameter. **c** Schematic illustration of Si–C and C–C ordering between bilayers. Reprinted with permission from [33]. Copyright 2010, American Physical Society

a binding energy per atom of 4.993 eV, whereas those are 5.989 and 4.812 eV for C-capped and Si-capped nanotubes, respectively.

The quasiparticle band structure and optical properties of single-walled zigzag and armchair SiC nanotubes were investigated by using ab initio many-body calculations using the GW and the GW plus Bethe–Salpeter equation (BSE) methods [35]. GW calculations show significant quasiparticle corrections of more than 1.0 eV to the Kohn–Sham bandgaps from the LDA calculations. The quasiparticle bandgaps of SiC nanotubes with different chirality behave very distinctively as a function of tube diameter, attributed to the difference in the curvature-induced orbital re-hybridization among different chiral nanotubes. The discrete exciton peaks dominate the optical absorption spectra because of the high exciton binding energy up to 2.0 eV of the SiC nanotube (Fig. 6.22). The formation of strongly bound excitons is ascribed to the enhanced electron–hole interaction in these nanotubes. The bound-exciton wavefunctions are more localized and also strongly cylindrically asymmetric in armchair SiC nanotubes. The armchair SiC nanotube has a high excitation energy (~ 3.0 eV) of the first bright exciton, with no dark exciton below it. The zigzag SiC nanotubes have many dark excitons with excitation energies below that of the first bright exciton.

The stability of the SiC nanotubes in the static buckling case was investigated [36]. Buckling behaviors of the SiC nanotubes were discussed using the continuum model based on the Euler–Bernoulli beam theory and employing different boundary conditions. The maximum buckling load corresponds to the clamped case and highest average diameter/length ratio with surface effect. The minimum buckling load corresponds to the clamped-free case and lowest average diameter/length ratio without surface effect. Surface effect affects buckling, especially for the case of the highest average diameter/length ratio. The surface elasticity parameters have more influence on buckling for the case of higher modes and C–C type boundary. The differential quadrature method by employing eleven grids for all types of boundary conditions yields good results. DFT–PBE calculations show that AB or Bernal-type stacked armchair and AB′ zigzag bilayered SiC nanoribbons can separately be converted into defect-free zigzag and armchair single-walled SiC nanotubes [37]. The edge dynamics is involved in the formation of the SiC nanotubes, and growing zigzag SiC nanotubes is more feasible as compared to armchair SiC nanotubes. Both types of nanotubes are semiconducting in nature, and the edge states of the nanoribbons have fully disappeared. The AA stacked zigzag bilayered SiC nanoribbons are converted into defective armchair SiC nanotubes, whereas conducting edge states are preserved owing to Si–Si and C–C bonds. The freestanding and defect-free SiC nanotube can form from the nanoribbon with a diameter up to 1.2 nm, above which the closed edge structure may evolve.

Impurities present in a semiconductor can strongly affect its electronic structure and properties. The nitrogen and boron atoms are common contaminations in bulk SiC. Nitrogen preferentially substitutes carbon, resulting in n-type conductivity in bulk SiC, whereas boron substitutes carbon (silicon), leading to a deep (shallow) acceptor in bulk SiC. The effects of the boron and nitrogen impurities on the electronic structures of the armchair and zigzag SiC nanotubes have been investigated by using ab initio supercell calculations [38]. The calculation shows that nitrogen can only act

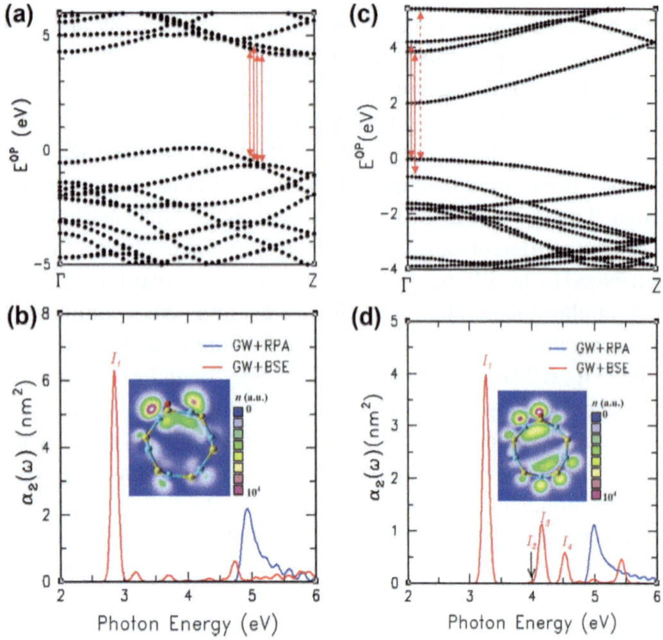

Fig. 6.22 **a**, **c** Quasiparticle band structure and **b**, **d** optical absorption spectra (broadened with a Gaussian width of 0.15 eV) of the armchair (3,3) (left panel) and zigzag (5,0) (right panel) SiC nanotubes. Optical spectra from GW + RPA and GW + BSE calculations are shown as dashed (blue) and solid (red) curves, respectively. The lowest energy bright exciton (I_1) is mainly related to the interband transition indicated as the solid (red) arrows in (**a**, **c**). The transition by the dashed (red) arrow in (**c**) corresponds to the third (I_3) bright exciton in (**d**). Insets in (**b**, **d**) show the squared electron amplitude of the I_1 exciton on a cross-sectional tubular plane with the hole fixed at the position of the red sphere. Si and C atoms are represented by cyan and yellow spheres, respectively. Reprinted with permission from [35]. Copyright 2011, American Physical Society

as an *n*-type dopant in the small bandgap (< 2.3 eV) SiC nanotube, whereas it is rather a compensating center for both holes and electrons in the wider bandgap SiC nanotube. When nitrogen acts as an *n*-type dopant, the corresponding ionization energy is about 0.3 eV, about 0.15–0.2 eV deeper than in bulk SiC polytypes. Similar to the case of bulk SiC, in SiC nanotubes, boron that substitutes for silicon (B_{Si}) acts as a relatively shallow acceptor and B_C acts as a relatively deep acceptor, and the energy levels are similar to that in bulk SiC. The properties of the boron substitutional defect show no significant difference between armchair and zigzag SiC nanotubes. The oxygen plays an important role in the properties of SiC nanotubes. The effects of oxygen on the properties of the SiC nanotubes must be seriously considered. Density functional calculations show that the doped boron atom in SiC nanotubes can substitute a carbon or silicon atom, generating two types of structures [39]. The transport properties of these B-doped SiC nanotubes were studied by using combined nonequilibrium Green's function and DFT methods. The negative differential resistance effect arises when the bias increases from 0.8 to 1.0 V. Both B-doped SiC nanotubes exhibit similar transport properties. The DFT–PBE calculations show the group-V atom doping effects on light absorption of SiC nanotubes [40]. The results show that the light absorption of the doped SiC nanotubes behaves differently within 600–1500 nm. The doping effects on the photoconductivity fall mainly in the spectral region above 620 nm, and the effects on the dielectric functions fall mainly in the region above 500 nm. The doping affects the band structures, thereby changing the dielectric and light absorption properties. The calculated formation energies suggest that the SiC nanotubes are more stable after substitution of Si atoms by group-V atoms, except for the case of N atoms. The influence of doping of group VIIIB transition metal atoms and adsorption of hydrogen molecules on SiC nanotubes were investigated using DFT–B3LYP/LanL2DZ calculations [41]. The Os atom doping on the SiC nanotubes is the most stable. The hydrogen molecule shows weak interaction with pristine SiC nanotubes, whereas it has strong interaction with transition metal-doped SiC nanotubes, particularly for the Os-doped nanotubes. The enhancement of the adsorption ability lies in the protruding structure of the doped nanotubes and induced charge transfer between the doped nanotube and the hydrogen molecule. The first layer of a doped SiC nanotube can store at most 3 hydrogen molecules.

The density functional calculation shows that the formation energies of the interstitial oxygen and O_C defects are negative [42]; thus in many actual growth processes of the SiC nanotubes, the oxygen atoms may readily be incorporated into the nanotubes, especially at elevated temperatures. The calculation shows that pure single-walled SiC nanotubes tend to be oxidized in the presence of ambient oxygen. At low temperatures, the basic structure of the SiC nanotubes is not destroyed by oxidation, but the interstitial oxygen defects form. At elevated temperatures, the electrically active O_C defects may form too. Although electrically inactive, the interstitial oxygen atoms diffuse fast in the SiC nanotubes even at room temperature. Spin-polarized density functional theory has been employed to investigate the structural and electronic properties of vacancies and antisites in zigzag, armchair, and chiral SiC nanotubes [43]. Antisites have lower formation energies relative to vacancies, resulting in an empty energy level near the conduction band minimum for both Si_C and C_{Si}. A carbon vacancy generates a pair of bonding and antibonding energy levels within the bandgap. A silicon vacancy has the largest formation energy and generates one spin-up occupied level within the valence band and three nearly degenerate (one spin-up and two spin-down) spin-polarized levels within the bandgap.

The stability of single-walled SiC nanotubes relative to SiC nanowires and against heating has been investigated using classical molecular dynamics simulations [44]. The simulation indicates that the SiC nanotubes are more stable than the SiC nanowires for diameters below about 1.6 nm, and the SiC nanowires are more stable than the nanotubes for greater diameters. This partially explains the difficulty in synthesizing actual SiC nanotubes. The SiC nanotubes with free ends will reconstruct to form a closed cap containing nonhexagonal defects at 1100 K, which acts as the nucleation site for heterogeneous melting as the temperature increases to 1620 K. The SiC nanotubes without free ends begin to melt at about 1820 K by homogeneous nucleation within the wall starting from thermally activated defects.

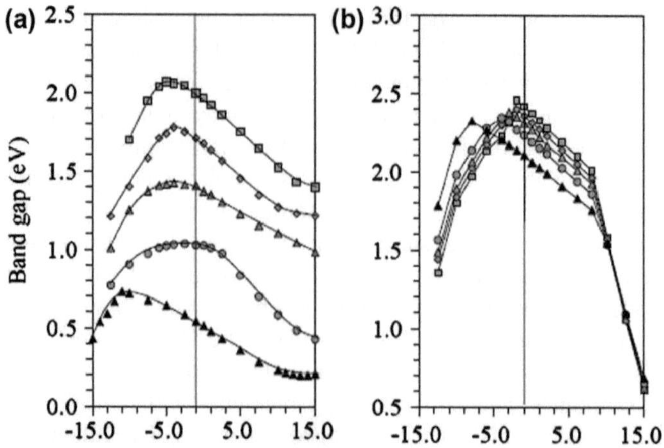

Fig. 6.23 Bandgap of zigzag (**a**) and armchair (**b**) SiC nanotubes as a function of uniaxial strain (% in unit, > 0 for tension and < 0 for compression). Reprinted with permission from [45]. Copyright 2008, American Institute of Physics

The first-principles calculation reveals that uniaxial strain applied on the single-walled SiC nanotubes alters their electronic band structures [45]. The electronic structure can be changed by mechanical strain in a wide energy range. The bandgap of the nanotube decreases with increasing uniaxial tensile strain, whereas it increases first and decreases then with increasing uniaxial compressive strain (Fig. 6.23).

The electronic band structure and optical dielectric function of single-walled zigzag, armchair, and chiral SiC nanotubes as well as single honeycomb SiC sheets have been investigated by using DFT–LDA calculations [46]. Figure 6.24a, b shows the calculated band structure and density of states for the graphitic hexagonal SiC sheet. It is apparently a semiconductor with a direct bandgap of 2.58 eV. The valence band maximum and the conduction band minimum both occur at the point *K* in the Brillouin zone. All the SiC nanotubes are semiconductors except for the ultrasmall (3,0) and (4,0) zigzag tubes which are metallic. The zigzag SiC nanotubes have direct bandgaps. The bandgap of the small SiC nanotube increases with increasing diameter, and it gradually approaches the bandgap of the SiC sheet for diameters over 2 nm (Fig. 6.24c). The static dielectric constants for all the SiC nanotubes are almost independent of diameter and chirality. First-principles calculations indicate that the optical spectra of SiC nanotubes depend on the diameter and chirality [47]. For a given chirality of SiC nanotubes, red or blue shift of the dielectric function peak with increasing tube diameter is possible due to the competition between the size effect and *p* orbitals overlapping. The shift decreases with increasing tube diameter.

The structural properties and energetics of nanotip and nanocone materials based on SiC nanotubes have been investigated using density functional calculations [48]. The calculation shows that edges made of multiple atoms are more stable than those made of a single atom. Tips with edges containing pentagon–heptagon defects are energetically more favorable than edges containing four-member rings and are

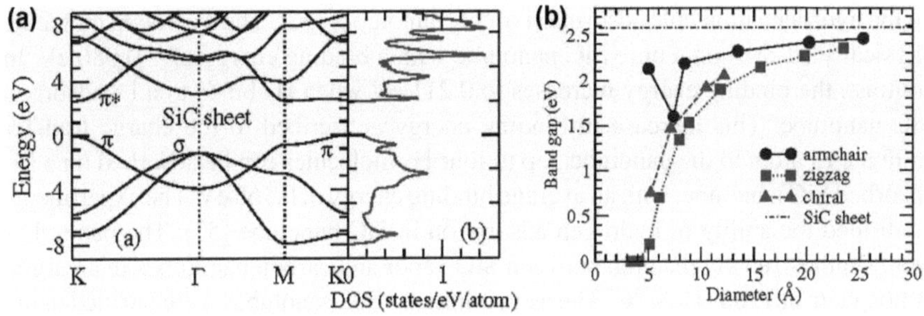

Fig. 6.24 **a** Band structure and **b** density of states of the SiC sheet. The zero energy is at the top of the valence band. **c** Calculated bandgap of the SiC nanotube versus diameter. The bandgap of the SiC sheet is also shown as the dashed-dotted line. Reprinted with permission from [46]. Copyright 2007, American Physical Society

thus suitable as tip materials for atomic force microscopy and scanning tunneling microscopy applications.

6.3.2 Adsorption of Small Molecules and Applications

Hydrogen can be readily incorporated into SiC nanotubes as an impurity during their growth. The theoretical study shows that silicon carbide nanotubes are very promising materials for hydrogen storages [49]. The density functional calculation reveals an increase of 20% in the binding energy of H_2 in SiC nanotubes compared with carbon nanotubes (Fig. 6.25), which may result from the alternative charges that exist in the SiC nanotube wall. The classical Monte Carlo simulation of nanotube bundles reveals an even larger increase in the storage capacity in SiC nanotubes, especially at low temperatures and at high pressures. Therefore, the SiC nanotubes seem to be more favorable than carbon nanotubes in hydrogen storage. The effect of one or two hydrogen impurity atoms in armchair and zigzag SiC nanotubes on their properties has been investigated using ab initio supercell calculations beyond the standard density functional theory [50]. Hydrogen atoms can be adsorbed onto the SiC nanotubes when they are present in the proximity of the nanotube. The calculation shows that a single hydrogen atom can act either as a donor or as an acceptor, depending on its adsorption site. The hydrogen atoms adsorbed on both sites are approximately equally stable, and thus these defects compensate each other in SiC nanotubes. Molecular dynamics simulations were carried out to examine the adsorption and diffusion of N_2, H_2, CO_2, CH_4, and n-C_4H_{10} in the SiC nanotubes, as a function of pressure and nanotube (zigzag, armchair, and chiral) type [51]. The results show strong effect of the nanotubes' chirality and curvature on the pressure dependence of the adsorption isotherms and self-diffusivity. The SiC nanotubes show higher hydrogen adsorption capacity compared to the carbon nanotubes under the adopted conditions. The DFT–PBE calculations show the effect of lithium adsorption on the hydrogen molecule adsorption of SiC nanotubes [52]. The hydrogen molecule physically adsorbs on a pure SiC nanotube with a binding energy of ~ 0.086 eV. In contrast, the binding energy increases to 0.211 eV when H_2 binds to a Li-adsorbed SiC nanotube. This increase in binding energy is ascribed to the charge transfer from the Li atom to the nanotube. Up to four H_2 molecules can be adsorbed on a Li-adsorbed SiC nanotube, with an average binding energy of 0.165 eV. The experiments confirmed the ability of hydrogen adsorption in SiC nanotube [53]. The nanotubes were synthesized via reaction between SiO vapor and carbon nanotubes in an argon atmosphere at 1200–1500 °C. The as-synthesized SiC nanotubes were purified using a sodium hydroxide solution to eliminate the by-products. The hydrogen adsorption characteristics were measured at 25 °C under a pressure of up to 100 bars. The results reveal hysteresis between the adsorption and desorption isotherms owing to the presence of metal impurities and/or multilayer nanotubes. The hydrogen storage capacity of the as-synthesized SiC nanotubes is close to that of carbon nanotubes, whereas that of the purified SiC nanotubes is 50% higher. The hydrogen uptake rate in the SiC nanotubes is five times faster than that in the carbon nanotubes, and the desorption from SiC nanotubes is completely reversible under vacuum.

The first-principles calculation indicates that CO and HCN molecules can be absorbed to Si atoms on the wall of SiC nanotubes with binding energies as high as 0.70 eV and can attract finite charges from SiC nanotubes [54]. Similar to oxygen molecules, the CO and HCN molecules adsorbed on SiC nanotubes may induce significant change in the conductivity of the nanotubes; hence, the SiC nanotubes are potential gas sensors for CO and HCN detection. The first-principles calculation

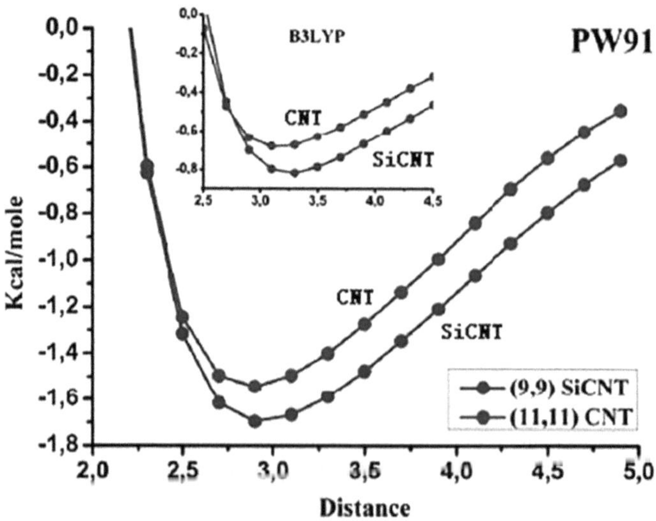

Fig. 6.25 Potential energy curves of an H_2 molecule that perpendicularly approaches the middle part of a hexagon of the (9,9) SiC nanotube's and (11,11) carbon nanotube's outer walls. Reprinted with permission from [49]. Copyright 2006, American Chemical Society

reveals that NO and NNO molecules can be chemisorbed on SiC nanotubes with appreciable binding energies [55]. The SiC nanotube–NO complex has large adsorption energy of – 0.7 eV. This complex has magnetism and even ferromagnetism in the case of more than one adsorbed NO molecules. This property may be used for sensing the amount of adsorbed NO molecules. In contrast, the SiC nanotube–NNO complex is nonmagnetic. The DFT–PBE calculations show the potential of zigzag $(n,0)$ $(n = 6, 8, 10, 12$, and $18)$ single-walled SiC nanotubes for CO_2 [56]. Tube diameter and CO_2 coverage play important roles in tube–CO_2 interaction. A single CO_2 can be chemisorbed to the Si–C bonds of the SiC nanotube with appreciable adsorption energy. The adsorption energy decreases with increasing tube diameter. CO_2 molecules prefer to be as far from each other as possible when they are adsorbed on the exemplified (8,0) tube. The increase in CO_2 coverage results in weakened interaction between CO_2 molecules and the nanotube. Up to eight CO_2 molecules can be adsorbed on a nanotube. The adsorption of CO_2 molecules lowers the bandgap.

The interaction between formaldehyde (HCOH) and infinitely long zigzag single-walled SiC nanotubes has been investigated using density functional theory [57]. In comparison with the weak adsorption on carbon nanotubes, HCOH molecule is chemisorbed to the Si–C bond of the SiC nanotube with appreciable adsorption energy. The Si–C pair is also very sensitive to the adsorption of the second HCOH molecule. The bandgap of the SiC nanotube decreases gradually with increasing adsorbed HCOH molecules, resulting in an increase of conductivity of the nanotube. Density functional theory calculation shows that the SiC nanotube can effectively break the N–H bond of ammonia and the OH bond of H–OX (X = H, CH_3, and C_2H_5) [58]. This process contains two steps, first, the NH_3 (or H–OX) molecule is adsorbed on the Si atom of the SiC nanotube, and then the adsorbed molecule is cleaved to an atomic H and a NH_2 (or OX) group. Adsorption of hydrogen peroxide (H_2O_2) on the BN and SiC nanotubes was compared through DFT calculations [59]. The molecule is chemically adsorbed on both nanotubes. It interacts with the SiC nanotube (adsorption energy – 0.97 eV) much more strongly than with the BN nanotube (adsorption energy – 0.47 eV). The H_2O_2 adsorption on BN nanotube results in slight decrease in the BN work function and increase in its field electron emission. In contrast, it may not affect those of the SiC nanotubes. On the contrary, the H_2O_2 adsorption improves the electrical conductivity of the SiC nanotube, but not significantly for the BN nanotube.

The adsorption of a series of transition metal atoms on single-walled SiC nanotubes have been investigated using density functional theory calculations [60]. Many transition metal atoms are chemically adsorbed on the outer surface of the single-walled SiC nanotube, with binding energies ranging from 1.17 eV (for Cu) to 3.18 eV (for Pt). The adsorbed metal atoms change the electronic structures of the single-walled SiC nanotubes and this change is nearly independent of the adsorption site. One exception is Ti, the (8,0) single-walled SiC nanotube exhibits characteristic of metal as Ti is adsorbed onto H site and characteristic of semiconductor as Ti is adsorbed onto C site.

6.4 Two-Dimensional SiC Sheets

6.4.1 Synthesis and Properties

Different from graphene or other 2D materials, there are much more theoretical studies on 2D SiC, whereas the experimental advances are very few and actually very challenging. DFT calculations had predicted that 2D graphene-like honeycomb-structured monolayer SiC is able to exist in nature, having a stable planar structure and behaving as a direct-bandgap semiconducting material. However, the synthesis or fabrication of truly 2D honeycomb-structured SiC has been extremely challenging, and mainly because unlike other 2D materials, bulk SiC crystal is not a van der Waals layered material. In a 3D SiC crystal, the atoms in the Si–C bonds take the form of sp^3 hybridization, and the corresponding Si–C bond is much stronger than the van der Waals bonding in layered materials.

Some signs of existence of two-dimensional SiC nanoflakes with thicknesses down to 0.5–1.5 nm have been presented by exfoliation of bulk SiC in polar solutions [61]. The nanosheets were prepared by sonication of the SiC powder in n-methylpyrrolidone or isopropyl alcohol. Figure 6.26 shows the high-resolution TEM image of the nanosheet on graphene. The lattice fringe spacing is around 0.28 nm. The fast Fourier transformation pattern reveals sixfold symmetry of this sheet. The measurement of more nanoflakes reveals sixfold-symmetry lattice fringes with spacings ranging from 0.24 to 0.28 nm. These values are bigger than the lattice parameter of graphene, 0.21 nm. The energy-dispersive X-ray spectroscopy shows that the nanosheet is mainly composed of silicon and carbon elements. The combined TEM, XRD, and XPS measurements indicate that the as-prepared ultrathin layers probably consist of graphitic SiC, graphene, and embedded ultrathin wurtzite SiC.

Quasi-two-dimensional SiC was produced by reaction between graphene and a silicon source [62], which was guided by the molecular dynamics simulation. The

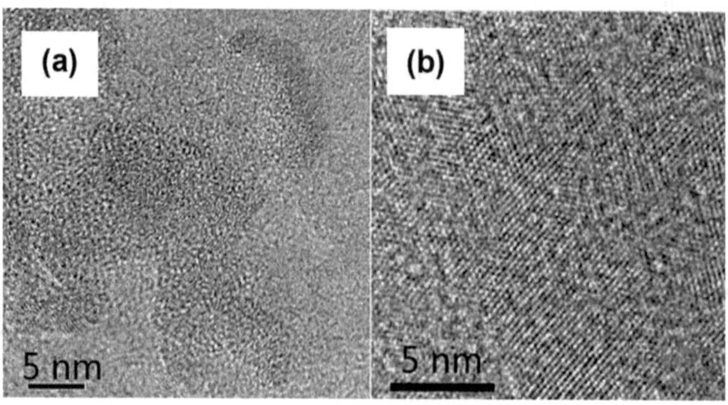

Fig. 6.26 **a** HRTEM image of two-dimensional SiC nanoflake supported by an ultrathin graphene layer. **b** HRTEM image of large SiC nanosheets. The lattice fringe spacing is 0.28 nm. Reprinted with permission from [61]. Copyright 2012, American Chemical Society

lateral length of the synthesized quasi-2D SiC lies in 0.3–5 μm, and the thickness is below 10 nm. The quasi-2D SiC_2 as a by-product is stable over 3 months in air. The quasi-2D multilayer SiC has a bandgap of around 3.72 eV and an exciton binding energy of 0.23 eV. In situ TEM confirmed the synthesis of quasi-2D SiC through the reaction between graphene and silicon QDs. An ultrathin SiC structure was prepared by a catalyst-free carbothermal and post-sonication method [63]. Combining ultra-light 3D graphene foam and SiO together at high temperature results in the formation of a complex SiC structure consisting of 3D SiC foam covered with 1D nanowires. The 3D SiC is made of 2D nanosheets. AFM reveals an average thickness of 2–3 nm and an average size of 2 μm for the flakes. The nanosheet has an interlayer spacing of 0.255 nm. The atomically thin and hexagonally bonded nanosized SiC grains assembling temporarily in graphene oxide pores were observed during an atomic resolution scanning TEM experiment [64]. Simulations indicate that the electronic structure of such small grains approaches that of 2D SiC. There is doubt about whether Si atoms take the sp^3 hybridization. The calculated chemical reactivity of 2D SiC suggests it is unstable against molecular adsorption or interlayer linking, like the case of silicone, but it can form stable van der Waals bilayers with either graphene or hexagonal BN. Carbothermal reduction between graphene oxide and silicon powder can also yield SiC nanosheets [65]. The high-resolution TEM image (Fig. 6.27) reveals an interplanar spacing of a typical nanosheet of 0.251 nm, corresponding to the (111) planes of 3C-SiC. There are stacking faults in the nanosheets. An impedance-type nonoxide humidity sensor fabricated on the basis of such nanosheets has a high surface area of 90.1 $m^2\ g^{-1}$. The SiC nanosheets have excellent sensing properties, with a high sensitivity to water in the ambient atmosphere (16,991.1 at 95% relative humidity), a wide range of relative humidity response (11–95%), short response time (3 s) and recovery time (3 s), and good linear response toward humidity. A later study shows that the surface hydroxyls on these SiC nanosheets can offer local protons, stabilize intermediates, and localize photogenerated electrons in the photoreduction of CO_2, promoting the efficiency and selectivity of CH_4 yield [66]. Ultrasmall 2D SiC QDs with blue photoluminescence were fabricated by a hydrothermal route using a SiC powder as the precursor [67]. Their sizes and thicknesses are tunable by changing the reaction time. The SiC QDs with an average lateral size of 3.0 nm were fabricated for a hydrothermal reaction time of 12 h. The ultrasmall 2D SiC QDs exhibited a long fluorescence lifetime of 2.59 μs. They were applied to the fluorescence labeling of A549, HeLa, and NHDF cells. SiC nanosheets synthesized via a catalyst-free carbothermal reaction of graphene oxide and Si powder were used for gas sensing [68]. The SiC nanosheets were utilized for sensing hazardous gases (acetone, ethanol, methanol, and ammonia) at a high temperature of 500 °C. The SiC nanosheet sensors exhibited a short response time (8–39 s) and a short recovery time (12–69 s) toward detecting gases.

Pyrolysis of polycarbosilane at 700 °C in a NaCl powder template followed by thermal treatment at 1200 °C can produce 3C-SiC nanoflakes [69]. The salt was removed by washing with water. The morphology of the SiC particles depends on the NaCl/polycarbosilane weight ratio. A NaCl/polycarbosilane ratio of 40:1 is critical for the formation of 3C-SiC nanoflakes. These nanoflakes have thicknesses below

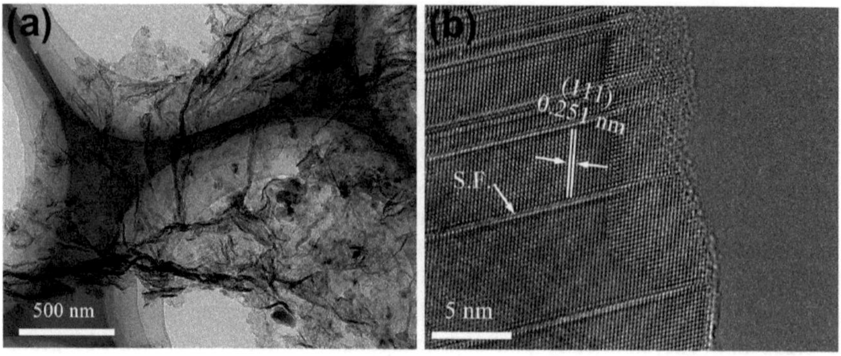

Fig. 6.27 **a** TEM and **b** high-resolution TEM images of the SiC nanosheet. Reprinted with permission from [65]. Copyright 2018, John Wiley & Sons, Inc.

5 nm, a high specific surface area of 169 m^2 g^{-1}, and a high total pore volume of up to 0.75 cm^3 g^{-1}. Bottom-up synthesis of colloidal 2D layered 3C-SiC QDs with lateral sizes of 5–10 nm from the liquid precursor tetramethylsilane was reported [70]. The plate-like nanocrystals arise from ⟨110⟩ exfoliation, with 1–6 atomic layers. They were surface-passivated with 1-dodecene for colloidal stability in organic solvents, which allow for size separation through density-gradient ultracentrifugation as a characterization tool. The samples shielded from oxygen and plasma-annealed exhibited narrow blue PL with a quantum yield (QY) above 60%; unshielded nanocrystals exhibited broad blue/green/white PL with 10–15% QY. The shape of the bright, narrow blue PL varies little with particle size. Figure 6.28 shows the PL spectra of the colloidal SiC NCs from across the postspin (density-gradient ultracentrifugation) gradient of a high-QY parent sample. The PL color and spectral shape show little variation with size, but with a slightly smaller FWHM for the best fractions (~ 35 nm). The maximum QY occurs near four atomic layers, with a drop in QY on moving from mesitylene to the gradient solvents hexane/toluene. The lifetime measured for the high-QY fractions reaches a minimum near the maximum QY. Upon drying under heat, the disk-like colloidal nanocrystals aggregated into microscale domains and further self-assembled into planar superlattice domains with common crystalline orientation.

Truly two-dimensional SiC was successfully isolated from bulk hexagonal SiC via a wet exfoliation method (Fig. 6.29a, b) [71]. Figure 6.29c shows the AFM image of two SiC nanosheets. The average heights of the two flakes are 0.28 and 0.25 nm, respectively (Fig. 6.29d). Considering the interlayer spacing in bulk SiC is about 0.25 nm, this suggests that the shown nanosheets are monolayers. The high-resolution TEM image of the 2D SiC indicates the highly ordered crystalline structure of the exfoliated SiC nanosheets (Fig. 6.29f). The graphene-like hexagonal structures shown in Fig. 6.29g, h confirm the creation of 2D SiC, where the honeycomb structure is apparent. 2D SiC shows Raman peaks that are different from that of bulk SiC (Fig. 6.30a, b). 2D SiC has an intense TO peak at around 790 cm^{-1}. But it has a

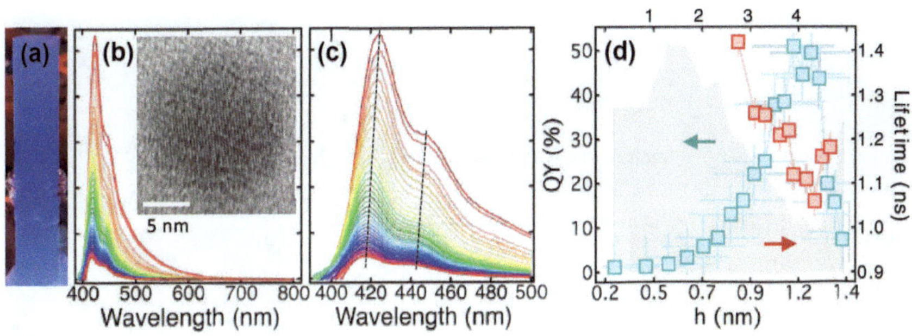

Fig. 6.28 **a** PL from a postspin gradient of a high-quantum yield parent. **b** PL spectra from across the gradient, with nanocrystal thickness increasing with depth. Inset shows a TEM image of a disk-like nanocrystal from a high-QY fraction ($h = 1.11$ nm, 5 nm scale). **c** Expanded view of the size-resolved spectra in (**b**), with the two dominant emission channels showing little dependence on nanocrystal size/thickness. **d** QY (blue) and lifetime (red) versus nanocrystal thickness, where the light-gray profile is the optical absorption. The upper horizontal scale is the layer number. Reprinted with permission from [70]. Copyright 2024, American Chemical Society

much weaker and broader LO peak (at around 930 cm^{-1}) compared to that of the SiC precursor. This asymmetric broadening and downshifting of the LO peak in 2D SiC is attributed to the quantum confinement effect. The absorption spectrum of the 2D SiC solution comprises two peaks at around 2.14 and 2.58 eV as well as an additional small peak at around 2.3 eV (Fig. 6.30c). All solutions of 2D SiC have discrete peaks in the range of 2.13–2.7 eV, with the most pronounced peak at around 2.13 eV. These discrete peaks might correlate with the strongly bound excitons in SiC nanosheets, quantum confinement effect, and their surface defects. Unlike bulk SiC, 2D SiC has strong PL. Figure 6.29d shows the PL (excitation: 266 nm) of the drop-casted SiC nanosheets on a sapphire substrate. It comprises an emission band at ~ 2.58 eV (or 480 nm). Its position matches that of one of the absorption peaks (at 2.58 eV), implying a probable origin of direct-bandgap transition. Unlike silicene, the produced 2D SiC nanosheets are environmentally stable, showing no signs of degradation.

As aforementioned, disordered SiC nanoflakes can indeed be fabricated with various exploited methods. Considering that sp^2 bonding between silicon and carbon is energetically favorable, so is it probable to explore a more controllable method to fabricate 2D honeycomb-structured SiC monolayers? A bottom–up method was developed to prepare large-area, high-quality honeycomb-structured SiC monolayers through annealing of ultrathin thin films of transition metal carbides growing on 4H-SiC(0001) [72]. The lattice-matched rocksalt monocarbides TaC and NbC have been demonstrated as favorable epitaxial substrates (Fig. 6.31a–d). For the fabrication, 3 nm thick film of TaC or NbC was sputter deposited onto Si-terminated 4H-SiC(0001) substrate, with a lateral extent of approximately 5 mm × 5 mm. The deposition was followed by 1700 °C anneal for 10 min in an Ar atmosphere. The atomistic detail of formation of this honeycomb SiC layer is not fully understood. For an isolated SiC substrate, when the temperature reaches ≈ 1400 °C, silicon atoms

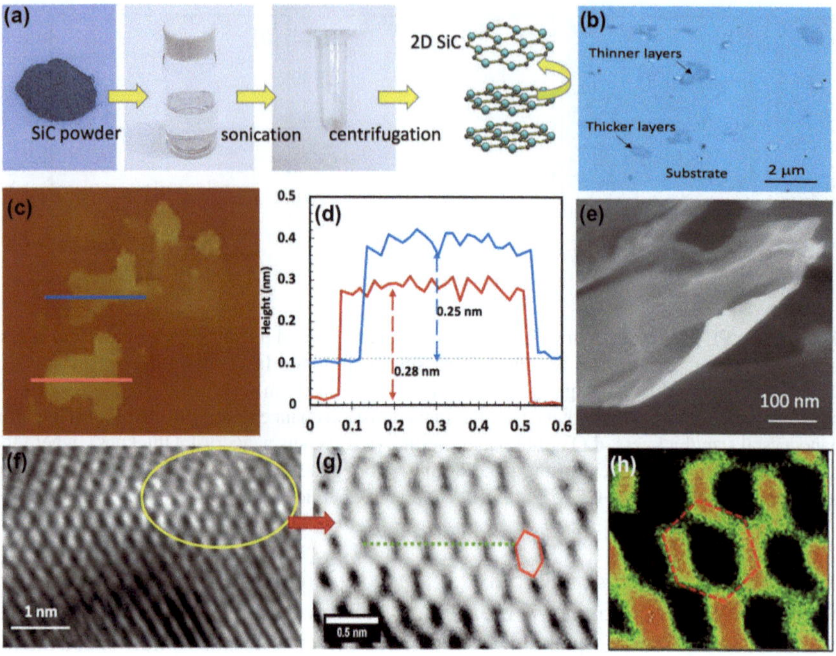

Fig. 6.29 **a** Photographs of exfoliation process and dispersed SiC. **b** Optical microscopy image. **c** AFM image. **d** Height profiles of the SiC nanosheets marked in (**c**). **e** SEM image of SiC nanosheets. **f–h** High-resolution TEM images of the ordered structure of SiC nanosheet. Reprinted with permission from [71]. Copyright 2021, MDPI

sublimate from the surface and carbon atoms do not, leaving a carbon-rich layer that would recrystallize as graphene. TaC is much more stable, and decomposing it on a timescale of minutes requires high temperatures above 3000 °C. These suggest that the formation of the SiC monolayer may be driven by decomposition of the underlying SiC, leaving the TaC film intact. It is likely that during the post-deposition anneal of the TaC film at 1700 °C, the underlying Si and C atoms diffuse through the TaC to accumulate and crystallize on the surface. Band structure calculations show good agreement with angle-resolved photoelectron spectroscopy (ARPES) measurements (Fig. 6.31e–h). They indicate that the π band of the SiC layer hybridizes with the underlying TaC surface states to give rise to a strongly spin–orbit split Dirac-like feature at the $\overline{K}$ point. For this approach of fabrication of the SiC monolayers, one difficulty remains, that is, how to decouple the SiC monolayer from the substrate through such as intercalation or adsorption scheme to give rise to free-standing SiC monolayers, so as to alter the band splitting or even bring the Dirac-like feature toward the Fermi energy.

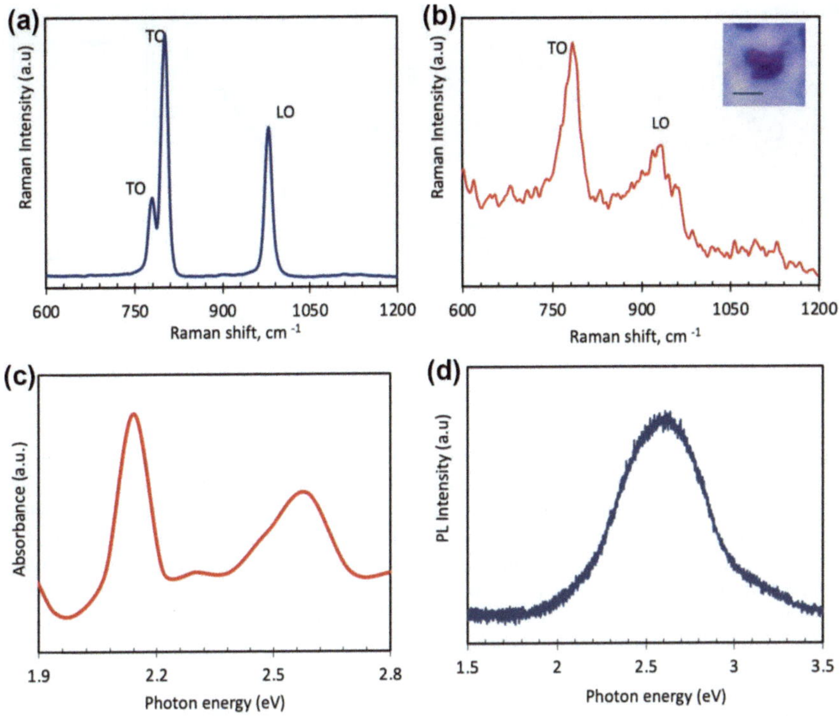

Fig. 6.30 Raman spectra of **a** SiC precursor and **b** exfoliated SiC nanosheets. The inset of (**b**) shows an optical image of the SiC nanosheet, with a scale bar of 500 nm. **c** Optical absorption spectrum of 2D SiC dispersion. **d** PL spectrum of drop-casted SiC nanosheets on a sapphire substrate. Reprinted with permission from [71]. Copyright 2021, MDPI

6.4.2 *Theoretical Perspectives*

There have been tremendous works devoted to theoretical explorations of two-dimensional SiC; major concerns include structural stability, electronics structures as well as related dielectric and light absorption properties, defects and adsorption, nanoribbons, and probable nonstoichiometric structures. The theoretical results predict the critical parameters of 2D SiC to guide the laboratory explorations and identification of these challenging and unique structures.

The quasiparticle band structure and optical properties of a single SiC sheet have been investigated by using ab initio many-body calculations using GW and GW plus BSE methods [35]. The more accurate GW quasiparticle calculations show significantly (bigger) correction by over 1.0 eV to the Kohn–Sham bandgaps from the LDA calculations. It is well known that the DFT calculation employing usual LDA or GGA exchange–correlation functional underestimates the bandgaps of semiconductors. Remarkably, the GW self-energy corrections transform the SiC sheet from

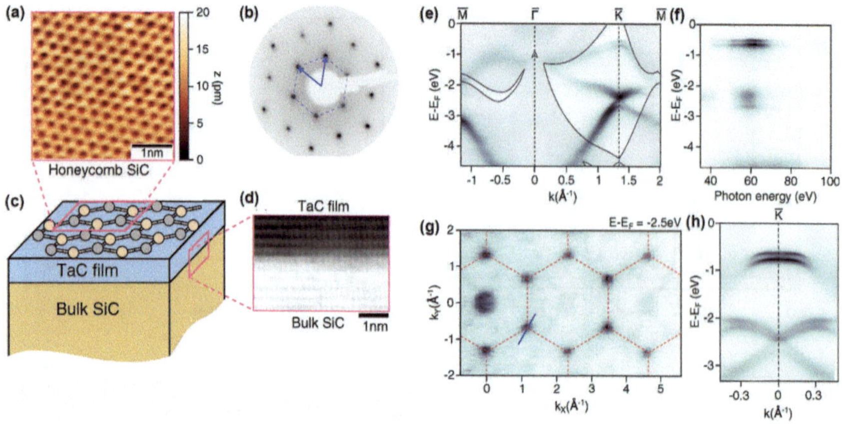

Fig. 6.31 **a** Topographic STM image of the monolayer honeycomb SiC layer terminating a TaC(111) surface. **b** LEED image; 2.32 $Å^{-1}$ unit cell reciprocal vectors are indicated. **c** Schematic sample structure. **d** Cross-sectional HAADF STEM image of the TaC–SiC interface. **e** ARPES measurement along denoted path, overlaid with surface-projected bulk band structure of TaC. **f** $h\nu$ resolved cut at $\overline{K}$ ($T = 80$ K). **g** Constant-energy surface through Dirac-like crossing of the surface state, with the surface Brillouin zone overlaid ($T = 80$ K). **h** High-resolution ARPES of the surface states at $\overline{K}$, direction indicated by the blue line in (**g**) ($T = 19$ K). Reprinted with permission from [72]. Copyright 2023, American Physical Society

an indirect LDA bandgap to a direct-bandgap material. The evaluated light absorption spectra are dominated by discrete exciton peaks (Fig. 6.32) as a result of the high binding energy (~ 2.0 eV) of the exciton state in the SiC sheet. The enhanced electron–hole interaction in the low-dimensional SiC sheet is responsible for the formation of strongly bound excitons.

The electronic and optical properties of two-dimensional honeycomb-structured SiC nanosheets were investigated by using DFT–GGA calculations [73]. The sheet thickness and geometric configuration strongly affect their properties. The calculation predicts an indirect bandgap for the 2D multilayer SiC sheet, whereas monolayer 2D SiC exhibits a direct bandgap (~ 2.5 eV) (Fig. 6.33). The monolayer 2D SiC converts from a direct to an indirect-bandgap semiconductor when the bond length is smaller than 1.74 Å. When the bond length varies within the range of 1.74–2.0 Å, the monolayer 2D SiC retains a direct bandgap. When the bond length increases above 2.0 Å, the monolayer becomes an indirect-bandgap material again. The optical conductivity of 2D multilayer SiC sheet is sensitive to the interlayer spacing. The authors further indicate that a controllable direct bandgap of bilayer/trilayer 2D SiC can be realized through interlayer oriented misalignment, and the direct-bandgap character is retained for most rotation angles [74].

The DFT–PBE calculations combined with the cluster expansion method give insights into the thermodynamic stability and electronic properties of 2D Si_xC_{1-x} monolayers with $0 \leq x \leq 1$ [75]. All the revealed ground state structures of 2D Si_xC_{1-x} show high thermodynamic stability and strong dependence on x. The 2D Si_xC_{1-x}

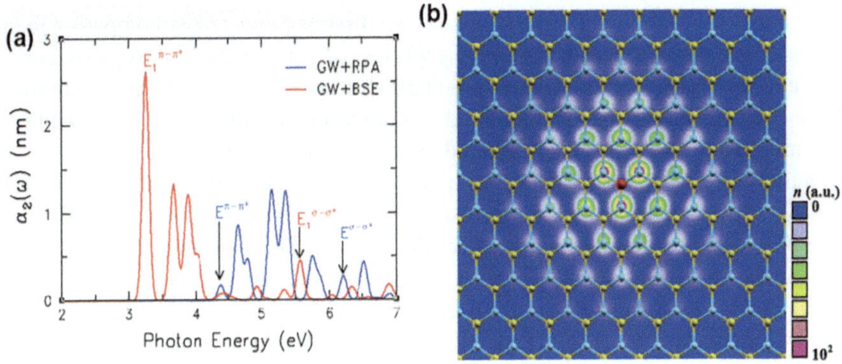

Fig. 6.32 **a** Optical polarizability per unit area from both GW + RPA and GW + BSE calculations. Theoretical spectra are broadened with a Gaussian smearing width of 0.15 eV. **b** $E_1^{\pi-\pi*}$ exciton wavefunction of the SiC sheet. The hole position (red sphere) is fixed at the top of a C atom (yellow sphere), and the squared electron amplitude [$n = |\Psi_S(\mathbf{r}_e, \mathbf{r}_h = 0)|^2$ in arbitrary units] lies mainly on Si atoms (cyan spheres) next to the hole. Reprinted with permission from [35]. Copyright 2011, American Physical Society

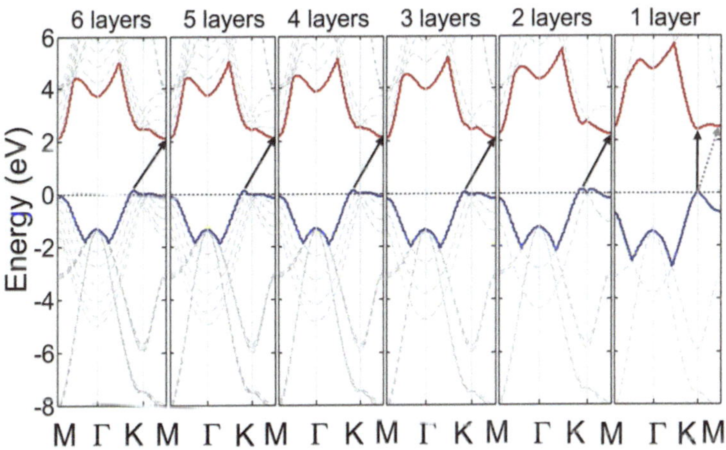

Fig. 6.33 Band structure evolution of 2D SiC from 1 to 6 layers. The conversion from indirect to direct-bandgap transition is apparent when the layer number decreases to one. Reprinted with permission from [73]. Copyright 2013, Royal Society of Chemistry

shows two distinct structural phases, a homogeneous phase with well-dispersed Si or C atoms and an in-plane hybrid phase consisting of SiC and graphene (or silicene for $x > 0.5$) domains (Fig. 6.34a–d). The in-plane hybrid structure shows uniform semiconducting properties with a tunable bandgap from 0 to 2.87 eV due to quantum confinement effect imposed by the SiC domains. The homogeneous structures can be semiconducting or semimetallic, depending on a superlattice vector connecting two adjacent doping sites, with a rule similar to the gap–chirality relationship in carbon nanotubes. The stability of these structures can first be evaluated by comparing their cohesive energies and molar formation energies (δF) (Fig. 6.34e, f). Negative cohesive energies indicate that all the studied 30 structures are energetically favorable. Their relative stability can be measured by comparing the altitude δF from the line. SiC monolayer with $x = 0.5$ is thermodynamically stable in that its δF is below the straight line with a negative δF of -0.016 eV. All the remaining Si_xC_{1-x} structures have δF above this line and are metastable with respect to graphene and silicene. The δF values of these structures fall within a narrow range of 0.057–0.329 eV, suggesting their considerable stability. These 2D SiC sheets exhibit fruitful electronic properties such as indirect and direct gaps as well as semimetallic behaviors while x changes from 0 to 1 (Fig. 6.35). Most of the Si_xC_{1-x} sheets have semiconducting properties; only two homogeneous structures, $Si_{0.17}C_{0.83}$ and $Si_{0.17}C_{0.83}$, are semimetallic, similar to graphene and silicene. The bandgap first increases with increasing x, reaches a peak value of 2.87 eV at $x = 0.5$ (2D SiC), beyond which it decreases to 0 eV when x reaches 1. Therefore, the bandgap of the Si_xC_{1-x} sheet is proportional to the Si or C doping concentration in graphene or silicene.

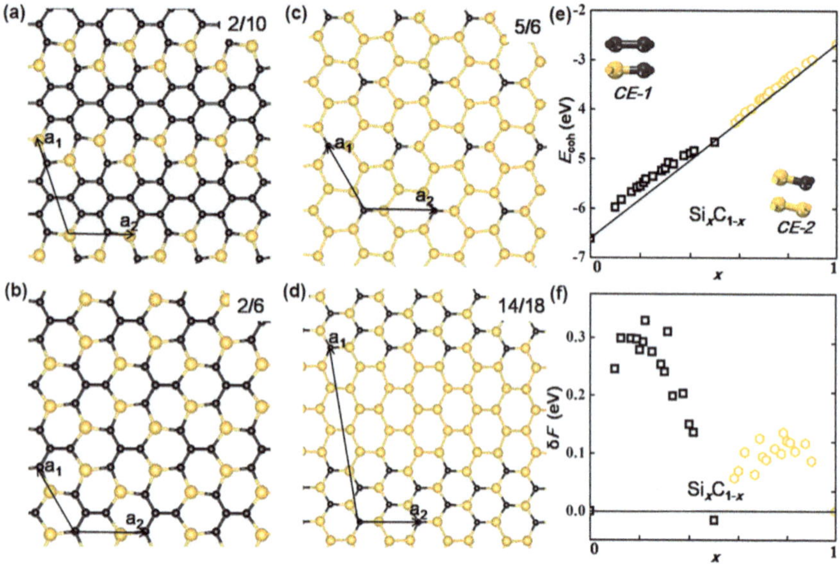

Fig. 6.34 **a–d** Atomic structures of four typical Si_xC_{1-x} sheets found by the cluster expansion (CE) method. The label-fraction represents the ratio of Si and total atoms in a unit cell. **a, d** Are hybrid structures, **b, c** Are homogeneous structures with well-dispersed dopants. **e, f** Cohesive energy and molar formation energy for the ground state Si_xC_{1-x} sheet as the function of x. Straight line connects cohesive energies and molar formation energies of graphene and silicene, respectively. Insets of (**e**) show primitive cell adopted in CE-1 and CE-2 calculations. Gray and yellow balls represent carbon and silicon atoms. Gray squares and yellow hexagons separately represent predicted structures from CE-1 and CE-2 calculations. Reprinted with permission from [75]. Copyright 2015, American Chemical Society

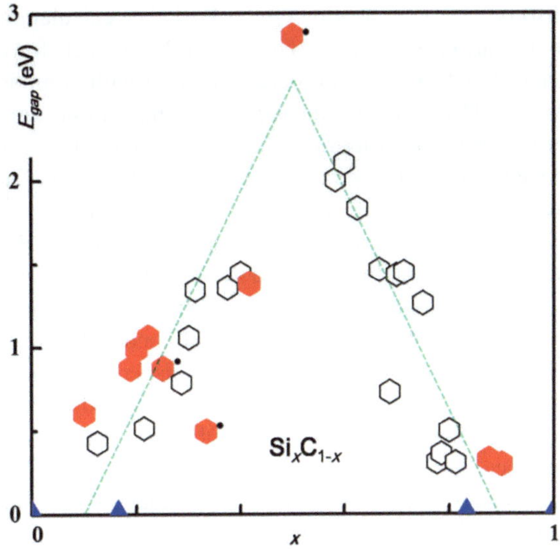

Fig. 6.35 Calculated bandgaps of the ground state Si_xC_{1-x} sheets as a function of x. Blue solid triangles, black hollow hexagons, and red solid hexagons designate semimetallic, indirect-, and direct-bandgap structures, respectively. Reprinted with permission from [75]. Copyright 2015, American Chemical Society

A van der Waals-induced DFT–PBE calculation shows that a planar honeycomb monolayer of siligraphene (SiC_7) decorated with various light metals might be useful for clean energy storage [76]. The binding energies of lithium (Li), sodium (Na), potassium (K), magnesium (Mg), calcium (Ca), scandium (Sc), and titanium (Ti) dopants on a SiC_7 monolayer are big enough to counteract the metal clustering effect. Bader charge analysis shows that upon adsorption, all the metal adatoms donate a fraction of electronic charges to the SiC_7 sheet. Each metal center on the SiC_7 sheet can bind maximally 4–5 H_2 molecules. A high H_2 gravimetric density is achievable for several types of dopants with a doping concentration of 12.50%. The H_2 binding energies fall within 0.2–0.6 eV. The DFT calculation shows that the mechanical parameters of g-SiC_x ($x = 7, 5, 3, 2, 1$) monolayers including ideal strength, Young's modulus, shear modulus, Poisson's ratio, and fracture toughness are related to the ratio of Si–C and C–C bonds [77]. The microscopic failure of different g-SiC_x differs, depending on the complex structural deformation modes introduced by Si substitution. The structural transformation of g-SiC_3 and g-SiC is ascribed to their singular structural symmetry. All the g-SiC_x monolayers preserve isotropic Young's moduli and Poisson's ratios. Another DFT–PBE calculation reveals that the Si_9C_{15} monolayer is an auxetic material with a negative Poisson ratio of 0.175 [78]. The DFT–HSE calculation shows that SiC monolayer is a semiconductor with a direct bandgap of 2.54 eV. The GW–BSE calculation shows that the monolayer has an optical bandgap of 2.73 eV and an exciton binding energy of 1.14 eV, implying a strongly bound exciton. The lattice thermal conductance of the monolayer at room temperature is 0.85 nW K^{-1}, and the maximal figure of merit is 1.2 at 800 K.

The energetics and thermodynamic properties of 2D binary graphene-like alloys made from graphene, silicene, or germanene have been investigated by combining first-principles total energy calculations and a statistical approach to account for disorder and composition effects [79]. The calculations of the electronic properties were carried out within the GGA-1/2 approach for an approximate quasiparticle band. For Si_xC_{1-x}, an ordered structure with composition $x = 0.5$ is stable at up to ~ 1000 K, whereas intermediate compositions are mainly realized by graphene or silicene. The ordering and decomposition effects strongly affect the composition dependence of the average fundamental energy gap. Large gaps that arise for Si_xC_{1-x} systems almost disappear for Ge_xSi_{1-x} and Ge_xC_{1-x}. The stable arrangements and band structures of graphitic silicon carbide were studied [80]. The proposed models were experimentally examined by preparing and characterizing a number of graphitic features. Graphitic silicon carbide shows a wide spectrum of optoelectronic properties (360–690 nm) and an unusual band structure with highly anisotropic transport properties arising from its nondispersive band near the K point. The effect of the number of layers on the electronic and optical properties of few-layer SiC nanosheets was studied by using DFT–PBE calculations [81]. With increasing layer number, the interlayer coupling becomes stronger, and the bandgap decreases. The nanosheets exhibit three absorption peaks at 4.35, 7.83, and 11.68 eV along their vertical direction, whereas there exists a strong absorption band within 2–10 eV along the parallel direction. The dielectric response increases with increasing layer number along the parallel direction, but decreases along the vertical direction. It shows metallic features within 1000–2250 THz along the parallel direction and only metallic features at around 2830 THz along the perpendicular direction.

Little has been known about the defect properties of 2D SiC. A first-principles calculation reveals that the two-dimensional graphitic SiC sheet acquires net magnetic moment when a Si-vacancy defect is present or the As, P, B, or N substitutional atom is present [82]. In contrast, the single C-vacancy, Si and C divacancy, and antisite defects do not induce magnetic moment in the SiC sheet. Reverse nonequilibrium molecular dynamics simulations show that various types of defects such as bi-, point, and mixed vacancies affect the thermal conductivity of 2D SiC [83]. The thermal conductivity decreases significantly with increasing vacancy concentration, and the effects of the point vacancies are more quickly compared to bi-vacancies and mixed vacancy defects. Increase in temperature results in further decrease in thermal conductivity due to strong phonon–vacancy scattering. The introduction of vacancies affects the acoustic phonon density of states and increase in temperature causes frequency shrinking in the curves of phonon density of states, both of which contribute to the diminished thermal conductivity.

The SiC monolayers show promise as potential single-photon sources, as revealed by DFT–HSE06 calculations [84]. A few (negative, positive, and neutral) point defects were taken into account. The formation energy calculations reveal that neutral and positive charge states of the Stone–Wales defects with and without combination of antisite defects are stable. The calculated Huang–Rhys factors and related Debye–Waller factors suggest the promise of the Stone–Wales defects as single-photon emitters. The calculations suggest that the negatively charged carbon antisite–vacancy defect is a promising candidate for realizing near-infrared single-photon luminescence [85], where the negative charge state may be generated by nitrogen doping. The neutral carbon vacancy with a spin-triplet ground state possesses broad visible emission and might be used for quantum sensing. The DFT–HSE06 calculation predicts an indirect bandgap of 3.38 eV and a direct bandgap of 3.43 eV for the pristine SiC monolayer (Fig. 6.36) [86]. The SiC monolayer shows a very low formation energy of 0.57 eV and can be stabilized on TaC{111} film. Nitrogen doping is the most favorable defect in the SiC monolayer due to its very low formation energy. The defects generate additional states in the forbidden gap, leading to significant reduction in the bandgap. The NC and AlSi defective systems show transition from semiconducting to metallic states. The conduction band of the pristine SiC monolayer is nearly flat in the $M \rightarrow K$ direction, leading to a big effective mass of 3.48 m_e. The defect induces significant red shift in the absorption edge and occurrence of additional absorption peaks. The 585 extended line defects in hexagonal SiC monolayer were investigated considering three types of interstitial atom pairs (SiSi-, SiC-, and CC-extended line defects) and using computational methods like DFT–PBE, Born–Oppenheimer molecular dynamics, and kinetic Monte Carlo [87]. The formation of all extended line defect systems is endothermic, and the structure with CC-extended line defects has the highest stability at 300 K. The simulated scanning tunneling microscopy allows for identifying and distinguishing of the three types of extended line defects. The extended line defects introduce mid-gap states derived from the p_z orbitals at the defect sites. The extended line defect region shows enhanced reactivity toward hydrogen adsorption, as confirmed by the kinetic Monte Carlo simulation.

Electronic structures of SiC nanoribbons have been investigated by using spin-polarized first-principles calculations [88]. The calculation shows that the two-dimensional SiC sheet has a direct energy gap of 2.55 eV at the K point in the hexagonal Brillouin zone (Fig. 6.37a–d). This value is smaller than the gap obtained by self-interaction corrected density functional calculation, suggesting that the calculation method employed in this study can give rise to qualitatively correct but quantitatively inaccurate results. All dangling σ bonds at the ribbon edges are saturated by hydrogen atoms for the calculation of the armchair and zigzag SiC nanoribbons. The armchair SiC nanoribbons with different widths have similar band structures, and they

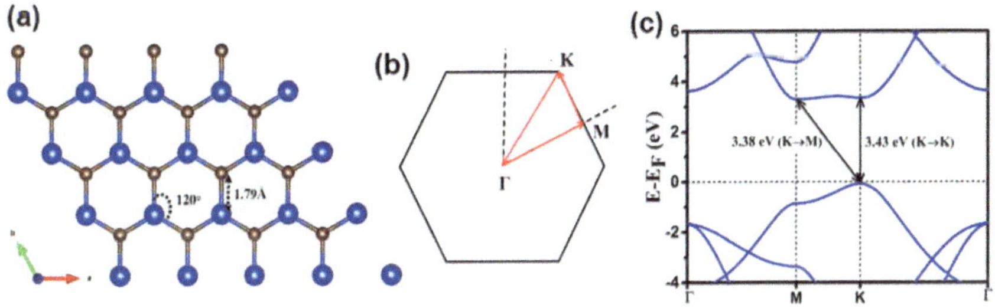

Fig. 6.36 **a** Optimized structure of 4 × 4 supercell of SiC monolayer with Si–C bond length of 1.79 Å, and Si–C–Si (and C–Si–C) bond angles 120° [blue (brown) spheres denote Si (C) atoms, the monolayer lies in the *xy* plane]. **b** High-symmetry path in the reciprocal space and **c** calculated HSE06 electronic band structure of a primitive cell of 2D SiC monolayer. Reprinted with permission from [86]. Copyright 2023, American Physical Society

are nonmagnetic semiconductors with a direct bandgap at the Γ point (Fig. 6.37e–g). The energy gap versus ribbon width curve exhibits a three-family behavior, similar to the case of armchair graphene nanoribbons. However, the energy gap increases with ribbon width for SiC nanoribbons, in comparison with the decrease for graphene nanoribbons. The zigzag nanoribbons are magnetic metals, and the spin polarization stems from the unpaired electrons localized on the ribbon edges. The zigzag nanoribbons narrower than 4 nm have half-metallic characters. The theoretical calculation based on the tight-binding model reveals that in the SiC nanoribbons, the charge redistribution at the ribbon edge is responsible for the bandgap oscillation with ribbon width [89].

The electronic properties of the zigzag edge and armchair edge SiC nanoribbons were studied using DFT–GGA calculations [90]. The zigzag nanoribbon is metallic except for the thinner ribbons (2–4 chains) with small direct bandgaps. The direct bandgaps of armchair nanoribbons exhibit sawtooth-like periodic oscillation features and approach 2.359 eV when the width increases. The projected density of states shows a sharp peak at the Fermi level for the broader zigzag nanoribbon, which arises from the edge C and Si atoms with H terminations. The valence charges are accumulated at around C atoms, indicating electron transfer from Si atoms to C atoms. The Si–H bond is ionic, and the C–H bond is covalent. The dangling bonds yield one (two) flat extra band at the Fermi level for the zigzag nanoribbon with either bare C or bare Si edge; for armchair nanoribbons with bare C and Si edges,

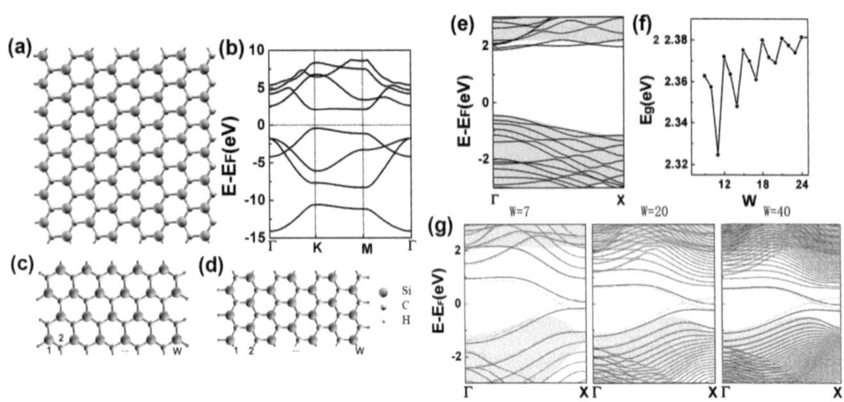

Fig. 6.37 **a** Geometry and **b** electronic band structure of the two-dimensional SiC sheet. Geometric structures of **c** armchair and **d** zigzag SiC nanoribbons with a width *W*. The nanoribbon width *W* is defined as the number of dimer lines (lines along the ribbon direction) for an armchair nanoribbon and the number of zigzag chains for a zigzag nanoribbon. **e** Band structure of armchair SiC nanoribbon with $W = 11$. Shaded areas show the projected band structure of a two-dimensional graphitic SiC sheet. **f** Bandgap of armchair SiC nanoribbons as a function of ribbon width *W*. **g** Spin-polarized band structures of zigzag SiC nanoribbons with $W = 7$, 20, and 40. Dotted red lines are spin-up bands, and solid blue lines are spin-down bands. Shaded areas show the projected band structure of a two-dimensional graphitic SiC sheet. Reprinted with permission from [88]. Copyright 2008, American Institute of Physics

two nearly flat extra bands arise up and below the Fermi level. The tight-binding model was applied to study the electronic structures of SiC and BN nanoribbons [89]. The edge effects can be represented by the modification of on-site energies and hopping integrals of the atoms on two edges. Considering the edge effects, this model reproduces the bandgaps and electronic band dispersion of the highest valence band and lowest conduction band. The charge redistribution at the edges causes bandgap oscillation of the SiC and BN nanoribbons. The modification of on-site energies and hopping integrals correlates with the charge redistribution of p electrons and can be achieved by decorating the nanoribbon edges with functional groups. DFT–HSE calculations show that the asymmetric hydrogen termination breaks the magnetic degeneracy and favors ferromagnetic state in the zigzag SiC nanoribbons [91]. The nanoribbons are transformed to half-metals by the bare Si edge, whereas the saturated Si/C edge brings quasi-half-metallic character, for which two spin-polarized bands intersect at the Fermi level with tiny bandgaps. The gap opening depends on the molecular origin of bands and ribbon width. All the wide zigzag SiC nanoribbons with asymmetric hydrogen termination are converted to half-metals. A GW–BSE calculation of unpassivated and hydrogen-passivated armchair SiC nanoribbons indicates that the quasiparticle bandgaps are up to 2 eV bigger than their Kohn–Sham calculation counterparts [92]. Inclusion of the electron–hole interaction modifies the absorption spectra, leading to strongly bound excitons. The GW–BSE calculation predicts an exciton binding energy of 1.07–1.93 eV. The analysis of the formation energies suggests that hydrogen-saturated ribbons are much more stable compared to bare ones. Another GW–BSE calculation reveals electronic and optical properties of nanoribbons with widths of 0.6–2.2 nm [93]. Zigzag SiC nanoribbons with widths larger than 1 nm exhibit half-metallicity at the mean-field level. The self-energy corrections bring substantial increase in bandgaps, converting the half-metallic zigzag SiC nanoribbons into narrow gap spin-polarized semiconductors. The narrowest nanoribbon exhibits strongly bound excitons, with a binding energy of 2.1 eV. The complex interplay of van der Waals, Coulomb, and direct edge-to-edge covalent interactions plays a key role in determining the stability of bilayer zigzag SiC nanoribbons [94]. The Si–Si edge of the homo-AA-stacked zigzag SiC nanoribbon shows a unique bond angle, indicative of sp^3-like covalent bonds. The C–C edge exhibits a flat structure, suggesting formation of the pseudochemical bonding between p_z orbitals at the edge. The homo-AA-stacked nanoribbon is nonmagnetic and metallic, although the monolayer nanoribbon shows magnetic features owing to spin-polarized p_z orbitals at the edge Si and C atoms. The hetero-AA-stacked nanoribbon is more stable than the homo one and has a finite bandgap.

A DFT–PBE calculation indicates that N doping at the carbon edge of SiC nanoribbons can selectively close the edge state channel and achieve 100% spin filtering, suggesting promise of SiC nanoribbons for achieving stable and completely spin-polarized transport [95]. Calculations predict perfect half-metal quantum transport in SiC nanoribbons [96]. When two zigzag SiC nanoribbons are connected by a C–Si–C–Si tetramer where the right ribbon is turned by 180° around the transport direction, 100% spin polarization with nearly perfect transmission is realized due to perfect momentum matching across the transport junction. Such a transport is independent of the magnetic configurations of the two nanoribbons. Quantum transport and spin current in a zigzag SiC nanoribbon device under a thermal gradient were theoretically investigated within the framework of the Landauer–Buttiker formalism using first-principles calculations [97]. Specific edge doping can turn off or keep open of the edge state transport channels. Different spin channels can be separately controlled. By substituting a B atom for an edge C atom and a P atom for an edge Si atom in the scattering region, a Seebeck thermopower with different signs for different spins and a finite conductance for both spins are achievable in the linear response regime. The thermoelectric field drives electrons of different spin channels in opposite directions, generating a spin current. Pure spin current is achievable through tuning the chemical potential and working temperature. Spin-related current generated by the photogalvanic effect in monolayer zigzag SiC nanoribbons was investigated using nonequilibrium Green's function combined with DFT [98]. 100% spin-polarized photocurrent is attainable in a wide range of photon energies by shining linearly/circularly polarized light when SiC nanoribbons are in the antiferromagnetic state. When they are in the ferromagnetic state, the spin polarization of photocurrent changes from 0 to 100% by altering the photon energy or polarization angle. Changing the circular polarization angle in the ferromagnetic state yields pure spin current.

The strain dependence of phonon transport in the monolayer SiC was investigated by solving the linearized phonon Boltzmann equation [99]. The room-temperature lattice thermal conductivity of the monolayer SiC is two orders of magnitude smaller than that of graphene owing to small group velocities and short phonon lifetimes. The SiC monolayer is mechanically and dynamically stable. At low strain (< 8%), the phonon lifetime enhancement causes increase in lattice thermal conductivity; at high strain (> 8%), the reduction of group velocities and the decrease in phonon lifetimes account for diminishing thermal conductivity. Reverse nonequilibrium molecular dynamics simulations reveal thermal properties of 2D SiC [100]. 2D SiC shows a thermal conductivity of ~ 313 W mK^{-1} at room temperature, one order bigger than that of silicene. The thermal conductivity deviates from the normal $1/T$ law above room temperature and shows an anomalous slowly decreasing behavior. The shrinking and softening of the peaks of the phonon density of states at high temperatures cause this anomaly. Molecular dynamics simulations show a highly anisotropic mechanical behavior of pristine and vacancy defected 2D SiC at 300–700 K and for all considered (point and bi-vacancies) defect conditions [101]. At 300 K, a fracture stress and an elastic modulus of 71.02 and 637.26 GPa are achievable along the armchair direction, which are ~ 24.42 and ~ 14.38% higher compared to those along the zigzag direction. In both armchair and zigzag directions, there is decrease in fracture stress, fracture strain, and elastic modulus with increasing temperature and defect concentration.

Catalytic activity of 2H-SiC monolayers and thin films toward CO oxidation was investigated using first-principles calculations [102]. 2H-SiC monolayer is planar and quite stable. It has moderate binding with O_2, while CO interacts physically. The Eley–Rideal mechanism prevails over the Langmuir–Hinshelwood mechanism.

2H-SiC thin film exhibits a nonplanar structure, allowing for moderate binding of both CO and O_2 on its surface, favoring the Langmuir–Hinshelwood mechanism. The adsorption of various gaseous molecules (NO, NO_2, NH_3, SO_2, CO, HCN, O_2, H_2, N_2, CO_2, and H_2O) on the graphitic SiC monolayer and bilayer has been investigated by using DFT calculations [103]. The SiC monolayer strongly adsorbs NO_2 and SO_2. The nitride gases (NO, NO_2, and NH_3) are chemisorbed on the SiC bilayer with moderate adsorption energies and apparent charge transfer; the other molecules are physically adsorbed. The bilayer only weakly adsorbs NO_2 and SO_2. The adsorption of NH_3 molecules dramatically changes the work functions of the SiC monolayer and bilayer.

Van der Waals coupling is powerful toward engineering atomically thin semiconductor materials with unique properties, bringing new hopes for catalysis and solar energy conversion. SiC can form heterostructures with various other materials. First-principles calculations give information about the SiC–graphene, SiC–MoS_2, and graphene–SiC–MoS_2 vdW heterostructures [104]. The SiC–graphene interface forms a *p*-type Schottky barrier, and an applied electric field can convert it into *n*-type Schottky contact. A conversion from a Schottky to an Ohmic contact at the interface is possible by changing the interlayer distance or applying an electric field. The SiC–MoS_2 interface is a type-II heterostructure, and a biaxial strain changes it into ion type-III. It has excellent optical absorption ability in the visible and far-infrared region. 2D SiC–MoS_2 van der Waals heterostructures have been investigated using DFT–HSE calculations [105]. The SiC monolayer–MoS_2 monolayer shows a narrower bandgap compared to corresponding pure materials with an energy gap of 1.46 and 1.36 eV. Electron transition mainly occurs between different constituents for these vdW heterostructures, facilitating effective separation of photogenerated carriers.

Photocatalytic properties of P, BSe, and SiC monolayers and their van der Waals heterostructures have been investigated by using DFT–HSE06 calculations [106]. The P–BSe (BSe–SiC) heterostructure behaves like an indirect type-II (type-I) bandgap semiconductor. The tensile strain of 10% (2%) converts P–BSe (BSe–SiC) into type-I (type-II) direct-bandgap nature. The P–SiC heterostructure behaves like a direct bandgap (type-II) semiconductor. The electron and hole carrier mobilities of these heterostructures lie in the range of 1.2×10^4–68.56×10^4 cm^2 V^{-1} s^{-1}.

The DFT–HSE06 calculation predicts a stable phase of 2D SiC in the popular biphenylene network, called SiC biphenylene [107]. This structure has a high melting point of 3475 K and a direct bandgap of 2.16 eV. Its quasiparticle bandgap is 2.89 eV at the G_0W_0 level. The G_0W_0–BSE calculation shows an optically active exciton peak at 2.07 eV, with a binding energy of 0.82 eV. The effective mass and Bohr radius of the exciton are 1.01 m_0 and 2.14 Å. Another calculation predicts a series of 2D SiC materials with a tetragonal lattice [108]. The band structure of SiC with tetragonal carbon rings and silicon rings exhibits Dirac cones. The Dirac cone of tetragonal SiC stems from "ring coupling". This mechanism refers to mutual coupling between four carbon atoms in the tetragonal C ring and the coupling in the tetragonal Si ring.

References

1. Pham-Huu C, Keller N, Ehret G, Ledoux MJ (2001) The first preparation of silicon carbide nanotubes by shape memory synthesis and their catalytic potential. J Catal 200:400–410
2. Keller N, Pham-Huu C, Ehret G, Keller V, Ledoux MJ (2003) Synthesis and characterisation of medium surface area silicon carbide nanotubes. Carbon 41:2131–2139
3. Nhut J-M, Vieira R, Pesant L, Tessonnier J-P, Keller N, Ehret G, Pham-Huu C, Ledoux MJ (2002) Synthesis and catalytic uses of carbon and silicon carbide nanostructures. Catal Today 76:11–32
4. Sun X-H, Li C-P, Wong W-K, Wong N-B, Lee C-S, Lee S-T, Teo B-K (2002) Formation of silicon carbide nanotubes and nanowires via reaction of silicon (from disproportionation of silicon monoxide) with carbon nanotubes. J Am Chem Soc 124:14464–14471
5. Taguchiw T, Igawa N, Yamamoto H, Jitsukawa S (2005) Synthesis of silicon carbide nanotubes. J Am Ceram Soc 88:459–461
6. Taguchi T, Igawa N, Yamamoto H, Shamoto S, Jitsukawa S (2005) Preparation and characterization of single-phase SiC nanotubes and C-SiC coaxial nanotubes. Phys E 28:431–438
7. Borowiak-Palen E, Ruemmeli MH, Gemming T, Knupfer M, Biedermann K, Leonhardt A, Pichler T (2005) Bulk synthesis of carbon-filled silicon carbide nanotubes with a narrow diameter distribution. J Appl Phys 97:056102
8. Li BS, Wu RB, Pan Y, Wu LL, Yang GY, Chen JJ, Zhu QM (2008) J Alloy Compd 462:446–451
9. Tony VCS, Voon CH, Lee CC, Lim BY, Gopinath SCB, Foo KL, Arshad MKM, Ruslinda AR, Hashim U, Nashaain MN, Al-Douri Y (2017) Effective synthesis of silicon carbide nanotubes by microwave heating of blended silicon dioxide and multi-walled carbon nanotube. Mater Res 20:1658–1668
10. Taguchi T, Tsubakiyama R, Miyajima K, Yamamoto S, Ohba H (2017) Effect of surface treatment on photoluminescence of silicon carbide nanotubes. Appl Surf Sci 403:308–313
11. Taguchi T, Yamamoto S, Ohba H (2018) Ion irradiation-induced novel microstructural change in silicon carbide nanotubes. Acta Mater 154:90–99
12. Hu JQ, Bando Y, Zhan JH, Golberg D (2004) Fabrication of ZnS/SiC nanocables, SiC-shelled ZnS nanoribbons (and sheets), and SiC nanotubes (and tubes). Appl Phys Lett 85:2932–2934
13. Zhou J, Liu J, Yang R, Lao C, Gao P, Tummala R, Xu NS, Wang ZL (2006) SiC-shell nanostructures fabricated by replicating ZnO nano-objects: a technique for producing hollow nanostructures of desired shape. Small 2:1344–1347
14. Zhou J, Zhou M, Chen Z, Zhang Z, Chen C, Li R, Gao X, Xie E (2009) SiC nanotubes arrays fabricated by sputtering using electrospun PVP nanofiber as templates. Surf Coat Technol 203:3219–3223
15. Drínek V, Šubrt J, Klementová M, Rieder M, Fajgar R (2009) From shelled Ge nanowires to SiC nanotubes. Nanotechnology 20:035606
16. Latu-Romain L, Ollivier M, Mantoux A, Auvert G, Chaix-Pluchery O, Sarigiannidou E, Bano E, Pelissier B, Roukoss C, Roussel H, Dhalluin F, Salem B, Jegenyes N, Ferro G, Chaussende D, Baron T (2011) From Si nanowire to SiC nanotube. J Nanopart Res 13:5425–5433
17. Wang H, Li X-D, Kim T-S, Kim D-P (2005) Inorganic polymer-derived tubular SiC arrays from sacrificial alumina templates. Appl Phys Lett 86:173104
18. Cheng Q-M, Interrante L, Lienhard M, Shen Q, Wu Z (2005) Methylene-bridged carbosilanes and polycarbosilanes as precursors to silicon carbide-from ceramic composites to SiC nanomaterials. J Eur Ceram Soc 25:233–241
19. Yang Z, Xia Y, Mokaya R (2004) High surface area silicon carbide whiskers and nanotubes nanocast using mesoporous silica. Chem Mater 16:3877–3884
20. Huang GS, Mei YF, Cavallo F, Baunack S, Coric E, Gemming T, Bertram F, Christen J, Fu RKY, Chu PK, Schmidt OG (2009) Fabrication and optical properties of C/β-SiC/Si hybrid rolled-up microtubes. J Appl Phys 105:016103
21. Pei LZ, Tang YH, Chen YW, Guo C, Li XX, Yuan Y, Zhang Y (2006) Preparation of silicon carbide nanotubes by hydrothermal method. J Appl Phys 99:114306

22. Xie ZF, Tao DL, Wang JQ (2007) Synthesis of silicon carbide nanotubes by chemical vapor deposition. J Nanosci Nanotechnol 7:647–652
23. Cui H, Sun Y, Yang GZ, Chen J, Jiang D, Wang CX (2009) Template- and catalyst-free synthesis, growth mechanism and excellent field emission properties of large scale single-crystalline tubular β-SiC. Chem Commun 6243–6245
24. Miyamoto Y, Yu BD (2002) Computational designing of graphitic silicon carbide and its tubular forms. Appl Phys Lett 80:586–588
25. Menon M, Richter E, Andriotis AN (2004) Structure and stability of SiC nanotubes. Phys Rev B 69:115322
26. Zhao M, Xia Y, Li F, Zhang RQ, Lee S-T (2005) Strain energy and electronic structures of silicon carbide nanotubes: density functional calculations. Phys Rev B 71:085312
27. Zhao M, Xia Y, Zhang RQ, Lee S-T (2005) Manipulating the electronic structures of silicon carbide nanotubes by selected hydrogenation. J Chem Phys 122:214707
28. Li F, Xi Y-Y, Zhao M-W, Liu X-D, Huang B-D, Yang Z-H, Ji Y-J, Song C (2005) Density-functional theory calculations of XH_3-decorated SiC nanotubes (X={C, Si}): structures, energetics, and electronic structures. J Appl Phys 97:104311
29. Baumeier B, Krüger P, Pollmann J (2007) Structural, elastic, and electronic properties of SiC, BN, and BeO nanotubes. Phys Rev B 76:085407
30. Alam KM, Ray AK (2008) Hybrid density functional study of armchair SiC nanotubes. Phys Rev B 77:035436
31. Alam KM, Ray AK (2007) A hybrid density functional study of zigzag SiC nanotubes. Nanotechnology 18:495706
32. Alfieri G, Kimoto T (2009) The structural and electronic properties of chiral SiC nanotubes: a hybrid density functional study. Nanotechnology 20:285703
33. Yu M, Jayanthi CS, Wu SY (2010) Geometric and electronic structures of graphitic-like and tubular silicon carbides: ab-initio studies. Phys Rev B 82:075407
34. Adhikari K, Ray AK (2011) Carbon- and silicon-capped silicon carbide nanotubes: an ab initio study. Phys Lett A 375:1817–1823
35. Hsueh HC, Guo GY, Louie SG (2011) Excitonic effects in the optical properties of a SiC sheet and nanotubes. Phys Rev B 84:085404
36. Mercan K, Civalek Ö (2017) Buckling analysis of silicon carbide nanotubes (SiCNTs) with surface effect and nonlocal elasticity using the method of HDQ. Compos B 114:34–45
37. Mahendiran D, Murugan P (2023) First-principles study of the formation of single-walled SiC nanotubes from bilayered SiC ribbons. Phys Rev B 107:205402
38. Gali A (2006) Ab initio study of nitrogen and boron substitutional impurities in single-wall SiC nanotubes. Phys Rev B 73:245415
39. Yang YT, Ding RX, Song JX (2011) Transport properties of boron-doped single-walled silicon carbide nanotubes. Phys B 406:216–219
40. Yang YY, Gong P, Ma WD, Hao R, Fang XY (2021) Effects of substitution of group V atoms for carbon or silicon atoms on optical properties of silicon carbide nanotubes. Chin Phys B 30:067803
41. Tabtimsai C, Ruangpornvisuti V, Tontapha S, Wanno B (2018) A DFT investigation on group 8B transition metal-doped silicon carbide nanotubes for hydrogen storage application. Appl Surf Sci 439:494–505
42. Szabó Á, Gali A (2009) Effect of oxygen on single-wall silicon carbide nanotubes studied by first-principles calculations. Phys Rev B 80:075425
43. Baierle RJ, Piquini P, Neves LP, Miwa RH (2006) Ab initio study of native defects in SiC nanotubes. Phys Rev B 74:155425
44. Zhang Y, Huang H (2008) Stability of single-wall silicon carbide nanotubes—molecular dynamics simulations. Comput Mater Sci 43:664–669
45. Wang Z, Zu X, Xiao H, Gao F, Weber WJ (2008) Tuning the band structures of single walled silicon carbide nanotubes with uniaxial strain: a first principles study. Appl Phys Lett 92:183116

46. Wu IJ, Guo GY (2007) Optical properties of SiC nanotubes: an *ab initio* study. Phys Rev B 76:035343
47. Huang SP, Wu DS, Hu JM, Zhang H, Xie Z, Hu H, Cheng WD (2007) First-principles study: size-dependent optical properties for semiconducting silicon carbide nanotubes. Opt Express 15:10947–10957
48. Mavrandonakis A, Froudakis GE, Andriotis A, Menon M (2006) Silicon carbide nanotube tips: promising materials for atomic force microscopy and/or scanning tunneling microscopy. Appl Phys Lett 89:123126
49. Mpourmpakis G, Froudakis GE, Lithoxoos GP, Samios J (2006) SiC nanotubes: a novel material for hydrogen storage. Nano Lett 6:1581–1583
50. Gali A (2007) Ab initio theoretical study of hydrogen and its interaction with boron acceptors and nitrogen donors in single-wall silicon carbide nanotubes. Phys Rev B 75:085416
51. Malek K, Sahimi M (2010) Molecular dynamics simulations of adsorption and diffusion of gases in silicon-carbide nanotubes. J Chem Phys 132:014310
52. Wang X, Liew KM (2011) Hydrogen storage in silicon carbide nanotubes by lithium doping. J Phys Chem C 115:3491–3496
53. Barghi SH, Tsotsis TT, Sahimi M (2014) Hydrogen sorption hysteresis and superior storage capacity of silicon-carbide nanotubes over their carbon counterparts. Int J Hydrogen Energy 39:21107–21115
54. Wu RQ, Yang M, Lu YH, Feng YP, Huang ZG, Wu QY (2008) Silicon carbide nanotubes as potential gas sensors for CO and HCN detection. J Phys Chem C 112:15985–15988
55. Gao G, Kang HS (2008) First principles study of NO and NNO chemisorption on silicon carbide nanotubes and other nanotubes. J Chem Theory Comput 4:1690–1697
56. Zhao JX, Ding YH (2009) Can silicon carbide nanotubes sense carbon dioxide? J Chem Theory Comput 5:1099–1105
57. Wang X, Liew KM (2011) Silicon carbide nanotubes serving as a highly sensitive gas chemical sensor for formaldehyde. J Phys Chem C 115:10388–10393
58. Zhao J-X, Xiao B, Ding Y-H (2009) Theoretical prediction of the N–H and O–H bonds cleavage catalyzed by the single-walled silicon carbide nanotube. J Phys Chem C 113:16736–16740
59. Soltani A, Peyghan AA, Bagheri Z (2013) H_2O_2 adsorption on the BN and SiC nanotubes: a DFT study. Phys E 48:176–180
60. Zhao J-X, Ding Y-H (2008) Silicon carbide nanotubes functionalized by transition metal atoms: a density-functional study. J Phys Chem C 112:2558–2564
61. Lin SS (2012) Light-emitting two-dimensional ultrathin silicon carbide. J Phys Chem C 116:3951–3955
62. Lin SS, Zhang SJ, Li XQ, Xu WL, Pi XD, Liu XY, Wang FC, Wu HA, Chen HS (2015) Quasi-two-dimensional SiC and SiC_2: interaction of silicon and carbon at atomic thin lattice plane. J Phys Chem C 119:19772–19779
63. Chabi S, Chang H, Xia YD, Zhu YQ (2016) From graphene to silicon carbide: ultrathin silicon carbide flakes. Nanotechnology 27:075602
64. Susi T, Skákalová V, Mittelberger A, Kotrusz P, Hulman M, Pennycook TJ, Mangler C, Kotakoski J, Meyer JC (2017) Computational insights and the observation of SiC nanograin assembly: towards 2D silicon carbide. Sci Rep 7:4399
65. Sun L, Wang B, Wang YD (2018) A novel silicon carbide nanosheet for high-performance humidity sensor. Adv Mater Interfaces 5:1701300
66. Han C, Lei YP, Wang B, Wu CZ, Zhang XS, Shen SJ, Sun L, Tian Q, Feng QG, Wang YD (2019) The functionality of surface hydroxyls on selective CH_4 generation from photoreduction of CO_2 over SiC nanosheets. Chem Commun 55:1572
67. Cao Y, Dong HF, Pu ST, Zhang XJ (2018) Photoluminescent two-dimensional SiC quantum dots for cellular imaging and transport. Nano Res 11:4074–4081
68. Sun L, Han C, Wu N, Wang B, Wang YD (2018) High temperature gas sensing performances of silicon carbide nanosheets with an n-p conductivity transition. RSC Adv 8:13697

69. Dragomir M, Valant M, Fanetti M, Mozharivskyj Y (2016) A facile chemical method for the synthesis of 3C-SiC nanoflakes. RSC Adv 6:21795
70. Thomas SA, Alharthi NS, Petersen RJ, Aldrees A, Tani S, Anderson KJ, Granlie J, Pringle TA, Payne SA, Choi Y (2024) Colloidal 2D layered SiC quantum dots from a liquid precursor: surface passivation, bright photoluminescence, and planar self-assembly. ACS Nano 18:26848–26857
71. Chabi S, Guler Z, Brearley AJ, Benavidez AD, Luk TS (2021) The creation of true two-dimensional silicon carbide. Nanomaterials 11:1799
72. Polley CM, Fedderwitz H, Balasubramanian T, Zakharov AA, Yakimova R, Bäcke O, Ekman J, Dash SP, Kubatkin S, Lara-Avila S (2023) Bottom-up growth of monolayer honeycomb SiC. Phys Rev Lett 130:076203
73. Lin X, Lin SS, Xu Y, Hakro AA, Hasan T, Zhang BL, Yu B, Luo JK, Li E, Chen HS (2013) Ab initio study of electronic and optical behavior of two-dimensional silicon carbide. J Mater Chem C 1:2131
74. Lin X, Lin SS, Xu Y, Chen HS (2015) Electronic structures of multilayer two-dimensional silicon carbide with oriented misalignment. J Mater Chem C 3:9057
75. Shi ZM, Zhang ZH, Kutana A, Yakobson BI (2015) Predicting two-dimensional silicon carbide monolayers. ACS Nano 9:9802–9809
76. Naqvi SR, Hussain T, Luo W, Ahuja R (2018) Metallized siligraphene nanosheets (SiC_7) as high capacity hydrogen storage materials. Nano Res 11:3802–3813
77. Lu XK, Xin TY, Zhang Q, Xu Q, Wei TH, Wang YX (2018) Versatile mechanical properties of novel g-SiC_x monolayers from graphene to silicene: a first-principles study. Nanotechnology 29:315701
78. Tagani MB (2023) Si_9C_{15} monolayer: a silicon carbide allotrope with remarkable physical properties. Phys Rev B 107:085114
79. Guilhon I, Teles LK, Marques M, Pela RR, Bechstedt F (2015) Influence of structure and thermodynamic stability on electronic properties of two-dimensional SiC, SiGe, and GeC alloys. Phys Rev B 92:075435
80. Yaghoubi A, Masenelli-Varlot K, Boisron O, Ramesh S, Melinon P (2018) Is graphitic silicon carbide (silagraphene) stable? Chem Mater 30:7234–7244
81. Liu WK, Kong SS, Yu XX, Li YL, Yang LZ, Ma Y, Fang XY (2023) Interlayer coupling, electronic and optical properties of few-layer silicon carbide nanosheets. Mater Today Commun 34:105030
82. Bekaroglu E, Topsakal M, Cahangirov S, Ciraci S (2010) First-principles study of defects and adatoms in silicon carbide honeycomb structures. Phys Rev B 81:075433
83. Islam ASMJ, Islam MS, Ferdous N, Park J, Hashimoto A (2020) Vacancy-induced thermal transport in two-dimensional silicon carbide: a reverse non-equilibrium molecular dynamics study. Phys Chem Chem Phys 22:13592
84. Hassanzada Q, Sarsari IA, Hashemi A, Ghojavand A, Gali A, Abdi M (2020) Theoretical study of quantum emitters in two-dimensional silicon carbide monolayers. Phys Rev B 102:134103
85. Mohseni M, Sarsari IA, Karbasizadeh S, Udvarhelyi P, Hassanzada Q, Ala-Nissila T, Gali A (2024) Vacancy-related color centers in two-dimensional silicon carbide monolayers. Phys Rev Mater 8:056201
86. Singh A, Mahamiya V, Shukla A (2023) Defect-driven tunable electronic and optical properties of two-dimensional silicon carbide. Phys Rev B 108:235311
87. Morais WP, Inacio GJ, Amorim RG, Paz WS, Pansini FNN, de Souza FAL (2023) Topological line defects in hexagonal SiC monolayer. Phys Chem Chem Phys 25:33048
88. Sun L, Li Y, Li Z, Li Q, Zhou Z, Chen Z, Yang J, Hou JG (2008) Electronic structures of SiC nanoribbons. J Chem Phys 129:174114
89. Zhao K, Zhao M, Wang Z, Fan Y (2010) Tight-binding model for the electronic structures of SiC and BN nanoribbons. Phys E 43:440–445
90. Zhang JM, Zheng FL, Zhang Y, Ji V (2010) First-principles study on electronic properties of SiC nanoribbon. J Mater Sci 45:3259–3265

91. Ding Y, Wang YL (2012) Electronic structures of zigzag SiC nanoribbons with asymmetric hydrogen-terminations. Appl Phys Lett 101:013102
92. Alaal N, Loganathan V, Medhekar N, Shukla A (2016) First principles many-body calculations of electronic structure and optical properties of SiC nanoribbons. J Phys D Appl Phys 49:105306
93. Alaal N, Loganathan V, Medhekar N, Shukla A (2017) From half-metal to semiconductor: electron-correlation effects in zigzag SiC nanoribbons from first principles. Phys Rev Appl 7:064009
94. Sharifi JA, Sun RY, Nakamura J (2024) First-principles study on edge-edge interactions of bilayer zigzag SiC nanoribbons. Jpn J Appl Phys 63:055001
95. Tao XX, Hao H, Wang XL, Zheng XH, Zeng Z (2016) Realizing stable fully spin polarized transport in SiC nanoribbons with dopant. Appl Phys Lett 108:233106
96. Zheng XH, Chen XB, Zhang L, Xiao LT, Jia ST, Zeng Z, Guo H (2017) Perfect spin and valley polarized quantum transport in twisted SiC nanoribbons. 2D Mater 4:025013
97. Jiang P, Tao XX, Hao H, Song LL, Zheng XH, Zhang L, Zeng Z (2017) Tuning a zigzag SiC nanoribbon as a thermal spin current generator. 2D Mater 4:035001
98. Chen J, Zhang LW, Zhang L, Zheng XH, Xiao LT, Jia ST, Wang J (2018) Photogalvanic effect induced fully spin polarized current and pure spin current in zigzag SiC nanoribbons. Phys Chem Chem Phys 20:26744
99. Guo SD, Dong J, Liu JT (2018) Nonmonotonic strain dependence of lattice thermal conductivity in monolayer SiC: a first-principles study. Phys Chem Chem Phys 20:22038
100. Islam ASMJ, Islam MS, Ferdousi N, Park J, Bhuiyan AG, Hashimoto A (2019) Anomalous temperature dependent thermal conductivity of two-dimensional silicon carbide. Nanotechnology 30:445707
101. Islam ASMJ, Islam S, Ferdous N, Park J, Bhuiyan AG, Hashimoto A (2019) Anisotropic mechanical behavior of two dimensional silicon carbide: effect of temperature and vacancy defects. Mater Res Express 6:125073
102. Sinthika S, Vala ST, Kawazoe Y, Thapa R (2016) CO oxidation prefers the Eley–Rideal or Langmuir–Hinshelwood pathway: monolayer vs thin film of SiC. ACS Appl Mater Interfaces 8:5290–5299
103. Zhao ZJ, Yong YL, Zhou QX, Kuang YM, Li XH (2020) Gas-sensing properties of the SiC monolayer and bilayer: a density functional theory study. ACS Omega 5:12364–12373
104. Li S, Sun ML, Chou JP, Wei J, Xing HZ, Hu A (2018) First-principles calculations of the electronic properties of SiC-based bilayer and trilayer heterostructures. Phys Chem Chem Phys 20:24726
105. Li Q, Xu L, Luo KW, Wang LL, Li XF (2018) SiC/MoS_2 layered heterostructures: promising photocatalysts revealed by a first-principles study. Mater Chem Phys 216:64–71
106. Idrees M, Din HU, Khan SA, Ahmad I, Gan LY, Nguyen CV, Amin B (2019) Van der Waals heterostructures of P, BSe, and SiC monolayers. J Appl Phys 125:094301
107. Singh A, Mahamiya V, Shukla A (2024) Characterization of silicon carbide biphenylene network through G_0W_0-BSE calculations. Phys Rev B 110:195305
108. Kong WX, Xiao XL, Xu WP, Wang R, Gan LY, Wei J, Fan J, Wu XZ (2021) The Dirac cone in two-dimensional tetragonal silicon carbides: a ring coupling mechanism. Nanoscale 13:18267

Chapter 7
SiC Nanostructured Films

Silicon carbide films have been intensively investigated concerning their fabrications and properties. In general, the thicknesses of the fabricated silicon carbide films can range from many micrometers to many nanometers. In this chapter, however, we shall focus on the nanostructured SiC films including the films with thicknesses on the nanoscale and the films comprising SiC nanocrystals. We shall summarize the fabrication, structures, and properties of these nanostructured SiC films.

7.1 Synthesis by Chemical Vapor Deposition

Chemical vapor deposition is an important method for the preparation of the silicon carbide films. Internal stress in 3C-SiC epilayers grown on Si substrates by using CVD using SiH_4, C_3H_8, and H_2 as starting materials at 1350 °C has been investigated by using Raman scattering [1]. Raman scattering from the 3C-SiC epitaxial layers grown on Si(001) and Si(111) substrates shows that the LO phonon frequencies shift to the lower-frequency side compared with stress-free 3C-SiC films. The frequency shift gives an estimate of the tensile stress of the 3C-SiC epilayer on Si(001) to be 5.4×10^9 dyn cm^{-2}. The stress difference between the 3C-SiC films on Si(001) and on Si(111) is smaller than that expected from the elastic deformation theory. A set of formulas have been derived [2, 3] to describe the Raman shifts of diamond and zinc-blende semiconductors under axial stress. The Raman shifts under hydrostatic pressure and uniaxial and biaxial stresses only act as special cases of the general formulas. Raman scattering measurements from a series of CVD 3C-SiC films on (100) Si with thickness between 60 nm and 17 μm (Fig. 7.1) show that the Raman spectra in samples with SiC thicknesses greater than 4 μm comprise a sharp and strong feature that obeys the selection rule for the 3C-SiC LO(Γ) phonon line. The stresses in 3C-SiC extracted from these data are between 0.3 and 1.1 GPa. 3C-SiC and Si have a large lattice mismatch of 20%, but the difference in Raman shift between

J. Fan and P. K. Chu, *Silicon Carbide Nanostructures*, Engineering Materials and Processes, https://doi.org/10.1007/978-3-032-10986-6_7

3C-SiC/Si and free 3C-SiC films is smaller than 2 cm^{-1}, corresponding to strain of 0.1–0.2%. The Raman signals from the SiC film and the Si substrate exhibited identical polarization behavior, confirming their same crystalline orientations. The 3C-SiC LO(Γ) phonon signal was enhanced by a factor of 2 or 3 after removal of the Si substrate owing to multiple reflections in the free 3C-SiC films.

Photoluminescence at 4.2 K in as-grown and ion-implanted 3C-SiC films grown by CVD on Si(100) substrates has been investigated [4]. The $D1$ and $D2$ defect PL bands previously observed in ion-implanted Lely-grown SiC were observed in the as-grown un-implanted films (Fig. 7.2), and they possess higher defect concentrations than the vapor transport Lely-grown 3C-SiC crystals. The ion implantation in the SiC films caused significant enhancement in the intensities of the $D1$ and $D2$ defect bands. Their intensities in both as-grown and ion-implanted films increased after annealing at temperatures up to 1600 °C and above this temperature, the intensity of the $D2$ zero-phonon line decreased sharply, suggesting possible lattice recovery which quenched this defect emission band. The intensity of the $D1$ emission band remained nearly constant after annealing at temperatures above 1600 °C and up to 1800 °C. Similar to the Lely-grown SiC, the spectral details of these defects-related PL bands and their annealing behaviors in the SiC films are independent of the implanted-ion species.

The photoluminescence at 2 K in 26 3C-SiC films with thicknesses of 60 nm to 25 μm grown on (100) Si by CVD has been investigated [5]. The bound-exciton PL spectrum (Fig. 7.3a) for the 3C-SiC free film comprises the nitrogen-bound-exciton zero-phonon line N_0 along with its (one, two, and three) phonon replicas. It also

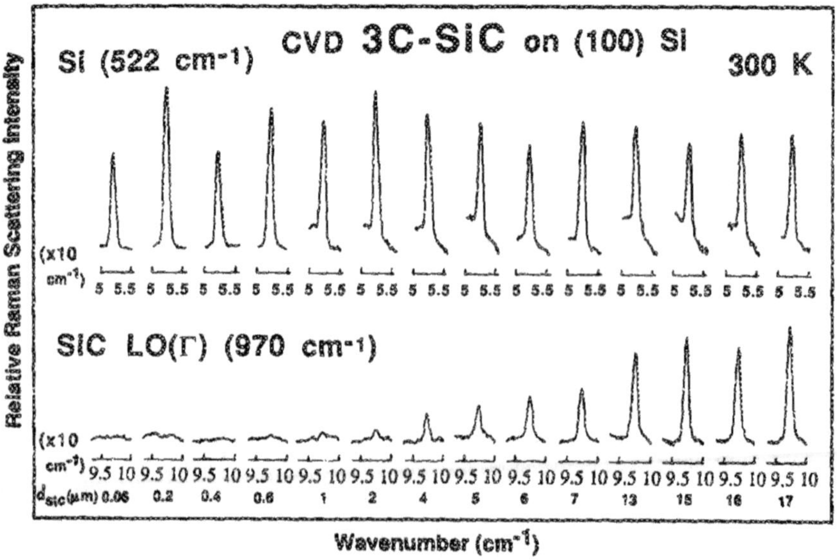

Fig. 7.1 Raman spectra of 14 CVD 3C-SiC/Si samples with the SiC film thickness varying from 60 nm to 17 μm. Reprinted with permission from [3]. Copyright 1988, American Institute of Physics

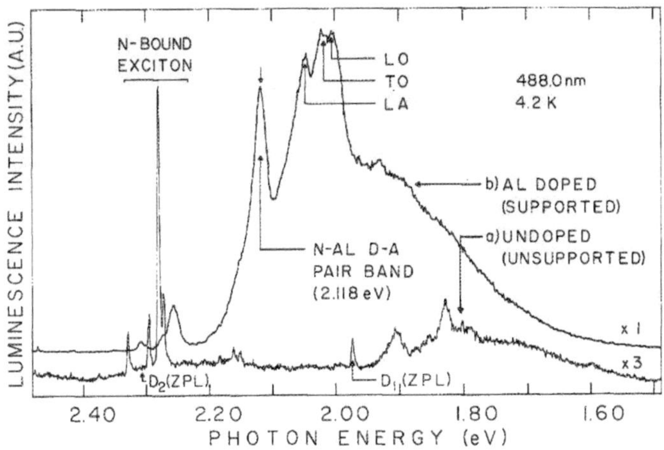

Fig. 7.2 PL spectra obtained from as-grown CVD 3C-SiC films. Spectrum (**a**) is from an undoped film that has been removed from its Si substrate. Spectrum (**b**) is from an Al-doped film on its Si substrate. Reprinted with permission from [4]. Copyright 1987, American Institute of Physics

contains a *B*-series bound-exciton line (B_0) and its phonon replicas that are related to some unidentified impurity. The B_0 line shifts to lower-energy side by 7 meV relative to the nitrogen-bound-exciton zero-phonon line. The bound-exciton lines have a full width at half maximum smaller than 2 meV. The PL spectra (Fig. 7.3b) obtained from the front and back sides of a 3C-SiC film with a thickness of 15 μm show much stronger bound-exciton lines in the near-surface region than the near-interface region. The D_I line is related to vacancies. The *G* band (near 1.90–1.92 eV) and its phonon replicas G_1 and G_2 bands as well as a background (below 1.95 eV) are all stronger in the interface layer than in the surface layer. This *G* band is associated with dislocations and extended defects. The very thin SiC films possess another defect-related *W* band near 2.15 eV.

Hydrogenated nanocrystalline silicon carbide films deposited on Si or Corning glass substrates have been prepared by using CVD using a mixture of methane, silane, and hydrogen as source gases [6]. Their infrared absorption spectrum comprises a 3C-SiC TO phonon absorption band at around 794 cm^{-1}. The Raman spectrum (Fig. 7.4a) consists of three main peaks at 785.3, 986.4, and 1606.5 cm^{-1} and a weak shoulder peak at 1390.2 cm^{-1}. The first two bands are assigned to TO and LO phonon modes of SiC and the peaks at 1606.5 and 1390.2 cm^{-1} are attributed to disordered graphite. The sample showed strong room-temperature photoluminescence centered at around 2.64 eV under 351 nm excitation, but the PL shifted to 2.2 eV under 325 nm excitation [7]. Figure 7.4b shows the PL and absorption spectra of two samples prepared under different source gas ratios [8], and the flow ratio of H_2:SiH_4:CH_4 is 100:10:3 for sample D and 100:5:1 for sample E. Their time-resolved PL spectra (Fig. 7.4c) can be fit by double-exponential functions with two decay time constants of 166 and 772 ps for sample D and 246 and 974 ps for sample E, respectively. These values are at least two orders of magnitude smaller than that of the bound-exciton transitions in

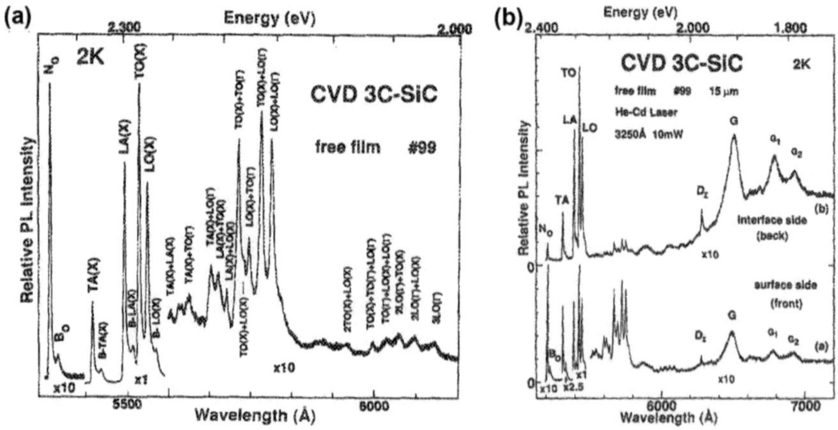

Fig. 7.3 **a** PL spectrum at 2 K showing the nitrogen-bound-exciton lines in a high-quality CVD free 3C-SiC film. **b** PL spectra at 2 K from both sides of a 15 μm CVD 3C-SiC film. Reprinted with permission from [5]. Copyright 1968, American Institute of Physics

bulk 3C-SiC at low temperatures. The double-exponential decay law suggests that multiple states were involved in the radiative recombination.

Hydrogenated amorphous silicon carbon films (a-SiC:H) prepared from a gas mixture of silane and acetylene by using catalytic CVD [9] have optical bandgaps that increase with increasing acetylene to silane ratio (Fig. 7.5). The bandgap reached 3.6 eV for a ratio over 0.8. The deposition rate decreased drastically with increasing acetylene fraction. These silicon carbide films have visible photoluminescence at room temperature.

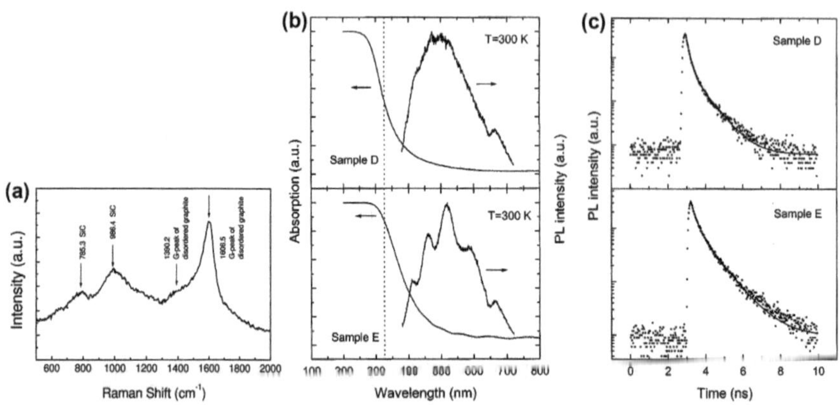

Fig. 7.4 **a** Raman spectrum of the CVD SiC nanocrystalline film [6]. **b** Room-temperature optical absorption and PL spectra (excited at 351 nm) and **c** time-resolved PL spectra (excited at 400 nm) of two CVD SiC films. The solid lines show the double-exponential fit to the experimental data. Reprinted with permission from [8]. Copyright 2000, American Institute of Physics

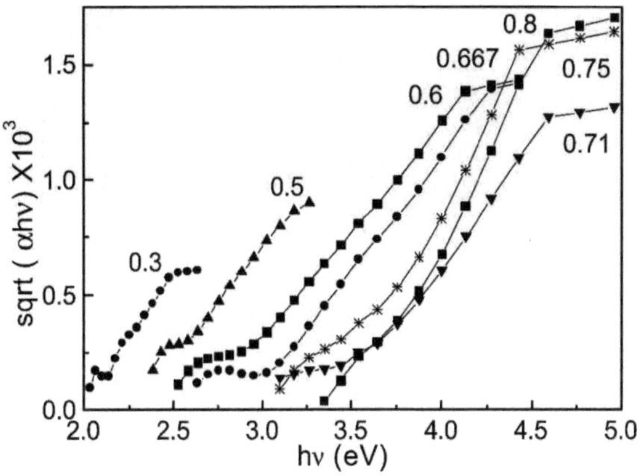

Fig. 7.5 Absorption (square root of absorbance) as a function of photon energy in a-SiC:H films that were prepared with different acetylene fractions. Reprinted with permission from [9]. Copyright 2001, Elsevier B.V.

Nanocrystalline SiC films deposited by using plasma-enhanced CVD at different gas flow ratios and at different deposition temperatures from 80 to 575 °C have been investigated [10]. Diethylsilane ($C_4H_{12}Si$) is the source gas, and a specific concentration of hydrogen, argon, or helium acts as the dilution gas. The infrared spectral analysis indicates that the SiC nanocrystals embedded in the amorphous film began to form at 300 °C. The crystallization degree increased with increasing temperature, and the crystalline fraction reached 65% at 575 °C. The hydrogen dilution gas caused the highest crystalline fraction. The high-resolution TEM measurement reveals the existence of the SiC nanocrystallites in the amorphous matrix, with their sizes ranging from 2 to 10 nm (Fig. 7.6). The refractive index of the film increased roughly linearly with temperature. The optical bandgap of the film can be derived from the absorption spectrum using $\alpha h\upsilon = A\left(h\upsilon - E_g\right)^2$, where α is the absorption coefficient, A is a constant, $h\upsilon$ is the photon energy, and E_g is the bandgap. Another characteristic value E_{04} is defined as the energy at which α equals 10^4 cm^{-1}. As shown in Fig. 7.6, both E_g and E_{04} increase monotonically with increasing deposition temperature. The bandgap varied from 2.6 to 4.47 eV as the temperature increased from 80 to 575 °C, showing a large blue shift relative to the bulk value. The micro-Raman measurement reveals an average crystallite size of 4 nm. The TEM observation shows that the crystallite sizes are 2–10 nm and that only the concentration of the nanocrystallites in the amorphous matrix increased with deposition temperature, whereas their sizes did not increase significantly with deposition temperature. Considering that the bandgap is determined by both the amorphous SiC and the SiC nanocrystallites, hence the origin of the large bandgaps was ascribed to the presence of nanocrystallites in the films.

Nanocrystalline silicon carbide films deposited on molybdenum substrates by using thermal plasma CVD at substrate temperatures of 750–1250 °C [11] are mainly composed of β-SiC nanocrystallites. Characterization of the films by use of nanoindentation shows that their average grain size, crystalline fraction in the film, and film hardness all increased with increasing substrate temperature (Fig. 7.7). The average grain sizes are between 10 and 20 nm, the crystalline fraction is 80–85%, and the

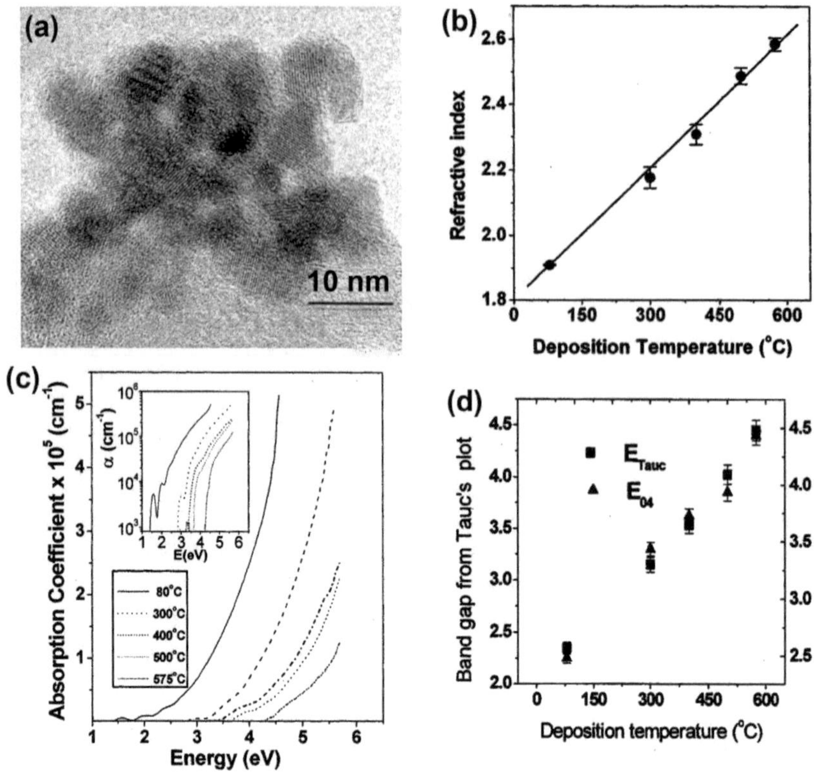

Fig. 7.6 **a** HRTEM image of the SiC film deposited with hydrogen dilution at 575 °C. **b** Refractive index versus deposition temperature. **c** Optical absorption spectra for the films deposited at different temperatures. The inset shows the same spectra in logarithmic scale. **d** Bandgap E_g and E_{04} as a function of deposition temperature. Reprinted with permission from [10]. Copyright 2003, American Institute of Physics

film hardness is about 50 GPa in the film obtained with substrate temperatures above 1200 °C.

Undoped and *n*- and *p*-type hydrogenated nanocrystalline 3C-SiC films have been deposited on glass and silicon substrates at a substrate temperature of 300 °C by using hot-wire CVD [12]. Monomethylsilane and hydrogen are the source gases for deposition of undoped films. Phosphine and hexamethyldisilazane are phosphorous and nitrogen sources for *n*-type doping, and dimethylaluminum hydride is aluminum source for *p*-type doping. Figure 7.8 shows the cross-sectional TEM images of the undoped films deposited at different H_2/monomethylsilane ratios. The corresponding ring electron diffraction pattern corresponds to the crystalline phases of 3C-SiC. The XRD analysis indicates that the average grain size in the film increased from 6.4 to 16.6 nm with increasing deposition temperature. The grain size affects the dark conductivity and the dielectric functions of the films. The former increased from 5.8×10^{-11} to 6.2×10^{-6} S cm^{-1} as the grain size increased from 6.4 to

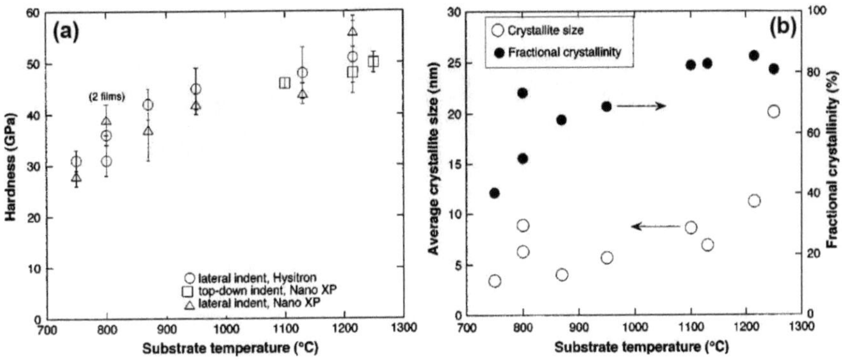

Fig. 7.7 **a** Film hardness as a function of substrate temperature during deposition of SiC film measured by nanoindenter. **b** Average crystallite size and crystalline fraction as a function of substrate temperature. Reprinted with permission from [11]. Copyright 2005, American Institute of Physics

16.6 nm. The dark conductivities are 5.32 and 7.67 $\times$ 10^{-4} S cm^{-1} for the *n*- and *p*-type films, respectively. The absorption data reveals that *p*-type doping significantly affects absorption coefficients above the bandgap of the doped film, whereas the absorption coefficients below the bandgap of the *n*-type film are affected by free carrier absorption and localized states within the bandgap.

Silicon carbide films with different carbon concentrations have been synthesized by using inductively coupled plasma CVD using a $SiH_4/CH_4/H_2$ gas mixture at a substrate temperature of 500 °C [13]. The film obtained with a carbon concentration of 49 at% is composed of nanocrystalline cubic silicon carbide, without any phase of silicon, graphite, or diamond crystallites (Fig. 7.9). The average size of SiC crystallites is about 6 nm. Polycrystalline silicon and amorphous silicon carbide coexist in the films obtained with lower carbon concentrations, and amorphous carbon and silicon carbide coexist in the films obtained with higher carbon concentrations.

Nanocrystalline SiC film has been deposited by using helicon wave plasma-enhanced CVD on Corning glass and single-crystal silicon substrates using a mixture of H_2, SiH_4, and CH_4 gases [14]. HRTEM observation (Fig. 7.10) reveals that the film contains many 3C-SiC nanocrystals embedded in the amorphous SiC matrix, with an average size of 3.96 nm. The SiC film showed photoluminescence with a peak wavelength ranging from 475 to 545 nm as the excitation wavelength increased from 360 to 500 nm. This shifting emission implies that the photoluminescence may be governed by quantum confinement effect. The time-resolved photoluminescence showed an average lifetime on the order of nanosecond. A thin film complex consisting of 3C-SiC nanocrystals as well as amorphous silicon and graphite was realized using an atmospheric pressure thermal plasma-enhanced CVD by employing $SiCl_4$ and C_2H_2 as silicon and carbon sources [15]. The synthesized SiC nanocrystals have an average diameter of 18–30 nm. They have red photoluminescence. Pure SiC nanocrystals can be achieved by a post-treatment of calcining and etching.

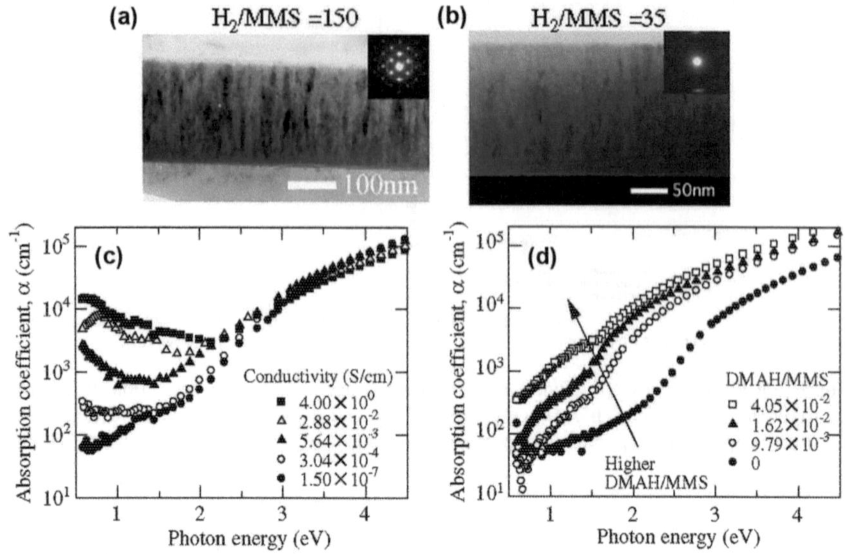

Fig. 7.8 **a**, **b** Cross-sectional TEM images of hydrogenated nanocrystalline 3C-SiC films deposited at different H_2/monomethylsilane ratios. **c** Absorption spectra of *n*-type SiC films with different dark conductivities. **d** Absorption spectra of *p*-type SiC films deposited at different dimethylaluminum hydride/monomethylsilane ratios. Reprinted with permission from [12]. Copyright 2007, IOP Publishing

7.2 Fabrication by Ions Implantation

Ions implantation in combination with annealing is another widely used method for fabricating SiC films containing SiC nanocrystals. The SiC microcrystallites in Si single crystals have been fabricated by carbon ion implantation at a fluence of 10^{17} cm^{-2} followed by annealing [16]. Infrared absorption examination reveals a broad absorption band centered at 700–725 cm^{-1} (Fig. 7.11) in the as-implanted sample, and it persisted after annealing at temperatures up to 825 °C. The band peak lies between the local mode of carbon in silicon (608 cm^{-1}) and the TO phonon mode of SiC (796 cm^{-1}) and was assigned to the former; the band's broadening and shifting to higher frequencies was ascribed to the high carbon concentration and the high disorder of the implanted layer. The SiC microcrystallites began to form at temperatures around 850 °C, as demonstrated by arising of the strong TO phonon absorption band of SiC at around 800 cm^{-1}. The annealing effect on the implanted film was completed by 875 °C, and there was no further change in the absorption band upon further anneal up to 1000 °C. The annealed implanted and un-implanted samples all possess an absorption band at around 1080 cm^{-1} attributed to the SiO_2 surface layers formed by annealing, and it vanished after removal of the surface layers by HF etching.

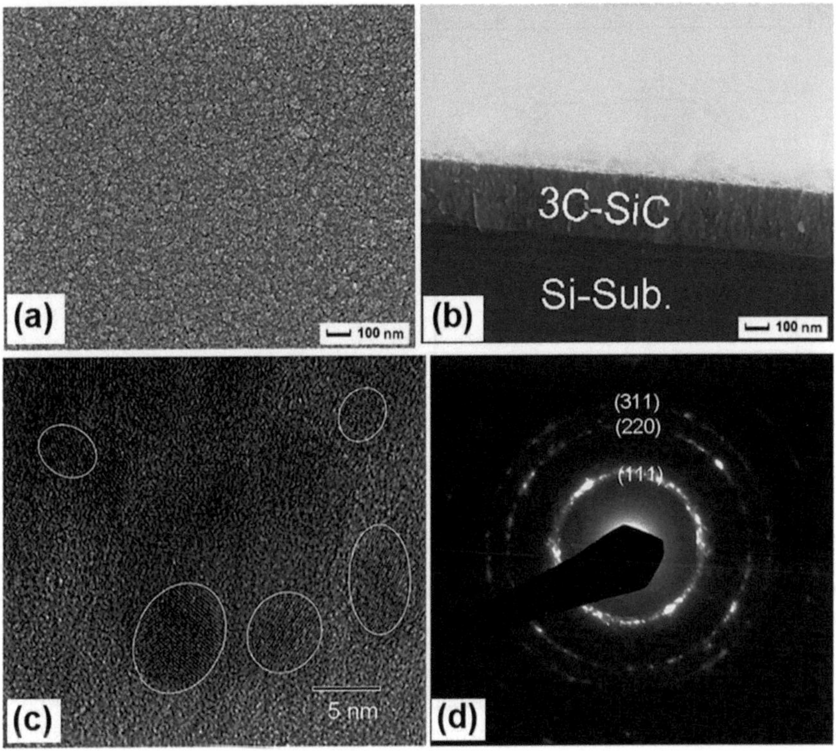

Fig. 7.9 **a, b** Surface morphology and fracture cross section of SiC film with a carbon concentration of 49 at%. **c** HRTEM image and **d** electron diffraction pattern of the same SiC film. Reprinted with permission from [13]. Copyright 2007, IOP Publishing

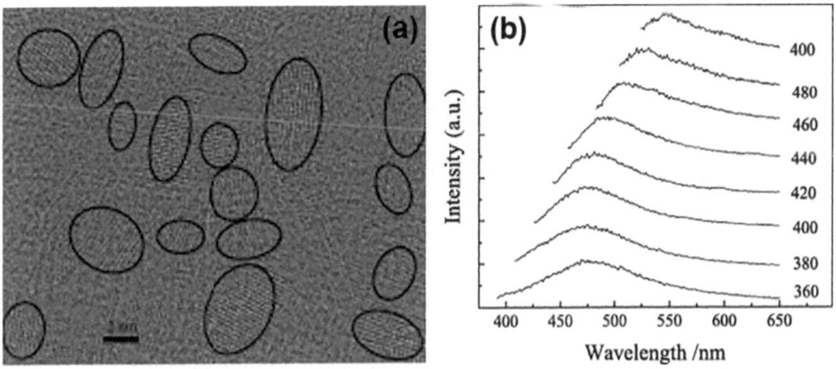

Fig. 7.10 **a** HRTEM image of the SiC film and **b** its PL spectra excited at different wavelengths. Reprinted with permission from [14]. Copyright 2011, Elsevier B.V.

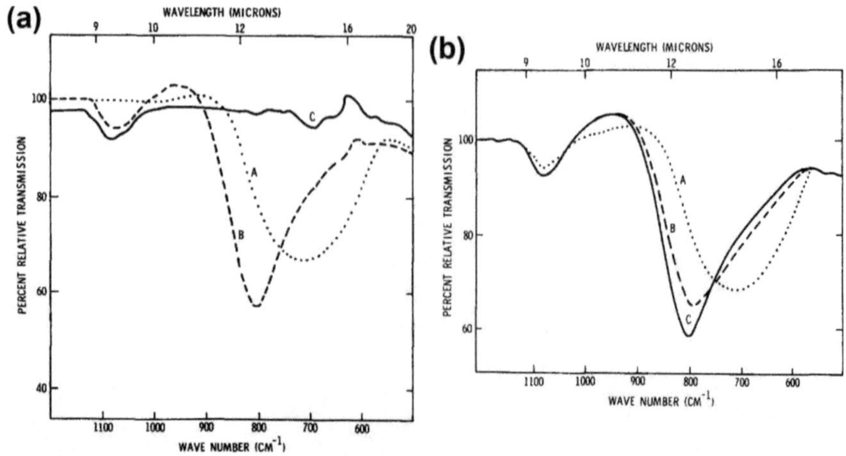

Fig. 7.11 **a** Infrared transmission spectra of (A) Si wafer implanted by 10^{17} carbon atoms cm^{-2} before annealing; (B) same sample after annealing at 1000 °C for 20 min; (C) un-implanted Si annealed at 1000 °C for 20 min. All spectra were taken with matched un-implanted unannealed Si as the reference sample. **b** Infrared transmission spectra of a Si wafer implanted by 10^{17} carbon atoms cm^{-2} followed by annealing for 20 min at (A) 800 and 825 °C, (B) 850 °C, and (C) 900 °C. Reprinted with permission from [16]. Copyright 1971, American Institute of Physics

The microcrystalline SiC in the Si matrix formed by carbon implantation has been investigated by using TEM, XRD, infrared absorption, Auger electron spectroscopy, Rutherford backscattering spectrometry, and nuclear reaction analysis [17]. A thick buried layer of β-SiC with a thickness of 300 nm formed after multiple implantations at 860 °C without further annealing. This buried layer consists of nanocrystalline SiC grains with an average size of 7 nm, which show nearly perfect epitaxial relation with the Si matrix. The grains are only slightly misoriented by 3.5° and are significantly twinned on {111} planes.

The β-SiC in Si crystal matrix has been fabricated by using carbon ion implantation into the Si wafer for an ion energy of 50 keV and a dose of 10^{17} cm^{-2} followed by annealing in nitrogen at 950 °C [18]. The TEM observation reveals the formation of a layer of polycrystalline β-SiC beneath the sample surface (Fig. 7.12), and the electron diffraction pattern shows features in accordance with polycrystalline β-SiC. The infrared spectrum has a strong absorption band at around 800 cm^{-1} that is assigned to the TO phonon mode of β-SiC. The porous SiC was achieved by etching the implanted and annealed sample in HF–ethanol (2:1 by volume) solution at a current density of 40 mA cm^{-2} with assistance of UV-light illumination. The resultant porous sample showed photoluminescence with two peaks at around 445 nm (2.79 eV) and 560 nm under 260-nm excitation. The former peak photon energy is higher than the bandgap of bulk β-SiC (2.2 eV), and this band may stem from the interband transition in the β-SiC nanocrystallites. The latter emission band was assigned to the porous Si. The distribution of the β-SiC nanocrystallites in the Si matrix has been investigated for samples achieved by carbon ion implantation into Si wafers at an energy 50 keV and a dose 5×10^{17} cm^{-2} at 600 °C followed by anneal in nitrogen at 1100 °C [19]. The cross-sectional TEM image (Fig. 7.13) of the as-implanted sample reveals three layers from the implanted surface down to the substrate; they are the upper damaged Si layer A (85 nm thick), the amorphous layer B (70 nm thick), and the lower damaged Si layer C (50 nm thick). After anneal, three layers have distinct concentrations of SiC nanocrystals due to the distribution of implanted ions, the concentration is low in layers A and C and reaches a maximum at the B/C interface. The embedded β-SiC crystallites have sizes of 2–8 nm.

The β-SiC nanocrystals have been synthesized by carbon implantation into Si crystals at 35 keV with various doses followed by annealing at 1200 °C in argon atmosphere and subsequent thermal oxidation at 1050 °C [20]. The infrared spectrum (Fig. 7.14) comprises two strong absorption peaks, one shifted from 820 to 795 cm^{-1} as the implantation dose increased from 1.2×10^{16} to 3.2×10^{17} cm^{-1} and it is assigned to the TO phonon mode of β-SiC, another at around 1080 cm^{-1} is assigned to the Si–O–Si vibration of the SiO_2 film formed on the sample surface. The sample showed photoluminescence, with two bands at around 460 and 535 nm under 300-nm excitation. Time-resolved photoluminescence spectrum of the sample comprises two decay components with lifetimes of around 3 and 0.35 ns, respectively.

The embedded SiC nanocrystallites can also be prepared by ion implantation into SiO_2 films instead of Si films. The quasi-white photoluminescence have been observed from the silica wafer implanted by Si^+ and C^+ ions or only by C^+ ions followed by thermal annealing in argon atmosphere [21]. The TEM measurement (Fig. 7.15a) reveals that the annealed Si^+/C^+ co-implanted silica wafer has three distinct layers beneath the surface. There are many Si and SiC nanocrystals in regions 1 and 3. The sample showed wide photoluminescence band (Fig. 7.15b) in the range

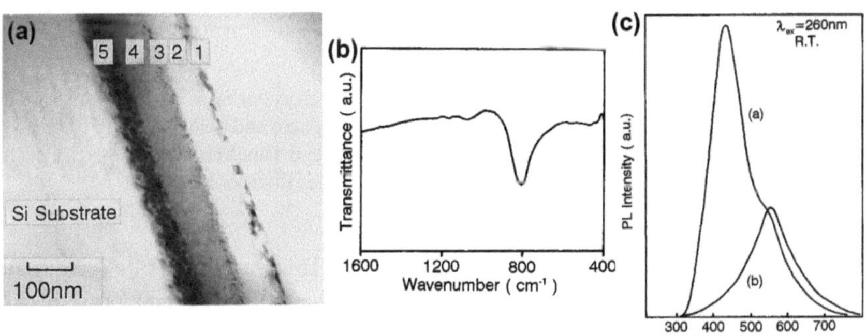

Fig. 7.12 **a** TEM image of the C^+-implanted Si crystal after anneal in N_2 at 950 °C. Regions 1–5 are successively upper damaged Si layer, β-SiC precipitate, closure boundary, β-SiC precipitate, and lower damaged Si layer. **b** FTIR spectrum of this sample. **c** PL spectra of (**a**) annealed C^+-implanted sample and (**b**) reference sample without implantation. Both samples were anodized in a HF-ethanol solution at 40 mA cm^{-2}. Reprinted with permission from [18]. Copyright 1995, American Institute of Physics

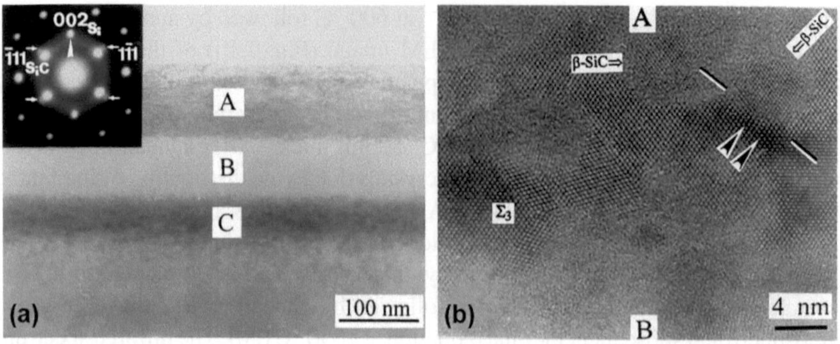

Fig. 7.13 **a** Cross-sectional TEM image showing three layers on the surface of carbon-implanted Si crystal without anneal. The inset is the electron diffraction pattern of this layered structure showing spots from Si matrix and β-SiC particles and a halo from the amorphous phase. **b** HRTEM image of the A/B interface in the annealed sample showing epitaxial growth and amorphous interface between Si and β-SiC nanocrystals. Reprinted with permission from [19]. Copyright 1997, Springer Nature Limited

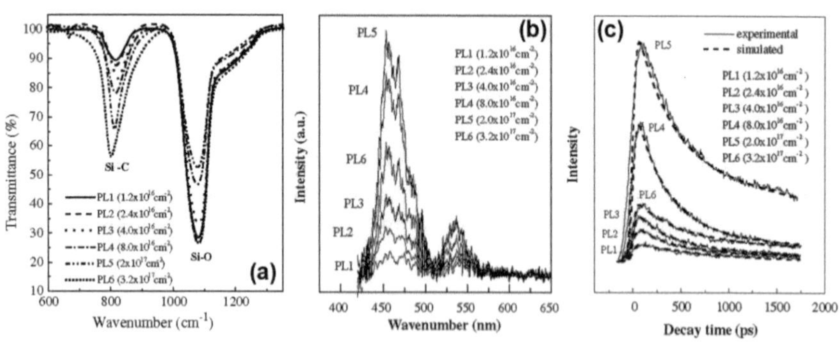

Fig. 7.14 **a** FTIR spectra of the carbon ion-implanted single-crystal Si at 35 keV with various doses. The samples were annealed at 1200 °C in argon atmosphere and oxidized at 1050 °C. **b** PL spectra of the samples excited at 300 nm at room temperature. **c** Time-resolved PL spectra of the samples. Reprinted with permission from [20]. Copyright 2003, Elsevier B.V.

of 350–800 nm with a maximum at around 550 nm. This quasi-white light emission arises from the graphitic carbon clusters as well as the Si and SiC nanocrystals.

Diamond was also employed as the implantation substrate for the synthesis of the embedded SiC crystallites. The natural IIa diamond has been used as the substrate to produce SiC by implantation of 90 keV N^+ at a dose of 1×10^{15} cm^{-2} followed by implantation of 150 keV Si^+ at a dose of 3×10^{17} cm^{-2} at 900 °C [22]. Cross-sectional transmission electron microscopy (Fig. 7.16) reveals a 70 nm thick damaged diamond layer (zone A) and a 150 nm thick buried layer (zone B) in which crystalline 3C-SiC domains are in perfect epitaxial relation to the weakly damaged diamond matrix. No amorphous or graphitic domains are present in the implanted layer owing to the high implantation temperature. There is an array of ordered misfit dislocations at

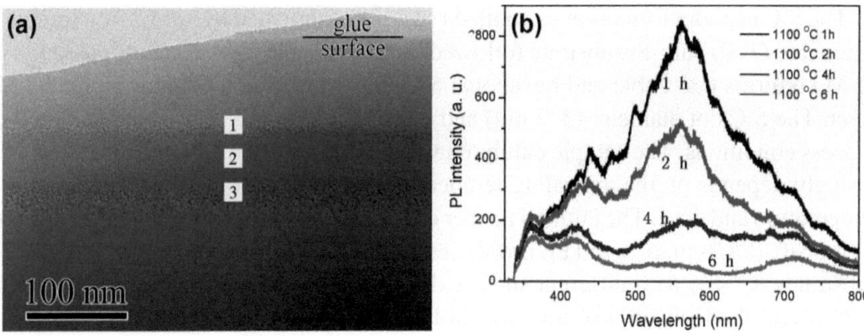

Fig. 7.15 **a** Cross-sectional TEM image of the Si^+/C^+ co-implanted silica wafer after anneal at 1100 °C showing three layers. There are many Si and SiC nanocrystals in regions 1 and 3. **b** PL spectra of the Si^+/C^+ co-implanted silica wafer before anneal and after anneal at different temperatures. Reprinted with permission from [21]. Copyright 2012, American Institute of Physics

the 3C-SiC/diamond interface. The Raman spectrum comprises a diamond signal at 1332 cm^{-1} and a weak peak at 797 cm^{-1} that is characteristic of 3C-SiC TO phonon mode. The infrared spectroscopy reveals 3C-SiC TO phonon absorption at around 796 cm^{-1}. The implantation temperature affects the structural properties of the Si-implanted diamond [23]. Implantation with a high Si fluence severely damaged the diamond matrix and the formed SiC crystallites. Raising the implantation temperature from 900 to 1000 °C reduced damage and increased the SiC crystallite concentration, size, and epitaxial alignment. Further increase of the implantation temperature did not improve the quality of the SiC-rich layer, and the damaged diamond transformed into graphite.

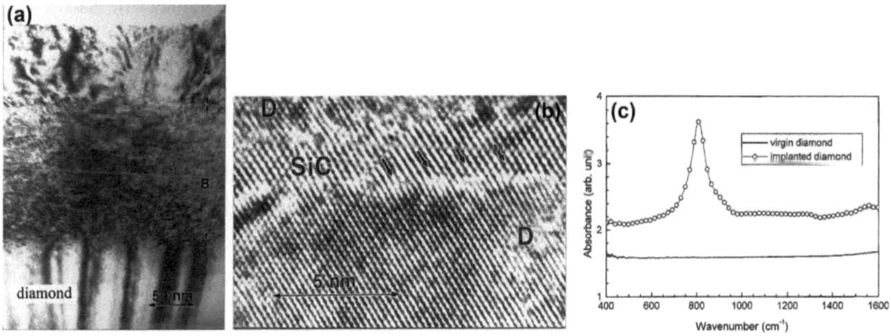

Fig. 7.16 **a** Cross-sectional TEM image of diamond doped successively by N^+ and Si^+. Zone A (upper layer) is the damaged crystalline diamond, and zone B (middle layer) is where the implanted Si atoms lie. **b** High-resolution micrograph of a 3C-SiC domain embedded in the diamond matrix of zone B. Letter D denotes diamond, and the arrows indicate the misfit dislocations. **c** Infrared absorption spectra of pristine and implanted diamond. Reprinted with permission from [22]. Copyright 2000, American Institute of Physics

The SiC nanodot formation in a bulk-Si substrate fabricated by hot-C^+-ion implantation into (100) bulk-Si substrate followed by annealing in N_2 was investigated [24]. TEM confirms that cubic and hexagonal SiC dots formed in a C^+-ion-implanted Si layer. The SiC dot diameter (3–7 nm) and density ($1–2 \times 10^{12}$ cm^{-2}) depend on the process conditions. The sample exhibited very strong PL after N_2-annealing. The PL strongly depends on the annealing temperature as well as hot-C^+-ion implantation temperature and dose. The authors further examined the Si surface orientation effects on the SiC-dot formation and PL properties in three (100)-, (110)-, and (111)-bulk-Si substrates (C^+-Si) with different surface densities of Si atoms (N_S) [25]. TEM and XPS reveal the formation of SiC dots in the (110)- and (111)-C^+-Si, in addition to (100)-C^+-Si. The diameter and surface density of the SiC dots are dependent on the Si surface orientation. The average diameter of the SiC dots in three surface-oriented C^+-Si decreased from 5 to 3 nm with increasing N_S. The Raman peak intensity of the TO Si–C vibration is nearly independent of N_S.

7.3 Fabrication by Sputtering

Sputtering is another common method to prepare silicon carbide nanostructured films. Silicon carbide thin films have been deposited by reactive magnetron co-sputtering of silicon and carbon on silicon or quartz substrates in pure hydrogen plasma at substrate temperatures of 100–600 °C [26]. Infrared absorption and TEM measurements (Fig. 7.17) indicate that the crystalline SiC began to form at a substrate temperature of 300 °C. The crystalline fraction increased with substrate temperature as demonstrated by the increase in intensity of the infrared absorption at 800 cm^{-1} that is assigned to the stretching vibration of Si–C bonds. The weaker absorption peaks at about 900 and 1000 cm^{-1} are separately assigned to the Si–H_2 bending and C–H wagging modes, and their respective stretching modes are present at 2100 and 2900 cm^{-1}. The electron diffraction ring stemming from the β-SiC (111) lattice planes became more clear with increasing substrate temperature above 300 °C, confirming the improvement of crystalline fraction. The refractive index of the layers increased from 1.9 to 2.4, and the room-temperature dark conductivity increased by six orders of magnitude as the substrate temperature increased from 100 to 600 °C. High-resolution transmission electron microscopy reveals the formation of randomly oriented β-SiC nanocrystals with an average size of a few nanometers for a substrate temperature over 400 °C [27].

The deposition temperature influences the growth of sputtered nanocrystalline SiC [28]. The sputtered film contains SiC nanocrystals with an average size of about 5 nm (Fig. 7.18a), and the number density increased with increasing deposition temperature. The sputtered film contains also cubic SiC crystallites possessing 6H-SiC faults. The optical transmission measurement performed on the sputtered SiC layers gave their static refractive index and pseudobandgap $E_{5.4}$, which is the energy corresponding to the absorption coefficient of 5×10^4 cm^{-1}, and both quantities

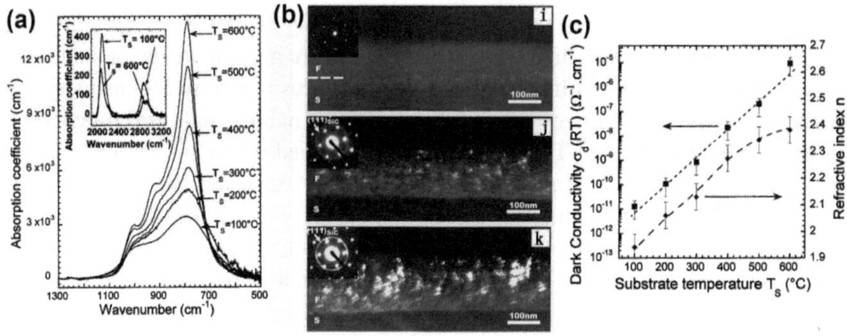

Fig. 7.17 **a** Infrared absorption spectra in the 500–1300 cm^{-1} region for the SiC layers obtained by co-sputtering of silicon and carbon at different substrate temperatures. The inset shows two typical spectra in the 1900–3300 cm^{-1} range. **b** Cross-sectional TEM images and corresponding electron diffraction patterns of the SiC films grown at (i) 200 °C, (j) 300 °C, and (k) 600 °C. **c** Refractive index and room-temperature dark conductivity as a function of substrate temperature. Reprinted with permission from [26]. Copyright 2000, American Institute of Physics

increased with deposition temperature (Fig. 7.18b). The slight increase of the pseudobandgap results from variation of the Si/C ratio and decrease of the hydrogen content in the SiC film with increasing deposition temperature.

The Si nanocrystals may coexist with the SiC nanocrystals in the Si and C co-sputtered films. The three-layer amorphous Si/C/amorphous Si films containing nanocrystalline SiC have been prepared by ion beaming sputtering of silicon and

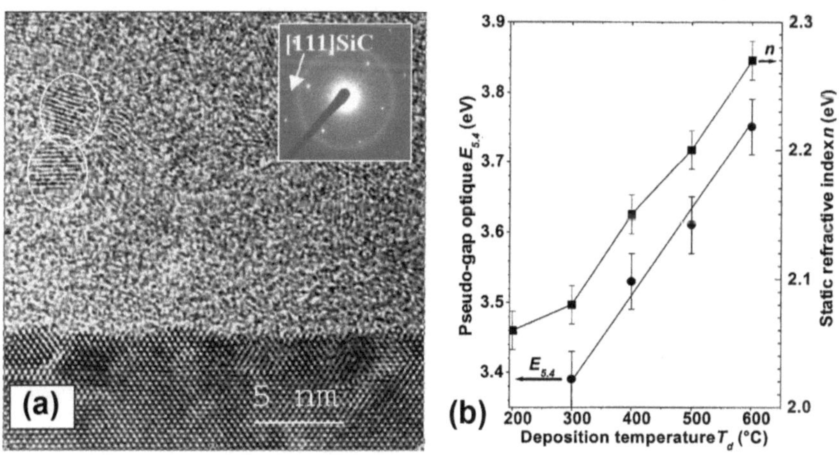

Fig. 7.18 **a** Cross-sectional TEM image for SiC film grown by co-sputtering of silicon and carbon at deposition temperature of 300 °C. The inset shows corresponding selected-area electron diffraction pattern. **b** Pseudogap $E_{5.4}$ obtained from absorption spectra and static refractive index as a function of deposition temperature.. The solid lines are the guide for the eyes. Reprinted with permission from [28]. Copyright 2005, American Institute of Physics

carbon on the Si substrates and subsequent annealing at 750 °C in vacuum [29]. The SiC nanocrystals with diameters of 20–120 nm are present on the top and at the bottom of the three-layer film. The two Si/C interfaces comprise SiC nanocrystals and Si nanocrystals with sizes below 5 nm, and these nanocrystals are embedded in an amorphous SiC matrix. The three-layer film showed photoluminescence with a single peak centered at around 480 nm under 325-nm excitation. $Si_{1-x}C_x$ films with varying carbon-to-silicon ratios have been fabricated by magnetron co-sputtering of C and Si targets [30]. The composition of the films changed with varying ratio of two targets. The as-sputtered films have been annealed at various temperatures from 800 to 1100 °C in nitrogen atmosphere. The infrared absorption spectrum comprises a Si–C stretching peak that shifted from 737 to 800 cm^{-1} as annealing temperature increased. Raman scattering (Fig. 7.19), XRD, and transmission electron microscopy reveal that the dominant nanocrystals in the sputtered film changed from Si to SiC in the films annealed at 1100 °C when the C/Si atomic ratio increased from 0.33 to 1.02. Co-sputtering of silicon and carbon targets on silica or silicon substrates followed by annealing can also yield silicon nanocrystals embedded in an amorphous SiC matrix [31].

The β-SiC nanocrystals can be fabricated by relaxation of $Si_{1-y}C_y$ ($0.005 > y > 0.05$) random alloys grown epitaxially on Si [32]. At temperatures over 900 °C the C atoms that are located in substitutional sites at lower temperatures will precipitate out of the lattice, leading to formation of β-SiC nanocrystals with the same lattice orientation as that of the mother Si lattice. The nanocrystals have uniform diameters, and they are between 2 and 4 nm for $y = 0.005$ $Si_{1-y}C_y$ material. The β-SiC materials with high surface areas and porosity have been prepared by reaction between mesoporous carbon and gaseous silicon coming from sublimation of the Si powder at temperatures of 1200–1300 °C [33]. The synthesized SiC sample has mesoporosity dimension of 5–40 nm.

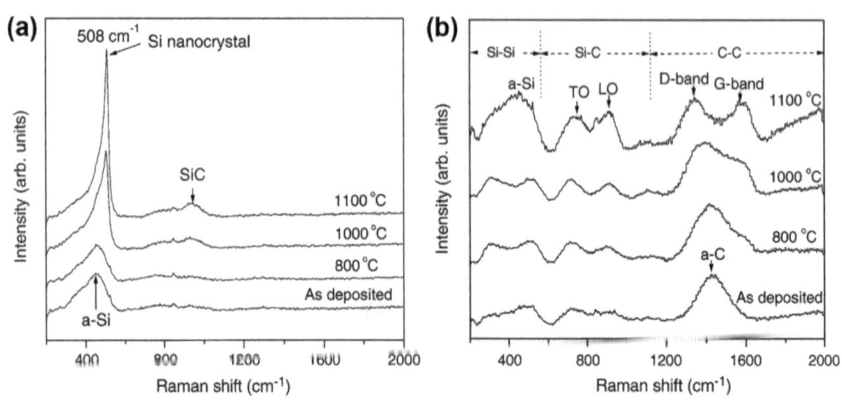

Fig. 7.19 Raman spectra of the Si and C co-sputtered films obtained with C/Si ratios of 0.33 (**a**) and 0.71 (**b**). The major peaks and annealing temperatures are indicated. Reprinted with permission from [30]. Copyright 2008, Elsevier B.V.

Deposition and annealing of the 200-nm-thick C_{60} film on (100) silicon substrate can give rise to silicon carbide films [34]. The reaction between C_{60} and silicon started at the interface and continued by silicon diffusion. Silicon diffusion and SiC formation require annealing at 800 °C for $t > 100$ min and at 900 °C for $t > 25$ min. The synthesized stoichiometric SiC film is uniform with a grain size of 20–40 nm (Fig. 7.20a). SiC nanocrystals can also be prepared by annealing C_{60}-covered porous Si [35]. X-ray diffraction reveals the existence of β-SiC particles along with Si, SiO_2, and graphite. The annealed sample exhibited photoluminescence (Fig. 7.20b) with an asymmetric band, and the double-Gaussian fit of the curve gives two peaks at 380 nm (3.26 eV) and 454 nm (2.73 eV). The 380 nm band has an excitation peak at 308 nm (Fig. 7.20c), and this luminescence was suggested to arise from the oxygen-vacancy defects in the SiO_2 matrix; the 454 nm band was related to the β-SiC particles.

The high-temperature reaction between Si wafer and carbon source material has been employed to synthesize the nanocrystalline SiC films [36]. The SiC nanocrystals have been epitaxially and void-free grown on single-crystal Si substrates through the reactive annealing process in CO at 1100–1190 °C. Crystal orientation shows influence on the nucleation density, size, and morphology of the nanocrystals. Irregular

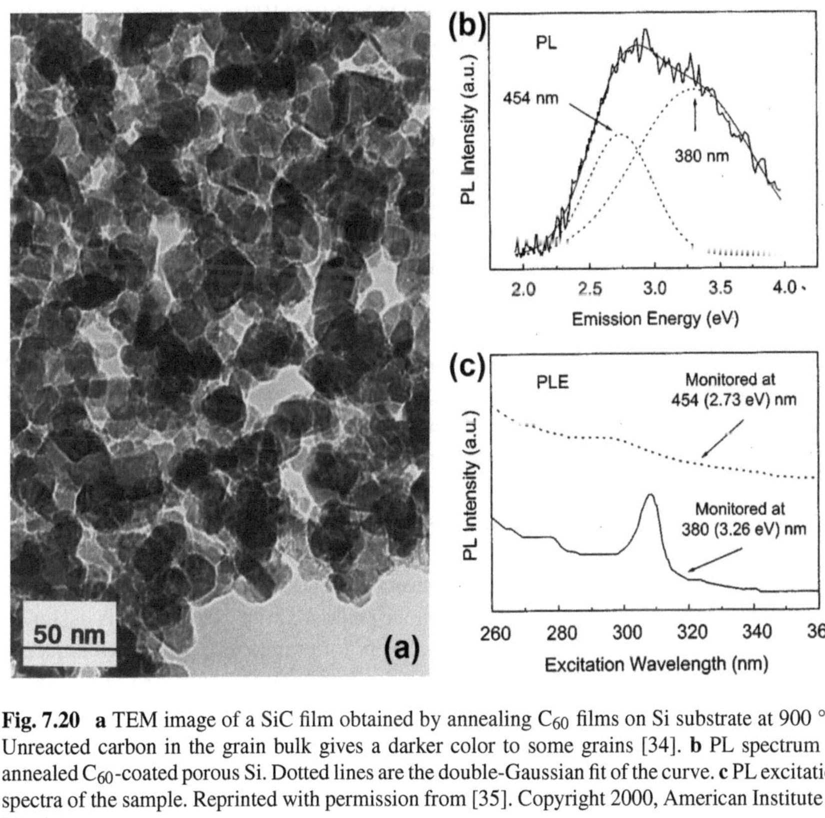

Fig. 7.20 **a** TEM image of a SiC film obtained by annealing C_{60} films on Si substrate at 900 °C. Unreacted carbon in the grain bulk gives a darker color to some grains [34]. **b** PL spectrum of annealed C_{60}-coated porous Si. Dotted lines are the double-Gaussian fit of the curve. **c** PL excitation spectra of the sample. Reprinted with permission from [35]. Copyright 2000, American Institute of Physics

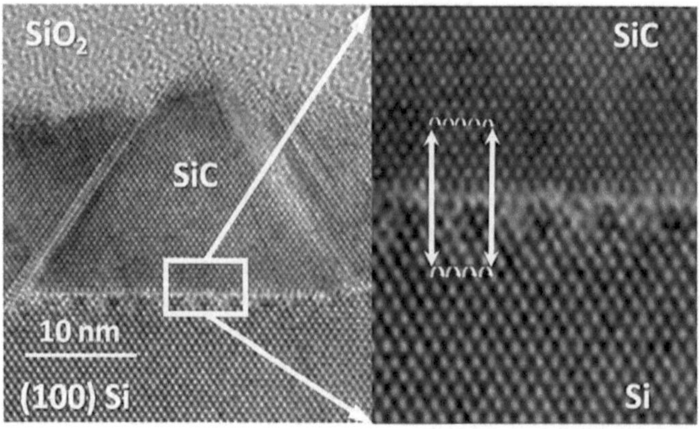

Fig. 7.21 HRTEM image of a SiC nanocrystal growing on Si(100) surface by annealing at 1100 °C in 5% CO. The nanocrystals are located at the SiO_2/Si interface. The SiC/Si interface is void-free and abrupt. Reprinted with permission from [36]. Copyright 2012, Elsevier B.V.

and large crystallites (63 nm) were prepared on the Si(111) surface for annealing in diluted CO (5%). The SiC nanocrystals growing on the (100) surfaces are a few tens of nanometers in diameter (Fig. 7.21). Much smaller grains were prepared on the Si(110) planes. The substrate orientation affects only the morphology for annealing in pure CO.

Direct sputtering of SiC targets can produce the SiC nanocrystals. The SiC–SiO_2 composite film has been fabricated by co-sputtering of SiC and SiO_2 targets on the Si wafer [37]. The as-deposited sample experienced post-anneal at 400–1200 °C. High-resolution TEM observation indicates that the as-deposited SiC–SiO_2 film is amorphous, but the SiC nanocrystals with diameters of about 5 nm are present in the sample annealed at 740 °C. The annealed SiC–SiO_2 composite film exhibited photoluminescence at around 460 nm (2.7 eV) (Fig. 7.22). The luminescence intensity is strongly dependent on the annealing temperature, and it reached a maximum at 950 °C. The infrared transmission measurement reveals two strong absorption bands at around 816 and 1055 cm^{-1}; the former is the optical vibration mode of SiC, and the latter is attributed to the Si–O–Si vibration.

The β-SiC films have been fabricated on the Si substrates by using pulsed laser deposition using a polycrystalline SiC target [38]. High-resolution TEM observation (Fig. 7.23) reveals that the deposited film contains both β-SiC and Si nanocrystallites. The infrared spectrum has a strong absorption band at around 780 cm^{-1}, which is the TO phonon mode of β-SiC. Another shoulder peak at about 1020 cm^{-1} is assigned to the Si–O–Si vibration. The film exhibited photoluminescence with two adjacent peaks at 416 and 435 nm. The corresponding photoluminescence excitation spectrum comprises a strong peak at 370 nm. The luminescence arises from the surface defect. The nanocrystalline SiC films comprising uniform-sized nanoislands have been prepared by plasma-assisted radio-frequency magnetron sputtering deposition

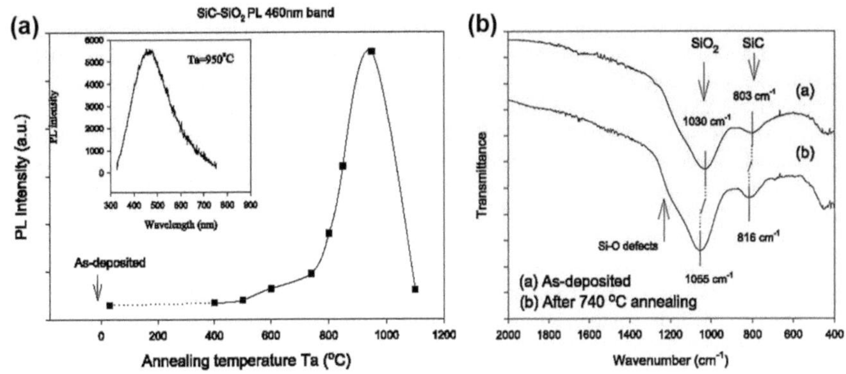

Fig. 7.22 **a** Intensity of PL at 460 nm in $SiC–SiO_2$ composite film under 325-nm excitation as a function of annealing temperature. The inset shows the PL spectrum of the $SiC–SiO_2$ film annealed at 950 °C. **b** FTIR spectra of the as-deposited and annealed $SiC–SiO_2$ composite film. Reprinted with permission from [37]. Copyright 2001, Elsevier B.V.

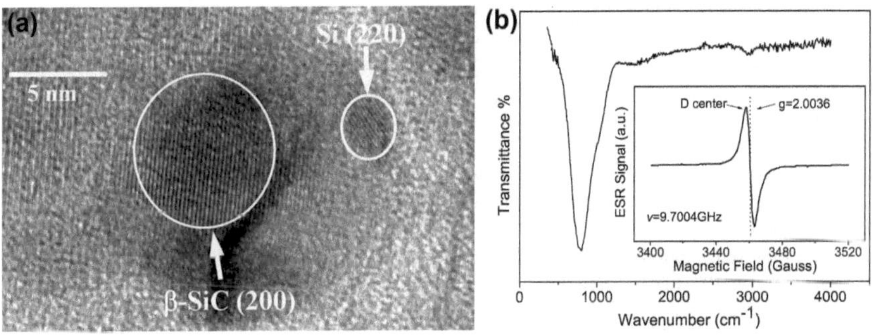

Fig. 7.23 **a** HRTEM image of the β-SiC film deposited on the (100) Si wafer showing Si and β-SiC nanocrystallites. **b** FTIR spectrum of the β-SiC film. The inset shows the electron paramagnetic resonance spectrum. Reprinted with permission from [38]. Copyright 2003, Elsevier B.V.

on the Si(100) surface using a SiC target [39]. The nanoislands have size of 20–35 nm and contain small (5 nm) nanocrystalline inclusions.

References

1. Mukaida H, Okumura H, Lee JH, Daimon H, Sakuma E, Misawa S, Endo K, Yoshida S (1987) Raman scattering of SiC: estimation of the internal stress in 3C-SiC on Si. J Appl Phys 62:254–257
2. Feng ZC, Choyke WJ, Powell JA (1988) Raman determination of layer stresses and strains for heterostructures and its application to the cubic SiC/Si system. J Appl Phys 64:6827–6835
3. Feng ZC, Mascarenhas AJ, Choyke WJ, Powell JA (1988) Raman scattering studies of chemical-vapor-deposited cubic SiC films of (100) Si. J Appl Phys 64:3176–3186

4. Freitas JA, Bishop SG, Edmond JA, Ryu J, Davis RF (1987) Photoluminescence spectroscopy of ion-implanted 3C-SiC grown by chemical vapor deposition. J Appl Phys 61:2011–2016
5. Choyke WJ, Feng ZC, Powell JA (1968) Low-temperature photoluminescence studies of chemical-vapor-deposition-grown 3C-SiC on Si. J Appl Phys 64:3163–3175
6. Yu MB, Rusli U, Yoon SF, Xu SJ, Chew K, Cui J, Ahn J, Zhang Q (2000) Hydrogenated nanocrystalline silicon carbide films synthesized by ECR-CVD and its intense visible photoluminescence at room temperature. Thin Solid Films 377–378:177–181
7. Yu MB, Rusli, Yoon SF, Chen ZM, Ahn J, Zhang Q, Chew K, Cui J (2000) Deposition of nanocrystalline cubic silicon carbide films using the hot-filament chemical-vapor-deposition method. J Appl Phys 87:8155–8158
8. Xu SJ, Yu MB, Rusli, Yoon SF, Che CM (2000) Time-resolved photoluminescence spectra of strong visible light-emitting SiC nanocrystalline films on Si deposited by electron-cyclotron-resonance chemical-vapor deposition. Appl Phys Lett 76:2550–2552
9. Kumbhar A, Patil SB, Kumar S, Lal R, Dusane RO (2001) Photoluminescent, wide-bandgap a-SiC:H alloy films deposited by Cat-CVD using acetylene. Thin Solid Films 395:244–248
10. Rajagopalan T, Wang X, Lahlouh B, Ramkumar C, Dutta P, Gangopadhyay S (2003) Low temperature deposition of nanocrystalline silicon carbide films by plasma enhanced chemical vapor deposition and their structural and optical characterization. J Appl Phys 94:5252–5260
11. Liao F, Girshick SL, Mook WM, Gerberich WW, Zachariah MR (2005) Superhard nanocrystalline silicon carbide films. Appl Phys Lett 86:171913
12. Miyajima S, Yamada A, Konagai M (2007) Characterization of undoped, n- and p-type hydrogenated nanocrystalline silicon carbide films deposited by hot-wire chemical vapor deposition at low temperatures. Jpn J Appl Phys 46:1415–1426
13. Cheng Q, Xu S, Long J, Huang S, Guo J (2007) Homogeneous nanocrystalline cubic silicon carbide films prepared by inductively coupled plasma chemical vapor deposition. Nanotechnology 18:465601
14. Yu W, Wang X, Geng C, Lve X, Lu W, Fu G (2011) Decay processes of photoluminescence in a nanocrystalline SiC thin film. Appl Surf Sci 258:1733–1737
15. Cao TF, Cheng Y, Zhang HB, Yan BH, Cheng Y (2015) High rate fabrication of room temperature red photoluminescent SiC nanocrystals. J Mater Chem C 3:4876
16. Borders JA, Picraux ST, Beezhold W (1971) Formation of SiC in silicon by ion implantation. Appl Phys Lett 18:509–511
17. Martin P, Daudin B, Dupuy M, Ermolieff A, Olivier M, Papon AM, Rolland G (1990) High-temperature ion beam synthesis of cubic SiC. J Appl Phys 67:2908–2912
18. Liao L-S, Bao X-M, Yang Z-F, Min N-B (1995) Intense blue emission from porous β-SiC formed on C^+-implanted silicon. Appl Phys Lett 66:2382–2384
19. Gao YH, Zhang Z, Liao LS, Bao XM (1997) A high-resolution electron microscopy study of blue-light emitting β-SiC nanoparticles in C^+-implanted silicon. J Mater Res 12:1640–1645
20. Chen D, Liao ZM, Wang L, Wang HZ, Zhao F, Cheung WY, Wong SP (2003) Photoluminescence from β-SiC nanocrystals embedded in SiO_2 films prepared by ion implantation. Opt Mater 23:65–69
21. Zhou XD, Ren F, Xiao XH, Xu JX, Dai ZG, Cai GX, Jiang CZ (2012) Origin of white light luminescence from Si^+/C^+ sequentially implanted and annealed silica. J Appl Phys 111:084304
22. Heera V, Fontaine F, Skorupa W, Pécz B, Barna Á (2000) Ion-beam synthesis of epitaxial silicon carbide in nitrogen-implanted diamond. Appl Phys Lett 77:226–228
23. Weishart H, Eichhorn F, Heera V, Pécz B, Barna Á, Skorupa W (2005) High-fluence Si-implanted diamond: optimum implantation temperature for SiC formation. J Appl Phys 98:043503
24. Mizuno T, Yamamoto M, Nakada S, Irie S, Aoki T, Sameshima T (2019) SiC nano-dot formation in bulk-Si substrate using hot-C^+-ion implantation process. Jpn J Appl Phys 58:081004
25. Mizuno T, Aoki T, Sameshima T (2022) Si surface orientation dependence of SiC-dot formation in bulk-Si using hot-C^+-ion implantation technique. J Appl Phys 131:074301
26. Kerdiles S, Berthelot A, Gourbilleau F, Rizk R (2000) Low temperature deposition of nanocrystalline silicon carbide thin films. Appl Phys Lett 76:2373–2375

27. Kerdiles S, Rizk R, Gourbilleau F, Pérez-Rodríguez A, Garrido B, González-Varona O, Morante JR (2000) Low temperature direct growth of nanocrystalline silicon carbide films. Mater Sci Eng B 69–70:530–535
28. Colder H, Rizk R, Morales M, Marie P, Vicens J, Vickridge I (2005) Influence of substrate temperature on growth of nanocrystalline silicon carbide by reactive magnetron sputtering. J Appl Phys 98:024313
29. Chung CK, Chen TY, Lai CW (2011) Low-temperature formation of nanocrystalline SiC particles and composite from three-layer Si/C/Si film for the novel enhanced white photoluminescence. J Nanopart Res 13:4821–4828
30. Song D, Cho E-C, Cho Y-H, Conibeer G, Huang Y, Huang S, Green MA (2008) Evolution of Si (and SiC) nanocrystal precipitation in SiC matrix. Thin Solid Films 516:3824–3830
31. Chang G-R, Ma F, Ma D-Y, Xu K-W (2010) Multi-band silicon quantum dots embedded in an amorphous matrix of silicon carbide. Nanotechnology 21:465605
32. Powell AR, LeGoues FK, Lyer SS (1994) Formation of β-SiC nanocrystals by the relaxation of $Si_{1-y}C_y$ random alloy layers. Appl Phys Lett 64:324–326
33. Liu Z, Shen W, Bu W, Chen H, Hua Z, Zhang L, Li L, Shi J, Tan S (2005) Low-temperature formation of nanocrystalline β-SiC with high surface area and mesoporosity via reaction of mesoporous carbon and silicon powder. Micropor Mesopor Mater 82:137–145
34. Moro L, Paul A, Lorents DC, Malhotra R, Ruoff RS, Lazzeri P, Vanzetti L, Lui A, Subramoney S (1997) Silicon carbide formation by annealing C_{60} films on silicon. J Appl Phys 81:6141–6146
35. Wu XL, Siu GG, Stokes MJ, Fan DL, Gu Y, Bao XM (2000) Blue-emitting β-SiC fabricated by annealing C_{60} coupled on porous silicon. Appl Phys Lett 77:1292–1294
36. Battistig G (2012) Orientation dependent growth of SiC nanocrystals at the SiO_2/Si interface. Thin Solid Films 520:1973–1977
37. Guo YP, Zheng JC, Wee ATS, Huan CHA, Li K, Pan JS, Feng ZC, Chua SJ (2001) Photoluminescence studies of SiC nanocrystals embedded in a SiO_2 matrix. Chem Phys Lett 339:319–322
38. Tan C, Wu XL, Deng SS, Huang GS, Liu XN, Bao XM (2003) Blue emission from silicon-based β-SiC films. Phys Lett A 310:236–240
39. Cheng Q, Xu S, Long J, Ostrikov K (2007) Deterministic plasma-aided synthesis of high-quality nanoislanded nc-SiC films. Appl Phys Lett 90:173112

Chapter 8
Biological Applications

Silicon carbide is a well-known semiconductor with excellent biocompatibility and at least two factors contribute to this favorable characteristic. The first one is that the compound does not contain heavy metals which tend to be detrimental to the human body, and the second one is that neither silicon nor carbon causes deleterious effects such as cytotoxicity. Therefore, silicon carbide is very promising in the biomedical field having many potential cardiovascular and orthopedic applications. In addition, luminescent silicon carbide nanocrystals may be good light emitters in biological imaging.

8.1 Biomorphic SiC Ceramics

Biomorphic SiC ceramics with a wood-like microstructure have been synthesized by infiltrating charcoal with tetraethyl orthosilicate in vacuum [1]. Tetraethyl orthosilicate was hydrolyzed in an ammonia solution to form the SiO_2 gel and then the charcoal covered with SiO_{2v} was heated to 1400 °C in Ar to produce SiC. X-ray diffraction (Fig. 8.1) reveals that with increasing infiltration and heating time, the carbon signal disappeared gradually but that of α-SiC emerged. After the residual carbon was burned off at 700 °C in air, almost the whole preform turned into porous α-SiC that retained the original wood structure. High-resolution SEM shows that the replicated vessels in the product are composed of polycrystalline SiC ceramics with grain sizes smaller than one micrometer.

Greil and colleagues have systematically investigated the processing and microstructures of cellular β-SiC ceramics with the anisotropic porous structure synthesized by infiltration of liquid silicon into carbonized wood and subsequent high-temperature treatment [2]. Various types of natural wood such as ebony, beech, oak, maple, pine, and balsa were carbonized at 800–1800 °C in an inert atmosphere to

J. Fan and P. K. Chu, *Silicon Carbide Nanostructures*, Engineering Materials and Processes, https://doi.org/10.1007/978-3-032-10986-6_8

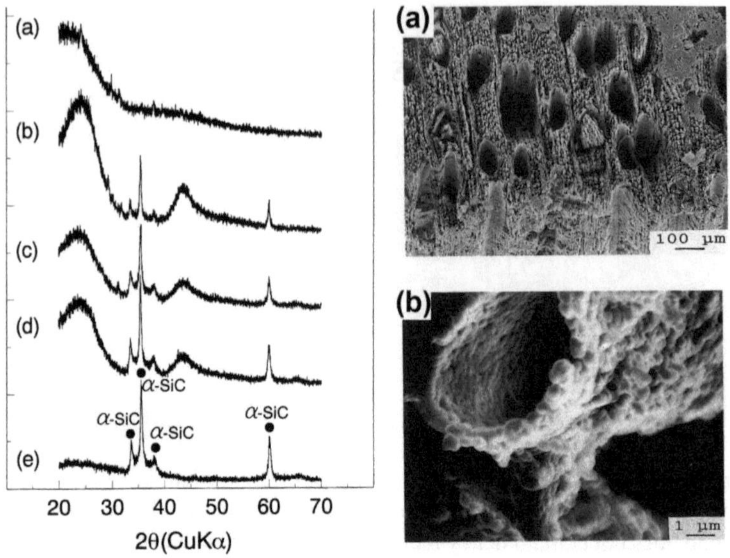

Fig. 8.1 (Left panel) X-ray diffraction patterns of **a** original charcoal, **b–d** specimens after tetraethyl orthosilicate infiltration and annealing at 1400 °C for 2 h **b** once, **c** three times, and **d** six times, and **e** specimen with residual carbon burned off at 700 °C in air after the treatment was repeated six times. (Right panel) SEM image of the specimen (α-SiC) with residual burned off at 700 °C in air after the treatment was repeated six times. Reprinted with permission from [1]. Copyright 1995, John Wiley & Sons, Inc.

produce wood resembling SiC. A rapid liquid Si infiltration-based solid–liquid reaction occurring on the pore surface at 1600 °C transformed the carbon template into β-SiC ceramics (Fig. 8.2). X-ray diffraction indicates that the carbon template was converted completely into β-SiC at a sufficient high temperature. The β-SiC shows distinct morphologies depending on the cellular microstructures of the starting wood materials. The mechanical properties of the SiC ceramics with an anisotropic porous microstructure were generally improved with fractional density, and the strength and strain-to-failure in the axial direction were substantially higher than that in the radial and tangential directions [3].

Pine wood was transformed into carbon preforms by high-temperature pyrolysis and then into biomorphic SiC ceramics by Si vapor infiltration in an inert atmosphere, with the morphology of the initial wood structure retained [4]. X-ray diffraction (Fig. 8.3) discloses that the final product has the β-SiC structure. As the reaction proceeded, the intensity of the (0002) peak of carbonized wood decreased, but that of the (111) peak of β-SiC increased, and no Si-peaks were detected. Only the strong β-SiC signals were observed from the sample infiltrated at 1600 °C for 4 h. SEM shows that the original cellular morphology of wood was retained as manifested by a nearly homogeneous strut with a thickness of about 2 µm. The specific surface area on the SiC ceramic was reduced to 3.3 $m^2\ g^{-1}$ from the original 46 $m^2\ g^{-1}$ of the carbon template, whereas the density increased from 0.34 to 1 $g\ cm^{-3}$.

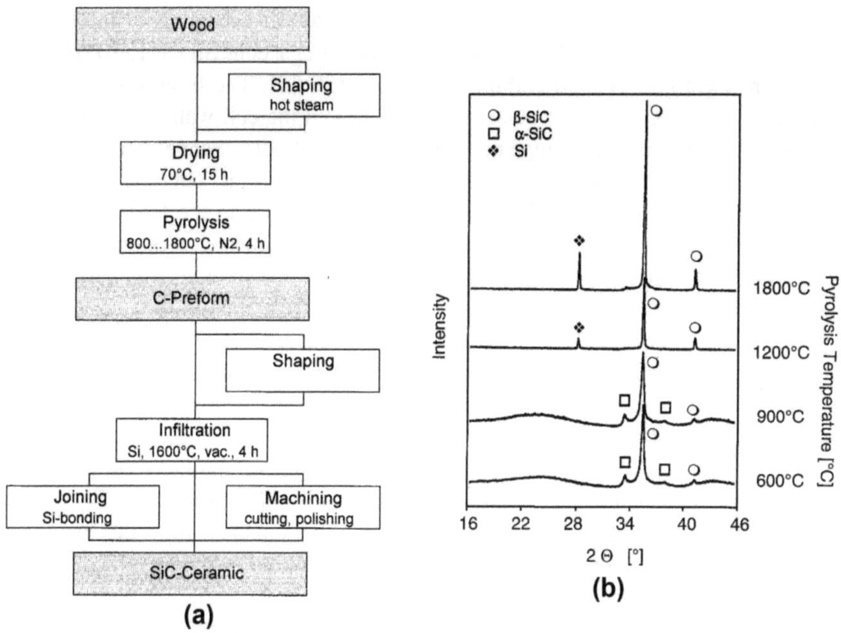

Fig. 8.2 **a** Processing scheme of preparing cellular β-SiC ceramics from wood. **b** XRD spectra of Si infiltrated ceramic reaction products of oak. Reprinted with permission from [2]. Copyright 1998, Elsevier B.V.

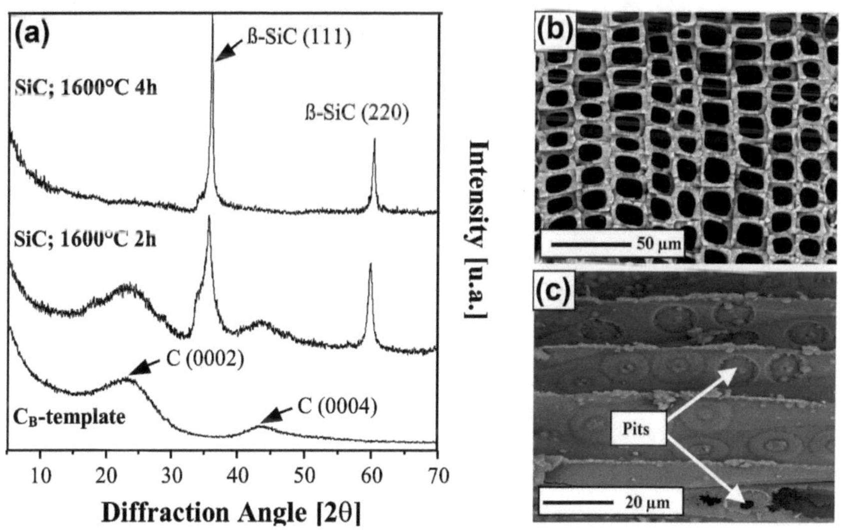

Fig. 8.3 **a** XRD spectra of the carbon template and Si infiltrated ceramic reaction products of pine. SEM images of β-SiC ceramic prepared by Si vapor infiltration of paralyzed pine wood in axial (**b**) and tangential (**c**) directions at 1600 °C for 4 h. Reprinted with permission from [4]. Copyright 2002, Elsevier B.V.

Oak can be converted to porous carbon template via annealing in inert atmosphere above 800 °C, and subsequent infiltration with gaseous SiO at 1550–1600 °C under flowing argon yielded cellular β-SiC ceramic [5]. The biomorphic cellular morphology of oak was again retained. The pores in the cell walls with diameters less than one micrometer disappeared, but two distinct pore channel maxima representing tracheidal cells and large vessels remained in the SiC ceramics (Fig. 8.4). X-ray diffraction shows that β-SiC is the dominant phase in the SiC ceramics in addition to a minor fraction of α-SiC whose amount diminished with reaction time. The carbon template can be completely converted to β-SiC ceramic after a long reaction. The β-SiC has an average size of 0.4 μm after reaction at 1600 °C for 16 h. In another study, oak was converted to highly porous SiC ceramic with wood-like microstructure by carbothermal reduction of oak charcoal/silica composites in static argon atmosphere at 1400–1600 °C [6]. The composites were fabricated by infiltrating silica sol into a porous biocarbon template from oak wood using a vacuum/pressure infiltration treatment. The biomorphic cellular morphology of oak wood charcoal was remained in the porous SiC ceramic with high precision. This porous SiC ceramic contains 3C-SiC and traces of α-SiC. Silica in the composites existed in the cellular pores as fibers and rods. The SiC strut material formed by the gas–solid reaction between SiO (g) and C (s).

The combined sol–gel and carbothermal reduction method using tetraethyl orthosilicate and *Tilia amurensis* wood as the starting materials conducted at 1600 °C

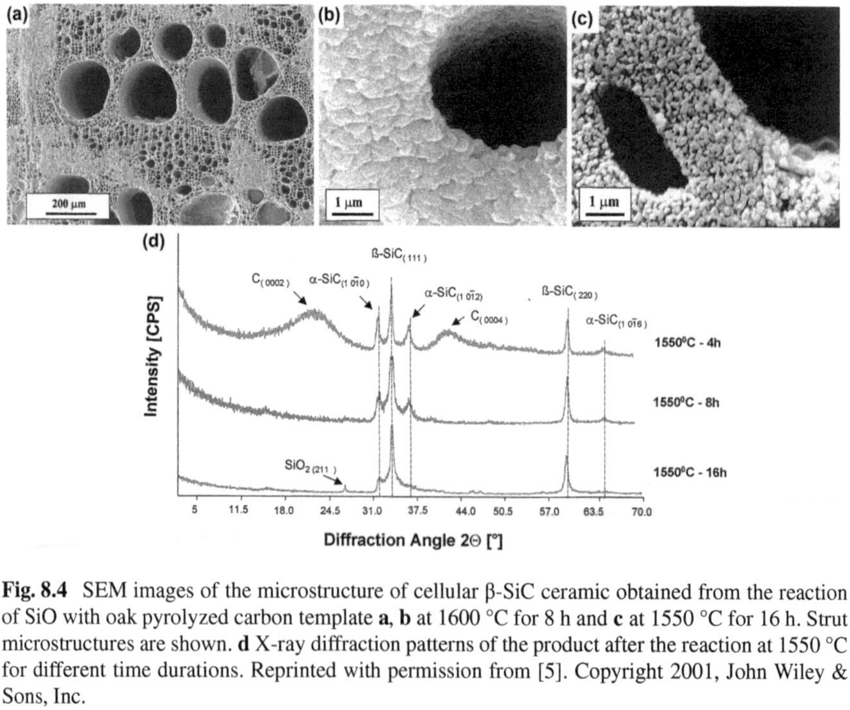

Fig. 8.4 SEM images of the microstructure of cellular β-SiC ceramic obtained from the reaction of SiO with oak pyrolyzed carbon template **a**, **b** at 1600 °C for 8 h and **c** at 1550 °C for 16 h. Strut microstructures are shown. **d** X-ray diffraction patterns of the product after the reaction at 1550 °C for different time durations. Reprinted with permission from [5]. Copyright 2001, John Wiley & Sons, Inc.

in argon atmosphere yielded biomorphic SiC ceramics [7]. The materials comprise mainly β-SiC and traces of α-SiC. The thickness of the ceramic strut became thinner with increasing cycle number of impregnation. The authors further prepared biocarbon template (charcoal) through carbonizing pine wood at 1200 °C under vacuum. Porous SiC ceramics with hybrid pore structures, a combination of tubular pores and network SiC struts in the tubular pores, were prepared via sol–gel conversion, carbonization, and carbothermal reduction reaction at elevated temperatures in argon atmosphere [8]. The resulting SiC ceramic contains 3C-SiC and a small amount of α-SiC. The conversion from pine charcoal to porous SiC ceramic with the hybrid pore structure improved bending strength from 16.4 to 42.2 MPa and reduced porosity from 76.1 to 48.3%. The SiC ceramics synthesized by carbothermal reduction of mineralized wood (pine and poplar) with silica in argon atmosphere were investigated [9]. The materials retained the cellular structure, and the wood cell walls are composed of crystalline SiC nanoparticles 200–700 nm in size. Some β-SiC whiskers 150–200 nm in diameter and 5–20 μm in length formed on the cellular structure.

The mechanical properties of the biomorphic SiC ceramics obtained by infiltration of molten silicon into carbonized mango wood at high temperature were investigated [10]. The materials contain residual silicon with a volume ratio of 35%. The cellular structure has very high strength even at a high temperature close to the melting point of silicon. The microstructure and strength anisotropy reveal that the biomorphic SiC ceramics have structures resembling a honeycomb. The authors further fabricated low-density biomorphic SiC ceramics through melt infiltration of pine carbon templates [11]. The resulting porous microstructure is anisotropic and presents a high level of connectivity. The material has been tested at a constant compression with a strain rate of 2×10^{-5} s^{-1} at temperatures of 1150–1350 °C. They exhibited high compressive strength up to 241 MPa in the axial direction. The elastic modulus and fracture toughness of the SiC biomorphic ceramics depend on the properties of the starting wood preforms and extent of molten silicon infiltration [12]. The materials contain both SiC and Si regions. The ceramics prepared from maple wood have flexural strength of about 344 MPa at room temperature and 230 MPa at 1350 °C. The fracture toughness is 2.6 MPa $m^{1/2}$ for the maple-derived ceramic and 2.0 MPa $m^{1/2}$ for the mahogany-derived one. The strength and fracture toughness increased with the density of the ceramic.

Biomorphic SiC ceramic materials are light, tough, and strong, and the excellent mechanical properties along with good biocompatibility of SiC make them ideal artificial materials in orthopedic implants. Biomorphic SiC coated with a uniform and adherent bioactive glass film by pulsed laser ablation has been investigated [13]. The rough SiC surface with the bioactive glass coating did not show spallation or detachment. The bioactivity was assessed in vitro by soaking in the simulated body fluid. A dense apatite layer formed even in the interconnecting pores of the ceramic after 72 h, suggesting good bioactivity. The biomorphic SiC ceramics with bioactive glass coatings can improve the mechanical and biochemical properties of the dental and orthopedic implants.

The thermal conductivity of biomorphic SiC fabricated from five different precursors was experimentally determined using flash diffusivity and specific heat measurements at temperatures ranging from room temperature to 1100 °C [14]. The axial thermal conductivity of all the materials decreased with increasing temperature. The thermal conductivity lies within 15–42 W mK^{-1} at room temperature, and the transverse thermal conductivity falls within 9–41 W mK^{-1}. The object-oriented finite element analysis was used to model and understand the heat-flow paths through the complex biomorphic SiC microstructures.

Permeability and mechanical integrity of porous biomorphic SiC ceramics from five different wood precursors, with porosities between 45 and 72%, were investigated [15]. Gas permeability and mechanical stability show dependence on the microstructure of the starting wood precursor. Air-permeation performance at room temperature was measured. The permeability constants were assessed by fitting the experimental data with Forchheimer's equation. The acquired Darcian permeability for biomorphic SiC samples fabricated from different precursors lies in the range of 10^{-11}–10^{-12} m^2 (Fig. 8.5), and the inertial terms fall in the range of 10^{-7}–10^{-8} m, indicating a correlation with the average pore size and orientation of the larger channels. The maximum compressive strength values lie in the range of 3–115 MPa.

Bioactivation by chemical treatment of biomorphous SiC was investigated in order to accelerate osseointegration and improve bone bonding ability [16]. Biomorphous SiC was obtained from sipo wood by heating in an inert atmosphere and infiltrating the resulting carbon replica with liquid silicon melt at 1450 °C. After removal of excess silicon by leaching in HF/HNO_3, the biomorphous preform consists of 3C-SIC with a small amount (6 wt%) of residual carbon. The preform was leached again in HCl/HNO_3 and then exposed to $CaCl_2$ solution. Oxidation of the residual carbon at the surface caused formation of carboxyl groups, which adsorbed Ca^{2+}. A local

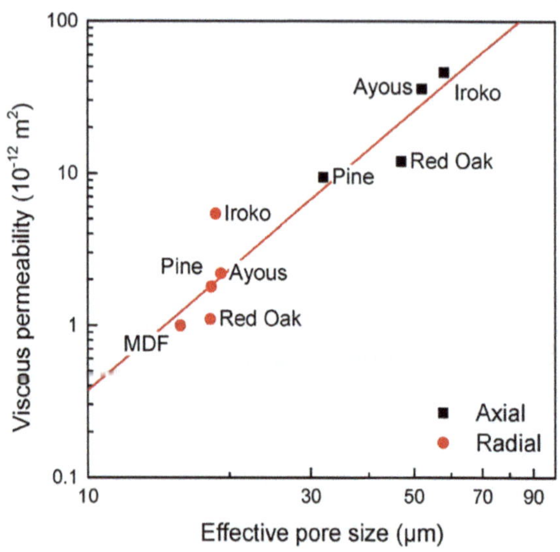

Fig. 8.5 Permeability (along two orientations) versus effective pore size for biomorphic SiC samples fabricated from different precursors. Reprinted with permission from [15]. Copyright 2016, Elsevier B.V.

increase in Ca^{2+} concentration stimulated in vitro precipitation of $Ca_5(PO_4)_3OH$ on the SiC preform surface upon exposure to simulated body fluid, indicating dramatically improved bone bonding activity compared with SiC. Biomorphic SiC skeletons embedded in NaCl–KCl molten salts can apply to solar and thermal energy storage [17]. A thermal conductivity of 116 W mK^{-1} was achieved by replicating cellular structure of oak wood, indicating an ultrafast thermal energy storage rate compared with molten salts alone. The solar absorptance was enhanced to 95.63% by further decorating TiN nanoparticles on the SiC skeletons, as a result of exciting broadband plasmonic resonances.

Ultrastrong, biomorphic cellular SiC ceramics were fabricated by the carbothermal reduction of gaseous SiO with a pyrolytic carbon template [18]. The anisotropic wood tissue microstructure was replicated in the biomorphic SiC materials, in which dense cell walls were assembled by directional columnar SiC nanocrystalline structures 100–200 nm in diameter and 100–500 nm in length (Fig. 8.6). The XRD patterns and Raman spectra demonstrate that the biomorphic ceramic is 3C-SiC. These ceramics have high porosity (76–86%), superior flexural strength ($\sigma_{\parallel}$: ~ 38.3 MPa), high thermal dissipation characteristics ($k_{\perp}$: 10.3 W m^{-1} K^{-1} at room temperature), and high anisotropic thermal conductivity ($k_{\perp}/k_{\parallel}$: ~ 1.6). The biomorphic SiC ceramic is a promising high-temperature filter [19]. The biomorphic SiC ceramic has good thermal stability at ~ 900 °C. It exhibited a high gas permeability ($k_1 = 11.83 \times 10^{-12}$ m^2, $k_2 = 1.86 \times 10^{-6}$ m) with ~ 81.9% porosity and a good filtration efficiency of 89.81% for PM > 2.5 μm under a pressure drop of 0.69 kPa.

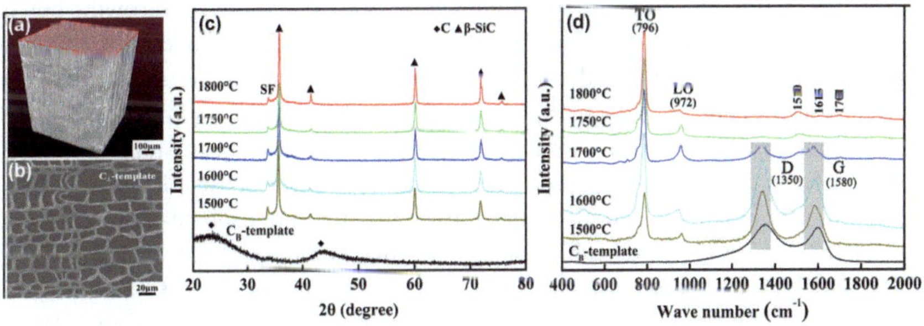

Fig. 8.6 **a** Micro-CT image of biomorphic SiC ceramics sintered at 1800 °C for 4 h, where the red area marks the cross section to radial direction. **b** Corresponding top-view SEM image. **c** XRD patterns, and **d** Raman spectra of pyrolytic biocarbon template and biomorphic SiC ceramics sintered at different temperatures. Reprinted with permission from [18]. Copyright 2019, American Chemical Society

8.2 SiC as Implant Materials

Diamond coatings can improve the tribological properties of the femoral head components in total hip replacements to prolong the lifetime, and the use of the SiC whisker–diamond composite can overcome the brittleness of the thin diamond coatings [20]. The reactions of regenerating bone tissue to diamond and SiC particles have been studied using implanted bone harvest chambers in rabbits. The particles dispersed in hyaluronan were introduced into a canal in the implant. After three weeks, both diamond and SiC particles did not mitigate bone formation, demonstrating that they have favorable biocompatibility. The authors further compared particles of diamond, SiC, and hydroxyapatite in serum-free cultures of human monocytes [21]. All particles were phagocytozed, and monocyte morphology changed except after the ingestion of diamond. Interleukin-1 beta production increased on average 30-fold and 38-fold in cultures separately exposed to hydroxyapatite and SiC, compared to control and diamond cultures. The introduced phagocytosis inhibitor cytochalasin B inhibited the morphological changes of the monocytes and reduced interleukin-1 beta production. For the experiments with introduced polymethylmethacrylate particles, the interleukin-1 beta stimulation fell in the same range as after hydroxyapatite and SiC stimulation.

Silicon carbide is a promising ceramic coating on titanium-based total hip replacements [22]. The SiC coating can prevent the formation of wear debris from the relatively soft titanium surface because SiC is a hard, tightly bonded, and not readily degraded ceramic material. The cytotoxicity test performed with JCRB0603 cells shows that 5 μm SiC particles prohibited one-third of colony outgrowth, but in contrast, the SiC-coated pins showed no inhibition, similar to uncoated titanium pins. This study demonstrates that SiC is a promising coating material on a total hip replacement implant.

Amorphous silicon carbide films prepared by CVD have been studied as insulating coatings on implantable microelectrodes [23]. The electrical leakage current of the amorphous SiC-coated platinum or iridium wire in phosphate-buffered saline is smaller than 10^{-11} A over a bias of $\pm$ 5 V. The resistivity of the amorphous SiC is 3×10^{13} Ω cm. The amorphous SiC showed a dissolution rate of 0.1 nm h^{-1} at 90 °C and no measurable dissolution at 37 °C in phosphate-buffered saline. When the amorphous SiC-coated quartz disk was implanted in the subcutaneous space of the rabbit, no chronic inflammatory response was observed. The amorphous SiC-coated microelectrode was implanted in the parietal cortex for up to 150 days, and the cortical response shows that the amorphous SiC is more stable in physiological saline than Si_3N_4 and well tolerated in the cortex. Micro/nanomechanical and tribological characterizations of SiC were carried out [24]. SiC, CoCrMo, Ti-6Al-4V, and stainless steel were examined for comparison. SiC exhibited higher hardness, elastic modulus, and scratch resistance as well as lower friction compared to other materials (Table 8.1). Fewer and smaller debris particles were generated for SiC.

The impact of amorphous SiC coating on Si probes by immunohistochemical characterization of key markers implicated in tissue-device response was examined [25].

Table 8.1 Mechanical and tribological properties of SiC, CoCrMo, Ti-6Al-4V, and stainless steel

Material	Hardness (GPa)	Elastic modulus (GPa)	Critical load (mN)	Wear coefficient of friction (beginning)	Wear coefficient of friction (end)	Size (μm)	Index
SiC	45.8	370	27	0.17	0.2	0.4	1
CoCrMo	10.1	300	13	0.17	0.23	1.3	2
Ti-6Al-4V	5.6	127	12	0.23	0.26	7.9	4
Stainless steel	6.2	223	9	0.45	0.40	8.1	3

After 4 and 8 weeks of implantation, device capture immunohistochemical labeling of neurons, astrocytes, and activated microglia/macrophages was evaluated. Neuron loss and microglia activation were similar between Si and amorphous SiC-coated probes, whereas the tissue implanted with amorphous SiC showed a reduction in astrocytes adjacent to the probe. Therefore, amorphous SiC shows similar biocompatibility profile as Si. SiC nanoparticles were introduced as an effective adhesive to acquire strong bonding (0.39 ± 0.03 kPa) between hybrid hydrogel films composed of alginate and collagen [26]. The authors fabricated multilayered heterogeneous constructs with embedded high-resolution microchannels (150 μm^{-1} mm) that were precisely interspaced (500–600 μm). Hydrogel layers were bonded with SiC nanoparticles, without blocking the hollow microchannels, and high cell viability (90.61% ± 3.28%) was maintained within the scaffold. Nanosilica caused clogging of smaller microchannels when used for interlayer bonding but was used to attach synthetic polymers such as Tygon to the hydrogels (32.5 ± 2.12 mN bond strength). This allows formation of inlet and outlet interconnections to the gel constructs. Biocompatibility of Si and 3C-SiC was evaluated using an in vitro model of human neuronal stem cells derived from dental pulp and mouse olfactory ensheathing cells [27]. Both dental pulp and olfactory ensheathing cells cultured on 3C-SiC did not experience consistent oxidative stress events and did not exhibit morphological modifications or adverse reactions in mitochondrial membrane potential.

3C-SiC nanomembranes that were grown on silicon wafers, released, and physically transferred to a final device substrate (e.g., polyimide) can apply to implantable flexible electronic devices [28]. SiC nanomembranes with a thickness of 230 nm showed no hydrolysis process (the etching rate is 0 nm per day at 96 °C in phosphate-buffered saline). There was no sign of water permeability for at least 60 days in phosphate-buffered saline at 96 °C and non-Na^+ ion diffusion detected at a thickness of 50 nm after being soaked in phosphate-buffered saline for 12 days. Prior to cell culture, the SiC nanomembranes were cleaned with 80% ethanol followed by surface UV exposure for 30 min. The cells had been seeded on the surface of SiC followed by storage at 37 °C for 3 days. Figure 8.7a, b shows fluorescent microscopy images of HDF cells after 24 h. The high optical transmittance of the free-standing SiC nanomembranes in the visible range enabled cell morphology observation and

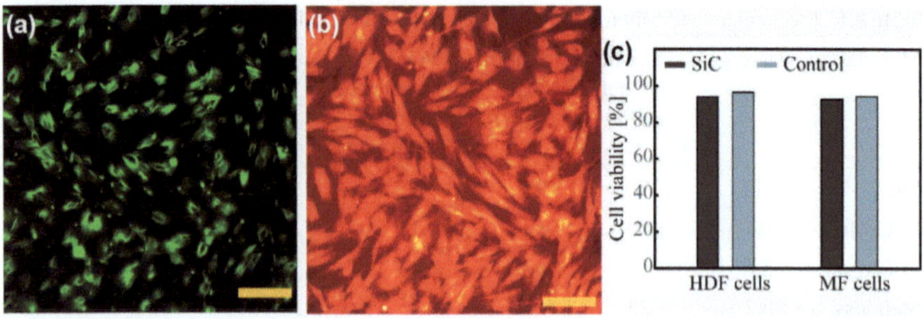

Fig. 8.7 Biocompatibility of SiC nanomembranes tested with mouse fibroblast cells and human dermal fibroblast cells (for practical health care applications). The cells were seeded on the SiC surface followed by storage at 37 °C for 3 days. **a** HDF cells show functional actin fiber formation. **b** Mitochondria distribution throughout the HDF cell was detected by using mitochondria-specific dye MitoTracker Deep Red. **c** Trypan blue exclusion test shows no significant difference in percentage of viable cells. Reprinted with permission from [28]. Copyright 2019, American Chemical Society

imaging using inverted fluorescent microscopes. The increase in the number of cells after 24 h indicates a good proliferation in the human dermal fibroblast (HDF) cells. The formation of functional actin fibers and cell nuclei captures the extensive stress of the HDF cytoskeleton, confirming that the cells well anchored on the SiC surface. The proper distribution of mitochondria throughout all cells reveals their healthiness and regular growth. Good proliferation was also observed in mouse fibroblast (MF) cells, with the formation of cytoskeleton and nucleic structures. Figure 8.7c shows the viability of the HDF and MF cells after culture on SiC nanomembranes for 72 h. Apparently, the viability is comparable to that of the control samples. Using a similar cell culture protocol, low-doped SiC also exhibited good cell proliferation, indicating that the doping concentration did not affect the biocompatibility of SiC nanomembranes.

Bioactive tantalum (Ta) metal deposited on porous SiC scaffolds by CVD was investigated to accelerate osseointegration and improve the bonding to bones [29]. The Ta coating evenly covered the whole scaffold structure. The bonding strength between the Ta coating and the SiC substrate is 88.4 MPa. The yield strength of porous SiC with a Ta coating (pTa) is 45.8 ± 2.9 MPa, the compressive strength is 61.4 ± 3.2 MPa, and the elastic modulus is ~ 4.8 GPa. When MG-63 human osteoblasts were co-cultured with pTa, osteoblasts exhibited good adhesion and spreading on the pTa surface and its porous structure, implying its excellent bioactivity and cytocompatibility. PTa and porous titanium were separately implanted into the femoral neck of goats for 12 weeks. The pTa group showed better osseointegration than the porous titanium group. The corrosion performance of SiC on anodized titanium dioxide nanotubes fabricated by using plasma-enhanced CVD was investigated through a potentiodynamic polarization test and bacterial incubation for 30 days [30]. These materials exhibited improved corrosion resistance, implying they are promising implant coating materials. SiC coated on carbon nanowall (CNW) scaffolds using CVD with a vinylsilane precursor at 700 °C was investigated [31]. The wall edge width ranges from 10 nm to those filling the wall-to-wall space without disrupting the CNW morphology. The SiC-coated CNWs were used as scaffolds for cell culture, indicating increase in cell viability until the edge area ratio reached 40%, and after that, cell viability turned to saturation, comparable to the cases of flat surfaces including SiC films on the Si substrate and control samples prepared using polystyrene. Calcification was suppressed in the CNWs, SiC/CNWs, and SiC/Si scaffolds compared to polystyrene.

The magnetic resonance imaging (MRI) compliance of epitaxial SiC films grown on various Si wafers used to implement a monolithic neural implant was studied by using finite element and Fourier-based simulations [32]. The specific absorption rate, induced heating, and image artifacts caused by the portion of the implant within a brain tissue phantom located in a 7 T small animal MRI machine were calculated and measured. The simulation and phantom experiments performed in the same 7 T MRI system reveal that free-standing 3C-SiC films have little to no image artifacts compared to silicon and platinum reference materials inside the MRI at 7 T. The simulations predict ~ 30% specific absorption rate reduction for 3C-SiC compared to Pt. Therefore, an all-SiC invasive neural interface may effectively reduce MRI induced heating and image artifacts in high field MRI.

8.3 Biological Applications and Toxicity of SiC Nanostructures

Nanostructures, including nanoparticles, nanowires, and nanotubes, are widely used in biodiagnostics such as electrical, electrochemical, optical, and magnetic detection and sensing [33]. The detected targets include proteins, nucleic acids, viruses, and biologically relevant small molecules. In particular, the use of QDs in in vitro and in vivo imaging, labeling, and sensing is very promising [34, 35]. Conventional organic fluorophores have broad absorption/emission bands and suffer from easy photo-bleaching, which limit their applications to long-term biological imaging. In contrast, QDs can have large absorption coefficients, high quantum yields, broad absorption bands, and narrow emission bands with peaks spanning the near-UV to the near-infrared region, large Stokes shifts, and high resistance to photo-bleaching. In particular, the emission color of QDs can be tuned by varying the size and surface functional groups. This property combined with the narrow emission bands renders them favorable for multicolor biological labeling and imaging at one excitation wavelength. However, the applications of traditional semiconductor (such as CdSe) QDs are limited by their cytotoxicity arising from leaking of heavy metal ions from the QDs [36–39]. In this regard, benign group IV nanoparticles including silicon, carbon, silicon carbide, and germanium nanoparticles fare better than those containing heavy metal atoms in vitro and in vivo [40]. Their luminescence covers the near-infrared to near-UV regime. Group IV nanoparticles generally have good resistance to photo-bleaching and have fluorescence lifetimes on the order of nanoseconds to microseconds boding well for biological labeling and imaging. SiC QDs have additional virtues such as ultrahigh chemical and thermal stability, low density, high hardness, and high photostability. Therefore, SiC nanostructures especially SiC QDs show great promise as biomaterials in imaging, sensing, drug delivery, and related applications.

8.3.1 SiC Nanoparticles

Biological imaging was demonstrated using 3C-SiC QDs having diameters of 1–7 nm prepared by electrochemical etching and ultrasonic stirring [41]. The quantum yield of these SiC QD dispersed in water was measured to be 17% by taking quinine sulfate in 0.1 N sulfuric acid as a reference standard and adopting 360 nm light excitation. After entering human fetal osteoblast cells, these QDs continue to show strong photoluminescence (Fig. 8.8). The intracellular green–yellow spots suggest that the nanocrystals penetrate the cell membrane by endocytosis. Fluorescence was observed throughout the cells including the nuclei, showing the presence of nanocrystals. The small dimension of 3C-SiC nanocrystals comparable to that of biological molecules is an advantage. The luminescence stability of 3C-SiC nanocrystals was compared to that of the organic dye propidium iodide (PI). The intensity of the luminescence from the PI labeled cells decreased to half the initial value after 5 min, but that from the 3C-SiC nanocrystal labeled cells diminished only slightly in the first several minutes and held for at least 40 min. This behavior is similar to that of conventional QDs which are photostable for minutes to hours. The methyl thiazolyl tetrazolium (MTT) assay was conducted on the HeLa cells to evaluate the biocompatibility of the 3C-SiC nanocrystals. The MTT assay is a colorimetric test which estimates the cell number by monitoring the mitochondrial function [a crucial role in maintaining the cellular structure and function via aerobic adenosine triphosphate (ATP) production] of the cultured cells. The cells incubated with 3C-SiC nanocrystals with a concentration up to 100 mg mL^{-1} (about 1.0×10^{15} particles) for as long as 48 h showed no remarkable difference in the MTT colorimetric reaction compared to the control cells. No cell morphology change or increase in the incidence of dead cells was observed even from cells incubated with a large dose of nanocrystals for a long time. These results indicate the 3C-SiC nanocrystals are benign to cells due to their extreme crystal structural stability, hydrophilicity of (hydroxyl and carboxylic) surfaces, and chemical inertness (particularly in water) as well as innate biocompatibility of silicon carbide.

The in vitro responses of eukaryotic cells exposed to SiC and TiO_2 nanoparticles as well as multiwalled carbon nanotubes were compared [42]. The primary organ cells (lung: A549 alveolar epithelial cells) and secondary organ cells (liver: WIF-B9, Can-10, and kidneys: NRK-52E, LLC-PK1 proximal cells) were employed for the evaluation because these organs were exposed when the nanoparticles were translocated

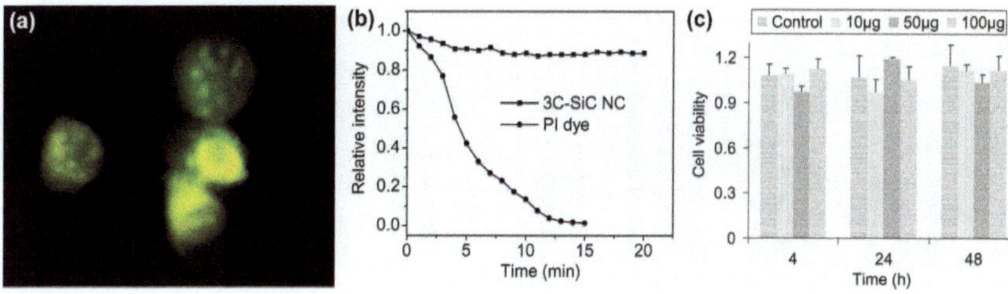

Fig. 8.8 **a** Dark-field fluorescence image of human fetal osteoblast (hFOB) cells after uptake of 3C-SiC nanocrystals (3.9 nm in average diameter) examined by fluorescence microscopy under 420–490 nm excitation. Image width: 93 μm. **b** Photostability test on 3C-SiC nanocrystal labeled and PI dye labeled hFOB cells excited at 420–490 nm and at 545 nm, respectively. **c** Cytotoxicity evaluation with HeLa cells incubated with different concentrations (in unit μg mL^{-1}) of 3C-SiC nanocrystals for 4, 24, and 48 h and analyzed by the standard methyl thiazolyl tetrazolium assay. Error bars indicate the standard deviation of three independent experiments. Reprinted with permission from [41]. Copyright 2008, John Wiley & Sons, Inc.

through the epithelial barriers. The toxicity of these nanoparticles showed dependence on their size, morphology, and chemical composition. The smallest spherical TiO_2 nanoparticles showed more cytotoxicity to NRK-52E cells, whereas the SiC nanoparticles showed nearly no cytotoxicity. Nanoparticle cytotoxicity showed also dependence on the cell line. All the nanoparticles uptaken by cells generated intracellular reactive oxygen species (ROS), and the ROS production depends on the nanoparticle chemical composition (Fig. 8.9). The 12–25 nm metal oxide nanoparticles generated the most ROS, while the 17 nm SiC nanoparticles did not increase intracellular ROS generation relative to the control. DNA damage was evaluated by the alkaline comet assay. Long carbon nanotubes with low concentration caused small DNA damage; the small spherical TiO_2 nanoparticles caused higher DNA damage. In contrast, short carbon nanotubes and SiC nanoparticles did not affect the DNA integrity. The effects of exposure to SiC nanoparticles on A549 lung epithelial cells were studied in vitro [43]. These SiC nanoparticles having Si/C ratios 0.8–1.3 were synthesized through laser pyrolysis of silane and acetylene. Note that they have much greater diameters of 13–68 nm compared to those SiC nanoparticles adopted in the aforementioned experiment. These SiC nanoparticles internalized in cells did not cause strong cell mortality but induced major redox disturbance and DNA damage. The cellular response showed dependence on the dimensions and Si/C ratio of the SiC nanoparticles. The biological effects of six SiC nanoparticle samples were investigated using a macrophage cell line in vitro [44]. The authors synthesized six different samples of SiC nanoparticles with different sizes (15–60 nm), crystallographic structures (3C-SiC with or without the presence of 6H-SiC), C/Si atomic ratio (excess silicon or carbon), surface oxidation (formation of a silica or an oxycarbide layer around the SiC grain), or chemical surface impurities (quantification of the amount of iron per unit surface) at the laboratory scale by using laser pyrolysis and sol–gel processes. A sample of pure α-SiC synthesized by mechanical wet-milling was also

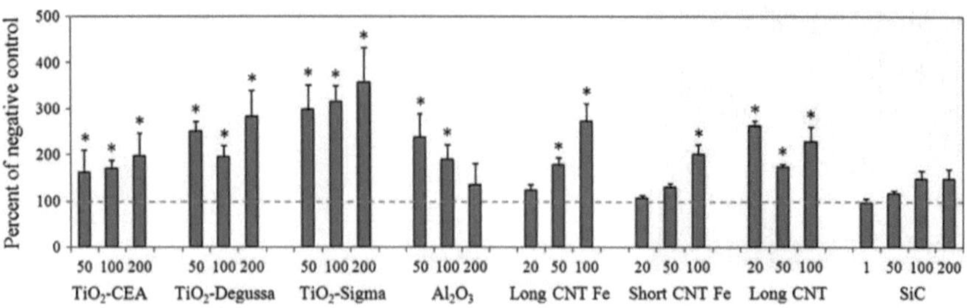

Fig. 8.9 Reactive oxygen species production induced by different types of nanoparticles in NRK-52E cells. The cells were exposed to 1–200 μg mL^{-1} of nanoparticles for 24 h, and reactive oxygen species production was evaluated by the H2DCF-DA assay. Luminescence is normalized by the total protein content and expressed as percentage of negative control, that is, unexposed cells. Reprinted with permission from [42]. Copyright 2010, Springer Nature Limited

employed. The cytotoxic issues studied include the cell membrane damage, inflammatory effects, and reactive oxygen species generation. All the SiC samples showed no cytotoxicity, whereas dramatic pro-oxidative reactions and inflammatory response were observed, and they showed dependence on the physico-chemical features of the SiC nanoparticles. The surface area and crystalline phase (α or β phase) of the SiC nanoparticles showed influence on the inflammatory effects.

SiC nanoparticles with a 2.5 nm most probable size dispersed in a solution can penetrate into the surrounding air environment during natural evaporation of the solution, resulting in labeling of adjacent cells via the vapor phase (Fig. 8.10) [45]. The cells placed over the SiC nanoparticle solution showed different colors after absorption of evaporated nanoparticles of different sizes, depending on the distance from the solution. The biodistribution, toxicity, and elemental composition changes in feces and organs after oral administration of SiC nanoparticles to a rat were studied [46]. Most of the SiC nanoparticles were excreted in the feces, and traces were retrieved in the urine, suggesting that SiC can cross the intestinal barrier. The SiC retention rate decreased with time through the gastrointestinal tract, and a single dose was mostly eliminated on the first day of administration. The presence of Ca, P, and Mg indicates that the SiC nanoparticles did not accumulate in the liver, kidneys, and spleen. The nanoparticles did not induce any morphological alteration in the organs studied including the esophagus, stomach, intestines, liver, spleen, and kidney or increase in the amount of granulomas. Eliminations of urinary sodium and potassium and urine osmolarity were not changed by the nanoparticles, showing that the SiC nanoparticles have no toxicity in rats for the studied doses.

Covalent grafting of amino groups onto the carboxylic acid functionalized fluorescent SiC nanoparticles via the reaction between the surface carboxylic acid groups and ethylenediamine changed the nanoparticle surface charge from negative to positive [47]. The sign of the surface charge showed influence on the distribution of the nanoparticles in the cells, as demonstrated by incubating 3T3-L1 fibroblast cells with differently charged SiC nanoparticles. Negatively charged SiC nanoparticles

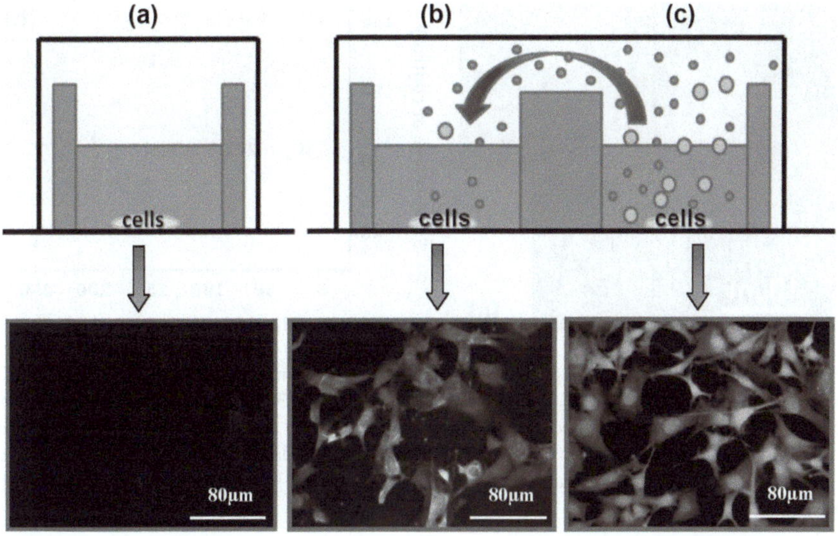

Fig. 8.10 Fluorescence images of the 3T3-L1 fibroblast cells. Isolated (**a**) and nonisolated (**b**) in the air environment of a cell culture (**c**). The cells in the right part were cultured with fluorescent SiC nanoparticles. Reprinted with permission from [45]. Copyright 2012, Springer Nature Limited

concentrated in the cell nuclei and positively charged nanoparticles in the cytoplasm, whereas weakly positively charged SiC nanoparticles were observed from both the cytoplasm and nuclei (Fig. 8.11). Cell proliferation affects the SiC nanoparticle uptake and its concentration in the cell nuclei, and cell division plays a key role in the penetration of the SiC nanoparticles into the cell nuclei [48].

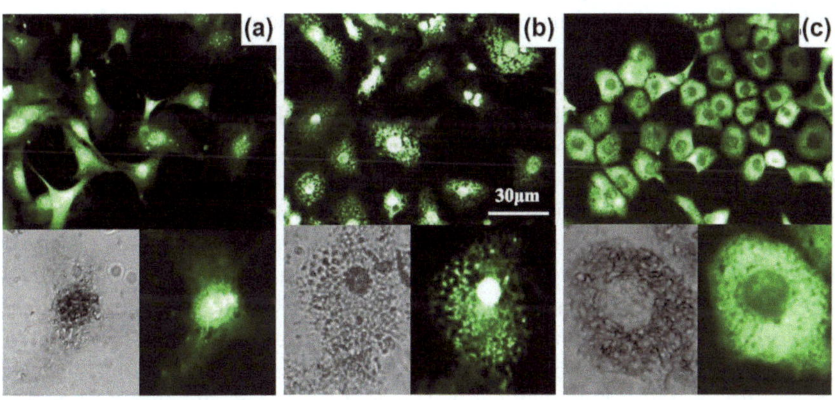

Fig. 8.11 Fluorescence and optical microscopy images of 3T3-L1 cells: **a** labeled with negatively charged SiC–CO_2H nanoparticles (zeta potential: − 30 mV); **b** labeled with quasi-neutral SiC–NH_2 nanoparticles (zeta potential: + 4 mV); and **c** labeled with positively charged SiC–NH_2 nanoparticles (zeta potential: + 100 mV). Reprinted with permission from [47]. Copyright 2012, IOP Publishing

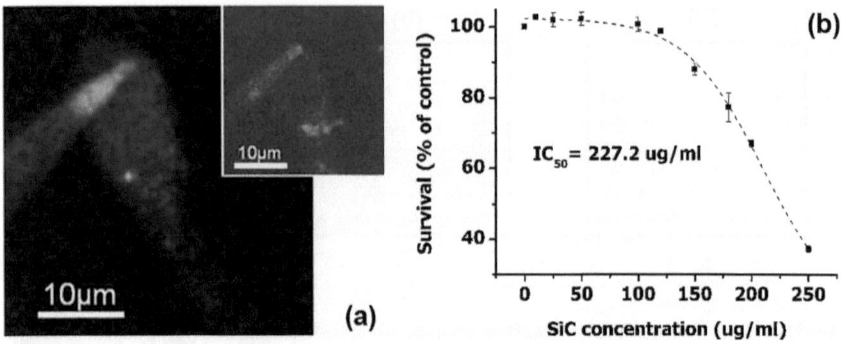

Fig. 8.12 **a** Two-photon imaging of a CA1 hippocampal pyramidal neuron cell labeled by SiC nanocrystals. Red fluorescence (600–700 nm) was generated by 830 nm excitation. The inset shows the same neuron cell with fluorescence collected from 425 to 525 nm. **b** Cell viability as a function of SiC nanoparticle concentration by alamarBlue assay. The calculated IC_{50} value (SiC nanoparticle concentration to allow for half-maximum inhibitory effect on cell growth) is shown. Reprinted with permission from [49]. Copyright 2013, Springer Nature Limited

Two-photon biological imaging based on SiC nanocrystals with diameters of around 3 nm fabricated by electroless chemical etching was investigated [49]. These nanocrystals showed violet–blue emission in the 410–450 nm region. Significant fluorescence from the CA1 hippocampal pyramidal cells labeled by the SiC nanocrystals was observed under two-photon excitation at 830 nm (Fig. 8.12). The cytotoxicity of the SiC nanoparticles was evaluated based on the alamarBlue^{TM} assay, showing that the SiC nanoparticles have only a small effect on cell growth in a wide concentration range up to 100 μg mL^{-1} for as long as 48 h.

Folate-modified SiC nanoparticles were used for cancer-cell-specific labeling [50]. An aminosilane was grafted onto the surface of the SiC nanoparticle, which was further modified with the folic acid using an isothiocyanate-based coupling method. Covalent grafting of silane and folic acid to the nanoparticle surface was confirmed. The average cancer-cell labeling percentage for these SiC nanoparticles is ~ 48%, much higher than those of 10–15% for controls (healthy cells, competition assay, and poly(ethylene glycol) modified SiC nanoparticles). The biological behavior of SiC nanoparticles of sizes ~ 1.5 or 9 nm in a medium modeling a living organism was investigated [51]. Transient nanoparticle–protein interaction owing to restricted diffusion behavior of the nanoparticles in the vicinity of a protein was observed. A single-molecule localization microscopic observation revealed the photoluminescence properties of the point defects in 3C- and 4H-SiC bulk materials, QDs, and nanoparticles of different sizes using intra-bandgap excitation [52]. The excitation light with a wavelength of 561 nm is ideal for observing sufficient blinking behavior, whereas the excitation light with a wavelength of 638 nm mostly quenched the PL. The incubation of the 4H-SiC nanoparticles with MCF10A cells in vitro was realized, with observation of super-resolved images of the nanoparticles in cells by combining

561 and 638 nm light excitation. Carbon-defective SiC nanoparticles having near-infrared absorption features can be applied to photothermal therapy [53]. The SiC nanoparticles-mediated photothermal therapy treatment could remove solid OSCC tumor, showing low hematotoxicity, cytotoxicity, and tissue toxicity. The low invasion of photothermal therapy treatment could prevent the invasion of bacteria and generate an antibacterial effect on the wound. These nanoparticles can be excreted from the body after the treatment.

8.3.2 SiC Whiskers and Nanoporous Membranes

The toxicity of SiC whiskers with sub-micrometer diameters has been investigated [54]. The cytotoxicity of SiC whiskers, alumina fibers, SiC powders, and crocidolite asbestos was evaluated and compared by monitoring the V79-4 Chinese hamster lung cell survival rate after exposure to the respective particles (Fig. 8.13). SiC whiskers and crocidolite asbestos exhibited comparable and severe toxicity to cells, but in contrast, SiC powders (with an average size of 400 nm) and alumina fibers showed rather low toxicity.

The short-term effect of SiC whiskers was investigated in vivo by instillation and inhalation to the rat lung [55]. The whiskers have a mean length of 5.1 μm and a mean width of 0.3 μm. Exposure to SiC whiskers caused pulmonary fibrotic changes, suggesting that SiC whiskers may have the fibrogenic potential. The Clara cell secretory protein was supposed to play a key role in regulating the acute inflammatory

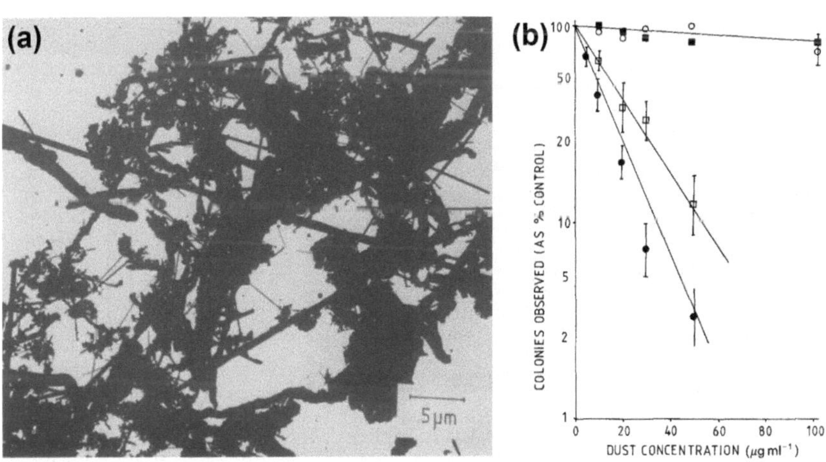

Fig. 8.13 **a** Transmission electron micrograph of SiC whiskers. **b** Cytotoxicity of alumina fiber (solid square), crocidolite asbestos (solid circle), SiC powders (empty circle), and SiC whiskers (empty square) expressed as percentage survival of V79-4 Chinese hamster lung cells exposed to the respective particles. Reprinted with permission from [54]. Copyright 1988, Springer Nature Limited

response in the lung and was involved in not only acute injury but also chronic injury of the lung exposed to SiC whiskers [56]. Silicon carbide nanowires are generally insoluble in water and when they are released into an aqueous environment, they can accumulate as sediments. The toxicity of SiC nanowires to freshwater sediment dwelling organisms such as amphipods, midges, oligochaetes, and mussels was studied [57]. The SiC nanowires have diameters ranging from 40 to 500 nm and lengths from 5 to 65 μm. Unsonicated SiC nanowires in water were not acutely toxic to amphipods but sonicated ones were acutely toxic to amphipods but not to midges, oligochaetes, or mussels. The cellular signal transduction processes and toxicity mechanisms of SiC nanowires have been investigated [58]. The Chinese hamster ovary cells in contact with the SiC nanowires showed genomic instability and significantly reduced reproduction rates. Activation of the mitogen activated protein kinases cellular signaling pathway and over-expression of cyclooxygenase-2 may be the major toxicity mechanisms in SiC nanowires irritation.

Porous SiC membranes with ultrasmall pores may act as semi-permeable materials in biomedical applications [59]. The free-standing *n*- and *p*-type SiC nanoporous membranes were fabricated by electrochemical etching of bulk 6H-SiC crystals in a 5% aqueous hydrofluoric acid electrolyte mixed with 5% ethanol. The *p*-type materials were anodized in darkness, and the *n*-type materials were anodized under UV-light irradiation from a 350 W mercury arc lamp. The porous film was peeled from the substrate by applying an additional high current density up to 100 mA cm^{-2}. The free-standing porous SiC membranes (Fig. 8.14) have thicknesses between 20 and 110 μm and permitted diffusion of proteins up to 29,000 Daltons excluding larger ones. Using the radioactive I^{125}-labeled albumin, porous SiC was demonstrated to have very low protein adsorption comparable to that of the best commercially available polymer nanoporous membrane.

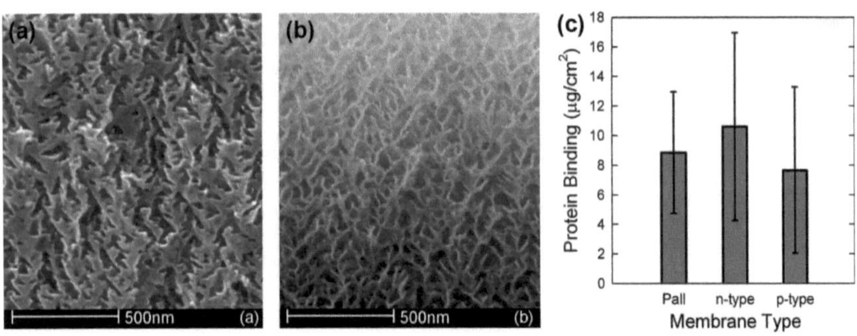

Fig. 8.14 Cross-sectional SEM images taken from pieces of porous membranes used in the tests: **a** *n*-type 6H-SiC and **b** *p*-type 6H-SiC. The *p*-type porous structure has a more feathery appearance and is more fragile than the *n*-type porous structure. **c** Comparison of binding of I^{125}-labeled albumin to porous *n*- and *p*-type SiC membranes and the best commercially available low adsorption polymer membrane. Error bars represent the standard deviation of three experiments. Reprinted with permission from [59]. Copyright 2004, Springer Nature Limited

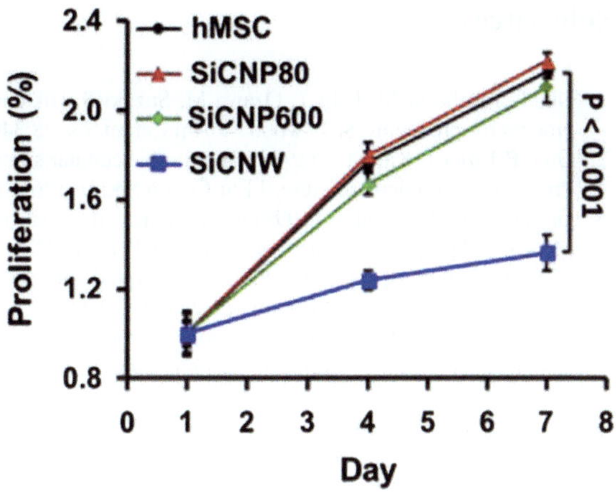

Fig. 8.15 SiCNWs greatly suppressed proliferation of human mesenchymal stem cells (hMSCs). Reprinted with permission from [62]. Copyright 2018, Elsevier B.V.

A study shows in vitro cytocompatibility of SiC/SiO_2 core/shell nanowires [60]. Different internalization mechanisms by adenocarcinomic alveolar basal epithelial cells, monocytic cell line derived from an acute monocytic leukemia, breast cancer cells, and normal human dermal fibroblasts were observed. The internalization occurred mainly for macropinocytosis and sporadically by direct penetration in all cells. The internalization occurred for phagocytosis only in monocytic leukemia cells. The experiments of cell proliferation, cell cycle progression, and oxidative stress on the cells treated with nanowires as compared to controls confirmed the cytocompatibility of the SiC/SiO_2 nanowires. The reactive oxygen species generation behaved as an early event and quickly ran out with a rapid decrease in adenocarcinomic alveolar basal epithelial and human dermal fibroblasts cells. The intracellular presence of nanowires caused identical molecular events in all the cell lines, but to a different extent: peroxidation of membrane lipids and oxidation of proteins. The nanowires did not exhibit midterm or long-term (10 days) cytotoxic activity. These nanowires conjugated via click-chemistry procedures with a tetracarboxyphenyl porphyrin derivative were demonstrated to be an efficient source of singlet oxygen for cell oxidative stress when irradiated with 6 MV X-rays at low doses (0.4–2 Gy) [61]. The in vitro clono genic survival assay on lung adenocarcinoma cells indicates that the cell population would decrease by ~ 75% with respect to control cells at the twelfth day after irradiation at a dose of 2 Gy. There has been a comparative evaluation of the toxicity of SiC nanowires and nanoparticles on human mesenchymal stem cells in terms of metabolism, viability, adhesion, proliferation, migration, oxidative stress, and differentiation ability [62]. TEM shows that the SiC nanoparticles have sizes of 78.8 ± 18.0 nm and 603.2 ± 97.6 nm. The widths of the SiC nanowires are 82.3 ± 12.4 nm. The results indicate that these SiC nanowires are toxic to human mesenchymal stem cells (Fig. 8.15) but not to cancer cell lines at the concentration of 0.1 mg mL^{-1}. The SiC nanoparticles were biocompatible to human mesenchymal stem cells at 0.1 mg mL^{-1}.

References

1. Ota T, Takahashi M, Hibi T, Ozawa M, Suzuki S, Hikichi Y, Suzuki H (1995) Biomimetic process for producing SiC "wood." J Am Ceram Soc 78:3409–3411
2. Greil P, Lifka T, Kaindl A (1998) Biomorphic cellular silicon carbide ceramics from wood: I. Processing and microstructure. J Eur Ceram Soc 18:1961–1973
3. Greil P, Lifka T, Kaindl A (1998) Biomorphic cellular silicon carbide ceramics from wood: II. Mechanical properties. J Eur Ceram Soc 18:1975–1983
4. Vogli E, Sieber H, Greil P (2002) Biomorphic SiC-ceramic prepared by Si-vapor phase infiltration of wood. J Eur Ceram Soc 22:2663–2668
5. Vogli E, Mukerji J, Hoffman C, Kladny R, Sieber H, Greil P (2001) Conversion of oak to cellular silicon carbide ceramic by gas-phase reaction with silicon monoxide. J Am Ceram Soc 84:1236–1240
6. Qian JM, Wang JP, Jin ZH (2004) Preparation of biomorphic SiC ceramic by carbothermal reduction of oak wood charcoal. Mater Sci Eng A 371:229–235
7. Qian J-M, Wang J-P, Qiao G-J, Jin Z-H (2004) Preparation of porous SiC ceramic with a woodlike microstructure by sol-gel and carbothermal reduction processing. J Eur Ceram Soc 24:3251–3259
8. Qian JM, Jin ZH (2006) Preparation and characterization of porous, biomorphic SiC ceramic with hybrid pore structure. J Eur Ceram Soc 26:1311–1316
9. Shin Y, Wang C, Exarhos GJ (2005) Synthesis of SiC ceramics by the carbothermal reduction of mineralized wood with silica. Adv Mater 17:73–77
10. Martínez-Fernández J, Valera-Feria FM, Singh M (2000) High temperature compressive mechanical behavior of joined biomorphic silicon carbide ceramics. Scripta Mater 43:813–818
11. Varela-Feria FM, Martínez-Fernández J, de Arellano-López AR, Singh M (2002) Low density biomorphic silicon carbide: microstructure and mechanical properties. J Eur Ceram Soc 22:2719–2725
12. Singh M, Salem JA (2002) Mechanical properties and microstructure of biomorphic silicon carbide ceramics fabricated from wood precursors. J Eur Ceram Soc 22:2709–2717
13. González P, Serra J, Liste S, Chiussi S, León B, Pérez-Amor M, Martínez-Fernández J, de Arellano-López AR, Varela-Feria FM (2003) New biomorphic SiC ceramics coated with bioactive glass for biomedical applications. Biomaterials 24:4827–4832
14. Pappacena KE, Faber KT, Wang H, Porter WD (2007) Thermal conductivity of porous silicon carbide derived from wood precursors. J Am Ceram Soc 90:2855–2862
15. Gómez-Martín A, Orihuela MP, Becerra JA, Martínez-Fernández J, Ramírez-Rico J (2016) Permeability and mechanical integrity of porous biomorphic SiC ceramics for application as hot-gas filters. Mater Des 107:450–460
16. Will J, Hoppe A, Müller FA, Raya CT, Fernández JM, Greil P (2010) Bioactivation of biomorphous silicon carbide bone implants. Acta Biomater 6:4488–4494
17. Xu Q, Liu X, Luo Q, Song Y, Wang H, Chen M, Xuan Y, Li Y, Ding Y (2021) Bifunctional biomorphic SiC ceramics embedded molten salts for ultrafast thermal and solar energy storage. Mater Today Energy 21:100764
18. Zhang JF, Zhou XN, Huang X, Hao LC, Zhi Q, Li ZX, Hou BQ, Yang JF, Wang B, Ishizaki K (2019) Biomorphic cellular silicon carbide nanocrystal-based ceramics derived from wood for use as thermally stable and lightweight structural materials. ACS Appl Nano Mater 2:7051–7060
19. Zhou XN, Hao X, Xu JQ, Ren Y, Zhi Q, Zhang NL, Wei ZL, Yang JF, Wang B, Ishizaki K (2024) Light-weight, wood-derived, biomorphic SiC ceramics by carbothermal reduction. Ceram Int 50:23135–23149
20. Aspenberg P, Anttila A, Konttinen YT, Lappalainen R, Goodman SB, Nordsletten L, Santavirta S (1996) Benign response to particles of diamond and SiC: bone chamber studies of new joint replacement coating materials in rabbits. Biomaterials 17:807–812

21. Nordsletten L, Hogasen AKM, Konttinen YT, Santavirta S, Aspenberg P (1996) Human monocytes stimulation by particles of hydroxyapatite, silicon carbide and diamond: in vitro studies of new prosthesis coatings. Biomaterials 17:1521–1527
22. Santavirta S, Takagi M, Nordsletten L, Anttila A, Lappalainen R, Konttinen YT (1998) Biocompatibility of silicon carbide in colony formation test in vitro. a promising new ceramic THR implant coating material. Arch Orthop Trauma Surg 118:89–91
23. Cogan SF, Edell DJ, Guzelian AA, Liu YP, Edell R (2003) Plasma-enhanced chemical vapor deposited silicon carbide as an implantable dielectric coating. J Biomed Mater Res A 67:856–867
24. Li XD, Wang XN, Bondokov R, Morris J, An YHH, Sudarshan TS (2005) Micro/nanoscale mechanical and tribological characterization of SiC for orthopedic applications. J Biomed Mater Res B 72:353–361
25. Knaack GL, McHail DG, Borda G, Koo B, Peixoto N, Cogan SF, Dumas TC, Pancrazio JJ (2016) In vivo characterization of amorphous silicon carbide as a biomaterial for chronic neural interfaces. Front Neurosci 10:301
26. Attalla R, Ling CSN, Selvaganapathy PR (2018) Silicon carbide nanoparticles as an effective bioadhesive to bond collagen containing composite gel layers for tissue engineering applications. Adv Healthcare Mater 7:1701385
27. Bonaventura G, Iemmolo R, La Cognata V, Zimbone M, La Via F, Fragalà ME, Barcellona ML, Pellitteri R, Cavallaro S (2019) Biocompatibility between silicon or silicon carbide surface and neural stem cells. Sci Rep 9:11540
28. Phan HP, Zhong YS, Nguyen TK, Park Y, Dinh T, Song EM, Vadivelu RK, Masud MK, Li JH, Shiddiky MJA, Dao DZ, Yamauchi Y, Rogers JA, Nguyen NT (2019) Long-lived, transferred crystalline silicon carbide nanomembranes for implantable flexible electronics. ACS Nano 13:11572–11581
29. Ma ZJ, Li JY, Cao F, Yang JH, Liu R, Zhao DW (2020) Porous silicon carbide coated with tantalum as potential material for bone implants. Regen Biomater 7:453–459
30. Hsu SM, Fares C, Xia XY, Rasel MA, Ketter J, Camargo SEA, Haque MA, Ren F, Esquivel-Upshaw JF (2021) In vitro corrosion of SiC-coated anodized Ti nano-tubular surfaces. J Funct Biomater 12:52
31. Ono K, Koide T, Ishikawa K, Tanaka H, Kondo H, Sugawara-Narutaki A, Jin Y, Yasuhara S, Hori M, Takeuchi W (2023) Biocompatibility of conformal silicon carbide on carbon nanowall scaffolds. Jpn J Appl Phys 62:SA1017
32. Beygi M, Dominguez-Viqueira W, Feng CY, Mumcu G, Frewin CL, La Via F, Saddow SE (2021) Silicon carbide and MRI: towards developing a MRI safe neural interface. Micromachines 12:126
33. Rosi NL, Mirkin CA (2005) Nanostructures in biodiagnostics. Chem Rev 105:1547–1562
34. Michalet X, Pinaud FF, Bentolila LA, Tsay JM, Doose S, Li JJ, Sundaresan G, Wu AM, Gambhir SS, Weiss S (2005) Quantum dots for live cells, in vivo imaging, and diagnostics. Science 307:538–544
35. Medintz IL, Uyeda HT, Goldman ER, Mattoussi H (2005) Quantum dot bioconjugates for imaging, labelling and sensing. Nat Mater 4:435–446
36. Derfus AM, Chan WCW, Bhatia SN (2004) Probing the cytotoxicity of semiconductor quantum dots. Nano Lett 4:11–18
37. Nel A, Xia T, Mädler L, Li N (2006) Toxic potential of materials at the nanolevel. Science 311:622–627
38. Oberdörster G, Oberdörster E, Oberdörster J (2005) Nanotoxicology: an emerging discipline evolving from studies of ultrafine particles. Environ Health Perspect 113:823–839
39. Lewinski N, Colvin V, Drezek R (2008) Cytotoxicity of nanoparticles. Small 4:26–49
40. Fan J, Chu PK (2010) Group IV nanoparticles: synthesis, properties, and biological applications. Small 6:2080–2098
41. Fan J, Li H, Jiang J, So LKY, Lam YW, Chu PK (2008) 3C-SiC nanocrystals as fluorescent biological labels. Small 4:1058–1062

42. Barillet S, Simon-Deckers A, Herlin-Boime N, Mayne-L'Hermite M, Reynaud C, Cassio D, Gouget B, Carrière M (2010) Toxicological consequences of TiO_2, SiC nanoparticles and multi-walled carbon nanotubes exposure in several mammalian cell types: an in vitro study. J Nanopart Res 12:61–73
43. Barillet S, Jugan M-L, Laye M, Leconte Y, Herlin-Boime N, Reynaud C, Carrière M (2010) In vitro evaluation of SiC nanoparticles impact on A549 pulmonary cells: cyto-, genotoxicity and oxidative stress. Toxicol Lett 198:324–330
44. Pourchez J, Forest V, Boumahdi N, Boudard D, Tomatis M, Fubini B, Herlin-Boime N, Leconte Y, Guilhot B, Cottier M, Grosseau P (2012) In vitro cellular responses to silicon carbide nanoparticles: impact of physico-chemical features on pro-inflammatory and pro-oxidative effects. J Nanopart Res 14:1143
45. Serdiuk T, Lysenko V, Skryshevsky VA, Géloën A (2012) Vapor phase mediated cellular uptake of sub 5 nm nanoparticles. Nanoscale Res Lett 7:212
46. Lozano O, Laloy J, Alpan L, Mejia J, Rolin S, Toussaint O, Dogné J-M, Lucas S, Masereel B (2012) Effects of SiC nanoparticles orally administered in a rat model: biodistribution, toxicity and elemental composition changes in feces and organs. Toxicol Appl Pharm 264:232–245
47. Serdiuk T, Alekseev SA, Lysenko V, Skryshevsky VA, Géloën A (2012) Charge-driven selective localization of fluorescent nanoparticles in live cells. Nanotechnology 23:315101
48. Serdiuk T, Lysenko V, Mognetti B, Skryshevsky V, Géloën A (2013) Impact of cell division on intracellular uptake and nuclear targeting with fluorescent SiC-based nanoparticles. J Biophotonics 6:291–297
49. Beke D, Szekrényes Z, Pálfi D, Róna G, Balogh I, Maák PA, Katona G, Czigány Z, Kamarás K, Rózsa B, Buday L, Vértessy B, Gali A (2013) Silicon carbide quantum dots for bioimaging. J Mater Res 28:205–209
50. Boksebeld M, Kilin V, Géloën A, Ceccone G, Jaffal A, Schmidt C, Alekseev S, Lysenko V, Wolf JP, Bonacina L, Souteyrand E, Chevolot Y, Monnier V (2017) Folate-modified silicon carbide nanoparticles as multiphoton imaging nanoprobes for cancer-cell-specific labeling. RSC Adv 7:27361
51. Dravecz G, Jánosi TZ, Beke D, Major DA, Károlyházy G, Erostyák J, Kamarás K, Gali A (2018) Identification of the binding site between bovine serum albumin and ultrasmall SiC fluorescent biomarkers. Phys Chem Chem Phys 20:13419
52. Castelletto S, Barbiero M, Charnley M, Boretti A, Gu M (2020) Imaging with nanometer resolution using optically active defects in silicon carbide. Phys Rev Appl 14:034021
53. Mu HB, Pang HY, Zheng C, Wang KX, Hu N, Zhang B (2021) Photothermal treatment of oropharyngeal cancer with carbon-defective silicon carbide. J Mater Chem B 9:5284
54. Birchall JD, Stanley DR, Mockford MJ, Pigott GH, Pinto PJ (1988) Toxicity of silicon carbide whiskers. J Mater Sci Lett 7:350–352
55. Ogami A, Morimoto Y, Yamato H, Oyabu T, Akiyama I, Tanaka I (2001) Short term effect of silicon carbide whisker to the rat lung. Ind Health 39:175–182
56. Morimoto Y, Ding L, Oyabu T, Hirohashi M, Kim H, Ogami A, Yamato H, Akiyama I, Hori H, Higashi T, Tanak I (2003) Expression of Clara cell secretory protein in the lungs of rats exposed to silicon carbide whisker in vivo. Toxicol Lett 145:273–279
57. Mwangi JN, Wang N, Ritts A, Kunz JL, Ingersoll CG, Li H, Deng B (2011) Toxicity of silicon carbide nanowires to sediment-dwelling invertebrates in water or sediment exposures. Environ Toxicol Chem 30:981–987
58. Jiang J, Wang J, Zhang X, Huo K, Wong HM, Yeung KWK, Zhang W, Hu T, Chu PK (2010) Activation of mitogen-activated protein kinases cellular signal transduction pathway in mammalian cells induced by silicon carbide nanowires. Biomaterials 31:7856–7862
59. Rosenbloom AJ, Sipe DM, Shishkin Y, Ke Y, Devaty RP, Choyke WJ (2004) Nanoporous SiC: a candidate semi-permeable material for biomedical applications. Biomed Microdevices 6:261–267
60. Cacchioli A, Ravanetti F, Alinovi R, Pinelli S, Rossi F, Negri M, Bedogni E, Campanini M, Galetti M, Goldoni M, Lagonegro P, Alfieri R, Bigi F, Salviati G (2014) Cytocompatibility and cellular internalization mechanisms of SiC/SiO_2 nanowires. Nano Lett 14:4368–4375

61. Rossi F, Bedogni E, Bigi F, Rimoldi T, Cristofolini L, Pinelli S, Alinovi R, Negri M, Dhanabalan SC, Attolini G, Fabbri F, Goldoni M, Mutti A, Benecchi G, Ghetti C, Iannotta S, Salviati G (2015) Porphyrin conjugated SiC/SiO_x nanowires for X-ray-excited photodynamic therapy. Sci Rep 5:7606
62. Chen F, Li GY, Zhao ER, Li JT, Hableel G, Lemaster JE, Bai YT, Sen GL, Jokerst JV (2018) Cellular toxicity of silicon carbide nanomaterials as a function of morphology. Biomaterials 179:60–70